Dynamics of Planetary Systems

PRINCETON SERIES IN ASTROPHYSICS

EDITED BY DAVID N. SPERGEL

Theory of Rotating Stars, *by Jean-Louis Tassoul*

Theory of Stellar Pulsation, *by John P. Cox*

Galactic Dynamics, Second Edition, *by James Binney and Scott Tremaine*

Dynamical Evolution of Globular Clusters, *by Lyman S. Spitzer, Jr.*

Supernovae and Nucleosynthesis: An Investigation of the History of Matter, from the Big Bang to the Present, *by David Arnett*

Unsolved Problems in Astrophysics, *edited by John N. Bahcall and Jeremiah P. Ostriker*

Galactic Astronomy, *by James Binney and Michael Merrifield*

Active Galactic Nuclei: From the Central Black Hole to the Galactic Environment, *by Julian H. Krolik*

Plasma Physics for Astrophysics, *by Russell M. Kulsrud*

Electromagnetic Processes, *by Robert J. Gould*

Conversations on Electric and Magnetic Fields in the Cosmos, *by Eugene N. Parker*

High-Energy Astrophysics, *by Fulvio Melia*

Stellar Spectral Classification, *by Richard O. Gray and Christopher J. Corbally*

Exoplanet Atmospheres: Physical Processes, *by Sara Seager*

Physics of the Interstellar and Intergalactic Medium, *by Bruce T. Draine*

The First Galaxies in the Universe, *by Abraham Loeb and Steven R. Furlanetto*

Exoplanetary Atmospheres: Theoretical Concepts and Foundations, *by Kevin Heng*

Magnetic Reconnection: A Modern Synthesis of Theory, Experiment, and Observations, *by Masaaki Yamada*

Physics of Binary Star Evolution: From Stars to X-ray Binaries and Gravitational Wave Sources, *by Thomas M. Tauris and Edward P.J. van den Heuvel*

Dynamics of Planetary Systems, *by Scott Tremaine*

Dynamics of Planetary Systems

Scott Tremaine

PRINCETON UNIVERSITY PRESS
PRINCETON AND OXFORD

Published by Princeton University Press
41 William Street, Princeton, New Jersey 08540
99 Banbury Road, Oxford OX2 6JX

press.princeton.edu

ISBN 9780691207124
ISBN (pbk.) 9780691207117
ISBN (e-book) 9780691244228

British Library Cataloging-in-Publication Data is available

Editorial: Ingrid Gnerlich & Whitney Rauenhorst
Production Editorial: Ali Parrington
Jacket/Cover Design: Wanda España
Production: Jacqueline Poirier
Publicity: Matthew Taylor & Charlotte Coyne

Cover image: The innermost Galilean satellite, Io, casts its shadow onto the cloud decks of Jupiter in this Voyager 1 mosaic, taken 4 March 1979, at a planet-spacecraft distance of 1 million km. Photograph by Ian Regan (source images courtesy of NASA/JPL, via the PDS Ring-Moon Systems Node's OPUS service).

This book has been composed in LaTeX. The publisher would like to acknowledge the author of this volume for providing the print-ready files from which this book was printed.

Printed on acid-free paper. ∞

Printed in the United States of America

10 9 8 7 6 5 4 3 2 1

Contents

Preface

The subject of this book, traditionally called celestial mechanics, is the oldest branch of theoretical physics. The publication in 1687 of the *Principia*, Newton's masterpiece on celestial mechanics, is widely regarded as the capstone of the Scientific Revolution. Since then, celestial mechanics has attracted the attention of many of the greatest physicists and mathematicians of the past several centuries, including Lagrange, Laplace, Gauss, Poincaré, Kolmogorov, and others. Concepts first explored in celestial mechanics are central to many if not most branches of physics, and its successful high-precision predictions of the motions of the planets have impacted disciplines as diverse as navigation and philosophy.

Celestial mechanics experienced a renaissance in the second half of the twentieth century. Starting in 1957, space flight created a demand for accurate and rapid orbit calculations as well as a need to understand the qualitative behavior of a wide variety of orbits. The development of high-speed digital computation enabled the study of classic problems in celestial mechanics with new tools. Advances in nonlinear dynamics and chaos theory provided new insights into the long-term behavior of orbits. Spacecraft visited every planet in the solar system and sent back data that dramatically expanded our understanding of the rich dynamics of their orbits, spins, and satellites. Finally, we have discovered thousands of planets outside the solar system, and celestial mechanics plays a central role in the analysis of the observations and the interpretation of their implications for the formation and evolution of planetary systems.

The primary goal of this book is to provide an introduction to celestial mechanics that reflects these developments. The reader is assumed to have an undergraduate background in classical mechanics and methods of mathematical physics (vectors, matrices, special functions, and so on), and much of what is needed is summarized in Appendixes B, C and D. The book contains most of the material that is needed for the reader to carry out research in the dynamics of planetary systems.

A book is defined in large part by what is left out, and a lot has been left out of this one. There is no analysis of spacecraft dynamics, except for a few examples and problems. There is almost no discussion of planet formation, since the tools that are needed to study this subject are mostly different from those of celestial mechanics. For similar reasons there is no discussion of planetary rings. Although general relativity offers a more accurate description of planetary motions than does Newtonian mechanics, its main use is in compiling high-accuracy planetary ephemerides and so it is only described briefly, in Appendix J. Perturbation theory for planets and satellites on nearly circular, nearly coplanar orbits was the main focus of celestial mechanics in the nineteenth and early twentieth centuries, but many of the problems for which this theory was needed can now be solved using computer algebra or numerical orbit integration; thus the topic is described in much less detail than in earlier books at this level. There is only limited discussion of the rich phenomenology of extrasolar planets, since this is a large and rapidly growing subject that deserves a book of its own.

There are problems at the end of the book, many of which are intended to elaborate on topics that are not covered fully in the main text. Some of the problems are more easily done using a computer algebra system.

The notation in the book is mostly standard. We regularly use the notation $f(x) = \mathrm{O}(x)$ to indicate that $|f(x)/x|$ is no larger than a constant value as $|x| \to \infty$. We assume that $0^0 = 1$, although most mathematical and scientific software treats it as undefined. The symbols $\simeq$ and $\sim$ are used to indicate approximate equality, with $\simeq$ suggesting higher accuracy than $\sim$. Vectors and matrices are denoted by boldface type ($\mathbf{a}$, $\mathbf{A}$) and operators by boldface sans-serif type ($\boldsymbol{\mathsf{A}}$). We usually do not distinguish row vectors from column vectors; thus we write $\mathbf{a} = (a_1, a_2, a_3)$, in which $\mathbf{a}$ is a row vector, as well as $\mathbf{Aa}$, in which the matrix $\mathbf{A}$ multiplies the column vector $\mathbf{a}$.

We are all indebted to the Smithsonian/NASA Astrophysics Data System, https://ui.adsabs.harvard.edu, and the arXiv e-print service, https://arxiv.org, which have revolutionized access to the astronomy literature. In large part thanks to their efforts, most of the literature referenced here can easily and freely be accessed on the web.

All of the plots were prepared using Matplotlib (Hunter 2007), and most of the orbit integrations were done using REBOUND (Rein & Liu 2012).

I have learned this subject largely through my colleagues, collaborators and students, including Eugene Chiang, Luke Dones, Subo Dong, Martin Duncan, Wyn Evans, Dan Fabrycky, Eric Ford, Jean-Baptiste Fouvry, Adrian Hamers, Julia Heisler, Kevin Heng, Matthew Holman, Mario Jurić, Boaz Katz, Jacques Laskar, Renu Malhotra, Norman Murray, Fathi Namouni, Annika Peter, Cristobal Petrovich, Gerald Quinlan, Thomas Quinn, Roman Rafikov, Nicole Rappaport, Hanno Rein, Prasenjit Saha, Kedron Silsbee, Aristotle Socrates, Serge Tabachnik, Dan Tamayo, Alar Toomre, Jihad Touma, Paul Wiegert, Jack Wisdom, Qingjuan Yu and Nadia Zakamska. I thank Hanno Rein, Renu Malhotra and her students, and especially Alar Toomre, who read and commented on large parts of the manuscript, as well as Alysa Obertas and David Vokrouhlický, who contributed their research results for the figures. Above all, I am indebted to Peter Goldreich, who introduced me to this subject. My long collaboration with him was one of the highlights of my research career.

Much of this book was completed at home during the pandemic that began in 2020. I am grateful to my wife Marilyn for her unswerving support for this project, without which it would neither have been started nor finished.

Dynamics of Planetary Systems

Chapter 1

The two-body problem

1.1 Introduction

The roots of celestial mechanics are two fundamental discoveries by Isaac Newton. First, in any inertial frame the acceleration of a body of mass m subjected to a force $\mathbf{F}$ is

$$\frac{\mathrm{d}^2\mathbf{r}}{\mathrm{d}t^2} = \frac{\mathbf{F}}{m}. \tag{1.1}$$

Second, the gravitational force exerted by a point mass m_1 at position $\mathbf{r}_1$ on a point mass m_0 at $\mathbf{r}_0$ is

$$\mathbf{F} = \frac{\mathbb{G}m_0 m_1(\mathbf{r}_1 - \mathbf{r}_0)}{|\mathbf{r}_1 - \mathbf{r}_0|^3}, \tag{1.2}$$

with $\mathbb{G}$ the gravitational constant.[1] Newton's laws have now been superseded by the equations of general relativity but remain accurate enough to describe all observable phenomena in planetary systems when they are supplemented by small relativistic corrections. A summary of the relevant effects of general relativity is given in Appendix J.

The simplest problem in celestial mechanics, solved by Newton but known as the **two-body problem** or the **Kepler problem**, is to determine

[1] For values of this and other constants, see Appendix A.

the orbits of two point masses ("particles") under the influence of their mutual gravitational attraction. This is the subject of the current chapter.[2]

The equations of motion for the particles labeled 0 and 1 are found by combining (1.1) and (1.2),

$$\frac{\mathrm{d}^2\mathbf{r}_0}{\mathrm{d}t^2} = \frac{\mathbb{G}m_1(\mathbf{r}_1 - \mathbf{r}_0)}{|\mathbf{r}_1 - \mathbf{r}_0|^3}, \quad \frac{\mathrm{d}^2\mathbf{r}_1}{\mathrm{d}t^2} = \frac{\mathbb{G}m_0(\mathbf{r}_0 - \mathbf{r}_1)}{|\mathbf{r}_0 - \mathbf{r}_1|^3}. \tag{1.3}$$

The total energy and angular momentum of the particles are

$$\begin{aligned} E_{\text{tot}} &= \tfrac{1}{2}m_0|\dot{\mathbf{r}}_0|^2 + \tfrac{1}{2}m_1|\dot{\mathbf{r}}_1|^2 - \frac{\mathbb{G}m_0m_1}{|\mathbf{r}_1 - \mathbf{r}_0|}, \\ \mathbf{L}_{\text{tot}} &= m_0\mathbf{r}_0 \times \dot{\mathbf{r}}_0 + m_1\mathbf{r}_1 \times \dot{\mathbf{r}}_1, \end{aligned} \tag{1.4}$$

in which we have introduced the notation $\dot{\mathbf{r}} = \mathrm{d}\mathbf{r}/\mathrm{d}t$. Using equations (1.3) it is straightforward to show that the total energy and angular momentum are conserved, that is, $\mathrm{d}E_{\text{tot}}/\mathrm{d}t = 0$ and $\mathrm{d}\mathbf{L}_{\text{tot}}/\mathrm{d}t = 0$.

We now change variables from $\mathbf{r}_0$ and $\mathbf{r}_1$ to

$$\mathbf{r}_{\text{cm}} \equiv \frac{m_0\mathbf{r}_0 + m_1\mathbf{r}_1}{m_0 + m_1}, \quad \mathbf{r} \equiv \mathbf{r}_1 - \mathbf{r}_0; \tag{1.5}$$

here $\mathbf{r}_{\text{cm}}$ is the **center of mass** or **barycenter** of the two particles and $\mathbf{r}$ is the **relative position**. These equations can be solved for $\mathbf{r}_0$ and $\mathbf{r}_1$ to yield

$$\mathbf{r}_0 = \mathbf{r}_{\text{cm}} - \frac{m_1}{m_0 + m_1}\mathbf{r}, \quad \mathbf{r}_1 = \mathbf{r}_{\text{cm}} + \frac{m_0}{m_0 + m_1}\mathbf{r}. \tag{1.6}$$

Taking two time derivatives of the first of equations (1.5) and using equations (1.3), we obtain

$$\frac{\mathrm{d}^2\mathbf{r}_{\text{cm}}}{\mathrm{d}t^2} = \mathbf{0}; \tag{1.7}$$

[2] Most of the basic material in the first part of this chapter can be found in textbooks on classical mechanics. The more advanced material in later sections and chapters has been treated in many books over more than two centuries. The most influential of these include Laplace (1799–1825), Tisserand (1889–1896), Poincaré (1892–1897), Plummer (1918), Brouwer & Clemence (1961) and Murray & Dermott (1999).

thus the center of mass travels at uniform velocity, a consequence of the absence of any external forces.

In these variables the total energy and angular momentum can be written

$$E_{\mathrm{tot}} = E_{\mathrm{cm}} + E_{\mathrm{rel}}, \quad \mathbf{L}_{\mathrm{tot}} = \mathbf{L}_{\mathrm{cm}} + \mathbf{L}_{\mathrm{rel}}, \tag{1.8}$$

where

$$\begin{aligned} E_{\mathrm{cm}} &= \tfrac{1}{2}M|\dot{\mathbf{r}}_{\mathrm{cm}}|^2, & \mathbf{L}_{\mathrm{cm}} &= M\mathbf{r}_{\mathrm{cm}} \times \dot{\mathbf{r}}_{\mathrm{cm}}, \\ E_{\mathrm{rel}} &= \tfrac{1}{2}\mu|\dot{\mathbf{r}}|^2 - \frac{\mathbb{G}\mu M}{|\mathbf{r}|}, & \mathbf{L}_{\mathrm{rel}} &= \mu\,\mathbf{r} \times \dot{\mathbf{r}}; \end{aligned} \tag{1.9}$$

here we have introduced the **reduced mass** and **total mass**

$$\mu \equiv \frac{m_0 m_1}{m_0 + m_1}, \quad M \equiv m_0 + m_1. \tag{1.10}$$

The terms E_{cm} and $\mathbf{L}_{\mathrm{cm}}$ are the kinetic energy and angular momentum associated with the motion of the center of mass. These are zero if we choose a reference frame in which the velocity of the center of mass $\dot{\mathbf{r}}_{\mathrm{cm}} = \mathbf{0}$. The terms E_{rel} and $\mathbf{L}_{\mathrm{rel}}$ are the energy and angular momentum associated with the relative motion of the two particles around the center of mass. These are the same as the energy and angular momentum of a particle of mass μ orbiting around a mass M (the "central body") that is fixed at the origin of the vector $\mathbf{r}$.

Taking two time derivatives of the second of equations (1.5) yields

$$\frac{\mathrm{d}^2\mathbf{r}}{\mathrm{d}t^2} = -\frac{\mathbb{G}M}{r^3}\mathbf{r} = -\frac{\mathbb{G}M}{r^2}\hat{\mathbf{r}}, \tag{1.11}$$

where $r = |\mathbf{r}|$ and the unit vector $\hat{\mathbf{r}} = \mathbf{r}/r$. Equation (1.11) describes any one of the following:

(i) the motion of a particle of arbitrary mass subject to the gravitational attraction of a central body of mass M that is fixed at the origin;

(ii) the motion of a particle of negligible mass (a **test particle**) under the influence of a freely moving central body of mass M;

(iii) the motion of a particle with mass equal to the reduced mass μ around a fixed central body that attracts it with the force $\mathbf{F}$ of equation (1.2).

Whatever the interpretation, the two-body problem has been reduced to a one-body problem.

Equation (1.11) can be derived from a Hamiltonian, as described in §1.4. It can also be written

$$\ddot{\mathbf{r}} = -\nabla\Phi_K, \tag{1.12}$$

where we have introduced the notation $\nabla f(\mathbf{r})$ for the gradient of the scalar function $f(\mathbf{r})$ (see §B.3 for a review of vector calculus). The function $\Phi_K(r) = -\mathbb{G}M/r$ is the **Kepler potential**. The solution of equations (1.11) or (1.12) is known as the **Kepler orbit**.

We begin the solution of equation (1.11) by evaluating the rate of change of the relative angular momentum $\mathbf{L}_{\text{rel}}$ from equation (1.9):

$$\frac{1}{\mu}\frac{\mathrm{d}\mathbf{L}_{\text{rel}}}{\mathrm{d}t} = \frac{\mathrm{d}\mathbf{r}}{\mathrm{d}t} \times \frac{\mathrm{d}\mathbf{r}}{\mathrm{d}t} + \mathbf{r} \times \frac{\mathrm{d}^2\mathbf{r}}{\mathrm{d}t^2} = -\frac{\mathbb{G}M}{r^2}\mathbf{r} \times \hat{\mathbf{r}} = \mathbf{0}. \tag{1.13}$$

Thus the relative angular momentum is conserved. Moreover, since $\mathbf{L}_{\text{rel}}$ is normal to the plane containing the test particle's position and velocity vectors, the position vector must remain in a fixed plane, the **orbital plane**. The plane of the Earth's orbit around the Sun is called the **ecliptic**, and the directions perpendicular to this plane are called the north and south ecliptic poles.

We now simplify our notation. Since we can always choose an inertial reference frame in which the center-of-mass angular momentum $\mathbf{L}_{\text{cm}} = \mathbf{0}$ for all time, we usually shorten "relative angular momentum" to "angular momentum." Similarly the "relative energy" E_{rel} is shortened to "energy." We usually work with the angular momentum per unit mass $\mathbf{L}_{\text{rel}}/\mu = \mathbf{r} \times \dot{\mathbf{r}}$ and the energy per unit mass $\frac{1}{2}|\dot{\mathbf{r}}|^2 - \mathbb{G}M/|\mathbf{r}|$. These may be called "specific angular momentum" and "specific energy," but we shall just write "angular momentum" or "energy" when the intended meaning is clear. Moreover we typically use the same symbol—$\mathbf{L}$ for angular momentum and E for energy—whether we are referring to the total quantity or the quantity per unit mass. This casual use of the same notation for two different physical quantities is less dangerous than it may seem, because the intended meaning can always be deduced from the units of the equations.

1.2 The shape of the Kepler orbit

We let (r, ψ) denote polar coordinates in the orbital plane, with ψ increasing in the direction of motion of the orbit. If $\mathbf{r}$ is a vector in the orbital plane, then $\mathbf{r} = r\hat{\mathbf{r}}$ where $(\hat{\mathbf{r}}, \hat{\boldsymbol{\psi}})$ are unit vectors in the radial and azimuthal directions. The acceleration vector lies in the orbital plane and is given by equation (B.18),

$$\ddot{\mathbf{r}} = (\ddot{r} - r\dot{\psi}^2)\hat{\mathbf{r}} + (2\dot{r}\dot{\psi} + r\ddot{\psi})\hat{\boldsymbol{\psi}}, \tag{1.14}$$

so the equations of motion (1.12) become

$$\ddot{r} - r\dot{\psi}^2 = -\frac{\mathrm{d}\Phi_\mathrm{K}(r)}{\mathrm{d}r}, \quad 2\dot{r}\dot{\psi} + r\ddot{\psi} = 0. \tag{1.15}$$

The second equation may be multiplied by r and integrated to yield

$$r^2\dot{\psi} = \text{constant} = L, \tag{1.16}$$

where $L = |\mathbf{L}|$. This is just a restatement of the conservation of angular momentum, equation (1.13).

We may use equation (1.16) to eliminate $\dot{\psi}$ from the first of equations (1.15),

$$\ddot{r} - \frac{L^2}{r^3} = -\frac{\mathrm{d}\Phi_\mathrm{K}}{\mathrm{d}r}. \tag{1.17}$$

Multiplying by $\dot{r}$ and integrating yields

$$\tfrac{1}{2}\dot{r}^2 + \frac{L^2}{2r^2} + \Phi_\mathrm{K}(r) = E, \tag{1.18}$$

where E is a constant that is equal to the energy per unit mass of the test particle. Equation (1.18) can be rewritten as

$$\tfrac{1}{2}v^2 - \frac{\mathbb{G}M}{r} = E, \tag{1.19}$$

where $v = (\dot{r}^2 + r^2\dot{\psi}^2)^{1/2}$ is the speed of the test particle.

Equation (1.18) implies that

$$\dot{r}^2 = 2E + \frac{2\mathbb{G}M}{r} - \frac{L^2}{r^2}. \tag{1.20}$$

As $r \to 0$, the right side approaches $-L^2/r^2$, which is negative, while the left side is positive. Thus there must be a point of closest approach of the test particle to the central body, known as the **periapsis** or **pericenter**.[3] In the opposite limit, $r \to \infty$, the right side of equation (1.20) approaches $2E$. Since the left side is positive, when $E < 0$ there is a maximum distance that the particle can achieve, known as the **apoapsis** or **apocenter**. Orbits with $E < 0$ are referred to as **bound** orbits since there is an upper limit to their distance from the central body. Orbits with $E > 0$ are **unbound** or **escape** orbits; they have no apoapsis, and particles on such orbits eventually travel arbitrarily far from the central body, never to return.[4]

The periapsis distance q and apoapsis distance Q of an orbit are determined by setting $\dot{r} = 0$ in equation (1.20), which yields the quadratic equation

$$2Er^2 + 2\mathbb{G}Mr - L^2 = 0. \tag{1.22}$$

For bound orbits, $E < 0$, there are two roots on the positive real axis,

$$q = \frac{\mathbb{G}M - \left[(\mathbb{G}M)^2 + 2EL^2\right]^{1/2}}{2|E|}, \quad Q = \frac{\mathbb{G}M + \left[(\mathbb{G}M)^2 + 2EL^2\right]^{1/2}}{2|E|}. \tag{1.23}$$

For unbound orbits, $E > 0$, there is only one root on the positive real axis,

$$q = \frac{\left[(\mathbb{G}M)^2 + 2EL^2\right]^{1/2} - \mathbb{G}M}{2E}. \tag{1.24}$$

[3] For specific central bodies other names are used, such as perihelion (Sun), perigee (Earth), periastron (a star), and so forth. "Periapse" is incorrect—an apse is not an apsis.

[4] The **escape speed** $v_{\rm esc}$ from an object is the minimum speed needed for a test particle to escape from its surface; if the object is spherical, with mass M and radius R, equation (1.19) implies that

$$v_{\rm esc} = \left(\frac{2\mathbb{G}M}{R}\right)^{1/2}. \tag{1.21}$$

To solve the differential equation (1.17) we introduce the variable $u \equiv 1/r$, and change the independent variable from t to ψ using the relation

$$\frac{\mathrm{d}}{\mathrm{d}t} = \dot{\psi}\frac{\mathrm{d}}{\mathrm{d}\psi} = Lu^2\frac{\mathrm{d}}{\mathrm{d}\psi}. \tag{1.25}$$

With these substitutions, $\dot{r} = -L\mathrm{d}u/\mathrm{d}\psi$ and $\ddot{r} = -L^2u^2\mathrm{d}^2u/\mathrm{d}\psi^2$, so equation (1.17) becomes

$$\frac{\mathrm{d}^2u}{\mathrm{d}\psi^2} + u = -\frac{1}{L^2}\frac{\mathrm{d}\Phi_\mathrm{K}}{\mathrm{d}u}. \tag{1.26}$$

Since $\Phi_\mathrm{K}(r) = -\mathbb{G}M/r = -\mathbb{G}Mu$ the right side is equal to a constant, $\mathbb{G}M/L^2$, and the equation is easily solved to yield

$$u = \frac{1}{r} = \frac{\mathbb{G}M}{L^2}[1 + e\cos(\psi - \varpi)], \tag{1.27}$$

where $e \geq 0$ and ϖ are constants of integration.[5] We replace the angular momentum L by another constant of integration, a, defined by the relation

$$L^2 = \mathbb{G}Ma(1 - e^2), \tag{1.28}$$

so the shape of the orbit is given by

$$r = \frac{a(1 - e^2)}{1 + e\cos f}, \tag{1.29}$$

where $f = \psi - \varpi$ is known as the **true anomaly**.[6]

The closest approach of the two bodies occurs at $f = 0$ or azimuth $\psi = \varpi$ and hence ϖ is known as the **longitude of periapsis**. The periapsis distance is $r(f = 0)$ or

$$q = a(1 - e). \tag{1.30}$$

[5] The symbol ϖ is a variant of the symbol for the Greek letter π, even though it looks more like the symbol for the letter ω; hence it is sometimes informally called "pomega."

[6] In a subject as old as this, there is a rich specialized vocabulary. The term "anomaly" refers to any angular variable that is zero at periapsis and increases by 2π as the particle travels from periapsis to apoapsis and back. There are also several old terms we shall not use: "semilatus rectum" for the combination $a(1 - e^2)$, "vis viva" for kinetic energy, and so on.

When the eccentricity is zero, the longitude of periapsis ϖ is undefined. This indeterminacy can drastically slow or halt numerical calculations that follow the evolution of the orbital elements, and can be avoided by replacing e and ϖ by two new elements, the **eccentricity components** or h **and** k **variables**

$$k \equiv e \cos \varpi, \quad h \equiv e \sin \varpi, \tag{1.31}$$

which are well defined even for $e = 0$. The generalization to nonzero inclination is given in equations (1.71).

Substituting q for r in equation (1.22) and replacing L^2 using equation (1.28) reveals that the energy per unit mass is simply related to the constant a:

$$E = -\frac{\mathbb{G}M}{2a}. \tag{1.32}$$

First consider bound orbits, which have $E < 0$. Then $a > 0$ by equation (1.32) and hence $e < 1$ by equation (1.28). A circular orbit has $e = 0$ and angular momentum per unit mass $L = (\mathbb{G}Ma)^{1/2}$. The circular orbit has the largest possible angular momentum for a given semimajor axis or energy, so we sometimes write

$$\mathbf{j} \equiv \frac{\mathbf{L}}{(\mathbb{G}Ma)^{1/2}}, \quad \text{where} \quad j = |\mathbf{j}| = (1 - e^2)^{1/2} \tag{1.33}$$

ranges from 0 to 1 and represents a dimensionless angular momentum at a given semimajor axis.

The apoapsis distance, obtained from equation (1.29) with $f = \pi$, is

$$Q = a(1 + e). \tag{1.34}$$

The periapsis and the apoapsis are joined by a straight line known as the **line of apsides**. Equation (1.29) describes an ellipse with one focus at the origin (**Kepler's first law**). Its major axis is the line of apsides and has length $q + Q = 2a$; thus the constant a is known as the **semimajor axis**. The **semiminor axis** of the ellipse is the maximum perpendicular distance of the orbit from the line of apsides, $b = \max_f[a(1 - e^2)\sin f/(1 + e\cos f)] = a(1 - e^2)^{1/2}$. The **eccentricity** of the ellipse, $(1 - b^2/a^2)^{1/2}$, is therefore equal to the constant e.

Box 1.1: The eccentricity vector

The **eccentricity vector** offers a more elegant but less transparent derivation of the equation for the shape of a Kepler orbit. Take the cross product of $\mathbf{L}$ with equation (1.11),

$$\mathbf{L} \times \ddot{\mathbf{r}} = -\frac{\mathbb{G}M}{r^3}\mathbf{L} \times \mathbf{r}. \tag{a}$$

Using the vector identity (B.9b), $\mathbf{L} \times \mathbf{r} = -\mathbf{r} \times \mathbf{L} = -\mathbf{r} \times (\mathbf{r} \times \dot{\mathbf{r}}) = r^2\dot{\mathbf{r}} - (\mathbf{r} \cdot \dot{\mathbf{r}})\mathbf{r}$, which is equal to $r^3 \mathrm{d}\hat{\mathbf{r}}/\mathrm{d}t$. Thus

$$\mathbf{L} \times \ddot{\mathbf{r}} = -\mathbb{G}M\frac{\mathrm{d}\hat{\mathbf{r}}}{\mathrm{d}t}. \tag{b}$$

Since $\mathbf{L}$ is constant, we may integrate to obtain

$$\mathbf{L} \times \dot{\mathbf{r}} = -\mathbb{G}M(\hat{\mathbf{r}} + \mathbf{e}), \tag{c}$$

where $\mathbf{e}$ is a vector constant of motion, the **eccentricity vector**. Rearranging equation (c), we have

$$\mathbf{e} = \frac{\dot{\mathbf{r}} \times (\mathbf{r} \times \dot{\mathbf{r}})}{\mathbb{G}M} - \frac{\mathbf{r}}{r}. \tag{d}$$

To derive the shape of the orbit, we take the dot product of (c) with $\hat{\mathbf{r}}$ and use the vector identity (B.9a) to show that $\hat{\mathbf{r}} \cdot (\mathbf{L} \times \dot{\mathbf{r}}) = -L^2/r$. The resulting formula is

$$r = \frac{L^2}{\mathbb{G}M}\frac{1}{1 + \mathbf{e} \cdot \hat{\mathbf{r}}} = \frac{a(1-e^2)}{1 + \mathbf{e} \cdot \hat{\mathbf{r}}}; \tag{e}$$

in the last equation we have eliminated L^2 using equation (1.28). This result is the same as equation (1.29) if the magnitude of the eccentricity vector equals the eccentricity, $|\mathbf{e}| = e$, and the eccentricity vector points toward periapsis.

The eccentricity vector is often called the **Runge–Lenz vector**, although its history can be traced back at least to Laplace (Goldstein 1975–1976). Runge and Lenz appear to have taken their derivation from Gibbs & Wilson (1901), the classic text that introduced modern vector notation.

Unbound orbits have $E > 0$, $a < 0$ and $e > 1$. In this case equation (1.29) describes a hyperbola with focus at the origin and asymptotes at azimuth

$$\psi = \varpi \pm f_\infty, \quad \text{where} \quad f_\infty \equiv \cos^{-1}(-1/e) \tag{1.35}$$

is the **asymptotic true anomaly**, which varies between π (for $e = 1$) and $\frac{1}{2}\pi$ (for $e \to \infty$). The constants a and e are still commonly referred to as semimajor axis and eccentricity even though these terms have no direct geometric interpretation.

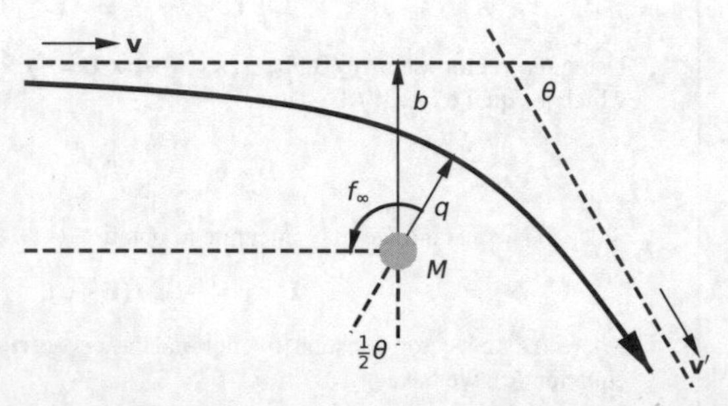

Figure 1.1: The geometry of an unbound or hyperbolic orbit around mass M. The impact parameter is b, the deflection angle is θ, the asymptotic true anomaly is f_∞, and the periapsis is located at the tip of the vector $\mathbf{q}$.

Suppose that a particle is on an unbound orbit around a mass M. Long before the particle approaches M, it travels at a constant velocity which we denote by $\mathbf{v}$ (Figure 1.1). If there were no gravitational forces, the particle would continue to travel in a straight line that makes its closest approach to M at a point $\mathbf{b}$ called the **impact parameter vector**. Long after the particle passes M, it again travels at a constant velocity $\mathbf{v}'$, where $v \equiv |\mathbf{v}| = |\mathbf{v}'|$ because of energy conservation. The deflection angle θ is the angle between $\mathbf{v}$ and $\mathbf{v}'$, given by $\cos\theta = \mathbf{v}\cdot\mathbf{v}'/v^2$. The deflection angle is related to the asymptotic true anomaly f_∞ by $\theta = 2f_\infty - \pi$; then

$$\tan\tfrac{1}{2}\theta = -\frac{\cos f_\infty}{\sin f_\infty} = \frac{1}{(e^2-1)^{1/2}}. \tag{1.36}$$

The relation between the pre- and post-encounter velocities can be written

$$\mathbf{v}' = \mathbf{v}\cos\theta - \hat{\mathbf{b}}v\sin\theta. \tag{1.37}$$

In many cases the properties of unbound orbits are best described by the asymptotic speed v and the impact parameter $b = |\mathbf{b}|$, rather than by orbital elements such as a and e. It is straightforward to show that the angular momentum and energy of the orbit per unit mass are $L = bv$ and $E = \frac{1}{2}v^2$. From equations (1.28) and (1.32) it follows that

$$a = -\frac{\mathbb{G}M}{v^2}, \quad e^2 = 1 + \frac{b^2v^4}{(\mathbb{G}M)^2}. \tag{1.38}$$

Then from equation (1.36),

$$\tan\tfrac{1}{2}\theta = \frac{\mathbb{G}M}{bv^2}. \tag{1.39}$$

The periapsis distance $q = a(1-e)$ is related to the impact parameter b by

$$q = \frac{\mathbb{G}M}{v^2}\left[\left(1 + \frac{b^2v^4}{\mathbb{G}^2M^2}\right)^{1/2} - 1\right] \quad \text{or} \quad b^2 = q^2 + \frac{2\,\mathbb{G}Mq}{v^2}. \tag{1.40}$$

Thus, for example, if the central body has radius R, the particle will collide with it if

$$b^2 \le b_{\text{coll}}^2 \equiv R^2 + \frac{2\,\mathbb{G}MR}{v^2}. \tag{1.41}$$

The corresponding cross section is πb_{coll}^2. If the central body has zero mass the cross section is just πR^2; the enhancement arising from the second term in equation (1.41) is said to be due to **gravitational focusing**.

In the special case $E = 0$, a is infinite and $e = 1$, so equation (1.29) is undefined; however, in this case equation (1.22) implies that the periapsis distance $q = L^2/(2\,\mathbb{G}M)$, so equation (1.27) implies

$$r = \frac{2q}{1 + \cos f}, \tag{1.42}$$

which describes a parabola. This result can also be derived from equation (1.29) by replacing $a(1-e^2)$ by $q(1+e)$ and letting $e \to 1$.

1.3 Motion in the Kepler orbit

The **period** P of a bound orbit is the time taken to travel from periapsis to apoapsis and back. Since $\mathrm{d}\psi/\mathrm{d}t = L/r^2$, we have $\int_{t_1}^{t_2} \mathrm{d}t = L^{-1}\int_{\psi_1}^{\psi_2} r^2 \mathrm{d}\psi$; the integral on the right side is twice the area contained in the ellipse between azimuths ψ_1 and ψ_2, so the radius vector to the particle sweeps out equal areas in equal times (**Kepler's second law**). Thus $P = 2A/L$, where the area of the ellipse is $A = \pi ab$ with a and $b = a(1-e^2)^{1/2}$ the semimajor and semiminor axes of the ellipse. Combining these results with equation (1.28), we find

$$P = 2\pi\left(\frac{a^3}{\mathbb{G}M}\right)^{1/2}. \tag{1.43}$$

The period, like the energy, depends only on the semimajor axis. The **mean motion** or mean rate of change of azimuth, usually written n and equal to $2\pi/P$, thus satisfies[7]

$$n^2a^3 = \mathbb{G}M, \tag{1.44}$$

which is **Kepler's third law** or simply **Kepler's law**. If the particle passes through periapsis at $t = t_0$, the dimensionless variable

$$\ell = 2\pi\frac{t-t_0}{P} = n(t-t_0) \tag{1.45}$$

is called the **mean anomaly**. Notice that the mean anomaly equals the true anomaly f when $f = 0, \pi, 2\pi, \ldots$ but not at other phases unless the orbit is circular; similarly, ℓ and f always lie in the same semicircle (0 to π, π to 2π, and so on).

[7] The relation $n = 2\pi/P$ holds because Kepler orbits are closed—that is, they return to the same point once per orbit. In more general spherical potentials we must distinguish the **radial period**, the time between successive periapsis passages, from the **azimuthal period** $2\pi/n$. For example, in a harmonic potential $\Phi(r) = \frac{1}{2}\omega^2r^2$ the radial period is π/ω but the azimuthal period is $2\pi/\omega$. Smaller differences between the radial and azimuthal period arise in perturbed Kepler systems such as multi-planet systems or satellites orbiting a flattened planet (§1.8.3). For the Earth the radial period is called the **anomalistic year**, while the azimuthal period of 365.256 363 d is the **sidereal year**. The anomalistic year is longer than the sidereal year by 0.003 27 d. When we use the term "year" in this book, we refer to the Julian year of exactly 365.25 d (§1.5).

The position and velocity of a particle in the orbital plane at a given time are determined by four **orbital elements**: two specify the size and shape of the orbit, which we can take to be e and a (or e and n, q and Q, L and E, and so forth); one specifies the orientation of the line of apsides (ϖ); and one specifies the location or phase of the particle in its orbit (f, ℓ, or t_0).

The trajectory $[r(t), \psi(t)]$ can be derived by solving the differential equation (1.20) for $r(t)$, then (1.16) for $\psi(t)$. However, there is a simpler method.

First consider bound orbits. Since the radius of a bound orbit oscillates between $a(1 - e)$ and $a(1 + e)$, it is natural to define a variable $u(t)$, the **eccentric anomaly**, by

$$r = a(1 - e\cos u); \tag{1.46}$$

since the cosine is multivalued, we must add the supplemental condition that u and f always lie in the same semicircle (0 to π, π to 2π, and so on). Thus u increases from 0 to 2π as the particle travels from periapsis to apoapsis and back. The true, eccentric and mean anomalies f, u and ℓ are all equal for circular orbits, and for any bound orbit $f = u = \ell = 0$ at periapsis and π at apoapsis.

Substituting equation (1.46) into the energy equation (1.20) and using equations (1.28) and (1.32) for L^2 and E, we obtain

$$\dot{r}^2 = a^2e^2\sin^2 u\,\dot{u}^2 = -\frac{\mathbb{G}M}{a} + \frac{2\mathbb{G}M}{a(1 - e\cos u)} - \frac{\mathbb{G}M(1 - e^2)}{a(1 - e\cos u)^2}, \tag{1.47}$$

which simplifies to

$$(1 - e\cos u)^2\dot{u}^2 = \frac{\mathbb{G}M}{a^3} = n^2 = \dot{\ell}^2. \tag{1.48}$$

Since $\dot{u}, \dot{\ell} > 0$ and $u = \ell = 0$ at periapsis, we may take the square root of this equation and then integrate to obtain **Kepler's equation**

$$\ell = u - e\sin u. \tag{1.49}$$

Kepler's equation cannot be solved analytically for u, but many efficient numerical methods of solution are available.

The relation between the true and eccentric anomalies is found by eliminating r from equations (1.29) and (1.46):

$$\cos f = \frac{\cos u - e}{1 - e\cos u}, \quad \cos u = \frac{\cos f + e}{1 + e\cos f}, \tag{1.50}$$

with the understanding that f and u always lie in the same semicircle. Similarly,

$$\sin f = \frac{(1-e^2)^{1/2}\sin u}{1 - e\cos u}, \quad \sin u = \frac{(1-e^2)^{1/2}\sin f}{1 + e\cos f}, \tag{1.51a}$$

$$\tan\tfrac{1}{2}f = \left(\frac{1+e}{1-e}\right)^{1/2}\tan\tfrac{1}{2}u, \tag{1.51b}$$

$$\exp(\mathrm{i}f) = \frac{\exp(\mathrm{i}u) - \beta}{1 - \beta\exp(\mathrm{i}u)}, \quad \exp(\mathrm{i}u) = \frac{\exp(\mathrm{i}f) + \beta}{1 + \beta\exp(\mathrm{i}f)}, \tag{1.51c}$$

where

$$\beta \equiv \frac{1 - (1-e^2)^{1/2}}{e}. \tag{1.52}$$

If we assume that the periapsis lies on the x-axis of a rectangular coordinate system in the orbital plane, the coordinates of the particle are

$$x = r\cos f = a(\cos u - e), \quad y = r\sin f = a(1-e^2)^{1/2}\sin u. \tag{1.53}$$

The position and velocity of a bound particle at a given time t can be determined from the orbital elements a, e, ϖ and t_0 by the following steps. First compute the mean motion n from Kepler's third law (1.44), then find the mean anomaly ℓ from (1.45). Solve Kepler's equation for the eccentric anomaly u. The radius r is then given by equation (1.46); the true anomaly f is given by equation (1.50); and the azimuth $\psi = f + \varpi$. The radial velocity is

$$v_r = \dot{r} = n\frac{\mathrm{d}r}{\mathrm{d}\ell} = n\frac{\mathrm{d}r/\mathrm{d}u}{\mathrm{d}\ell/\mathrm{d}u} = \frac{nae\sin u}{1 - e\cos u} = \frac{nae\sin f}{(1-e^2)^{1/2}}, \tag{1.54}$$

and the azimuthal velocity is

$$v_\psi = r\dot{\psi} = \frac{L}{r} = na\frac{(1-e^2)^{1/2}}{1 - e\cos u} = na\frac{1 + e\cos f}{(1-e^2)^{1/2}}, \tag{1.55}$$

$$\langle (r/a)^2 \sin^2 f \rangle = \tfrac{1}{2} - \tfrac{1}{2}e^2, \tag{1.65e}$$

$$\langle (r/a)^2 \cos f \sin f \rangle = 0. \tag{1.65f}$$

Equation (1.64) gives

$$\langle (a/r)^2 \rangle = (1 - e^2)^{-1/2}, \tag{1.66a}$$

$$\langle (a/r)^3 \rangle = (1 - e^2)^{-3/2}, \tag{1.66b}$$

$$\langle (a/r)^3 \cos^2 f \rangle = \tfrac{1}{2}(1 - e^2)^{-3/2}, \tag{1.66c}$$

$$\langle (a/r)^3 \sin^2 f \rangle = \tfrac{1}{2}(1 - e^2)^{-3/2}, \tag{1.66d}$$

$$\langle (a/r)^3 \sin f \cos f \rangle = 0. \tag{1.66e}$$

Additional orbit averages are given in Problems 1.2 and 1.3.

1.3.2 Motion in three dimensions

So far we have described the motion of a particle in its orbital plane. To characterize the orbit fully we must also specify the spatial orientation of the orbital plane, as shown in Figure 1.2.

We work with the usual Cartesian coordinates (x, y, z) and spherical coordinates (r, θ, ϕ) (see Appendix B.2). We call the plane $z = 0$, corresponding to $\theta = \frac{1}{2}\pi$, the **reference plane**. The **inclination** of the orbital plane to the reference plane is denoted I, with $0 \le I \le \pi$; thus $\cos I = \hat{\mathbf{z}} \cdot \hat{\mathbf{L}}$, where $\hat{\mathbf{z}}$ and $\hat{\mathbf{L}}$ are unit vectors in the direction of the z-axis and the angular-momentum vector. Orbits with $0 \le I \le \frac{1}{2}\pi$ are **direct** or **prograde**; orbits with $\frac{1}{2}\pi < I < \pi$ are **retrograde**.

Any bound Kepler orbit pierces the reference plane at two points known as the **nodes** of the orbit. The particle travels upward ($\dot{z} > 0$) at the **ascending node** and downward at the **descending node**. The azimuthal angle ϕ of the ascending node is denoted Ω and is called the **longitude of the ascending node**. The angle from ascending node to periapsis, measured in the direction of motion of the particle in the orbital plane, is denoted ω and is called the **argument of periapsis**.

An unfortunate feature of these elements is that neither ω nor Ω is defined for orbits in the reference plane ($I = 0$). Partly for this reason, the

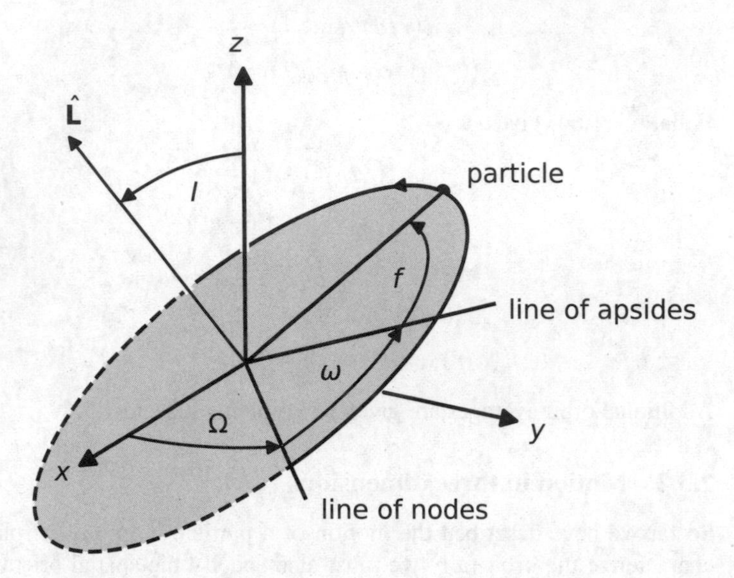

Figure 1.2: The angular elements of a Kepler orbit. The usual Cartesian coordinate axes are denoted by (x, y, z), the reference plane is $z = 0$, and the orbital plane is denoted by a solid curve above the equatorial plane ($z > 0$) and a dashed curve below. The plot shows the inclination I, the longitude of the ascending node Ω, the argument of periapsis ω and the true anomaly f.

argument of periapsis is often replaced by a variable called the **longitude of periapsis** which is defined as

$$\varpi \equiv \Omega + \omega. \tag{1.67}$$

For orbits with zero inclination, the longitude of periapsis has a simple interpretation—it is the azimuthal angle between the x-axis and the periapsis, consistent with our earlier definition of the same symbol following equation (1.29)—but if the inclination is nonzero, it is the sum of two angles measured in different planes (the reference plane and the orbital plane).[8] Despite this awkwardness, for most purposes the three elements (Ω, ϖ, I) provide the most convenient way to specify the orientation of a Kepler orbit.

The **mean longitude** is

$$\lambda \equiv \varpi + \ell = \Omega + \omega + \ell, \tag{1.68}$$

where ℓ is the mean anomaly; like the longitude of perihelion, the mean longitude is the sum of angles measured in the reference plane (Ω) and the orbital plane ($\omega + \ell$).

Some of these elements are closely related to the Euler angles that describe the rotation of one coordinate frame into another (Appendix B.6). Let (x', y', z') be Cartesian coordinates in the **orbital reference frame**, defined such that the z'-axis points along the angular-momentum vector $\mathbf{L}$ and the x'-axis points toward periapsis, along the eccentricity vector $\mathbf{e}$. Then the rotation from the (x, y, z) reference frame to the orbital reference frame is described by the Euler angles

$$(\alpha, \beta, \gamma) = (\Omega, I, \omega). \tag{1.69}$$

The position and velocity of a particle in space at a given time t are specified by six orbital elements: two specify the size and shape of the orbit (e and a); three specify the orientation of the orbit (I, Ω and ω), and one specifies the location of the particle in the orbit (f, u, ℓ, λ, or t_0). Thus, for example, to find the Cartesian coordinates (x, y, z) in terms of the orbital elements, we write the position in the orbital reference frame as $(x', y', z') = r(\cos f, \sin f, 0)$ and use equation (1.69) and the rotation matrix for the transformation from primed to unprimed coordinates (eq. B.61):

$$\begin{aligned}
\frac{x}{r} &= \cos\Omega\cos(f+\omega) - \cos I \sin\Omega\sin(f+\omega), \\
\frac{y}{r} &= \sin\Omega\cos(f+\omega) + \cos I \cos\Omega\sin(f+\omega), \\
\frac{z}{r} &= \sin I \sin(f+\omega);
\end{aligned} \tag{1.70}$$

[8] Thus "longitude of periapsis" is a misnomer, since ϖ is *not* equal to the azimuthal angle of the eccentricity vector, except for orbits of zero inclination.

here r is given in terms of the orbital elements by equation (1.29).

When the eccentricity or inclination is small, the polar coordinate pairs e–ϖ and I–Ω are sometimes replaced by the eccentricity and inclination components[9]

$$k \equiv e\cos\varpi, \quad h \equiv e\sin\varpi, \quad q \equiv \tan I\cos\Omega, \quad p \equiv \tan I\sin\Omega. \tag{1.71}$$

The first two equations are the same as equations (1.31).

For some purposes the shape, size and orientation of an orbit can be described most efficiently using the angular-momentum and eccentricity vectors, $\mathbf{L}$ and $\mathbf{e}$. The two vectors describe five of the six orbital elements: the missing element is the one specifying the location of the particle in its orbit, f, u, ℓ, λ or t_0 (the six components of the two vectors determine only five elements, because $\mathbf{e}$ is restricted to the plane normal to $\mathbf{L}$).

Note that ω and Ω are undefined for orbits with zero inclination; ω and ϖ are undefined for circular orbits; and ϖ, Ω and I are undefined for radial orbits ($e \to 1$). In contrast the angular-momentum and eccentricity vectors are well defined for *all* orbits. The cost of avoiding indeterminacy is redundancy: instead of five orbital elements we need six vector components.

1.3.3 Gauss's f and g functions

A common task is to determine the position and velocity, $\mathbf{r}(t)$ and $\mathbf{v}(t)$, of a particle in a Kepler orbit given its position and velocity $\mathbf{r}_0$ and $\mathbf{v}_0$ at some initial time t_0. This can be done by converting $\mathbf{r}_0$ and $\mathbf{v}_0$ into the orbital elements $a, e, I, \omega, \Omega, \ell_0$, replacing ℓ_0 by $\ell = \ell_0 + n(t - t_0)$ and then reversing the conversion to determine the position and velocity from the new orbital elements. But there is a simpler method, due to Gauss.

Since the particle is confined to the orbital plane, and $\mathbf{r}_0, \mathbf{v}_0$ are vectors lying in this plane, we can write

$$\mathbf{r}(t) = f(t, t_0)\mathbf{r}_0 + g(t, t_0)\mathbf{v}_0, \tag{1.72}$$

[9] The function $\tan I$ in the elements q and p can be replaced by any function that is proportional to I as $I \to 0$. Various authors use I, $\sin\frac{1}{2}I$, and so forth. The function $\sin I$ is not used because it has the same value for I and $\pi - I$.

which defines **Gauss's f and g functions**. This expression also gives the velocity of the particle,

$$\mathbf{v}(t) = \frac{\partial f(t,t_0)}{\partial t}\mathbf{r}_0 + \frac{\partial g(t,t_0)}{\partial t}\mathbf{v}_0. \tag{1.73}$$

To evaluate f and g for bound orbits we use polar coordinates (r,ψ) and Cartesian coordinates (x,y) in the orbital plane, and assume that $\mathbf{r}_0$ lies along the positive x-axis ($\psi_0 = 0$). Then the components of equation (1.72) along the x- and y-axes are:

$$\begin{aligned} r(t)\cos\psi(t) &= f(t,t_0)r_0 + g(t,t_0)v_r(t_0), \\ r(t)\sin\psi(t) &= g(t,t_0)v_\psi(t_0), \end{aligned} \tag{1.74}$$

where v_r and v_ψ are the radial and azimuthal velocities. These equations can be solved for f and g:

$$\begin{aligned} f(t,t_0) &= \frac{r(t)}{r_0}\left[\cos\psi(t) - \frac{v_r(t_0)}{v_\psi(t_0)}\sin\psi(t)\right], \\ g(t,t_0) &= \frac{r(t)}{v_\psi(t_0)}\sin\psi(t). \end{aligned} \tag{1.75}$$

We use equations (1.16), (1.28), (1.29), (1.54) and the relation $\psi = f - f_0$ to replace the quantities on the right sides by orbital elements (unfortunately f is used to denote both true anomaly and one of Gauss's functions). The result is

$$\begin{aligned} f(t,t_0) &= \frac{\cos(f-f_0) + e\cos f}{1+e\cos f}, \\ g(t,t_0) &= \frac{(1-e^2)^{3/2}\sin(f-f_0)}{n(1+e\cos f)(1+e\cos f_0)}. \end{aligned} \tag{1.76}$$

Since these expressions contain only the orbital elements n, e and f, they are valid in any coordinate system, not just the one we used for the derivation. For deriving velocities from equation (1.73), we need

$$\frac{\partial f(t,t_0)}{\partial t} = n\frac{e\sin f_0 - e\sin f - \sin(f-f_0)}{(1-e^2)^{3/2}},$$

$$\frac{\partial g(t,t_0)}{\partial t} = \frac{e\cos f_0 + \cos(f - f_0)}{1 + e\cos f_0}. \tag{1.77}$$

The f and g functions can also be expressed in terms of the eccentric anomaly, using equations (1.50) and (1.51a):

$$\begin{aligned}
f(t,t_0) &= \frac{\cos(u-u_0) - e\cos u_0}{1 - e\cos u_0}, \\
g(t,t_0) &= \frac{1}{n}[\sin(u-u_0) - e\sin u + e\sin u_0], \\
\frac{\partial f(t,t_0)}{\partial t} &= -\frac{n\sin(u-u_0)}{(1-e\cos u)(1-e\cos u_0)}, \\
\frac{\partial g(t,t_0)}{\partial t} &= \frac{\cos(u-u_0) - e\cos u}{1 - e\cos u}.
\end{aligned} \tag{1.78}$$

To compute $\mathbf{r}(t)$, $\mathbf{v}(t)$ from $\mathbf{r}_0 \equiv \mathbf{r}(t_0)$, $\mathbf{v}_0 = \mathbf{v}(t_0)$ we use the following procedure. From equations (1.19) and (1.32) we have

$$\frac{1}{a} = \frac{2}{r} - \frac{v^2}{\mathbb{G}M}; \tag{1.79}$$

so we can compute the semimajor axis a from $r_0 = |\mathbf{r}_0|$ and $v_0 = |\mathbf{v}_0|$. Then Kepler's law (1.44) yields the mean motion n. The total angular momentum is $L = |\mathbf{r}_0 \times \mathbf{v}_0|$ and this yields the eccentricity e through equation (1.28). To determine the eccentric anomaly at t_0, we use equation (1.46) which determines $\cos u_0$, and then determine the quadrant of u_0 by observing that the radial velocity $\dot{r}$ is positive when $0 < u_0 < \pi$ and negative when $\pi < u_0 < 2\pi$. From Kepler's equation (1.49) we then find the mean anomaly ℓ_0 at $t = t_0$.

The mean anomaly at t is then $\ell = \ell_0 + n(t - t_0)$. By solving Kepler's equation numerically we can find the eccentric anomaly u. We may then evaluate the f and g functions using equations (1.78) and the position and velocity at t from equations (1.72) and (1.73).

1.4 Canonical orbital elements

The powerful tools of Lagrangian and Hamiltonian dynamics are essential for solving many of the problems addressed later in this book. A summary of the relevant aspects of this subject is given in Appendix D. In this section we show how Hamiltonian methods are applied to the two-body problem.

The Hamiltonian that describes the trajectory of a test particle around a point mass M at the origin is

$$H_{\rm K}(\mathbf{r},\mathbf{v}) = \tfrac{1}{2}\mathbf{v}^2 - \frac{\mathbb{G}M}{|\mathbf{r}|}. \tag{1.80}$$

Here $\mathbf{r}$ and $\mathbf{v}$ are the position and velocity, which together determine the position of the test particle in 6-dimensional phase space. The vectors $\mathbf{r}$ and $\mathbf{v}$ are a canonical coordinate-momentum pair.[10] Hamilton's equations read

$$\frac{\mathrm{d}\mathbf{r}}{\mathrm{d}t} = \frac{\partial H_{\rm K}}{\partial \mathbf{v}} = \mathbf{v}, \qquad \frac{\mathrm{d}\mathbf{v}}{\mathrm{d}t} = -\frac{\partial H_{\rm K}}{\partial \mathbf{r}} = -\frac{\mathbb{G}M}{|\mathbf{r}|^3}\mathbf{r}. \tag{1.81}$$

These are equivalent to the usual equations of motion (1.11).

The advantage of Hamiltonian methods is that the equations of motion are the same in any set of phase-space coordinates $\mathbf{z} = (\mathbf{q},\mathbf{p})$ that are obtained from $(\mathbf{r},\mathbf{v})$ by a canonical transformation (Appendix D.6). For example, suppose that the test particle is also subject to an additional potential $\Phi(\mathbf{r},t)$ arising from some external mass distribution, such as another planet. Then the Hamiltonian and the equations of motion in the original variables are

$$H(\mathbf{r},\mathbf{v},t) = H_{\rm K}(\mathbf{r},\mathbf{v}) + \Phi(\mathbf{r},t), \qquad \frac{\mathrm{d}\mathbf{r}}{\mathrm{d}t} = \frac{\partial H}{\partial \mathbf{v}}, \qquad \frac{\mathrm{d}\mathbf{v}}{\mathrm{d}t} = -\frac{\partial H}{\partial \mathbf{r}}. \tag{1.82}$$

[10] We usually—but not always—adopt the convention that the canonical momentum $\mathbf{p}$ that is conjugate to the position $\mathbf{r}$ is velocity $\mathbf{v}$ rather than Newtonian momentum $m\mathbf{v}$. Velocity is often more convenient than Newtonian momentum in gravitational dynamics since the acceleration of a body in a gravitational potential is independent of mass. If necessary, the convention used in a particular set of equations can be verified by dimensional analysis.

In the new canonical variables,[11]

$$H(\mathbf{z},t) = H_{\mathrm{K}}(\mathbf{z}) + \Phi(\mathbf{z},t), \qquad \frac{\mathrm{d}\mathbf{q}}{\mathrm{d}t} = \frac{\partial H}{\partial \mathbf{p}}, \quad \frac{\mathrm{d}\mathbf{p}}{\mathrm{d}t} = -\frac{\partial H}{\partial \mathbf{q}}. \tag{1.83}$$

If the additional potential is small compared to the Kepler potential, $|\phi(\mathbf{r},t)| \ll \mathbb{G}M/r$, then the trajectory will be close to a Kepler ellipse. Therefore the analysis can be much easier if we use new coordinates and momenta $\mathbf{z}$ in which Kepler motion is simple.[12] The six orbital elements—semimajor axis a, eccentricity e, inclination I, longitude of the ascending node Ω, argument of periapsis ω and mean anomaly ℓ—satisfy this requirement as all of the elements are constant except for ℓ, which increases linearly with time. This set of orbital elements is not canonical, but they can be rearranged to form a canonical set called the **Delaunay variables**, in which the coordinate-momentum pairs are:

$$\begin{aligned} &\ell, && \Lambda \equiv (\mathbb{G}Ma)^{1/2}, \\ &\omega, && L = [\mathbb{G}Ma(1-e^2)]^{1/2}, \\ &\Omega, && L_z = L\cos I. \end{aligned} \tag{1.84}$$

Here L_z is the z-component of the angular-momentum vector $\mathbf{L}$ (see Figure 1.2); $L = |\mathbf{L}|$ (eq. 1.28); and Λ is sometimes called the **circular angular momentum** since it equals the angular momentum for a circular orbit. The proof that the Delaunay variables are canonical is given in Appendix E.

The Kepler Hamiltonian (1.80) is equal to the energy per unit mass, which is related to the semimajor axis by equation (1.32); thus

$$H_{\mathrm{K}} = -\frac{\mathbb{G}M}{2a} = -\frac{(\mathbb{G}M)^2}{2\Lambda^2}. \tag{1.85}$$

[11] For notational simplicity, we usually adopt the convention that the Hamiltonian and the potential are functions of position, velocity, or position in phase space rather than functions of the coordinates. Thus $H(\mathbf{r},\mathbf{v},t)$ and $H(\mathbf{z},t)$ have the same value if $(\mathbf{r},\mathbf{v})$ and $\mathbf{z}$ are coordinates of the same phase-space point in different coordinate systems.

[12] However, the additional potential $\Phi(\mathbf{z},t)$ is often much more complicated in the new variables; for a start, it generally depends on all six phase-space coordinates rather than just the three components of $\mathbf{r}$. Since dynamics is more difficult than potential theory, the tradeoff—simpler dynamics at the cost of more complicated potential theory—is generally worthwhile.

Since the Kepler Hamiltonian is independent of the coordinates, the momenta Λ, L and L_z are all constants along a trajectory in the absence of additional forces; such variables are called **integrals of motion**. Because the Hamiltonian is independent of the momenta L and L_z their conjugate coordinates ω and Ω are also constant, and $\mathrm{d}\ell/\mathrm{d}t = \partial H_\mathrm{K}/\partial\Lambda = (\mathbb{G}M)^2\Lambda^{-3} = (\mathbb{G}M/a^3)^{1/2} = n$, where n is the mean motion defined by Kepler's law (1.44). Of course, all of these conclusions are consistent with what we already know about Kepler orbits.

Because the momenta are integrals of motion in the Kepler Hamiltonian and the coordinates are angular variables that range from 0 to 2π, the Delaunay variables are also angle-action variables for the Kepler Hamiltonian (Appendix D.7). For an application of this property, see Box 1.2.

One shortcoming of the Delaunay variables is that they have coordinate singularities at zero eccentricity, where ω is ill-defined, and zero inclination, where Ω and ω are ill-defined. Even if the eccentricity or inclination of an orbit is small but nonzero, these elements can vary rapidly in the presence of small perturbing forces, so numerical integrations that follow the evolution of the Delaunay variables can grind to a near-halt.

To address this problem we introduce other sets of canonical variables derived from the Delaunay variables. We write $\mathbf{q} = (\ell, \omega, \Omega)$, $\mathbf{p} = (\Lambda, L, L_z)$ and introduce a generating function $S_2(\mathbf{q}, \mathbf{P})$ as described in Appendix D.6.1. From equations (D.63)

$$\mathbf{p} = \frac{\partial S_2}{\partial \mathbf{q}}, \quad \mathbf{Q} = \frac{\partial S_2}{\partial \mathbf{P}}, \tag{1.86}$$

and these equations can be solved for the new variables $\mathbf{Q}$ and $\mathbf{P}$. For example, if $S_2(\mathbf{q}, \mathbf{P}) = (\ell + \omega + \Omega)P_1 + (\omega + \Omega)P_2 + \Omega P_3$ then the new coordinate-momentum pairs are

$$\begin{aligned} \lambda = \ell + \omega + \Omega, &\quad \Lambda, \\ \varpi = \omega + \Omega, &\quad L - \Lambda = (\mathbb{G}Ma)^{1/2}\left[(1-e^2)^{1/2} - 1\right], \\ \Omega, &\quad L_z - L = (\mathbb{G}Ma)^{1/2}(1-e^2)^{1/2}(\cos I - 1). \end{aligned} \tag{1.87}$$

Here we have reintroduced the mean longitude λ (eq. 1.68) and the longitude of periapsis ϖ (eq. 1.67). Since λ and ϖ are well defined for orbits of zero inclination, these variables are better suited for describing nearly equatorial prograde orbits. The longitude of the node Ω is still ill-defined when the inclination is zero, although if the motion is known or assumed to be restricted to the equatorial plane the first two coordinate-momentum pairs are sufficient to describe the motion completely.

With the variables (1.87) two of the momenta $L - \Lambda$ and $L_z - L$ are always negative. For this reason some authors prefer to use the generating function $S_2(\mathbf{q}, \mathbf{P}) = (\ell + \omega + \Omega)P_1 - (\omega + \Omega)P_2 - \Omega P_3$, which yields new coordinates and momenta

$$\begin{aligned} \lambda = \ell + \omega + \Omega, &\quad \Lambda, \\ -\varpi = -\omega - \Omega, &\quad \Lambda - L = (\mathbb{G}Ma)^{1/2}[1 - (1 - e^2)^{1/2}], \\ -\Omega, &\quad L - L_z = (\mathbb{G}Ma)^{1/2}(1 - e^2)^{1/2}(1 - \cos I). \end{aligned} \tag{1.88}$$

Another set is given by the generating function $S_2(\mathbf{q}, \mathbf{P}) = \ell P_1 + (\ell + \omega)P_2 + \Omega P_3$, which yields coordinates and momenta

$$\begin{aligned} \ell, &\quad \Lambda - L = (\mathbb{G}Ma)^{1/2}[1 - (1 - e^2)^{1/2}], \\ \ell + \omega, &\quad L = (\mathbb{G}Ma)^{1/2}(1 - e^2)^{1/2}, \\ \Omega, &\quad L_z = (\mathbb{G}Ma)^{1/2}(1 - e^2)^{1/2}\cos I. \end{aligned} \tag{1.89}$$

The action $\Lambda - L$ that appears in (1.88) and (1.89) has a simple physical interpretation. At a given angular momentum L, the radial motion in the Kepler orbit is governed by the Hamiltonian $H(r, p_r) = \frac{1}{2}p_r^2 + \frac{1}{2}L^2/r^2 - \mathbb{G}M/r$ (cf. eq. 1.18). The corresponding action is $J_r = \oint \mathrm{d}r\, p_r/(2\pi)$ (eq. D.72). The radial momentum $p_r = \dot{r}$ by Hamilton's equations; writing r and $\dot{r}$ in terms of the eccentric anomaly u using equations (1.46) and (1.54) gives

$$J_r = \frac{na^2e^2}{2\pi}\int_0^{2\pi} \mathrm{d}u\, \frac{\sin^2 u}{1 - e\cos u} = na^2[1 - (1 - e^2)^{1/2}] = \Lambda - L. \tag{1.90}$$

Thus $\Lambda - L$ is the action associated with the radial coordinate, sometimes called the **radial action**. The radial action is zero for circular orbits and equal to $\frac{1}{2}(\mathbb{G}Ma)^{1/2}e^2$ when $e \ll 1$.

Box 1.2: The effect of slow mass loss on a Kepler orbit

If the mass of the central object is changing, the constant M in equations like (1.11) must be replaced by a variable $M(t)$. We assume that the evolution of the mass is (i) due to some spherically symmetric process (e.g., a spherical wind from the surface of a star), so there is no recoil force on the central object; (ii) slow, in the sense that $|\mathrm{d}M/\mathrm{d}t| \ll M/P$, where P is the orbital period of a planet.

Since the gravitational potential remains spherically symmetric, the angular momentum $L = (\mathbb{G}Ma)^{1/2}(1-e^2)^{1/2}$ (eq. 1.28) is conserved.

Moreover, actions are adiabatic invariants (Appendix D.10), so during slow mass loss the actions remain almost constant. The Delaunay variable $\Lambda = (\mathbb{G}Ma)^{1/2}$ (eq. 1.84) is an action. Since Λ and L are distinct functions of Ma and e, and both are conserved—one adiabatically and one exactly—then both Ma and e are also conserved. In words, during slow mass loss the orbit expands, with $a(t) \propto 1/M(t)$, but its eccentricity remains constant. The accuracy of this approximate conservation law is explored in Problem 2.8.

At present the Sun is losing mass at a rate $\dot{M}/M = -(1.1 \pm 0.3) \times 10^{-13}\,\mathrm{yr}^{-1}$ (Pitjeva et al. 2021). Near the end of its life, the Sun will become a red-giant star and expand dramatically in radius and luminosity. At the tip of the red-giant branch, about 7.6 Gyr from now, the solar radius will be about 250 times its present value or 1.2 au and its luminosity will be 2 700 times its current value (Schröder & Connon Smith 2008). During its evolution up the red-giant branch the Sun will lose about 30% of its mass, and according to the arguments above the Earth's orbit will expand by the same fraction. Whether or not the Earth escapes being engulfed by the Sun depends on the uncertain relative rates of the Sun's future expansion and its mass loss.

Finally, consider the generating function $S_2(\mathbf{q}, \mathbf{P}) = P_1(\ell + \omega + \Omega) + \frac{1}{2}{P_2}^2 \cot(\omega + \Omega) + \frac{1}{2}{P_3}^2 \cot \Omega$, which yields the **Poincaré variables**

$$
\begin{aligned}
&\lambda = \ell + \omega + \Omega, && \Lambda, \\
&[2(\Lambda - L)]^{1/2} \cos \varpi, && [2(\Lambda - L)]^{1/2} \sin \varpi, \\
&[2(L - L_z)]^{1/2} \cos \Omega, && [2(L - L_z)]^{1/2} \sin \Omega.
\end{aligned} \tag{1.91}
$$

These are well defined even when $e = 0$ or $I = 0$. In particular, in the limit of small eccentricity and inclination the Poincaré variables simplify to

$$\begin{aligned} &\lambda, && \Lambda, \\ &(\mathbb{G}Ma)^{1/4} e \cos\varpi, && (\mathbb{G}Ma)^{1/4} e \sin\varpi, \\ &(\mathbb{G}Ma)^{1/4} I \cos\Omega, && (\mathbb{G}Ma)^{1/4} I \sin\Omega. \end{aligned} \tag{1.92}$$

Apart from the constant of proportionality $(\mathbb{G}Ma)^{1/4}$ these are just the Cartesian elements defined in equations (1.71).

All of these sets of orbital elements remain ill-defined when the inclination $I = \pi$ (retrograde orbits in the reference plane) or $e = 1$ (orbits with zero angular momentum); however, such orbits are relatively rare in planetary systems.[13]

1.5 Units and reference frames

Measurements of the trajectories of solar-system bodies are some of the most accurate in any science, and provide exquisitely precise tests of physical theories such as general relativity. Precision of this kind demands careful definitions of units and reference frames. These will only be treated briefly in this book, since our focus is on understanding rather than measuring the behavior of celestial bodies.

Tables of physical, astronomical and solar-system constants are given in Appendix A.

1.5.1 Time

The unit of time is the Système Internationale or SI second (s), which is defined by a fixed value for the frequency of a particular transition of cesium atoms. Measurements from several cesium frequency standards are combined to form a timescale known as **International Atomic Time** (TAI).

[13] A set of canonical coordinates and momenta that is well defined for orbits with zero angular momentum is described by Tremaine (2001). Alternatively, the orbit can be described using the angular-momentum and eccentricity vectors, which are well defined for any Kepler orbit; see §5.3 or Allan & Ward (1963).

In contrast, **Universal Time** (UT) employs the Earth's rotation on its axis as a clock. UT is not tied precisely to this clock because the Earth's angular speed is not constant. The most important nonuniformity is that the length of the day increases by about 2 milliseconds per century because of the combined effects of tidal friction and post-glacial rebound. There are also annual and semiannual variations of a few tenths of a millisecond. Despite these irregularities, a timescale based approximately on the Earth's rotation is essential for everyday life: for example, we would like noon to occur close to the middle of the day. Therefore all civil timekeeping is based on **Coordinated Universal Time** (UTC), which is an atomic timescale that is kept in close agreement with UT by adding extra seconds ("leap seconds") at regular intervals.[14] Thus UTC is a discontinuous timescale composed of segments that follow TAI apart from a constant offset.

An inconvenient feature of TAI for high-precision work is that it measures the rate of clocks at sea level on the Earth; general relativity implies that the clock rate depends on the gravitational potential and hence the rate of TAI is different from the rate measured by the same clock outside the solar system. For example, the rate of TAI varies with a period of one year and an amplitude of 1.7 milliseconds because of the eccentricity of the Earth's orbit. **Barycentric Coordinate Time** (TCB) measures the proper time experienced by a clock that co-moves with the center of mass of the solar system but is far outside it. TCB ticks faster than TAI by 0.49 seconds per year, corresponding to a fractional speedup of 1.55×10^{-8}.

The times of astronomical events are usually measured by the **Julian date**, denoted by the prefix JD. The Julian date is expressed in days and decimals of a day. Each day has 86 400 seconds. The Julian year consists of exactly 365.25 days and is denoted by the prefix J. For example, the initial conditions of orbits are often specified at a standard epoch, such as

$$\text{J2000.0} = \text{JD}\,2\,451\,545.0, \tag{1.93}$$

which corresponds roughly to noon in England on January 1, 2000. The modified Julian day is defined as

$$\text{MJD} = \text{JD} - 2\,400\,000.5, \tag{1.94}$$

[14] The utility of leap seconds is controversial, and their future is uncertain.

the integer offset reduces the length of the number specifying relatively recent dates, and the half-integer offset ensures that the MJD begins at midnight rather than noon.

In contrast to SI seconds (s) and days (1 d = 86 400 s) there is no unique definition of "year": most astronomers use the Julian year but there is also the anomalistic year, sidereal year, and the like (see footnote 7). For this reason the use of "year" as a precise unit of time is deprecated. However, we shall occasionally use years, megayears and gigayears (abbreviated yr, Myr, Gyr) to denote 1, 10^6 and 10^9 Julian years. The age of the solar system is 4.567 Gyr and the age of the Universe is 13.79 Gyr. The future lifetime of the solar system as we know it is about 7.6 Gyr (see Box 1.2).

The SI unit of length is defined in terms of the second, such that the speed of light is exactly

$$c \equiv 299\,792\,458\,\mathrm{m\,s^{-1}}. \tag{1.95}$$

1.5.2 Units for the solar system

The history of the determination of the scale of the solar system and the mass of the Sun is worth a brief description. Until the mid-twentieth century virtually all of our data on the orbits of the Sun and planets came from tracking their positions on the sky as functions of time. This information could be combined with the theory of Kepler orbits developed earlier in this chapter (plus small corrections arising from mutual interactions between the planets, which are handled by the methods of Chapter 4) to determine all of the orbital elements of the planets including the Earth, except for the overall scale of the system. Thus, for example, the ratio of semimajor axes of any two planets was known to high accuracy, but the values of the semimajor axes in meters were not.[15] To reflect this uncertainty, astronomers introduced the concept of the **astronomical unit** (abbreviated au), which was originally defined to be the semimajor axis of the Earth's orbit. Thus the semimajor axes of the planets were known in astronomical units long before the value of the astronomical unit was known to comparable accuracy.

[15] This indeterminacy follows from dimensional analysis: measurements of angles and times cannot be combined to find a quantity with dimensions of length.

Since Kepler's third law (1.44) is $\mathbb{G}M = 4\pi^2 a^3/P^2$, and orbital periods P can be determined so long as we have accurate clocks, any fractional uncertainty ϵ in the astronomical unit implies a fractional uncertainty of 3ϵ in the **solar mass parameter** $\mathbb{G}M_\odot$.

Over the centuries, the astronomical unit was measured by many different techniques, including transits of Venus, parallaxes of nearby solar-system objects over Earth-sized baselines and stellar aberration. Nevertheless, even in the 1950s the astronomical unit was only known with a fractional accuracy of about 10^{-3}. Soon after, radar observations of Venus and Mars and ranging data from interplanetary spacecraft reduced the uncertainty by several orders of magnitude. The current uncertainty is much smaller than variations in the Earth's semimajor axis due to perturbations from the other planets, so in 2012 the International Astronomical Union (IAU) re-defined the astronomical unit to be an exact unit of length,

$$1\,\text{au} \equiv 149\,597\,870\,700\,\text{m}. \tag{1.96}$$

Distances to other stars are measured in units of **parsecs** (abbreviated pc), the distance at which 1 au subtends one second of arc. Thus the parsec is also an exact unit of length, though an irrational number of meters:

$$1\,\text{pc} = \frac{648\,000}{\pi}\,\text{au} \simeq 3.085\,677\,6 \times 10^{16}\,\text{m}. \tag{1.97}$$

The determination of the scale of the solar system allowed the determination of $\mathbb{G}M_\odot$ to comparable accuracy. In contrast, the gravitational constant $\mathbb{G}$, determined by laboratory experiments, is only known to a fractional accuracy of 2×10^{-5} (see Appendix A). Therefore the masses of the Sun and solar-system planets are much less well known than $\mathbb{G}$ times the masses, and for accurate work they should always be quoted along with the assumed value of $\mathbb{G}$.

In 2015 the IAU recommended that orbit calculations should be based on the nominal value of the solar mass parameter

$$\mathbb{G}M_\odot \equiv 1.327\,124\,4 \times 10^{20}\,\text{m}^3\,\text{s}^{-2}. \tag{1.98}$$

The adjective "nominal" means that this should be understood as a standard conversion factor that is close to the "actual" value (probably with a fractional error of less than 1×10^{-9}). For most dynamical problems it is better to use a consistent set of constants that is common to the whole community rather than the best current estimate of each constant.

1.5.3 The solar system reference frame

The **Barycentric Celestial Reference System** (BCRS) is a coordinate system created in 2000 by the IAU. The system uses harmonic coordinates (eq. J.6), with origin at the solar system barycenter and time given by TCB. This is the reference system appropriate for solving the equations of motion of solar system bodies. The orientation of the BCRS coordinate system coincides with that of the International Celestial Reference System (ICRS), which is defined by the adopted angular coordinates of a set of extragalactic radio sources. For more detail see Kaplan (2005) and Urban & Seidelmann (2013).

These definitions are based on the assumption that the local inertial reference frame (the BCRS) is not rotating relative to the distant universe (the ICRS), sometimes called **Mach's principle**. This assumption is testable: the relative rotation of these frames is consistent with zero and less than 4×10^{-5} arcsec yr^{-1} (Folkner 2010).

1.6 Orbital elements for exoplanets

The orbital elements of extrasolar planets ("exoplanets") are much more difficult to determine accurately than the elements of solar-system bodies. In most cases we only know some of the six orbital elements, depending on the detection method.

Here we describe three methods of planet detection based on the classical observational techniques of spectroscopy, photometry, astrometry and imaging. We do not discuss a further important technique, gravitational microlensing, because it measures only the mass of the planet and its projected separation from the host star and thus provides only limited constraints on the orbital elements and dynamics (Gaudi 2011).

1.6.1 Radial-velocity planets

One of the most powerful methods to detect and characterize exoplanets is through periodic variations in the velocity of their host star, which arise as both star and planet orbit around their common center of mass.[16] These variations can be detected through small Doppler shifts in the stellar spectrum.[17]

To illustrate the analysis, we consider a system containing a single planet. The star is at $\mathbf{r}_0$ and the planet is at $\mathbf{r}_1$. The velocity of the star is given by the time derivative of equation (1.6),

$$\mathbf{v}_0 = \mathbf{v}_{\rm cm} - \frac{m_1}{m_0 + m_1}\mathbf{v}, \tag{1.99}$$

where $\mathbf{v}$ is the velocity of the planet relative to the star. The velocity of the center of mass $\mathbf{v}_{\rm cm}$ is constant (eq. 1.7). We may choose our coordinates such that the positive z-axis is parallel to the line of sight from the observer to the system and pointing away from the observer; thus edge-on orbits have $I = 90°$, face-on orbits have $I = 0$, and positive line-of-sight velocity implies that the star is receding from us. Then the line-of-sight velocity of the star relative to the center of mass is

$$v_{\rm los} \equiv (\mathbf{v}_0 - \mathbf{v}_{\rm cm}) \cdot \hat{\mathbf{z}} = -\frac{m_1}{m_0 + m_1}\mathbf{v} \cdot \hat{\mathbf{z}}. \tag{1.100}$$

From equation (1.70), $\mathbf{v} \cdot \hat{\mathbf{z}} = \dot{z} = \sin I[\dot{r}\sin(f + \omega) + r\cos(f + \omega)\dot{f}] = \sin I[v_r \sin(f+\omega)+v_\psi \cos(f+\omega)]$. Then using equations (1.54) and (1.55),

$$v_{\rm los} = -\frac{m_1}{m_0 + m_1}\frac{na}{(1-e^2)^{1/2}}\sin I\big[\cos(f + \omega) + e\cos\omega\big]. \tag{1.101}$$

[16] The possibility of detecting planets by radial-velocity variations and by transits was first discussed in a prescient short paper by Struve (1952).

[17] Unfortunately the term "radial velocity" is commonly used to denote two different quantities: (i) the component of the planet's velocity relative to the host star along the line joining them, and (ii) the component of the star's velocity relative to the observer along the line joining them. In practice the meaning is usually clear from the context, but when there is the possibility of confusion we shall use the term "line-of-sight velocity" as an unambiguous replacement for (ii).

Since the orbital period $P = 2\pi a^{3/2}/[\,\mathbb{G}(m_0+m_1)]^{1/2}$ is directly observable while the semimajor axis is not, we eliminate a in favor of P to obtain

$$v_{\rm los} = -\frac{m_1}{m_0+m_1}\left[\frac{2\pi\,\mathbb{G}(m_0+m_1)}{P}\right]^{1/3}\frac{\sin I}{(1-e^2)^{1/2}}\big[\cos(f+\omega)+e\cos\omega\big]. \tag{1.102}$$

Using equations (1.50) and (1.51a), this result can also be expressed in terms of the eccentric anomaly,

$$\begin{aligned} v_{\rm los} = &-\frac{m_1}{m_0+m_1}\left[\frac{2\pi\,\mathbb{G}(m_0+m_1)}{P}\right]^{1/3}\sin I \\ &\times\frac{(1-e^2)^{1/2}\cos u\cos\omega-\sin u\sin\omega}{1-e\cos u}. \end{aligned} \tag{1.103}$$

To obtain $v_{\rm los}(t)$, the line-of-sight velocity as a function of time (the **velocity curve**), we write the mean anomaly as $\ell = 2\pi(t-t_0)/P$ where t_0 is the time of periapsis passage, solve Kepler's equation (1.49) for u, and then substitute u into equation (1.103). The velocity curve is not sinusoidal unless the orbit is circular, but it is still useful to define the **semi-amplitude** K as half the difference between the maximum and minimum velocity. From equation (1.102) the extrema of $v_{\rm los}$ occur at $f=-\omega$ and $f=\pi-\omega$, so

$$K = \frac{m_1}{m_0+m_1}\left[\frac{2\pi\,\mathbb{G}(m_0+m_1)}{P}\right]^{1/3}\frac{\sin I}{(1-e^2)^{1/2}}. \tag{1.104}$$

These results tell us what can and cannot be determined from the velocity curve. The orbital period P is equal to the period of the velocity curve, and the eccentricity e and argument of periapsis ω can be determined from the shape of the curve. The longitude of the node Ω is not constrained. The masses of the star and planet, m_0 and m_1, and the inclination I cannot be individually determined, only the combination

$$\mu \equiv \frac{m_1^3\sin^3 I}{(m_0+m_1)^2}, \tag{1.105}$$

known as the **mass function**. The mass function is related to the semi-amplitude and period by

$$\mu = \frac{P}{2\pi \mathbb{G}} K^3 (1 - e^2)^{3/2}. \tag{1.106}$$

Since exoplanet masses are usually much smaller than the mass of their host star, and the mass of the host star can usually be determined from its spectral properties, the mass function yields a combination of the planetary mass and orbital inclination, $m_1 \sin I$.

The semi-amplitude K varies as $a^{-1/2}$ for planets of a given mass, so radial-velocity searches are most sensitive to planets orbiting close to the host star. Planets whose orbital periods are much larger than the survey duration will contribute a constant acceleration or linear drift to the line-of-sight velocity of the host star, and this signal provides evidence for the existence of a distant planet but only weak constraints on its properties.

The most precisely measured radial-velocity planets are found orbiting pulsars. The pulsar emits pulsed radio signals at regular intervals Δt. The pulse emitted at $t_n = n\Delta t + \text{const}$ arrives at $t'_n = t_n + r(t_n)/c$ where $r(t_n)$ is the distance of the pulsar at t_n and c is the speed of light. Now write $r(t) = \text{const} + v_{\text{los}} t$ where v_{los} is the line-of-sight velocity of the pulsar, and we have $\Delta t'_n = t'_{n+1} - t'_n = \Delta t(1 + v_{\text{los}}/c)$. Thus measuring the intervals between pulses yields the line-of-sight velocity (up to an undetermined constant, since the rest-frame pulse interval Δt is unknown), and as usual periodic variations in the line-of-sight velocity are the signature of a planet.

Pulsar planets are rare, presumably because planets cannot survive the supernova explosion that creates the pulsar, and only a handful are known. The prototype is the system of three planets discovered around the pulsar PSR B1257+12 (Wolszczan & Frail 1992).

1.6.2 Transiting planets

In a small fraction of cases, a planetary system is oriented such that one or more of its planets crosses the face of the host star as seen from Earth, an event known as a **transit**.[18] During the transit, there is a characteristic dip in the stellar flux, which repeats with a period equal to the planet's orbital period.

Suppose that the planet has radius R_p and the star has radius R_*. In most cases $R_\mathrm{p} \ll R_*$; for example, the radii of Earth and Jupiter relative to the Sun are $R_\oplus/R_\odot = 0.009\,153$ and $R_\mathrm{J}/R_\odot = 0.099\,37$.[19] During a transit the visible area of the stellar disk is reduced to a fraction $1 - f$ of its unobscured value, where

$$f = \frac{R_\mathrm{p}^2}{R_*^2}, \tag{1.107}$$

and the flux from the star is reduced by a similar amount (depending on limb darkening, to be discussed later in this subsection). An observer watching Earth or Jupiter transit the Sun would find $f = 8.377 \times 10^{-5}$ and $f = 0.009\,88$ respectively. With current technology, Jupiter-like transits can be detected from the ground but Earth-like transits can only be detected by space-based observatories.

The probability that a planet will transit depends strongly on its semimajor axis. To determine this probability, we again use a coordinate system in which the z-axis is parallel to the line of sight. Then the planet transits if and only if the minimum value of $x^2 + y^2$ is less than $(R_* + R_\mathrm{p})^2$. From equations (1.70), $x^2 + y^2 = r^2 - z^2 = r^2[1 - \sin^2 I \sin^2(f + \omega)]$ so the minimum value of $x^2 + y^2$ is $r^2 \cos^2 I$. Therefore if the planet is on a circular orbit with semimajor axis a, it transits if and only if $|\cos I| < (R_* + R_\mathrm{p})/a$. If the distribution of orientations of the planetary orbits is random—an untested

[18] Transits and occultations are usually distinguished from eclipses. In an eclipse (e.g., an eclipse of the Sun by the Moon) both bodies have similar angular size. In a transit (e.g., a transit of Venus across the Sun) a small body passes in front of a large one, and in an occultation a small body passes behind a large one.

[19] Planets are not perfect spheres: in general, the polar radius R_pol of a rotating planet is smaller than its equatorial radius R_eq, and the planet is said to have an **equatorial bulge** (Box 1.3). If we assume that the spin and orbital axes of the planet are aligned, then both are normal to the line of sight if the planet transits the star. Approximating the shape of the planet as an ellipse, its area on the plane normal to the line of sight is $\pi R_\mathrm{eq} R_\mathrm{pol}$ so the effective radius for computing the transit depth is $R_\mathrm{eff} = (R_\mathrm{eq} R_\mathrm{pol})^{1/2}$. For the Earth and Jupiter the effective radii are $R_{\oplus,\mathrm{eff}} = 6\,367.4\,\mathrm{km}$ and $R_{\mathrm{J,eff}} = 69\,134\,\mathrm{km}$. In contrast the Sun is nearly spherical, with a fractional difference in the polar and equatorial radii $\lesssim 10^{-5}$.

but extremely plausible assumption—then $|\cos I|$ is uniformly distributed between 0 and 1, so the probability of transit is

$$p = \frac{R_* + R_\mathrm{p}}{a}. \tag{1.108}$$

A useful reference time for the duration of the transit is

$$\tau_0 = \frac{2R_*}{v} = 2R_* \left(\frac{a}{\mathbb{G}M_*}\right)^{1/2} = 12.98\,\text{hours}\,\frac{R_*}{R_\odot}\left(\frac{a}{\text{au}}\frac{M_\odot}{M_*}\right)^{1/2}. \tag{1.109}$$

Here v is the planet's orbital velocity, M_* is the stellar mass, and a is the planet's semimajor axis; in deriving these equations we have assumed that the planet's orbit is circular. The reference time equals the actual transit time only if the planet radius $R_\mathrm{p} \ll R_*$, the stellar radius $R_* \ll a$, and the transit passes through the center of the star. The actual transit time is usually shorter than τ_0 since the planet travels along a chord across the star rather than through its center.

The interval between transits equals the orbital period (eq. 1.43),

$$P = 2\pi \left(\frac{a^3}{\mathbb{G}M_*}\right)^{1/2}. \tag{1.110}$$

The shape and duration of the transit event can be described more accurately using Figure 1.3. The point of closest approach of the planet to the center of the star is bR_* where the **impact parameter** b is a dimensionless number in the range 0 to ~ 1. There are four milestones during the transit event: first contact, where the projected planetary disk first touches the edge of the star; second contact, where the entire planetary disk first obscures the star, third contact, the last time at which the entire planetary disk obscures the star, and fourth contact, when the transit ends. Between first and second contact the flux from the star is steadily decreasing as more and more of the stellar disk is obscured; between second and third contact the flux is constant; and between third and fourth contact the flux is steadily returning to its original value. If the closest approach to the center of the star is at $t = 0$, then straightforward trigonometry shows that the times associated

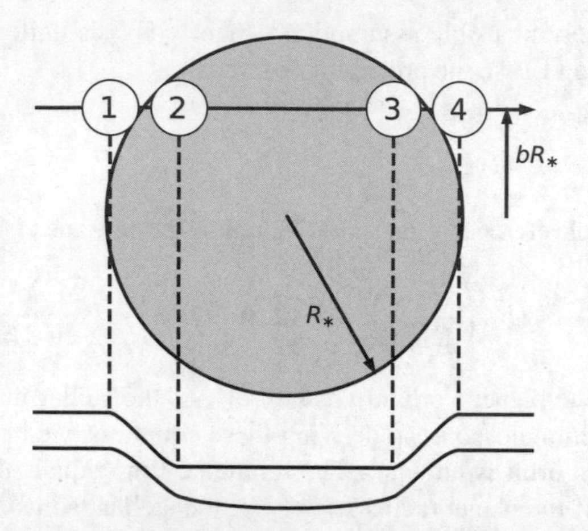

Figure 1.3: The geometry of a planetary transit. The large shaded circle of radius R_* shows the disk of the host star, and the unshaded circles of radius R_p show the position of the planetary disk at first, second, third and fourth contact. The minimum distance between the centers of the planet and the star is bR_*, where b is the impact parameter. In this image $b = 0.6$ and $R_\mathrm{p}/R_* = 0.15$. The curves at the bottom of the figure show the stellar flux as a function of time in two cases: no limb darkening (top), and solar limb darkening (bottom) as described in the paragraph containing equation (1.114). Analytic expressions for transit light curves are given by Sackett (1999), Mandel & Agol (2002) and Seager & Mallén-Ornelas (2003).

with these events are

$$\begin{aligned} t_4 = -t_1 &= \frac{1}{v}\left[(R_* + R_\mathrm{p})^2 - b^2R_*^2\right]^{1/2} = \tfrac{1}{2}\tau_0[(1 + R_\mathrm{p}/R_*)^2 - b^2]^{1/2}, \\ t_3 = -t_2 &= \frac{1}{v}\left[(R_* - R_\mathrm{p})^2 - b^2R_*^2\right]^{1/2} = \tfrac{1}{2}\tau_0[(1 - R_\mathrm{p}/R_*)^2 - b^2]^{1/2}. \end{aligned} \tag{1.111}$$

Here we have assumed that $R_* \ll a$ so the planet travels across the star at nearly constant velocity v; an equivalent constraint is that the transit duration is much less than the orbital period, $\tau_0 \ll P$. The total duration of the transit is

$$t_4 - t_1 = \frac{2}{v}\left[(R_* + R_\mathrm{p})^2 - b^2 R_*^2\right]^{1/2} = \tau_0\left[(1 + R_\mathrm{p}/R_*)^2 - b^2\right]^{1/2}, \quad (1.112)$$

and the duration of the flat part of the transit, between second and third contact, is

$$t_3 - t_2 = \tau_0\left[(1 - R_\mathrm{p}/R_*)^2 - b^2\right]^{1/2}. \quad (1.113)$$

What can we measure from the transit depth, duration and shape? The fractional depth f of the transit determines the ratio of the planetary and stellar radii R_p/R_* through equation (1.107). Once this is known, the total duration $t_4 - t_1$ (eq. 1.112) and the duration of the flat part of the transit $t_3 - t_2$ (eq. 1.113) give two constraints on the impact parameter b and the reference time τ_0, so both can be determined. If the stellar mass M_* and radius R_* can be determined from the star's luminosity, colors and spectrum then equations (1.109) for the reference time and (1.43) for the orbital period give two constraints on the semimajor axis: if these agree then the planetary orbit is likely to be circular, and if not it must be eccentric.

This simple model predicts that the flux from the star is constant between second and third contact, which requires that the surface brightness of the star is uniform. In practice the surface brightness of the stellar disk is usually higher near the center, a phenomenon called **limb darkening**. One common parametrization of limb darkening is that the surface brightness at distance R from the center of the stellar disk of radius R_* is given by

$$\frac{I(R)}{I(0)} = 1 - a(1-\mu) - b(1-\mu)^2, \quad \text{where} \quad \mu = (1 - R^2/R_*^2)^{1/2}. \quad (1.114)$$

The limb-darkening coefficients a and b depend on the spectral type of the star and the wavelength range in which the surface brightness is measured. For a solar-type star measured in the Kepler wavelength band, $a \simeq 0.41$ and $b \simeq 0.26$.[20]

The depth of a transit (eq. 1.107) is independent of the semimajor axis a of the planet, but the probability that a planet will transit varies as a^{-1} (eq.

[20] Limb-darkening models for a wide range of stars are described in Claret & Bloemen (2011).

1.108), so transit searches are most sensitive to planets close to the host star. Planets whose orbital periods are larger than the survey duration are difficult to verify: a useful rule of thumb is that at least three transits are needed for a secure detection.

1.6.3 Astrometric planets

Planets can be detected by the periodic variations in the position of their host star as the star orbits around the center of mass of the star and planet.

The Kepler ellipse described by the star is projected onto an ellipse on the sky plane perpendicular to the line of sight. However, the semimajor axis and eccentricity of the projected ellipse are generally different from those of the original ellipse, and the focus of the projected ellipse differs from the projection of the focus of the original ellipse. Nevertheless all of the orbital elements, with some minor degeneracies, can be deduced from these measurements.

We consider a system containing a single planet of mass m_1 orbiting a star of mass m_0. We choose coordinates such that the positive z-axis is parallel to the line of sight from the observer to the system and pointing toward the observer.[21] The position of the star is $\mathbf{r}_0 = \mathbf{r}_{\text{cm}} - m_1\mathbf{r}/(m_0+m_1)$ (eq. 1.6), where $\mathbf{r}_{\text{cm}}$ is the position of the center of mass and $\mathbf{r} = \mathbf{r}_1 - \mathbf{r}_0$ is the vector from the star to the planet. Using equations (1.29) and (1.70) the position of the star on the sky, in the Cartesian coordinates x and y, is

$$\begin{aligned} x_0 &= x_{\text{cm}} - \frac{1-e^2}{1+e\cos f}(A\cos f + F\sin f), \\ y_0 &= y_{\text{cm}} - \frac{1-e^2}{1+e\cos f}(B\cos f + G\sin f), \end{aligned} \tag{1.115}$$

[21] Unfortunately this orientation is opposite to the orientation of the coordinate system in §1.6.1. The line-of-sight velocity is always defined to be positive if the star is receding from the observer, which implies that the positive z-axis points *away* from the observer. For astrometric binaries the x-y coordinate system on the sky is assumed to be right-handed (the positive y-axis is 90° counterclockwise from the positive x-axis), which requires that the positive z-axis points *toward* the observer.

where the **Thiele–Innes elements** are

$$\begin{aligned} A &= \frac{m_1 a}{m_0 + m_1}(\cos\Omega\cos\omega - \cos I\sin\Omega\sin\omega), \\ B &= \frac{m_1 a}{m_0 + m_1}(\sin\Omega\cos\omega + \cos I\cos\Omega\sin\omega), \\ F &= \frac{m_1 a}{m_0 + m_1}(-\cos\Omega\sin\omega - \cos I\sin\Omega\cos\omega), \\ G &= \frac{m_1 a}{m_0 + m_1}(-\sin\Omega\sin\omega + \cos I\cos\Omega\cos\omega); \end{aligned} \tag{1.116}$$

as usual a and e are the semimajor axis and eccentricity of the relative orbit, and f, I, ω and Ω are the true anomaly, inclination, argument of periapsis and longitude of the ascending node. The four Thiele-Innes elements replace a, I, Ω and ω; their advantage is that the positions are linear functions of these elements, which simplifies orbit fitting.

Equations (1.115) are simpler when written in terms of the eccentric anomaly, using equations (1.46), (1.50) and (1.51a):

$$\begin{aligned} x_0 &= x_{\rm cm} - A(\cos u - e) - F(1-e^2)^{1/2}\sin u, \\ y_0 &= y_{\rm cm} - B(\cos u - e) - G(1-e^2)^{1/2}\sin u. \end{aligned} \tag{1.117}$$

The eccentric anomaly is related to the time t through Kepler's equation (1.49), and with equation (1.45) this reads $n(t - t_0) = u - e\sin u$. Using these results we can fit the observations of x_0 and y_0 as a function of time to equations (1.117) to determine $x_{\rm cm}$, $y_{\rm cm}$, A, B, F and G, the eccentricity e, the mean motion n and the epoch of periapsis t_0.

The usual orbital elements are straightforward to derive from the Thiele–Innes elements. First,

$$\tan(\Omega + \omega) = \frac{B - F}{A + G}, \quad \tan(\Omega - \omega) = \frac{B + F}{A - G}, \tag{1.118}$$

and these equations can be solved for Ω and ω. If these are solutions then so are $\Omega + k_1\pi$ and $\omega + k_2\pi$, where k_1 and k_2 are integers. All but one of these solutions can be discarded because we also require that (i) $\sin(\Omega + \omega)$ has the same sign as $B - F$; (ii) $\sin(\Omega - \omega)$ has the same sign as $B + F$;

(iii) $0 \leq \omega < 2\pi$; and (iv) $0 \leq \Omega < \pi$. The last of these is a convention that is imposed because astrometric observations alone cannot distinguish the solutions (Ω, ω) and $(\Omega + \pi, \omega + \pi)$.

Next define

$$q_1 = \frac{A + G}{\cos(\Omega + \omega)}, \quad q_2 = \frac{A - G}{\cos(\Omega - \omega)}. \tag{1.119}$$

Then

$$I = 2\tan^{-1}(q_2/q_1)^{1/2}, \quad \frac{m_1 a}{m_0 + m_1} = \tfrac{1}{2}(q_1 + q_2). \tag{1.120}$$

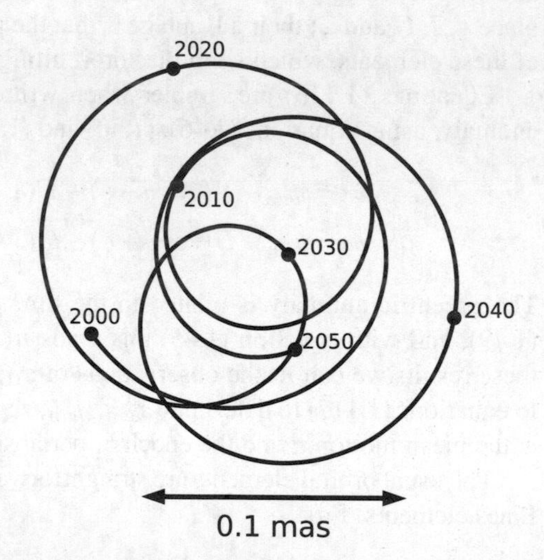

Figure 1.4: The astrometric signal from the solar system over the 50-year period from 2000 to 2050, as viewed from a star 100 parsecs away in the direction of the north ecliptic pole. The arrows mark an angular distance of 0.1 milliarcseconds.

The fit to the observations also yields the mean motion n, which constrains the semimajor axis and masses through Kepler's third law, $n^2 a^3 = \mathbb{G}(m_0 + m_1)$ (eq. 1.44). Combining this relation with the last of equations (1.120), we have

$$\frac{m_1^3}{(m_0 + m_1)^2} = \frac{(q_1 + q_2)^3 n^2}{8\,\mathbb{G}}; \tag{1.121}$$

the quantities on the right are observables and the left side is the **mass function** for astrometric planets. The mass m_0 of the host star can usually be determined from its spectral properties, so the mass function determines the planetary mass m_1.

The astrometric signal from a planet is proportional to its semimajor axis, so planets on larger orbits are easier to detect astrometrically. However, a reliable determination of the orbital elements usually requires data over at least one orbit, unless the data are extremely accurate. Thus the easiest planets to detect astrometrically are those with an orbital period smaller than the span of observations, but not by too much.

Astrometric data from multi-planet systems are hard to interpret if *any* of the massive planets in the system has an orbital period longer than the span of the observations. As an example, the astrometric signal arising from the motion of the Sun around the barycenter of the solar system is shown in Figure 1.4, as seen from a star 100 parsecs away. The figure shows that determining the masses and orbits of the giant planets in a planetary system like our own, even with an astrometric baseline of 1–2 decades, would be quite difficult.

1.6.4 Imaged planets

Imaging planets is difficult because the host star is so much brighter than the planet. For example, the luminosity of the Earth at visible wavelengths is only about 10^{-10} times the luminosity of the Sun. The contrast ratio is more favorable for young, massive planets at infrared wavelengths, in part because such planets are self-luminous, emitting thermal energy as they contract (Burrows et al. 1997). Even Jupiter emits roughly as much energy per unit time from contraction as it reflects from the Sun.

Most planets that have been successfully imaged are in orbits with large semimajor axes, where they are not swallowed in the glare from their host star: the median estimated semimajor axis of planets detected by direct imaging is well over 100 au. For a solar-mass host star the orbital period at 100 au is 1 000 yr, so the motion of most imaged planets relative to their host star has not been detected at all. What motion has been detected covers only a small fraction of the orbit, so the uncertainties in the orbital elements are large. Nevertheless, it is worth examining briefly what elements can be detected in principle for imaged planets.

In contrast to astrometric planets, where the position of the host star relative to the center of mass is measured on the sky plane, we measure the position of an imaged planet at $\mathbf{r}_1$ relative to the host star at $\mathbf{r}_0$. By analogy with equations (1.117) we may write the Cartesian coordinates of this relative position on the sky plane as

$$x = x_1 - x_0 = A'(\cos u - e) + F'(1-e^2)^{1/2}\sin u,$$
$$y = y_1 - y_0 = B'(\cos u - e) + G'(1-e^2)^{1/2}\sin u, \qquad (1.122)$$

where u is the eccentric anomaly, e is the eccentricity, and the Thiele–Innes elements are

$$A' = a(\cos\Omega\cos\omega - \cos I\sin\Omega\sin\omega),$$
$$B' = a(\sin\Omega\cos\omega + \cos I\cos\Omega\sin\omega),$$
$$F' = a(-\cos\Omega\sin\omega - \cos I\sin\Omega\cos\omega),$$
$$G' = a(-\sin\Omega\sin\omega + \cos I\cos\Omega\cos\omega). \qquad (1.123)$$

As usual a, ω and Ω are the semimajor axis, argument of periapsis and longitude of the ascending node. The eccentric anomaly is related to the time t through Kepler's equation (1.49), which reads $n(t-t_0) = u - e\sin u$ where n is the mean motion. We can fit the observations of x and y as a function of time to equations (1.122) to determine A', B', F', G', e, n and t_0. Then we can follow the procedure in equations (1.118)–(1.120) to determine the other orbital elements. Like astrometry, imaging cannot distinguish the solutions (Ω, ω) and $(\Omega + \pi, \omega + \pi)$. A check of the results comes from Kepler's third law (1.44): this determines the mass of the host star from the mean motion and the semimajor axis, and this mass can be determined independently from the spectral properties of the star.

1.7 Multipole expansion of a potential

In most cases the distance between a planet and its host star, or a satellite and its host planet, is large enough that both can be treated as point masses.

However, accurate dynamical calculations must sometimes account for the distribution of mass within one or both of these bodies. Examples include tracking artificial satellites of the Earth, measuring the relativistic precession of Mercury's perihelion, or determining the precession rate of a planet's spin axis.

Let $\rho(\mathbf{r})$ denote the density of a planet at position $\mathbf{r}$. The total mass of the planet is M and we assume that the origin is the center of mass of the planet. Then

$$\int \mathrm{d}\mathbf{r}\,\rho(\mathbf{r}) = M, \qquad \int \mathrm{d}\mathbf{r}\,\rho(\mathbf{r})\mathbf{r} = \mathbf{0}. \tag{1.124}$$

Using equations (C.44) and (C.55), the gravitational potential can be written in spherical coordinates $\mathbf{r} = (r, \theta, \phi)$ as

$$\begin{aligned}
\Phi(r,\theta,\phi) &= -\mathbb{G}\int \frac{\mathrm{d}\mathbf{r}'\,\rho(\mathbf{r}')}{|\mathbf{r}-\mathbf{r}'|} \qquad (1.125)\\
&= -\mathbb{G}\sum_{l=0}^{\infty}\int \mathrm{d}\mathbf{r}'\,\rho(\mathbf{r}')\frac{r_<^l}{r_>^{l+1}}\,\mathsf{P}_l(\cos\gamma)\\
&= -\sum_{l=0}^{\infty}\frac{4\pi\mathbb{G}}{2l+1}\sum_{m=-l}^{l}\int \mathrm{d}\mathbf{r}'\,\rho(\mathbf{r}')\frac{r_<^l}{r_>^{l+1}}\,Y_{lm}^*(\theta',\phi')Y_{lm}(\theta,\phi).
\end{aligned}$$

Here $\mathsf{P}_l(\cos\gamma)$ and $Y_{lm}(\theta,\phi)$ are a Legendre polynomial and a spherical harmonic (Appendices C.6 and C.7), $r_<$ and $r_>$ are the smaller and larger of r and r', $\cos\gamma = \mathbf{r}'\cdot\mathbf{r}/(r'r)$ is the cosine of the angle between the vectors $\mathbf{r}$ and $\mathbf{r}'$, and the asterisk denotes the complex conjugate. Any satellite must orbit outside all of the planetary mass, so the potential seen by the satellite simplifies to

$$\Phi(r,\theta,\phi) \equiv \sum_{l=0}^{\infty}\Phi_l(r,\theta,\phi), \tag{1.126}$$

where

$$\begin{aligned}
\Phi_l(r,\theta,\phi) &= -\frac{\mathbb{G}}{r^{l+1}}\int \mathrm{d}\mathbf{r}'\,\rho(\mathbf{r}')r'^l\mathsf{P}_l(\cos\gamma) \qquad (1.127)\\
&= -\frac{4\pi\mathbb{G}}{(2l+1)r^{l+1}}\sum_{m=-l}^{l}Y_{lm}(\theta,\phi)\int \mathrm{d}\mathbf{r}'\,\rho(\mathbf{r}')r'^lY_{lm}^*(\theta',\phi').
\end{aligned}$$

We examine the first three of these terms:

Monopole ($l = 0$) Since $\mathsf{P}_0(\cos\gamma) = 1$ (eq. C.45) and $\int d\mathbf{r}'\,\rho(\mathbf{r}') = M$ (eq. 1.124), we have $\Phi_0(r,\theta,\phi) = -\mathbb{G}M/r$, the same as if all the mass of the planet were concentrated in a point at the origin.

Dipole ($l = 1$) Since $\mathsf{P}_1(\cos\gamma) = \cos\gamma = \mathbf{r}'\cdot\mathbf{r}/(r'r)$, the combination $r'\mathsf{P}_1(\cos\gamma)$ is a linear function of $\mathbf{r}'$ at fixed $\mathbf{r}$ and zero at $\mathbf{r}' = \mathbf{0}$. Then the second of equations (1.124) implies that the integral in the first line of equation (1.127) is zero. Thus $\Phi_1(r,\theta,\phi) = 0$.

Quadrupole ($l = 2$) Since $\mathsf{P}_2(\cos\gamma) = \frac{3}{2}\cos^2\gamma - \frac{1}{2}$, the combination $r'^2\mathsf{P}_2(\cos\gamma) = \frac{3}{2}(\mathbf{r}'\cdot\mathbf{r})^2/r^2 - \frac{1}{2}r'^2$. Therefore the quadrupole potential can be written

$$\Phi_2(r,\theta,\phi) = \frac{\mathbb{G}}{2r^5}\int d\mathbf{r}'\,\rho(\mathbf{r}')\big[r'^2r^2 - 3(\mathbf{r}'\cdot\mathbf{r})^2\big]. \tag{1.128}$$

When written in terms of the inertia tensor $\mathbf{I}$ of the planet (eq. D.85), this yields **MacCullagh's formula**

$$\Phi_2(r,\theta,\phi) = \frac{3\,\mathbb{G}}{2r^5}\sum_{ij=1}^{3} r_i I_{ij} r_j - \frac{\mathbb{G}}{2r^3}\sum_{i=1}^{3} I_{ii} = \frac{3\,\mathbb{G}}{2r^5}\mathbf{r}^{\mathrm{T}}\,\mathbf{I}\mathbf{r} - \frac{\mathbb{G}}{2r^3}\mathrm{Tr}(\mathbf{I}); \tag{1.129}$$

here $\mathbf{r}^{\mathrm{T}}$ is the row vector that is the transpose of the column vector $\mathbf{r}$, and $\mathrm{Tr}\,(\mathbf{I})$ is the trace of the inertia tensor.

Since $\Phi_l(r,\theta,\phi)$ in equation (1.127) falls off with distance $\propto r^{-l-1}$, at large distances from the host planet the potential is dominated by the monopole potential ($\propto r^{-1}$) and quadrupole potential ($\propto r^{-3}$).

1.7.1 The gravitational potential of rotating fluid bodies

Small bodies, such as rocks, comets and most asteroids, are irregularly shaped. Larger astronomical bodies are nearly spherical, because the forces due to gravity overwhelm the ability of any solid material to maintain other shapes (a brief quantitative discussion of this transition is given at the end of §8.6). Stars and planets are large enough that they can usually be treated as a fluid. In this case the distribution of the matter is determined by a balance between gravity, pressure and centrifugal force due to rotation. Models of stellar and planetary interiors show that the resulting density distribution is always axisymmetric around the spin axis.[22]

Axisymmetry allows us to simplify the spherical-harmonic expansion (1.127) for the gravitational potential of the planet. If the axis of symmetry of the planet is chosen to be the polar axis ($\theta = 0$), the second line of equation (1.127) vanishes when $m \neq 0$ since $\int d\phi' Y_{lm}(\theta', \phi') \propto \int d\phi' \exp(im\phi') = 0$ when $m \neq 0$. Using the definition (C.46) of spherical harmonics in terms of associated Legendre functions, equations (1.126) and (1.127) can be rewritten as

$$\Phi(r, \theta) = -\frac{\mathbb{G}M}{r}\left[1 - \sum_{l=2}^{\infty} J_l \left(\frac{R_\text{p}}{r}\right)^l \mathsf{P}_l(\cos\theta)\right], \tag{1.130}$$

where the dimensionless **multipole moments** J_l are given by

$$J_l \equiv -\frac{1}{MR_\text{p}^l}\int d\mathbf{r}'\, \rho(\mathbf{r}')\mathsf{P}_l(\cos\theta')r'^l. \tag{1.131}$$

The quantity R_p is an arbitrary reference radius that is introduced so that J_l is dimensionless; conventionally it is chosen to be close to the planetary radius.

Since $\mathsf{P}_2(\cos\theta) = \frac{1}{2}(3\cos^2\theta - 1)$ (eq. C.45), the **quadrupole moment** J_2 can be written in Cartesian coordinates as

$$J_2 = \frac{1}{MR_\text{p}^2}\int d\mathbf{r}\, \rho(\mathbf{r})(\tfrac{1}{2}x^2 + \tfrac{1}{2}y^2 - z^2). \tag{1.132}$$

For an axisymmetric body we define the moments of inertia of the planet around the equatorial and polar axes as (cf. eqs. D.87)

$$A = \int d\mathbf{r}\, \rho(\mathbf{r})(y^2 + z^2) = I_{xx} = I_{yy},$$

[22] Non-axisymmetric equilibrium bodies of self-gravitating fluid do exist. The first and most famous example is the sequence of Jacobi ellipsoids (Chandrasekhar 1969), which are uniformly rotating masses of homogeneous, incompressible fluid. However, only axisymmetric equilibria exist for typical planets, in which the material is compressible so the mass is concentrated toward the center.

$$C = \int d\mathbf{r}\, \rho(\mathbf{r})(x^2 + y^2) = I_{zz}, \tag{1.133}$$

which implies that

$$J_2 = \frac{C - A}{MR_\mathrm{p}^2}. \tag{1.134}$$

Then either MacCullagh's formula (1.129) or equation (1.130) yields[23]

$$\begin{aligned}\Phi(r,\theta) &= -\frac{\mathbb{G}M}{r} + \frac{\mathbb{G}MJ_2R_\mathrm{p}^2}{2r^3}(3\cos^2\theta - 1) + \mathrm{O}(r^{-4}) \\ &= -\frac{\mathbb{G}M}{r} + \mathbb{G}\frac{C - A}{2r^3}(3\cos^2\theta - 1) + \mathrm{O}(r^{-4}).\end{aligned} \tag{1.135}$$

Notice that measurements of the potential external to the planet allow us to determine the *difference* between the moments of inertia A and C but not the moments themselves. The rate of precession of the spin axis due to the torque from an external body, such as the Sun, yields the dynamical ellipticity$(C - A)/C$ (cf. eq. 7.10), so measurements of both the external gravitational field and the precession are needed to determine both moments of inertia C and A.

We also expect that rotating planets or stars are symmetric about the equatorial plane (the plane normal to the polar axis that passes through their center of mass),[24] so $\rho(r,\theta)$ is an even function of $\cos\theta$ if the center of mass coincides with the origin. Since $\mathsf{P}_l(-\cos\theta) = (-1)^l\mathsf{P}_l(\cos\theta)$ (eq. C.38), all multipole moments J_l with odd values of l vanish. In this case there is a sharper limit on the error in equation (1.135): $\mathrm{O}(r^{-5})$ rather than $\mathrm{O}(r^{-4})$.

Rotation flattens the density distribution of a planet (i.e., the planet becomes **oblate**), so the moment of inertia C around the polar axis is larger than the moment A around an equatorial axis, which in turn implies through equation (1.134) that the quadrupole moment J_2 is positive. In general the

---

[23] A function $f(r)$ is $\mathrm{O}(r^{-p})$ if $r^p f(r)$ is less than some constant when r is large enough.

[24] This result can be proved analytically in simple models of a planetary interior. In particular, if the planet is uniformly rotating (i.e., the fluid has zero velocity in a frame rotating at a constant angular speed $\boldsymbol{\Omega}$) and the equation of state is barotropic (i.e., the pressure is a function only of the density), then **Lichtenstein's theorem** states that in equilibrium the fluid has reflection symmetry around a plane perpendicular to $\boldsymbol{\Omega}$ (e.g., Lindblom 1992).

Box 1.3: Rotation, quadrupole moment and flattening

If the quadrupole moment J_2 is much larger than all of the J_n with $n > 2$, equation (1.135) implies that the gravitational potential outside the planet is

$$\Phi(r,\theta) = -\frac{\mathbb{G}M}{r}\left[1 - \frac{J_2 R_\mathrm{p}^2}{2r^2}(3\cos^2\theta - 1)\right]. \tag{a}$$

We assume that the planet is rotating uniformly with angular speed Ω around its polar axis. Then the centrifugal potential is (eq. D.21)

$$\Phi_\mathrm{cent}(r,\theta) = -\tfrac{1}{2}\Omega^2(x^2 + y^2) = -\tfrac{1}{2}\Omega^2 r^2 \sin^2\theta. \tag{b}$$

If the surface of the planet can be treated as a fluid—that is, if it has an atmosphere or is large enough that the strength of the material at its surface is negligible—then the effective potential $\Phi_\mathrm{eff}(r,\theta) \equiv \Phi(r,\theta) + \Phi_\mathrm{cent}(r,\theta)$ must be constant on the surface.[a] Let the surface be $r = R_\mathrm{p} + \Delta R(\theta)$; we assume that the reference radius R_p is close enough to the mean radius of the surface that $|\Delta R(\theta)| \ll R_\mathrm{p}$. Then we may expand the effective potential to first order in $\Delta R(\theta)$, Ω^2 and J_2:

$$\Phi_\mathrm{eff}(R,\theta) = \text{constant} + \frac{\mathbb{G}M}{R_\mathrm{p}^2}\Delta R(\theta) + \frac{3\mathbb{G}M}{2R_\mathrm{p}} J_2 \cos^2\theta + \tfrac{1}{2}\Omega^2 R_\mathrm{p}^2 \cos^2\theta. \tag{c}$$

If this is to be independent of the polar angle θ on the surface, we require

$$\frac{\Delta R(\theta)}{R_\mathrm{p}} = -\left(\tfrac{3}{2}J_2 + \frac{\Omega^2 R_\mathrm{p}^3}{2\mathbb{G}M}\right)\cos^2\theta + \text{constant}. \tag{d}$$

Thus the difference between the equatorial radius $R_\mathrm{eq} = R_\mathrm{p} + \Delta R(\tfrac{1}{2}\pi)$ and the polar radius $R_\mathrm{pol} = R_\mathrm{p} + \Delta R(0)$ is

$$\frac{R_\mathrm{eq} - R_\mathrm{pol}}{R_\mathrm{p}} = \tfrac{3}{2}J_2 + \frac{\Omega^2 R_\mathrm{p}^3}{2\mathbb{G}M}. \tag{e}$$

This simple relation connects three observables: the flattening or oblateness of the planet, the rotation rate and the quadrupole moment.

---

[a] Hydrostatic equilibrium in the rotating frame requires $\nabla p = -\rho\nabla\Phi_\mathrm{eff}$ where $p(\mathbf{r})$ is the pressure and $\rho(\mathbf{r})$ is the density. Since $\nabla \times \nabla p = \mathbf{0}$ for any scalar field $p(\mathbf{r})$ (eq. B.36a), we must have $\nabla\rho \times \nabla\Phi_\mathrm{eff} = \mathbf{0}$. This result implies that the gradient of the density must be parallel to the gradient of the effective potential, so surfaces of constant density and effective potential coincide.

multipole moments with even values of l decrease rapidly as l grows, so the non-spherical part of the potential is dominated by the quadrupole term even at the surface of the planet. Given this, there is a simple relation between the rotation rate, the quadrupole moment and the flattening of the planetary surface (Box 1.3).

1.8 Nearly circular orbits

1.8.1 Expansions for small eccentricity

Most planet and satellite orbits are nearly circular, so expansions of the trajectory in powers of the eccentricity e were an essential tool for studying orbits in the days when all algebra was done by hand. Such expansions continue to provide insight in many problems of celestial mechanics. Here we illustrate the derivations of these expansions, which are given to $\mathrm{O}(e^3)$. Expansions for other variables, or higher order expansions, can easily be derived by computer algebra.

(a) True anomaly in terms of eccentric anomaly Take the log of the first of equations (1.51c),

$$f = u - \mathrm{i}\log[1 - \beta\exp(-\mathrm{i}u)] + \mathrm{i}\log[1 - \beta\exp(\mathrm{i}u)], \tag{1.136}$$

and replace β by its expression (1.52) in terms of the eccentricity e. Then expand as a Taylor series in e:

$$f = u + e\sin u + \tfrac{1}{4}e^2\sin 2u + e^3(\tfrac{1}{4}\sin u + \tfrac{1}{12}\sin 3u) + \mathrm{O}(e^4). \tag{1.137}$$

(b) Eccentric anomaly in terms of true anomaly Similarly, using the second of equations (1.51c),

$$u = f - e\sin f + \tfrac{1}{4}e^2\sin 2f - e^3(\tfrac{1}{4}\sin f + \tfrac{1}{12}\sin 3f) + \mathrm{O}(e^4). \tag{1.138}$$

(c) Mean anomaly in terms of eccentric anomaly This is simply Kepler's equation (1.49),

$$\ell = u - e\sin u. \tag{1.139}$$

(d) Mean anomaly in terms of true anomaly Combining Kepler's equation with equation (1.138) and expanding as a Taylor series in e yields

$$\ell = f - 2e\sin f + \tfrac{3}{4}e^2\sin 2f - \tfrac{1}{3}e^3\sin 3f + \mathrm{O}(e^4). \tag{1.140}$$

The most important expansions are those in terms of the mean anomaly, since time is the natural independent variable for a trajectory and mean anomaly is a linear function of time.

(e) Eccentric anomaly in terms of mean anomaly Kepler's equation implies that the eccentric anomaly u changes by 2π when the mean anomaly ℓ changes by 2π. Thus any function $g(u)$ is a periodic function of the mean anomaly, which can be expanded in a Fourier series (see Appendix B.4). In particular, setting $g(u) = \exp(\mathrm{i}ju)$ with j an integer, we may write

$$\exp(\mathrm{i}ju) = \sum_{m=-\infty}^{\infty} c_m(j)\exp(\mathrm{i}m\ell), \tag{1.141}$$

where (eq. B.48)

$$c_m(j) = \frac{1}{2\pi}\int_0^{2\pi} \mathrm{d}\ell\, \exp[\mathrm{i}(ju - m\ell)]. \tag{1.142}$$

Eliminating ℓ using Kepler's equation $\ell = u - e\sin u$,

$$c_m(j) = \frac{1}{2\pi}\int_0^{2\pi} \mathrm{d}u\,(1 - e\cos u)\exp[\mathrm{i}(j-m)u + \mathrm{i}me\sin u]. \tag{1.143}$$

For $m = 0$ it is straightforward to show that

$$c_0(j) = \delta_{j0} - \tfrac{1}{2}e\delta_{j1} - \tfrac{1}{2}e\delta_{j,-1}, \tag{1.144}$$

where δ_{mn} is the Kronecker delta (eq. C.1). For $m \neq 0$ we write (eq. C.29)

$$\exp(\mathrm{i}me\sin u) = \sum_{k=-\infty}^{\infty} J_k(me)\exp(\mathrm{i}ku), \tag{1.145}$$

where $J_k(z)$ is a Bessel function (Appendix C.5). Using the first of the identities (C.28), equation (1.143) simplifies to

$$c_m(j) = \frac{j}{m} J_{m-j}(me), \quad m \neq 0. \tag{1.146}$$

Now set $j = 1$ and take the imaginary part of equation (1.141):

$$\sin u = \sum_{\substack{m=-\infty \\ m\neq 0}}^{\infty} \frac{J_{m-1}(me)}{m} \sin m\ell. \tag{1.147}$$

Using relations (C.26) and (C.28), this result simplifies to

$$\sin u = 2 \sum_{m=1}^{\infty} \frac{J_m(me)}{me} \sin m\ell, \tag{1.148}$$

which may be combined with Kepler's equation to yield

$$u = \ell + 2 \sum_{m=1}^{\infty} \frac{J_m(me)}{m} \sin m\ell. \tag{1.149}$$

Finally, the power series for Bessel functions (C.24) can be used to convert equation (1.149) into a power series in eccentricity:

$$u = \ell + e \sin \ell + \tfrac{1}{2} e^2 \sin 2\ell + \tfrac{1}{8} e^3 (3 \sin 3\ell - \sin \ell) + \mathrm{O}(e^4). \tag{1.150}$$

(f) True anomaly in terms of mean anomaly Inserting the series (1.150) into equation (1.137) and expanding the result as a power series in eccentricity, we find

$$f = \ell + 2e \sin \ell + \tfrac{5}{4} e^2 \sin 2\ell + \tfrac{1}{12} e^3 (13 \sin 3\ell - 3 \sin \ell) + \mathrm{O}(e^4). \tag{1.151}$$

(g) Radius in terms of mean anomaly Take the real part of equation (1.141) with $j = 1$. We find

$$\cos u = -\tfrac{1}{2} e + \sum_{\substack{m=-\infty \\ m\neq 0}}^{\infty} \frac{J_{m-1}(me)}{m} \cos m\ell; \tag{1.152}$$

using equations (C.26) and (C.28), this result simplifies to

$$\cos u = -\tfrac{1}{2}e + 2\sum_{m=1}^{\infty} \frac{J'_m(me)}{m} \cos m\ell. \tag{1.153}$$

Since $r/a = 1 - e\cos u$ (eq. 1.46) we have

$$\frac{r}{a} = 1 + \tfrac{1}{2}e^2 - 2e\sum_{m=1}^{\infty} \frac{J'_m(me)}{m} \cos m\ell. \tag{1.154}$$

Finally, using the power series (C.24), we obtain a power series in eccentricity,

$$\frac{r}{a} = 1 - e\cos\ell + \tfrac{1}{2}e^2(1 - \cos 2\ell) + \tfrac{3}{8}e^3(\cos\ell - \cos 3\ell) + \mathrm{O}(e^4). \tag{1.155}$$

All of these expansions share the following important property. Consider a term of the form $e^k \cos nx$ or $e^k \sin nx$, where k and n are non-negative integers and x is any of the three anomalies u, f, or ℓ. Then k is always at least as large as n; for example, a term proportional to $\cos 3\ell$ is always multiplied at least by e^3. This behavior, which is also seen in expansions of the Hamiltonian in powers of the eccentricity and inclination (§4.3), is sometimes called the **d'Alembert property**.

1.8.2 The epicycle approximation

The equation of motion for orbits with small eccentricities and inclinations can be solved in more general axisymmetric potentials than the Kepler potential $\Phi_\mathrm{K}(r) = -\mathbb{G}M/r$. Applications of such solutions include the study of satellites orbiting an oblate planet, planets in a massive circumstellar disk, and planets orbiting close enough to the host star that relativistic corrections are important.

We consider an axisymmetric potential $\Phi(R, z)$ in cylindrical coordinates (R, ϕ, z), and assume that the potential is symmetric about the equatorial plane $z = 0$, so $\Phi(R, -z) = \Phi(R, z)$. The equations of motion for a test particle, $\ddot{\mathbf{r}} = -\nabla\Phi$, can be written (eq. B.18):

$$\ddot{R} - R\dot{\phi}^2 = -\frac{\partial\Phi(R, z)}{\partial R}, \quad 2\dot{R}\dot{\phi} + R\ddot{\phi} = 0, \quad \ddot{z} = -\frac{\partial\Phi(R, z)}{\partial z}. \tag{1.156}$$

The second equation may be multiplied by R and integrated to yield

$$R^2\dot{\phi} = \text{constant} = L_z, \tag{1.157}$$

which states that the z-component of the angular momentum is conserved, a consequence of the axisymmetry of the potential. The first of equations (1.156) can then be rewritten as

$$\ddot{R} - \frac{L_z^2}{R^3} = -\frac{\partial\Phi(R,z)}{\partial R}. \tag{1.158}$$

We first examine a circular orbit in the equatorial plane, $R(t) \equiv R_g =$ constant, $z(t) = 0$; we assume the orbit is prograde so $\dot{\phi} > 0$. The third of equations (1.156) is trivially satisfied because the potential is even in z, so $\partial\Phi/\partial z$ must vanish at $z = 0$. Equation (1.158) yields

$$L_z^2 = R_g^3\left(\frac{\partial\Phi}{\partial R}\right)_{(R_g,0)}, \tag{1.159}$$

which relates the orbital radius to the angular momentum. Equation (1.157) can then be solved,

$$\phi(t) = \kappa_\phi(R_g)t + \phi_0, \tag{1.160}$$

where ϕ_0 is an integration constant and the **azimuthal frequency** is

$$\kappa_\phi(R_g) \equiv \frac{L_z}{R_g^2} = \left(\frac{1}{R}\frac{\partial\Phi}{\partial R}\right)^{1/2}_{(R_g,0)}. \tag{1.161}$$

The **azimuthal period** $P_\phi \equiv 2\pi/\kappa_\phi$.

Now consider a nearly circular orbit with the same z-component of angular momentum L_z as in equation (1.159). We let

$$x \equiv R - R_g, \tag{1.162}$$

and expand the potential in a Taylor series around $(R,z) = (R_g,0)$:

$$\Phi(R,z) = \left(\frac{\partial\Phi}{\partial R}\right)_{(R_g,0)} x + \tfrac{1}{2}\left(\frac{\partial^2\Phi}{\partial R^2}\right)_{(R_g,0)} x^2 + \tfrac{1}{2}\left(\frac{\partial^2\Phi}{\partial z^2}\right)_{(R_g,0)} z^2, \tag{1.163}$$

plus terms that are $O(x^3, xz^2, z^4)$ and an unimportant constant. The terms in the expansion proportional to $\partial\Phi/\partial z$ and $\partial^2\Phi/\partial R\partial z$ have vanished because the potential is even in z. The **epicycle approximation** consists of neglecting all terms in this expansion that are higher than second order in x and z, which corresponds to ignoring all terms in the equations of motion (which involve $\nabla\Phi$) that are higher than first order in x and z.

We first examine motion in the z-direction, substituting the Taylor series (1.163) into the third of equations (1.156) to obtain

$$\ddot{z} + \kappa_z^2 z = 0, \tag{1.164}$$

where the **vertical frequency** is

$$\kappa_z(R_g) = \left(\frac{\partial^2\Phi}{\partial z^2}\right)^{1/2}_{(R_g,0)}. \tag{1.165}$$

Thus, in the epicycle approximation the vertical motion is decoupled from the horizontal motion and described by the solution to the harmonic-oscillator equation (1.164),

$$z(t) = z_0 \cos(\kappa_z t + \zeta), \tag{1.166}$$

with integration constants $z_0 \geq 0$ and ζ.

We next turn to the radial equation of motion (1.158), replacing $\ddot{R}$ by $\ddot{x}$, the potential $\Phi(R, z)$ by its Taylor expansion (1.163), and L_z^2/R^3 by its Taylor expansion $L_z^2/(R_g + x)^3 = L_z^2/R_g^3 - 3(L_z^2/R_g^4)x + O(x^2)$. The terms independent of x cancel because of equation (1.159), and discarding all terms that are higher than first order in x or z, we obtain

$$\ddot{x} + \kappa_R^2 x = 0, \tag{1.167}$$

where the **radial** or **epicycle frequency** is

$$\kappa_R(R_g) = \left(\frac{3L_z^2}{R^4} + \frac{\partial^2\Phi}{\partial R^2}\right)^{1/2}_{(R_g,0)} = \left(\frac{3}{R}\frac{\partial\Phi}{\partial R} + \frac{\partial^2\Phi}{\partial R^2}\right)^{1/2}_{(R_g,0)}, \tag{1.168}$$

and the **radial period** is $P_R \equiv 2\pi/\kappa_R$.

Like the vertical motion, the radial motion is described by the solution to a harmonic-oscillator equation,

$$x(t) = x_0 \cos(\kappa_R t + \eta), \tag{1.169}$$

with integration constants $x_0 \geq 0$ and η.

Finally, we solve for the azimuthal motion by writing equation (1.157) in the form $\dot{\phi} = L_z/(R_g + x)^2 = \kappa_\phi(1 - 2x/R_g) + \mathrm{O}(x^2)$; dropping terms higher than $\mathrm{O}(x)$ and using equation (1.169), we find

$$\phi(t) = \kappa_\phi t + \phi_0 - \frac{2x_0}{R_g}\frac{\kappa_\phi}{\kappa_R}\sin(\kappa_R t + \eta). \tag{1.170}$$

Thus, although the radial and vertical motions are decoupled, the radial and azimuthal motions are not. In particular, motion in the orbital plane is the superposition of (i) uniform circular motion of a **guiding center** with coordinates $(R, \phi) = (R_g, \kappa_\phi t + \phi_0)$, and (ii) motion around an ellipse (the **epicycle**) centered on the guiding center. The motion around the epicycle is retrograde, that is, clockwise if the motion of the guiding center is counterclockwise. The semi-axes of the ellipse are x_0 in the radial direction and $2x_0\kappa_\phi/\kappa_R$ in the azimuthal direction, so the axis ratio of the ellipse is

$$\frac{\text{radial axis}}{\text{azimuthal axis}} = \frac{\kappa_R}{2\kappa_\phi}. \tag{1.171}$$

Since the motion around the epicycle has fixed frequency κ_R, this is also the ratio of the root-mean-square velocities relative to the guiding center in the radial and azimuthal directions.

For example, in the Kepler potential $\Phi_{\mathrm{K}}(R, z) = -\mathbb{G}M/(R^2 + z^2)^{1/2}$,

$$\kappa_R = \kappa_\phi = \kappa_z = n = \left(\frac{\mathbb{G}M}{R_g^3}\right)^{1/2}, \tag{1.172}$$

where n is the usual mean motion. The periapsis and apoapsis are $R_g - x_0$ and $R_g + x_0$ and since these equal $a(1 - e)$ and $a(1 + e)$ in the usual orbital elements, we conclude that $R_g = a$ and $x_0 = ae$.

We can also describe the shape of the orbit. Since we are only working to first order in the displacement from a circular orbit, the time t appearing in the oscillatory first-order terms of equations (1.166) and (1.169) can be replaced by the azimuth ϕ using the zero-order part of equation (1.170), $\phi(t) = \kappa_\phi t + \phi_0$. Thus we find

$$x = x_0 \cos\left[\frac{\kappa_R}{\kappa_\phi}(\phi - \phi_R)\right], \qquad z = z_0 \cos\left[\frac{\kappa_z}{\kappa_\phi}(\phi - \phi_z)\right], \tag{1.173}$$

where the constants $\phi_R \equiv \phi_0 - (\kappa_\phi/\kappa_R)\eta$ and $\phi_z \equiv \phi_0 - (\kappa_\phi/\kappa_z)\zeta$. Unless κ_R/κ_ϕ and κ_z/κ_ϕ are rational numbers (as in eq. 1.172), the orbit is not closed; eventually the particle passes arbitrarily close to every point in the square $|x| \le x_0$, $|z| \le z_0$.

The longitude of periapsis ϖ is the azimuth at which the orbital radius is smallest[25] and is determined by setting the argument of the cosine in the first of equations (1.173) to $\pi, 3\pi, 5\pi, \ldots$. Thus $\varpi = \phi_R + \pi\kappa_\phi/\kappa_R, \phi_R + 3\pi\kappa_\phi/\kappa_R, \ldots$. For Kepler orbits, with $\kappa_\phi = \kappa_R$, these angles are all the same modulo 2π so the longitude of periapsis is fixed. If the potential is close to a Kepler potential, the longitude of periapsis will appear to change slowly, by an amount $\Delta\varpi = 2\pi[(\kappa_\phi/\kappa_R) - 1]$ between successive periapsis passages. Since the time between such passages is $\Delta t = 2\pi/\kappa_R$, the longitude of periapsis advances at an average rate

$$\frac{\mathrm{d}\varpi}{\mathrm{d}t} = \frac{\Delta\varpi}{\Delta t} = \kappa_\phi - \kappa_R. \tag{1.174}$$

Similarly, the longitude of the ascending node Ω is the azimuth at which the particle pierces the equatorial plane traveling upward, and is determined by setting the argument of the cosine in the second of equations (1.173) to $\frac{3}{2}\pi, \frac{7}{2}\pi, \ldots$. The longitude of the ascending node advances at a rate

$$\frac{\mathrm{d}\Omega}{\mathrm{d}t} = \kappa_\phi - \kappa_z. \tag{1.175}$$

[25] This definition differs slightly from the one given in §1.3.2, which measures the longitude of periapsis as the sum of two angles in different planes. However, the difference between the two definitions is $\mathrm{O}(z^2)$ and hence is negligible in the context of the epicycle approximation.

Box 1.4: Osculating elements

Orbital elements such as semimajor axis and eccentricity are defined in the Kepler potential. For these to be useful when additional forces are present we must agree on the prescription to be used for calculating the elements from the position and velocity. The **osculating elements** at time t are defined to be the elements that the orbit would have if the perturbing forces were switched off instantaneously at time t. In other words the position and velocity are converted to orbital elements using the same formulas that would apply if there were no forces other than the Kepler gravitational force.

This definition sounds obvious, but some of its consequences are not. For example, consider a particle on a circular orbit of radius R in the equatorial plane of an oblate planet, in which the quadrupole moment $J_2 \neq 0$ but all the higher multipoles vanish. Then the semimajor axis a is not equal to the radius, and the eccentricity e is not equal to zero, even though the orbit is circular. Quantitatively, the radial velocity $v_R = 0$ and from equation (1.176) the azimuthal velocity is $v_\phi = R\kappa_\phi(R) = (\mathbb{G}M/R)^{1/2}(1 + \frac{3}{2}J_2R_\mathrm{p}^2/R^2)^{1/2}$. In a Kepler potential the semimajor axis and eccentricity are related to the position and velocity by $\frac{1}{2}(v_R^2 + v_\phi^2) - \mathbb{G}M/R = -\frac{1}{2}\mathbb{G}M/a$, and $(Rv_\phi)^2 = \mathbb{G}Ma(1-e^2)$. Therefore

$$a = \frac{R}{1 - \frac{3}{2}J_2R_\mathrm{p}^2/R^2}, \quad e = \frac{3J_2R_\mathrm{p}^2}{2R^2}. \tag{a}$$

An example of non-osculating elements is given in §7.1.1.

1.8.3 Orbits and the multipole expansion

As an example of the use of the epicycle approximation, we compute the apsidal and nodal precession rates $\dot{\varpi}$ and $\dot{\Omega}$ for low-eccentricity, low-inclination orbits in the gravitational field of a oblate, axisymmetric planet.

We expand the gravitational potential using equation (1.130). In the $z = 0$ plane we have $\theta = \frac{1}{2}\pi$ and $r = R$. As in the preceding subsection, we assume that the potential $\Phi(R, z)$ is an even function of z, which implies that the multipole moments J_l vanish for odd l. Then equations (1.161) and

(1.168) yield

$$\kappa_\phi^2(R) = \frac{\mathbb{G}M}{R^3}\left[1 - \sum_{l=2}^{\infty}(l+1)\mathsf{P}_l(0)J_l\left(\frac{R_\mathrm{p}}{R}\right)^l\right], \tag{1.176}$$

$$\kappa_R^2(R) = \frac{\mathbb{G}M}{R^3}\left[1 + \sum_{l=2}^{\infty}(l^2-1)\mathsf{P}_l(0)J_l\left(\frac{R_\mathrm{p}}{R}\right)^l\right]. \tag{1.177}$$

Here $\mathsf{P}_l(0)$ is given by equation (C.42) and the sums can be restricted to even values of l.

To derive κ_z^2 we write $r = (R^2 + z^2)^{1/2}$, $\cos\theta = z/(R^2 + z^2)^{1/2}$, and note that

$$\left.\frac{\partial^2 r^{-p}}{\partial z^2}\right|_{z=0} = -\frac{p}{R^{p+2}}, \quad \left.\frac{\partial^2 \mathsf{P}_l(\cos\theta)}{\partial z^2}\right|_{z=0} = \frac{\mathsf{P}_l''(0)}{R^2} = -\frac{l(l+1)}{R^2}\mathsf{P}_l(0), \tag{1.178}$$

where the last equality follows from equation (C.36). Thus equation (1.165) yields

$$\kappa_z^2(R) = \frac{\mathbb{G}M}{R^3}\left[1 - \sum_{l=2}^{\infty}(l+1)^2\mathsf{P}_l(0)J_l\left(\frac{R_\mathrm{p}}{R}\right)^l\right]. \tag{1.179}$$

These expressions for κ_ϕ, κ_R and κ_z can be employed in equations (1.174) and (1.175) to derive exact expressions for the apsidal and nodal precession rates $\dot\varpi$ and $\dot\Omega$ in the limit of very small eccentricity and inclination. However, simpler expressions are usually sufficient, because the corrections to the Kepler potential from the multipole potentials are generally small even for the first term, $l = 2$, and decrease rapidly with increasing l. For most purposes, expressions of sufficient accuracy can be derived by (i) neglecting all multipole moments J_l with $l > 4$; and (ii) taking the square root of equations (1.176), (1.177) and (1.179), expanding the result as a series in J_2 and J_4, and dropping all terms of order J_2^3 or higher and J_4^2 and higher. With these approximations,

$$\frac{\mathrm{d}\varpi}{\mathrm{d}t} = \left(\frac{\mathbb{G}M}{R^3}\right)^{1/2}\left[\frac{3J_2R_\mathrm{p}^2}{2R^2} - \frac{15J_4R_\mathrm{p}^4}{4R^4} + \mathrm{O}(J_2^3, J_4^2, J_6)\right], \tag{1.180a}$$

$$\frac{\mathrm{d}\Omega}{\mathrm{d}t} = \left(\frac{\mathbb{G}M}{R^3}\right)^{1/2}\left[-\frac{3J_2R_\mathrm{p}^2}{2R^2} + \frac{15J_4R_\mathrm{p}^4}{4R^4} + \frac{9J_2^2R_\mathrm{p}^4}{4R^4} + \mathrm{O}(J_2^3, J_4^2, J_6)\right]. \tag{1.180b}$$

If the terms of order J_2^2 are neglected as well, we have $\mathrm{d}\varpi/\mathrm{d}t = -\mathrm{d}\Omega/\mathrm{d}t$; at this level of approximation, the nodes regress at the same rate that the apsides advance. The generalization to orbits of arbitrary eccentricity and inclination is given in Problem 5.3.

1.9 Response of an orbit to an external force

The evolution of a planetary trajectory under the influence of a force other than the attraction of the host star (an **external force**) is described by the equation

$$\frac{\mathrm{d}^2\mathbf{r}}{\mathrm{d}t^2} = -\frac{\mathbb{G}M}{r^3}\mathbf{r} + \mathbf{F}_\mathrm{ext}, \tag{1.181}$$

where $\mathbf{r}$ is the position of the planet relative to the host star, M is the sum of the masses of the planet and star, and $\mathbf{F}_\mathrm{ext}$ is the external force per unit mass on the planet (for an explicit derivation of this result, follow the steps leading to eq. 1.11).[26]

If the external force is weak, the trajectory is approximately a Kepler ellipse, and therefore can be described more economically in terms of the time evolution of the orbital elements rather than the time evolution of the position. The main goal of this section is to derive the equations that describe this evolution.

[26] The interpretation of the external force $\mathbf{F}_\mathrm{ext}$ requires care in more general cases. If there are forces per unit mass $\mathbf{F}_\mathrm{p}$ and $\mathbf{F}_*$ on the planet and star, then $\mathbf{F}_\mathrm{ext} = \mathbf{F}_\mathrm{p} - \mathbf{F}_*$. If the external force arises from the negative gradient of a potential of the form $m_\mathrm{p}m_*\phi(\mathbf{r}_\mathrm{p} - \mathbf{r}_*)$, say because one or both of the bodies is not a point mass, then $\mathbf{F}_\mathrm{ext} = -(m_* + m_\mathrm{p})\partial\phi(\mathbf{r})/\partial\mathbf{r}$. Notice that in this case $\mathbf{F}_\mathrm{ext}$ is larger than the actual force per unit mass on the planet by a factor $1 + m_\mathrm{p}/m_*$. Of course, in most practical cases $m_\mathrm{p} \ll m_*$ so this distinction is not important.

1.9.1 Lagrange's equations

When the external force can be derived from a Hamiltonian H_{ext}, we can use the tools of Hamiltonian dynamics to find the evolution of the orbital elements. Since the Delaunay variables $\mathbf{q} \equiv (\ell, \omega, \Omega)$, $\mathbf{p} \equiv (\Lambda, L, L_z)$ (eq. 1.84) are canonical, the equation of motion is simply (eq. D.13)

$$\frac{\mathrm{d}\mathbf{z}}{\mathrm{d}t} = \mathbf{J}\frac{\partial}{\partial \mathbf{z}}(H_{\text{K}} + H_{\text{ext}}), \tag{1.182}$$

where $\mathbf{z} = (\mathbf{q}, \mathbf{p})$, $\mathbf{J}$ is the symplectic matrix (eq. D.14), and the Kepler Hamiltonian $H_{\text{K}} = -\frac{1}{2}(\mathbb{G}M)^2/\Lambda^2$ (eq. 1.85).

Despite the simplicity of the Delaunay variables, it is often easier to work with the non-canonical elements $\mathbf{E} \equiv (\lambda, \varpi, \Omega, a, e, I)$—mean longitude, longitude of periapsis, longitude of the ascending node, semimajor axis, eccentricity and inclination. Then from equations (D.49)

$$\frac{\mathrm{d}\mathbf{E}}{\mathrm{d}t} = \mathbf{G}\mathbf{J}\mathbf{G}^{\text{T}}\frac{\partial}{\partial \mathbf{E}}(H_{\text{K}} + H_{\text{ext}}), \tag{1.183}$$

where $G_{ij} \equiv \partial E_i/\partial z_j$ is the Jacobian matrix relating $\mathbf{z}$ and $\mathbf{E}$.

The Jacobian matrix is straightforward to evaluate using the relations

$$\lambda = \ell + \omega + \Omega, \quad \varpi = \omega + \Omega, \quad a = \frac{\Lambda^2}{\mathbb{G}M}, \quad e = (1 - L^2/\Lambda^2)^{1/2}, \quad I = \cos^{-1} L_z/L. \tag{1.184}$$

We find

$$\mathbf{G} = \begin{bmatrix} 1 & 1 & 1 & 0 & 0 & 0 \\ 0 & 1 & 1 & 0 & 0 & 0 \\ 0 & 0 & 1 & 0 & 0 & 0 \\ 0 & 0 & 0 & \dfrac{2\Lambda}{\mathbb{G}M} & 0 & 0 \\ 0 & 0 & 0 & \dfrac{L^2}{\Lambda^2(\Lambda^2 - L^2)^{1/2}} & -\dfrac{L}{\Lambda(\Lambda^2 - L^2)^{1/2}} & 0 \\ 0 & 0 & 0 & 0 & \dfrac{L_z}{L(L^2 - L_z^2)^{1/2}} & -\dfrac{1}{(L^2 - L_z^2)^{1/2}} \end{bmatrix}. \tag{1.185}$$

Then

$$\mathbf{GJG}^{\mathrm{T}} = \frac{1}{na^2}\begin{bmatrix} 0 & 0 & 0 & 2a & -\dfrac{j(1-j)}{e} & -\dfrac{\tan\frac{1}{2}I}{j} \\ 0 & 0 & 0 & 0 & -\dfrac{j}{e} & -\dfrac{\tan\frac{1}{2}I}{j} \\ 0 & 0 & 0 & 0 & 0 & -\dfrac{1}{j\sin I} \\ -2a & 0 & 0 & 0 & 0 & 0 \\ \dfrac{j(1-j)}{e} & \dfrac{j}{e} & 0 & 0 & 0 & 0 \\ \dfrac{\tan\frac{1}{2}I}{j} & \dfrac{\tan\frac{1}{2}I}{j} & \dfrac{1}{j\sin I} & 0 & 0 & 0 \end{bmatrix}, \tag{1.186}$$

where $j \equiv (1-e^2)^{1/2}$ and the mean motion $n = (\mathbb{G}M/a^3)^{1/2}$. Inserting this matrix into (1.183), we obtain **Lagrange's equations**:[27]

$$\begin{aligned}
\frac{\mathrm{d}\lambda}{\mathrm{d}t} &= n + \frac{2}{na}\frac{\partial H_{\mathrm{ext}}}{\partial a} - \frac{j(1-j)}{na^2e}\frac{\partial H_{\mathrm{ext}}}{\partial e} - \frac{\tan\frac{1}{2}I}{na^2j}\frac{\partial H_{\mathrm{ext}}}{\partial I}, \\
\frac{\mathrm{d}\varpi}{\mathrm{d}t} &= -\frac{j}{na^2e}\frac{\partial H_{\mathrm{ext}}}{\partial e} - \frac{\tan\frac{1}{2}I}{na^2j}\frac{\partial H_{\mathrm{ext}}}{\partial I}, \\
\frac{\mathrm{d}\Omega}{\mathrm{d}t} &= -\frac{1}{na^2j\sin I}\frac{\partial H_{\mathrm{ext}}}{\partial I}, \\
\frac{\mathrm{d}a}{\mathrm{d}t} &= -\frac{2}{na}\frac{\partial H_{\mathrm{ext}}}{\partial \lambda}, \\
\frac{\mathrm{d}e}{\mathrm{d}t} &= \frac{j(1-j)}{na^2e}\frac{\partial H_{\mathrm{ext}}}{\partial \lambda} + \frac{j}{na^2e}\frac{\partial H_{\mathrm{ext}}}{\partial \varpi}, \\
\frac{\mathrm{d}I}{\mathrm{d}t} &= \frac{\tan\frac{1}{2}I}{na^2j}\frac{\partial H_{\mathrm{ext}}}{\partial \lambda} + \frac{\tan\frac{1}{2}I}{na^2j}\frac{\partial H_{\mathrm{ext}}}{\partial \varpi} + \frac{1}{na^2j\sin I}\frac{\partial H_{\mathrm{ext}}}{\partial \Omega}.
\end{aligned} \tag{1.187}$$

When the eccentricity and inclination are small, we can simplify Lagrange's equations by evaluating the factors multiplying the partial derivatives of

[27] Traditionally, Lagrange's equations have been written using a function $R = -H_{\mathrm{ext}}$ on the right side, so all of the signs are reversed. Probably this convention arose because R is positive when the external forces arise from a gravitational potential.

$H_{\rm ext}$ as power series in e and I and dropping all terms that are $\mathrm{O}(e, I)$ or higher:

$$\begin{aligned}
\frac{\mathrm{d}\lambda}{\mathrm{d}t} &= n + \frac{2}{na}\frac{\partial H_{\rm ext}}{\partial a}, & \frac{\mathrm{d}a}{\mathrm{d}t} &= -\frac{2}{na}\frac{\partial H_{\rm ext}}{\partial \lambda}, \\
\frac{\mathrm{d}\varpi}{\mathrm{d}t} &= -\frac{1}{na^2 e}\frac{\partial H_{\rm ext}}{\partial e}, & \frac{\mathrm{d}e}{\mathrm{d}t} &= \frac{1}{na^2 e}\frac{\partial H_{\rm ext}}{\partial \varpi}, \\
\frac{\mathrm{d}\Omega}{\mathrm{d}t} &= -\frac{1}{na^2 I}\frac{\partial H_{\rm ext}}{\partial I}, & \frac{\mathrm{d}I}{\mathrm{d}t} &= \frac{1}{na^2 I}\frac{\partial H_{\rm ext}}{\partial \Omega}.
\end{aligned} \tag{1.188}$$

Several of Lagrange's equations are ill-defined when the eccentricity e or inclination I is zero, and as a result the equations are difficult to integrate numerically or analytically when e or I is small. In these situations it is better to work with the orbital elements $\mathbf{E}' = (\lambda, k, q, a, h, p)$. From equations (1.71) we have

$$\begin{aligned}
\lambda &= \ell + \omega + \Omega, \\
k &= e\cos\varpi = (1 - L^2/\Lambda^2)^{1/2}\cos(\omega + \Omega), \\
q &= \tan I \cos\Omega = \frac{(L^2 - L_z^2)^{1/2}}{L_z}\cos\Omega, \\
a &= \frac{\Lambda^2}{\mathbb{G}M}, \\
h &= e\sin\varpi = (1 - L^2/\Lambda^2)^{1/2}\sin(\omega + \Omega), \\
p &= \tan I \sin\Omega = \frac{(L^2 - L_z^2)^{1/2}}{L_z}\sin\Omega.
\end{aligned} \tag{1.189}$$

The Jacobian matrix $G'_{ij} = \partial E'_i/\partial z_j$ is

$$\mathbf{G}' = \begin{bmatrix} 1 & 1 & 1 & 0 & 0 & 0 \\ 0 & -h & -h & \dfrac{L^2}{\Lambda(\Lambda^2 - L^2)}k & -\dfrac{L}{\Lambda^2 - L^2}k & 0 \\ 0 & 0 & -p & 0 & \dfrac{L}{L^2 - L_z^2}q & -\dfrac{L^2}{L_z(L^2 - L_z^2)}q \\ 0 & 0 & 0 & \dfrac{2\Lambda}{\mathbb{G}M} & 0 & 0 \\ 0 & k & k & \dfrac{L^2}{\Lambda(\Lambda^2 - L^2)}h & -\dfrac{L}{\Lambda^2 - L^2}h & 0 \\ 0 & 0 & q & 0 & \dfrac{L}{L^2 - L_z^2}p & -\dfrac{L^2}{L_z(L^2 - L_z^2)}p \end{bmatrix}. \tag{1.190}$$

Then

$$\mathbf{G}'\mathbf{J}\mathbf{G}'^{\mathrm{T}} = \frac{1}{na^2}\begin{bmatrix} 0 & -\dfrac{jk}{1+j} & -\dfrac{q}{jc} & 2a & -\dfrac{jh}{1+j} & -\dfrac{p}{jc} \\ \dfrac{jk}{1+j} & 0 & \dfrac{hq}{jc} & 0 & j & \dfrac{hp}{jc} \\ \dfrac{q}{jc} & -\dfrac{hq}{jc} & 0 & 0 & \dfrac{kq}{jc} & \dfrac{1}{j\cos^3 I} \\ -2a & 0 & 0 & 0 & 0 & 0 \\ \dfrac{jh}{1+j} & -j & -\dfrac{kq}{jc} & 0 & 0 & -\dfrac{kp}{jc} \\ \dfrac{p}{jc} & -\dfrac{hp}{jc} & -\dfrac{1}{j\cos^3 I} & 0 & \dfrac{kp}{jc} & 0 \end{bmatrix}, \tag{1.191}$$

where as usual $j = (1 - e^2)^{1/2} = (1 - k^2 - h^2)^{1/2}$, and $c \equiv \cos I + \cos^2 I$. The analog to equation (1.183) for the primed elements gives

$$\begin{aligned}
\frac{\mathrm{d}\lambda}{\mathrm{d}t} &= n + \frac{2}{na}\frac{\partial H_{\mathrm{ext}}}{\partial a} - \frac{j}{na^2(1+j)}\left(k\frac{\partial H_{\mathrm{ext}}}{\partial k} + h\frac{\partial H_{\mathrm{ext}}}{\partial h}\right) \\
&\quad - \frac{1}{na^2 jc}\left(q\frac{\partial H_{\mathrm{ext}}}{\partial q} + p\frac{\partial H_{\mathrm{ext}}}{\partial p}\right), \\
\frac{\mathrm{d}k}{\mathrm{d}t} &= \frac{jk}{na^2(1+j)}\frac{\partial H_{\mathrm{ext}}}{\partial \lambda} + \frac{j}{na^2}\frac{\partial H_{\mathrm{ext}}}{\partial h} + \frac{h}{na^2 jc}\left(q\frac{\partial H_{\mathrm{ext}}}{\partial q} + p\frac{\partial H_{\mathrm{ext}}}{\partial p}\right), \\
\frac{\mathrm{d}q}{\mathrm{d}t} &= \frac{q}{na^2 jc}\frac{\partial H_{\mathrm{ext}}}{\partial \lambda} + \frac{q}{na^2 jc}\left(k\frac{\partial H_{\mathrm{ext}}}{\partial h} - h\frac{\partial H_{\mathrm{ext}}}{\partial k}\right) + \frac{1}{na^2 j\cos^3 I}\frac{\partial H_{\mathrm{ext}}}{\partial p},
\end{aligned}$$

$$\frac{\mathrm{d}a}{\mathrm{d}t} = -\frac{2}{na}\frac{\partial H_{\mathrm{ext}}}{\partial \lambda}, \tag{1.192}$$
$$\frac{\mathrm{d}h}{\mathrm{d}t} = \frac{jh}{na^2(1+j)}\frac{\partial H_{\mathrm{ext}}}{\partial \lambda} - \frac{j}{na^2}\frac{\partial H_{\mathrm{ext}}}{\partial k} - \frac{k}{na^2jc}\left(q\frac{\partial H_{\mathrm{ext}}}{\partial q} + p\frac{\partial H_{\mathrm{ext}}}{\partial p}\right),$$
$$\frac{\mathrm{d}p}{\mathrm{d}t} = \frac{p}{na^2jc}\frac{\partial H_{\mathrm{ext}}}{\partial \lambda} + \frac{p}{na^2jc}\left(k\frac{\partial H_{\mathrm{ext}}}{\partial h} - h\frac{\partial H_{\mathrm{ext}}}{\partial k}\right) - \frac{1}{na^2j\cos^3 I}\frac{\partial H_{\mathrm{ext}}}{\partial q}.$$

When the eccentricity and inclination are small, we can simplify these equations using the same approximations that we used to derive equations (1.188):

$$\frac{\mathrm{d}\lambda}{\mathrm{d}t} = n + \frac{2}{na}\frac{\partial H_{\mathrm{ext}}}{\partial a}, \qquad \frac{\mathrm{d}a}{\mathrm{d}t} = -\frac{2}{na}\frac{\partial H_{\mathrm{ext}}}{\partial \lambda},$$
$$\frac{\mathrm{d}k}{\mathrm{d}t} = \frac{1}{na^2}\frac{\partial H_{\mathrm{ext}}}{\partial h}, \qquad \frac{\mathrm{d}h}{\mathrm{d}t} = -\frac{1}{na^2}\frac{\partial H_{\mathrm{ext}}}{\partial k},$$
$$\frac{\mathrm{d}q}{\mathrm{d}t} = \frac{1}{na^2}\frac{\partial H_{\mathrm{ext}}}{\partial p}, \qquad \frac{\mathrm{d}p}{\mathrm{d}t} = -\frac{1}{na^2}\frac{\partial H_{\mathrm{ext}}}{\partial q}. \tag{1.193}$$

1.9.2 Gauss's equations

An alternative approach is to work directly with the external force per unit mass $\mathbf{F}_{\mathrm{ext}}$ rather than the corresponding Hamiltonian H_{ext}. To do so, we first introduce the orbital elements $\overline{\mathbf{E}} \equiv (f, \omega, \Omega, a, e, I)$ where f is the true anomaly and $\omega = \varpi - \Omega$ is the argument of periapsis. In Cartesian coordinates, the position of the planet is given by equations (1.70), which can be written

$$\mathbf{r} = \mathbf{R}^{\mathrm{T}}(\Omega, I, f + \omega)\begin{bmatrix} r \\ 0 \\ 0 \end{bmatrix}, \tag{1.194}$$

in which $\mathbf{R}^{\mathrm{T}}$ is the transpose of the rotation matrix (B.60), given explicitly by equation (B.61). The radius $r = a(1 - e^2)/(1 + e\cos f)$ (eq. 1.29), so equation (1.194) expresses $\mathbf{r}$ as a function of the elements $\overline{\mathbf{E}}$.

We assume that the Hamiltonian H_{ext} is written as a function of the orbital elements $\overline{\mathbf{E}}$ but that it is derived from a potential that depends only

Box 1.5: Radiation pressure and Poynting–Robertson drag

Small bodies orbiting a star experience forces from the radiation field of the star.

We first assume that the small body is spherical, with radius R, and perfectly absorbing. If the luminosity of the star is L, the flux of radiation (energy per unit time crossing unit area on a spherical surface at a distance r from the star) is $F = L/(4\pi r^2)$. A stationary body at this distance absorbs energy at a rate $\dot{E} = \pi R^2 F$. If the body is moving at velocity $\mathbf{v}$, the rate of energy absorption is modified by the Doppler effect to $\dot{E} = \pi R^2 F(1-\mathbf{v}\cdot\hat{\mathbf{r}}/c)$ where $\hat{\mathbf{r}}$ is the unit vector from the star to the body. Since a photon of energy ϵ carries momentum ϵ/c, the corresponding rate of absorption of linear momentum is $\dot{\mathbf{p}} = \dot{E}\hat{\mathbf{r}}/c$. The rate of change of momentum equals the force, so the force due to **radiation pressure** is

$$\mathbf{F}_{\rm rad} = \hat{\mathbf{r}}\frac{LR^2}{4r^2c}\left(1 - \frac{\mathbf{v}\cdot\hat{\mathbf{r}}}{c}\right). \qquad \text{(a)}$$

In thermal equilibrium, the body re-radiates all the energy that it absorbs. Since a photon of energy ϵ has mass ϵ/c^2, this re-radiation implies a mass-loss rate $\dot{M} = -\dot{E}/c^2$. If the re-radiation is isotropic in the body's rest frame, the net rate of momentum change associated with this mass loss is $\dot{\mathbf{p}} = \dot{M}\mathbf{v} = -\mathbf{v}\dot{E}/c^2$ in the inertial frame, equivalent to a **Poynting–Robertson drag** force[a]

$$\mathbf{F}_{\rm PR} = -\mathbf{v}\frac{LR^2}{4r^2c^2}; \qquad \text{(b)}$$

here we have dropped a term that is smaller by $\mathrm{O}(v/c)$. The total force is

$$\mathbf{F} = \mathbf{F}_{\rm rad} + \mathbf{F}_{\rm PR} = \frac{LR^2}{4r^2c}\left[\left(1 - \frac{\mathbf{v}\cdot\hat{\mathbf{r}}}{c}\right)\hat{\mathbf{r}} - \frac{\mathbf{v}}{c}\right] + \mathrm{O}(v^2/c^2). \qquad \text{(c)}$$

In practice this result should be multiplied by an efficiency factor Q that accounts for scattering, diffraction and incomplete absorption (Burns et al. 1979).

A useful reference number is the ratio of the radiation force on a stationary body to the gravitational force from the host star, $F_g = \mathbb{G}M_*m/r^2$, where $m = \frac{4}{3}\pi\rho R^3$ is the mass of the body and ρ is its density:

$$\beta \equiv \frac{F_{\rm rad}(\mathbf{v}=0)}{F_g} = \frac{3LQ}{16\pi\,\mathbb{G}M_*c\rho R} = 0.191\,Q\frac{L}{L_\odot}\frac{M_\odot}{M_*}\frac{3\,\mathrm{g\,cm^{-3}}}{\rho}\frac{1\,\mu}{R}; \qquad \text{(d)}$$

here we have written the radius in units of microns ($1\,\mu = 10^{-4}\,\mathrm{cm} = 10^{-6}\,\mathrm{m}$). Note that β is independent of the distance from the star, r.

In general $Q \ll 1$ for particles orbiting the Sun with sizes much smaller than the wavelength at the peak of the solar spectrum, around $0.5\,\mu$. Thus β peaks for most materials at $R \sim 0.1\,\mu$. When $\beta > 1$ the outward force from radiation pressure exceeds the inward force from gravity, and the body is unbound.

[a] Some authors label the velocity-dependent term in equation (a) as part of the Poynting–Robertson drag rather than the radiation pressure, while others label the radial component of equation (b) as part of the radiation pressure.

on $\mathbf{r}$; then the external force per unit mass is $\mathbf{F}_{\text{ext}} = -\partial H_{\text{ext}}/\partial \mathbf{r}$ and we have

$$\frac{\partial H_{\text{ext}}}{\partial \overline{E}_j} = \sum_{k=1}^{3} \frac{\partial H_{\text{ext}}}{\partial r_k}\frac{\partial r_k}{\partial \overline{E}_j} = -\sum_{k=1}^{3} F_{\text{ext},k}\frac{\partial r_k}{\partial \overline{E}_j} = -\sum_{k=1}^{3} F_{\text{ext},k}\frac{\partial}{\partial \overline{E}_j} r R_{1k}. \tag{1.195}$$

Then the derivatives of the Hamiltonian in terms of the orbital elements $\mathbf{E} = (\lambda, \varpi, \Omega, a, e, I)$ are

$$\frac{\partial H_{\text{ext}}}{\partial E_m} = \sum_{j=1}^{6} \frac{\partial H_{\text{ext}}}{\partial \overline{E}_j} C_{jm} = -\sum_{j=1}^{6} C_{jm} \sum_{k=1}^{3} F_{\text{ext},k}\frac{\partial}{\partial \overline{E}_j} r R_{1k}, \tag{1.196}$$

where $\mathbf{C}$ is the Jacobian matrix[28]

$$\mathbf{C} = \left[\frac{\partial \overline{E}_j}{\partial E_m}\right] = \frac{\partial(f, \omega, \Omega, a, e, I)}{\partial(\lambda, \varpi, \Omega, a, e, I)} \tag{1.197}$$

$$= \begin{bmatrix} \frac{(1+e\cos f)^2}{j^3} & -\frac{(1+e\cos f)^2}{j^3} & 0 & 0 & \frac{\sin f(2+e\cos f)}{j^2} & 0 \\ 0 & 1 & -1 & 0 & 0 & 0 \\ 0 & 0 & 1 & 0 & 0 & 0 \\ 0 & 0 & 0 & 1 & 0 & 0 \\ 0 & 0 & 0 & 0 & 1 & 0 \\ 0 & 0 & 0 & 0 & 0 & 1 \end{bmatrix}.$$

The expression (1.196) gives the derivatives of the Hamiltonian as functions of the external force components along the three coordinate axes, $\mathbf{F}_{\text{ext}} = \sum_{k=1}^{3} F_{\text{ext},k}\hat{\mathbf{n}}_k$ with $(\hat{\mathbf{n}}_1, \hat{\mathbf{n}}_2, \hat{\mathbf{n}}_3) = (\hat{\mathbf{x}}, \hat{\mathbf{y}}, \hat{\mathbf{z}})$. The results are much simpler if we use new coordinates, $\mathbf{F}_{\text{ext}} = \sum_{k=1}^{3} F'_{\text{ext},k}\hat{\mathbf{n}}'_k$, with $\hat{\mathbf{n}}'_1$ along the outward radial direction through the planet; $\hat{\mathbf{n}}'_2$ in the orbital plane, perpendicular to the radius vector and in the direction of orbital motion; and $\hat{\mathbf{n}}'_3$ normal to the orbital plane, positive in the direction from which the orbital motion appears counterclockwise. Thus $(\hat{\mathbf{n}}'_1, \hat{\mathbf{n}}'_2, \hat{\mathbf{n}}'_3)$ form a right-handed

[28] Evaluating this matrix is tedious but straightforward using Kepler's equation (1.49) in the form $\lambda = u - e\sin u + \varpi$, the relation $\varpi = \omega + \Omega$ (eq. 1.67), and the relations (1.50) and (1.51a) between the eccentric anomaly u and the true anomaly f.

triad of unit vectors. The relation between these components is given by equation (B.61), $F_{\text{ext},k} = \sum_{p=1}^{3} R^{\text{T}}_{kn} F'_{\text{ext},n}$ Then equation (1.196) becomes

$$\frac{\partial H_{\text{ext}}}{\partial E_m} = -\sum_{j=1}^{6} C_{jm} \sum_{k,n=1}^{3} F'_{\text{ext},n} R_{nk} \frac{\partial}{\partial E_j} r R_{1k}. \tag{1.198}$$

For brevity, write $(F'_{\text{ext},1}, F'_{\text{ext},2}, F'_{\text{ext},3}) = (R, T, N)$; thus R is the external force per unit mass along the radial direction, T is the azimuthal or tangential force per unit mass in the orbital plane, and N is the force per unit mass normal to the orbital plane. Evaluating (1.198) gives

$$\begin{aligned}
\frac{\partial H_{\text{ext}}}{\partial \lambda} &= -R\frac{ae\sin f}{j} - T\frac{a^2 j}{r}, \\
\frac{\partial H_{\text{ext}}}{\partial \varpi} &= R\frac{ae\sin f}{j} - T\left(r - \frac{a^2 j}{r}\right), \\
\frac{\partial H_{\text{ext}}}{\partial \Omega} &= Tr(1-\cos I) + Nr\cos(f+\omega)\sin I, \\
\frac{\partial H_{\text{ext}}}{\partial a} &= -\frac{Rr}{a}, \\
\frac{\partial H_{\text{ext}}}{\partial e} &= Ra\cos f - T\frac{r(2+e\cos f)\sin f}{j^2}, \\
\frac{\partial H_{\text{ext}}}{\partial I} &= -Nr\sin(f+\omega).
\end{aligned} \tag{1.199}$$

Inserting these results in equations (1.187), we obtain **Gauss's equations**,

$$\begin{aligned}
\frac{\mathrm{d}\lambda}{\mathrm{d}t} &= n - R\frac{2r(1+j) + aej\cos f}{na^2(1+j)} + T\frac{r(1-j)(2+e\cos f)\sin f}{na^2 je} \\
&\quad + N\frac{r\tan\frac{1}{2}I\sin(f+\omega)}{na^2 j}, \\
\frac{\mathrm{d}\varpi}{\mathrm{d}t} &= -R\frac{j\cos f}{nae} + T\frac{r(2+e\cos f)\sin f}{na^2 je} + N\frac{r\tan\frac{1}{2}I\sin(f+\omega)}{na^2 j}, \\
\frac{\mathrm{d}\Omega}{\mathrm{d}t} &= N\frac{r\sin(f+\omega)}{na^2 j\sin I},
\end{aligned}$$

$$\frac{\mathrm{d}a}{\mathrm{d}t} = R\,\frac{2e\sin f}{nj} + T\,\frac{2aj}{nr},$$
$$\frac{\mathrm{d}e}{\mathrm{d}t} = R\,\frac{j\sin f}{na} + T\,\frac{j(\cos u + \cos f)}{na},$$
$$\frac{\mathrm{d}I}{\mathrm{d}t} = N\,\frac{r\cos(f+\omega)}{na^2 j}. \tag{1.200}$$

Although we derived Gauss's equations by assuming that the forces were derived from a Hamiltonian, they remain valid for any forces.

Alternative derivations of Gauss's equations are given by Brouwer & Clemence (1961) and Burns (1976); see also Problem 1.21.

As an illustration we compute the orbital evolution of a body subjected to radiation forces from its host star. From equation (c) of Box 1.5 the radial, tangential and normal radiation forces per unit mass are

$$R = \frac{k_{\mathrm{rad}}}{r^2}\left(1 - 2\frac{v_r}{c}\right), \quad T = -\frac{k_{\mathrm{rad}} v_\psi}{r^2 c}, \quad N = 0, \quad \text{where} \quad k_{\mathrm{rad}} \equiv \frac{LR^2}{4mc}, \tag{1.201}$$

v_r and v_ψ are the radial and azimuthal velocities, and m is the mass of the body. We substitute expressions for v_r and v_ψ from equations (1.54) and (1.55) and insert the results in Gauss's equations for the evolution of the semimajor axis and eccentricity. Eliminating the radius using equation (1.29) and the eccentric anomaly using equation (1.50), we find

$$\frac{\mathrm{d}a}{\mathrm{d}t} = \frac{2k_{\mathrm{rad}}\, e\sin f(1+e\cos f)^2}{na^2 j^5} - \frac{4k_{\mathrm{rad}}\, e^2\sin^2 f(1+e\cos f)^2}{acj^6} - \frac{2k_{\mathrm{rad}}(1+e\cos f)^4}{acj^6},$$
$$\frac{\mathrm{d}e}{\mathrm{d}t} = \frac{k_{\mathrm{rad}}\sin f(1+e\cos f)^2}{na^3 j^3} - \frac{2k_{\mathrm{rad}}\, e\sin^2 f(1+e\cos f)^2}{a^2 cj^4} - \frac{k_{\mathrm{rad}}(e + 2\cos f + e\cos^2 f)(1+e\cos f)^2}{a^2 cj^4}. \tag{1.202}$$

If the radiation forces are weak, the orbital elements will be relatively constant over a single orbit, so we can determine their long-term evolution by averaging these equations over an orbit. Using equation (1.64) we have

$$\left\langle \frac{\mathrm{d}a}{\mathrm{d}t} \right\rangle = -\frac{k_{\mathrm{rad}}(2 + 3e^2)}{ac(1 - e^2)^{3/2}}, \quad \left\langle \frac{\mathrm{d}e}{\mathrm{d}t} \right\rangle = -\frac{5k_{\mathrm{rad}}e}{2a^2c(1 - e^2)^{1/2}}. \tag{1.203}$$

The exact solution of these equations is described in Problem 1.22, or see Wyatt & Whipple (1950). For circular orbits, the equation for $\langle \mathrm{d}a/\mathrm{d}t \rangle$ is easily integrated to give

$$a(t) = a_0 \left(1 - \frac{t}{t_{\mathrm{rad}}}\right)^{1/2}, \quad \text{where} \quad t_{\mathrm{rad}} = \frac{mc^2 a_0^2}{LR^2}. \tag{1.204}$$

Here a_0 is the semimajor axis at the initial time $t = 0$. In terms of β, the ratio of the radiation pressure to the gravitational attraction from the host star (eq. d of Box 1.5),

$$t_{\mathrm{rad}} = \frac{a_0^2 c}{4\,\mathbb{G}M_* \beta} = \frac{400.5\,\mathrm{yr}}{\beta}\left(\frac{a_0}{1\,\mathrm{au}}\right)^2 \frac{M_\odot}{M_*}. \tag{1.205}$$

Since we have orbit-averaged the equations of motion, these results are only valid if the evolution is slow, that is, if $|\mathrm{d}a/\mathrm{d}t| \ll a/P$, where P is the orbital period.

Chapter 2

Numerical orbit integration

2.1 Introduction

The trajectories in any system containing more than one planet cannot be determined analytically, except in special cases. Therefore numerical orbit integration is indispensable for celestial mechanics.

A brief and readable introduction to numerical integration of differential equations is given by Press et al. (2007). For more comprehensive treatments see Hairer et al. (1993, 2006) and Blanes & Casas (2016).

The equation of motion for a planetary system can be written in the general form

$$\frac{\mathrm{d}\mathbf{z}}{\mathrm{d}t} = \mathbf{f}(\mathbf{z}, t), \tag{2.1}$$

where t is the time and $\mathbf{z}$ is a vector representing the coordinates of the system in phase space. The trajectory or orbit of a system is the curve $\mathbf{z}(t)$ that is determined by equation (2.1). If we know a single point on the curve, say $\mathbf{z}(t_0)$, the entire trajectory can be determined by solving this equation.

The simplest example is a single test particle with position $\mathbf{r}$ and velocity $\mathbf{v}$ that orbits in a gravitational potential $\Phi(\mathbf{r}, t)$. Its phase-space position can be written[1] as a 6-dimensional vector $\mathbf{z} \equiv (\mathbf{r}, \mathbf{v})$. The right side of equation

[1] This differs from the conventional definition of phase space, in which the momentum is $m\mathbf{v}$

(2.1) is then

$$\mathbf{f}(\mathbf{z},t) = [\mathbf{v}, -\nabla\Phi(\mathbf{r},t)]. \tag{2.2}$$

This equation of motion can equally well be written as a second-order differential equation

$$\frac{\mathrm{d}^2\mathbf{r}}{\mathrm{d}t^2} = -\nabla\Phi(\mathbf{r},t); \tag{2.3}$$

we shall call this the **Newtonian form** of the equation of motion, even though the concept of force as the gradient of a scalar potential was developed after Newton's death.

A more general example is motion in a dynamical system governed by a Hamiltonian $H(\mathbf{q},\mathbf{p},t)$ where $\mathbf{q}$ and $\mathbf{p}$ are vectors of canonical coordinates and momenta. Then $\mathbf{z} = (\mathbf{q},\mathbf{p})$, and according to Hamilton's equations (D.12) the right side of equation (2.1) is

$$\mathbf{f}(\mathbf{z},t) = \left(\frac{\partial H}{\partial \mathbf{p}}, -\frac{\partial H}{\partial \mathbf{q}}\right). \tag{2.4}$$

The equation of motion (2.2) is a special case in which $\mathbf{q} = \mathbf{r}$, $\mathbf{p} = \mathbf{v}$ and $H(\mathbf{q},\mathbf{p},t) = \frac{1}{2}p^2 + \Phi(\mathbf{q},t)$.

The demands of celestial mechanics are varied enough that no single numerical method for solving the differential equation (2.1) always works well. Consider the following example problems:

Solar-system ephemeris Fit the trajectories of the Sun, Moon, Earth, the other solar-system planets and satellites, and the most massive asteroids to determine their orbital elements and masses. The relevant data include radio ranges to interplanetary spacecraft, laser ranges to the Moon, radar ranges to Mercury and Venus, and optical observations of asteroids, the outer planets, and their satellites. The numerical errors in the integrations should be less than a centimeter (cm) over 100 years or more, to ensure that they are much smaller than the residuals from the fits (currently a few cm for the Moon and a few tens of meters for the terrestrial planets Mercury, Venus and Mars). For this task, the integration algorithm must be extremely precise; on the

rather than $\mathbf{v}$. See footnote 10 of Chapter 1.

Box 2.1: Extended phase space

The differential equation (2.1) is said to be **autonomous** if the right side is independent of the time t. Any equation such as (2.1) can be converted to an autonomous one. To do this, define a vector $\mathbf{Z} \equiv (\mathbf{z}, t)$ and a **fictitious time** τ, related to the time t by $\mathrm{d}t = g(\mathbf{z})\mathrm{d}\tau$ (note that the fictitious time may be different for different trajectories at the same instant of real time t). Then the equation

$$\frac{\mathrm{d}\mathbf{Z}}{\mathrm{d}\tau} = \mathbf{F}(\mathbf{Z}), \quad \text{where} \quad \mathbf{F}(\mathbf{Z}) \equiv g(\mathbf{z})\,[\mathbf{f}(\mathbf{Z}), 1] \tag{a}$$

has the same solution as (2.1) but is autonomous (because t is now a component of $\mathbf{Z}$ rather than the independent variable). We are free to choose any function for $g(\mathbf{z})$, and if $g(\mathbf{z}) = 1$ the fictitious time is the same as the true time.

Similarly, a time-dependent Hamiltonian $H(\mathbf{q}, \mathbf{p}, t)$ can be converted to an autonomous Hamiltonian $\Gamma(\mathbf{Q}, \mathbf{P})$ in an extended phase space with fictitious time τ. The extended phase space has coordinates and momenta $\mathbf{Q} \equiv (Q_0, \mathbf{q})$, $\mathbf{P} \equiv (P_0, \mathbf{p})$, where $Q_0 \equiv t$. If we set

$$\Gamma(\mathbf{Q}, \mathbf{P}) \equiv g(\mathbf{q}, \mathbf{p})[H(\mathbf{q}, \mathbf{p}, Q_0) + P_0], \tag{b}$$

then Hamilton's equations for the evolution of $\mathbf{q}$, $\mathbf{p}$ and Q_0 are

$$\frac{\mathrm{d}\mathbf{q}}{\mathrm{d}\tau} = \frac{\partial \Gamma}{\partial \mathbf{p}} = g\frac{\partial H}{\partial \mathbf{p}} + \frac{\partial g}{\partial \mathbf{p}}(H + P_0), \qquad \frac{\mathrm{d}Q_0}{\mathrm{d}\tau} = \frac{\partial \Gamma}{\partial P_0} = g, \tag{c}$$

$$\frac{\mathrm{d}\mathbf{p}}{\mathrm{d}\tau} = -\frac{\partial \Gamma}{\partial \mathbf{q}} = -g\frac{\partial H}{\partial \mathbf{q}} - \frac{\partial g}{\partial \mathbf{q}}(H + P_0), \qquad \frac{\mathrm{d}P_0}{\mathrm{d}\tau} = -\frac{\partial \Gamma}{\partial Q_0} = -g\frac{\partial H}{\partial Q_0}. \tag{d}$$

The equation for $\mathrm{d}Q_0/\mathrm{d}\tau$ says that

$$\mathrm{d}t = g(\mathbf{q}, \mathbf{p})\mathrm{d}\tau. \tag{e}$$

Using this result to eliminate $\mathrm{d}\tau$ in favor of $\mathrm{d}t$, the remaining equations become

$$\frac{\mathrm{d}\mathbf{q}}{\mathrm{d}t} = \frac{\partial H}{\partial \mathbf{p}} + \frac{1}{g}\frac{\partial g}{\partial \mathbf{p}}(H+P_0), \quad \frac{\mathrm{d}\mathbf{p}}{\mathrm{d}t} = -\frac{\partial H}{\partial \mathbf{q}} - \frac{1}{g}\frac{\partial g}{\partial \mathbf{q}}(H+P_0), \quad \frac{\mathrm{d}P_0}{\mathrm{d}t} = -\frac{\partial H}{\partial t}. \tag{f}$$

There is a simple interpretation of the momentum P_0. Since the Hamiltonian $\Gamma(\mathbf{Q}, \mathbf{P})$ is independent of τ it is conserved along a trajectory. Let $E(t)$ be the energy on that trajectory, $E(t) = H[\mathbf{q}(t), \mathbf{p}(t), t]$. If we choose $P_0 = -E$ at the initial point of the trajectory, then $\Gamma = 0$ at that point so Γ vanishes on the whole trajectory. Therefore $P_0(t) = -E(t)$ at all times. In words, the momentum P_0 conjugate to the time coordinate Q_0 is minus the energy.

If $P_0 = -E$ so $\Gamma = 0$, or if $g(\mathbf{q}, \mathbf{p}) = \text{const}$, then the first two of equations (f) reduce to the original Hamilton's equations (2.1) and (2.4).

other hand the speed of the calculation is not a major consideration, since the integration interval is only a few thousand orbital periods for the Moon and much less for the other bodies.[2]

Long-term stability of planetary systems Determine the stability of an exoplanet system over the lifetime of the Galaxy, $\sim 10\,\text{Gyr}$. Exoplanets can have orbital periods as short as a few hours, so we must follow up to 10^{13} orbits of a planet, which requires the fastest possible integration algorithm. On the other hand the masses and orbital parameters of exoplanets are not well known, so accuracy is not so important as long as the qualitative features of the orbital evolution are correct.

Evolution of cometary orbits Follow the orbits of thousands of comets from the Oort cloud as they pass through the Sun's planetary system. A typical Oort-cloud comet has a semimajor axis of $\sim 30\,000\,\text{au}$ and thus an orbital period of $5\,\text{Myr}$ (§9.5). However, if it has a close encounter with Jupiter, its orbital elements can change dramatically within a few hours. This task requires a fast and accurate algorithm that can follow occasional changes in an orbit on timescales as short as 10^{-10} orbital periods.

Evolution of the spin of Mars The current obliquity of Mars—the angle between the spin and orbital angular-momentum vectors—is $25.19°$, and the spin angular momentum precesses with a period of $1.70 \times 10^5\,\text{yr}$, mostly due to torques from the Sun. If there were no other planets in the solar system, the obliquity and precession rates would be constant. However, gravitational torques from other planets cause the obliquity to vary chaotically between nearly zero and almost $65°$ (see §7.1.2). To follow the history of the Martian obliquity requires integrating both the equations of motion of the planets and the rigid-body equations of motion for Mars for the age of the solar system, $4.57\,\text{Gyr}$.

[2] High-precision ephemerides are currently available from several sources: the Harvard–Smithsonian Center for Astrophysics in Cambridge, Massachusetts (Chandler et al. 2021), the Institute of Geodesy in Hannover (Müller et al. 2019), the Jet Propulsion Laboratory in Pasadena (Park et al. 2021), the Paris Observatory (Viswanathan et al. 2018) and the Institute of Applied Astronomy in St. Petersburg (Pitjeva & Pitjev 2014).

Over the last several decades the capabilities of scientific computing have grown exponentially, mostly through two trends: declining cost of computer memory and improvements in hardware and software for parallel computing. Unfortunately, neither of these trends has a big impact on most problems in celestial mechanics: the number of bodies is small enough that the memory requirements are small, and following a trajectory is an intrinsically serial calculation that is difficult to parallelize.

Many of the integrators described in this chapter are implemented in open-source software packages, such as SciPy for python. The most sophisticated general-purpose software for orbit integrations is REBOUND (Rein & Liu 2012) at https://rebound.readthedocs.io/en/latest/#. REBOUND was used for all the long orbit integrations in this book.

2.1.1 Order of an integrator

Numerical integration produces a sequence of phase-space positions $\mathbf{z}_n$ at times t_n. To keep the exposition simple, we assume at first that the **timestep** h is fixed, so $t_n = t_0 + nh$.

Given the position $\mathbf{z}_n$, and possibly information from earlier positions $\mathbf{z}_{n-1}, \mathbf{z}_{n-2}$, and so on, we are seeking a formula—an **integrator**—that generates a new phase-space position $\mathbf{z}_{n+1}$ that approximates the trajectory at t_{n+1}. The **local error**[3] of the integrator is the difference between $\mathbf{z}_{n+1}$ and the phase-space position that would be found by an exact solution of the equation of motion, starting from the initial condition $\mathbf{z}_n$ at t_n. The **order** of the integrator is k if the local error varies with timestep as $\mathrm{O}(h^{k+1})$. It is prudent to assume the worst-case scenario in which the error accumulates at every step, in which case the error after a fixed time interval $\Delta t = Nh$ (the **global error**) can be as large as $\mathrm{O}(Nh^{k+1}) = \Delta t\,\mathrm{O}(h^k)$. An integrator is only useful if the global error approaches zero as the timestep approaches zero, which requires $k > 0$.

High order does not necessarily imply high accuracy. The local error in a kth-order integrator typically involves derivatives of order $k + 1$ in $\mathbf{z}(t)$.

[3] Sometimes called the local **truncation error** in contradistinction to **roundoff error**, which arises because arithmetic operations cannot be carried out exactly in a computer for most real numbers. We discuss roundoff error in §2.7, assuming for now that it is negligible.

Thus, if the solution $\mathbf{z}(t)$ varies rapidly, its high-order derivatives can be large enough that the local error actually grows as the order of the integrator is increased beyond some critical value.

The relative performance of integrators is often measured by comparing the local error at a given timestep h. This may not be a fair comparison, in part because the computing time needed for a single timestep can be very different for different integrators. Since the most expensive part of the calculation is usually the evaluation of the force $-\nabla\Phi$ (eq. 2.2), it is fairer to compare the local error at an "effective" timestep, which is the mean interval between force evaluations. For example, the classical Runge–Kutta integrator in equation (2.65) has four force evaluations per timestep, so its effective timestep is $\frac{1}{4}h$.

Numerical methods for integrating ordinary differential equations are simpler in celestial mechanics than in most other subjects in one respect: many important problems can be done with a fixed timestep, such as integrations of multi-planet systems like the solar system. Nevertheless, some applications require an adaptive timestep, which shrinks or grows depending on the changing behavior of the orbit. A challenging task requiring an adaptive timestep is to follow a particle on a highly eccentric orbit; for example, the typical eccentricity of comets coming from the Oort cloud is $e = 0.9999$ (Problem 2.9).

2.1.2 The Euler method

The simplest integrator is the **Euler method**, in which the equation of motion (2.1) is approximated by

$$\mathbf{z}_{n+1} = \mathbf{z}_n + h\mathbf{f}(\mathbf{z}_n, t_n). \tag{2.5}$$

The method is **explicit**, which means that $\mathbf{z}_{n+1}$ can be computed directly from $\mathbf{z}_n$. Later we will encounter **implicit** methods, in which a nonlinear equation involving both $\mathbf{z}_n$ and $\mathbf{z}_{n+1}$ has to be solved at each step.

We now determine the order of the Euler method. The exact trajectory $\mathbf{z}(t)$ that passes through $\mathbf{z}_n$ can be expanded in a Taylor series around t_n,

$$\mathbf{z}(t) = \mathbf{z}_n + \dot{\mathbf{z}}(t_n)(t - t_n) + \tfrac{1}{2}\ddot{\mathbf{z}}(t_n)(t - t_n)^2 + \mathrm{O}[(t - t_n)^3]. \tag{2.6}$$

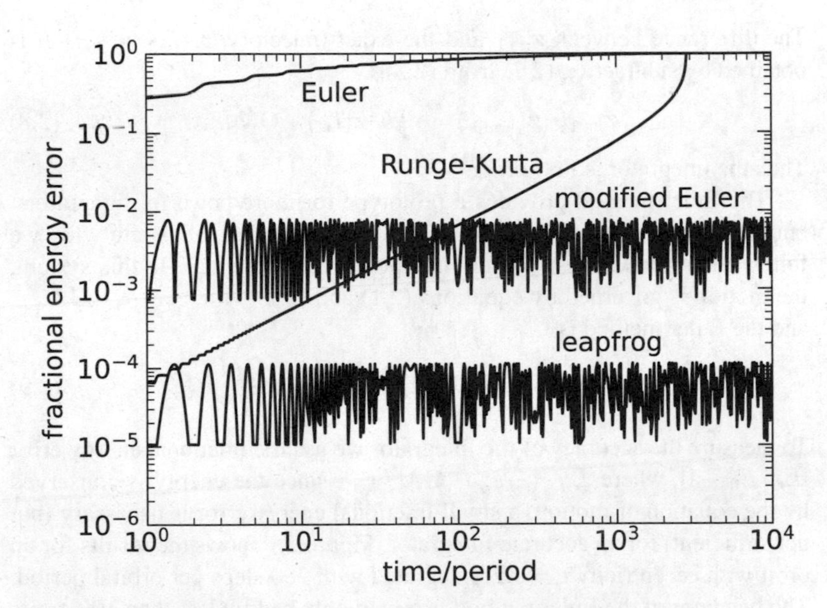

Figure 2.1: The absolute value of the fractional energy error as a function of time during integrations of a Kepler orbit with eccentricity $e = 0.2$. Each integrator is allowed 200 force evaluations per orbit. The integrations are carried out using the Euler and modified Euler methods (eqs. 2.9 and 2.21), which are first-order; leapfrog, which is second-order (eq. 2.29); and the classical fourth-order Runge–Kutta method (eq. 2.65). The energy error for modified Euler and leapfrog oscillates between positive and negative values; to reduce clutter these curves are truncated below 10% of the maximum error. Note the poor performance of the Euler method. In contrast, the modified Euler method and leapfrog exhibit no long-term growth in the error.

Since this is a solution of equation (2.1) we must have $\dot{\mathbf{z}}(t_n) = \mathbf{f}(\mathbf{z}_n, t_n)$ so

$$\mathbf{z}(t) = \mathbf{z}_n + \mathbf{f}(\mathbf{z}_n, t_n)(t - t_n) + \tfrac{1}{2}\ddot{\mathbf{z}}(t_n)(t - t_n)^2 + \mathrm{O}[(t - t_n)^3]. \quad (2.7)$$

The difference between $\mathbf{z}_{n+1}$ and the exact trajectory at $t_{n+1} = t_n + h$ is obtained by subtracting (2.7) from (2.5):

$$\mathbf{z}_{n+1} - \mathbf{z}(t_{n+1}) = -\tfrac{1}{2}h^2\ddot{\mathbf{z}}(t_n) + \mathrm{O}(h^3). \quad (2.8)$$

Thus the integrator is first-order.

The Euler method provides a prototype for more powerful integrators, but should *never* be used for practical calculations. To illustrate why, we follow the orbit of a test particle in a point-mass potential. In this system, the motion is governed by equations (2.1) and (2.2) with $\Phi(\mathbf{r}) = -\mathbb{G}M/r$, and the Euler method is

$$\mathbf{r}_{n+1} = \mathbf{r}_n + h\mathbf{v}_n, \quad \mathbf{v}_{n+1} = \mathbf{v}_n - h\frac{\mathbb{G}M}{|\mathbf{r}_n|^3}\mathbf{r}_n. \quad (2.9)$$

To measure the accuracy of the integrator we use the fractional energy error $|E_n/E_0 - 1|$, where $E_n = \frac{1}{2}v_n^2 - \mathbb{G}M/|\mathbf{r}_n|$; since the energy is conserved by the equation of motion, a small fractional energy error is necessary (but not sufficient) for an accurate integrator. Figure 2.1 shows the results for an orbit with eccentricity $e = 0.2$, integrated with 200 steps per orbital period. The behavior of the Euler method is remarkably bad: in less than 100 orbits the fractional energy error is of order unity.[4]

The behavior of the Euler method can also be investigated analytically using the simple harmonic potential, $\Phi(\mathbf{r}) = \frac{1}{2}\omega^2 r^2$—although this is less realistic than the Kepler potential for our purposes, the performance of most integrators in the harmonic and Kepler potentials is qualitatively similar. If we treat the vector $\mathbf{z} = (\mathbf{r}, \mathbf{v})$ as a column matrix,[5] the equation of motion can be written

$$\dot{\mathbf{z}} = \mathbf{f}(\mathbf{z}) = \mathbf{A}\mathbf{z}, \quad \text{where} \quad \mathbf{A} \equiv \begin{bmatrix} \mathbf{0} & \mathbf{I} \\ -\omega^2\mathbf{I} & \mathbf{0} \end{bmatrix}. \quad (2.10)$$

[4] The fractional error $E_n/E_0 - 1$ asymptotically approaches -1 because the semimajor axis a_n of the orbit grows without limit, so the energy $E_n = -\frac{1}{2}\mathbb{G}M/a_n \to 0$.

[5] Our notation does not distinguish whether vectors such as $\mathbf{z}$ are $1 \times N$ column matrices or $N \times 1$ row matrices if the meaning is clear from the context.

Here $\mathbf{0}$ and $\mathbf{I}$ are the 3×3 zero and identity matrices. The general solution of this matrix differential equation is

$$\begin{aligned}\mathbf{r}(t) &= \tfrac{1}{2}\mathbf{r}_0\big(\mathrm{e}^{\mathrm{i}\omega t} + \mathrm{e}^{-\mathrm{i}\omega t}\big) - \tfrac{1}{2}\mathrm{i}\omega^{-1}\mathbf{v}_0\big(\mathrm{e}^{\mathrm{i}\omega t} - \mathrm{e}^{-\mathrm{i}\omega t}\big),\\ \mathbf{v}(t) &= \tfrac{1}{2}\mathrm{i}\omega\mathbf{r}_0\big(\mathrm{e}^{\mathrm{i}\omega t} - \mathrm{e}^{-\mathrm{i}\omega t}\big) + \tfrac{1}{2}\mathbf{v}_0\big(\mathrm{e}^{\mathrm{i}\omega t} + \mathrm{e}^{-\mathrm{i}\omega t}\big),\end{aligned} \tag{2.11}$$

where $\mathbf{r}_0$ and $\mathbf{v}_0$ are the position and velocity at $t = 0$. Of course, despite the complex numbers in equations (2.11), $\mathbf{r}(t)$ and $\mathbf{v}(t)$ are always real.

Using this notation, Euler's method is

$$\mathbf{z}_{n+1} = \mathbf{z}_n + h\mathbf{A}\mathbf{z}_n. \tag{2.12}$$

The general solution of this matrix equation is a linear combination of sequences of the form

$$\mathbf{z}_n = \kappa^n \mathbf{a}, \tag{2.13}$$

where the scale factor κ is a nonzero (possibly complex) constant. Substituting (2.13) into (2.12), we have

$$\frac{\kappa - 1}{h}\mathbf{a} = \mathbf{A}\mathbf{a}. \tag{2.14}$$

Thus $\mathbf{a}$ must be an eigenvector of $\mathbf{A}$, with $(\kappa - 1)/h$ the corresponding eigenvalue. The eigenvalues λ are the solutions of $\det(\mathbf{A} - \lambda\mathbf{I}) = 0$, where "det" is shorthand for the determinant. It is simple to show that there are two eigenvalues, $\lambda_\pm = \pm\mathrm{i}\omega$. Then $\mathbf{z}_n = \kappa_+^n\mathbf{a}_+ + \kappa_-^n\mathbf{a}_-$, where

$$\kappa_\pm = 1 \pm \mathrm{i}\omega h, \quad \mathbf{a}_\pm = \begin{bmatrix} \mathbf{c}_\pm \\ \pm\mathrm{i}\omega\, \mathbf{c}_\pm \end{bmatrix}, \tag{2.15}$$

and $\mathbf{c}_\pm$ is a 1×3 column vector determined by the initial conditions.

These results can be rewritten to give the position and velocity at step n in terms of their values at step 0,

$$\begin{aligned}\mathbf{r}_n &= \tfrac{1}{2}\mathbf{r}_0\big(\kappa_+^n + \kappa_-^n\big) - \tfrac{1}{2}\mathrm{i}\omega^{-1}\mathbf{v}_0\big(\kappa_+^n - \kappa_-^n\big),\\ \mathbf{v}_n &= \tfrac{1}{2}\mathrm{i}\omega\mathbf{r}_0\big(\kappa_+^n - \kappa_-^n\big) + \tfrac{1}{2}\mathbf{v}_0\big(\kappa_+^n + \kappa_-^n\big).\end{aligned} \tag{2.16}$$

Comparing equations (2.11) and (2.16), we see that the Euler method simply replaces the exponential $\exp(\pm i\omega h)$ by the scale factor $\kappa_\pm = 1 \pm i\omega h$. The difference between the two is $O(h^2)$, as expected since the Euler method is first-order. A less obvious but equally important difference is that $|\exp(\pm i\omega h)| = 1$ but $|\kappa_\pm| = (1 + \omega^2 h^2)^{1/2} > 1$. Equation (2.16) shows that $\mathbf{r}_n$ and $\mathbf{v}_n$ are the sum of terms that vary in magnitude as $|\kappa_\pm|^n$, so on average the radius $|\mathbf{r}_n|$ and speed $|\mathbf{v}_n|$ grow as $(1 + \omega^2 h^2)^{n/2}$. Setting $n = (t - t_0)/h$, we find that the radius and speed tend to grow *exponentially*, as $\exp[\gamma(t - t_0)]$ with

$$\gamma = \frac{1}{h} \log |\kappa_\pm| = \frac{1}{2h} \log(1 + \omega^2 h^2) = \tfrac{1}{2}\omega^2 h + O(h^3). \tag{2.17}$$

This exponential growth is the cause of the poor performance of the Euler method in Figure 2.1.

The **implicit** or **backward Euler method** differs from the explicit Euler method (2.5) in that the force is evaluated at the end of the timestep rather than the start:

$$\mathbf{z}_{n+1} = \mathbf{z}_n + h\mathbf{f}(\mathbf{z}_{n+1}, t_{n+1}). \tag{2.18}$$

This is an implicit equation for $\mathbf{z}_{n+1}$, which is solved iteratively starting from a first guess for $\mathbf{z}_{n+1}$ that can be provided by the explicit Euler method. The qualitative behavior of implicit Euler can be found without any new calculations. We simply rewrite equation (2.18) as

$$\mathbf{z}_n = \mathbf{z}_{n+1} - h\mathbf{f}(\mathbf{z}_{n+1}, t_{n+1}), \tag{2.19}$$

which shows that implicit Euler with timestep h is equivalent to explicit Euler with timestep $-h$. Thus if the radius or energy grows rapidly with explicit Euler in a given system, it will decay just as fast with implicit Euler. Needless to say, growth or decay of this kind is unacceptable numerical behavior in a conservative system such as the two-body problem. For dissipative systems, implicit integrators are usually more reliable than explicit ones, but this is not true for orbit integrations.

2.1.3 The modified Euler method

We can do better—*much* better—than the Euler method in a large class of dynamical systems. The equations describing the motion of a test particle in a potential $\Phi(\mathbf{r}, t)$ are (2.1) and (2.2), and for these equations the Euler method is

$$\mathbf{r}_{n+1} = \mathbf{r}_n + h\mathbf{v}_n, \quad \mathbf{v}_{n+1} = \mathbf{v}_n - h\nabla\Phi(\mathbf{r}_n, t_n), \tag{2.20}$$

with $t_{n+1} = t_n + h$. The **modified Euler method** is

$$\mathbf{r}_{n+1} = \mathbf{r}_n + h\mathbf{v}_n, \quad \mathbf{v}_{n+1} = \mathbf{v}_n - h\nabla\Phi(\mathbf{r}_{n+1}, t_n + h). \tag{2.21}$$

The only change is that the force is evaluated at the new position and time $\mathbf{r}_{n+1}$ and t_{n+1} rather than the old position and time $\mathbf{r}_n$ and t_n. The key feature that allows us to make this modification without solving a nonlinear equation at each step is that the time derivative of the position $\mathbf{r}$ depends only on the velocity $\mathbf{v}$, while the derivative of $\mathbf{v}$ depends only on $\mathbf{r}$ (and the time).

The performance of the modified Euler method for a test particle in a Kepler potential is shown in Figure 2.1. Although modified Euler is still only a first-order method, the rapid growth in energy error seen in the Euler method is *completely* absent: the energy error oscillates rather than growing. The sharp downward cusps in the error arise because we are plotting the logarithm of the absolute value of the error, which diverges to $-\infty$ as the oscillations pass through zero. With a larger timestep, the amplitude of the oscillations is larger but there is still no growth in the maximum error, no matter how long the integration continues.

The secret of this success is illuminated by examining the harmonic oscillator, as we did in equations (2.10)–(2.17). Using the same notation, the modified Euler method can be written

$$\mathbf{z}_{n+1} = \mathbf{z}_n + h\mathbf{A}_{\mathrm{mod}}\mathbf{z}_n, \quad \text{where} \quad \mathbf{A}_{\mathrm{mod}} \equiv \begin{bmatrix} \mathbf{0} & \mathbf{I} \\ -\omega^2\mathbf{I} & -\omega^2 h\mathbf{I} \end{bmatrix}. \tag{2.22}$$

The solution of this equation is a linear combination of sequences of the form (2.13), with scale factor

$$\kappa_\pm = 1 - \tfrac{1}{2}\omega^2h^2 \pm \mathrm{i}\omega h\left(1 - \tfrac{1}{4}\omega^2h^2\right)^{1/2}. \tag{2.23}$$

In this case it is simple to show that $|\kappa_\pm| = 1$, independent of the timestep h so long as $h \le 2/\omega$ (which is much larger than the timesteps used in practice). Thus, remarkably, the exponential growth in $|\mathbf{r}_n|$ and $|\mathbf{v}_n|$ that arises with the Euler method is eliminated completely in the modified Euler method.

It is useful to think of a single timestep of the modified Euler method as the composition of two separate steps. First we advance the position at constant velocity, which we represent by an operator

$$\mathsf{D}_h(\mathbf{r}, \mathbf{v}) = (\mathbf{r} + h\mathbf{v}, \mathbf{v}); \tag{2.24}$$

this is called the **drift operator** since the particle drifts without accelerating. The drift operator also advances the time by the timestep h. Then we advance the velocity at constant position and time, as represented by the operator

$$\mathsf{K}_h(\mathbf{r}, \mathbf{v}) = [\mathbf{r}, \mathbf{v} - h\nabla\Phi(\mathbf{r}, t)], \tag{2.25}$$

called the **kick operator** since it gives an instantaneous kick or impulse to the velocity without affecting the position. Note that

$$\mathsf{D}_{h/2}\mathsf{D}_{h/2} = \mathsf{D}^2_{h/2} = \mathsf{D}_h, \quad \mathsf{D}_{-h} = \mathsf{D}^{-1}_h; \tag{2.26}$$

with similar relations for the kick operator K_h.

The modified Euler method (2.21) can be written

$$\mathbf{z}_{n+1} = (\mathbf{r}_{n+1}, \mathbf{v}_{n+1}) = \mathsf{K}_h\mathsf{D}_h\mathbf{z}_n; \tag{2.27}$$

here the operators are applied sequentially, starting at the right. More precisely, (2.27) is the "drift-kick" version of modified Euler. Equally good is the "kick-drift" integrator, in which a single timestep has the form

$$\mathbf{z}_{n+1} = \mathsf{D}_h\mathsf{K}_h\,\mathbf{z}_n. \tag{2.28}$$

2.1.4 Leapfrog

A single step of the **leapfrog integrator** can be written[6]

$$\begin{aligned}
\mathbf{r}' &= \mathbf{r}_n + \tfrac{1}{2}h\mathbf{v}_n, \\
\mathbf{v}_{n+1} &= \mathbf{v}_n - h\nabla\Phi(\mathbf{r}', t_n + \tfrac{1}{2}h), \\
\mathbf{r}_{n+1} &= \mathbf{r}' + \tfrac{1}{2}h\mathbf{v}_{n+1}.
\end{aligned} \tag{2.29}$$

These equations can be written in terms of the drift and kick operators as

$$\mathbf{z}_{n+1} = \mathsf{D}_{h/2}\mathsf{K}_h\mathsf{D}_{h/2}\,\mathbf{z}_n, \tag{2.30}$$

so this method is also called "drift-kick-drift" leapfrog. An equally good alternative is "kick-drift-kick" leapfrog,

$$\mathbf{z}_{n+1} = \mathsf{K}_{h/2}\mathsf{D}_h\mathsf{K}_{h/2}\mathbf{z}_n. \tag{2.31}$$

Leapfrog is a second-order method, as can be shown using an approach analogous to equations (2.6)–(2.8).

The performance of leapfrog for a test particle in a Kepler potential is shown in Figure 2.1. As with the modified Euler method, the energy error is oscillatory rather than growing. The maximum error of leapfrog is much smaller than the maximum error of modified Euler, as expected since leapfrog is second-order rather than first-order.

An integration of P steps of drift-kick-drift leapfrog can be written

$$\mathbf{z}_P = \left(\mathsf{D}_{h/2}\mathsf{K}_h\mathsf{D}_{h/2}\right)^P\mathbf{z}_0. \tag{2.32}$$

Using equations (2.26), this result can be rewritten in several equivalent forms (recall that the operators are applied from right to left):

$$\begin{aligned}
\mathbf{z}_P &= \mathsf{D}_{-h/2}(\mathsf{D}_h\mathsf{K}_h)^P\mathsf{D}_{h/2}\,\mathbf{z}_0 \\
&= \mathsf{D}_{h/2}(\mathsf{K}_h\mathsf{D}_h)^P\mathsf{D}_{-h/2}\,\mathbf{z}_0
\end{aligned}$$

[6] Leapfrog is also known as the Störmer or the Verlet method, although the concepts can be traced back to Newton (Hairer et al. 2006).

$$= \mathsf{D}_{h/2}\mathsf{K}_{h/2}(\mathsf{K}_{h/2}\mathsf{D}_h\mathsf{K}_{h/2})^{P-1}\mathsf{K}_{h/2}\mathsf{D}_{h/2}\,\mathbf{z}_0. \tag{2.33}$$

These formulas show that apart from one or two steps at the beginning and end of the integration, the drift-kick and kick-drift forms of the modified Euler method and the drift-kick-drift and kick-drift-kick forms of leapfrog are all identical. They also show, remarkably, that a multistep integration using modified Euler can be improved in accuracy from first-order to second-order simply by adding two drift steps of opposite sign before and after the integration.[7]

Despite its simplicity, the leapfrog integrator is widely used to study the N-body problem in molecular dynamics, cosmology, galaxy formation and evolution, and so forth. One of its advantages for large N, where computer memory is a limiting factor, is that the phase-space positions can be updated at each timestep without any temporary additional storage: following equations (2.29), the position $\mathbf{r}_n$ is replaced by $\mathbf{r}'$, then the velocities $\mathbf{v}_n$ are replaced by $\mathbf{v}_{n+1}$, then $\mathbf{r}'$ is replaced by $\mathbf{r}_{n+1}$.

These results prompt an obvious but deep question: what properties of the modified Euler and leapfrog integrators lead to oscillatory energy errors rather than growing ones, and how can we design more accurate integrators with similar properties? These issues are the subject of the next section.

2.2 Geometric integration methods

The goal of general-purpose integrators is to minimize the local error—the difference accrued over a single timestep between the true phase-space position and the position predicted by the integrator. However, not all errors are equally important.

To illustrate this point, suppose that two integrators A and B having the same timestep h are used to follow a circular orbit with semimajor axis a in the gravitational field of a point mass M. Integrator A makes an error ϵ in the azimuthal coordinate or orbital phase at each timestep, but is otherwise exact. Then the phase error after an integration time T, requiring N =

[7] The application of these extra steps to improve the order of an integrator is called **symplectic correction** (Wisdom et al. 1996).

Box 2.2: Taylor-series methods

The Taylor series for the trajectory $\mathbf{z}(t+h)$ is

$$\mathbf{z}(t+h) = \mathbf{z}(t) + h\frac{\mathrm{d}}{\mathrm{d}t}\mathbf{z}(t) + \frac{h^2}{2}\frac{\mathrm{d}^2}{\mathrm{d}t^2}\mathbf{z}(t) + \cdots + \frac{h^n}{n!}\frac{\mathrm{d}^n}{\mathrm{d}t^n}\mathbf{z}(t) + \mathrm{O}(h^{n+1}). \quad \text{(a)}$$

If we discard all terms in the Taylor series that are $\mathrm{O}(h^2)$ or higher and use (2.1) to eliminate $\mathrm{d}\mathbf{z}/\mathrm{d}t$ then $\mathbf{z}(t+h) = \mathbf{z}(t) + h\mathbf{f}[\mathbf{z}(t), t]$, which is simply the Euler method (2.5). A natural approach to constructing more accurate integrators is to include more terms in the series. For example, including the next term yields

$$z_j(t+h) = z_j(t) + hf_j + \tfrac{1}{2}h^2\left(\sum_k f_k\frac{\partial f_j}{\partial z_k} + \frac{\partial f_k}{\partial t}\right), \quad \text{(b)}$$

where $\mathbf{f}$ and its derivatives are all evaluated at $[\mathbf{z}(t), t]$.

Similar formulas can be derived to any desired order. However, the functions on the right side rapidly become quite complicated, particularly if the phase space has many degrees of freedom. Because the other integrators described in this chapter are usually much simpler at a given order, Taylor-series methods have only seen limited use in celestial mechanics. This situation is changing as computer algebra now enables Taylor-series expansions of the gravitational N-body problem to arbitrarily high order (e.g., Hayes 2008; Biscani & Izzo 2021).

T/h steps, is $(\Delta\phi)_A = N\epsilon = T\epsilon/h$. In contrast, integrator B makes a fractional error ϵ in the semimajor axis at each timestep. Since the mean motion n is equal to $(\mathbb{G}M)^{1/2}a^{-3/2}$ by Kepler's law, this semimajor axis error corresponds to an error in n of $-\frac{3}{2}n\epsilon$ per step or $-\frac{3}{2}n\epsilon/h$ per unit time. The orbital phase grows in time as $\mathrm{d}\phi/\mathrm{d}t = n$, so the error in phase for integrator B grows as $\mathrm{d}^2(\Delta\phi)_B/\mathrm{d}t^2 = -\frac{3}{2}n\epsilon/h$. So long as ϵ is small enough that n is approximately constant, this is easily integrated to give $(\Delta\phi)_B = -\frac{3}{4}n\epsilon T^2/h$. These arguments show that local phase errors lead to a global phase error that grows linearly with the integration time T, while local semimajor axis errors lead to much worse behavior: a global phase error that grows quadratically with T. For example, with an integration time $T = 10^{10}$ years, an orbital period of 1 year and a timestep $h = 0.01$

years these estimates yield $(\Delta\phi)_A = 1 \times 10^{12}\epsilon$ and $(\Delta\phi)_B = -5 \times 10^{22}\epsilon$, larger by a factor of more than 10^{10}.

As this example illustrates, semimajor axis or energy errors are far more dangerous than phase errors in long orbit integrations, but general-purpose integrators do not distinguish between the two. Similarly they do not distinguish cumulative errors from oscillatory errors, although the former are far more damaging in long integrations.

How do we use this insight to design integrators? One powerful approach is through **geometric integrators**, which preserve (some of) the geometric properties of the phase-space flow described by the original equation of motion. Geometric integrators accept a larger local error at the end of one timestep to ensure that the geometrical properties of the flow in phase space are the same in the numerical trajectory and the true trajectory.

A simple example of a geometric integrator would be one that conserves the total energy and angular momentum of an N-body system. In practice, such integrators have not proved to be very useful in celestial mechanics. One reason is that smaller planets may contribute very little to the total energy and angular momentum. For example, because of its high eccentricity and short orbital period Mercury is the most difficult planet to follow accurately in numerical integrations of the solar system, yet it contains less than 0.2% of the orbital energy in the solar system and an even smaller fraction of the angular momentum.

Most geometric integrators are designed to inherit one or both of two specific properties of the phase-space flow: in reversible integrators, a particle returns to its exact starting point in phase space if its velocity is reversed; while in symplectic integrators, the transformation from initial to final phase-space position is symplectic or canonical.

For book-length treatments of geometric integrators see Hairer et al. (2006) and Blanes & Casas (2016).

2.2.1 Reversible integrators

In an autonomous dynamical system, the equation of motion (2.1) reads $\mathrm{d}\mathbf{z}/\mathrm{d}t = \mathbf{f}(\mathbf{z})$; that is, the right side has no explicit time dependence (see Box 2.1). In an inertial reference frame, isolated gravitational N-body systems are autonomous. They are also reversible, by which we mean the following. Suppose the initial position and velocity of a particle are $(\mathbf{r}_0, \mathbf{v}_0)$ and we integrate the trajectory for an interval h, at which point its phase-space coordinates are $(\mathbf{r}_1, \mathbf{v}_1)$. Now reverse the velocity to obtain phase-space coordinates $(\mathbf{r}_1, -\mathbf{v}_1)$ and integrate again for an interval h. Then the final position and velocity will be $(\mathbf{r}_2, -\mathbf{v}_2)$. We now reverse the velocity again, so the particle is at $(\mathbf{r}_2, \mathbf{v}_2)$. The integrator is reversible if the particle has now returned to its original phase-space position, that is, if $(\mathbf{r}_2, \mathbf{v}_2) = (\mathbf{r}_0, \mathbf{v}_0)$.

We now restate this concept more generally. The trajectory of a dynamical system is its position $\mathbf{z}(t)$ in phase space as a function of time t. For any autonomous dynamical system we may define a nonlinear operator or **propagator** G_h that maps $\mathbf{z}(t)$ to $\mathbf{z}(t+h)$. It follows from the definition that $\mathsf{G}_0 = \mathsf{I}$, the identity operator, and that

$$\mathsf{G}_{h+k} = \mathsf{G}_h \mathsf{G}_k. \tag{2.34}$$

Setting $k = -h$, we conclude that

$$\mathsf{G}_{-h} = \mathsf{G}_h^{-1}. \tag{2.35}$$

We define the time-reversal operator T such that $\mathsf{T}\mathbf{z}$ is the phase-space position that corresponds to $\mathbf{z}$ if the direction of time is reversed. For example if we use phase-space coordinates in which $\mathbf{z} = (\mathbf{z}_1, \mathbf{z}_2)$, where $\mathbf{z}_1$ represents position and $\mathbf{z}_2$ represents velocity, then $\mathsf{T}(\mathbf{z}_1, \mathbf{z}_2) = (\mathbf{z}_1, -\mathbf{z}_2)$. Since T is linear in these coordinates, it can be written as a matrix:

$$\mathsf{T}\begin{bmatrix} \mathbf{z}_1 \\ \mathbf{z}_2 \end{bmatrix} = \begin{bmatrix} \mathbf{I} & \mathbf{0} \\ \mathbf{0} & -\mathbf{I} \end{bmatrix}\begin{bmatrix} \mathbf{z}_1 \\ \mathbf{z}_2 \end{bmatrix}. \tag{2.36}$$

To keep the discussion general, we shall make the weaker assumption that T is a *linear* operator, that is, $\mathsf{T}(\mathbf{z}_1 + \mathbf{z}_2) = \mathsf{T}\mathbf{z}_1 + \mathsf{T}\mathbf{z}_2$ and $\mathsf{T}(c\mathbf{z}) = c\mathsf{T}\mathbf{z}$. This assumption is not valid in all phase-space coordinates. For example, in Delaunay variables

$$\mathsf{T}(\Lambda, L, L_z, \ell, \omega, \Omega) = (\Lambda, L, -L_z, -\ell, \pi - \omega, \pi + \Omega). \tag{2.37}$$

Reversing the direction of time twice has no effect, so in *any* coordinates $\mathsf{T}^2 = \mathsf{I}$.

An autonomous dynamical system is **time-reversible** or just **reversible** if (Arnold 1984; Roberts & Quispel 1992)

$$\mathsf{G}_h \mathsf{T} \mathsf{G}_h = \mathsf{T}. \tag{2.38}$$

Because of equation (2.35), an equivalent statement is

$$\mathsf{T} \mathsf{G}_h = \mathsf{G}_{-h} \mathsf{T}. \tag{2.39}$$

What properties of the equation of motion are required for reversibility? By applying equation (2.34) repeatedly, the operator G_h can be rewritten as $(\mathsf{G}_{h/N})^N$ for any integer N and it is straightforward to show by induction that if $\mathsf{G}_{h/N}$ is reversible, then G_h must be as well. By letting $N \to \infty$ we conclude that it is sufficient to show that G_h is reversible for very small timesteps h. If the equation of motion is $\mathrm{d}\mathbf{z}/\mathrm{d}t = \mathbf{f}(\mathbf{z})$, then

$$\mathsf{G}_h = \mathsf{I} + h\mathsf{F} + \mathrm{O}(h^2), \tag{2.40}$$

where F is the nonlinear operator defined by $\mathsf{F}\mathbf{z} = \mathbf{f}(\mathbf{z})$. The system is reversible if and only if equation (2.38) is satisfied up to terms of order h; since T is linear this requires

$$\mathsf{T}\mathsf{F} + \mathsf{F}\mathsf{T} = \mathbf{0} \quad \text{or} \quad \mathsf{T}\mathbf{f}(\mathbf{z}) = -\mathbf{f}(\mathsf{T}\mathbf{z}). \tag{2.41}$$

For example, in the differential equations (2.1) and (2.2) we have $\mathbf{z} = (\mathbf{r}, \mathbf{v})$ and $\mathbf{f}(\mathbf{z}) = [\mathbf{v}, -\nabla\Phi(\mathbf{r})]$. Using equation (2.36), $\mathsf{T}\mathbf{z} = (\mathbf{r}, -\mathbf{v})$ and $\mathsf{T}\mathbf{f}(\mathbf{z}) = [\mathbf{v}, \nabla\Phi(\mathbf{r}, t)]$, while $\mathbf{f}(\mathsf{T}\mathbf{z}) = [-\mathbf{v}, -\nabla\Phi(\mathbf{r})]$. Thus equation (2.41) is satisfied and the system is reversible.

We now apply these concepts to integrators. Let $\mathsf{\Gamma}_h$ be the operator corresponding to an integrator with timestep h, that is, if the phase-space position at time t is $\mathbf{z}$ then the position at the next timestep $t + h$ is $\mathbf{z}' = \mathsf{\Gamma}_h \mathbf{z}$.

The definition of an integrator with negative timestep needs some attention. In principle there need be no relation between $\mathsf{\Gamma}_h$ and $\mathsf{\Gamma}_{-h}$; for example, $\mathsf{\Gamma}_h$ could be leapfrog for $h > 0$ and the modified Euler method for $h < 0$. However, it is natural to assume that $\mathsf{\Gamma}_h \mathbf{z}$ is a smooth (analytic) function of h, that is, it has a Taylor-series expansion in h that is valid for both positive and negative h. Moreover, integrating a reversible dynamical system with timestep $h > 0$ should give the same result as integrating the time-reversed system with timestep $-h$, that is,

$$\mathbf{T}\boldsymbol{\Gamma}_h = \boldsymbol{\Gamma}_{-h}\mathbf{T} \quad \text{or} \quad \boldsymbol{\Gamma}_{-h}^{-1}\mathbf{T}\boldsymbol{\Gamma}_h = \mathbf{T}. \tag{2.42}$$

We say that an integrator is **normal** if it satisfies this constraint when applied to a reversible dynamical system, as defined by equation (2.41). For example, the Euler method (eq. 2.5) is $\boldsymbol{\Gamma}_h = \mathbf{I} + h\mathbf{F}$ so $\mathbf{T}\boldsymbol{\Gamma}_h = \mathbf{T} + h\mathbf{T}\mathbf{F}$ and $\boldsymbol{\Gamma}_{-h}\mathbf{T} = (\mathbf{I} - h\mathbf{F})\mathbf{T} = \mathbf{T} - h\mathbf{F}\mathbf{T}$; then the relation (2.41) shows that it is normal. Similarly, it is straightforward to show that the drift and kick operators (2.24) and (2.25) are normal. Moreover, if $\mathbf{A}_h$ and $\mathbf{B}_h$ are normal then so is $\mathbf{A}_h\mathbf{B}_h$. Thus the modified Euler method and leapfrog are both normal, since they are composed of kick and drift operators. In fact almost all of the integrators that we encounter in this chapter are normal.

An integrator is said to be **symmetric** if

$$\boldsymbol{\Gamma}_{-h} = \boldsymbol{\Gamma}_h^{-1}. \tag{2.43}$$

The analogous equation (2.35) for autonomous systems with continuous time holds automatically. However, (2.43) does not hold automatically for integrators. For example, the drift-kick modified Euler method (2.27) has $\boldsymbol{\Gamma}_h = \mathbf{K}_h\mathbf{D}_h$. From the second of equations (2.26) and its analog for the kick operator we have $\boldsymbol{\Gamma}_{-h}^{-1} = \mathbf{D}_{-h}^{-1}\mathbf{K}_{-h}^{-1} = \mathbf{D}_h\mathbf{K}_h$, which is the kick-drift integrator (2.28), not drift-kick. Thus the modified Euler method is not symmetric. Another example is the Euler method (2.5), which has $\boldsymbol{\Gamma}_h\mathbf{z} = \mathbf{z} + h\mathbf{f}(\mathbf{z})$; in this case $\boldsymbol{\Gamma}_{-h}^{-1}$ is the *backward* Euler method (2.18). Thus the Euler method is also not symmetric. The simplest symmetric method is leapfrog.

We now show that any symmetric integrator must have even order. If an integrator $\boldsymbol{\Gamma}_h$ has order k, then it must be related to the propagator $\mathbf{G}_h$ that describes the exact flow by

$$\boldsymbol{\Gamma}_h = \mathbf{G}_h + h^{k+1}\mathbf{E} + \mathrm{O}(h^{k+2}), \tag{2.44}$$

where $h^{k+1}\mathbf{E}\mathbf{z}$ is the dominant error term for a single timestep h starting at $\mathbf{z}$. Similarly,

$$\boldsymbol{\Gamma}_{-h} = \mathbf{G}_{-h} + (-h)^{k+1}\mathbf{E} + \mathrm{O}(h^{k+2}). \tag{2.45}$$

Therefore

$$\mathbf{\Gamma}_{-h}\mathbf{\Gamma}_h = \mathbf{I} + (-h)^{k+1}\mathbf{E}\mathbf{G}_h + h^{k+1}\mathbf{G}_{-h}\mathbf{E} + \mathrm{O}(h^{k+2}). \tag{2.46}$$

Since $\mathbf{G}_h = \mathbf{I} + \mathrm{O}(h)$, we have

$$\mathbf{\Gamma}_{-h}\mathbf{\Gamma}_h = \mathbf{I} + [(-h)^{k+1} + h^{k+1}]\mathbf{E} + \mathrm{O}(h^{k+2}). \tag{2.47}$$

If the integrator is symmetric, the right side must equal the identity, so the quantity in square brackets must vanish, which requires that the order k is even.

By analogy with equation (2.38), an integrator is reversible if

$$\mathbf{\Gamma}_h\mathbf{T}\mathbf{\Gamma}_h = \mathbf{T}. \tag{2.48}$$

Comparing this condition with equations (2.42) and (2.43), we conclude that a normal integrator is reversible if and only if it is symmetric. Since leapfrog is symmetric and normal it is also reversible.

Of course, reversibility of an integrator is only a useful property if the underlying dynamical system is reversible, that is, if it satisfies (2.38) or (2.41).

Any normal, non-symmetric integrator can be used to construct a time-reversible one: it is straightforward to show from equation (2.42) that if $\mathbf{\Gamma}_h$ is normal, then $\mathbf{\Gamma}_{-h}^{-1}\mathbf{\Gamma}_h$ is a reversible integrator with timestep $2h$. For example, let $\mathbf{\Gamma}_h = \mathbf{K}_h\mathbf{D}_h$, the drift-kick version of modified Euler. Just below equation (2.43) we showed that $\mathbf{\Gamma}_{-h}^{-1} = \mathbf{D}_h\mathbf{K}_h$, so $\mathbf{\Gamma}_{-h}^{-1}\mathbf{\Gamma}_h = \mathbf{D}_h\mathbf{K}_{2h}\mathbf{D}_h$, which is the drift-kick-drift leapfrog operator with timestep $2h$ (this is another proof that leapfrog is reversible). Similarly, If $\mathbf{E}_h$ denotes the Euler method, then $\mathbf{E}_{-h}^{-1}\mathbf{E}_h$ is the trapezoidal rule (2.75), an implicit integrator that is reversible. See Problem 2.4 for another example.

2.2.2 Symplectic integrators

The motion of most systems relevant to celestial mechanics is governed by a Hamiltonian $H(\mathbf{q}, \mathbf{p}, t)$. In Hamiltonian systems the flow of trajectories through phase space is strongly constrained; for example, Liouville's theorem tells us that phase-space volumes are conserved by the flow (see the discussion surrounding eq. D.47). Thus integrators derived from Hamiltonians, by the methods described below, may be able to follow Hamiltonian systems more accurately than general-purpose integrators.

Let $\mathsf{\Gamma}_h$ be an integrator with timestep h, so a single timestep of the integrator is $\mathbf{z}' = \mathsf{\Gamma}_h \mathbf{z}$. Its Jacobian matrix $\mathbf{\Gamma}_h(\mathbf{z})$ is defined by (cf. eq. D.42)

$$\Gamma_{h,ij}(h,\mathbf{z}) \equiv \frac{\partial z_i'}{\partial z_j} = \frac{\partial}{\partial z_j}[\mathsf{\Gamma}_h(\mathbf{z})]_i. \tag{2.49}$$

If the integrator is derived from a Hamiltonian then its Jacobian matrix must be symplectic, that is, it must satisfy the symplectic condition (D.46),

$$\mathbf{\Gamma}_h^{\mathrm{T}} \mathbf{J} \mathbf{\Gamma}_h = \mathbf{J}, \tag{2.50}$$

where $\mathbf{J}$ is the symplectic matrix (D.14) and "T" denotes the transpose. Integrators satisfying this condition are known as **symplectic integrators.**[8]

The symplectic condition (2.50) superficially resembles the reversibility condition (2.48), but there are important differences. In particular, reversibility is a feature of an *individual* orbit specified by the operator $\mathsf{\Gamma}_h$, while symplecticity is a feature of a *family* of nearby orbits because it depends on the Jacobian matrix of $\mathsf{\Gamma}_h$.

For an example of a symplectic integrator, consider a system in which the canonical coordinates and momenta are $\mathbf{q} = \mathbf{r}$ and $\mathbf{p} = \mathbf{v}$, the position and velocity, and the Hamiltonian is $H(\mathbf{r}, \mathbf{v}, t) = \frac{1}{2}v^2 + \Phi(\mathbf{r}, t)$. To construct an integrator with timestep h, we introduce the periodic delta function defined by equation (C.9),

$$\delta_h(t) = h \sum_{k=-\infty}^{\infty} \delta(t - kh). \tag{2.51}$$

Over time intervals much longer than h, the average value of the periodic delta function is nearly unity, which suggests that so long as the orbital period is much larger than the timestep we can approximate the Hamiltonian by

$$H_{\mathrm{num}}(\mathbf{r}, \mathbf{v}, t) = \tfrac{1}{2}v^2 + \delta_h(t - t_0 - fh)\Phi(\mathbf{r}, t), \tag{2.52}$$

8 Yoshida (1993) gives a clear review of early work on symplectic integrators.

where t_0 is the initial time of the integration and f is a constant between 0 and 1. We call this the **numerical Hamiltonian**. The corresponding equations of motion are

$$\frac{\mathrm{d}\mathbf{r}}{\mathrm{d}t} = \frac{\partial H_{\mathrm{num}}}{\partial \mathbf{v}} = \mathbf{v}, \quad \frac{\mathrm{d}\mathbf{v}}{\mathrm{d}t} = -\frac{\partial H_{\mathrm{num}}}{\partial \mathbf{r}} = -\delta_h(t - t_0 - fh)\nabla\Phi(\mathbf{r}, t). \quad (2.53)$$

We now solve these to determine $\mathbf{r}(t)$ and $\mathbf{v}(t)$ over the interval from $t_n = t_0 + nh$ to $t_{n+1} = t_0 + (n+1)h$. Let $t' = t_n + fh$; t' must lie between t_n and t_{n+1}. Let t'_- and t'_+ denote times slightly before and after t'. Then from $t = t_n$ to $t = t'_-$ we have $\delta_h(t - t_0 - fh) = 0$, so the velocity is constant at $\mathbf{v}_n$ and the position advances to $\mathbf{r}' \equiv \mathbf{r}_n + fh\mathbf{v}_n$. From $t = t'_-$ to $t = t'_+$ the change in position is negligible but the velocity is subject to an impulse $\Delta\mathbf{v} = -\nabla\Phi(\mathbf{r}', t') \int_{t'_-}^{t'_+} \mathrm{d}t\, \delta_h(t - t') = -h\nabla\Phi(\mathbf{r}, t')$. Thus $\mathbf{r}(t'_+) = \mathbf{r}(t'_-) = \mathbf{r}'$ and $\mathbf{v}(t'_+) = \mathbf{v}(t'_-) - h\nabla\Phi(\mathbf{r}', t')$. Finally, between t'_+ and t_{n+1} the position advances to $\mathbf{r}_{n+1} = \mathbf{r}' + (1 - f)h\mathbf{v}(t'_+)$ and the velocity is constant, so $\mathbf{v}_{n+1} = \mathbf{v}(t'_+)$. Summarizing, the position and velocity at $t_{n+1} = t_n + h$ are given by

$$\begin{aligned} \mathbf{r}' &= \mathbf{r}_n + fh\mathbf{v}_n, \\ \mathbf{v}_{n+1} &= \mathbf{v}_n - h\nabla\Phi(\mathbf{r}', t_n + fh), \\ \mathbf{r}_{n+1} &= \mathbf{r}' + (1 - f)h\mathbf{v}_{n+1}. \end{aligned} \quad (2.54)$$

By letting $f \to 1$ we recover the drift-kick modified Euler method of equation (2.27). If $f \to 0$ we obtain the kick-drift modified Euler method (2.28), and if $f = \frac{1}{2}$ we obtain the drift-kick-drift leapfrog integrator (2.29). Therefore all of these integrators and operators can be derived from the Hamiltonian (2.52), and thus all are symplectic. Moreover the composition of symplectic operators is symplectic (see discussion in the paragraph following eq. D.47), so other compositions of the kick and drift operators such as kick-drift-kick leapfrog are symplectic as well.

An alternative proof that these integrators are symplectic is based on the Jacobian matrices of the drift and kick operators D_h (eq. 2.24) and K_h (eq. 2.25). In N-dimensional space, these are the $2N \times 2N$ matrices

$$\mathbf{D}(h) = \begin{bmatrix} \mathbf{I} & h\mathbf{I} \\ \mathbf{0} & \mathbf{I} \end{bmatrix}, \quad \mathbf{K}(h, \mathbf{r}, t) = \begin{bmatrix} \mathbf{I} & \mathbf{0} \\ -h\mathbf{\Phi} & \mathbf{I} \end{bmatrix}. \quad (2.55)$$

Here $\mathbf{0}$ and $\mathbf{I}$ are the $N \times N$ zero and identity matrices, and $\boldsymbol{\Phi}(\mathbf{r}, t)$ is the Hessian of the potential $\Phi(\mathbf{r}, t)$, that is, the $N \times N$ matrix with entries $\Phi_{ij}(\mathbf{r}, t) = \partial^2\Phi/\partial r_i \partial r_j$. It is straightforward to show that $\mathbf{D}(h)$ and $\mathbf{K}(h, \mathbf{r}, t)$ satisfy the symplectic condition (2.50). Therefore the operators D_h and K_h are symplectic, and so are compositions of these operators such as the modified Euler and leapfrog integrators.

If the original Hamiltonian is time-independent, then it is conserved along a trajectory and equal to the energy. A symplectic integrator does not conserve the energy. However, experiments such as those reported in Figure 2.1 show that in most cases, the energy oscillates around a mean that is close to the conserved energy of the exact trajectory.

We have derived explicit symplectic integrators for Hamiltonians such as $H(\mathbf{r}, \mathbf{v}, t) = \frac{1}{2}v^2 + \Phi(\mathbf{r}, t)$. Symplectic integrators also exist for general Hamiltonians $H(\mathbf{q}, \mathbf{p}, t)$ but these are usually implicit. The simplest first-order symplectic integrator with timestep h is

$$\mathbf{q}_{n+1} = \mathbf{q}_n + h\frac{\partial H}{\partial \mathbf{p}_{n+1}}(\mathbf{q}_n, \mathbf{p}_{n+1}, t), \quad \mathbf{p}_{n+1} = \mathbf{p}_n - h\frac{\partial H}{\partial \mathbf{q}_n}(\mathbf{q}_n, \mathbf{p}_{n+1}, t). \tag{2.56}$$

Informally, the superior performance of symplectic integrators over long integration times arises because the geometrical constraints on Hamiltonian flows in phase space are so strong that systematic errors (in, say, the energy, phase-space volume, or other conserved quantities) cannot accumulate. The properties of symplectic integrators are discussed further in §2.5.1.

2.2.3 Variable timestep

One serious limitation of symplectic integrators is that they work well only with fixed timesteps, as the following example shows. Suppose the timestep depends on phase-space position, $h = \tau(\mathbf{r}, \mathbf{v})$. The Hamiltonian (2.52) becomes

$$H_{\mathrm{num}}(\mathbf{r}, \mathbf{v}, t) = \tfrac{1}{2}v^2 + \delta_{\tau(\mathbf{r},\mathbf{v})}[t - t_0 - f\tau(\mathbf{r}, \mathbf{v})]\Phi(\mathbf{r}, t). \tag{2.57}$$

Since Hamilton's equations (D.37) require derivatives of H_{num} they now involve derivatives of delta functions, which means that there are no simple

Box 2.3: How do geometric integrators fail?

The usual criterion for the success of a numerical integration is that the global error—the difference between the numerical solution and the true solution at the end of the integration—is sufficiently small. Geometric integrators, however, preserve the properties of the phase-space flow and therefore can give qualitatively correct results even when the global error is relatively large. For example, a geometric integrator could fail to predict the orbital phases in a multi-planet system after a 1 Gyr integration, but still correctly predict whether the system is stable.

To illustrate this behavior, we integrate the equation of motion for a simple pendulum using the modified Euler method. The pendulum Hamiltonian for a particle of unit mass is (eq. 6.1)

$$H(q,p) = \tfrac{1}{2}p^2 - \omega^2 \cos q, \tag{a}$$

where ω is the frequency of small-amplitude oscillations. The equations of motion are

$$\frac{\mathrm{d}q}{\mathrm{d}t} = \frac{\partial H}{\partial p} = p, \qquad \frac{\mathrm{d}p}{\mathrm{d}t} = -\frac{\partial H}{\partial q} = -\omega^2 \sin q. \tag{b}$$

The kick-drift modified Euler integrator (2.28) with timestep h is

$$p_{n+1} = p_n - \omega^2 h \sin q_n, \quad q_{n+1} = q_n + hp_{n+1}. \tag{c}$$

We set $y_n \equiv hp_n$ and $x_n \equiv q_n + \pi$ to derive a simpler form,

$$y_{n+1} = y_n + K \sin x_n, \quad x_{n+1} = x_n + y_{n+1} \tag{d}$$

with $K \equiv \omega^2 h^2$. This is the Chirikov–Taylor map described in Appendix F.

Plots of this map (modulo 2π) are shown in Figure F.1 for $K = 0.1, 0.5, 1.0$, and 2.0. For small K, the trajectories in the map closely match the level surfaces of the Hamiltonian (a), as they must since the modified Euler method is a well-behaved integrator for sufficiently small timestep. As K increases, the trajectories become more distorted and a significant fraction of phase space becomes chaotic. This example illustrates that symplectic integrators with too large a timestep can fail by introducing spurious structure, such as chaotic regions, into the phase space of a system described by a regular Hamiltonian.

analytic operators corresponding to this Hamiltonian. In words, a symplectic integrator with fixed timestep is generally no longer symplectic once the timestep is varied.[9]

Fortunately, the geometric constraints on the phase-space flow imposed by time reversibility are also strong, so the leapfrog integrator retains its good behavior if the timestep is adjusted in a time-reversible manner, even though the resulting integrator is no longer symplectic. Here is one simple way to do this: we modify equations (2.29) to

$$\begin{aligned}
\mathbf{r}' &= \mathbf{r}_n + \tfrac{1}{2}h\mathbf{v}_n, \\
\mathbf{v}' &= \mathbf{v}_n - \tfrac{1}{2}h\nabla\Phi(\mathbf{r}', t_n + \tfrac{1}{2}h), \\
t_{n+1} &= t_n + \tfrac{1}{2}(h + h'), \\
\mathbf{v}_{n+1} &= \mathbf{v}' - \tfrac{1}{2}h'\nabla\Phi(\mathbf{r}', t_{n+1} - \tfrac{1}{2}h'), \\
\mathbf{r}_{n+1} &= \mathbf{r}' + \tfrac{1}{2}h'\mathbf{v}_{n+1}.
\end{aligned} \tag{2.58}$$

Here h' is determined from h by solving the equation $u(h, h') = \tau(\mathbf{r}', \mathbf{v}')$, where $u(h, h')$ is a symmetric function of h and h' such that $u(h, h) = h$; for example, $u(h, h') = \frac{1}{2}(h + h')$ or $u(h, h') = 2hh'/(h + h')$. Inspection of these equations shows that they are reversible, and like leapfrog the integrator is explicit and requires no auxiliary storage.

This result can be generalized to any normal integrator $\mathbf{\Gamma}_h$ for which there is an explicit inverse $\mathbf{\Gamma}_h^{-1}$. A single step of a reversible integrator is given by

$$\mathbf{z}' = \mathbf{\Gamma}_{h/2}\mathbf{z}_n, \quad u(h, h') = \tau(\mathbf{z}'), \quad \mathbf{z}_{n+1} = \mathbf{\Gamma}_{-h'/2}^{-1}\mathbf{z}'. \tag{2.59}$$

A different approach to developing symplectic integrators with variable timestep is based on the extended phase space described in Box 2.1. Suppose that the optimum timestep for an integrator at the phase-space position $(\mathbf{q}, \mathbf{p})$ is $g(\mathbf{q}, \mathbf{p})$. We introduce the fictitious time τ defined by equation

[9] A symplectic integrator *does* remain symplectic if the timesteps are varied in some fixed pattern that does not depend on the phase-space coordinates, but this sort of behavior of the timestep has little practical use.

(e) of Box 2.1; then a constant timestep of 1 in τ corresponds to a variable timestep $g(\mathbf{q},\mathbf{p})$ in the real time t, so long as we replace the Hamiltonian $H(\mathbf{q},\mathbf{p},t)$ by the Hamiltonian $\Gamma(\mathbf{Q},\mathbf{P})$ defined in equation (b) in the box. The principal limitation of this approach is that symplectic integrators for the Hamiltonian $\Gamma(\mathbf{P},\mathbf{Q})$ are usually implicit (cf. eq. 2.56), in contrast to the explicit integrators (such as leapfrog) that can be used on a simpler Hamiltonian such as $H(\mathbf{r},\mathbf{v}) = \frac{1}{2}v^2 + \Phi(\mathbf{r},t)$ (Mikkola & Tanikawa 1999; Preto & Tremaine 1999).

2.3 Runge–Kutta and collocation integrators

2.3.1 Runge–Kutta methods

This is a broad class of integrators in which the function $\mathbf{f}(\mathbf{z},t)$ on the right side of the differential equation (2.1) is evaluated at several intermediate times ("stages") between t_n and t_{n+1}, and the evaluations are combined to match a Taylor-series expansion of the trajectory to as high an order as possible.

We illustrate this process for explicit second-order methods. We write

$$\begin{aligned} \mathbf{z}' &= \mathbf{z}_n + \alpha h\mathbf{f}(\mathbf{z}_n,t_n), \\ \mathbf{z}_{n+1} &= \mathbf{z}_n + \beta h\mathbf{f}(\mathbf{z}_n,t_n) + \gamma h\mathbf{f}(\mathbf{z}',t_n+\delta h), \end{aligned} \tag{2.60}$$

where the Greek letters α, β, γ, δ denote four coefficients that are to be determined.

To keep the next few equations simpler, we temporarily replace the vectors $\mathbf{z}$ and $\mathbf{f}$ by scalars z and f. This restriction does not affect any of our conclusions. Combining equations (2.60) and expanding them in powers of the timestep h, we have

$$z_{n+1} = z_n + (\beta+\gamma)hf + \alpha\gamma h^2 f_z f + \gamma\delta h^2 f_t + \mathrm{O}(h^3); \tag{2.61}$$

here $f_z \equiv \partial f/\partial z$, $f_t \equiv \partial f/\partial t$, and all of the functions on the right side of the equation are evaluated at (z_n,t_n). Using the same notation, the Taylor series for the solution of $z_t = f(z,t)$ (eq. 2.1) is

$$z_{n+1} = z_n + hz_t + \tfrac{1}{2}h^2 z_{tt} + \mathrm{O}(h^3)$$

$$= z_n + hf + \tfrac{1}{2}h^2 f_z f + \tfrac{1}{2}h^2 f_t + \mathrm{O}(h^3). \tag{2.62}$$

If the integrator (2.61) is to match the Taylor series (2.62) up to and including terms of order h^2, we must have

$$\beta + \gamma = 1, \quad \alpha = \delta = \frac{1}{2\gamma}. \tag{2.63}$$

Since there are three equations for four unknowns, there is one free parameter for second-order methods of this class, which we can choose to be γ. The most common choice is $\gamma = 1$, which implies $\beta = 0$ and $\alpha = \delta = \frac{1}{2}$ and gives the **explicit midpoint integrator**,

$$\mathbf{z}_{n+1} = \mathbf{z}_n + h\mathbf{f}[\mathbf{z}_n + \tfrac{1}{2}h\mathbf{f}(\mathbf{z}_n, t_n), t_n + \tfrac{1}{2}h]. \tag{2.64}$$

Runge–Kutta methods with more stages and higher orders can be generated similarly. The most popular is a fourth-order, four-stage integrator known as RK4 or the "classical" Runge–Kutta method, defined by

$$\begin{aligned}
\mathbf{k}_1 &= \mathbf{f}(\mathbf{z}_n, t_n),\\
\mathbf{k}_2 &= \mathbf{f}(\mathbf{z}_n + \tfrac{1}{2}h\mathbf{k}_1, t_n + \tfrac{1}{2}h),\\
\mathbf{k}_3 &= \mathbf{f}(\mathbf{z}_n + \tfrac{1}{2}h\mathbf{k}_2, t_n + \tfrac{1}{2}h),\\
\mathbf{k}_4 &= \mathbf{f}(\mathbf{z}_n + h\mathbf{k}_3, t_n + h),\\
\mathbf{z}_{n+1} &= \mathbf{z}_n + \tfrac{1}{6}h(\mathbf{k}_1 + 2\mathbf{k}_2 + 2\mathbf{k}_3 + \mathbf{k}_4).
\end{aligned} \tag{2.65}$$

There are four evaluations of $\mathbf{f}(\mathbf{z})$, so the effective timestep is $\frac{1}{4}h$.

When RK4 is applied to the harmonic oscillator the solution is again a sum of sequences of the form $\mathbf{z}_n = \kappa^n \mathbf{a}$ (eq. 2.13), with two solutions for the scale factor given by

$$\kappa_\pm = 1 - \tfrac{1}{2}\omega^2 h^2 + \tfrac{1}{24}\omega^4 h^4 \pm \mathrm{i}(\omega h - \tfrac{1}{6}\omega^3 h^3). \tag{2.66}$$

The radius and speed tend to decay exponentially as $\exp[\gamma(t - t_0)]$ with

$$\gamma = \frac{1}{h}\log|\kappa_\pm| = \frac{1}{2h}\log\left(1 - \tfrac{1}{72}\omega^6 h^6 + \tfrac{1}{576}\omega^8 h^8\right) = -\tfrac{1}{144}\omega^6 h^5 + \mathrm{O}(h^8). \tag{2.67}$$

The rate of decay is far smaller than the rate of growth that we found with the Euler method in equation (2.17). The numerical experiment shown in Figure 2.1 confirms that RK4 is a much more accurate integrator, although with 200 force evaluations per orbit it still fails after less than 1 000 orbits.

The general form of an s-stage Runge–Kutta method is

$$\begin{aligned} \mathbf{k}_i &= \mathbf{f}(\mathbf{z}_n + h\sum_{j=1}^{s} A_{ij}\mathbf{k}_j, t_n + hc_i), \quad i = 1, \ldots, s, \\ \mathbf{z}_{n+1} &= \mathbf{z}_n + h\sum_{j=1}^{s} w_j\mathbf{k}_j. \end{aligned} \tag{2.68}$$

The method is explicit if $A_{ij} = 0$ for $j \geq i$. For example, RK4 has

$$\mathbf{A} = \begin{bmatrix} 0 & 0 & 0 & 0 \\ \frac{1}{2} & 0 & 0 & 0 \\ 0 & \frac{1}{2} & 0 & 0 \\ 0 & 0 & 1 & 0 \end{bmatrix}, \quad \mathbf{c} = \left(0\ \tfrac{1}{2}\ \tfrac{1}{2}\ 1\right), \quad \mathbf{w} = \left(\tfrac{1}{6}\ \tfrac{1}{3}\ \tfrac{1}{3}\ \tfrac{1}{6}\right). \tag{2.69}$$

Runge–Kutta methods are good choices for short integrations—hundreds or thousands of orbits—but not for the Gyr integrations needed to investigate the long-term stability of planetary systems. They are also well matched to problems that require a variable timestep, such as following the evolution of highly eccentric orbits.

Current practice is to estimate the timestep required for a given accuracy using an **embedded Runge–Kutta integrator**. These are integrators designed such that the same function evaluations can be used with two different weights to give Runge–Kutta methods of different orders. Then the local error can be estimated from the difference between the two methods, so the timestep can be adjusted if the local error is too large or too small compared to some pre-set accuracy criterion.

For example, the **Dormand–Prince method** (Shampine 1986; Hairer et al. 1993) is an explicit seven-stage Runge–Kutta integrator with

$$\mathbf{A} = \begin{bmatrix} 0 & 0 & 0 & 0 & 0 & 0 & 0 \\ \frac{1}{5} & 0 & 0 & 0 & 0 & 0 & 0 \\ \frac{3}{40} & \frac{9}{40} & 0 & 0 & 0 & 0 & 0 \\ \frac{44}{45} & -\frac{56}{15} & \frac{32}{9} & 0 & 0 & 0 & 0 \\ \frac{19\,372}{6\,561} & -\frac{25\,360}{2\,187} & \frac{64\,448}{6\,561} & -\frac{212}{729} & 0 & 0 & 0 \\ \frac{9\,017}{3\,168} & -\frac{355}{33} & \frac{46\,732}{5\,247} & \frac{49}{176} & -\frac{5\,103}{18\,656} & 0 & 0 \\ \frac{35}{384} & 0 & \frac{500}{1\,113} & \frac{125}{192} & -\frac{2\,187}{6\,784} & \frac{11}{84} & 0 \end{bmatrix}, \tag{2.70}$$

and

$$\mathbf{c} = \left(0\ \tfrac{1}{5}\ \tfrac{3}{10}\ \tfrac{4}{5}\ \tfrac{8}{9}\ 1\ 1\right). \tag{2.71}$$

This yields a fifth-order integrator if we choose

$$\mathbf{w} = \left(\tfrac{35}{384}\ 0\ \tfrac{500}{1\,113}\ \tfrac{125}{192}\ -\tfrac{2\,187}{6\,784}\ \tfrac{11}{84}\ 0\right). \tag{2.72}$$

Although this is a seven-stage method, it only requires six evaluations of $\mathbf{f}(\mathbf{z}, t)$ per step because the last stage of step n is evaluated at the same location as the first stage of step $n+1$ (because the last row of the matrix $\mathbf{A}$ is the same as the vector $\mathbf{w}$). Equations (2.70) and (2.71) also produce an embedded fourth-order integrator if we replace $\mathbf{w}$ by

$$\mathbf{w}' = \left(\tfrac{5\,179}{57\,600}\ 0\ \tfrac{7\,571}{16\,695}\ \tfrac{393}{640}\ -\tfrac{92\,097}{339\,200}\ \tfrac{187}{2\,100}\ \tfrac{1}{40}\right). \tag{2.73}$$

Let the new phase-space position obtained by equations (2.68) and (2.70)–(2.72) be $\mathbf{z}_{n+1}$, while the position obtained by replacing (2.72) by (2.73) is $\mathbf{z}'_{n+1}$. The local error in $\mathbf{z}_{n+1}$ should be much smaller than the local error in $\mathbf{z}'_{n+1}$ since the former is derived by a fifth-order method and the latter by a fourth-order method. Therefore the local error in $\mathbf{z}'_{n+1}$ is $\mathbf{\Delta} \simeq \mathbf{z}'_{n+1} - \mathbf{z}_{n+1}$ and scales as $\mathrm{O}(h^5)$. If we change the timestep to some new value h_{new}, the error should be $\mathbf{\Delta}(h_{\mathrm{new}}/h)^5$. If we want the error to be less than some pre-set value ϵ in all coordinates of the vector $\mathbf{z}'_{n+1}$ then the new timestep should be

$$h_{\mathrm{new}} = h\left[\frac{\epsilon}{|\max(\Delta_i)|}\right]^{1/5}. \tag{2.74}$$

This formula allows the timestep to be expanded if $|\max(\Delta_i)| < \epsilon$ or shrunk if $|\max(\Delta_i)| > \epsilon$; of course, in the latter case the trial timestep has failed to give the desired accuracy and so must be re-taken with the smaller timestep. The error estimate applies to $\mathbf{z}'_{n+1}$, but in practice we use $\mathbf{z}_{n+1}$ as the predicted position at t_{n+1} since it is more accurate.

The Dormand–Prince integrator is implemented in many software packages, including the Python-based SciPy package.

There are geometric Runge–Kutta methods. The simplest of these are the **trapezoidal rule**,

$$\mathbf{z}_{n+1} = \mathbf{z}_n + \tfrac{1}{2}h\mathbf{f}(\mathbf{z}_n, t_n) + \tfrac{1}{2}h\mathbf{f}(\mathbf{z}_{n+1}, t_{n+1}), \tag{2.75}$$

and the **implicit midpoint method**,

$$\mathbf{z}_{n+1} = \mathbf{z}_n + h\mathbf{f}\big[\tfrac{1}{2}(\mathbf{z}_n + \mathbf{z}_{n+1}), t_n + \tfrac{1}{2}h\big]. \tag{2.76}$$

These are both second-order methods. They are implicit integrators since the new position $\mathbf{z}_{n+1}$ appears as an argument of $\mathbf{f}(\mathbf{z}, t)$. Both methods are symmetric (eq. 2.43), and when applied to reversible systems they are normal (eq. 2.42) and therefore reversible (eq. 2.48). The implicit midpoint method is also symplectic when applied to Hamiltonian systems, but the trapezoidal rule is not. Both methods are closely related to Euler's method, as described at the end of §2.2.1 and in Problem 2.4.

In terms of the general formula (2.68), the trapezoidal rule can be written

$$\mathbf{A} = \begin{bmatrix} 0 & 0 \\ \frac{1}{2} & \frac{1}{2} \end{bmatrix}, \quad \mathbf{c} = \big(0\;1\big), \quad \mathbf{w} = \big(\tfrac{1}{2}\;\tfrac{1}{2}\big), \tag{2.77}$$

and the implicit midpoint method is

$$A_1 = \tfrac{1}{2}, \quad c_1 = \tfrac{1}{2}, \quad w_1 = 1. \tag{2.78}$$

Unfortunately, all Runge–Kutta integrators described by equation (2.68) that are reversible or symplectic are also implicit (Hairer et al. 2006) and so require several iterations per step to converge.

An alternative approach, which we now describe, is to develop explicit Runge–Kutta integrators of sufficiently high order that the local truncation error is smaller than the roundoff error. Then if the original dynamical system is reversible or symplectic the integrator will be as well, at least to within roundoff error.

2.3.2 Collocation methods

Integrators for ordinary differential equations are closely related to numerical methods for evaluating integrals, since integration of the differential equation $\mathrm{d}z/\mathrm{d}t = f(t)$ is equivalent to finding the integral $\int \mathrm{d}t\, f(t)$. To reduce confusion in this subsection, we shall always use the term **quadrature** to denote the evaluation of integrals, in contradistinction to **integration**, which denotes the solution of differential equations.

For example, if $b = a+h$ then the Euler method (2.5) yields the following approximation to the integral $I = \int_a^b \mathrm{d}t\, f(t)$:

$$I = hf_0, \tag{2.79}$$

in which $f_\alpha \equiv f(a+\alpha h)$. This quadrature rule requires one function evaluation per timestep h and is first-order, that is, the error over an interval h is $\mathrm{O}(h^2)$. The explicit midpoint method (2.64) gives the approximation

$$I = hf_{1/2}. \tag{2.80}$$

The classical Runge–Kutta or RK4 method (2.65) gives

$$I = h\big(\tfrac{1}{6}f_0 + \tfrac{2}{3}f_{1/2} + \tfrac{1}{6}f_1\big), \tag{2.81}$$

which is Simpson's quadrature rule. The trapezoidal rule (2.75) gives

$$I = h\big(\tfrac{1}{2}f_0 + \tfrac{1}{2}f_1\big). \tag{2.82}$$

These observations motivate integrators known as **collocation methods**. Let $c_1, \ldots, c_s$ be distinct real numbers, usually between 0 and 1, called the **nodes**. To integrate the differential equation (2.1), we seek a polynomial $\mathbf{u}(t)$ of degree s such that

$$\mathbf{u}(t_n) = \mathbf{z}_n, \quad \dot{\mathbf{u}}(t_n + hc_j) = \mathbf{f}[\mathbf{u}(t_n + hc_j), t_n + hc_j], \quad j = 1, \ldots, s. \tag{2.83}$$

For each dimension of the vector $\mathbf{u}(t)$, these equations give $s+1$ constraints on the $s+1$ coefficients of the polynomial, so usually the degree-s polynomial satisfying (2.83) is unique. Once the polynomial is found, the next step of the integration is $\mathbf{z}_{n+1} = \mathbf{u}(t_n+h)$. The Euler method (2.5), the backward Euler method (2.18), and the implicit midpoint method (2.76) are collocation methods with $s = 1$ and $c_1 = 0, 1, \frac{1}{2}$, respectively. The trapezoidal rule (2.75) is a collocation method with $s = 2$ and $c_1 = 0$, $c_2 = 1$. Collocation methods are special cases of Runge–Kutta methods (see Box 2.4).

All collocation methods other than the Euler method are implicit, and they are reversible if and only if the nodes c_i are symmetrically distributed around $\frac{1}{2}$, that is, $c_{s-i+1} = 1 - c_i$ for $i \leq 1 \leq s$.

The general quadrature rule that underlies equations (2.79)–(2.82) is

$$I = h \sum_{j=1}^{s} w_j f(t_0 + hc_j). \tag{2.84}$$

Not surprisingly, if the error in the quadrature rule is $\mathrm{O}(h^{k+1})$ then the error in the collocation method with the same nodes is also $\mathrm{O}(h^{k+1})$. Therefore high-order quadrature rules can be used to generate high-order integrators.

The principle of **Gaussian quadrature** is that a wise choice of the nodes c_i yields a quadrature rule whose order k can be as large as twice the number s of nodes per timestep. In practice, most collocation methods use either the **Gauss–Legendre** or the **Gauss–Radau** rule for choosing the nodes.

The Gauss–Legendre rule has order $k = 2s$, which is the theoretical maximum. Moreover it can be shown that collocation methods using this rule are symplectic (Sanz–Serna 1988). The nodes are given by the roots of the Legendre polynomials of Appendix C.6, $\mathsf{P}_s(2c_i - 1) = 0$, and the weights are

$$w_i = \frac{1}{4(c_i - c_i^2)[\mathsf{P}_s'(2c_i - 1)]^2}, \quad i = 1, \ldots, s. \tag{2.85}$$

For example, the two-stage, fourth-order Gauss–Legendre rule has

$$c_1 = \tfrac{1}{2} - \tfrac{1}{2} \cdot 3^{-1/2}, \quad c_2 = \tfrac{1}{2} + \tfrac{1}{2} \cdot 3^{-1/2}, \quad w_1 = w_2 = \tfrac{1}{2}. \tag{2.86}$$

Box 2.4: Collocation and Runge–Kutta methods

To explore the relation between these two types of integrator, write

$$\dot{\mathbf{u}}(t) = \sum_{j=1}^{s} \mathbf{k}_j P_j\left(\frac{t - t_n}{h}\right), \tag{a}$$

where $P_j(\tau)$ is the **Lagrange interpolating polynomial**

$$P_j(x) \equiv \prod_{\substack{l=1 \\ l \neq j}}^{s} \frac{x - c_l}{c_j - c_l}, \tag{b}$$

which equals 1 at $x = c_j$ and zero at $x = c_l$ if $l \neq j$, and has degree $s - 1$. The second of equations (2.83) implies that for a collocation method

$$\mathbf{k}_i = \mathbf{f}[\mathbf{u}(t_n + hc_i), t_n + hc_i]. \tag{c}$$

Integrating equation (a) from t_n to $t_n + hc_i$ and using the first of equations (2.83) we find

$$\mathbf{u}(t_n + hc_i) = \mathbf{z}_n + h \sum_{j=1}^{s} A_{ij} \mathbf{k}_j, \quad \text{where} \quad A_{ij} \equiv \int_0^{c_i} \mathrm{d}\tau\, P_j(\tau), \tag{d}$$

and integrating from t_n to $t_n + h$ gives

$$\mathbf{z}_{n+1} = \mathbf{z}_n + h \sum_{j=1}^{s} w_j \mathbf{k}_j \quad \text{where} \quad w_j \equiv \int_0^1 \mathrm{d}\tau\, P_j(\tau). \tag{e}$$

Equations (c)–(e) are equivalent to the definition (2.68) of an s-stage Runge–Kutta method. Thus all collocation methods are Runge–Kutta methods. The converse is not true, because not all choices of the coefficients A_{ij} and w_j satisfy equations (d) and (e).

The Gauss–Radau rule has one node $c_1 = 0$ at the start of the timestep and order $k = 2s - 1$. The order is one less than the Gauss–Legendre rule, but the force at the first node is just $\mathbf{f}(\mathbf{z}_n, t_n)$ and can be computed once and for all at each step without iterating to convergence. The nodes and weights for the Gauss–Radau rule are given by

$$c_1 = 0, \quad w_1 = \frac{1}{s^2}, \tag{2.87}$$
$$\mathsf{P}_{s-1}(2c_i - 1) + \mathsf{P}_s(2c_i - 1) = 0, \; w_i = \frac{1 - c_i}{s^2[\mathsf{P}_{s-1}(2c_i - 1)]^2}, \; i = 2, \ldots, s.$$

For example the two-stage, third-order Gauss–Radau rule has

$$c_1 = 0, \quad c_2 = \tfrac{2}{3}, \quad w_1 = \tfrac{1}{4}, \quad w_2 = \tfrac{3}{4}. \tag{2.88}$$

This can be written in the notation of equation (2.68) as

$$\mathbf{A} = \begin{bmatrix} 0 & 0 \\ \frac{1}{3} & \frac{1}{3} \end{bmatrix}, \quad \mathbf{c} = \left(0 \; \tfrac{2}{3}\right), \quad \mathbf{w} = \left(\tfrac{1}{4} \; \tfrac{3}{4}\right). \tag{2.89}$$

Gauss–Legendre or Gauss–Radau integrators can also be designed for the Newtonian form (2.3) of the equation of motion. The most popular of these (Everhart 1985; Rein & Spiegel 2015) is a Gauss–Radau integrator with order $k = 15$ and requires $s = 8n$ function evaluations per step, where n is the number of iterations that the implicit integrator needs to converge (typically only 2–3 if the differential equations are well behaved and the timestep is well chosen). These integrators are particularly well suited for high-accuracy integrations of orbits with rapidly changing accelerations that require a variable timestep. For typical planetary orbits, high-order integrators of this kind can achieve local errors smaller than the roundoff error (§2.7) with timesteps of a few percent of the orbital period.

2.4 Multistep integrators

2.4.1 Multistep methods for first-order differential equations

The integrators that we have described so far are **one-step** methods, that is, the only information used to predict $\mathbf{z}_{n+1}$ is $\mathbf{z}_n$, and all the history contained in the earlier steps of the trajectory is discarded. In contrast, **multistep methods** also use the positions $\mathbf{z}_{n-m}$ and derivatives $\mathbf{f}(\mathbf{z}_{n-m}, t_{n-m})$ from several earlier steps $m = 1, 2, \ldots$ to enhance the accuracy of the predicted position (Henrici 1962; Gear 1971; Hairer et al. 1993, 2006).

The most common multistep methods for the differential equation (2.1) are defined by the formula

$$\sum_{m=0}^{M} \alpha_m \mathbf{z}_{n+1-m} + h \sum_{m=0}^{M} \beta_m \mathbf{f}(\mathbf{z}_{n+1-m}, t_{n+1-m}) = \mathbf{0}, \tag{2.90}$$

where h is the timestep and at least one of α_M and β_M is nonzero. Without loss of generality we can set $\alpha_0 = -1$, and rewrite the formula as

$$\mathbf{z}_{n+1} = \sum_{m=1}^{M} \alpha_m \mathbf{z}_{n+1-m} + h \sum_{m=0}^{M} \beta_m \mathbf{f}(\mathbf{z}_{n+1-m}, t_{n+1-m}), \tag{2.91}$$

which shows how $\mathbf{z}_{n+1}$ is predicted from up to M earlier positions and $M + 1$ derivatives. The method is explicit if $\beta_0 = 0$; otherwise it depends on the derivative at the predicted point, $\mathbf{f}(\mathbf{z}_{n+1}, t_{n+1})$, and thus is implicit. A method can be specified compactly by its characteristic polynomials

$$\rho(x) \equiv \sum_{m=0}^{M} \alpha_m x^m, \quad \sigma(x) \equiv \sum_{m=0}^{M} \beta_m x^m. \tag{2.92}$$

We define the order of a multistep integrator as follows. If $\mathbf{z}^\star(t)$ is an exact solution of the differential equation $\dot{\mathbf{z}} = \mathbf{f}(\mathbf{z}, t)$, and $\mathbf{z}_n^\star \equiv \mathbf{z}^\star(t_n)$, then a method of order k satisfies

$$\sum_{m=0}^{M} \alpha_m \mathbf{z}^\star_{n+1-m} + h \sum_{m=0}^{M} \beta_m \mathbf{f}(\mathbf{z}^\star_{n+1-m}, t_{n+1-m}) = \mathrm{O}(h^{k+1}). \tag{2.93}$$

We justify this definition and describe its relation to the definition of order for one-step methods later in this section.

We now determine the coefficients α_m and β_m. Any solution of the differential equation $\dot{\mathbf{z}} = \mathbf{f}(\mathbf{z}, t)$ can be written as a Taylor series expansion around t_{n+1}, $\mathbf{z}^\star(t) = \sum_{j=0}^\infty \mathbf{a}_j (t - t_{n+1})^j$. Then $\mathbf{z}^\star_{n+1-m} = \sum_{j=0}^\infty \mathbf{a}_j (-mh)^j$

and $\mathbf{f}(\mathbf{z}_{n+1-m}, t_{n+1-m}) = \dot{\mathbf{z}}^\star(t_{n+1-m}) = \sum_{j=0}^\infty j\mathbf{a}_j(-mh)^{j-1}$. Equation (2.93) becomes

$$\sum_{j=0}^\infty (-1)^j h^j \mathbf{a}_j \left[\sum_{m=0}^M m^j \alpha_m - j\sum_{m=0}^M m^{j-1}\beta_m\right] = \mathrm{O}(h^{k+1}). \tag{2.94}$$

Therefore the integrator has order k if and only if

$$E_j \equiv \sum_{m=0}^M m^j \alpha_m - j\sum_{m=0}^M m^{j-1}\beta_m = 0 \quad \text{for } j = 0, 1, \ldots, k \tag{2.95}$$

and $E_{k+1} \neq 0$. The quantities E_j can be written in terms of the characteristic polynomials: $E_0 = \rho(1)$, $E_1 = \rho'(1) - \sigma(1)$, and so forth.

For an integrator of order k there are $k+1$ constraints, $E_0 = E_1 = \cdots = E_k = 0$. In an M-step method there are are $2(M+1)$ coefficients α_m and β_m, but we have set $\alpha_0 = -1$, and $\beta_0 = 0$ if the method is explicit. Thus there are $2M$ or $2M+1$ unknowns, depending on whether the integrator is explicit or implicit. Since linear equations usually have a solution if the number of equations is less than or equal to the number of unknowns, we might hope to choose the coefficients such that the integrator has order $2M-1$ (explicit) or $2M$ (implicit). This can be done, but such integrators have little or no practical value because of the **Dahlquist barrier**, a theorem showing that the maximum order of a stable M-step method is M if the method is explicit, $M+1$ if the method is implicit and M is odd, or $M+2$ if the method is implicit and M is even.

Given the limitations on order imposed by the Dahlquist barrier, it is simpler to construct integrators by choosing the coefficients α_m and then solving for the β_m. The α_m can be arbitrary except that (i) $\alpha_0 = -1$ by assumption, and (ii) the condition $E_0 = 0$ requires $\sum_{m=0}^M \alpha_m = 0$. Once the α_m are fixed, we have k constraints on the β_m from $E_1 = \cdots = E_k = 0$. For explicit methods $\beta_0 = 0$, so we have M variables β_m to satisfy these constraints. Typically this can be done if $M \geq k$, so the maximum order of an M-step method of this kind is $k = M$, as high as can be expected given the Dahlquist barrier.

The simplest polynomial consistent with the constraints $\alpha_0 = -1$ and $\sum \alpha_m = 0$ has $\alpha_1 = 1$ and $\alpha_2 = \alpha_3 = \cdots = \alpha_M = 0$. The corresponding characteristic polynomial is $\rho(x) = x - 1$. This choice yields the explicit **Adams–Bashforth** methods; for orders $k = 1, 2, 3, 4$ these are

$$\begin{aligned}
\mathbf{z}_{n+1} &= \mathbf{z}_n + h\mathbf{f}_n, \\
\mathbf{z}_{n+1} &= \mathbf{z}_n + \tfrac{1}{2}h\big(3\mathbf{f}_n - \mathbf{f}_{n-1}\big), \\
\mathbf{z}_{n+1} &= \mathbf{z}_n + \tfrac{1}{12}h\big(23\mathbf{f}_n - 16\mathbf{f}_{n-1} + 5\mathbf{f}_{n-2}\big), \\
\mathbf{z}_{n+1} &= \mathbf{z}_n + \tfrac{1}{24}h\big(55\mathbf{f}_n - 59\mathbf{f}_{n-1} + 37\mathbf{f}_{n-2} - 9\mathbf{f}_{n-3}\big).
\end{aligned} \tag{2.96}$$

Here $\mathbf{f}_n \equiv \mathbf{f}(\mathbf{z}_n, t_n)$. The first line is just the Euler method.

For implicit methods we have $M + 1$ variables β_m and k equations, so the maximum order is $k = M + 1$, again consistent with the Dahlquist barrier. The same choice of characteristic polynomial $\rho(x) = x - 1$ yields the implicit **Adams–Moulton** methods: for orders $1, 2, 3, 4$ these are

$$\begin{aligned}
\mathbf{z}_{n+1} &= \mathbf{z}_n + h\mathbf{f}_{n+1}, \\
\mathbf{z}_{n+1} &= \mathbf{z}_n + \tfrac{1}{2}h\big(\mathbf{f}_{n+1} + \mathbf{f}_n\big), \\
\mathbf{z}_{n+1} &= \mathbf{z}_n + \tfrac{1}{12}h\big(5\mathbf{f}_{n+1} + 8\mathbf{f}_n - \mathbf{f}_{n-1}\big), \\
\mathbf{z}_{n+1} &= \mathbf{z}_n + \tfrac{1}{24}h\big(9\mathbf{f}_{n+1} + 19\mathbf{f}_n - 5\mathbf{f}_{n-1} + \mathbf{f}_{n-2}\big).
\end{aligned} \tag{2.97}$$

The first line is the implicit Euler method (2.18) and the second is the trapezoidal rule (2.75).

In **predictor-corrector** methods an explicit integrator is used to estimate $\mathbf{z}_{n+1}$, which then provides the starting estimate of $\mathbf{f}_{n+1}$ for an implicit integrator. The Adams–Bashforth and Adams–Moulton pair produce a successful predictor-corrector method for many problems.

The behavior of multistep methods can be explored further by following a test particle in a harmonic potential, along the lines of the discussion in §2.1.2. Substituting $\dot{\mathbf{z}} = \mathbf{f}(\mathbf{z}) = \mathbf{A}\mathbf{z}$ (eq. 2.10) into equation (2.90), we have

$$\sum_{m=0}^{M} \alpha_m \mathbf{z}_{n+1-m} + h \sum_{m=0}^{M} \beta_m \mathbf{A}\mathbf{z}_{n+1-m} = \mathbf{0}, \tag{2.98}$$

where $\mathbf{z} \equiv (\mathbf{r}, \mathbf{v})$ and

$$\mathbf{A} \equiv \begin{bmatrix} \mathbf{0} & \mathbf{I} \\ -\omega^2\mathbf{I} & \mathbf{0} \end{bmatrix}. \tag{2.99}$$

The solution of this difference equation is a linear combination of sequences of the form $\mathbf{z}_n = \kappa^n \mathbf{a}$. Substituting this expression in equation (2.98), we have

$$\rho(\kappa^{-1})\mathbf{a} + h\sigma(\kappa^{-1})\mathbf{A}\mathbf{a} = \mathbf{0}. \tag{2.100}$$

Thus $\mathbf{a}$ must be an eigenvector of $\mathbf{A}$. The eigenvalues of $\mathbf{A}$ are $\lambda_\pm = \pm\mathrm{i}\omega$ so the solutions of equation (2.100) are $\kappa_\pm$ where

$$\rho(\kappa_\pm^{-1}) \pm \mathrm{i}\omega h\sigma(\kappa_\pm^{-1}) = 0. \tag{2.101}$$

In general there are multiple solutions of each of these equations. Assuming that the coefficients α_m and β_m are real, the solutions of equations (2.101) are related by $\kappa_- = \kappa_+^*$ where the asterisk denotes a complex conjugate.

We now use these results to provide a heuristic justification of the definition (2.93) of the order of a multistep method. The general solution of the difference equation (2.98) at timestep n is a sum of terms of the form $c_\pm\kappa_\pm^n\mathbf{a}_\pm$ where $c_\pm$ are constants, $\mathbf{a}_\pm$ are eigenvectors of $\mathbf{A}$, and $\kappa_\pm$ is a solution of (2.101). The exact solution at $t_1 = t_0 + h$ with initial condition $c_\pm\mathbf{a}_\pm$ at t_0 is $c_\pm\exp(\pm\mathrm{i}\omega h)\mathbf{a}_\pm$. Thus the one-step error between t_0 and t_1 is a sum of terms of the form $c_\pm\epsilon_\pm\mathbf{a}_+$ where $\epsilon_\pm \equiv \kappa_\pm - \exp(\pm\mathrm{i}\omega h)$. Eliminating $\kappa_\pm$ from equation (2.101) in favor of $\epsilon_\pm$, we have

$$\rho[(\mathrm{e}^{\pm\mathrm{i}\omega h} + \epsilon_\pm)^{-1}] \pm \mathrm{i}\omega h\sigma[(\mathrm{e}^{\pm\mathrm{i}\omega h} + \epsilon_\pm)^{-1}] = 0; \tag{2.102}$$

and keeping only terms up to first order in the small quantity $\epsilon_\pm$,

$$\rho(\mathrm{e}^{\mp\mathrm{i}\omega h}) \pm \mathrm{i}\omega h\sigma(\mathrm{e}^{\mp\mathrm{i}\omega h}) - \epsilon_\pm\mathrm{e}^{\mp 2\mathrm{i}\omega h}\big[\rho'(\mathrm{e}^{\mp\mathrm{i}\omega h}) \pm \mathrm{i}\omega h\sigma'(\mathrm{e}^{\mp\mathrm{i}\omega h})\big] = 0. \tag{2.103}$$

The sum of the first two terms must be $\mathrm{O}(h^{k+1})$ according to the definition (2.93). Since the quantity in square brackets in (2.103) is of order unity, $\epsilon_\pm$ must be $\mathrm{O}(h^{k+1})$, so the one-step error is also $\mathrm{O}(h^{k+1})$, consistent with the definition of order for one-step methods in §2.1.1.

In the limit $h \to 0$, the solutions of (2.101) reduce to the roots of $\rho(\kappa^{-1})$; one of these is $\kappa = 1$ because of the constraint $E_0 = 0$. However, since $\rho(x)$ is a polynomial of order $\leq M$, there are up to $M - 1$ additional ("parasitic") roots, and the behavior of the integrator depends on these as well. In particular, if any of these roots have $|\kappa| > 1$ then small perturbations (e.g., due to roundoff error) grow exponentially, no matter how small the timestep may be. If there are roots with $|\kappa| = 1$ with multiplicity greater than 1 (say, p) then small perturbations grow as n^{p-1}. Whether the growth is exponential or polynomial, the behavior of the numerical trajectory is eventually dominated by these parasitic roots. Therefore any useful multistep integrator must be **zero-stable**, by which we mean that there are no roots of $\rho(\kappa^{-1})$ with $|\kappa| > 1$ and all roots with $|\kappa| = 1$ have multiplicity 1. For example, the Adams–Bashforth and Adams–Moulton methods have $\rho(x) = x - 1$, so the only root is $x = 1$ and the methods are zero-stable. Zero-stability is a necessary condition for a practical multistep integrator but it is not sufficient. For example, the Euler method is zero-stable but performs badly, as we saw in §2.1.2.

2.4.2 Multistep methods for Newtonian differential equations

The Newtonian form (2.3) of the equation of motion is

$$\frac{\mathrm{d}^2\mathbf{r}}{\mathrm{d}t^2} = \mathbf{F}(\mathbf{r}, t), \tag{2.104}$$

where $\mathbf{F} = -\nabla\Phi$ is the force per unit mass. A linear multistep method for this equation can be written (cf. eq. 2.90)

$$\sum_{m=0}^{M} \alpha_m \mathbf{r}_{n+1-m} + h^2 \sum_{m=0}^{M} \beta_m \mathbf{F}(\mathbf{r}_{n+1-m}, t_{n+1-m}) = \mathbf{0}, \tag{2.105}$$

and if we set $\alpha_0 = -1$,

$$\mathbf{r}_{n+1} = \sum_{m=1}^{M} \alpha_m \mathbf{r}_{n+1-m} + h^2 \sum_{m=0}^{M} \beta_m \mathbf{F}(\mathbf{r}_{n+1-m}, t_{n+1-m}). \tag{2.106}$$

The method is explicit if $\beta_0 = 0$ and otherwise implicit.

We define the method to have order k if

$$\sum_{m=0}^{M} \alpha_m \mathbf{r}^\star_{n+1-m} + h^2 \sum_{m=0}^{M} \beta_m \mathbf{F}(\mathbf{r}^\star_{n+1-m}, t_{n+1-m}) = \mathrm{O}(h^{k+2}), \tag{2.107}$$

where $\mathbf{r}^\star_n(t)$ is an exact solution of the differential equation (2.104). Note the exponent $k+2$ on the right side compared to $k+1$ on the right side of the analogous definition in equation (2.93);[10] the reasons for this choice are described after equation (2.113).

We derive the coefficients α_m and β_m using the same arguments that follow equation (2.93). Any solution of the differential equation can be written as a Taylor series expansion around t_{n+1}, $\mathbf{r}^\star(t) = \sum_{j=0}^\infty \mathbf{a}_j(t-t_{n+1})^j$. Then $\mathbf{r}^\star_{n+1-m} = \sum_{j=0}^\infty \mathbf{a}_j(-mh)^j$ and $\mathbf{F}(\mathbf{r}_{n+1-m}, t_{n+1-m}) = \ddot{\mathbf{r}}^\star(t_{n+1-m}) = \sum_{j=0}^\infty j(j-1)\mathbf{a}_j(-mh)^{j-2}$. Equation (2.107) becomes

$$\sum_{j=0}^{\infty}(-1)^j h^j \mathbf{a}_j\left[\sum_{m=0}^{M} m^j\alpha_m + j(j-1)\sum_{m=0}^{M} m^{j-2}\beta_m\right] = \mathrm{O}(h^{k+2}). \quad (2.108)$$

Therefore the integrator has order k if and only if

$$\overline{E}_j \equiv \sum_{m=0}^{M} m^j\alpha_m + j(j-1)\sum_{m=0}^{M} m^{j-2}\beta_m = 0 \quad \text{for } j = 0, 1, \ldots, k+1 \quad (2.109)$$

and $\overline{E}_{k+2} \neq 0$. The quantities $\overline{E}_j$ can be written in terms of the characteristic polynomials: $\overline{E}_0 = \rho(1)$, $\overline{E}_1 = \rho'(1)$, and so forth.

Once again the simplest way to construct integrators is to choose the coefficients α_m and then solve equations (2.109) for the β_m. We have assumed that $\alpha_0 = -1$; in addition, the constraints $\overline{E}_0 = \overline{E}_1 = 0$ imply that $\sum\alpha_m = 0$ and $\sum m\alpha_m = 0$, independent of the choice of β_m. The simplest choice consistent with these constraints is $\alpha_1 = 2$, $\alpha_2 = -1$, with $\alpha_3 = \cdots = \alpha_M = 0$. The corresponding characteristic polynomial is $\rho(x) = -(x-1)^2$. This process yields the explicit **Störmer** multistep methods of orders $k = 2, 3, 4, 5$:

$$\begin{aligned}
\mathbf{r}_{n+1} &= 2\mathbf{r}_n - \mathbf{r}_{n-1} + h^2\mathbf{F}_n,\\
\mathbf{r}_{n+1} &= 2\mathbf{r}_n - \mathbf{r}_{n-1} + \tfrac{1}{12}h^2\big(13\mathbf{F}_n - 2\mathbf{F}_{n-1} + \mathbf{F}_{n-2}\big), \qquad (2.110)\\
\mathbf{r}_{n+1} &= 2\mathbf{r}_n - \mathbf{r}_{n-1} + \tfrac{1}{12}h^2\big(14\mathbf{F}_n - 5\mathbf{F}_{n-1} + 4\mathbf{F}_{n-2} - \mathbf{F}_{n-3}\big),\\
\mathbf{r}_{n+1} &= 2\mathbf{r}_n - \mathbf{r}_{n-1}
\end{aligned}$$

[10] Not all authors subscribe to the same definition: some use an exponent $k+1$ in both (2.93) and (2.107).

$$+ \tfrac{1}{240}h^2\big(299\mathbf{F}_n - 176\mathbf{F}_{n-1} + 194\mathbf{F}_{n-2} - 96\mathbf{F}_{n-3} + 19\mathbf{F}_{n-4}\big).$$

Here $\mathbf{F}_n = \mathbf{F}(\mathbf{r}_n, t_n)$. The first of these is closely related to leapfrog (see Box 2.5).

The first two implicit **Cowell** methods have orders 4 and 5:

$$\mathbf{r}_{n+1} = 2\mathbf{r}_n - \mathbf{r}_{n-1} + \tfrac{1}{12}h^2\big(\mathbf{F}_{n+1} + 10\mathbf{F}_n + \mathbf{F}_{n-1}\big), \quad (2.111)$$
$$\mathbf{r}_{n+1} = 2\mathbf{r}_n - \mathbf{r}_{n-1} + \tfrac{1}{240}h^2\big(19\mathbf{F}_{n+1} + 204\mathbf{F}_n + 14\mathbf{F}_{n-1} + 4\mathbf{F}_{n-2} - \mathbf{F}_{n-3}\big).$$

Box 2.5: Leapfrog and the Störmer method

The simplest Störmer multistep method, from the first line of equation (2.110), is

$$\mathbf{r}_{n+1} = 2\mathbf{r}_n - \mathbf{r}_{n-1} + h^2\mathbf{F}_n, \quad \text{(a)}$$

where $\mathbf{r}_n$ is the position at step n, $\mathbf{F}_n$ is the force at $\mathbf{r}_n$, and h is the timestep. The average velocities between steps $n - 1$ and n and between steps n and $n + 1$ are

$$\mathbf{v}_{n-1/2} = \frac{\mathbf{r}_n - \mathbf{r}_{n-1}}{h}, \quad \mathbf{v}_{n+1/2} = \frac{\mathbf{r}_{n+1} - \mathbf{r}_n}{h}. \quad \text{(b)}$$

The positions at timesteps $n \pm \frac{1}{2}$ can be estimated as

$$\mathbf{r}_{n-1/2} = \frac{\mathbf{r}_n + \mathbf{r}_{n-1}}{2}, \quad \mathbf{r}_{n+1/2} = \frac{\mathbf{r}_{n+1} + \mathbf{r}_n}{2}. \quad \text{(c)}$$

Now solve equations (b) for $\mathbf{r}_{n-1}$ and $\mathbf{r}_{n+1}$ and eliminate these variables from equations (a) and (c). We find

$$\mathbf{r}_n = \mathbf{r}_{n-1/2} + \tfrac{1}{2}h\mathbf{v}_{n-1/2},$$
$$\mathbf{v}_{n+1/2} = \mathbf{v}_{n-1/2} + h\mathbf{F}_n,$$
$$\mathbf{r}_{n+1/2} = \mathbf{r}_n + \tfrac{1}{2}h\mathbf{v}_{n+1/2}. \quad \text{(d)}$$

Apart from minor differences in notation, this is the drift-kick-drift leapfrog integrator of equation (2.29).

The behavior of these methods can be explored by following a test particle orbiting in a harmonic potential with force law $\mathbf{F}(\mathbf{r}) = -\omega^2\mathbf{r}$. Substituting this force law into equation (2.105), we have

$$\sum_{m=0}^{M} (\alpha_m - \beta_m \omega^2 h^2)\mathbf{r}_{n+1-m} = \mathbf{0}. \tag{2.112}$$

The solution of this difference equation is a linear combination of sequences of the form $\mathbf{r}_n = \kappa^n \mathbf{a}$. Substituting this expression in equation (2.112), we have

$$\rho(\kappa^{-1}) - \omega^2 h^2 \sigma(\kappa^{-1}) = 0. \tag{2.113}$$

We now provide a heuristic justification for the definition (2.107) of the order of these integrators. Following arguments similar to those leading to equation (2.103), we find that the error between timesteps n and $n+1$ is of order $\epsilon_\pm$ where to first order in $\epsilon_\pm$

$$\rho(\mathrm{e}^{\mp \mathrm{i}\omega h}) - \omega^2 h^2 \sigma(\mathrm{e}^{\mp \mathrm{i}\omega h}) - \epsilon_\pm \mathrm{e}^{\mp 2\mathrm{i}\omega h}\big[\rho'(\mathrm{e}^{\mp \mathrm{i}\omega h}) - \omega^2 h^2 \sigma'(\mathrm{e}^{\mp \mathrm{i}\omega h})\big] = 0. \tag{2.114}$$

From equation (2.107) the sum of the first two terms is $\mathrm{O}(h^{k+2})$. The quantity in square brackets is equal to $\rho'(1) \mp \mathrm{i}\omega h \rho''(1) + \mathrm{O}(h^2)$ which equals $\mp \mathrm{i}\omega h \rho''(1) + \mathrm{O}(h^2)$ because of the condition $\overline{E}_1 = \rho'(1) = 0$. Therefore the quantity in square brackets is $\mathrm{O}(h)$, which implies that the one-step error is $\mathrm{O}(h^{k+1})$, consistent with the definition of order for one-step methods.

As we discussed after equation (2.103), a practical integrator must be zero-stable, that is, all roots of $\rho(\kappa^{-1})$ must lie inside or on the unit circle in the complex plane, and roots on the unit circle must be simple.[11]

A minor disadvantage of all multistep methods is that they require a special procedure, usually employing some other integrator, to generate the $M-1$ initial positions needed to get the multistep integrator started. A more serious disadvantage is that changing the timestep is much more complicated than in one-step methods. For this reason, multistep integrators are mostly used for long orbit integrations in planetary systems like the solar system, in which all of the planets are on well separated, nearly circular and

[11] With one exception: the conditions $\overline{E}_0 = \overline{E}_1 = 0$ are equivalent to $\rho(1) = \rho'(1) = 0$, so there is always a double root at $\kappa = 1$. Because of the double root the effect of a small perturbation grows as n, corresponding physically to a small perturbation in the initial velocity.

nearly coplanar orbits. Almost all long solar-system integrations up to the 1990s (e.g., Cohen et al. 1973) used a Störmer method with $M = 13$ steps and order $k = 13$, having coefficients defined by the characteristic polynomials $\rho(x) = -(x-1)^2$ and[12]

$$\begin{aligned}\sigma(x) = (&4\,621\,155\,471\,343x - 13\,232\,841\,914\,856x^2 + 47\,013\,743\,726\,958x^3\\&-114\,321\,700\,672\,600x^4 + 202\,271\,967\,611\,865x^5 - 266\,609\,549\,584\,656x^6\\&+264\,429\,021\,895\,332x^7 - 197\,106\,808\,276\,656x^8 + 108\,982\,933\,333\,425x^9\\&-43\,427\,592\,828\,040x^{10} + 11\,807\,143\,978\,638x^{11} - 1\,962\,777\,574\,776x^{12}\\&+150\,653\,570\,023x^{13})/2\,615\,348\,736\,000.\end{aligned} \tag{2.115}$$

Polynomials of high degree such as this should be always be evaluated using a method that minimizes roundoff error, as described in §2.7.2.

Störmer and Cowell integrators are neither reversible nor symplectic. Therefore test-particle orbits in a fixed potential are subject to energy drift. However, with a high-order multistep method and a suitable timestep the energy drift can be negligible for a well behaved planetary system (Grazier et al. 2005).

2.4.3 Geometric multistep methods

The concept of symplectic integration (§2.2.2) is difficult to apply to multistep integrators because they map multiple times in the present and past to a single future time. In contrast, the condition for reversibility (§2.2.1) is easy to state: the coefficients α_m and β_m in equation (2.101) must satisfy

$$\alpha_{M-j} = \alpha_j, \quad \beta_{M-j} = \beta_j, \quad j = 0, 1, \ldots, M. \tag{2.116}$$

A multistep method satisfying this condition is said to be **symmetric**[13] (Lambert & Watson 1976). It is straightforward to show that if the method is

[12] These coefficients are straightforward to derive by solving the linear system of equations (2.109) using computer algebra. Tables of coefficients are given by Maury & Segal (1969).

[13] Here "symmetric" refers to the symmetry of the coefficients, but the usage is consistent with our earlier definition of a symmetric integrator in equation (2.43).

symmetric and κ is a root of equation (2.101) then so is $1/\kappa$; thus if there is any root inside the unit circle there must also be one outside. If the method is zero-stable then none of the roots of $\rho(\kappa^{-1})$ is outside the unit circle, so they must all lie *on* the circle.

Since they are reversible, all symmetric methods have even order (see discussion surrounding eq. 2.44). Of the Störmer methods shown in equations (2.110), only the first is symmetric and therefore reversible; similarly, only the first of the Cowell methods in (2.111) is symmetric.

As we have shown, when $h = 0$ all of the solutions κ of equation (2.101) for a zero-stable symmetric method lie on the unit circle. As h increases from zero, there comes a point h_0, the termination of the **interval of periodicity**, at which one pair of solutions moves off the unit circle, one inside and one outside. For example, the interval of periodicity for the $k = 2$ Störmer method on the first line of equation (2.110) terminates at $h_0 = 2/\omega$. In practice, a reliable integration of a Kepler orbit requires a timestep substantially smaller than h_0 because any small eccentricity adds higher frequencies that lead to instability.

High-order symmetric multistep methods exhibit a variety of resonances and instabilities—narrow ranges of timestep in which the errors are much larger than normal (Quinlan 1999). These appear only when integrating nonlinear systems and are unrelated to the instability that arises at the termination of the interval of periodicity.

To construct high-order symmetric multistep methods we must use a characteristic polynomial of higher degree than the simple $\rho(x) = -(x - 1)^2$ used in the Störmer and Cowell methods. For a method of order k we must satisfy equations (2.109) and (2.116), but these conditions still leave considerable freedom, so the design of good multistep methods is something of an art. In practice, a reasonable goal is to seek an integrator of a given order with a large interval of periodicity and weak resonances (Quinlan & Tremaine 1990; Quinlan 1999; Fukushima 1999).

2.5 Operator splitting

In many systems, the right side of the equation of motion (2.1) can be decomposed into a sum of terms arising from different physical effects; thus

$$\frac{\mathrm{d}\mathbf{z}}{\mathrm{d}t} = \mathbf{f}(\mathbf{z},t) = \mathbf{f}_A(\mathbf{z},t) + \mathbf{f}_B(\mathbf{z},t) + \cdots. \tag{2.117}$$

For example, the motion of a test particle with phase-space coordinates $\mathbf{z} \equiv (\mathbf{r},\mathbf{v})$ in a gravitational potential $\Phi(\mathbf{r},t)$ is governed by equation (2.117) with $\mathbf{f}(\mathbf{z},t) = [\mathbf{v}, -\nabla\Phi(\mathbf{r},t)]$, so we may choose

$$\mathbf{f}_A(\mathbf{z},t) = (\mathbf{v},\mathbf{0}), \quad \mathbf{f}_B(\mathbf{z},t) = -[\mathbf{0}, \nabla\Phi(\mathbf{r},t)]. \tag{2.118}$$

A second example is motion in a multi-planet system, where the acceleration of a planet is the sum of the accelerations from the central star and the other planets.

Usually the equation of motion resulting from a single term in this sum [$\dot{\mathbf{z}} = \mathbf{f}_A(\mathbf{z},t)$, $\dot{\mathbf{z}} = \mathbf{f}_B(\mathbf{z},t)$, and so forth] is simpler than the original equation of motion, and may even be analytically soluble. The concept of **operator splitting** (Glowinski et al. 2016) is that the full equation of motion (2.117) can be solved numerically by advancing the trajectory for a short time under the influence of $\mathbf{f}_A(\mathbf{z},t)$, then under the influence of $\mathbf{f}_B(\mathbf{z},t)$, and so on until all of the terms in the sum have contributed, and then repeating the process.

The most important applications of this approach involve splitting into two components, $\mathbf{f}(\mathbf{z},t) = \mathbf{f}_A(\mathbf{z},t) + \mathbf{f}_B(\mathbf{z},t)$, and we restrict ourselves to this case from now on. To simplify the presentation we also restrict ourselves to autonomous systems in which $\mathbf{f}(\mathbf{z})$ is independent of time, since non-autonomous systems can always be converted to this form as described in Box 2.1.

We shall use the operator notation of §2.2.1, in which the trajectory of the system in phase space is

$$\mathbf{z}(t+h) = \mathsf{G}_h\mathbf{z}(t). \tag{2.119}$$

The propagator G_h satisfies $\mathrm{d}\mathsf{G}_h/\mathrm{d}h = \mathsf{F}$, where F is the nonlinear operator defined by $\mathsf{F}\mathbf{z} = \mathbf{f}(\mathbf{z})$. Similarly, $\mathsf{G}_{A,h}$ satisfies $\mathrm{d}\mathsf{G}_{A,h}/\mathrm{d}h = \mathsf{F}_A$, where

$\mathsf{F}_A\mathbf{z} = \mathbf{f}_A(\mathbf{z})$, with analogous definitions for $\mathsf{G}_{B,h}$ and F_B. For example, the propagators for the splitting (2.118) are

$$\mathsf{G}_{A,h}(\mathbf{r},\mathbf{v}) = (\mathbf{r}+h\mathbf{v},\mathbf{v}), \quad \mathsf{G}_{B,h}(\mathbf{r},\mathbf{v}) = (\mathbf{r},\mathbf{v}-h\nabla\Phi). \tag{2.120}$$

These are simply the drift and kick operators D_h and K_h from equations (2.24) and (2.25).

Let $\mathsf{\Gamma}_h$ be the propagator corresponding to an integrator with timestep h, so $\mathbf{z}_{n+1} = \mathsf{\Gamma}_h\mathbf{z}_n$. The integrator defined by **Lie–Trotter splitting** is (as usual, operators are applied from right to left)

$$\mathsf{\Gamma}_h = \mathsf{G}_{B,h}\mathsf{G}_{A,h} \quad \text{or} \quad \mathsf{G}_{A,h}\mathsf{G}_{B,h}. \tag{2.121}$$

For the operators (2.120) these are $\mathsf{K}_h\mathsf{D}_h$ and $\mathsf{D}_h\mathsf{K}_h$, the first-order drift-kick and kick-drift modified Euler integrators of equations (2.27) and (2.28). In general, Lie–Trotter splitting produces a first-order integrator.

A more accurate integrator is defined by **Strang splitting** (Strang 1968),

$$\mathsf{\Gamma}_h = \mathsf{G}_{A,h/2}\mathsf{G}_{B,h}\mathsf{G}_{A,h/2} \quad \text{or} \quad \mathsf{G}_{B,h/2}\mathsf{G}_{A,h}\mathsf{G}_{B,h/2}. \tag{2.122}$$

For the operators (2.120) these are drift-kick-drift leapfrog $\mathsf{D}_{h/2}\mathsf{K}_h\mathsf{D}_{h/2}$ (eq. 2.30) and kick-drift-kick leapfrog $\mathsf{K}_{h/2}\mathsf{D}_h\mathsf{K}_{h/2}$ (eq. 2.31). Strang splitting produces a second-order integrator.

If one or both of the propagators G_A and G_B is not analytic, it can be evaluated numerically using one of the integration methods described already in this chapter. Naturally, there is no reason to use an integrator for G_A or G_B that is higher order than the splitting scheme.

2.5.1 Operator splitting for Hamiltonian systems

Most of the systems in celestial mechanics are Hamiltonian, and Hamiltonian dynamics provides a powerful tool for analyzing operator-splitting methods.

We consider a Hamiltonian that is the sum of two simpler ones, $H = H_A + H_B$. In Hamiltonian systems the propagator G_h (eq. 2.119) is given by equation (D.39),

$$\mathsf{G}_h \equiv \exp(h\mathsf{L}_H) \tag{2.123}$$

where $H(\mathbf{z})$ is the Hamiltonian and the Lie operator $\mathbf{L}_H f = \{f, H\}$ is the Poisson bracket of the Hamiltonian and any function $f(\mathbf{z})$.

Box 2.6: The Baker–Campbell–Hausdorff formula

This states that the product of the exponentials of operators $\mathbf{X}$ and $\mathbf{Y}$ is given by

$$\exp(\epsilon\mathbf{X})\exp(\epsilon\mathbf{Y}) = \exp(\epsilon\mathbf{Z}), \qquad \text{(a)}$$

where

$$\mathbf{Z} = \mathbf{X} + \mathbf{Y} + \tfrac{1}{2}\epsilon[\mathbf{X},\mathbf{Y}] + \tfrac{1}{12}\epsilon^2\left([\mathbf{X},[\mathbf{X},\mathbf{Y}]] + [\mathbf{Y},[\mathbf{Y},\mathbf{X}]]\right) - \tfrac{1}{24}\epsilon^3[\mathbf{Y},[\mathbf{X},[\mathbf{X},\mathbf{Y}]]] + \mathrm{O}(\epsilon^4). \qquad \text{(b)}$$

Here $[\mathbf{X},\mathbf{Y}] \equiv \mathbf{XY} - \mathbf{YX}$ is the commutator of $\mathbf{X}$ and $\mathbf{Y}$, and $\epsilon \ll 1$.

The series in equation (b) is not necessarily convergent. It should be regarded as an asymptotic series, that is, if the series is truncated at order ϵ^n, the error will be of order ϵ^{n+1} as $\epsilon \to 0$.

A related formula is

$$\exp(\epsilon\mathbf{X})\exp(\epsilon\mathbf{Y})\exp(\epsilon\mathbf{X}) = \exp(\epsilon\mathbf{Z}), \qquad \text{(c)}$$

where

$$\mathbf{Z} = 2\mathbf{X} + \mathbf{Y} + \tfrac{1}{6}\epsilon^2\left([\mathbf{Y},[\mathbf{Y},\mathbf{X}]] - [\mathbf{X},[\mathbf{X},\mathbf{Y}]]\right) + \mathrm{O}(\epsilon^4). \qquad \text{(d)}$$

The Baker–Campbell–Hausdorff formula requires that the operators have suitable definitions of addition ($\mathbf{X} + \mathbf{Y}$), multiplication ($\mathbf{XY}$) and multiplication by a complex number λ ($\lambda\mathbf{X}$). Addition and multiplication of operators are both associative [$(\mathbf{X}+\mathbf{Y})+\mathbf{Z} = \mathbf{X}+(\mathbf{Y}+\mathbf{Z})$; $\mathbf{X}(\mathbf{YZ}) = (\mathbf{XY})\mathbf{Z}$]. Moreover addition commutes ($\mathbf{X}+\mathbf{Y} = \mathbf{Y}+\mathbf{X}$), although multiplication need not do so (if it did, all commutators would be zero). These conditions are all satisfied by the Lie operator $\mathbf{L}_g$ (eq. D.33).

Using this result, the first of the integrators (2.122) is

$$\mathbf{S}_h \equiv \exp(\tfrac{1}{2}h\mathbf{L}_{H_A})\exp(h\mathbf{L}_{H_B})\exp(\tfrac{1}{2}h\mathbf{L}_{H_A}); \qquad (2.124)$$

here we have introduced the symbol $\mathbf{S}_h$ as a reminder that the integrator is based on Strang splitting. The integrator is symplectic since it is the composition of symplectic propagators (see the discussion near the end of

Appendix D.5). Moreover it is straightforward to show that if the dynamical systems governed by the Hamiltonians H_A and H_B are reversible (eq. 2.38), then the integrator $\mathbf{S}_h$ is also reversible.

Using the Baker–Campbell–Hausdorff formula (eq. d of Box 2.6), equation (2.124) can be written as $\mathbf{S}_h = \exp(h\mathbf{Z})$ where

$$\begin{aligned}\mathbf{Z} = {} & \mathbf{L}_{H_A} + \mathbf{L}_{H_B} \\ & + \tfrac{1}{24}h^2\big(2[\mathbf{L}_{H_B},[\mathbf{L}_{H_B},\mathbf{L}_{H_A}]] - [\mathbf{L}_{H_A},[\mathbf{L}_{H_A},\mathbf{L}_{H_B}]]\big) + \mathrm{O}(h^4).\end{aligned} \tag{2.125}$$

Equation (D.36) implies that any commutator such as $[\mathbf{L}_a,[\mathbf{L}_b,[\mathbf{L}_c,\cdots]]]$ is equal to $\mathbf{L}_g$ where $g \equiv \{\{\{\cdots,c\},b\},a\}$. Moreover $\mathbf{L}_a + \mathbf{L}_b + \mathbf{L}_c = \mathbf{L}_{a+b+c}$ because the Poisson bracket is linear. Therefore we can write $\mathbf{Z} = \mathbf{L}_H$ where the **numerical Hamiltonian** is

$$H_{\mathrm{num}} \equiv H_A + H_B + H_{\mathrm{err}}(h), \tag{2.126}$$

and the **error Hamiltonian** is

$$H_{\mathrm{err}}(h) = \tfrac{1}{24}h^2\big(2\{\{H_A,H_B\},H_B\} - \{\{H_B,H_A\},H_A\}\big) + \mathrm{O}(h^4). \tag{2.127}$$

In words, we have shown that the integrator (2.124) follows exactly the trajectory of a particle in a new Hamiltonian that differs from the original one by the error Hamiltonian, which is smaller that the original by $\mathrm{O}(h^2)$.

In most numerical analysis of Hamiltonian systems, we think of an integrator as yielding an approximate trajectory in the exact Hamiltonian. This result illuminates a quite different point of view: the integrator yields an *exact* trajectory in a Hamiltonian that differs from the exact one by an error Hamiltonian H_{err}. Integrators of higher order correspond to error Hamiltonians of higher order, and a method of order k has an error Hamiltonian that is $\mathrm{O}(h^k)$. One advantage of this viewpoint is that errors in the trajectory can be analyzed using Hamiltonian perturbation theory, which is simpler and more powerful than analyzing the errors in numerical integrators directly (of course this tool is only available if the integrator is symplectic, which is one of the reasons for the popularity of integrators of this kind). Ironically, numerical integration became the central tool in celestial mechanics because of the limitations of perturbation theory, yet perturbation theory turns out to be the best way to analyze the errors in numerical integrations.

As described in Box 2.6, series such as (2.127) are asymptotic rather than convergent; in practice this means that they describe the behavior of the integrator only when the timestep h is sufficiently small. To illustrate this limitation, consider a Hamiltonian $H_A + H_B$ that is autonomous and has one degree of freedom (i.e., is time-independent with two phase-space dimensions). Such Hamiltonians are always integrable (Appendix D.7) and therefore cannot exhibit chaos. Equation (2.127) implies that the numerical Hamiltonian $H_{\rm num}$ is also autonomous with one degree of freedom, and therefore should not exhibit chaos either; however, simple symplectic integrators for autonomous Hamiltonians with one degree of freedom *do* exhibit chaos when the timestep is large enough (see Box 2.3). The resolution of this apparent paradox lies in the lack of convergence of the series for the error Hamiltonian.

An alternative approach is to express the numerical Hamiltonian using the periodic delta function, as in equation (2.52). This approach avoids concerns about convergence but yields a time-dependent numerical Hamiltonian rather than an autonomous one.

2.5.2 Composition methods

Operator splitting can be generalized to construct symplectic and reversible integrators of arbitrarily high order. Consider the integrator

$$\mathbf{\Phi}_h \equiv \mathbf{S}_{ah}\mathbf{S}_{bh}\mathbf{S}_{ah}, \tag{2.128}$$

where $\mathbf{S}$ is given by equation (2.124). The integrator is symplectic since it is the composition of operators that are individually symplectic (see the discussion near the end of Appendix D.5). Moreover if the dynamical systems governed by H_A and H_B are reversible, then $\mathbf{S}_h$ is reversible and so $\mathbf{\Phi}_h$ is also reversible.

Equation (2.125) implies that

$$\mathbf{S}_h = \exp[h\mathbf{Z}_1 + h^3\mathbf{Z}_3 + \mathrm{O}(h^5)], \tag{2.129}$$

where $\mathbf{Z}_1 \equiv \mathbf{L}_{H_A} + \mathbf{L}_{H_B}$. Then

$$\begin{aligned}\mathbf{\Phi}_h &\equiv \exp\left[ah\mathbf{Z}_1 + a^3h^3\mathbf{Z}_3 + \mathrm{O}(h^5)\right]\exp\left[bh\mathbf{Z}_1 + b^3h^3\mathbf{Z}_3 + \mathrm{O}(h^5)\right] \\ &\quad\times \exp\left[ah\mathbf{Z}_1 + a^3h^3\mathbf{Z}_3 + \mathrm{O}(h^5)\right] \\ &= \exp\left[(2a+b)h\mathbf{Z}_1 + (2a^3+b^3)h^3\mathbf{Z}_3 + \mathrm{O}(h^5)\right];\end{aligned} \tag{2.130}$$

the last equation follows from equation (d) of Box 2.6. The exact propagator is $\exp(h\mathbf{L}_{H_A+H_B}) = \exp(-h\mathbf{L}_{H_A} - h\mathbf{L}_{H_B}) = \exp(-h\mathbf{Z}_1)$ so if we choose $2a + b = 1$, $2a^3 + b^3 = 0$, the one-step error will be $\mathrm{O}(h^5)$ and $\mathbf{\Phi}_h$ will be a fourth-order integrator. These constraints on a and b require that

$$a = \frac{1}{2 - 2^{1/3}} = 1.351\,21, \quad b = -\frac{2^{1/3}}{2 - 2^{1/3}} = -1.702\,41, \tag{2.131}$$

and this equation together with (2.128) define the **Forest–Ruth** integrator (Forest & Ruth 1990; Suzuki 1990; Yoshida 1990). Notice that $b < 0$; thus by taking one Strang step forward, one back and a third forward we have promoted the second-order Strang integrator to a fourth-order one.

This procedure can be generalized to construct sequences of Strang splittings that yield reversible, symplectic integrators of any even order (e.g., Yoshida 1993; Hairer et al. 2006; Blanes & Casas 2016). In general these have one or more backward timesteps and because of this the errors tend to be larger than the errors in other (non-symplectic) integrators of the same order.

2.5.3 Wisdom–Holman integrators

So far we have applied operator splitting to follow the motion of a test particle governed by the Hamiltonian $H = \frac{1}{2}v^2 + \Phi(\mathbf{r}, t)$, by splitting the Hamiltonian into $H_A = \frac{1}{2}v^2$, $H_B = \Phi(\mathbf{r}, t)$ (eqs. 2.118 and 2.120). However, other splits are possible. Suppose that we have a Hamiltonian of the form $\frac{1}{2}v^2 - \mathbb{G}M/|\mathbf{r}| + \epsilon\phi(\mathbf{r}, t)$, representing a test particle orbiting in a Kepler potential with an additional perturbing potential that is smaller by $\mathrm{O}(\epsilon)$. We can make the split $H_A = \frac{1}{2}v^2 - \mathbb{G}M/|\mathbf{r}|$ and $H_B = \epsilon\phi(\mathbf{r}, t)$. Then the operator $\mathbf{G}_{A,h}$ corresponds to advancing the particle on a Kepler orbit for a time h—most easily done using the Gauss f and g functions—and $\mathbf{G}_{B,h}$ is the usual kick operator (2.120) due to the potential $\epsilon\phi(\mathbf{q})$. The advantage of this approach is that the integration errors after one timestep are of order ϵh^3 rather than h^3 as in standard leapfrog (Wisdom & Holman 1991). The term "mixed-variable" is sometimes used to describe these methods because the integrator consists of alternating steps that are trivial in orbital elements and Cartesian coordinates.

The most important use of these methods is for long-term integrations of planetary systems. In this case, we must split the Hamiltonian for the N-body problem into a sum of Kepler Hamiltonians of the form $\frac{1}{2}p^2/m - \mathbb{G}Mm/r$, which will be H_A, and an interaction term that depends only on the coordinates, which will be H_B. Unfortunately this cannot be done in barycentric coordinates, because of the presence of the kinetic energy of the host star in the Hamiltonian (the term $\frac{1}{2}p_0^2/m_0$ in eq. 4.3). Astrocentric coordinates can also be used, but in this case the Hamiltonian (4.13) contains an additional term $\frac{1}{2}|\sum_j \mathbf{p}_j^\star|^2/m_0$ so the Hamiltonian has to be split three ways (Duncan et al. 1998; Wisdom 2006). The best choice is Jacobi coordinates, for which the Hamiltonian is given in equation (4.39). A comparison of the merits of different coordinate systems is given by Rein & Tamayo (2019), and Wisdom–Holman schemes of higher order are reviewed by Rein et al. (2019). Wisdom–Holman integrators are usually the method of choice for long integrations of planetary systems in which the planets are on nearly circular and coplanar orbits.

2.6 Regularization

High-eccentricity orbits are difficult for most integrators to handle, because the acceleration is very large for a small fraction of the orbit. The simplest, and most extreme, version of this problem arises for a **collision orbit**, an orbit with negligible angular momentum. In this case the radius of the orbit varies as $r(t) \propto |t-t_0|^{2/3}$ as $r(t) \to 0$ (see Problem 1.14), and no integrator can easily manage a non-differentiable trajectory of this kind.

This obstacle can be circumvented by transforming to a coordinate system in which the two-body problem has no singularity—a procedure called

regularization. Standard integrators can then be used to solve the equation of motion in the regularized coordinates.

2.6.1 Time regularization

The simplest form of regularization is time transformation. We write the equation of motion for the perturbed two-body problem as

$$\ddot{\mathbf{r}} = -\frac{\mathbb{G}M}{r^3}\mathbf{r} - \nabla\Phi_{\text{ext}}, \tag{2.132}$$

where $\Phi_{\text{ext}}(\mathbf{r}, t)$ is the gravitational potential from sources other than the central mass M. We change to a fictitious time τ defined by

$$\mathrm{d}t = r^p \,\mathrm{d}\tau. \tag{2.133}$$

If a collision occurs at $t = 0$, then the radius satisfies $r(t) \propto |t|^{2/3}$ so equation (2.133) implies that $|\tau| \propto |t|^{1-2p/3}$ and $r(\tau) \propto |\tau|^{2/(3-2p)}$. The simplest choice that makes the radius a smooth function of τ across the encounter is $p = 1$, which yields $r(\tau) \propto \tau^2$, and henceforth we restrict ourselves to this value of p.

Denoting derivatives with respect to τ by primes, we find for $p = 1$

$$\dot{\mathbf{r}} = \frac{\mathrm{d}\tau}{\mathrm{d}t}\frac{\mathrm{d}\mathbf{r}}{\mathrm{d}\tau} = \frac{\mathbf{r}'}{r}, \quad \ddot{\mathbf{r}} = \frac{\mathrm{d}\tau}{\mathrm{d}t}\frac{\mathrm{d}}{\mathrm{d}\tau}\frac{\mathbf{r}'}{r} = \frac{\mathbf{r}''}{r^2} - \frac{(\mathbf{r}\cdot\mathbf{r}')}{r^4}\mathbf{r}'. \tag{2.134}$$

Substituting these results into the equation of motion, we obtain

$$\mathbf{r}'' = \frac{(\mathbf{r}\cdot\mathbf{r}')}{r^2}\mathbf{r}' - \mathbb{G}M\frac{\mathbf{r}}{r} - r^2\nabla\Phi_{\text{ext}}. \tag{2.135}$$

The eccentricity vector $\mathbf{e}$ helps us to simplify this equation. From Box 1.1, using the vector identity (B.9b), we have

$$\mathbf{e} = \frac{\dot{\mathbf{r}}\times(\mathbf{r}\times\dot{\mathbf{r}})}{\mathbb{G}M} - \frac{\mathbf{r}}{r} = \frac{1}{\mathbb{G}M}\left[\frac{|\mathbf{r}'|^2}{r^2}\mathbf{r} - \frac{(\mathbf{r}\cdot\mathbf{r}')}{r^2}\mathbf{r}'\right] - \frac{\mathbf{r}}{r}. \tag{2.136}$$

By using this result to eliminate the term proportional to $(\mathbf{r}\cdot\mathbf{r}')\mathbf{r}'$, equation (2.135) can be written

$$\mathbf{r}'' = |\mathbf{r}'|^2\frac{\mathbf{r}}{r^2} - 2\mathbb{G}M\frac{\mathbf{r}}{r} - \mathbb{G}M\mathbf{e} - r^2\nabla\Phi_{\rm ext}. \tag{2.137}$$

The energy of the orbit is

$$E = \tfrac{1}{2}v^2 - \frac{\mathbb{G}M}{r} + \Phi_{\rm ext} = \frac{|\mathbf{r}'|^2}{2r^2} - \frac{\mathbb{G}M}{r} + \Phi_{\rm ext}, \tag{2.138}$$

so we may eliminate the term proportional to $|\mathbf{r}'|^2$, giving (Burdet 1967; Heggie 1973)

$$\mathbf{r}'' - 2E\mathbf{r} + \mathbb{G}M\mathbf{e} = -\nabla(r^2\Phi_{\rm ext}). \tag{2.139}$$

This equation of motion must be supplemented by equations for the rates of change of t, E and $\mathbf{e}$ with fictitious time τ,

$$\begin{aligned}
t' &= r,\\
E' &= r\frac{\partial\Phi_{\rm ext}}{\partial t},\\
\mathbf{e}' &= \frac{1}{\mathbb{G}M}\big[\mathbf{r}'(\mathbf{r}\cdot\nabla\Phi_{\rm ext}) + \nabla\Phi_{\rm ext}(\mathbf{r}\cdot\mathbf{r}') - 2\mathbf{r}(\mathbf{r}'\cdot\nabla\Phi_{\rm ext})\big].
\end{aligned} \tag{2.140}$$

When the external potential vanishes the energy E and the eccentricity vector $\mathbf{e}$ are constant, and (2.139) is the equation of motion of a harmonic oscillator with frequency $\omega = (-2E)^{1/2}$ that is subject to a constant force $-\mathbb{G}M\mathbf{e}$. Therefore $\mathbf{r}(\tau)$ is a smooth function, even for a collision orbit. Moreover it is straightforward to show from equations (1.46) and (1.49) that in this case the fictitious time τ is related to the eccentric anomaly u by $u = na\tau + \text{const}$, where n and a are the mean motion and semimajor axis. We therefore call this method **eccentric-anomaly regularization**.

Figure 2.2 shows the fractional energy error that arises in the integration of an orbit with eccentricity $e = 0.999$ using eccentric-anomaly regularization. Note that regularization by time transformation in this way is different from integrating in Cartesian coordinates with a variable timestep; in particular, regularization allows us to integrate collision orbits, while the use of a variable timestep does not.

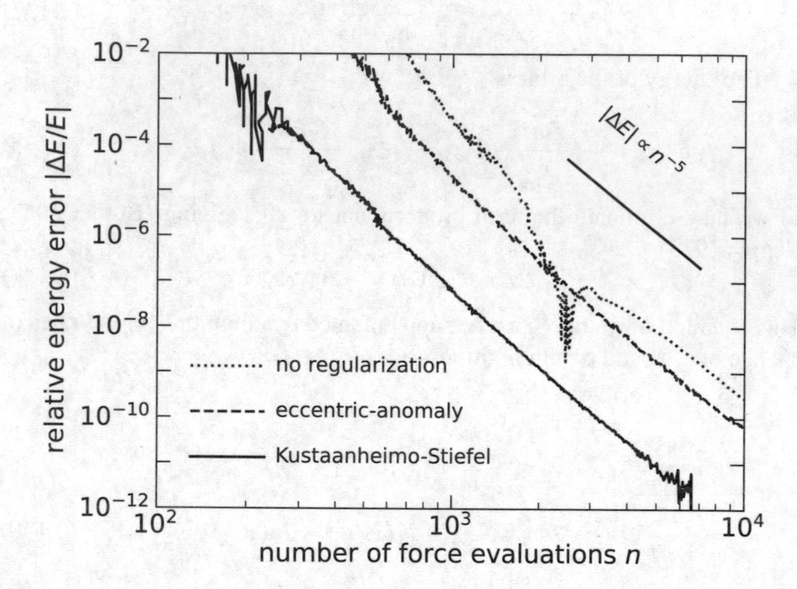

Figure 2.2: Relative energy error $|\Delta E/E|$ in the integration of a Kepler orbit with eccentricity $e = 0.999$ over one periapsis passage. The orbit is integrated from $r = \frac{1}{2}a, \dot{r} < 0$ to $r = \frac{1}{2}a, \dot{r} > 0$. The x-axis is the number of force evaluations n. All orbits are followed using the Dormand–Prince integrator defined in equations (2.70)–(2.73); since this is a fifth-order integrator we expect the energy error to scale as $|\Delta E| \propto n^{-5}$, shown by a short solid line. The curve labeled "no regularization" is computed in Cartesian coordinates; the curve labeled "eccentric-anomaly" is computed in Cartesian coordinates using eccentric-anomaly regularization; and the curve labeled "Kustaanheimo–Stiefel" is computed as described in §2.6.2.

2.6.2 Kustaanheimo–Stiefel regularization

This is an alternative regularization procedure that involves the transformation of both the coordinates and the time. The Kustaanheimo–Stiefel (KS) formulas can be derived using the symmetry group of the Kepler problem, the theory of quaternions and spinors, or several other methods (Stiefel & Scheifele 1971; Yoshida 1982; Waldvogel 2008; Saha 2009; Mikkola 2020), but since the derivation has limited relevance to other aspects of celestial mechanics we only give the results.

We define a 4-vector $\mathbf{u} = (u_1, u_2, u_3, u_4)$ that is related to the position $\mathbf{r} = (x, y, z)$ by

$$u_1 = \left[\tfrac{1}{2}(x+r)\right]^{1/2} \cos\psi, \qquad u_4 = \left[\tfrac{1}{2}(x+r)\right]^{1/2} \sin\psi,$$
$$u_2 = \frac{yu_1 + zu_4}{x+r}, \qquad u_3 = \frac{zu_1 - yu_4}{x+r}, \tag{2.141}$$

where ψ is an arbitrary parameter. The inverse relations are

$$x = u_1^2 - u_2^2 - u_3^2 + u_4^2, \quad y = 2(u_1u_2 - u_3u_4), \quad z = 2(u_1u_3 + u_2u_4). \tag{2.142}$$

Then

$$\dot{x} = 2(u_1\dot{u}_1 - u_2\dot{u}_2 - u_3\dot{u}_3 + u_4\dot{u}_4),$$
$$\dot{y} = 2(u_1\dot{u}_2 + u_2\dot{u}_1 - u_3\dot{u}_4 - u_4\dot{u}_3),$$
$$\dot{z} = 2(u_1\dot{u}_3 + u_3\dot{u}_1 + u_2\dot{u}_4 + u_4\dot{u}_2). \tag{2.143}$$

The equations of motion for $\mathbf{u}(t)$ are not uniquely determined by the equations of motion for $\mathbf{r}(t)$. We may therefore impose the additional constraint

$$u_4\dot{u}_1 - u_1\dot{u}_4 + u_2\dot{u}_3 - u_3\dot{u}_2 = 0. \tag{2.144}$$

The inverse relation for time derivatives is then

$$\dot{u}_1 = \frac{1}{2r}(u_1\dot{x} + u_2\dot{y} + u_3\dot{z}), \qquad \dot{u}_2 = \frac{1}{2r}(-u_2\dot{x} + u_1\dot{y} + u_4\dot{z}),$$
$$\dot{u}_3 = \frac{1}{2r}(-u_3\dot{x} - u_4\dot{y} + u_1\dot{z}), \qquad \dot{u}_4 = \frac{1}{2r}(u_4\dot{x} - u_3\dot{y} + u_2\dot{z}). \tag{2.145}$$

These relations imply that the radius r and velocity v are given by

$$r = |\mathbf{u}|^2 = u_1^2 + u_2^2 + u_3^2 + u_4^2, \quad v^2 = |\dot{\mathbf{r}}|^2 = 4|\mathbf{u}|^2|\dot{\mathbf{u}}|^2 = 4r(\dot{u}_1^2 + \dot{u}_2^2 + \dot{u}_3^2 + \dot{u}_4^2). \tag{2.146}$$

Now consider a test particle subject to the gravitational field of a point mass M and an additional potential $\Phi_{\rm ext}(\mathbf{r}, t)$. The Lagrangian is

$$L(\mathbf{r}, \dot{\mathbf{r}}, t) = \tfrac{1}{2}v^2 + \frac{\mathbb{G}M}{r} - \Phi_{\rm ext}(\mathbf{r}, t). \tag{2.147}$$

In the new variables

$$L(\mathbf{u}, \dot{\mathbf{u}}, t) = 2|\mathbf{u}|^2|\dot{\mathbf{u}}|^2 + \frac{\mathbb{G}M}{|\mathbf{u}|^2} - \Phi_{\rm ext}(\mathbf{u}, t). \tag{2.148}$$

The momentum conjugate to the coordinates $\mathbf{u}$ is $\mathbf{p}_u = \partial L/\partial\dot{\mathbf{u}} = 4|\mathbf{u}|^2\dot{\mathbf{u}}$, and the Hamiltonian is

$$H(\mathbf{u}, \mathbf{p}_u) = \mathbf{p}_u \cdot \dot{\mathbf{u}} - L = \frac{|\mathbf{p}_u|^2}{8|\mathbf{u}|^2} - \frac{\mathbb{G}M}{|\mathbf{u}|^2} + \Phi_{\rm ext}(\mathbf{u}, t). \tag{2.149}$$

We now introduce a transformation to an extended phase space as described in Box 2.1. The fictitious time is defined by equation (e) of that box, with $g(\mathbf{u}, \mathbf{p}_u) = r = |\mathbf{u}|^2$ (the same transformation used in eccentric-anomaly regularization). The new coordinates (now 5-vectors) are $\mathbf{U} = (U_0, \mathbf{u})$ where $U_0 = t$ and $\mathbf{P} \equiv (P_0, \mathbf{p}_u)$. The transformed Hamiltonian is given by equation (b) of Box 2.1,

$$\Gamma(\mathbf{U}, \mathbf{P}) = \tfrac{1}{8}|\mathbf{p}_u|^2 - \mathbb{G}M + |\mathbf{u}|^2\Phi_{\rm ext}(\mathbf{U}) + P_0|\mathbf{u}|^2. \tag{2.150}$$

Denoting derivatives with respect to the fictitious time τ by a prime, the regularized equations of motion are

$$\mathbf{u}'' + \tfrac{1}{2}P_0\mathbf{u} = -\tfrac{1}{4}\frac{\partial}{\partial\mathbf{u}}\left(|\mathbf{u}|^2\Phi_{\rm ext}\right), \quad P_0' = -|\mathbf{u}|^2\frac{\partial\Phi_{\rm ext}}{\partial t}, \quad t' = |\mathbf{u}|^2. \tag{2.151}$$

We may choose the initial condition for P_0 such that $\Gamma(\mathbf{U}, \mathbf{P}) = 0$ at the initial time. Since $\Gamma(\mathbf{U}, \mathbf{P})$ is independent of the fictitious time τ, it will remain zero everywhere on the trajectory, so $P_0 = -\frac{1}{8}\mathbf{p}_u^2/|\mathbf{u}|^2 - \Phi_{\rm ext}(\mathbf{U}) + \mathbb{G}M/|\mathbf{u}|^2$. Comparison with equation (2.149) shows that $P_0 = -E$, where E is the energy. Thus we can rewrite the regularized equations of motion as

$$\begin{aligned} t' &= |\mathbf{u}|^2, \\ E' &= |\mathbf{u}|^2 \frac{\partial \Phi_{\rm ext}}{\partial t}, \\ \mathbf{u}'' - \tfrac{1}{2}E\mathbf{u} &= -\tfrac{1}{4}\frac{\partial}{\partial \mathbf{u}}\left(|\mathbf{u}|^2 \Phi_{\rm ext}\right), \\ E &= \tfrac{1}{2}v^2 - \frac{\mathbb{G}M}{r} + \Phi_{\rm ext} = 2\frac{|\mathbf{u}'|^2}{|\mathbf{u}|^2} - \frac{\mathbb{G}M}{|\mathbf{u}|^2} + \Phi_{\rm ext}. \end{aligned} \tag{2.152}$$

When the external potential vanishes, the energy E is conserved and the equation of motion is that of a harmonic oscillator with frequency $\omega = (-\frac{1}{2}E)^{1/2}$.

The equations of motion (2.152) resemble equations (2.139) and (2.140) resulting from time regularization: both describe the trajectory of a particle in a harmonic potential, with squared frequency ω^2 proportional to $-E$. An important difference is that in eccentric-anomaly regularization the frequency is $\omega = (-2E)^{1/2}$, while in KS regularization $\omega = (-E/2)^{1/2}$, a factor of two smaller. Because the frequency is smaller, an integrator can follow the motion more accurately at a given timestep.

Figure 2.2 shows the fractional energy error that arises in the integration of an orbit with eccentricity $e = 0.999$ using KS regularization. This approach reduces the energy error by two orders of magnitude compared to eccentric-anomaly regularization and even more compared to integration in Cartesian coordinates.

2.7 Roundoff error

Computer arithmetic with real numbers is not exact. The integrators discussed in this chapter have been designed to minimize the **truncation error** that arises when a differential equation is approximated using finite timesteps, but even the most accurate integrators are subject to **roundoff error** arising from the limitations of computer arithmetic. Shrinking the timestep h of a kth-order integrator shrinks the global error over a fixed time interval as h^k (§2.1.1) but only so long as the roundoff error is less than the truncation error; if the timestep is reduced too far the error will be dominated by roundoff and will grow again, typically as $h^{-\kappa}$ where κ is between 0.5 and 1 (see §2.7.3).

Roundoff error is more difficult to study and to control than truncation error. Fortunately, with modern computers roundoff is usually unimportant except in the most demanding orbit integrations. The aim of this section is to provide a brief introduction to the properties of roundoff error and how it can be managed.[14]

In numerical calculations, reproducibility is almost as important as accuracy. Ideally, when the same code is run with different compilers or different machines it should give the same answer, down to the last bit (Rein & Tamayo 2017). To accomplish this goal, the computing community has agreed on a set of conventions for floating-point formats and arithmetic known as IEEE 754 (The Institute of Electrical and Electronics Engineers Standard for Floating-Point Arithmetic). Most modern computers are compliant with IEEE 754,[15] and we shall refer our discussion to this standard.

An alternative approach is to use integer arithmetic, which does not suffer from roundoff. Either the phase space can be discretized on a grid of integers (Rannou 1974; Earn & Tremaine 1992), or selected operations can be carried out in integer arithmetic (Levesque & Verlet 1993; Rein & Tamayo 2018).

---

[14] A classic description of the early history and basic algorithms of floating-point arithmetic is Knuth (1981). William Kahan is responsible for many of the advances in understanding, improving and standardizing floating-point arithmetic, and his many papers on the web are worth reading to learn about the history and the subtleties of this subject. A comprehensive reference is Muller et al. (2010).

[15] Graphics processing units (GPUs) can be an exception.

2.7.1 Floating-point numbers

Almost all computers store real numbers in floating-point format: a real number x is represented as a sequence $[k_0, k_2, \ldots, k_{p-1}; s, e]$ where

$$x = s \sum_{n=0}^{p-1} k_n b^{e-n}. \tag{2.153}$$

Here $s = \pm 1$ is the sign, p is the precision, e is the exponent, and b is the base. Almost always $b = 2$ (binary arithmetic). The set of numbers $(k_0, \ldots, k_{p-1})$ is called the **significand**; each k_n is an integer between 0 and $b - 1$ (0 or 1 in binary arithmetic). The exponent e is an integer in the range $e_{\min} \leq e \leq e_{\max}$. To ensure that the representation of a given number is unique, we require that the leading digit k_0 is nonzero unless $x = 0$ or $e = e_{\min}$.

Any number that has the form (2.153) is a **representable number**. All integers j with $|j| < b^p$ are representable. All representable numbers are rational numbers but not all rational numbers are representable; for example, $\frac{1}{3}$ is not representable in binary arithmetic. Irrational numbers such as π or $\sqrt{2}$ are never representable.

Numbers that are not representable are rounded to one of the two adjacent representable numbers. The best approach, and the default in the IEEE 754 standard, is to round to the nearest, and if the distances are equal round to the one whose least significant digit is even (**round to nearest, ties to even**). We denote the rounded value of a real number x by $\mathrm{rnd}\,(x)$.

The most common format for floating-point numbers in IEEE 754 is the **binary64** format. Here the precision $p = 53$, corresponding to a fractional difference between adjacent representable numbers of $2^{-53} \simeq 1 \times 10^{-16}$. The exponent range is from $e_{\min} = -1022$ to $e_{\max} = +1023$, which allows the significand, exponent and sign to be stored in 64 bits.

2.7.2 Floating-point arithmetic

When two representable numbers x and y are added, subtracted, multiplied or divided, the result is often not representable. The designer of the computer's arithmetic engine must decide what representable number to use to approximate the result of each of these operations, along with others such as square roots.

The most important principle guiding this design is **exact rounding**, which states that the result of a function operating on representable numbers should be the representable number that is closest to the exact result of the same function.

In mathematical notation, let $F(x_1, \ldots, x_N)$ be a function of N real numbers and let $\mathrm{fl}\,[F(x_1, \ldots, x_N)]$ be the representable number that results from evaluating this function in floating-point arithmetic.[16] Then rounding is exact for the function F if

$$\mathrm{rnd}\,[F(x_1, \ldots, x_N)] = \mathrm{fl}\,[F(x_1, \ldots, x_N)]. \tag{2.154}$$

In the IEEE 754 standard, addition, subtraction, multiplication, division and square roots are rounded exactly (the inclusion of square roots is a blessing for orbit integration). The current version, IEEE 754-2019, recommends but does not require exact rounding of exponentials, logarithms, trigonometric and other common transcendental functions as well.

Exact rounding enables floating-point arithmetic to preserve many of the properties of exact arithmetic. For example, the binary operations of addition and multiplication commute: $x + y = y + x$ and $x \times y = y \times x$ for all real x and y. These properties are preserved with exact rounding:

$$\mathrm{fl}\,(x + y) = \mathrm{fl}\,(y + x) \quad \text{and} \quad \mathrm{fl}\,(x \times y) = \mathrm{fl}\,(y \times x). \tag{2.155}$$

for all representable x and y. Many other familiar arithmetic properties are preserved with exact rounding; for example,

$$\begin{aligned} \mathrm{fl}\,(x + 0) &= x, \\ \mathrm{fl}\,(x - y) &= \mathrm{fl}\,[x + (-y)], \\ \mathrm{fl}\,(x/x) &= 1, \end{aligned}$$

[16] Note that fl is *not* a function of $F(x_1, \ldots, x_N)$. Mathematically, fl is a functional of F and $x_1, \ldots, x_N$ are parameters. Also note that the result of operating with fl on some functions such as $F(x_1, x_2, x_3) = x_1 + x_2 + x_3$ is ambiguous because we need to specify the order in which the additions are performed, e.g., $x_1 + (x_2 + x_3)$ or $(x_1 + x_2) + x_3$.

$$\begin{aligned}
\mathrm{fl}\,(x/1) &= x, \\
\mathrm{fl}\,(x+y) &= 0 \text{ if and only if } x = -y, \\
\mathrm{fl}\,(x \times y) &= 0 \text{ if and only if } x = 0 \text{ or } y = 0.
\end{aligned} \tag{2.156}$$

On the other hand, the associative $[x+(y+z) = (x+y)+z,\, x(yz) = (xy)z]$, and distributive $[z(x+y) = zx + zy]$ properties of normal arithmetic no longer hold.

The lack of an associative property means that sums of the form $\sum_{i=1}^{N} x_i$ depend on the order in which the summation is done. Worse still, there can be catastrophic loss of accuracy if there is near-cancellation between two or more terms in this sum. Exact rounding allows us to fix this problem by implementing a **compensated summation** algorithm, which dramatically reduces the error in the sum. The basic idea is to track an extra variable that keeps a running total of the roundoff errors. The pseudocode for compensated summation is

$s = 0$ {This is the running sum}
$c = 0$ {This is an estimate of the error in the sum}
for $i = 1$ **to** N **do**
  $y = x_i - c$ {c is initially zero}
  $t = s + y$
  $c = (t - s) - y$ {The brackets ensure that $t - s$ is evaluated first}
  $s = t$
end for
print s {This is the required sum}

A variant of this algorithm allows the roundoff error in an addition to be eliminated entirely, by a **sum-conserving transformation** that converts x and y into two floating-point numbers s and c with *exactly* the same sum. The first number s is the rounded sum, while c is the roundoff error. In the usual case that the base b is 2 (binary arithmetic), if rounding is exact and x and y are representable then (Dekker 1971; Knuth 1981)

$$x + y = s + c \quad \text{where} \quad s = \mathrm{fl}\,(x+y),\; c = \begin{cases} \mathrm{fl}\,[(x-s)+y] & \text{if } |x| \geq |y|, \\ \mathrm{fl}\,[(y-s)+x] & \text{if } |y| \geq |x|. \end{cases} \tag{2.157}$$

A product-conserving transformation also exists, but requires many more operations (Dekker 1971).

Many other small changes in the evaluation of algebraic expressions can reduce roundoff error. For example, the polynomial $p(x) = \sum_{j=0}^{N} a_j x^j$ should be evaluated by **Horner's rule**,

$p = a_N$
for $j = N - 1$ **to** 0 **do**
  $p = px + a_j$
end for
print p {This is $p(x)$}

2.7.3 Good and bad roundoff behavior

The equation of motion for a bound test particle in a time-independent potential $\Phi(\mathbf{r})$ conserves the energy $E = \frac{1}{2}v^2 + \Phi(\mathbf{r})$. If we follow the motion numerically the energy will not be conserved, because of both truncation error and roundoff error. For the current discussion we ignore truncation error, and assume that the integration is carried out in base $b = 2$ floating-point arithmetic with precision p. We can write the energy error accrued in the timestep from t_j to $t_j + h$ as $\Delta E_j = 2^{-p} f_j E$ where $|f_j|$ varies from step to step but is always of order unity (although for a complicated integrator, $|f_j|$ could be as large as a few hundred).

Since the roundoff error in a given timestep depends on the least significant bits of the phase-space coordinates at the start of the timestep, it is useful to think of each f_j as an independent random variable, sampled from a distribution with a mean $\overline{f}$ and a standard deviation σ. Now consider the total fractional energy error $\Delta E/E$ after an integration time T requiring $N = T/h \gg 1$ timesteps. So long as $|\Delta E/E| \ll 1$ we can write $\Delta E/E = 2^{-p} \sum_{j=1}^{N} f_j$. Since this is the sum of a large number of independent random variables, the central-limit theorem implies that $\Delta E/E$ will have a Gaussian probability distribution with mean and standard deviation given by

$$\frac{\overline{\Delta E}}{E} = \frac{T}{2^p h}\overline{f}, \quad \frac{\sigma_{\Delta E}}{|E|} = 2^{-p}\sigma N^{1/2} = \frac{T^{1/2}}{2^p h^{1/2}}\sigma. \tag{2.158}$$

The effect of roundoff on the orbital phase is much larger than the effect on the energy. By Kepler's law, the mean motion $n = (\mathbb{G}M/a^3)^{1/2}$ (eq. 1.44) and the semimajor axis $a = -\frac{1}{2}\mathbb{G}M/E$ (eq. 1.32) so $n \propto (-E)^{3/2}$. Therefore if the fractional energy error is $\Delta E/E$, the fractional error in mean motion will be $\Delta n/n \sim \Delta E/E$ and the error in orbital phase that accumulates over time T will be $\Delta\ell \sim T\Delta n \sim nT\Delta E/E$. Then so long as $|\Delta E/E| \ll 1$, equations (2.158) imply that $\overline{\Delta\ell} \sim nT^2\overline{f}/(2^p h)$ and $\sigma_{\Delta l} \sim nT^{3/2}\sigma/(2^p h^{1/2})$. We replace the mean motion by the orbital period $P = 2\pi/n$ and add numerical coefficients derived from analytic solutions of the drift and diffusion equations to obtain (Problem 2.10)

$$\overline{\Delta\ell} = \frac{3\pi T^2}{2^{p+1}Ph}\overline{f}, \quad \sigma_{\Delta l} = \frac{3^{1/2}\pi T^{3/2}}{2^p P h^{1/2}}\sigma. \tag{2.159}$$

This analysis shows that there are two contributions to the roundoff error in a long integration: *drift* yielding an energy error that grows as the integration time T and a phase error that grows as T^2, and *diffusion* yielding an energy error that grows as $T^{1/2}$ and a phase error that grows as $T^{3/2}$. Over long integrations the drift errors are far larger than the diffusion errors unless $\overline{f} = 0$.

As an illustration, suppose we integrate the motion of a planet using binary64 floating-point arithmetic ($p = 53$) and a timestep $h = 0.01\,\text{yr} = 3.65$ days. Then equations (2.158) and (2.159) yield

$$\begin{aligned}
\frac{\overline{\Delta E}}{E} &= 1.1 \times 10^{-6}\,\overline{f}\,\frac{T}{10^8\,\text{yr}}\frac{0.01\,\text{yr}}{h}, \\
\frac{\sigma_{\Delta E}}{|E|} &= 1.1 \times 10^{-11}\sigma\left(\frac{T}{10^8\,\text{yr}}\frac{0.01\,\text{yr}}{h}\right)^{1/2}, \\
\overline{\Delta\ell} &= 523\,\overline{f}\left(\frac{T}{10^8\,\text{yr}}\right)^2\frac{0.01\,\text{yr}}{h}\frac{1\,\text{yr}}{P}, \\
\sigma_{\Delta l} &= 0.006\,\sigma\left(\frac{T}{10^8\,\text{yr}}\right)^{3/2}\left(\frac{0.01\,\text{yr}}{h}\right)^{1/2}\frac{1\,\text{yr}}{P}.
\end{aligned} \tag{2.160}$$

This result implies that even after an integration time of only 100 Myr, only 2% of the age of the solar system, drift leads to errors in the planet's mean longitude of many radians, while the diffusion errors are almost 10^5 times smaller.

These arguments lead to a crude but useful characterization of roundoff error in long integrations: **bad roundoff** exhibits linear growth $\Delta E \propto T$ in the errors of conserved quantities such as energy, while **good roundoff** exhibits the much slower growth $\Delta E \propto T^{1/2}$, behavior sometimes referred to as **Brouwer's law** (Newcomb 1899; Brouwer 1937). Achieving good roundoff requires that the mean energy error per timestep vanishes, that is, $\overline{f} = 0$.

Good roundoff behavior is a necessary component of accurate long-term orbit integrations but requires unusual care in programming. The following practices help:

- Use only compilers that comply with the IEEE 754 standard for floating-point arithmetic, to ensure that arithmetic operations are exactly rounded and that the code is portable.

- Check that compiler optimization flags do not lead to unexpected behavior such as replacing $(x+y)+z$ by $x+(y+z)$ or x/y by $x \times (1/y)$. Fused multiply-add replaces $\mathrm{fl}\left[\mathrm{fl}\left(x \times y\right) + z\right]$ by $\mathrm{rnd}\left(x \times y + z\right)$; the latter is more accurate but portability demands that the same formula is used on all compilers.

- Use the "round to nearest, ties to even" rounding mode to eliminate one source of drift in the phase-space positions.

- At each timestep, an integrator increments the phase-space positions $\mathbf{z}_j$ by some amount $\Delta\mathbf{z}_j$ that is proportional to the timestep h. This addition should be carried out using compensated summation or some other algorithm that minimizes or eliminates roundoff error in addition.

- Use extended-precision arithmetic on critical operations if it is available and not too slow.

- Avoid using mathematical functions such as $\sin$ and $\cos$ since these often depend on the compiler and may yield biased results that lead to bad roundoff behavior.

- Avoid using any mathematical constants that are not representable. For example, in the classical Runge–Kutta integrator (2.65) the factor $\frac{1}{6}$ will be rounded to a slightly different number, leading to drift in the phase-space positions. This problem can be evaded by writing all multiplications of x by a fixed rational number p/q as $(x \times p)/q$ rather than $x \times (p/q)$; since the least significant bits of x vary randomly from step to step, the roundoff error from the multiplication and subsequent division should lead to diffusion rather than drift. Irrational constants such as π should be avoided in repetitive operations.

Chapter 3

The three-body problem

The three-body problem is to determine the trajectories of three points interacting through their mutual gravity. In contrast to the two-body problem described in Chapter 1, there is no general analytic solution to the three-body problem. The three-body problem was originally stated by Newton, and ever since then has driven much of our understanding of the physics and mathematics of dynamical systems. In his famous treatise on mechanics the mathematician E. T. Whittaker (1873–1956) called it "the most celebrated of all dynamical problems" (Whittaker 1937). Books focused on the three-body problem include Szebehely (1967), Marchal (1990), Valtonen & Karttunen (2005) and Musielak & Quarles (2014). See Barrow–Green (1997) for a historical review.

We label the three bodies by the subscripts 0, 1, 2, so their masses and trajectories are m_0, m_1, m_2 and $\mathbf{r}_0(t)$, $\mathbf{r}_1(t)$, $\mathbf{r}_2(t)$. In an inertial frame, the equations of motion are

$$\frac{\mathrm{d}^2\mathbf{r}_i}{\mathrm{d}t^2} = -\mathbb{G}\sum_{\substack{j=0,1,2\\ j\neq i}} m_j \frac{\mathbf{r}_i - \mathbf{r}_j}{|\mathbf{r}_i - \mathbf{r}_j|^3}, \quad i = 0, 1, 2. \tag{3.1}$$

The total energy is

$$E_{\rm tot} = \tfrac{1}{2}\sum_{i=0}^{2} m_i|\dot{\mathbf{r}}_i|^2 - \mathbb{G}\sum_{\substack{i,j=0\\ j>i}}^{2} \frac{m_i m_j}{|\mathbf{r}_i - \mathbf{r}_j|}. \tag{3.2}$$

The total momentum is

$$\mathbf{P}_{\rm tot} = \sum_{i=0}^{2} m_i \dot{\mathbf{r}}_i, \tag{3.3}$$

and the total angular momentum is

$$\mathbf{L}_{\rm tot} = \sum_{i=0}^{2} m_i \mathbf{r}_i \times \dot{\mathbf{r}}_i. \tag{3.4}$$

The total energy, momentum, and angular momentum are all conserved, as can be verified by substituting equation (3.1) into the equations for $\dot{E}_{\rm tot}$, $\dot{\mathbf{P}}_{\rm tot}$, and $\dot{\mathbf{L}}_{\rm tot}$. It is often convenient to work in the barycentric frame, the inertial frame in which the center of mass $\sum_i m_i \mathbf{r}_i / \sum_i m_i$ is fixed at the origin (§4.1).

The phase space for the three-body problem has 18 dimensions (6 for the positions and velocities of each of the three bodies). Using the 10 conserved quantities ($E_{\rm tot}$, $\mathbf{P}_{\rm tot}$, $\mathbf{L}_{\rm tot}$, and the position of the center of mass), the phase space can be reduced to a manifold of 8 dimensions, although this is still much too large for a comprehensive exploration of the trajectories.

The most important special case is the **restricted** three-body problem, in which one of the masses, say m_2, is set to zero (i.e., particle 2 is a test particle). Usually body 1 is then chosen to be the less massive of m_0 and m_1, so $m_1 < m_0$. In the restricted problem the massive bodies follow a Kepler orbit as described in Chapter 1, so we need only study the motion of m_2. In the **circular** restricted three-body problem, the orbit of the massive bodies m_0 and m_1 is further assumed to be circular.

3.1 The circular restricted three-body problem

In the restricted problem, it makes sense to drop the subscript "2" on the position and velocity of the test particle. Then the equation of motion for the test particle is

$$\frac{\mathrm{d}^2\mathbf{r}}{\mathrm{d}t^2} = \ddot{\mathbf{r}} = -\mathbb{G}m_0\frac{\mathbf{r}-\mathbf{r}_0}{|\mathbf{r}-\mathbf{r}_0|^3} - \mathbb{G}m_1\frac{\mathbf{r}-\mathbf{r}_1}{|\mathbf{r}-\mathbf{r}_1|^3} = -\frac{\partial\Phi}{\partial\mathbf{r}}, \tag{3.5}$$

where

$$\Phi(\mathbf{r},t) = -\frac{\mathbb{G}m_0}{|\mathbf{r}-\mathbf{r}_0(t)|} - \frac{\mathbb{G}m_1}{|\mathbf{r}-\mathbf{r}_1(t)|}. \tag{3.6}$$

The test particle's energy and angular momentum per unit mass[1] are

$$E = \tfrac{1}{2}|\dot{\mathbf{r}}|^2 + \Phi(\mathbf{r},t) = \tfrac{1}{2}|\dot{\mathbf{r}}|^2 - \frac{\mathbb{G}m_0}{|\mathbf{r}-\mathbf{r}_0(t)|} - \frac{\mathbb{G}m_1}{|\mathbf{r}-\mathbf{r}_1(t)|},$$
$$\mathbf{L} = \mathbf{r}\times\dot{\mathbf{r}}. \tag{3.7}$$

In general, neither E nor $\mathbf{L}$ is constant along a trajectory governed by the equation of motion (3.5), in contrast to the total energy and angular momentum E_{tot} and $\mathbf{L}_{\mathrm{tot}}$, which *are* conserved in a system with three nonzero masses.

In the circular restricted problem, the two massive bodies are separated by a fixed distance equal to their semimajor axis a, and orbit at a constant angular speed $\mathbf{\Omega}$, whose magnitude is given by Kepler's law (1.44),

$$\Omega^2 = \frac{\mathbb{G}(m_0+m_1)}{a^3}. \tag{3.8}$$

The equation of motion for the test particle is simplest in the rotating frame in which the massive bodies are stationary. We denote the position in this frame by $\mathbf{x} = (x_a, x_b, x_c)$. The origin is chosen at the center of mass with the positive x_c-axis parallel to $\mathbf{\Omega}$. The massive bodies labeled 0 and 1 are chosen to lie on the negative and positive x_a-axis at $\mathbf{x}_0$ and $\mathbf{x}_1$, so

$$\mathbf{x}_0 = -\frac{m_1}{m_0+m_1}a\,\hat{\mathbf{x}}_a, \quad \mathbf{x}_1 = +\frac{m_0}{m_0+m_1}a\,\hat{\mathbf{x}}_a. \tag{3.9}$$

[1] Since a test particle has zero mass, these are actually the ratio of the energy and angular momentum to the mass in the limit as the mass approaches zero.

The equation of motion of the test particle is (eq. D.20)

$$\ddot{\mathbf{x}} = -\mathbb{G}m_0\frac{\mathbf{x}-\mathbf{x}_0}{|\mathbf{x}-\mathbf{x}_0|^3} - \mathbb{G}m_1\frac{\mathbf{x}-\mathbf{x}_1}{|\mathbf{x}-\mathbf{x}_1|^3} - 2\boldsymbol{\Omega}\times\dot{\mathbf{x}} - \boldsymbol{\Omega}\times(\boldsymbol{\Omega}\times\mathbf{x}), \quad (3.10)$$

where the velocity $\dot{\mathbf{x}}$ of the test particle in the rotating frame is related to its velocity in the inertial frame, $\dot{\mathbf{r}}$, by equation (D.17), which reads in the current notation

$$\dot{\mathbf{x}} = \dot{\mathbf{r}} - \boldsymbol{\Omega}\times\mathbf{x}. \quad (3.11)$$

The last term in equation (3.10) is the centrifugal force, which is the negative gradient of the centrifugal potential $\Phi_{\text{cent}}(\mathbf{x}) \equiv -\frac{1}{2}|\boldsymbol{\Omega}\times\mathbf{x}|^2$ (eq. D.21). Then the **Jacobi constant** is

$$E_{\text{J}} \equiv \tfrac{1}{2}|\dot{\mathbf{x}}|^2 + \Phi(\mathbf{x}) + \Phi_{\text{cent}}(\mathbf{x}) = \tfrac{1}{2}|\dot{\mathbf{x}}|^2 + \Phi_{\text{eff}}(\mathbf{x}), \quad (3.12)$$

where the **effective potential** is the sum of the gravitational and centrifugal potentials,

$$\begin{aligned}\Phi_{\text{eff}}(\mathbf{x}) &\equiv -\frac{\mathbb{G}m_0}{|\mathbf{x}-\mathbf{x}_0|} - \frac{\mathbb{G}m_1}{|\mathbf{x}-\mathbf{x}_1|} - \tfrac{1}{2}|\boldsymbol{\Omega}\times\mathbf{x}|^2 \\ &= -\frac{\mathbb{G}m_0}{|\mathbf{x}-\mathbf{x}_0|} - \frac{\mathbb{G}m_1}{|\mathbf{x}-\mathbf{x}_1|} - \tfrac{1}{2}\Omega^2(x_a^2 + x_b^2). \end{aligned} \quad (3.13)$$

In terms of the effective potential, the equation of motion (3.10) reads

$$\ddot{\mathbf{x}} = -2\boldsymbol{\Omega}\times\dot{\mathbf{x}} - \nabla\Phi_{\text{eff}}; \quad (3.14)$$

in words, the acceleration of the test particle in the rotating frame is the sum of the Coriolis acceleration and the gradient of the effective potential.

Using equation (3.11) and the vector identity (B.9a) we can show that the Jacobi constant is related to the energy and angular momentum per unit mass in the non-rotating frame (eq. 3.7) by (cf. eq. D.24)

$$E_{\text{J}} = E - \boldsymbol{\Omega}\cdot\mathbf{L}. \quad (3.15)$$

Although neither E nor $\mathbf{L}$ are constants of motion, it is straightforward to show from equation (3.14) that E_{J} *is* a constant or integral of motion. The existence of this constant is peculiar to the circular, restricted three-body problem. If either $m_2 \neq 0$ or the eccentricity of the orbit of m_0 and m_1 is nonzero, then no such constant exists.

Since $|\dot{\mathbf{x}}|^2 \geq 0$, a particle with Jacobi constant E_{J} is restricted to the region where $\Phi_{\mathrm{eff}}(\mathbf{x}) \leq E_{\mathrm{J}}$. The surface on which $\Phi_{\mathrm{eff}}(\mathbf{x}) = E_{\mathrm{J}}$ is called the **zero-velocity surface** and separates regions forbidden to the motion from those allowed—though of course there is no guarantee that the test particle will explore all of the allowed region.

The contours of the effective potential in the plane of the orbit of the two massive bodies are shown in Figure 3.1 for a system with $m_1/m_0 = 0.1$. If the initial position and velocity vectors lie in the $z = 0$ plane (the orbital plane of the two massive bodies), they will remain so; this "planar, circular, restricted three-body problem" is the simplest version of the three-body problem, yet as we shall see it still yields rich and instructive dynamical behavior.

If the smaller mass $m_1 \ll m_0$ and the test particle is not close to the smaller mass, the Jacobi constant takes on a simplified form called the Tisserand parameter (Box 3.1).

3.1.1 The Lagrange points

The first step in exploring the circular restricted three-body problem is to look for trajectories that are stationary in the rotating frame. These are located at isolated positions called **Lagrange points**. Such trajectories must have zero velocity $\dot{\mathbf{x}}$ and zero acceleration $\ddot{\mathbf{x}}$, so equation (3.14) implies that $\nabla\Phi_{\mathrm{eff}} = 0$, which means that the Lagrange points are located at extrema—minima, maxima, or saddle points—of the effective potential.

Differentiating equation (3.13) with respect to the three coordinates and using equation (3.9) gives

$$\frac{\partial\Phi_{\mathrm{eff}}}{\partial x_a} = (\nu^2 - \Omega^2)x_a + \frac{\mathbb{G}m_0m_1a}{m_0+m_1}\left(\frac{1}{|\mathbf{x}-\mathbf{x}_0|^3} - \frac{1}{|\mathbf{x}-\mathbf{x}_1|^3}\right),$$

$$\frac{\partial\Phi_{\mathrm{eff}}}{\partial x_b} = (\nu^2 - \Omega^2)x_b,$$

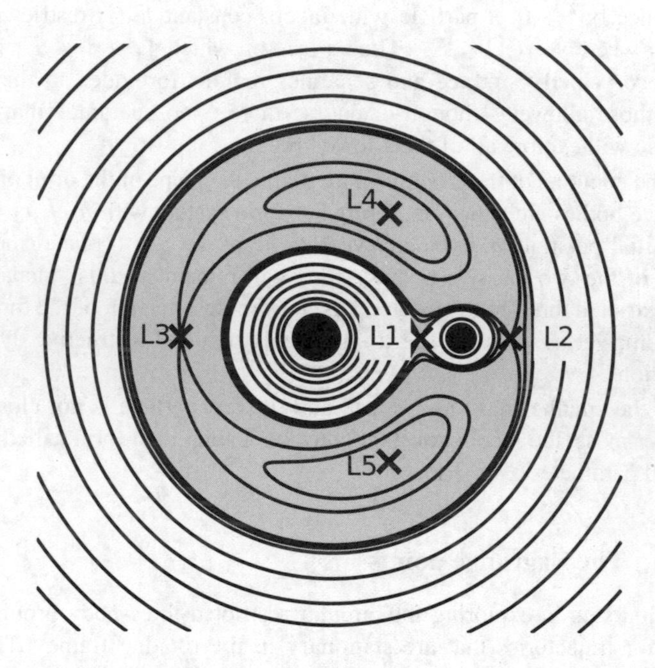

Figure 3.1: Contours of the effective potential (3.13), also known as zero-velocity curves, for the planar, circular, restricted three-body problem. The Lagrange points L1, . . . ,L5 are marked by crosses. Shading marks regions in which the effective potential is greater than the value at the L1 point. The massive bodies, located at the centers of the black circles, have mass ratio $m_1/m_0 = 0.1$. The smaller mass is on the right.

Box 3.1: The Tisserand parameter

In many planetary systems, the gravitational fields from the host star and one planet—typically the most massive one—dominate the motion of small bodies over a large region. For example, the dynamics of asteroids is determined mostly by the Sun and Jupiter.

Suppose that the star and planet have masses M and $m_{\mathrm{p}} \ll M$, and the planet travels on a circular orbit with semimajor axis a_{p}. In most of the volume of the planetary system the gravitational potential from the planet is small compared to the potential from the host star, so the Jacobi constant (3.15) can be written as $E_{\mathrm{J}} = E_{\mathrm{K}} - \mathbf{\Omega} \cdot \mathbf{L} + \mathrm{O}(m_{\mathrm{p}}/M)$ where E_{K} is the Kepler energy (1.19). We have $E_{\mathrm{K}} = -\frac{1}{2}\,\mathbb{G}M/a$ from equation (1.32) and $\mathbf{\Omega} \cdot \mathbf{L} = \Omega(\mathbb{G}Ma)^{1/2}(1-e^2)^{1/2}\cos I$ from equation (1.28), where e is the eccentricity and I is the inclination relative to the orbital plane of the planet. Thus

$$E_{\mathrm{J}} = -\frac{\mathbb{G}M}{2a} - \Omega(\mathbb{G}Ma)^{1/2}(1-e^2)^{1/2}\cos I + \mathrm{O}(m_{\mathrm{p}}/M). \qquad \text{(a)}$$

Furthermore the planet's angular speed $\Omega = (\mathbb{G}M/a_{\mathrm{p}}^3)^{1/2}$ so the Jacobi constant can be written

$$E_{\mathrm{J}} = -\frac{\mathbb{G}M}{2a_{\mathrm{p}}}T + \mathrm{O}(m_{\mathrm{p}}/M), \qquad \text{(b)}$$

where the **Tisserand parameter** is

$$T \equiv \frac{a_{\mathrm{p}}}{a} + 2\left(\frac{a}{a_{\mathrm{p}}}\right)^{1/2}(1-e^2)^{1/2}\cos I. \qquad \text{(c)}$$

One use of the Tisserand parameter is to connect fragmentary observations of small bodies such as asteroids and comets. Since the parameter is conserved during an encounter with a planet, it can be used to determine whether or not a newly discovered body is the same as a known one that recently suffered a close encounter that changed its orbital elements.

In the solar system, the Tisserand parameter is usually defined using Jupiter's semimajor axis, $a_{\mathrm{p}} = 5.203\,\mathrm{au}$. Classes of small bodies in the solar system are sometimes defined by a range of Tisserand parameters.

$$\frac{\partial \Phi_{\text{eff}}}{\partial x_c} = \nu^2 x_c, \tag{3.16}$$

where a is the orbital radius of the binary and

$$\nu^2 \equiv \frac{\mathbb{G}m_0}{|\mathbf{x}-\mathbf{x}_0|^3} + \frac{\mathbb{G}m_1}{|\mathbf{x}-\mathbf{x}_1|^3} \tag{3.17}$$

is positive-definite. The right side of the last of equations (3.16) is nonzero whenever $x_c \neq 0$. Therefore any extrema of Φ_{eff} must have $x_c = 0$, that is, they lie in the orbital plane of the two massive bodies. The second of equations (3.16) implies that at an extremum either $\nu^2 = \Omega^2$ or $x_b = 0$. We examine each of these possibilities in turn.

Triangular Lagrange points Consider first the case $x_b \neq 0$. Then the condition $\partial\Phi_{\text{eff}}/\partial x_b = 0$ requires that $\nu^2 = \Omega^2$, and $\partial\Phi_{\text{eff}}/\partial x_a = 0$ requires that

$$\frac{1}{|\mathbf{x}-\mathbf{x}_0|^3} - \frac{1}{|\mathbf{x}-\mathbf{x}_1|^3} = 0, \tag{3.18}$$

which in turn requires that $|\mathbf{x}-\mathbf{x}_0| = |\mathbf{x}-\mathbf{x}_1|$. Let us call this distance b; then (3.17) implies that $\nu^2 = \mathbb{G}(m_0 + m_1)/b^3$, so the condition $\nu^2 = \Omega^2$ together with (3.8) requires that $a = b$, which means that the triangle formed by m_0, m_1, and the test particle is equilateral, and since $x_c = 0$ the triangle must lie in the orbital plane. The positions occupied by the test particle in this configuration are the two **triangular Lagrange points**, labeled L4 and L5 in Figure 3.1. The L4 point leads m_1 by 60° in the direction of orbital motion, and L5 trails by the same angle.

Collinear Lagrange points We next look for solutions with $x_b = x_c = 0$. In this case we can use the first of equations (3.16) to write the condition $\partial\Phi_{\text{eff}}/\partial x_a = 0$ as

$$f(X) = \frac{1-\mu}{|X+\mu|^3}(X+\mu) + \frac{\mu}{|X-1+\mu|^3}(X-1+\mu) - X = 0, \tag{3.19}$$

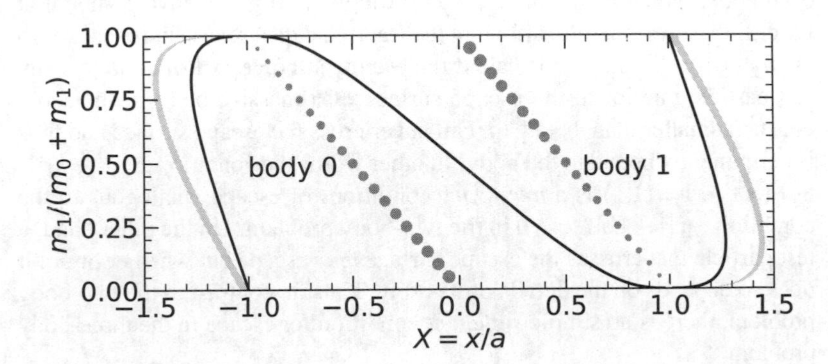

Figure 3.2: The locations of the three collinear Lagrange points as a function of $\mu = m_1/(m_0 + m_1)$ are shown as the solid line (eq. 3.19). The circles indicate the locations of m_0 and m_1, and the narrow shaded band shows the region where the equilibria would be stable (eq. 3.32). All three collinear points are always unstable.

where $X \equiv x_a/a$ and $\mu \equiv m_1/(m_0 + m_1)$. The function $f(X)$ is singular at the locations of the two masses, $X_0 = -\mu$ and $X_1 = 1 - \mu$. For $X < X_0$,

$$f(X) = \frac{1-\mu}{(X+\mu)^2} - \frac{\mu}{(X-1+\mu)^2} - X. \tag{3.20}$$

This function decreases monotonically from $f(X) \to \infty$ as $X \to -\infty$ to $f(X) \to -\infty$ as $X \to X_0 = -\mu$ from below. Therefore it has one and only one root for $X < X_0$. Similar arguments show that there is one root in the range $X_0 < X < X_1$ and one with $X > X_1$. These are the three **collinear Lagrange points**, shown by crosses in Figure 3.1. The usual (but not universal) conventions are that $m_1 \leq m_0$ and that L1, L2 and L3 are labeled as shown in Figure 3.1. In general the roots $f(X) = 0$ that determine the locations of the collinear Lagrange points must be determined numerically. These locations are plotted as a function of μ in Figure 3.2.

The zero-velocity surfaces close to each body are closed. As we move away, we eventually encounter the last closed zero-velocity surface around each body, which is the one passing through L1 (the effective potential at L2 or L3 is always larger, although the fractional difference approaches zero as $m_1 \to 0$); this surface is called the **escape surface** or **Roche lobe**. Any test particle interior to the escape surface of a massive body, with Jacobi constant smaller than $\Phi_{\text{eff}}(\text{L1})$, can never cross the escape surface and thus is permanently bound to the body. In other words the condition $E_\text{J} = \frac{1}{2}|\dot{\mathbf{x}}|^2 + \Phi_{\text{eff}}(\mathbf{x}) \geq \Phi_{\text{eff}}(\text{L1})$ is a *necessary* condition for escape, analogous to the condition $\frac{1}{2}|\dot{\mathbf{r}}|^2 - \mathbb{G}M/r \geq 0$ in the two-body problem. On the other hand, a test particle that crosses the escape surface *may* escape but whether or not it does so depends on the details of its orbit. Thus, in contrast to the two-body problem, there is no simple *sufficient* condition for escape in the three-body problem.

The most important special case of the collinear Lagrange points arises when $\mu \ll 1$, which occurs, for example, when m_0 and m_1 are respectively a star and a planet, or a planet and a satellite. As $\mu \to 0$, $f(X) \to \text{sgn}(X)/X^2 - X$ except near $X = 1$.[2] The only root of this function away from $X = 1$ is at $X = -1$, corresponding to the Lagrange point L3. The other roots of $f(X)$, corresponding to the Lagrange points L1 and L2, are close to m_1, so we write $X = X_1 + \delta X = 1 - \mu + \delta X$. Substituting this result into equation (3.19), the condition $f(X) = 0$ can be written

$$\frac{1-\mu}{(1+\delta X)^2} + \frac{\mu\,\text{sgn}(\delta X)}{(\delta X)^2} - 1 + \mu - \delta X = 0. \tag{3.21}$$

Expanding the left side as a Taylor series in δX and μ, we obtain

$$\frac{\mu\,\text{sgn}(\delta X)}{(\delta X)^2} - 3\delta X + \text{O}(\delta X^2, \mu\delta X) = 0, \tag{3.22}$$

and dropping the higher order terms,

$$|\delta X| = \left(\tfrac{1}{3}\mu\right)^{1/3}. \tag{3.23}$$

At this order of accuracy we can set $\mu = m_1/m_0$. We conclude that when $m_1 \ll m_0$, the collinear Lagrange points L1 and L2 are separated from the

[2] The function $\text{sgn}(X)$ is +1 if $X > 0$ and −1 if $X < 0$.

mass m_1 by the **Hill radius**,

$$r_{\mathrm{H}} \equiv a\left(\frac{m_1}{3m_0}\right)^{1/3}, \tag{3.24}$$

named after the American mathematician and astronomer George W. Hill (1838–1914).

The sphere centered on m_1 with radius equal to the Hill radius is sometimes called the **sphere of influence**.[3]

3.1.2 Stability of the Lagrange points

A casual observer might conclude that the Lagrange points are all unstable equilibria, because they lie at maxima or saddle points of the effective potential Φ_{eff}, so a particle at rest at the equilibrium can always slide downhill. This conclusion is wrong, because the Coriolis force can stabilize the motion.

Let $\mathbf{x}_{\mathrm{L}}$ be the location of one of the Lagrange points. To determine its stability to small perturbations we substitute $\mathbf{x} = \mathbf{x}_{\mathrm{L}} + \Delta\mathbf{x} = \mathbf{x}_{\mathrm{L}} + (\Delta x_a, \Delta x_b, \Delta x_c)$ in the equation of motion (3.14) and expand to first order in $|\Delta\mathbf{x}|$. Since $\boldsymbol{\Omega}$ is parallel to the positive x_c-axis, we have

$$\Delta\ddot{x}_a = 2\Omega\Delta\dot{x}_b - \Phi_{aa}\Delta x_a - \Phi_{ab}\Delta x_b - \Phi_{ac}\Delta x_c,$$

[3] Sometimes the radius of the sphere of influence is defined to be $r_{\mathrm{s}} = a(m_1/m_0)^{2/5}$, based on the following argument. Consider a test particle a distance r from m_1, which is much less massive than the central body m_0. The acceleration of the test particle due to m_1 is $\mathbb{G}m_1/r^2$ and the acceleration due to m_0 is approximately $\mathbb{G}m_0/a^2$, where a is the semimajor axis of the m_0-m_1 orbit. The ratio of these accelerations, which measures the relative strength of the perturbation from m_1, is $g_0 \sim m_1 a^2/(m_0 r^2)$. Now switch to the non-inertial frame centered on m_1. The acceleration due to m_1 is again $\mathbb{G}m_1/r^2$. The dominant perturbation is the acceleration due to the quadrupole potential from m_0 (eq. 3.71), $\mathbb{G}m_0 r/a^3$. The ratio of these accelerations, which measures the strength of the perturbation from m_0, is $g_1 \sim m_0 r^3/(m_1 a^3)$. The sphere of influence is defined by $g_0 = g_1$ which implies $r \sim a(m_1/m_0)^{2/5} = r_{\mathrm{s}}$. Inside the sphere of influence it is more accurate to consider the trajectory as a Kepler orbit around m_1 that is perturbed by m_0, while outside it is more accurate to treat the trajectory as a Kepler orbit around m_0 that is perturbed by m_1. Which of the two definitions of the sphere of influence is more appropriate depends on the dynamical context.

$$
\begin{aligned}
\Delta\ddot{x}_b &= -2\Omega\Delta\dot{x}_a - \Phi_{ba}\Delta x_a - \Phi_{bb}\Delta x_b - \Phi_{bc}\Delta x_c, \\
\Delta\ddot{x}_c &= -\Phi_{ca}\Delta x_a - \Phi_{cb}\Delta x_b - \Phi_{cc}\Delta x_c,
\end{aligned} \tag{3.25}
$$

where $\Phi_{aa} = (\partial^2\Phi_{\text{eff}}/\partial x_a^2)_{\mathbf{x}_\text{L}}$, and so forth. The Lagrange points lie in the $x_c = 0$ plane and the potential Φ_{eff} is even in x_c, so $\Phi_{ac} = \Phi_{bc} = \Phi_{ca} = \Phi_{cb} = 0$. Moreover $\Phi_{ba} = \Phi_{ab}$ from the properties of partial derivatives.

The solution of equations (3.25) is a sum of terms of the form $\mathbf{x} = \mathbf{a}\exp(\lambda t)$ where $\mathbf{a}$ and λ satisfy the matrix equation

$$
\mathbf{A}(\lambda)\mathbf{a} = \mathbf{0}, \quad \mathbf{A}(\lambda) \equiv \begin{bmatrix} \lambda^2 + \Phi_{aa} & -2\Omega\lambda + \Phi_{ab} & 0 \\ 2\Omega\lambda + \Phi_{ab} & \lambda^2 + \Phi_{bb} & 0 \\ 0 & 0 & \lambda^2 + \Phi_{cc} \end{bmatrix}. \tag{3.26}
$$

Solutions other than the trivial solution $\mathbf{a} = \mathbf{0}$ exist only if the determinant of $\mathbf{A}$ vanishes, which requires

$$
(\lambda^2 + \Phi_{cc})\left[(\lambda^2 + \Phi_{aa})(\lambda^2 + \Phi_{bb}) + 4\Omega^2\lambda^2 - \Phi_{ab}^2\right] = 0. \tag{3.27}
$$

One pair of roots of this equation is $\lambda = \pm\mathrm{i}\Phi_{cc}^{1/2}$. Using the last line of equations (3.16) $(\partial^2\Phi_{\text{eff}}/\partial x_c^2)_{x_c=0} = \nu^2$, so $\Phi_{cc} = \nu_\text{L}^2$ where $\nu_\text{L} \equiv \nu(\mathbf{x}_\text{L})$; since ν^2 is positive-definite, ν_L is real. Thus $\mathbf{x} = \mathbf{a}\exp(\pm\mathrm{i}\nu_\text{L} t)$, and since the motion is oscillatory these terms in the solution are stable.

The other solutions of equation (3.27) have

$$
(\lambda^2 + \Phi_{aa})(\lambda^2 + \Phi_{bb}) + 4\Omega^2\lambda^2 - \Phi_{ab}^2 = 0, \tag{3.28}
$$

and investigating these takes more work. We rewrite this equation as

$$
\lambda^4 + b\lambda^2 + c = 0, \quad \text{where} \quad b \equiv \Phi_{aa} + \Phi_{bb} + 4\Omega^2,\ c = \Phi_{aa}\Phi_{bb} - \Phi_{ab}^2. \tag{3.29}
$$

This is a polynomial containing only even powers of λ. Thus if λ is a root then so is $-\lambda$. A stable solution must have $\operatorname{Re}\lambda \le 0$ for all roots, but this can only be true for both λ and $-\lambda$ if $\operatorname{Re}\lambda = 0$. Therefore stability requires that $z \equiv \lambda^2$ is real and negative or zero for all roots. The variable z satisfies the quadratic equation $z^2 + bz + c = 0$, which has solutions $z = -\frac{1}{2}b \pm \frac{1}{2}(b^2 -$

$4c)^{1/2}$. These are real and negative or zero—implying that the Lagrange point is stable—if and only if

$$b^2 \geq 4c, \quad b \geq 0, \quad c \geq 0. \tag{3.30}$$

The parameter c is negative if and only if the Lagrange point is a saddle point of the effective potential. Therefore all saddle points are unstable, whereas maxima or minima of the potential may be either stable or unstable.

First consider the collinear Lagrange points. From equation (3.13) we have

$$\Phi_{aa} = -2\nu_{\rm L}^2 - \Omega^2, \quad \Phi_{bb} = \nu_{\rm L}^2 - \Omega^2, \quad \Phi_{ab} = 0, \tag{3.31}$$

where $\nu_{\rm L}^2 = \nu^2(\mathbf{x}_{\rm L})$ is given by equation (3.17) evaluated at the collinear Lagrange point $\mathbf{x}_{\rm L}$. Then $b = 2\Omega^2 - \nu_{\rm L}^2$ and $c = (\Omega^2 - \nu_{\rm L}^2)(\Omega^2 + 2\nu_{\rm L}^2)$. The stability constraints (3.30) then require

$$\tfrac{8}{9}\Omega^2 \leq \nu_{\rm L}^2 \leq \Omega^2. \tag{3.32}$$

The narrow regions where these inequalities are satisfied are marked in gray in Figure 3.2. They do not intersect the locations of the collinear Lagrange points, shown by a black line, for any value of the mass ratio $m_1/(m_0+m_1)$, so we conclude that the collinear points are always unstable.

At the triangular Lagrange points,

$$\Phi_{aa} = -\tfrac{3}{4}\Omega^2, \quad \Phi_{bb} = -\tfrac{9}{4}\Omega^2, \quad \Phi_{ab} = \mp\frac{3^{3/2}}{4}\frac{m_0 - m_1}{m_0 + m_1}\Omega^2; \tag{3.33}$$

in the last equation the minus and plus signs refer to the L4 and L5 points respectively. Then $b = \Omega^2$ and $c = \frac{27}{4}\Omega^4 m_0 m_1/(m_0 + m_1)^2$. Stability requires $b^2 \geq 4c$ or

$$\frac{m_0 m_1}{(m_0 + m_1)^2} \leq \frac{1}{27} \quad \text{or} \quad \frac{m_1}{m_0 + m_1} \leq \frac{1}{2} - \frac{1}{2}\left(\frac{23}{27}\right)^{1/2} = 0.0385, \tag{3.34}$$

assuming as usual that m_1 is the smaller mass. We conclude that the L4 and L5 Lagrange points are stable provided that the mass ratio $\mu = m_1/(m_0 + m_1)$ is sufficiently small. This is the case for all of the planets in the solar system—even for Jupiter, the most massive planet, $\mu = M_{\rm J}'/(M_\odot + M_{\rm J}') = 0.000\,953\,9$.[4] For the Earth–Moon system $M_{☽}/(M_\oplus + M_{☽}) = 0.012\,15$, so

[4] Here $M_{\rm J}'$ denotes the mass of Jupiter plus its satellites.

L4 and L5 are stable in this system as well. The triangular points are also stable for all the satellites of the giant planets.

In most cases of interest $m_1 \ll m_0$; then at the triangular points $b \gg c \simeq \frac{27}{4}\Omega^4 m_1/m_0$, and the quadratic equation $z^2 + bz + c = 0$ has solutions

$$z = \lambda^2 = -\Omega^2 + \mathrm{O}(m_1/m_0), \quad z = \lambda^2 = -\tfrac{27}{4}\Omega^2 \frac{m_1}{m_0} + \mathrm{O}(m_1/m_0)^2. \tag{3.35}$$

The first solution corresponds to displacements from the Lagrange points that oscillate as $\exp(\pm\mathrm{i}\Omega t)$; these are simply the epicyclic oscillations described in §1.8.2, which would be present even in the absence of the secondary mass m_1. The second solution describes slow oscillations with the much lower frequency $\pm\frac{1}{2}3^{3/2}\Omega(m_1/m_0)^{1/2}$, mainly in the azimuthal direction. Oscillations of this kind, in which the frequency approaches zero as the strength of the perturbation (in this case m_1) approaches zero, are called **librations**. A more complete description of librations around the triangular Lagrange points is given in §3.2.

The solar system contains a wide variety of objects at the triangular points. Thousands of asteroids orbit around the L4 and L5 points of the Sun–Jupiter system; these are called **Trojan asteroids** and by association the L4 and L5 points are sometimes called **Trojan points** in this and other systems. Although we have only proved that orbits near L4 and L5 are linearly stable in the circular restricted three-body problem, most of the Jupiter Trojans appear to orbit stably around the Lagrange points for the lifetime of the solar system despite perturbations from the other planets, eccentricities as large as 0.3, and inclinations as large as almost 60° (Problem 3.1).

The triangular Lagrange points also exist in some cases of the general three-body problem, in which all three masses are nonzero and the orbits all have the same eccentricity. For some values of the mass ratios and the eccentricity, motion around the triangular points is stable (Danby 1964a,b).

A handful of asteroids have also been discovered orbiting the triangular Lagrange points of the Sun–Neptune, Sun–Uranus, and Sun–Mars systems, and Earth has at least one known asteroid orbiting the L4 point. The absence of similar asteroids in the Sun–Saturn system is probably because of long-term instabilities induced by Jupiter. Saturn's satellites Tethys and Dione each have two smaller satellites orbiting their triangular Lagrange points.

The L1 and L2 points of the Sun–Earth system are popular destinations for spacecraft. Even though orbits near these Lagrange points are unstable, spacecraft can be kept close to them using occasional small "station-keeping" thruster burns (see Problem 3.3). Orbits near L1 are useful for observations of the Earth, since they always view the sunlit hemisphere, and for observations of the Sun and solar wind. Orbits near L2 are the best sites for many space observatories since the Sun, Earth and Moon are relatively close together in the sky, so the spacecraft optics can be shielded by a single sunshade. A spacecraft placed exactly at L1 or L2 has the undesirable property that the Sun is in line with the Earth and the spacecraft and interferes with radio transmissions. Thus most spacecraft are placed in orbits around the Lagrange points (Problem 3.4).

The unstable orbits leading from the L1 and L2 points visit much of the solar system (the **interplanetary transport network**), so trajectories starting from these points can reach distant targets with very little additional fuel if the mission designer is prepared to wait long enough. For example, NASA's ISEE-3/ICE spacecraft visited the L1 and L2 Sun–Earth Lagrange points and two comets over 8 years.

3.1.3 Surface of section

The simplest autonomous (time-independent) Hamiltonians have one degree of freedom. Understanding the geometry of their trajectories in phase space is straightforward because the phase space has only two dimensions, so the trajectories can be shown as curves on a plane surface (e.g., Figures 6.2 or 6.4). The simplest version of the three-body problem is the planar, circular, restricted three-body problem, which has four phase-space dimensions that can be taken to be the two components of the position and velocity of the test particle in the orbital plane of the two massive bodies. The conservation of the Jacobi constant (3.12) implies that the trajectory is restricted to a manifold of 3 dimensions in this 4-dimensional space, but even in 3 dimensions trajectories are difficult to visualize.

The **surface of section** or **Poincaré map** is a device invented by the mathematician Henri Poincaré (1854–1912) that enables us to study dynamical systems such as these. We consider the events when an orbit crosses a given curve in the orbital plane; the curve is in principle arbitrary but a simple choice is the line joining the two massive bodies, that is, the locus $x_b = 0$. We restrict ourselves to crossings from negative to positive x_b ($\dot{x}_b > 0$). From equation (3.12) the Jacobi constant can be written $E_\mathrm{J} = \frac{1}{2}\dot{x}_a^2 + \frac{1}{2}\dot{x}_b^2 + \Phi_\mathrm{eff}(x_a, x_b)$, so at one of these crossings

$$\dot{x}_b = [2E_\mathrm{J} - 2\Phi_\mathrm{eff}(x_a, 0) - \dot{x}_a^2]^{1/2}. \tag{3.36}$$

For a given value of E_J an orbit is completely defined by the two coordinates $(x_a, \dot{x}_a)$, since $x_b = 0$ by definition and $\dot{x}_b$ is given by (3.36). Therefore we can represent a crossing by a point in the $(x_a, \dot{x}_a)$ plane. Since this point defines the orbit, it also defines the coordinates of the next crossing $(x'_a, \dot{x}'_a)$. Thus the trajectory has been reduced to a mapping of the $(x_a, \dot{x}_a)$ plane into itself, the Poincaré map, which we may write as $\mathsf{P}(x_a, \dot{x}_a) = (x'_a, \dot{x}'_a)$.

Several properties of Poincaré map are worth noting:

- There is a different map for each value of the Jacobi constant E_J; thus a better notation for the map is $\mathsf{P}_{E_\mathrm{J}}$.

- The map does not cover the whole plane; it is only defined in the region where the argument of the square root in equation (3.36) is non-negative, or

$$\dot{x}_a^2 \leq 2E_\mathrm{J} - 2\Phi_\mathrm{eff}(x_a, 0). \tag{3.37}$$

- The map provides no information on orbits that do not cross the line $x_b = 0$ with $\dot{x}_b > 0$, such as orbits that remain close to the triangular Lagrange points.

- The map is area-preserving, that is, if $\mathbf{P}_{E_\mathrm{J}}$ is the Jacobian matrix of $\mathsf{P}_{E_\mathrm{J}}$ (cf. eq. 2.49), then the Jacobian determinant $\det(\mathbf{P}_{E_\mathrm{J}}) = 1$. This result is reminiscent of Liouville's theorem, which states that phase-space volume is conserved by a Hamiltonian flow (see discussion following eq. D.47), but Liouville's theorem relates the volumes at two successive times whereas the Poincaré map relates the areas at successive crossings of the line $x_b = 0$, which generally occur at different times for different trajectories. For proofs see Binney et al. (1985), Tabor (1989) or Lichtenberg & Lieberman (1992).

- Suppose the trajectory has a second constant or integral of motion in addition to the Jacobi constant, say $g(x_a, x_b, \dot{x}_a, \dot{x}_b) = \text{const}$. Then we can set $x_b = 0$ and eliminate $\dot{x}_b$ using (3.36) to obtain a relation between x_a and $\dot{x}_a$ that specifies a curve in the $(x_a, \dot{x}_a)$ plane. Therefore if a trajectory has a second integral, its successive images $\mathsf{P}, \mathsf{P}^2, \ldots, \mathsf{P}^N$ must lie on a curve in the surface of section. Of course, if the orbit is N-periodic, that is, if $\mathsf{P}^N(x_a, \dot{x}_a) = (x_a, \dot{x}_a)$, then the curve degenerates into N distinct points.

Surfaces of section are shown in Figures 3.3 and 3.4 for two values of the Jacobi constant. In the first figure, $E_{\rm J} = -2$ is smaller than the effective potential at L1, $\Phi_{\rm eff}(\text{L1}) = -1.7851$. In this case the test-particle trajectories are permanently confined either to the region around m_0, to the region around m_1, or to the region outside the Lagrange points L2 and L3 (cf. Figure 3.1); the last of these regions is outside the boundary of the plots here. The dots arrange themselves on well defined closed curves—usually there is one curve per orbit but occasionally, if one of the orbital frequencies is nearly resonant with the orbital frequency of the m_0-m_1 binary, a single orbit may appear as two or more distinct closed curves. Although the plot shows only 200 iterations of the Poincaré map, the trajectories would remain on the curves if we iterated the Poincaré map millions or billions of times (the justification for this claim comes from the application of the KAM theorem to autonomous Hamiltonians with 2 degrees of freedom, as described in Appendix D.8). Thus almost all of the orbits enjoy a second integral of motion in addition to the Jacobi constant.

In the second figure, 3.4, the Jacobi constant $E_{\rm J} = -1.75$ is larger than the effective potential at L1 but smaller than the effective potential at L2 and L3. There is a single allowed region that encloses both m_0 and m_1 but particles cannot escape to infinity. The surface of section exhibits some closed curves but most of the allowed region is filled by points that are scattered randomly over an area, rather than a curve. In fact almost all of the area not occupied by curves could be filled by a *single* orbit if we iterated the Poincaré map enough times. Similar behavior—"islands" of regular orbits surrounded by a chaotic "sea"—is seen in many Hamiltonian systems, as described in Appendix D.8.

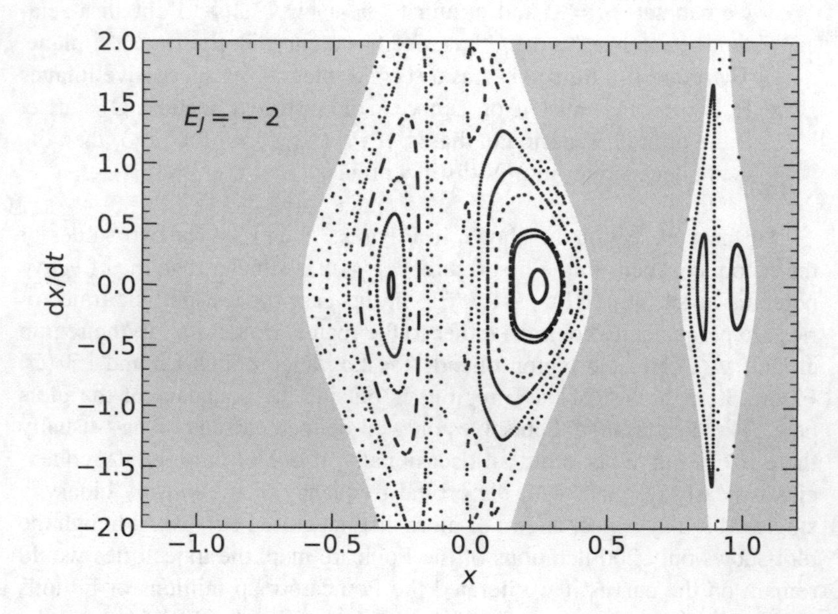

Figure 3.3: Surface of section for the planar, circular, restricted three-body problem. The mass ratio is $m_1/m_0 = 0.1$, as in Figure 3.1. The units are chosen such that $\mathbb{G} = 1$, $m_0 + m_1 = 1$, $a = 1$, which implies that the angular speed of the m_0–m_1 binary is $\Omega = [\mathbb{G}(m_0 + m_1)/a^3]^{1/2} = 1$. In these units the Jacobi constant is $E_J = -2$. For comparison, the effective potential at the Lagrange points L1, L2, and L3 is -1.7851, -1.7258, and -1.5453 respectively. The surface of section is defined by $x_b = 0$, $\dot{x}_b > 0$ and the plots show x_a and $\dot{x}_a$ on the horizontal and vertical axes. The trajectories are excluded from the shaded regions, where $\frac{1}{2}\dot{x}_b^2 + \Phi_{\text{eff}}(x_b, 0) > E_J$. Additional allowed regions outside the Lagrange points L2 and L3 are not shown.

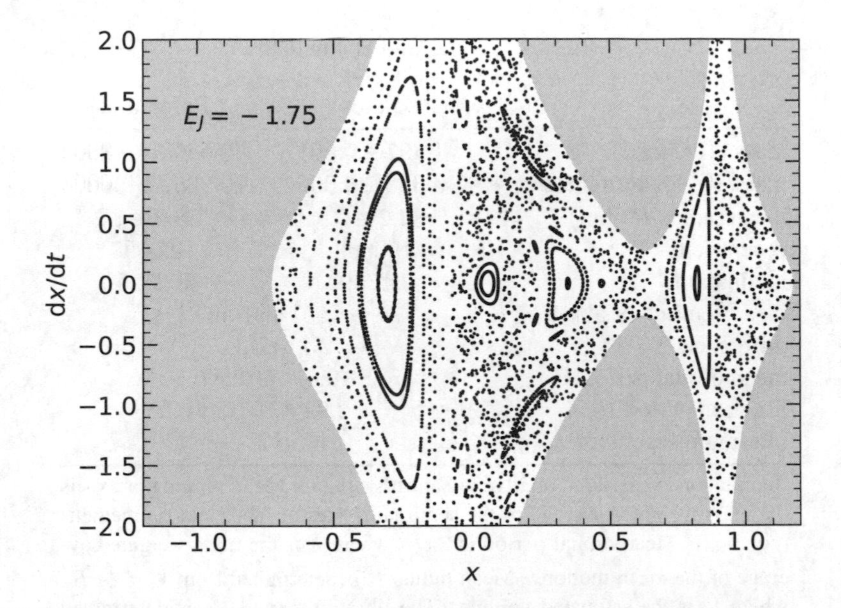

Figure 3.4: As in Figure 3.3, but for Jacobi constant $E_J = -1.75$.

A sequence of surfaces of section of this kind, for different values of the Jacobi constant, provides a complete picture of the behavior of orbits in the planar, circular, restricted three-body problem. Unfortunately it is not possible to generalize this approach to Hamiltonian systems with more than two degrees of freedom.

3.2 Co-orbital dynamics

Motion near the triangular Lagrange points is a special case of the general problem of co-orbital dynamics, the determination of the behavior of two or more bodies orbiting a common host with almost the same semimajor

Table 3.1: Janus and Epimetheus

	Janus	Epimetheus
mass (10^{18} kg)	1.897 ± 0.001	0.5263 ± 0.0003
mass/10^{-9} Saturn mass	3.338 ± 0.002	0.9262 ± 0.0005
mean radius $\overline{R}$ (km)	89.2 ± 0.8	58.2 ± 1.2
eccentricity	0.0068	0.0097
inclination	0.1639°	0.3525°
mean semimajor axis $\overline{a}$	151 450 km	
difference Δa	50 km	
mean orbital period P	0.694 59 days	
libration period P_L	2.92×10^3 days	
minimum separation $\phi_{\min}$	5.2°	

Inclination is measured relative to Saturn's equator. Mean semimajor axis is the mass-weighted average of the semimajor axes of Janus and Epimetheus (eq. 3.47). Mean orbital period is $2\pi/\overline{n}$ where $\overline{n}$ is the mass-weighted average of the mean motions. Mean radius $\overline{R}$ is determined from $V = \frac{4}{3}\pi\overline{R}^3$ where V is the estimated volume. The libration period P_L is determined from equation (3.52). Data from Nicholson et al. (1992), Spitale et al. (2006), Jacobson et al. (2008), Thomas et al. (2013), and JPL Solar System Dynamics at https://ssd.jpl.nasa.gov/. For Saturn's properties see Appendix A.

axes. This subject dates back to Maxwell's 1856 work on Saturn's rings[5] and was investigated long ago by Brown (1911). The study of co-orbital dynamics was re-kindled by the Voyager flybys of Saturn in 1980 and 1981, which provided close-up observations of Janus and Epimetheus, two co-orbital satellites of Saturn (Table 3.1).

We examine a three-body system consisting of a host mass M and two

[5] Maxwell examined the equilibria and stability of N small satellites with identical masses orbiting a massive central body. The satellites are assumed to have the same mean motion and are equally spaced in azimuth. Maxwell concluded that this configuration of satellites is stable for all N if the mass of the satellites is sufficiently small. Unfortunately, Maxwell assumed that Saturn was fixed and thereby neglected the indirect term of the potential (eq. 4.6). In fact, equally spaced satellites are only stable if $N \geq 7$ (Pendse 1935; Salo & Yoder 1988).

small satellites with masses $m_1, m_2 \ll M$. In the frame centered on the host, the gravitational potential experienced by satellite 1 due to satellite 2 is given by equation (4.6),

$$\Phi_1(\mathbf{r}_1) = -\frac{\mathbb{G}m_2}{|\mathbf{r}_1 - \mathbf{r}_2|} + \frac{\mathbb{G}m_2\,\mathbf{r}_2 \cdot \mathbf{r}_1}{|\mathbf{r}_2|^3}. \tag{3.38}$$

The second term is the indirect potential that arises because the frame centered on the host is not an inertial frame. The potential experienced by satellite 2 is obtained by exchanging the subscripts 1 and 2.

We assume that the satellite semimajor axes a_1 and a_2 are nearly the same, and that the eccentricities and inclinations are small (we shall show below that if the eccentricities and inclinations are initially small, they will remain so unless the two satellites have a close encounter). Since the orbits are nearly circular and coplanar and the semimajor axes are nearly equal, we can replace the orbital radii $r_1 = |\mathbf{r}_1|$ and $r_2 = |\mathbf{r}_2|$ in equation (3.38) by the average semimajor axis of the two satellites, which we denote by $\overline{a}$ (see eq. 3.47). We introduce polar coordinates (r, ϕ) in the common orbital plane, oriented such that the satellite orbits are prograde, $\dot{\phi}_1, \dot{\phi}_2 > 0$. In these coordinates the gravitational potential on satellite 1 due to satellite 2 becomes

$$\begin{aligned}\Phi_1(\mathbf{r}_1) &= -\frac{\mathbb{G}m_2}{\overline{a}[2 - 2\cos(\phi_1 - \phi_2)]^{1/2}} + \frac{\mathbb{G}m_2\cos(\phi_1 - \phi_2)}{\overline{a}} \\ &= -\frac{\mathbb{G}m_2}{2\overline{a}|\sin\frac{1}{2}(\phi_1 - \phi_2)|} + \frac{\mathbb{G}m_2\cos(\phi_1 - \phi_2)}{\overline{a}}.\end{aligned} \tag{3.39}$$

In this equation $\overline{a}$ should be regarded as a constant.

Since the orbits are nearly circular and coplanar, the primary effect of the mutual gravity of the two satellites is on their semimajor axes. The relation between semimajor axis and angular momentum for satellite 1 is $L_1 = m_1(\mathbb{G}Ma_1)^{1/2}$ (eq. 1.28). The rate of change of angular momentum is $\mathrm{d}L_1/\mathrm{d}t = N_1$, where $N_1 = -m_1\partial\Phi_1/\partial\phi_1$ is the torque due to the potential $\Phi_1(r_1, \phi_1)$. Thus

$$\frac{\mathrm{d}a_1}{\mathrm{d}t} = -\frac{2a_1^{1/2}}{(\mathbb{G}M)^{1/2}}\frac{\partial\Phi_1}{\partial\phi_1} \simeq -\frac{2\overline{a}^{1/2}}{(\mathbb{G}M)^{1/2}}\frac{\partial\Phi_1}{\partial\phi_1}. \tag{3.40}$$

Moreover from Kepler's law (1.44),

$$\frac{\mathrm{d}\phi_1}{\mathrm{d}t} = \frac{[\mathbb{G}(M+m_1)]^{1/2}}{a_1^{3/2}} \simeq \frac{(\mathbb{G}M)^{1/2}}{\overline{a}^{3/2}}\left[1 - \frac{3(a_1-\overline{a})}{2\overline{a}}\right]; \tag{3.41}$$

the last equation uses the first two terms of a Taylor expansion of a_1 around $\overline{a}$. Taking the time derivative and substituting equation (3.40) gives

$$\frac{\mathrm{d}^2\phi_1}{\mathrm{d}t^2} = -\frac{3(\mathbb{G}M)^{1/2}}{2\overline{a}^{5/2}}\frac{\mathrm{d}a_1}{\mathrm{d}t} = \frac{3}{\overline{a}^2}\frac{\partial\Phi_1}{\partial\phi_1}. \tag{3.42}$$

This simple expression can be compared to the equation of motion for a rotating rigid body. Consider a body of mass m at the end of a massless rod of length a that rotates in a plane around the origin, with azimuthal angle ϕ. The moment of inertia of the rod and body is $I = ma^2$ and its angular momentum is $L = I\dot{\phi}$. The torque on the body due to an external potential Φ is $N = -m\,\partial\Phi/\partial\phi$. The resulting angular acceleration is given by $\dot{L} = N$ so

$$\frac{\mathrm{d}^2\phi}{\mathrm{d}t^2} = \frac{N}{I} = -\frac{m}{I}\frac{\partial\Phi}{\partial\phi} = -\frac{1}{a^2}\frac{\partial\Phi}{\partial\phi}. \tag{3.43}$$

Comparing equations (3.42) and (3.43) we see that co-orbital satellites act as if they have a *negative* moment of inertia $I = -\frac{1}{3}m\overline{a}^2$, that is, an attractive torque tends to repel them. The basis of this counterintuitive behavior, sometimes called the **donkey principle**, is simple to describe physically. Suppose that satellite 2 leads satellite 1 in its orbit ($0 < \phi_2 - \phi_1 < \pi$). Then 2 exerts a positive torque on 1, which adds angular momentum to its orbit, which increases its semimajor axis. As its semimajor axis grows, the mean motion of satellite 1 shrinks and it orbits more slowly, thereby receding from satellite 2 as if it were being repelled.

Combining the potential (3.39) with the angular acceleration (3.42), we arrive at the equation of motion for a co-orbital satellite,

$$\frac{\mathrm{d}^2\phi_1}{\mathrm{d}t^2} = \frac{3\,\mathbb{G}m_2}{\overline{a}^3}\left[\frac{s\cos\frac{1}{2}(\phi_1-\phi_2)}{4\sin^2\frac{1}{2}(\phi_1-\phi_2)} - \sin(\phi_1-\phi_2)\right]. \tag{3.44}$$

Here $s = \mathrm{sgn}[\sin\frac{1}{2}(\phi_1 - \phi_2)]$. Similarly, the equation of motion for ϕ_2 is

$$\frac{\mathrm{d}^2\phi_2}{\mathrm{d}t^2} = -\frac{3\,\mathbb{G}m_1}{\overline{a}^3}\left[\frac{s\cos\frac{1}{2}(\phi_1 - \phi_2)}{4\sin^2\frac{1}{2}(\phi_1 - \phi_2)} - \sin(\phi_1 - \phi_2)\right]. \tag{3.45}$$

We now change variables from ϕ_1 and ϕ_2 to

$$\phi_{\mathrm{cm}} \equiv \frac{m_1\phi_1 + m_2\phi_2}{m_1 + m_2}, \qquad \phi \equiv \phi_1 - \phi_2. \tag{3.46}$$

The first of these can be thought of as the angular center of mass of the two bodies and the second is their relative angle or angular separation (cf. eq. 1.5). Using equations (3.44) and (3.45) we find that $\mathrm{d}^2\phi_{\mathrm{cm}}/\mathrm{d}t^2 = 0$, so the speed of the angular center of mass is constant. Moreover using this result together with equation (3.41) and its analog for $\mathrm{d}\phi_2/\mathrm{d}t$, we find that $\mathrm{d}(m_1a_1 + m_2a_2)/\mathrm{d}t = 0$ so the mass-weighted mean semimajor axis is constant. Therefore it makes sense to define the average semimajor axis as

$$\overline{a} = \frac{m_1a_1 + m_2a_2}{m_1 + m_2}. \tag{3.47}$$

Using equations (3.44) and (3.45) and the second of equations (3.46), we find

$$\frac{\mathrm{d}^2\phi}{\mathrm{d}t^2} = \frac{3\,\mathbb{G}(m_1 + m_2)}{\overline{a}^3}\left(\frac{s\cos\frac{1}{2}\phi}{4\sin^2\frac{1}{2}\phi} - \sin\phi\right). \tag{3.48}$$

We can multiply by $\mathrm{d}\phi/\mathrm{d}t$ and integrate to find

$$\frac{\overline{a}^2}{2}\left(\frac{\mathrm{d}\phi}{\mathrm{d}t}\right)^2 - \frac{3\,\mathbb{G}(m_1 + m_2)}{\overline{a}}\left(\cos\phi - \frac{1}{2|\sin\frac{1}{2}\phi|}\right) \equiv E_{\mathrm{c}} = \text{constant}. \tag{3.49}$$

The **corotation constant** E_{c} is an integral of motion with the dimensions of energy per unit mass that is reminiscent of the Jacobi constant (3.12), but the two are not the same. For example, equation (3.49) is valid for two arbitrary small masses m_1 and m_2 and contains the average semimajor axis $\overline{a}$ as a constant parameter, while the Jacobi constant is an integral of motion

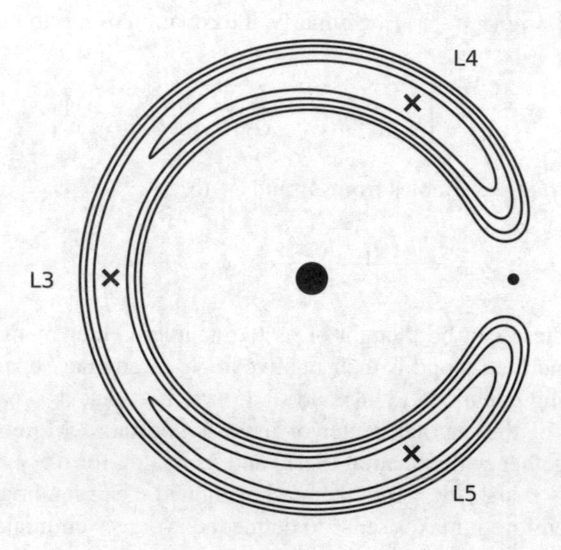

Figure 3.5: Trajectories of co-orbital satellites as described by equation (3.51). The radius is $\overline{a} + \Delta a$ and the azimuth is ϕ, where $\overline{a}$ is the mean semimajor axis and Δa and ϕ are the differences in semimajor axes and azimuth of the two satellites. The mass ratio $(m_1 + m_2)/M = 0.003$. The Lagrange points L3, L4, and L5 are marked by crosses (cf. Figure 3.1).

only when one of the masses is zero and does not contain the semimajor axis.

The corotation constant can also be written in terms of the difference in semimajor axes $\Delta a \equiv a_1 - a_2$: from equation (3.41) and its analog for $\mathrm{d}\phi_2/\mathrm{d}t$,

$$\frac{\mathrm{d}\phi}{\mathrm{d}t} = \frac{\mathrm{d}\phi_1}{\mathrm{d}t} - \frac{\mathrm{d}\phi_2}{\mathrm{d}t} \simeq -\frac{3(\mathbb{G}M)^{1/2}}{2\overline{a}^{5/2}}\Delta a, \tag{3.50}$$

so equation (3.49) becomes

$$\frac{9(\Delta a)^2}{8\overline{a}^2} - \frac{3(m_1 + m_2)}{M}\left(\cos\phi - \frac{1}{2|\sin\frac{1}{2}\phi|}\right) = \frac{\overline{a}E_\mathrm{c}}{\mathbb{G}M} = \text{constant.} \quad (3.51)$$

These contours are plotted in Figure 3.5, in which the radius is $\overline{a} + \Delta a$ and the azimuth is ϕ. Since the orbits are nearly circular, $\overline{a} + \Delta a$ is nearly equal to the radius so the contours show the actual shape of the orbits. The figure is reminiscent of Figure 3.1, but here the contours represent orbits rather than zero-velocity surfaces.

Figure 3.5 shows two types of orbit. **Tadpole orbits** librate around the Lagrange points L4 and L5 and never cross the line $\phi = \pi$, while **horseshoe orbits** librate around the collinear Lagrange point L3 and are symmetric around the line $\phi = \pi$. Tadpole orbits have corotation constant $E_\mathrm{c} < E_\mathrm{crit}$ where the critical value $E_\mathrm{crit} = \frac{9}{2}\mathbb{G}(m_1 + m_2)/\overline{a}$, while horseshoes have $E_\mathrm{c} > E_\mathrm{crit}$. The largest tadpole orbit has $E_\mathrm{c} = E_\mathrm{crit}$ and librates between $\phi = \pi$ and $\phi = \phi_\mathrm{min} = 0.417\,23 = 23.906°$, where ϕ_min is given by the solution of $\cos\phi_\mathrm{min} - \frac{1}{2}|\sin\frac{1}{2}\phi_\mathrm{min}|^{-1} + \frac{3}{2} = 0$.

The equilibrium solutions of (3.48) are found by setting $\mathrm{d}^2\phi/\mathrm{d}t^2 = 0$. By replacing $\sin\phi$ with $2\sin\frac{1}{2}\phi\cos\frac{1}{2}\phi$ we find that equilibrium requires either (i) $\cos\frac{1}{2}\phi = 0$ or (ii) $|\sin^3\frac{1}{2}\phi| = \frac{1}{8}$. Condition (i) implies $\phi = \pi = 180°$ and condition (ii) requires $\phi = \pm\frac{1}{3}\pi = \pm 60°$, which correspond respectively to the collinear Lagrange point L3 and the triangular Lagrange points L4 and L5. L4 and L5 are minima of the corotation constant (3.51).[6]

Small perturbations to ϕ around the Lagrange points have time dependence $\exp(\lambda t)$, where $\lambda^2 = \frac{21}{8}\mathbb{G}(m_1 + m_2)/\overline{a}^3$ at the collinear Lagrange point and $\lambda^2 = -\frac{27}{4}\mathbb{G}(m_1 + m_2)/\overline{a}^3$ at the triangular points. Thus the collinear Lagrange point is unstable but the triangular points are stable, extending the conclusions we already reached in §3.1.1 from the case $m_2 = 0$ to the case where $m_1, m_2 \ll M$.

To determine the period P_L of the librations, we can integrate equation

[6] But recall that L4 and L5 are *maxima* of the effective potential Φ_eff (3.13).

(3.49):

$$P_L = 2\int_{\phi_{\min}}^{\phi_{\max}} \mathrm{d}\phi \left[\frac{6\,\mathbb{G}(m_1+m_2)}{\overline{a}^3}\left(\cos\phi - \frac{1}{2|\sin\frac{1}{2}\phi|}\right) + \frac{2E_{\rm c}}{\overline{a}^2}\right]^{-1/2}$$
$$= \frac{\overline{P}}{\pi}\left[\frac{M}{6(m_1+m_2)}\right]^{1/2}\int_{\phi_{\min}}^{\phi_{\max}} \frac{\mathrm{d}\phi}{\left(\cos\phi - \frac{1}{2}|\sin\frac{1}{2}\phi|^{-1} + \frac{1}{3}\epsilon_{\rm c}\right)^{1/2}} \tag{3.52}$$

where

$$\epsilon_{\rm c} = \frac{\overline{a}E_{\rm c}}{\mathbb{G}(m_1+m_2)}. \tag{3.53}$$

Here $\overline{P} = 2\pi\overline{a}^{3/2}/(\mathbb{G}M)^{1/2}$ is the Kepler orbital period at semimajor axis $\overline{a}$ (eq. 1.43), and $\phi_{\min}$ and $\phi_{\max}$ are the two azimuths at which the square root in the denominator of the second of equations (3.52) vanishes. For $\epsilon_{\rm c} < \frac{9}{2}$ the orbits are tadpoles and both $\phi_{\min}$ and $\phi_{\max}$ are between 0 and π, while for $\epsilon_{\rm c} > \frac{9}{2}$ the orbits are horseshoes and $\phi_{\max} = 2\pi - \phi_{\min}$.

For Janus and Epimetheus the closest approach angle is $\phi_{\min} = 5.2°$ and the libration period is $P_L = 2.92 \times 10^3$ d $= 8.00$ yr. This is the only known pair of satellites in permanent horseshoe orbits, although some quasi-satellites of Earth spend part of their time on horseshoe orbits as described in the next subsection. Many small bodies in the solar system are found in tadpole orbits, such as the Trojan asteroids.

We now investigate the effect of the interactions between co-orbital satellites on their eccentricities and inclinations. For simplicity we consider a test particle that co-orbits with a massive satellite on a circular orbit, but the results apply equally well to satellites of comparable mass. We denote the mean motion of the massive satellite by n_0 and describe the test-particle orbit using the canonical angles $(\lambda, -\varpi, -\Omega)$ and actions $(\Lambda, \Lambda - L, L - L_z)$ (eq. 1.88). The Hamiltonian for the test particle is the sum of the Kepler Hamiltonian $-\frac{1}{2}(\mathbb{G}M)^2/\Lambda^2$ and the gravitational potential Φ due to the massive satellite. The latter depends on azimuth only through the difference in azimuth between the satellite and the test particle, so it must have the form $\Phi(\lambda - n_0 t, \varpi - n_0 t, \Omega - n_0 t, \Lambda, \Lambda - L, L - L_z)$. This form motivates a canonical transformation to new angles $\boldsymbol{\theta}$ and new actions $\mathbf{J}$ defined by the generating function

$$S_2(\mathbf{J}, \lambda, -\varpi, -\Omega, t) = J_1(\lambda - n_0 t) - J_2(\varpi - n_0 t) - J_3(\Omega - n_0 t). \quad (3.54)$$

Then from equations (D.63),

$$\begin{aligned} \Lambda &= J_1, & \Lambda - L &= J_2, & L - L_z &= J_3, \\ \theta_1 &= \lambda - n_0 t, & \theta_2 &= n_0 t - \varpi, & \theta_3 &= n_0 t - \Omega, \end{aligned} \quad (3.55)$$

and the new Hamiltonian is

$$H(\boldsymbol{\theta}, \mathbf{J}) = H_0(\mathbf{J}) + \Phi(\boldsymbol{\theta}, \mathbf{J}) \quad (3.56)$$

where

$$H_0(\mathbf{J}) = -\frac{(\mathbb{G}M)^2}{2J_1^2} - n_0(J_1 - J_2 - J_3). \quad (3.57)$$

The frequencies associated with the unperturbed Hamiltonian $H_0(\mathbf{J})$ are $\boldsymbol{\Omega} = \partial H_0/\partial \mathbf{J} = [(\mathbb{G}M)^2/J_1^3 - n_0, n_0, n_0] = (n - n_0, n_0, n_0)$, where $n = (\mathbb{G}M/a^3)^{1/2}$ is the mean motion of the test particle. Since $|n - n_0| \ll n_0$ for a co-orbiting particle, the angle θ_1 varies slowly while θ_2 and θ_3 vary rapidly. Then according to the averaging principle (Appendix D.9), we can average $\Phi(\boldsymbol{\theta}, \mathbf{J})$ over the fast angles θ_2 and θ_3. The averaged Hamiltonian is independent of the fast angles so the conjugate actions J_2 and J_3 are integrals of motion. Since $J_2 = (\mathbb{G}Ma)^{1/2}[1 - (1 - e^2)^{1/2}]$ and $J_3 = (\mathbb{G}Ma)^{1/2}(1 - e^2)^{1/2}(1 - \cos I)$ and the fractional variation of the semimajor axis a is small for a co-orbital satellite, we conclude that the eccentricity e and inclination I are almost constant in co-orbital dynamics. In particular, if the eccentricities and inclinations of the test particle are initially small or zero, they will remain so.

As the corotation constant increases above E_{crit}, the horseshoe orbits become wider and the minimum separation ϕ_{min} between the satellites becomes smaller. Eventually the assumption on which this averaging principle is based—that the changes in the azimuthal angle ϕ are much slower than the orbital frequency n_0—becomes invalid. We can use equation (3.48) to estimate when this occurs. When $|\phi| \ll 1$ this equation simplifies to

$\ddot{\phi} = 3\mathbb{G}(m_1 + m_2)s/(\overline{a}^3\phi^2)$. The averaging principle fails unless $|\ddot{\phi}| \ll |\phi|/P^2$ which requires $|\phi| \gg [(m_1 + m_2)/M]^{1/3}$, equivalent to the statement that the minimum separation of the two satellites must be much larger than the Hill radius $r_{\mathrm{H}} = \overline{a}[(m_1 + m_2)/(3M)]^{1/3}$ (eq. 3.112). For Janus and Epimetheus the minimum separation is roughly 90 Hill radii, so the averaging principle is safe by a large margin.

There is no consensus on how Janus and Epimetheus formed on or evolved into their current orbits.

3.2.1 Quasi-satellites

We showed in §3.1.1 that the circular restricted three-body problem admits a necessary condition for escape but has no simple sufficient condition. **Quasi-satellites** are an example of this distinction: they orbit a planet stably at distances much larger than the Hill radius (3.24), outside the escape surface of their host planet.

This behavior can be interpreted using the epicycle approximation of §1.8.2. In the simplest case the satellite is a test particle on an eccentric orbit, and its host planet is on a circular orbit and located close to the guiding center of the satellite's epicyclic motion. Even though the test particle is not bound to the planet, it can orbit around it permanently if the gravitational attraction from the planet is sufficiently strong to keep the guiding center bound to it.

Because the motion of the satellite around the epicycle is much faster than the motion of the guiding center relative to the planet, we can use the averaging principle again: in effect, we replace the satellite by a rigid, elliptical wire that has the same size, shape and mass distribution as the epicycle. Normally, a rigid wire or hoop centered on the planet would be unstable, because the gravitational potential of the wire has a maximum at its center.[7] However, in this case the donkey principle described earlier in this section suggests that the wire repels the planet instead of attracting it, so

[7] This instability was known to Laplace (1799–1825) and was used by Maxwell in 1856 to argue that Saturn's rings could not be composed of solid material. Laplace's argument is valid if and only if the radius of the ring is much less than the Hill radius (3.24).

the motion of the guiding center is stable when the planet is at a maximum of the wire potential.

We now provide a quantitative description of this phenomenon.

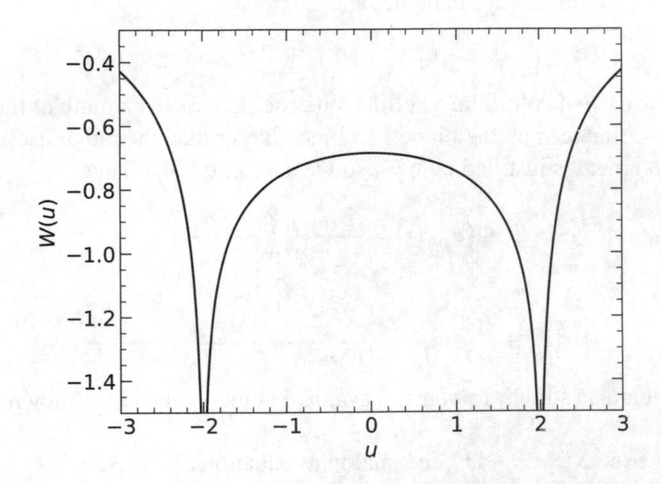

Figure 3.6: The quasi-satellite potential $W(u)$ defined in equation (3.61).

Since the planet is on a circular orbit, its polar coordinates can be written $(a_\mathrm{p}, \psi_\mathrm{p})$ with a_p = const and ψ_p increasing uniformly. The location of the guiding center of the satellite orbit can be written (a, ψ) where a is the semimajor axis. The satellite eccentricity is e; we assume for simplicity that the inclination is zero although the analysis is straightforward to generalize to nonzero inclination. We then average the gravitational potential due to the satellite's mass m over the epicycle orbit. To do this we use Cartesian coordinates with origin at the guiding center, x-axis pointing radially outward, and y-axis pointing in the direction of the planet's motion. In these coordinates the epicyclic motion is given by equations (1.169) and (1.170) as $x = x_0 \cos\tau$, $y = -2x_0 \sin\tau$, where τ is the epicycle phase and $x_0 = ae$. If $|a - a_\mathrm{p}| \ll a_\mathrm{p}$ and $|\psi - \psi_\mathrm{p}| \ll 1$, the position of the planet is approximately

$x = a_{\rm p} - a$, $y = a_{\rm p}(\psi_{\rm p} - \psi)$. Then the averaged potential is

$$\Phi(a, \psi) = -\frac{\mathbb{G}m}{2\pi}\int_0^{2\pi}\frac{\mathrm{d}\tau}{\Delta}, \tag{3.58}$$

where m is the mass of the planet and

$$\Delta^2 = (a + ae\cos\tau - a_{\rm p})^2 + (a_{\rm p}\psi - 2ae\sin\tau - a_{\rm p}\psi_{\rm p})^2. \tag{3.59}$$

As in the case of co-orbital satellites, the oscillations in azimuth of the guiding center induced by the planet are much larger than the radial oscillations (this statement is justified below), so we can set $a = a_{\rm p}$. Thus

$$\Phi(a_{\rm p}, \psi) = \frac{\mathbb{G}m}{a_{\rm p}e}W\left(\frac{\psi - \psi_{\rm p}}{e}\right) \tag{3.60}$$

where

$$W(u) \equiv -\frac{1}{2\pi}\int_0^{2\pi}\frac{\mathrm{d}\tau}{[\cos^2\tau + (u - 2\sin\tau)^2]^{1/2}}. \tag{3.61}$$

This integral is straightforward to evaluate numerically and is shown in Figure 3.6.

If we set $\Delta\psi \equiv \psi - \psi_{\rm p}$, the analog to equation (3.42) is

$$\frac{\mathrm{d}^2\Delta\psi}{\mathrm{d}t^2} = \frac{3}{a_{\rm p}^2}\frac{\partial\Phi}{\partial\Delta\psi}, \tag{3.62}$$

and this can be multiplied by $\mathrm{d}\Delta\psi/\mathrm{d}t$ and integrated to give the integral of motion

$$\frac{a_{\rm p}^2}{2}\left(\frac{\mathrm{d}\Delta\psi}{\mathrm{d}t}\right)^2 - \frac{3\,\mathbb{G}m}{a_{\rm p}e}W(\Delta\psi/e) \equiv E_q = \text{constant}. \tag{3.63}$$

The analog of equation (3.50) is

$$\frac{\mathrm{d}\Delta\psi}{\mathrm{d}t} = -\frac{3(\mathbb{G}M)^{1/2}}{2a_{\rm p}^{5/2}}(a - a_{\rm p}), \tag{3.64}$$

so equation (3.63) can be rewritten

$$\frac{9(a - a_{\rm p})^2}{8a_{\rm p}^2} - \frac{3m}{Me}W(\Delta\psi/e) \equiv \frac{a_{\rm p}E_q}{\mathbb{G}M} = \text{constant}. \tag{3.65}$$

Thus the guiding center of the quasi-satellite orbit undergoes coupled oscillations in semimajor axis and azimuth relative to the planet. Since the ratio of planet mass to stellar mass $m/M \ll 1$, the fractional amplitude of the oscillations in semimajor axis is much smaller than the amplitude in azimuth.

Equations (3.63) and (3.65) are the analogs of equations (3.49) and (3.51) for co-orbital satellites.

If the oscillations of the guiding center are small, $|\Delta\psi| \ll e$, we can expand the potential $W(\Delta\psi/e)$ in a Taylor series around the origin, and equation (3.62) becomes

$$\frac{\mathrm{d}^2\psi}{\mathrm{d}t^2} = \frac{3\,\mathbb{G}m}{a_\mathrm{p}^3 e^3} W''(0)\Delta\psi = -0.300\,95\frac{\mathbb{G}m}{a_\mathrm{p}^3 e^3}\Delta\psi, \tag{3.66}$$

corresponding to stable harmonic oscillations of the guiding center around the planet. This result is valid only for small oscillations; for larger excursions of the guiding center, we must use the full potential from equation (3.61). Eventually when $|\Delta\psi| > 2e$ there can be close encounters or collisions between the planet and the satellite, and the analysis here is no longer valid. The condition $|\Delta\psi| \lesssim 2e$ requires in turn by equation (3.65) that $|a - a_\mathrm{p}| \lesssim a_\mathrm{p}(m/M)^{1/2}e^{-1/2}$. Thus if the planet mass is small, the semimajor axis of the quasi-satellite must be nearly equal to that of the planet.

A typical quasi-satellite orbit is shown in Figure 3.7. All quasi-satellites are on retrograde orbits (i.e., clockwise if the planet's orbit is counterclockwise) since the motion around the epicyclic ellipse is retrograde. The properties of periodic quasi-satellite orbits are described further in §3.4.1.

Several small asteroids currently occupy horseshoe and quasi-satellite orbits around the Earth, although the estimated lifetimes of these orbits are far less than the age of the solar system. Most of these undergo multiple transitions between horseshoe and quasi-satellite orbits. Quasi-satellite orbits that are stable for the lifetime of the solar system exist around several of the outer planets (Wiegert et al. 2000; Shen & Tremaine 2008), but so far no objects have been found in these orbits.

These dynamical arguments suggest that retrograde satellites can orbit stably at distances from their host planet much larger than are possible for

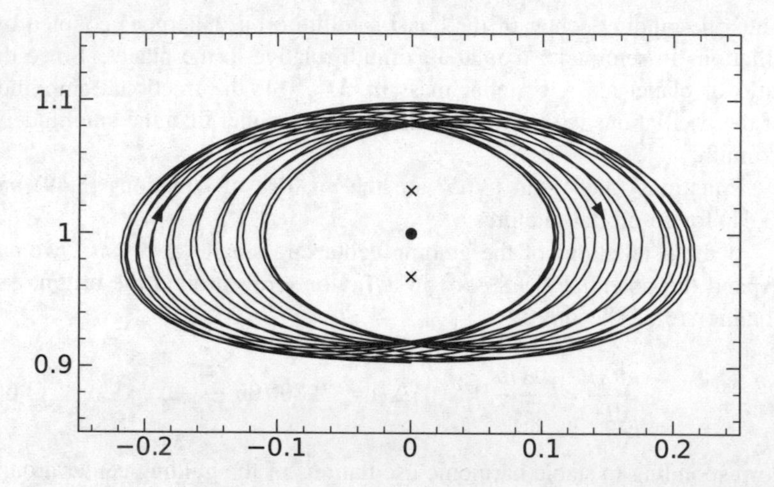

Figure 3.7: A quasi-satellite orbit, as seen in a frame rotating with the planet. The black circle marks the location of the planet, which travels on a counterclockwise circular orbit around a host star located at the origin, off the bottom of the plot. The crosses mark the locations of the Lagrange points L1 and L2. The planet mass is 10^{-4} times the mass of its host star. The quasi-satellite orbit is retrograde and has eccentricity $e = 0.1$.

prograde satellites. In fact almost all of the dozens of small satellites found at large distances ($\gtrsim 0.3$ Hill radii) from Jupiter and Saturn have retrograde orbits.

3.3 The hierarchical three-body problem

The co-orbiting systems that we described in the preceding two sections are interesting but rare. The vast majority of three-body systems in astrophysics are **hierarchical**, which means that they consist of two bodies orbiting one another (the "inner binary") plus a third at much larger distance that orbits the center of mass of the inner two (the "outer binary"). Such systems are generally stable if the ratio of the semimajor axes of the outer and inner binary is large enough. Examples include systems consisting of a planet, a satellite, and the planet's host star; systems of two planets with very different semimajor axes; a planet orbiting a star that belongs to a binary system; and triple star systems.

In this section we focus on hierarchical systems in which the orbits of the inner and outer binaries are nearly circular and coplanar. Some of the behavior of systems with large eccentricities and/or inclinations is described in §5.4.

Let m_1 and m_2 denote the masses of the bodies in the inner binary, with m_0 the mass of the distant body. Here we focus on the dynamics of the inner binary, treating the distant body as traveling around the center of mass of the inner two bodies in a fixed orbit. We work in a reference frame centered on one of the two bodies in the inner binary, which we take to be body 1. Then the equation of motion for body 2 has the form $\ddot{\mathbf{r}}_2 = -\partial\Phi_2(\mathbf{r}_2, t)/\partial\mathbf{r}_2$, where (eq. 4.6)[8]

$$\Phi_2(\mathbf{r}, t) = -\frac{\mathbb{G}(m_1 + m_2)}{|\mathbf{r}|} - \frac{\mathbb{G}m_0}{|\mathbf{r} - \mathbf{r}_0(t)|} + \frac{\mathbb{G}m_0\,\mathbf{r}(t)\cdot\mathbf{r}_0}{|\mathbf{r}_0(t)|^3}, \tag{3.67}$$

and the positions $\mathbf{r}_2$ and $\mathbf{r}_0$ of bodies 2 and 0 are measured relative to body 1.

In a hierarchical system $r = |\mathbf{r}|$ is much smaller than $r_0 = |\mathbf{r}_0|$, so we may expand the potential in powers of r/r_0. To keep track of the ordering it is helpful to replace $\mathbf{r}_0$ by $\lambda\mathbf{r}_0$ and then expand (3.67) as a power series in λ^{-1}:

$$\begin{aligned}\Phi_2(\mathbf{r}, t) = &-\frac{\mathbb{G}(m_1 + m_2)}{r} - \frac{\mathbb{G}m_0}{\lambda r_0} + \frac{\mathbb{G}m_0 r^2}{2\lambda^3 r_0^3} - \frac{3\,\mathbb{G}m_0(\mathbf{r}\cdot\mathbf{r}_0)^2}{2\lambda^3 r_0^5} \\ &+ \frac{3\,\mathbb{G}m_0 r^2(\mathbf{r}\cdot\mathbf{r}_0)}{2\lambda^4 r_0^5} - \frac{5\,\mathbb{G}m_0(\mathbf{r}\cdot\mathbf{r}_0)^3}{2\lambda^4 r_0^7} + \mathrm{O}(\lambda^{-5}).\end{aligned} \tag{3.68}$$

[8] There are notational differences between equations (3.67) and (4.6). Here the positions are measured relative to body 1, while in §4.1.2 the positions are measured relative to body 0. The reason for this difference is that in both cases we would like to attach the label "0" to the most massive body—the Sun in the Earth-Moon-Sun system to be examined in §3.3.1, and the central star in a multi-planet system.

Notice that the **dipole** term proportional to λ^{-2} has vanished (as it does in the discussion following eq. 1.127).

An alternative approach that leads to the same answer is to rewrite the term $-\mathbb{G}m_0|/|\mathbf{r}-\mathbf{r}_0(t)|$ in (3.67) using the expansion (C.44):

$$\Phi_2(\mathbf{r},t) = -\frac{\mathbb{G}(m_1+m_2)}{r} - \mathbb{G}m_0\sum_{l=0}^{\infty}\frac{r^l}{(\lambda r_0)^{l+1}}\mathsf{P}_l(\cos\gamma) + \frac{\mathbb{G}m_0\mathbf{r}\cdot\mathbf{r}_0}{\lambda^2 r_0^3}, \tag{3.69}$$

where $\mathsf{P}_l(\cos\gamma)$ is a Legendre polynomial, γ is the angle between $\mathbf{r}$ and $\mathbf{r}_0$, and $\cos\gamma = \mathbf{r}\cdot\mathbf{r}_0/(rr_0)$. Using the formulas for the Legendre polynomials in equations (C.45), it is straightforward to verify that the series in (3.69) yields (3.68).

The term of order λ^{-1} is called the **monopole** potential; since it is independent of $\mathbf{r}$ its gradient vanishes, so it exerts no force and can be dropped. Then equation (3.68) simplifies to

$$\begin{aligned}\Phi_2(\mathbf{r},t) = &-\frac{\mathbb{G}(m_1+m_2)}{r} + \frac{\mathbb{G}m_0 r^2}{2\lambda^3 r_0^3} - \frac{3\,\mathbb{G}m_0(\mathbf{r}\cdot\mathbf{r}_0)^2}{2\lambda^3 r_0^5}\\ &+\frac{3\,\mathbb{G}m_0 r^2(\mathbf{r}\cdot\mathbf{r}_0)}{2\lambda^4 r_0^5} - \frac{5\,\mathbb{G}m_0(\mathbf{r}\cdot\mathbf{r}_0)^3}{2\lambda^4 r_0^7} + \mathrm{O}(\lambda^{-5}).\end{aligned} \tag{3.70}$$

There are two terms proportional to λ^{-3} that represent the **quadrupole tidal potential** and two terms proportional to λ^{-4} for the **octopole tidal potential**. If we keep only the quadrupole terms and set $\lambda = 1$, we can rewrite this result as

$$\Phi_2(\mathbf{r},t) = -\frac{\mathbb{G}(m_1+m_2)}{r} + \frac{\mathbb{G}m_0 r^2}{2r_0^3} - \frac{3\,\mathbb{G}m_0(\mathbf{r}\cdot\mathbf{r}_0)^2}{2r_0^5} + \frac{\mathbb{G}m_0}{r_0}\mathrm{O}(r^3/r_0^3). \tag{3.71}$$

The corresponding equation of motion for body 2 is

$$\begin{aligned}\frac{\mathrm{d}^2\mathbf{r}_2}{\mathrm{d}t^2} &= -\frac{\partial\Phi_2(\mathbf{r}_2,t)}{\partial\mathbf{r}_2} \qquad (3.72)\\ &= -\frac{\mathbb{G}(m_1+m_2)}{r_2^3}\mathbf{r}_2 - \frac{\mathbb{G}m_0}{r_0^3}\mathbf{r}_2 + \frac{3\,\mathbb{G}m_0(\mathbf{r}_2\cdot\mathbf{r}_0)}{r_0^5}\mathbf{r}_0 + \frac{\mathbb{G}m_0}{r_0^2}\mathrm{O}(r^2/r_0^2).\end{aligned}$$

3.3.1 Lunar theory

The prototypical hierarchical three-body problem is the Earth–Moon–Sun system, which is hierarchical because the Sun is roughly 400 times more distant than the Moon. The development of "lunar theories"—analytic expressions for the trajectory of the Moon, usually based on expansions of the Hamiltonian representing the dynamical effects of the Sun as power series in the lunar eccentricity and inclination—was a centerpiece of solar-system dynamics until the mid-twentieth century (Gutzwiller 1998). Since then, the lunar orbit has been studied more simply and accurately by numerical integrations. Nevertheless, the most important features of analytic lunar theories still provide insight into the history of the lunar orbit and the properties of planetary and satellite orbits in exoplanet systems.

In our notation the Earth, Moon, and Sun are bodies 1, 2, and 0 respectively. The reference frame is centered on the Earth, so from this viewpoint both the Moon and Sun orbit the Earth. We work in Cartesian coordinates in which the z-axis is normal to the orbital plane of the solar orbit around the Earth (the ecliptic). Thus the coordinates of the Sun are $\mathbf{r}_0 = (x_0, y_0, 0) = r_0[\cos(f_0 + \varpi_0), \sin(f_0 + \varpi_0), 0]$, where f_0 and ϖ_0 are the true anomaly and longitude of periapsis of the Sun, and r_0 is its distance from Earth. The potential is then given by equation (3.71) as:

$$\Phi_2(\mathbf{r}, t) = H_\mathrm{K} + H_\odot, \quad \text{where} \quad H_\mathrm{K} = -\frac{\mathbb{G}(m_1 + m_2)}{r} \tag{3.73}$$

is the Kepler Hamiltonian and

$$H_\odot = \frac{\mathbb{G}m_0}{r_0}\left[\frac{r^2}{2r_0^2} - \frac{3[x\cos(f_0 + \varpi_0) + y\sin(f_0 + \varpi_0)]^2}{2r_0^2} + \mathrm{O}(r^3/r_0^3)\right] \tag{3.74}$$

is the Hamiltonian due to the Sun.

Perturbation theory requires that $H_\odot$ is "small" relative to the Hamiltonian H_K that describes the Kepler motion of the Earth-Moon two-body system. Many of the complications of lunar theory arise because "small" can have three distinct meanings in this context:

(i) As described earlier in this section, the solar Hamiltonian $H_\odot$ contains quadrupole, octopole, and higher multipole terms, each smaller than its predecessor by a factor $\sim a/a_0 = 0.002\,570$, the ratio of the semimajor axes of the Moon and Sun. In this section we keep only the quadrupole terms, which corresponds to dropping the terms $\mathrm{O}(r^3/r_0^3)$ from the Hamiltonian (3.74).

(ii) The eccentricity and inclination of the lunar orbit relative to the ecliptic, $e = 0.0549$ and $I = 5.145°$, are both small, as is the eccentricity of the Sun's orbit relative to the Earth, $e_0 = 0.0167$. Thus the Hamiltonian $H_\odot$ is simplified further by expanding it as a power series in e, I, and e_0, and truncating the power series at some maximum degree. In this section we keep terms up to $\mathrm{O}(e^2, e_0^2, ee_0, I^2)$.

(iii) The Kepler Hamiltonian H_K is of order $\mathbb{G}(m_1 + m_2)/a$, while the solar Hamiltonian $H_\odot$ is of order $\mathbb{G}m_0a^2/a_0^3$, so their ratio $H_\odot/H_\mathrm{K} \sim m_0a^3/[(m_1+m_2)a_0^3]$. The mean motion n_0 of the solar orbit is given by Kepler's law, $n_0^2a_0^3 = \mathbb{G}(m_0 + m_1 + m_2)$; the masses of the Earth and Moon, m_1 and m_2, are so much smaller than the solar mass m_0 that we can write $\mathbb{G}m_0 = n_0^2a_0^3$. Similarly the mean motion n of the lunar orbit is given by $\mathbb{G}(m_1 + m_2) = n^2a^3$. Thus $H_\odot/H_\mathrm{K} \sim n_0^2/n^2$, where $n_0/n = 0.0748$ is the ratio of the sidereal month to the sidereal year, that is, the ratio of the orbital periods of the Moon and Sun relative to the fixed stars. Solving Hamilton's equations to higher and higher order using perturbation theory yields expressions for the trajectory involving higher and higher powers of n_0/n. In this section we solve the equations of motion only to first order; that is, we use the unperturbed Kepler motion of the Moon on the right side of Hamilton's equations.

As the accuracy of a lunar theory is improved, both items (i) and (iii) give rise to series in powers of a/a_0, the first because the ratio of radii r/r_0 expressed in orbital elements is proportional to a/a_0, and the second because $n_0/n = [m_0/(m_1 + m_2)]^{1/2}(a/a_0)^{3/2}$. However, the origin of these expansions is quite different: the first is a series of better and better approximations to the Hamiltonian $H_\odot$, while the second is a series of better and better approximations to the solutions of Hamilton's equations for a given $H_\odot$. Note that the small parameter n_0/n is much larger than the small parameter a/a_0, so it may be appropriate to include higher orders in n_0/n than in a/a_0. The developer of any lunar theory must decide the maximum orders of e, I, e_0, a/a_0 and n_0/n that will be accurately represented in the solutions.

Using equations (1.70), (1.151), and (1.155) to rewrite equation (3.74) in terms of orbital elements, and truncating the expansion of the Hamiltonian as described in items (i) and (ii), we obtain

$$\begin{aligned}
H_\odot = \frac{\mathbb{G}m_0 a^2}{a_0^3}\Big[& -\tfrac{1}{4} - \tfrac{3}{8}e^2 - \tfrac{3}{8}e_0^2 + \tfrac{3}{8}I^2 \\
& + (-\tfrac{3}{4} + \tfrac{15}{8}e^2 + \tfrac{15}{8}e_0^2 + \tfrac{3}{8}I^2)\cos(2\lambda - 2\lambda_0) \\
& + \tfrac{1}{2}e\cos(\lambda - \varpi) - \tfrac{3}{4}e\cos(3\lambda - 2\lambda_0 - \varpi) + \tfrac{9}{4}e\cos(\lambda - 2\lambda_0 + \varpi) \\
& - \tfrac{21}{8}e_0\cos(2\lambda - 3\lambda_0 + \varpi_0) + \tfrac{3}{8}e_0\cos(2\lambda - \lambda_0 - \varpi_0) \\
& - \tfrac{3}{4}e_0\cos(\lambda_0 - \varpi_0) - \tfrac{9}{8}e_0^2\cos(2\lambda_0 - 2\varpi_0) \\
& + \tfrac{1}{8}e^2\cos(2\lambda - 2\varpi) - \tfrac{15}{8}e^2\cos(2\lambda_0 - 2\varpi) - \tfrac{3}{4}e^2\cos(4\lambda - 2\lambda_0 - 2\varpi) \\
& - \tfrac{3}{8}I^2\cos(2\lambda - 2\Omega) - \tfrac{3}{8}I^2\cos(2\lambda_0 - 2\Omega) - \tfrac{51}{8}e_0^2\cos(2\lambda - 4\lambda_0 + 2\varpi_0) \\
& + \tfrac{3}{8}ee_0\cos(3\lambda - \lambda_0 - \varpi - \varpi_0) + \tfrac{3}{4}ee_0\cos(\lambda + \lambda_0 - \varpi - \varpi_0) \\
& - \tfrac{9}{8}ee_0\cos(\lambda - \lambda_0 + \varpi - \varpi_0) - \tfrac{21}{8}ee_0\cos(3\lambda - 3\lambda_0 - \varpi + \varpi_0) \\
& + \tfrac{3}{4}ee_0\cos(\lambda - \lambda_0 - \varpi + \varpi_0) + \tfrac{63}{8}ee_0\cos(\lambda - 3\lambda_0 + \varpi + \varpi_0)\Big]. \qquad (3.75)
\end{aligned}$$

The main effects of solar perturbations on the lunar orbit are best explained by looking one by one at a few selected collections of these terms.

Secular terms These are independent of the mean longitudes of the Moon and Sun,[9] λ and λ_0:

$$H_{\rm sec} = \frac{\mathbb{G}m_0 a^2}{a_0^3}\big[-\tfrac{1}{4} - \tfrac{3}{8}e^2 - \tfrac{3}{8}e_0^2 + \tfrac{3}{8}I^2\big]. \qquad (3.76)$$

[9] The adjective "secular" is used in astronomy generally and celestial mechanics in particular to denote changes that are long-lasting rather than oscillating on short timescales.

Since we have already discarded terms in the disturbing function that are of order higher than $\mathrm{O}(e^2, e_0^2, ee_0, I^2)$, there is no additional loss of accuracy if we analyze the effects of these perturbations using the simplified Lagrange equations (1.188). From these we find that a, e, and I are all constants, which we call $\overline{a}$, $\overline{e}$, and $\overline{I}$. Then

$$\frac{\mathrm{d}\varpi}{\mathrm{d}t} = \frac{3\,\mathbb{G}m_0}{4\overline{n}a_0^3}, \quad \frac{\mathrm{d}\Omega}{\mathrm{d}t} = -\frac{3\,\mathbb{G}m_0}{4\overline{n}a_0^3}, \tag{3.77}$$

where $\overline{n}^2\overline{a}^3 = \mathbb{G}(m_0 + m_1)$. Thus the line of apsides precesses forward, while the line of nodes precesses backward at the same rate (compare the discussion following eqs. 1.180, and see Problem 1.20). Writing Kepler's law for the solar orbit as $\mathbb{G}m_0 = n_0^2 a_0^3$, equations (3.77) simplify to

$$\frac{\mathrm{d}\varpi}{\mathrm{d}t} = -\frac{\mathrm{d}\Omega}{\mathrm{d}t} = \frac{3n_0^2}{4\overline{n}}. \tag{3.78}$$

Of course these formulas are only a first approximation to the time-averaged apsidal and nodal precession rates of the Moon. As described earlier in this subsection, if we neglect the octopole and higher multipole moments and assume that the eccentricity and inclination of the lunar orbit are small, the precession rates are given by power series in $m \equiv n_0/\overline{n}$, of which equations (3.78) give the first terms. The next few terms are[10]

$$\begin{aligned} \frac{1}{\overline{n}}\frac{\mathrm{d}\varpi}{\mathrm{d}t} &= \tfrac{3}{4}m^2 + \tfrac{225}{32}m^3 + \tfrac{4\,071}{128}m^4 + \tfrac{265\,493}{2\,048}m^5 + \mathrm{O}(m^6), \\ \frac{1}{\overline{n}}\frac{\mathrm{d}\Omega}{\mathrm{d}t} &= -\tfrac{3}{4}m^2 + \tfrac{9}{32}m^3 + \tfrac{273}{128}m^4 + \tfrac{9\,797}{2\,048}m^5 + \mathrm{O}(m^6). \end{aligned} \tag{3.79}$$

The series for $\dot{\varpi}$ is given to m^{11} by Hill (1894) and the series for $\dot{\Omega}$ is given to m^6 by Delaunay (1860, 1867). The series for $\dot{\varpi}$ is notorious for its slow convergence. In the case of the Moon, with $m = 0.074\,80$, the value of $\dot{\varpi}$ obtained from the first term in the series is smaller than the exact result by a factor 2.042 57, illustrating the danger of using first-order perturbation

[10] The second terms in the series are derived in equation (5.111).

theory in hierarchical systems when the period ratio is not very small.[11] The series for $\dot{\Omega}$ converges more quickly. Numerical solutions for $\dot{\varpi}$ are described in §3.4.1 and shown in Figure 3.12.

The evection This is the term in the disturbing function (3.75)

$$H_{\rm ev} = -\frac{15\,\mathbb{G}m_0 a^2}{8a_0^3} e^2 \cos(2\lambda_0 - 2\varpi). \tag{3.80}$$

The simplified Lagrange equations (1.188) show that under the influence of this term a, I, and Ω are constants and

$$\frac{\mathrm{d}\lambda}{\mathrm{d}t} = \overline{n} - \frac{15n_0^2 e^2}{2\overline{n}} \cos(2\lambda_0 - 2\varpi),$$
$$\frac{\mathrm{d}\varpi}{\mathrm{d}t} = \frac{15n_0^2}{4\overline{n}} \cos(2\lambda_0 - 2\varpi), \quad \frac{\mathrm{d}e}{\mathrm{d}t} = -\frac{15n_0^2 e}{4\overline{n}} \sin(2\lambda_0 - 2\varpi). \tag{3.81}$$

The solar mean longitude advances at a uniform rate, $\lambda_0 = n_0 t + \text{const}$. We use first-order perturbation theory, which means that we integrate these equations assuming that the other orbital elements on the right sides are fixed, at $\overline{e}$ and $\overline{\varpi}$. Then

$$\lambda = \overline{n}t + \lambda_{\rm i} - \frac{15n_0\overline{e}^2}{4\overline{n}} \sin(2\lambda_0 - 2\overline{\varpi}), \tag{3.82}$$
$$\varpi = \overline{\varpi} + \frac{15n_0}{8\overline{n}} \sin(2\lambda_0 - 2\overline{\varpi}), \quad e = \overline{e} + \frac{15n_0\overline{e}}{8\overline{n}} \cos(2\lambda_0 - 2\overline{\varpi}).$$

The most obvious signature of these variations is in the longitude or azimuthal angle ϕ of the Moon, which is related to the mean longitude by equation (1.151),

$$\phi = \lambda + 2e\sin(\lambda - \varpi) + \mathrm{O}(e^2). \tag{3.83}$$

[11] This problem was recognized by Newton, who complained in the *Principia* that the rotation of "the lunar apsis is about twice as speedy" as his calculations implied. He is reported to have said that "his head never ached but with his studies of the Moon" (Whiteside 1976; Cook 2000). In the eighteenth century, the discrepancy between equation (3.78) and the observed apsidal precession of the Moon prompted speculation that Newton's law of gravity was incorrect.

Inserting equations (3.82) and keeping only the lowest order terms in the eccentricity and in the strength of the perturbation, we have

$$\phi = \overline{n}t + \lambda_{\rm i} + 2\overline{e}\sin(\overline{n}t + \lambda_{\rm i} - \overline{\varpi}) + \frac{15n_{\rm O}\overline{e}}{4\overline{n}}\sin(\overline{n}t + \lambda_{\rm i} - 2\lambda_{\rm O} + \overline{\varpi}). \quad (3.84)$$

The first three terms describe the unperturbed Kepler orbit, and the last term is the evection. This is the largest periodic perturbation in the Moon's azimuth, with amplitude $\frac{15}{4}n_{\rm O}\overline{e}/\overline{n} = 0.882°$ according to this calculation. A more accurate estimate of the coefficient of this term, according to the lunar theory of Brown (1897–1908), is 1.274°.

An **evection resonance** can occur in the three-body problem if the mean motion $n_{\rm O}$ of the distant body (the Sun) is equal to the apsidal precession rate $\dot{\varpi}$ of the satellite (the Moon), for example due to the quadrupole moment of the host body (the Earth). In resonance the longitude of periapsis ϖ librates around an azimuth that is ±90° from the azimuth of the distant body. An example is shown in Figure 3.8.

If the Moon formed close to the Earth as debris from a giant impact and subsequently evolved to its current orbit as a result of tidal friction between the Earth and Moon, then it likely passed through an evection resonance early in its history (at this time, the apsidal precession is mostly due to the equatorial bulge of the rapidly rotating Earth, not the tidal field from the Sun as is presently the case). The evection resonance can excite the lunar eccentricity, leading to substantial tidal heating and perhaps melting of the lunar interior, and it can drain orbital angular momentum from the Earth-Moon system and transfer it to the solar orbit.

There is an inclination-dependent term in the Hamiltonian (3.75) analogous to the evection,

$$H = -\frac{3\,\mathbb{G}m_{\rm O}a^2}{8a_{\rm O}^3}I^2\cos(2\lambda_{\rm O} - 2\Omega). \quad (3.85)$$

This term is responsible for the largest non-Kepler oscillations in the Moon's latitude. Nevertheless its effects are less significant than those of evection, in part because the numerical coefficient is five times smaller.

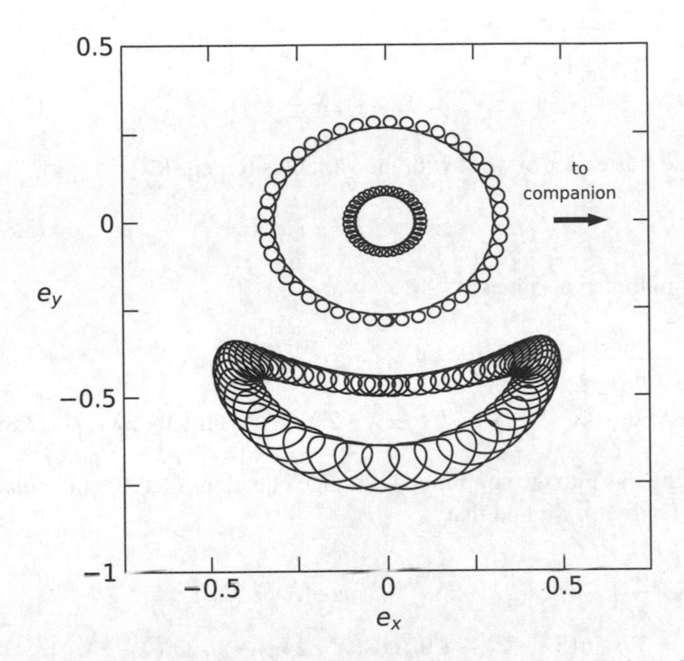

Figure 3.8: The evection resonance. The figure shows the evolution of the eccentricity vectors $\mathbf{e} = (e_x, e_y)$ of three test particles orbiting a host body that in turn is orbited by a distant companion. The host body and the companion have unit mass, the test particle has unit semimajor axis, and the companion is on a circular orbit with semimajor axis 10. The evolution is plotted in a frame that rotates with the orbital motion of the distant companion, and the positive x-axis points toward the companion. The host body has a quadrupole moment $J_2R^2 = 0.01$ (eq. 1.135). The three test particles have initial eccentricity $e = 0.1$, 0.3, and 0.5. The last of these is in an evection resonance, in which the mean precession rate due to the quadrupole moment $\dot{\varpi} = \dot{\Omega} + \dot{\omega}$ (eq. P.30) equals the mean motion of the distant companion, and the eccentricity vector librates around a direction perpendicular to the companion.

The variation We next examine the terms

$$H_{\rm var} = \frac{\mathbb{G}m_0 a^2}{a_0^3}\Big[-\tfrac{3}{4}\cos(2\lambda - 2\lambda_0) - \tfrac{3}{4}e\cos(3\lambda - 2\lambda_0 - \varpi) + \tfrac{9}{4}e\cos(\lambda - 2\lambda_0 + \varpi)\Big]. \tag{3.86}$$

In this case it is easier to work with the variables (cf. eq. 1.71)

$$k = e\cos\varpi, \quad h = e\sin\varpi, \tag{3.87}$$

so the Hamiltonian becomes

$$H_{\rm var} = \frac{\mathbb{G}m_0 a^2}{a_0^3}\Big[-\tfrac{3}{4}\cos(2\lambda - 2\lambda_0) - \tfrac{3}{4}k\cos(3\lambda - 2\lambda_0) - \tfrac{3}{4}h\sin(3\lambda - 2\lambda_0) + \tfrac{9}{4}k\cos(\lambda - 2\lambda_0) - \tfrac{9}{4}h\sin(\lambda - 2\lambda_0)\Big]. \tag{3.88}$$

Inserting this into the simplified Lagrange equations (1.193) and replacing $\mathbb{G}m_0/a_0^3$ by n_0^2 we find that

$$\begin{aligned}
\frac{\mathrm{d}\lambda}{\mathrm{d}t} &= n + \frac{n_0^2}{n}\Big[-3\cos(2\lambda - 2\lambda_0) - 3k\cos(3\lambda - 2\lambda_0) \\
&\qquad - 3h\sin(3\lambda - 2\lambda_0) + 9k\cos(\lambda - 2\lambda_0) - 9h\sin(\lambda - 2\lambda_0)\Big],\\
\frac{\mathrm{d}a}{\mathrm{d}t} &= \frac{n_0^2 a}{n}\Big[-3\sin(2\lambda - 2\lambda_0) - \tfrac{9}{2}k\sin(3\lambda - 2\lambda_0) + \tfrac{9}{2}h\cos(3\lambda - 2\lambda_0) \\
&\qquad + \tfrac{9}{2}k\sin(\lambda - 2\lambda_0) + \tfrac{9}{2}h\cos(\lambda - 2\lambda_0)\Big],\\
\frac{\mathrm{d}k}{\mathrm{d}t} &= \frac{n_0^2}{n}\Big[-\tfrac{3}{4}\sin(3\lambda - 2\lambda_0) - \tfrac{9}{4}\sin(\lambda - 2\lambda_0)\Big],\\
\frac{\mathrm{d}h}{\mathrm{d}t} &= \frac{n_0^2}{n}\Big[\tfrac{3}{4}\cos(3\lambda - 2\lambda_0) - \tfrac{9}{4}\cos(\lambda - 2\lambda_0)\Big].
\end{aligned} \tag{3.89}$$

In first-order perturbation theory, we solve the differential equations for $\mathrm{d}a/\mathrm{d}t$, $\mathrm{d}k/\mathrm{d}t$, and $\mathrm{d}h/\mathrm{d}t$ by replacing the orbital elements a, k, h on the right side with fixed quantities $\overline{a}$, $\overline{k}$, $\overline{h}$, and by replacing the mean longitude λ with $\overline{\lambda} = \overline{n}t$ + constant. The mean longitude of the Sun is λ_0 =

$n_0 t +$ constant; this variable is not perturbed, as we can ignore perturbations of the Sun by the Moon. We have

$$a = \overline{a} + \frac{n_0^2 \overline{a}}{\overline{n}} \left[\frac{3\cos(2\overline{\lambda} - 2\lambda_0)}{2\overline{n} - 2n_0} + \frac{9\overline{k}\cos(3\overline{\lambda} - 2\lambda_0)}{2(3\overline{n} - 2n_0)} + \frac{9\overline{h}\sin(3\overline{\lambda} - 2\lambda_0)}{2(3\overline{n} - 2n_0)} - \frac{9\overline{k}\cos(\overline{\lambda} - 2\lambda_0)}{2(\overline{n} - 2n_0)} + \frac{9\overline{h}\sin(\overline{\lambda} - 2\lambda_0)}{2(\overline{n} - 2n_0)} \right],$$

$$k = \overline{k} + \frac{n_0^2}{\overline{n}} \left[\frac{3\cos(3\overline{\lambda} - 2\lambda_0)}{4(3\overline{n} - 2n_0)} + \frac{9\cos(\overline{\lambda} - 2\lambda_0)}{4(\overline{n} - 2n_0)} \right],$$

$$h = \overline{h} + \frac{n_0^2}{\overline{n}} \left[\frac{3\sin(3\overline{\lambda} - 2\lambda_0)}{4(3\overline{n} - 2n_0)} - \frac{9\sin(\overline{\lambda} - 2\lambda_0)}{4(\overline{n} - 2n_0)} \right]. \tag{3.90}$$

The small parameter in this perturbation expansion is the ratio of the mean motion of the Sun to the mean motion of the Moon, $n_0/\overline{n}$, so at the accuracy to which we are working we can simplify these expressions by replacing denominators like $2\overline{n} - 2n_0$ by $2\overline{n}$. Furthermore, since the eccentricity is small we may focus on the case where $\overline{k} = \overline{h} = 0$, so these expressions simplify to

$$a = \overline{a} + \frac{3n_0^2 \overline{a}}{2\overline{n}^2} \cos(2\overline{\lambda} - 2\lambda_0),$$

$$k = \frac{n_0^2}{\overline{n}^2} \left[\tfrac{1}{4}\cos(3\overline{\lambda} - 2\lambda_0) + \tfrac{9}{4}\cos(\overline{\lambda} - 2\lambda_0) \right],$$

$$h = \frac{n_0^2}{\overline{n}^2} \left[\tfrac{1}{4}\sin(3\overline{\lambda} - 2\lambda_0) - \tfrac{9}{4}\sin(\overline{\lambda} - 2\lambda_0) \right]. \tag{3.91}$$

To integrate the equation in (3.89) for $\mathrm{d}\lambda/\mathrm{d}t$ at the same level of accuracy, we must use Kepler's law $n^2 a^3 = \mathbb{G}(m_1 + m_2)$ to write the term n as $\overline{n} - \frac{3}{2}(\overline{n}/\overline{a})(a - \overline{a})$ and substitute for $a - \overline{a}$ from the first of equations (3.91). Upon integrating the result we have

$$\lambda = \overline{\lambda} - \frac{21 n_0^2}{8\overline{n}^2} \sin(2\lambda - 2\lambda_0). \tag{3.92}$$

To interpret these expressions in terms of the shape of the lunar orbit, we use the expansions (1.155) and (1.151) for the radius r and azimuthal angle ϕ. Truncating these at first order in the eccentricity, we have

$$r = a(1 - k\cos\lambda - h\sin\lambda), \quad \phi = \lambda + 2k\sin\lambda - 2h\cos\lambda. \tag{3.93}$$

Then substituting from equations (3.91) and (3.92) and truncating at first order in the small parameter $n_0^2/\overline{n}^2$, we find

$$r = \overline{a} - \frac{n_0^2\overline{a}}{n^2}\cos(2\overline{\lambda} - 2\lambda_0), \quad \phi = \overline{\lambda} + \frac{11n_0^2}{8n^2}\sin(2\overline{\lambda} - 2\lambda_0). \tag{3.94}$$

The amplitude of the periodic variation in the azimuth is $\frac{11}{8}n_0^2/n^2 = 0.441°$; a more accurate estimate from Brown's lunar theory is 0.658°.

At the order to which we are working, eliminating $\overline{\lambda}$ in favor of ϕ gives the following polar equation:

$$r = \overline{a} - \frac{n_0^2\overline{a}}{n^2}\cos(2\phi - 2\lambda_0). \tag{3.95}$$

This is an approximate ellipse with short axis pointing toward the Sun and long axis 90° away from the Sun, sometimes called the **variational ellipse**. The variational ellipse is centered on body 1 (the Earth), in contrast to the Kepler ellipse which has one focus at body 1. A more direct derivation of equation (3.95) is described in Problem 3.7. The variational ellipse is a member of family "g" of periodic orbits in Hill's problem, as described in the following section.

Additional terms of the lunar Hamiltonian are analyzed in Problems 3.8 and 3.9.

3.4 Hill's problem

Hill's problem is a simplified version of the hierarchical three-body problem that can be used when the third body is much more massive and more distant than the other two, as in the case of the Earth–Moon–Sun system (the ratio of the solar mass to the mass of the Earth and Moon is 3.289×10^5 and the ratio of the Sun-Earth semimajor axis to the Earth–Moon semimajor axis is 389.17). Hill's problem is perhaps the simplest non-integrable case of the N-body problem, with *no* free parameters, yet it is accurate enough to reproduce most of the complex behavior of the lunar orbit.

As in the preceding section, m_1 and m_2 denote the masses of the bodies in the inner binary, and m_0 is the mass of the distant massive body. We initially work in a frame with origin at body 0. The equation of motion for m_1 is (cf. eq. 4.5)

$$\ddot{\mathbf{r}}_1 = -\frac{\mathbb{G}(m_0 + m_1)}{|\mathbf{r}_1|^3}\mathbf{r}_1 + \frac{\mathbb{G}m_2}{|\mathbf{r}_2 - \mathbf{r}_1|^3}(\mathbf{r}_2 - \mathbf{r}_1) - \frac{\mathbb{G}m_2}{|\mathbf{r}_2|^3}\mathbf{r}_2, \tag{3.96}$$

with a similar equation for $\mathbf{r}_2$.

In a hierarchical three-body system such as this one, the motion of the center of mass of m_1 and m_2 relative to m_0 is not far from the motion of a test particle on a Kepler orbit around m_0. Therefore we introduce a reference vector $\overline{\mathbf{a}}(t)$ from the origin at $\mathbf{r}_0$ to a point close to m_1 and m_2 that obeys the equation of motion

$$\frac{\mathrm{d}^2\overline{\mathbf{a}}}{\mathrm{d}t^2} = -\frac{\mathbb{G}m_0}{|\overline{\mathbf{a}}|^3}\overline{\mathbf{a}}, \tag{3.97}$$

and define a new coordinate $\Delta\mathbf{r}_1$ to be the difference between the position of m_1 and the tip of $\overline{\mathbf{a}}$; thus $\Delta\mathbf{r}_1 \equiv \mathbf{r}_1 - \overline{\mathbf{a}}$, with a similar definition for $\Delta\mathbf{r}_2$. The equation of motion becomes

$$\begin{aligned}\frac{\mathrm{d}^2\Delta\mathbf{r}_1}{\mathrm{d}t^2} = &\frac{\mathbb{G}m_2}{|\Delta\mathbf{r}_2 - \Delta\mathbf{r}_1|^3}(\Delta\mathbf{r}_2 - \Delta\mathbf{r}_1) - \frac{\mathbb{G}(m_0 + m_1)}{|\overline{\mathbf{a}} + \Delta\mathbf{r}_1|^3}(\overline{\mathbf{a}} + \Delta\mathbf{r}_1) \\ &- \frac{\mathbb{G}m_2}{|\overline{\mathbf{a}} + \Delta\mathbf{r}_2|^3}(\overline{\mathbf{a}} + \Delta\mathbf{r}_2) + \frac{\mathbb{G}m_0}{|\overline{\mathbf{a}}|^3}\overline{\mathbf{a}},\end{aligned} \tag{3.98}$$

with a similar equation for $\Delta\mathbf{r}_2$.

We are interested in the case where $m_0 \gg m_1, m_2$ and the distance to m_0 is much larger than the distance between m_1 and m_2. Hill's insight was that we can represent this case by replacing $\overline{\mathbf{a}}$ by $\lambda\overline{\mathbf{a}}$ and m_0 by $\lambda^3 m_0$ and letting $\lambda \to \infty$. To carry out this procedure, we use the identity

$$\frac{1}{|\lambda\overline{\mathbf{a}} + \Delta\mathbf{r}|^3} = \frac{1}{\lambda^3\overline{a}^3}\left[1 - \frac{3\overline{\mathbf{a}}\cdot\Delta\mathbf{r}}{\lambda\overline{a}^2} + \mathrm{O}(\lambda^{-2})\right]. \tag{3.99}$$

Keeping only the terms proportional to λ^k with $k \geq 0$ we find

$$\frac{\mathrm{d}^2\Delta\mathbf{r}_1}{\mathrm{d}t^2} = \frac{\mathbb{G}m_2}{|\Delta\mathbf{r}_2 - \Delta\mathbf{r}_1|^3}(\Delta\mathbf{r}_2 - \Delta\mathbf{r}_1) - \frac{\mathbb{G}m_0}{|\overline{\mathbf{a}}|^3}\Delta\mathbf{r}_1 + \frac{3\,\mathbb{G}m_0}{|\overline{\mathbf{a}}|^5}(\overline{\mathbf{a}}\cdot\Delta\mathbf{r}_1)\overline{\mathbf{a}}. \tag{3.100}$$

So long as the eccentricities and inclinations of the orbits of m_1 and m_2 around m_0 are small, we may assume that the reference vector $\overline{\mathbf{a}}$ traces out a circular orbit, so $\overline{a} = |\overline{\mathbf{a}}|$ is a constant and $\overline{\mathbf{a}}$ rotates at constant angular speed $\Omega = (\mathbb{G}m_0/\overline{a}^3)^{1/2}$. The equation of motion then simplifies to

$$\frac{\mathrm{d}^2\Delta\mathbf{r}_1}{\mathrm{d}t^2} = \frac{\mathbb{G}m_2}{|\Delta\mathbf{r}_2 - \Delta\mathbf{r}_1|^3}(\Delta\mathbf{r}_2 - \Delta\mathbf{r}_1) - \Omega^2\Delta\mathbf{r}_1 + 3\Omega^2\frac{\overline{\mathbf{a}}\cdot\Delta\mathbf{r}_1}{\overline{a}^2}\overline{\mathbf{a}}. \tag{3.101}$$

We now transform to a uniformly rotating reference frame in which the reference vector $\overline{\mathbf{a}}$ is fixed. In this frame we denote the vectors from $\overline{\mathbf{a}}$ to $\mathbf{r}_{1,2}$ by $\Delta\mathbf{x}_{1,2}$ rather than $\Delta\mathbf{r}_{1,2}$, and we must include the Coriolis force $-2\boldsymbol{\Omega}\times\mathrm{d}\Delta\mathbf{x}_1/\mathrm{d}t$ and the centrifugal force $-\boldsymbol{\Omega}\times(\boldsymbol{\Omega}\times\Delta\mathbf{x}_1) = \Omega^2\Delta\mathbf{x}_1 - \boldsymbol{\Omega}(\boldsymbol{\Omega}\cdot\Delta\mathbf{x}_1)$ (eq. D.20):

$$\begin{aligned}\frac{\mathrm{d}^2\Delta\mathbf{x}_1}{\mathrm{d}t^2} + 2\boldsymbol{\Omega}\times\frac{\mathrm{d}\Delta\mathbf{x}_1}{\mathrm{d}t} &= \frac{\mathbb{G}m_2}{|\Delta\mathbf{x}_2 - \Delta\mathbf{x}_1|^3}(\Delta\mathbf{x}_2 - \Delta\mathbf{x}_1) - \boldsymbol{\Omega}(\boldsymbol{\Omega}\cdot\Delta\mathbf{x}_1)\\ &\quad + 3\Omega^2\frac{\overline{\mathbf{a}}\cdot\Delta\mathbf{x}_1}{\overline{a}^2}\overline{\mathbf{a}}.\end{aligned} \tag{3.102}$$

We may choose the x-axis to be parallel to $\overline{\mathbf{a}}$ and the z-axis to be parallel to $\boldsymbol{\Omega}$, so the positive x-axis points radially outward and the positive y-axis points in the direction of orbital motion. Then we arrive at **Hill's equations**,

$$\begin{aligned}\frac{\mathrm{d}^2\Delta x_1}{\mathrm{d}t^2} - 2\Omega\frac{\mathrm{d}\Delta y_1}{\mathrm{d}t} &= \frac{\mathbb{G}m_2}{|\Delta\mathbf{x}_2 - \Delta\mathbf{x}_1|^3}(\Delta x_2 - \Delta x_1) + 3\Omega^2\Delta x_1,\\ \frac{\mathrm{d}^2\Delta y_1}{\mathrm{d}t^2} + 2\Omega\frac{\mathrm{d}\Delta x_1}{\mathrm{d}t} &= \frac{\mathbb{G}m_2}{|\Delta\mathbf{x}_2 - \Delta\mathbf{x}_1|^3}(\Delta y_2 - \Delta y_1),\\ \frac{\mathrm{d}^2\Delta z_1}{\mathrm{d}t^2} &= \frac{\mathbb{G}m_2}{|\Delta\mathbf{x}_2 - \Delta\mathbf{x}_1|^3}(\Delta z_2 - \Delta z_1) - \Omega^2\Delta z_1.\end{aligned} \tag{3.103}$$

Notice that $\overline{a}$ and m_0 have disappeared from the equations—they enter only through the orbital frequency Ω.

If $m_2 = 0$ or the distance $|\Delta\mathbf{x}_2 - \Delta\mathbf{x}_1|$ is large enough that the force from m_2 is negligible, the solution of Hill's equations of motion is analytic:

$$\begin{aligned}
\Delta x_1 &= \alpha_1 - \epsilon_1 \cos(\Omega t + \delta_1),\\
\Delta y_1 &= -\tfrac{3}{2}\alpha_1 \Omega t + \gamma_1 + 2\epsilon_1 \sin(\Omega t + \delta_1),\\
\Delta z_1 &= \epsilon_{z1} \cos(\Omega t + \delta_{z1}).
\end{aligned} \tag{3.104}$$

This solution corresponds to the epicyclic motion described in §1.8.2. The difference between the two treatments is that epicycle theory provides an *approximate* solution of the exact equations of motion for the two-body problem, while equations (3.104) are an *exact* solution of Hill's equations, which approximate the two-body problem. The variables α, ϵ and ϵ_z are closely related to the semimajor axis a, eccentricity e and inclination I in the original Kepler problem. In particular, for an assumed value of $\overline{a}$ we have $a \simeq \overline{a} + \alpha$, $e \simeq \epsilon/\overline{a}$, $I \simeq \epsilon_z/\overline{a}$ (of course, strictly speaking, Hill's equations are only valid in the limit $\overline{a} \to \infty$).

Just as in the two-body problem, we can change variables from $\Delta\mathbf{x}_1$ and $\Delta\mathbf{x}_2$ to

$$\mathbf{x}_{\rm cm} \equiv \frac{m_1 \Delta\mathbf{x}_1 + m_2 \Delta\mathbf{x}_2}{m_1 + m_2}, \qquad \mathbf{x} \equiv \Delta\mathbf{x}_2 - \Delta\mathbf{x}_1; \tag{3.105}$$

here $\mathbf{x}_{\rm cm}$ is the barycenter of m_1 and m_2 and $\mathbf{x}$ is the relative position. The barycenter satisfies the equations of motion

$$\begin{aligned}
\ddot{x}_{\rm cm} - 2\Omega \dot{y}_{\rm cm} &= 3\Omega^2 x_{\rm cm},\\
\ddot{y}_{\rm cm} + 2\Omega \dot{x}_{\rm cm} &= 0,\\
\ddot{z}_{\rm cm} &= -\Omega^2 z_{\rm cm},
\end{aligned} \tag{3.106}$$

which have solutions analogous to equations (3.104).

The relative position satisfies the equations of motion

$$\begin{aligned}
\frac{\mathrm{d}^2 x}{\mathrm{d}t^2} - 2\Omega \frac{\mathrm{d}y}{\mathrm{d}t} &= -\frac{\mathbb{G}(m_1 + m_2)}{|\mathbf{x}|^3} x + 3\Omega^2 x,\\
\frac{\mathrm{d}^2 y}{\mathrm{d}t^2} + 2\Omega \frac{\mathrm{d}x}{\mathrm{d}t} &= -\frac{\mathbb{G}(m_1 + m_2)}{|\mathbf{x}|^3} y,
\end{aligned}$$

$$\frac{\mathrm{d}^2 z}{\mathrm{d}t^2} = -\frac{\mathbb{G}(m_1 + m_2)}{|\mathbf{x}|^3} z - \Omega^2 z. \tag{3.107}$$

To study these we introduce a dimensionless time $\tau \equiv \Omega t$ and coordinates

$$(\xi, \eta, \zeta) \equiv \left[\frac{\Omega^2}{\mathbb{G}(m_1 + m_2)}\right]^{1/3} (x, y, z) = \frac{1}{a}\left(\frac{m_0}{m_1 + m_2}\right)^{1/3} (x, y, z), \tag{3.108}$$

and we arrive at a dimensionless version of Hill's equations,

$$\begin{aligned} \frac{\mathrm{d}^2\xi}{\mathrm{d}\tau^2} - 2\frac{\mathrm{d}\eta}{\mathrm{d}\tau} &= -\frac{\xi}{\rho^3} + 3\xi, \\ \frac{\mathrm{d}^2\eta}{\mathrm{d}\tau^2} + 2\frac{\mathrm{d}\xi}{\mathrm{d}\tau} &= -\frac{\eta}{\rho^3}, \\ \frac{\mathrm{d}^2\zeta}{\mathrm{d}\tau^2} &= -\frac{\zeta}{\rho^3} - \zeta, \end{aligned} \tag{3.109}$$

where $\rho^2 \equiv \xi^2 + \eta^2 + \zeta^2$. Notice that there are no free parameters in these equations. A Hamiltonian formulation of Hill's equations is described in Problem 3.12.

It is straightforward to verify that the dimensionless Hill's equations have an integral of motion

$$E_{\mathrm{H}} = \tfrac{1}{2}\left(\frac{\mathrm{d}\xi}{\mathrm{d}\tau}\right)^2 + \tfrac{1}{2}\left(\frac{\mathrm{d}\eta}{\mathrm{d}\tau}\right)^2 + \tfrac{1}{2}\left(\frac{\mathrm{d}\zeta}{\mathrm{d}\tau}\right)^2 - \frac{1}{\rho} - \tfrac{3}{2}\xi^2 + \tfrac{1}{2}\zeta^2, \tag{3.110}$$

which we call the **Jacobi–Hill constant**, analogous to the Jacobi constant.[12] It is also straightforward to show that the stationary solutions of Hill's equations are

$$\xi = \pm 3^{-1/3}, \quad \eta = 0, \quad \zeta = 0, \tag{3.111}$$

analogous to the collinear Lagrange points L2 and L1. In this approximation L2 and L1 are at the same distance, known as the **Hill radius**. Restoring the

[12] We use the term "analogous to" rather than "special case of," because Hill's problem is not a special case of the restricted three-body problem—the masses of all three bodies are nonzero, whereas in the restricted problem one mass must vanish.

dimensional factors, the Hill radius is

$$r_{\rm H} = \left[\frac{\mathbb{G}(m_1+m_2)}{3\Omega^2}\right]^{1/3} = \overline{a}\left(\frac{m_1+m_2}{3m_0}\right)^{1/3}, \tag{3.112}$$

where $\overline{a}$ is the distance from the center of mass of m_1 and m_2 to the distant body m_0. The distance $r_{\rm H}$ is sometimes called the **mutual Hill radius** since it depends on the masses of both small bodies. The special case of the restricted three-body problem, in which $m_2 = 0$, was already described by equation (3.24).

The Jacobi–Hill constant restricts the motion to the region

$$\Phi_{\rm eff}(\xi,\eta,\zeta) \leq E_{\rm H}, \quad \text{where} \quad \Phi_{\rm eff}(\xi,\eta) = -\frac{1}{(\xi^2+\eta^2+\zeta^2)^{1/2}} - \tfrac{3}{2}\xi^2 + \tfrac{1}{2}\zeta^2 \tag{3.113}$$

is the effective potential and the surface $\Phi_{\rm eff} = E_{\rm H}$ is the zero-velocity surface, concepts introduced in §3.1 in the context of the restricted three-body problem. Figure 3.9 shows the zero-velocity curves in the planar Hill's problem ($\zeta = {\rm d}\zeta/{\rm d}\tau = 0$), analogous to Figure 3.1.

3.4.1 Periodic orbits in Hill's problem

To study the behavior of trajectories in Hill's problem we focus on periodic orbits, which form the "skeleton" around which other orbits can be grouped. There is an infinite number of periodic orbits, so we examine only the simplest of them: we restrict our attention to orbits that (i) remain in the $\zeta = 0$ plane; (ii) are symmetric with respect to the radial or ξ-axis; and (iii) are simple-periodic, by which we mean that they intersect the ξ-axis at only two locations—one with ${\rm d}\eta/{\rm d}\tau > 0$ and the other with ${\rm d}\eta/{\rm d}\tau < 0$—before returning to their original position and velocity. These conditions imply that the orbits must cross the ξ-axis at right angles, that is, ${\rm d}\xi/{\rm d}\tau = 0$ when $\eta = 0$.

In Hill's problem, periodic orbits of this kind are organized in one-parameter families. The parameter is usually chosen to be the Jacobi–Hill constant $E_{\rm H}$ of equation (3.110). Since the orbit remains in the $\zeta = 0$ plane,

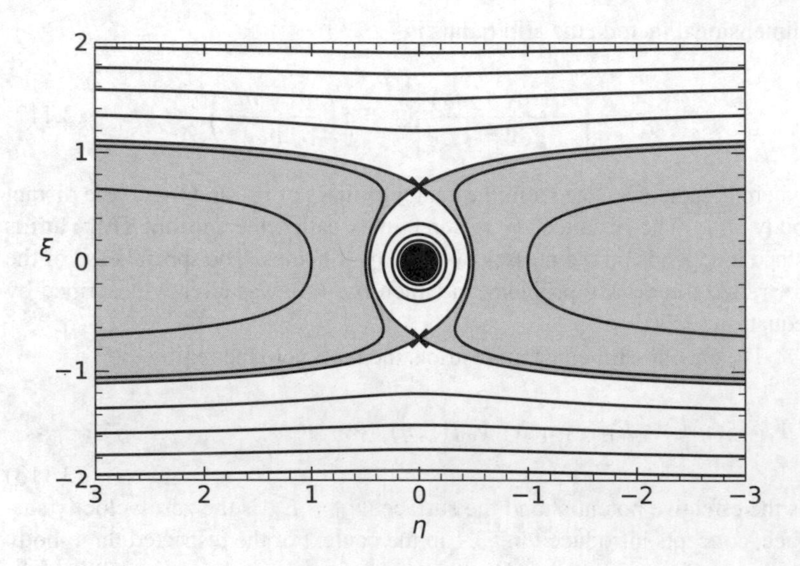

Figure 3.9: Contours of the effective potential (3.113), also known as zero-velocity curves, for the planar Hill's problem. The Lagrange points L1 and L2 are marked by crosses. Shading marks regions in which the effective potential is greater than the value at the L1 and L2 points, $\Phi_{\rm eff} = -3^{4/3}/2 = -2.163\,37$. Note that the horizontal or η-axis increases from right to left such that (ξ, η) is a right-handed coordinate system.

it must have $\zeta = {\rm d}\zeta/{\rm d}\tau = 0$. Thus

$$E_{\rm H} = \tfrac{1}{2}\left(\frac{{\rm d}\xi}{{\rm d}\tau}\right)^2 + \tfrac{1}{2}\left(\frac{{\rm d}\eta}{{\rm d}\tau}\right)^2 - \frac{1}{(\xi^2+\eta^2)^{1/2}} - \tfrac{3}{2}\xi^2. \tag{3.114}$$

The periodic orbit can be specified completely by $E_{\rm H}$ and the value $\xi = \xi_{\rm p}$ at which it crosses the ξ-axis in the direction of increasing η. To see this, note that at this point $\eta = {\rm d}\xi/{\rm d}\tau = 0$ since the orbit is simple-periodic. Then equation (3.114) can be solved to find $({\rm d}\eta/{\rm d}\tau)_{\xi=\xi_{\rm p},\eta=0}$, which we know to be positive by assumption:

$$\left(\frac{\mathrm{d}\eta}{\mathrm{d}\tau}\right)_{\xi=\xi_\mathrm{p},\eta=0} = \left(2E_\mathrm{H} + \frac{2}{\xi_\mathrm{p}} + 3\xi_\mathrm{p}^2\right)^{1/2}. \tag{3.115}$$

Thus all of the initial phase-space coordinates are specified, and the trajectory can be determined by numerical integration of the first two of equations (3.109).

Let us imagine integrating a trajectory numerically, starting at $\tau = 0$ when the particle crosses the ξ-axis at $\xi = \xi_\mathrm{p}$ in the direction of increasing η and continuing until the orbit crosses the ξ-axis again in the direction of decreasing η. Let $\xi'_{1/2}(E_\mathrm{H}, \xi_\mathrm{p})$ and $\tau_{1/2}(E_\mathrm{H}, \xi_\mathrm{p})$ be the values of $\mathrm{d}\xi/\mathrm{d}\tau$ and τ at the end of the integration. If $\xi'_{1/2}(E_\mathrm{H}, \xi_\mathrm{p}) = 0$ the orbit is periodic, since it will return to its starting point after an additional interval $\tau_{1/2}$. Thus to find a periodic orbit with a given Jacobi–Hill constant E_H, we simply solve numerically the equation $\xi'_{1/2}(E_\mathrm{H}, \xi_\mathrm{p}) = 0$ for ξ_p, using standard methods for finding the roots of nonlinear equations (e.g., Press et al. 2007). The period of the orbit in the rotating frame or **synodic period**[13] is $2\tau_{1/2}(E_\mathrm{H}, \xi_\mathrm{p})$.

Determining the stability of a periodic orbit requires only a simple extension of these arguments. Once again consider a particle—not necessarily on a periodic orbit—that crosses the ξ-axis in the direction of increasing η; at this point its coordinates are $\xi \equiv \xi_0$, $\eta = 0$, $\mathrm{d}\xi/\mathrm{d}\tau \equiv \xi'_0$ and $\mathrm{d}\eta/\mathrm{d}\tau > 0$, the last of which is determined by the given value of the Jacobi–Hill constant E_H. When the particle next crosses the ξ-axis in the direction of increasing η, its coordinates are ξ_1, $\eta = 0$, $\mathrm{d}\xi/\mathrm{d}\tau \equiv \xi'_1$ and $\mathrm{d}\eta/\mathrm{d}\tau$, which is again determined by the Jacobi–Hill constant. Thus we can write

$$\xi_1 = X(\xi_0, \xi'_0, E_\mathrm{H}), \quad \xi'_1 = Y(\xi_0, \xi'_0, E_\mathrm{H}). \tag{3.116}$$

[13] For a more general definition of the synodic period, suppose that bodies m_2 and m_0 are in coplanar orbits in the same direction around body m_1 (one may think of m_1 as the Earth, m_2 as the Moon and m_0 as the Sun). Then the synodic period is the time between successive conjunctions of m_2 and m_0 as viewed from m_1 (a **conjunction** occurs when the two bodies have the same azimuth). Quantitatively, if the orbital periods are P_2 and P_0 in an inertial frame, then the synodic period of m_2 is given by $P_\mathrm{syn}^{-1} = |P_2^{-1} - P_0^{-1}|$. The same concept can be applied to spins: the synodic period of the Earth's spin with respect to the Sun is 1 day, but the period of the Earth's spin in an inertial frame (the **sidereal period**) is 0.997 27 days.

A simple-periodic orbit is a fixed point of this transformation, with $\xi_1 = \xi_0 \equiv \xi_\mathrm{p}$ and $\xi_1' = \xi_0' = 0$. Now consider a neighboring orbit with initial conditions $\xi_0 = \xi_\mathrm{p} + \Delta\xi_0$, $\xi_0' = \Delta\xi_0'$, which is transformed to $\xi_1 = \xi_\mathrm{p} + \Delta\xi_1$, $\xi_1' = \Delta\xi_1'$. In the linear approximation

$$\Delta\xi_1 = a\Delta\xi_0 + b\Delta\xi_0', \quad \Delta\xi_1' = c\Delta\xi_0 + d\Delta\xi_0', \tag{3.117}$$

where

$$a = \frac{\partial X}{\partial \xi_0}, \quad b = \frac{\partial X}{\partial \xi_0'}, \quad c = \frac{\partial Y}{\partial \xi_0}, \quad d = \frac{\partial Y}{\partial \xi_0'}, \tag{3.118}$$

with the derivatives evaluated at $\xi_0 = \xi_\mathrm{p}$ and $\xi_0' = 0$. This map can be iterated to determine the linearized phase-space location at successive orbits, $\xi_\mathrm{p} + \Delta\xi_2$, $\Delta\xi_2'$, and so on. The map can be written in matrix notation as

$$\Delta\mathbf{z}_{n+1} = \mathbf{A}\Delta\mathbf{z}_n \quad \text{where} \quad \mathbf{A} = \begin{bmatrix} a & b \\ c & d \end{bmatrix}, \tag{3.119}$$

with $\Delta\mathbf{z}_n = [\Delta\xi_n \; \Delta\xi_n']$ (we do not distinguish whether vectors such as $\mathbf{z}$ are $1 \times N$ column matrices or $N \times 1$ row matrices, as the meaning is clear from the context). The general solution of this equation is a linear combination of sequences of the form

$$\Delta\mathbf{z}_n = k^n\mathbf{a}, \tag{3.120}$$

where $\mathbf{a}$ is an eigenvector of $\mathbf{A}$ and k is a (possibly complex) eigenvalue, given by

$$k = k_\pm \equiv \tfrac{1}{2}(a + d) \pm \tfrac{1}{2}\left[(a - d)^2 + 4bc\right]^{1/2}. \tag{3.121}$$

The orbit is stable if and only if $|k_+| \leq 1$ and $|k_-| \leq 1$.

To calculate the constant a numerically, we simply integrate two orbits with the same value of E_H and ξ_0' and values of ξ_0 differing by some small $\delta\xi_0$. After both orbits return to $\eta = 0$ with $\mathrm{d}\eta/\mathrm{d}\tau > 0$—which typically will occur at slightly different times—we calculate the difference in their ξ coordinates, $\delta\xi_1$, and then $a = \delta\xi_1/\delta\xi_0$. The constants b, c, and d are calculated similarly. Although we do not need these results except as a numerical check, it can be shown that (i) the map defined by equation (3.119) is area-preserving (Binney et al. 1985; Tabor 1989; Lichtenberg &

Lieberman 1992), which in turn requires that $ad - bc = 1$; (ii) for symmetric periodic orbits, $a = d$ (Hénon 1965). The first of these results implies that $k_- = 1/k_+$, so either (i) $(a - d)^2 + 4bc > 0$, which implies that k_+ and k_- are real, one of them exceeds unity in absolute value, and the orbit is unstable; or (ii) $(a - d)^2 + 4bc \leq 0$, in which case k_+ and k_- are complex conjugates, $|k_+| = |k_-| = 1$, and the orbit is stable.

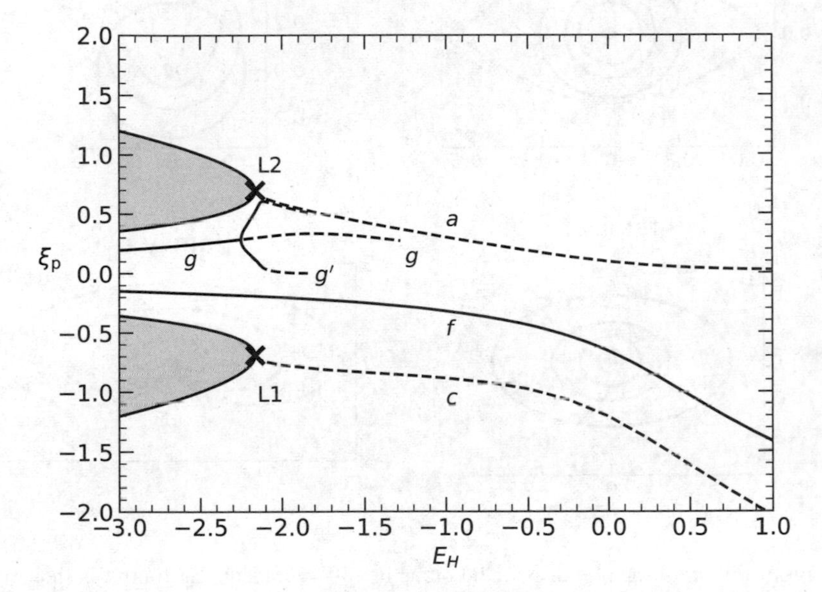

Figure 3.10: Periodic orbits in Hill's problem. Each family of periodic orbits is denoted by a line, which is solid if the orbits are stable and dashed if they are unstable. Each orbit is specified by the value of the Jacobi–Hill constant E_H (eq. 3.110) and the value of $\xi = \xi_\mathrm{p}$ when the orbit crosses the ξ-axis with $\mathrm{d}\eta/\mathrm{d}\tau > 0$. The diagram only shows periodic orbits that (i) lie in the $\zeta = 0$ plane; (ii) are symmetric with respect to the ξ-axis and (iii) cross the ξ-axis only twice. The shaded regions are forbidden, because the argument of the square root in equation (3.115) is negative. The Lagrange points of equation (3.111) are marked as L1 and L2. Prograde orbits have $\xi_\mathrm{p} > 0$ and retrograde orbits have $\xi_\mathrm{p} < 0$. For further detail see Hénon (1969).

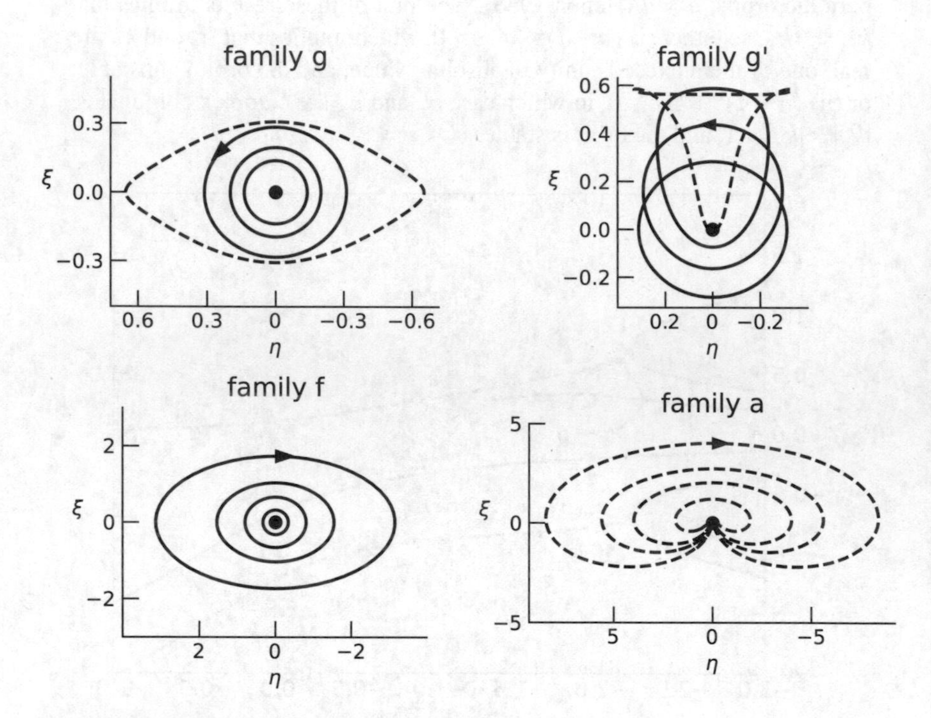

Figure 3.11: Examples of periodic orbits in Hill's problem. Each orbit is shown by a solid line if stable, and by a dashed line if unstable. The origin is marked by a black circle. Additional orbits in family g' can be derived by reflection in the line $\xi = 0$. Orbits in family c can be derived from orbits in family a by the same reflection. Note that the horizontal or η-axis increases from right to left, as in Figures 3.9 and 3.13. The values of the Jacobi–Hill constant $E_{\rm H}$ for each family are: family g, −1.5, −2.25, −3, −4; family g', −2.25, −2.2, −2.135 71, −2; family f, 1.5, 0.5, 0, −1; family a, 4.5, 2, 1, 0. The orbits in families g and g' with $E_{\rm H} = -2.25$ and $E_{\rm H} = -2.135\,71$ respectively are the last stable members of each family.

The main families of periodic orbits in Hill's problem are shown in Figure 3.10, taken from Hénon (1969). Stable orbits are shown by solid lines and unstable orbits by dashed lines.

Family f These are retrograde orbits (here "retrograde" means orbiting in the opposite direction to the orbit of m_2 around m_0). The orbits are symmetric around the ξ-axis by construction but they are also symmetric around the η-axis. Examples are shown in the bottom left panel of Figure 3.11.

As $E_{\mathrm{H}} \to -\infty$ and $\xi_{\mathrm{p}} \to 0$, the perturbing effects of the mass m_0 become negligible and the periodic orbits of family f approach circular orbits described by Kepler's laws. At the other extreme, as E_{H} becomes large, the orbits become quasi-satellite orbits, with dynamics described in §3.2.1. All orbits in family f are stable.

Family g These are prograde orbits, symmetric around both the ξ-axis and the η-axis. Examples are shown in the top left panel of Figure 3.11.

As $E_{\mathrm{H}} \to -\infty$ and $\xi_{\mathrm{p}} \to 0$, the periodic orbits of family g approach circular orbits described by Kepler's laws. As E_{H} grows from $-\infty$, so the perturbations from m_0 are small but not negligible, the shape of the periodic orbits is described by the variational ellipse of equation (3.95). As E_{H} continues to grow, the periodic orbit becomes more and more non-circular. Finally, at $E_{\mathrm{H}} = -2.250$ family g becomes unstable and branches to a stable family g' of orbits that are asymmetric around the η-axis.

The results we have derived can be used to determine numerically the apsidal precession rate of nearly circular orbits induced by the distant companion. To see how to do this, consider a test particle on a nearly circular Kepler orbit. To first order in the eccentricity, the orbit has radius $r = a + \Delta r$ and radial velocity Δv_r where (eqs. 1.29 and 1.54)

$$\Delta r = -ae\cos(\phi - \varpi), \quad \Delta v_r = nae\sin(\phi - \varpi). \tag{3.122}$$

We now transform to the frame rotating with the distant body, at constant angular speed Ω, but for the moment we ignore the gravitational effects of this body. As usual in this section we let $\tau = \Omega t$ be the dimensionless time, and we denote by P the synodic period of the test particle in units of the dimensionless time. Then the time of the jth conjunction of the test particle with the distant body, when their azimuths in the rotating frame are the same, is $\tau_j = jP + \tau_0$. Similarly the azimuth of the conjunction in the inertial frame is $\phi_j = jP + \phi_0$.

If the line of apsides precesses at a mean rate $\dot{\varpi}$ in an inertial frame, then the longitude of periapsis at conjunction j is $\varpi_j = \dot{\varpi} jP/\Omega + \varpi_0$. The radius and radial velocity at conjunction j are then given by

$$\begin{aligned}\Delta r_j &= -ae\cos(\phi_j - \varpi_j) = -ae\cos[jP(1-\dot{\varpi}/\Omega) + \phi_0 - \varpi_0],\\ \Delta v_{rj} &= nae\sin(\phi_j - \varpi_j) = nae\sin[jP(1-\dot{\varpi}/\Omega) + \phi_0 - \varpi_0].\end{aligned} \tag{3.123}$$

These equations can be rewritten as

$$\Delta r_{j+1} = \cos\psi\Delta r_j + \frac{1}{n}\sin\psi\Delta v_{rj}, \quad \Delta v_{r,j+1} = -n\sin\psi\Delta r_j + \cos\psi\Delta v_{rj} \tag{3.124}$$

with $\psi = P(1 - \dot{\varpi}/\Omega)$.

We now include the gravitational forces from the distant body. To do so, we replace the circular orbit by the periodic orbit from family g, and equations (3.124) by equations (3.117). In making this replacement, we can identify $a = d = \cos\psi$, $b = n^{-1}\sin\psi$ and $c = -n\sin\psi$. Thus for any stable orbit, the mean precession rate $\dot{\varpi}$ can be determined from the coefficients a, b, c, d and the synodic period P.

The properties of family g are plotted in Figure 3.12 as a function of the synodic period, in units where the mean motion of the distant body m_0 is $\Omega = 1$. The stable branch of family g terminates when the synodic period $P = 1.2259$, at which point the orbital radius—defined as the average of its maximum and minimum radii—is $\overline{r} = 0.297\,67$, about 43% of the Hill radius $3^{-1/3} = 0.693\,36$. For comparison, the Moon's synodic period is 29.531 d and the sidereal year—the time taken for the Earth to orbit the Sun once with respect to the fixed stars—is 365.256 d, so $P = 2\pi \times 29.531/365.256 = 0.508\,00$.

A comprehensive study of bound orbits in Hill's problem is given by Hénon (1969, 1970, 1974).

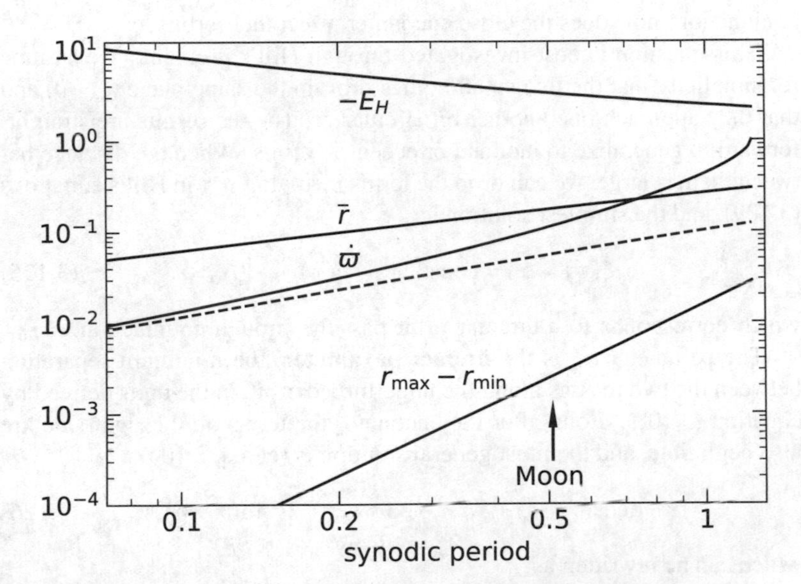

Figure 3.12: Stable orbits in family g in Hill's problem. The solid curves show the negative of the Jacobi–Hill constant, the mean apsidal precession rate $\dot{\varpi}$, $\overline{r} = \frac{1}{2}(r_{\rm max} + r_{\rm min})$ and $r_{\rm max} - r_{\rm min}$, where $r_{\rm max}$ and $r_{\rm min}$ are the maximum and minimum radius of the periodic orbit. All quantities are plotted as functions of the synodic period P, in units where the orbital period of the distant body is 2π. The dashed line shows the estimate of the apsidal precession rate from equation (3.77). The synodic period of the Moon is marked by an arrow at $P = 0.508\,00$. The plots terminate at $P = 1.2259$, $E_{\rm H} = -2.250$, where family g of periodic orbits becomes unstable.

3.4.2 Unbound orbits in Hill's problem

One of the fundamental problems in celestial mechanics is to understand the behavior of two small bodies on nearby circular orbits as they pass through conjunction: how does the close encounter affect their orbits?

This question is best investigated through Hill's problem. We assume for simplicity that the two small bodies orbit in the same plane ($\zeta = 0$) and that they approach one another on circular orbits—the results are straightforward to generalize to inclined or eccentric orbits. When the distance between them is large, we can drop the terms involving ρ^{-3} in Hill's equations (3.109), and the simplest solution is

$$\xi(\tau) = \alpha_0 = \text{constant}, \quad \eta(\tau) = -\tfrac{3}{2}\alpha_0\tau, \tag{3.125}$$

which corresponds to a circular orbit passing through conjunction at $\tau = 0$. The parameter α_0 is the **impact parameter**, the minimum separation between the two masses along the unperturbed orbit, in the units defined by equation (3.108). Long after the encounter the terms on the right side are also negligible, and the most general solution is (cf. eq. 3.104)

$$\xi = \alpha - \epsilon\cos(\tau + \delta), \quad \eta = -\tfrac{3}{2}\alpha\tau + \gamma + 2\epsilon\sin(\tau + \delta), \tag{3.126}$$

which can be rewritten as

$$\xi = \alpha - k_{\rm H}\cos\tau + h_{\rm H}\sin\tau, \quad \eta = -\tfrac{3}{2}\alpha\tau + \gamma + 2k_{\rm H}\sin\tau + 2h_{\rm H}\cos\tau, \tag{3.127}$$

where the Hill eccentricity components are

$$k_{\rm H} = \epsilon\cos\delta, \quad h_{\rm H} = \epsilon\sin\delta. \tag{3.128}$$

We are mostly interested in the constants α, $k_{\rm H}$, and $h_{\rm H}$, which determine the change in semimajor axis and the eccentricity excited during the encounter. These can be written in terms of the phase-space coordinates as

$$\begin{aligned} \alpha &= 4\xi + 2\frac{\mathrm{d}\eta}{\mathrm{d}\tau}, \\ k_{\rm H} &= 2\frac{\mathrm{d}\eta}{\mathrm{d}\tau}\cos\tau + 3\xi\cos\tau + \frac{\mathrm{d}\xi}{\mathrm{d}\tau}\sin\tau, \end{aligned}$$

$$h_{\rm H} = -2\frac{d\eta}{d\tau}\sin\tau - 3\xi\sin\tau + \frac{d\xi}{d\tau}\cos\tau. \tag{3.129}$$

Taking the time derivatives of these equations and simplifying the results using Hill's equations (3.109), we have

$$\begin{aligned}\frac{d\alpha}{d\tau} &= -2\frac{\eta}{\rho^3},\\ \frac{dk_{\rm H}}{d\tau} &= -\frac{1}{\rho^3}(2\eta\cos\tau + \xi\sin\tau),\\ \frac{dh_{\rm H}}{d\tau} &= \frac{1}{\rho^3}(2\eta\sin\tau - \xi\cos\tau).\end{aligned} \tag{3.130}$$

If the impact parameter is large enough, then the distance ρ between the two bodies is also large, so the changes in the orbit are small. In this case we can determine these changes in α, $k_{\rm H}$ and $h_{\rm H}$ by evaluating the right sides of equations (3.130) along the unperturbed circular orbit (3.125). Since $\eta(\tau)$ is an odd function of τ and $\rho(\tau) = |\alpha_0|(1 + \frac{9}{4}\tau^2)^{1/2}$ is even, the change in α integrates to zero at this level of approximation. For similar reasons the change in $k_{\rm H}$ integrates to zero. The value of $h_{\rm H}$ changes from zero to (Julian & Toomre 1966)

$$\begin{aligned}\Delta h_{\rm H} &= -\frac{3\,{\rm sgn}(\alpha_0)}{\alpha_0^2}\int_{-\infty}^{\infty}\frac{d\tau\,\tau\sin\tau}{(1+\frac{9}{4}\tau^2)^{3/2}} - \frac{{\rm sgn}(\alpha_0)}{\alpha_0^2}\int_{-\infty}^{\infty}\frac{d\tau\,\cos\tau}{(1+\frac{9}{4}\tau^2)^{3/2}}\\ &= -\frac{8\cdot 2.5195\,{\rm sgn}(\alpha_0)}{9\alpha_0^2} = -\frac{2.2396\,{\rm sgn}(\alpha_0)}{\alpha_0^2},\end{aligned} \tag{3.131}$$

where the factor $2.5195 = 2K_0(\frac{2}{3}) + K_1(\frac{2}{3})$, and $K_n(\cdot)$ is a Bessel function (Appendix C.5). This linearized approximation to the change in the orbit should be accurate if $|\Delta h_{\rm H}| \ll |\alpha_0|$ or $|\alpha_0| \gtrsim 3$.

We can use these results to determine the change in the constant α more accurately. The Jacobi–Hill constant $E_{\rm H}$ (eq. 3.110) is conserved exactly in Hill's problem. Its value long before the encounter is $E_{\rm H} = -\frac{3}{8}\alpha_0^2$, as can be seen by substituting equations (3.125) into equation (3.110) and letting $\tau \to -\infty$. Likewise, its value long after the encounter is obtained from equation (3.127), $E_{\rm H} = -\frac{3}{8}\alpha^2 + \frac{1}{2}k_{\rm H}^2 + \frac{1}{2}h_{\rm H}^2$. The two expressions for $E_{\rm H}$ must be equal, so we may conclude that

$$\alpha^2 = \alpha_0^2 + \tfrac{4}{3}(\Delta h_{\rm H})^2 = \alpha_0^2 + \frac{2^8 \cdot 2.5195^2}{3^5 \alpha_0^4}. \tag{3.132}$$

Writing $\alpha = \alpha_0 + \Delta\alpha$ and keeping only terms linear in $\Delta\alpha$, we obtain

$$\Delta\alpha = \frac{2^7 \cdot 2.5195^2}{3^5 \alpha_0^5} = \frac{3.3438}{\alpha_0^5}. \tag{3.133}$$

Rescaling to physical units using equation (3.108), the change in impact parameter due to the encounter is

$$\Delta(a_2 - a_1) = \frac{2^7 \cdot 2.5195^2\, \overline{a}^6}{3^5 (a_2 - a_1)^5} \frac{(m_1 + m_2)^2}{m_0^2}. \tag{3.134}$$

This result does not contradict our earlier finding that the change in α is zero: that result was accurate to first order in the maximum strength of the perturbation, which is proportional to α_0^{-2}, whereas equation (3.132) shows that the change in α^2 is proportional to α_0^{-4} and thus is second order in the maximum strength of the perturbation. Using the Jacobi–Hill constant has allowed us to extract a result accurate to second order from a first-order calculation.

Notice that equation (3.132) implies that α^2 is always greater than α_0^2 so the semimajor-axis difference between the two bodies is always larger in absolute value after the conjunction than it was before. In this sense the bodies *repel* one another despite the attractive force between them, an example of the donkey principle described after equation (3.43).

Figure 3.13 shows the trajectories in Hill's problem of particles that approach conjunction on circular orbits. For impact parameters $\alpha_0 \gtrsim 3$, the encounter simply excites the eccentricity, as described by the linear analysis leading to equation (3.131). For $\alpha_0 \lesssim 0.5$ the approaching body reverses course without significant eccentricity excitation, the same behavior described for co-orbital satellites in §3.2. At intermediate impact parameters, the encounter leads to complex behavior that can include close encounters or collisions between the two small bodies. See Hénon & Petit (1986)

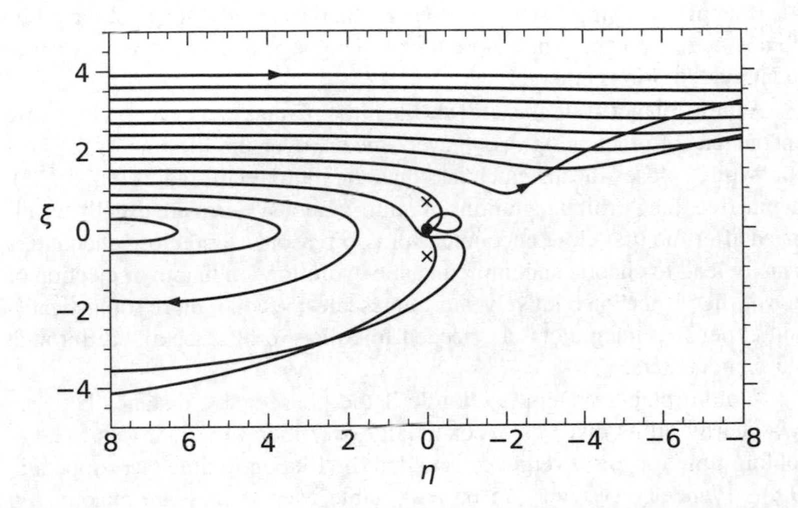

Figure 3.13: Unbound orbits in Hill's problem, as described by the differential equations (3.109) with $\zeta = \mathrm{d}\zeta/\mathrm{d}t = 0$. The bodies approach on circular orbits, coming from the upper left on an initial orbit described by equations (3.125). A symmetric set of orbits coming from the lower right is not shown. The Lagrange points (3.111) are marked by crosses. Note that the horizontal or η-axis increases from right to left, as in Figures 3.9 and 3.11.

and Petit & Hénon (1986) for a thorough description of encounters in Hill's problem.

3.5 Stability of two-planet systems

The configuration of an isolated planetary system is specified by the masses and orbits of its planets and the mass of its host star. An observed planetary system should normally be in a configuration that is stable over times comparable to the age of its host star, except in the unlikely case that it was born in the recent past or will change in the near future.

Roughly speaking, "stability" means that the size and shape of the orbits do not change much. When more precise language is used, several complementary definitions emerge.

A multi-planet system is **Hill stable** if the planets can never have a close encounter. In practical terms, "never" means over the lifetime of the host star, while "close" means much less than the mutual Hill radius (eq. 3.112). In practice, long orbit integrations of multi-planet systems are usually terminated after the first close encounter, for two reasons: first, close encounters usually lead to chaotic and unpredictable evolution, collision, or ejection of one of the planets on relatively short timescales; second, more sophisticated and expensive integrators are needed to follow orbits accurately through close encounters.

A multi-planet system is **chaotic** if the phase-space distance between two nearby orbits diverges exponentially over long time intervals. The e-folding time for this divergence is called the Liapunov time (see Appendix D.8). Planetary systems can be Hill stable even if they are chaotic. In particular, in many cases the exponential divergence is mainly in the mean longitudes of the two orbits and only affects the size and shape of the orbit over intervals much longer than the Liapunov time.

A multi-planet system is **Lagrange stable** if no planet can escape the system or collide with the host star. A two-planet system can be Hill stable but Lagrange unstable or vice versa.

We now describe the stability of two-planet systems using these categories.[14]

Hill stability The simplest example of a multi-planet system consists of two planets on nearly circular and coplanar orbits around the same host star. In this case the configuration is specified by the planet masses m_1 and m_2, the mass of the host star m_0, and the initial semimajor axes a_1 and a_2. The planetary masses are much less than the mass of the star, $m_1, m_2 \ll m_0$, so we expect that orbits that are well separated are likely to be stable. Thus we are mostly interested in the stability of orbits having $|a_1 - a_2| \ll \overline{a}$ where $\overline{a}$

[14] A closely related problem is the stability of a single planet that orbits in a binary-star system, as described briefly in Box 3.2.

Box 3.2: Stability of planetary orbits in binary-star systems

Many planets are found in binary-star systems. Planets that circle one of the two stars are said to have **S-type** orbits, while those that circle both have **P-type** or **circumbinary** orbits.

The possible orbits that such planets can occupy are constrained by the requirement that they be stable over long times. As a simple example, consider a system containing two stars of equal mass on an orbit with semimajor axis a_* and eccentricity e_*, along with a single zero-mass planet. Numerical orbit integrations show that a planet on an initially circular S-type orbit will survive if its semimajor axis satisfies (Holman & Wiegert 1999)

$$\frac{a}{a_*} \lesssim 0.27 - 0.34e_* + 0.05e_*^2. \quad \text{(a)}$$

A planet on an initially circular P-type orbit survives if

$$\frac{a}{a_*} \gtrsim 2.3 + 3.8e_* - 1.7e_*^2. \quad \text{(b)}$$

These results are based on simulations lasting $\sim 10^4$ binary periods—far less than the ages of typical binary stars but still long enough that the stability boundary is evolving only slowly.

is the mean semimajor axis. In this case we can approximate the three-body problem by Hill's problem.

Figure 3.9 shows the contours of the effective potential $\Phi_{\text{eff}}(\xi, \eta)$ in Hill's problem. The effective potential becomes large and negative close to the origin and at large values of $|\xi|$. These regions are separated by saddle points at the Lagrange points L1 and L2, where $\Phi_{\text{eff}} = \Phi_{\text{L}} \equiv -3^{4/3}/2 = -2.163\,37$. Orbits having Jacobi–Hill constant $E_{\text{H}} < \Phi_{\text{L}}$ cannot cross these saddle points. Therefore if the two planets have $E_{\text{H}} < \Phi_{\text{L}}$ and are initially at large separation, they can never have a close encounter. In particular if the two planets are initially far from conjunction, on circular orbits with impact parameter α_0 (eq. 3.125), their Jacobi–Hill constant is $E_{\text{H}} = -\frac{3}{8}\alpha_0^2$ so they can never have a close encounter if $|\alpha_0| > 2.401\,87$. Restoring the dimensional factors using equation (3.108), we conclude that the planets are

Hill stable if

$$|a_1 - a_2| > 2 \cdot 3^{1/6}\,\overline{a}\left(\frac{m_1 + m_2}{m_0}\right)^{1/3} = 2.40\,\overline{a}\left(\frac{m_1 + m_2}{m_0}\right)^{1/3} = 3.46\,r_{\rm H}, \tag{3.135}$$

where $r_{\rm H}$ is the mutual Hill radius defined in equation (3.112). This result is valid so long as $m_1, m_2 \ll m_0$ and the eccentricities and inclinations are small compared to $|a_1 - a_2|/\overline{a}$. This is a sufficient criterion for Hill stability, not a necessary one. For example the tadpole and horseshoe orbits shown in Figure 3.5 do not satisfy (3.135) but never have a close encounter.

Similarly, if the two planets start on eccentric orbits with Hill eccentricity components $(k_{\rm H}, h_{\rm H})$ and impact parameter α_0 (eq. 3.127), the Jacobi–Hill constant for the relative orbit is $E_{\rm H} = -\frac{3}{8}\alpha_0^2 + \frac{1}{2}k_{\rm H}^2 + \frac{1}{2}h_{\rm H}^2$. Restoring the dimensional factors, we find that the planets are Hill stable if

$$|a_1 - a_2| > 2 \cdot 3^{1/6}\,\overline{a}\left[\left(\frac{m_0 + m_1}{m_0}\right)^{2/3} + \frac{(\mathbf{e}_1 - \mathbf{e}_2)^2}{3^{4/3}}\right]^{1/2}, \tag{3.136}$$

where $\mathbf{e}_1$ and $\mathbf{e}_2$ are the eccentricity vectors of the two planets. If the planets are on circular orbits with mutual inclination I, a similar calculation yields

$$|a_1 - a_2| > 2 \cdot 3^{1/6}\,\overline{a}\left[\left(\frac{m_0 + m_1}{m_0}\right)^{2/3} + \frac{I^2}{3^{4/3}}\right]^{1/2}. \tag{3.137}$$

Similar arguments can be used to determine Hill stability in the case where the mass m_1 of planet 1 is comparable to the stellar mass m_0, so long as the mass of the second planet $m_2 = 0$ and m_1 and m_0 are on a circular orbit. This is the circular restricted three-body problem studied in §3.1, and the system is Hill stable if its Jacobi constant (3.12) is smaller than the value of the effective potential (3.13) at the Lagrange point L1 (if m_2 is initially interior to m_1), or L2 (if m_2 is exterior to m_1).

The most general criterion for Hill stability is based on the conservation of the total energy and angular momentum of the three-body system, and can be applied for arbitrary masses and initial conditions. The derivation of this criterion is described in Appendix G.

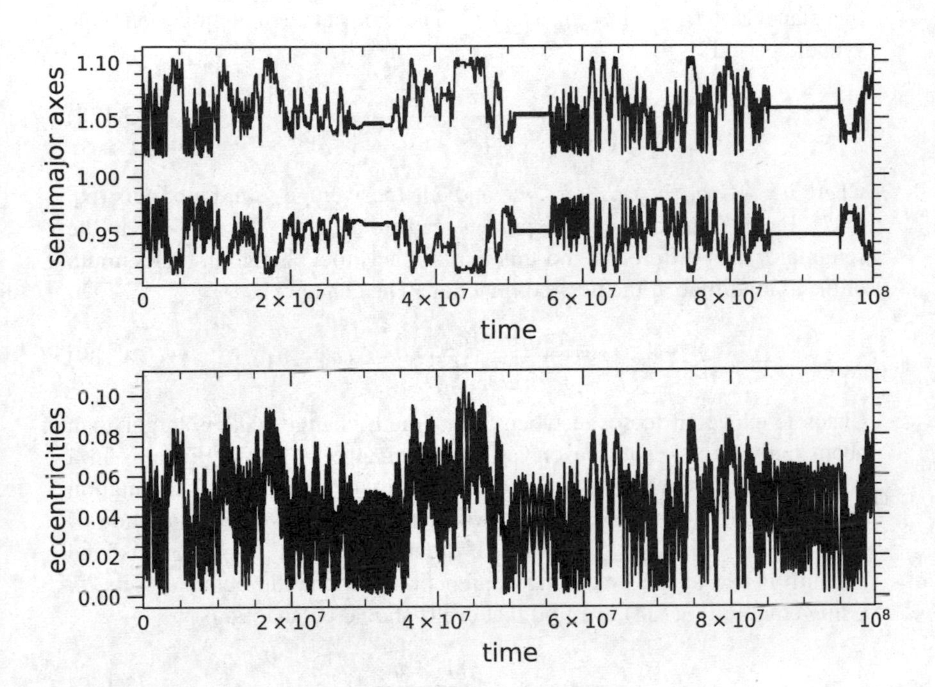

Figure 3.14: The chaotic evolution of a two-planet system. The host star has unit mass and the planet masses are $m_1 = m_2 = 10^{-6}$. The planets initially are on circular orbits with semimajor axes $1 \pm \frac{1}{2}\Delta$, $\Delta = 0.033$. We use units in which $\mathbb{G} = 1$ so the orbital period is about 2π. The top panel shows the semimajor axes of the two planets and the bottom panel shows the eccentricities. In the bottom panel the eccentricities are difficult to separate because they closely track each other, for reasons described in the text. The system is Hill stable according to equation (3.135), but chaotic according to (3.140).

Chaos We begin with a heuristic derivation of the size of the chaotic zone created by the interaction of two nearby planets. The mean motions of the two planets are $n_{1,2} = (\mathbb{G}m_0/a_{1,2}^3)^{1/2}$. The time between conjunctions (the synodic period) is

$$P_{\text{syn}} = \frac{2\pi}{|n_1 - n_2|} = \frac{2\pi}{(\mathbb{G}m_0)^{1/2}|a_1^{-3/2} - a_2^{-3/2}|} \simeq \frac{4\pi\overline{a}}{3n|a_2 - a_1|}, \tag{3.138}$$

where $n = (\mathbb{G}m_0/\overline{a}^3)^{1/2} \simeq n_1, n_2$, and $\overline{a}$ is the average semimajor axis (eq. 3.47). Each conjunction of the planets leads to a change $\Delta(a_2 - a_1)$ in the semimajor axis difference and this in turn leads to a change in the azimuth or mean longitude at the next conjunction. The change is

$$\Delta\phi = n\Delta P_{\text{syn}} \simeq -\frac{4\pi\overline{a}\,\text{sgn}(a_2 - a_1)}{3(a_2 - a_1)^2}\Delta(a_2 - a_1). \tag{3.139}$$

Chaos is expected to set in when the azimuth change $|\Delta\phi|$ is larger than about one radian. The reason is that the change in semimajor axis in a conjunction of two planets on eccentric orbits depends on the orientation of the eccentricity vectors of the two planets relative to the longitude of conjunction, so if $|\Delta\phi| \gtrsim c$, with c of order unity, the system "forgets" the azimuth it had at the preceding conjunction and chaotic behavior sets in. Using equation (3.134), we find that orbits should be chaotic if

$$\frac{|a_2 - a_1|}{\overline{a}} \lesssim f\left(\frac{m_1 + m_2}{m_0}\right)^{2/7}, \tag{3.140}$$

where $f = 1.46/c^{1/7}$. This estimate is consistent with an approximate estimate of the size of the chaotic zone using the resonance-overlap criterion, as described in §6.2.1, and with numerical orbit integrations, which yield $f \simeq 1.2$–1.5 (Wisdom 1980; Gladman 1993; Deck et al. 2013; Morrison & Malhotra 2015). This estimate excludes tadpole and horseshoe orbits. The boundary of the chaotic zone is not precisely defined, because there is a transition range of semimajor axes in which some orbits are chaotic and some are regular.

Notice that the exponent of equation (3.140) is slightly smaller than the exponent of (3.135)—$\frac{2}{7} = 0.2857$ versus $\frac{1}{3} = 0.3333$. Thus for sufficiently small planetary masses, there is a range of semimajor axes at which the orbits are Hill stable but chaotic. Despite the chaotic nature of the orbits, the planets can never suffer a close encounter.

Orbits in the chaotic region exhibit significant and irregular excursions in semimajor axis and eccentricity. An example is shown in Figure 3.14, which follows the evolution over $10^8/(2\pi)$ orbital periods of two planets with masses $m_1 = m_2 = 10^{-6}m_0$ on initially circular orbits in the chaotic zone. The figure exhibits several striking features. Despite the long integration time and the chaotic nature of the evolution there are no close encounters, and both the semimajor axes and the eccentricities remain bounded within narrow ranges. There are long quiescent intervals, lasting for up to $10^7/(2\pi)$ orbital periods, in which the semimajor axes remain nearly constant and there is no obvious sign of chaos. The top panel shows that the variations in the semimajor axes are out of phase, such that the average semimajor axis is nearly conserved. This is simply a consequence of energy conservation: in a two-planet system the total Kepler energy is $-\frac{1}{2}\mathbb{G}M_\star(m_1/a_1 + m_2/a_2)$, so if the two semimajor axes are nearly the same they must vary out of phase. In contrast, the bottom panel shows that the variations in the eccentricity are in phase, such that $e_1 \simeq e_2$. The reasons for this are explored in Problem 3.10.

Lagrange stability A two-planet system is Lagrange unstable if one planet either escapes or collides with the host star. To obtain some analytic insight, we examine a system containing two planets: planet 1 has mass $m_1 \ll m_0$ and follows a circular orbit with semimajor axis a_1, while planet 2 has zero mass and initially follows a nearly circular, coplanar orbit with semimajor axis $a_2 \simeq a_1$. The initial value of the Tisserand parameter of planet 2 (see Box 3.1) is $T \simeq 3$, and the Tisserand parameter is conserved, except for short intervals when the two planets are in close proximity. If planet 2 escapes, it is likely to do so with near-zero energy; in this case $a_2^{-1} \simeq 0$, $e_2 \simeq 1$ and $a_2(1 - e_2^2) \simeq 2q_2$ where q_2 is the periapsis distance, so $T = 2(2q_2/a_1)^{1/2}$. Equating the two values of T, we conclude that $q_2 = \frac{9}{8}a_1$; thus planet 2 escapes when its periapsis is only 12% larger than the semimajor axis of planet 1, so the two can still interact strongly. On the other hand, if the test particle collides with the host star, then $e_2 = 1$ and so $T = a_2/a_1$, and we may conclude that at collision the semimajor axis is $a_2 = \frac{1}{3}a_1$ and the apoapsis distance is $Q_2 = a_2(1 + e_2) = \frac{2}{3}a_1$, too far from planet 1 for strong gravitational interactions.

These considerations suggest that Lagrange instability usually leads to escape rather than collision with the host star, and this is confirmed by numerical orbit integrations (Deck et al. 2013; Morrison & Malhotra 2015). The simulations also show that (i) for typical planetary masses, the Lagrange stability boundary lies close to the boundaries for Hill stability and for chaos; and (ii) for systems containing two low-mass planets, the timescale for Lagrange instability is very long—roughly $0.02(m_2/m_0)^{-1.5}$ orbital periods according to Morrison & Malhotra (2015)—so orbits like the ones shown in Figure 3.14 may still escape in the future.

3.6 Disk-driven migration

Gaseous protoplanetary disks disperse a few Myr after their host star forms (e.g., Williams & Cieza 2011). Giant planets such as Jupiter and Saturn must have formed before the disk dispersed, since it is the only plausible source for the gas that comprises their massive atmospheres. Therefore any theory of planet formation and evolution must account for the dynamical interactions between a massive planet and a surrounding gaseous disk. Our understanding of this behavior is still incomplete, and this understanding requires analytic and numerical tools from fluid mechanics that are outside the scope of this book (see Baruteau et al. 2014 for a review), but some of the most important features of disk-planet interactions can be described with the tools we have already developed.

We examine the behavior of a small body of mass m on a circular orbit of semimajor axis a, passing through conjunction with a planet of mass M_p on a circular orbit with semimajor axis a_p. Both objects orbit a host star of mass M_*, with $m \ll M_\mathrm{p} \ll M_*$ and $|a - a_\mathrm{p}| \ll a_\mathrm{p}$. The orbital period of the planet is $P_\mathrm{p} = 2\pi/n_\mathrm{p}$ with $n_\mathrm{p} = (\mathbb{G}M_*/a_\mathrm{p}^3)^{1/2}$, and the orbital period of the small body is $P = 2\pi/n$ with $n = (\mathbb{G}M_*/a^3)^{1/2}$. The synodic period is given by equation (3.138), which in the present notation reads

$$P_{\rm syn} = \frac{4\pi a_{\rm p}}{3n_{\rm p}|a - a_{\rm p}|}. \tag{3.141}$$

We now take the small body m to be an element of gas in the protoplanetary disk. We assume that the interaction with the planet at conjunction excites the eccentricity of the gas element in the same way that it excites the eccentricity of a test particle. Thus, following conjunction the eccentricity of the gas element should be given by equation (3.131). Over the synodic period $P_{\rm syn}$, which is much longer than the time needed to pass through conjunction, viscous or other collective interactions within the gas disk are likely to damp the eccentricity, so the gas element will approach the next conjunction having returned to a circular orbit. Nevertheless, the orbits are not the same at the two conjunctions: the gas element also suffers a change in its semimajor axis according to equation (3.134), which is rewritten in the present context as

$$\Delta a = \frac{2^7 \cdot 2.5195^2\, \mathbb{G}^2 M_{\rm p}^2}{3^5 n_{\rm p}^4 (a - a_{\rm p})^5}. \tag{3.142}$$

The change in semimajor axis corresponds to a change in angular momentum, and this change accumulates as it is repeated at subsequent conjunctions. The long-term rate of change of the semimajor axis of the gas element is therefore given by

$$\frac{1}{a_{\rm p}}\frac{{\rm d}a}{{\rm d}t} = \frac{1}{a_{\rm p}}\frac{\Delta a}{P_{\rm syn}} = {\rm sgn}(a - a_{\rm p})\frac{2^5 \cdot 2.5195^2\, \mathbb{G}^2 M_{\rm p}^2}{3^4 \pi n_{\rm p}^3 a_{\rm p}^2 (a - a_{\rm p})^4}. \tag{3.143}$$

For a solar-mass host star,

$$\frac{1}{a_{\rm p}}\frac{{\rm d}a}{{\rm d}t} = \frac{{\rm sgn}(a - a_{\rm p})}{244\,{\rm yr}}\left(\frac{M_{\rm p}}{M_{\rm J}}\right)^2\left(\frac{0.1a_{\rm p}}{a - a_{\rm p}}\right)^4\left(\frac{5\,{\rm au}}{a_{\rm p}}\right)^{3/2} \tag{3.144}$$

where $M_{\rm J}$ is the mass of Jupiter. The characteristic timescale is only $\sim 10^{-4}$ of the lifetime of the gas disk, confirming that disk-planet gravitational interactions must play a central role in sculpting planetary systems.

Equation (3.143) shows that the planet repels the disk material, and this repulsion has two important consequences. First, if the interactions with the planet are stronger than viscous stresses or other mechanisms such as density waves that transport angular momentum within the disk, the planet will carve out an annular gap in the disk. Such gaps are seen in Saturn's rings (e.g., the Encke gap, caused by the satellite Pan). Gaps in the dust distribution are also common features in images of protoplanetary disks (Andrews et al. 2018), although at the present time there is little direct evidence that these gaps are caused by planets.

Second, just as the planet repels the disk, the disk repels the planet, causing the planetary orbit to spiral inward or outward, a process known as **disk-driven migration** of planets. The migration rate of the planet due to a ring of material of mass m_r at semimajor axis a is $\mathrm{d}a_\mathrm{p}/\mathrm{d}t = -(m_\mathrm{r}/M_\mathrm{p})\mathrm{d}a/\mathrm{d}t$, where $\mathrm{d}a/\mathrm{d}t$ is given by equation (3.143). If the planet is embedded in the disk, the repulsion from gas inside the planet is largely canceled by the repulsion from gas outside the planet. The cancellation is not exact, but whether the inner or outer disk wins cannot be determined from the analysis leading to equation (3.143), in part because the contribution from a ring of material diverges strongly as $|a - a_\mathrm{p}| \to 0$. In a real disk the divergence is suppressed when $|a - a_\mathrm{p}| < \max(r_\mathrm{H}, h)$, where r_H is the Hill radius (eq. 3.24) and h is the thickness of the disk.

Careful analyses of realistic protoplanetary disk models show that, in general, the repulsion from the outer disk is stronger and the planet migrates inward. The migration rate depends strongly on whether the planet is massive enough to open a gap in the disk, and the process is called **Type I migration** if there is no gap, and otherwise **Type II**. In either case the migration timescale is typically much shorter than the lifetime of the protoplanetary disk for planets of Earth mass or larger.

Perhaps the biggest surprise in the early history of exoplanet discoveries was that giant planets were found at semimajor axes much smaller than those of the giant planets in the solar system. The distribution of orbital periods of exoplanets discovered by radial-velocity surveys (§1.6.1) is shown in Figure 3.15. Not shown are planets discovered by transit surveys, to avoid the strong bias of transit surveys toward planets with small semimajor axes and orbital periods (see discussion at the end of §1.6.2); radial-velocity sur-

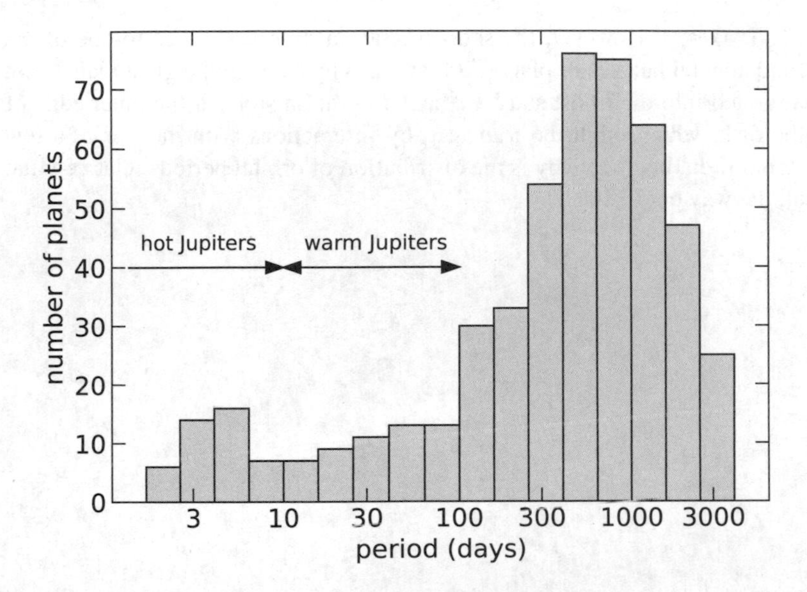

Figure 3.15: Orbital periods of 529 planets with $M_\mathrm{p} \sin I$ between 0.1 and 10 Jupiter masses. Only planets discovered by radial-velocity surveys are plotted. Data are from the NASA Exoplanet Archive, https://exoplanetarchive.ipac.caltech.edu/index.html.

veys are also biased toward planets with small semimajor axes, but the bias is weaker. Only planets with $M_\mathrm{p} \sin I$ between 0.1 and 10 Jupiter masses are plotted in the figure, which should mostly be giants with massive atmospheres. The plot shows that most known giant planets have orbital periods between about 300 and 3 000 days, but there is a significant tail at smaller periods containing about one-third of the planets. The most extreme examples are the **hot Jupiters**, typically defined as giant planets with orbital periods < 10 d. Similarly, **warm Jupiters** are defined to have orbital periods between 10 d and 100 d.

It is difficult to form hot Jupiters in situ and disk-driven migration offers a mechanism to explain why giant planets are found at such small semimajor axes.[15] However, the short migration timescales lead to one of the fundamental puzzles of planet formation: why have all the giant planets not migrated into their host star? Perhaps migration stops at the inner edge of the disk, which might be truncated by interactions with the star at a few stellar radii, but then why is the distribution of orbital periods relatively flat all the way out to 100 d?

[15] An alternative mechanism is high-eccentricity migration, described in §5.4.2.

Chapter 4

The N-body problem

Determining the trajectories of $N-1$ planets orbiting a common host star is an example of the **gravitational N-body problem**. The gravitational N-body problem also describes the behavior of many other astrophysical systems: multiple-star systems, stars in star clusters and galaxies, galaxies in clusters of galaxies, elementary particles in dark-matter halos, and so forth. Celestial mechanics is distinguished by its focus on systems in which (i) there is a dominant central mass; (ii) the eccentricities and inclinations are small; and (iii) a large number of orbital periods elapse in the lifetime of the system, typically 10^9–10^{11}.

4.1 Reference frames and coordinate systems

A host star with mass m_0 and position $\mathbf{r}_0$ is orbited by $N-1$ planets with masses m_j and positions $\mathbf{r}_j$, $j = 1, \ldots, N-1$. We assume that these coordinates are defined in an **inertial reference frame**: a frame in which an isolated body does not accelerate, or, equivalently, there are no fictitious forces.

Fictitious forces can arise from rotation of the frame around its origin or from linear acceleration of the origin of the frame. The rotation of any solar-system reference frame relative to an inertial frame can be determined dynamically—by finding the Coriolis and centrifugal forces required for a dynamical model to fit the observed trajectories of the Moon and planets—or cosmologically, by assuming that the positions of distant extragalactic objects such as quasars are fixed in an inertial frame (§1.5.3). Linear acceleration of the frame does not lead to fictitious forces so long as the frame is freely falling.

We can almost always assume that planetary systems are isolated, at least if we are investigating the trajectories of objects within a few tens of astronomical units of the host star. Tidal forces on planetary systems arise from the overall Galactic mass distribution and individual nearby stars, but these are far too small to be detectable—for example, the mean density of the solar system inside Neptune's orbit is more than $\sim 10^{12}$ times the mean density of the Galaxy in the solar neighborhood (see Problem 9.6). Tidal forces of this kind are only relevant for the dynamics of comets, as described in §9.4.

The equations of motion of an isolated N-body system in an inertial frame are

$$\ddot{\mathbf{r}}_j = \sum_{k=0,k\neq j}^{N-1} \frac{\mathbb{G}m_k(\mathbf{r}_k - \mathbf{r}_j)}{|\mathbf{r}_j - \mathbf{r}_k|^3}, \quad j = 0, \ldots, N-1. \tag{4.1}$$

In terms of a gravitational potential,

$$\ddot{\mathbf{r}}_j = -\frac{\partial \Phi_j(\mathbf{r}_j, t)}{\partial \mathbf{r}_j}, \quad \text{where} \quad \Phi_j(\mathbf{r}, t) = -\sum_{k=0,k\neq j}^{N-1} \frac{\mathbb{G}m_k}{|\mathbf{r} - \mathbf{r}_k(t)|}. \tag{4.2}$$

These equations of motion result from the Hamiltonian

$$H(\mathbf{r}_0, \ldots, \mathbf{r}_{N-1}, \mathbf{p}_0, \ldots, \mathbf{p}_{N-1}) \equiv \sum_{j=0}^{N-1} \frac{\mathbf{p}_j^2}{2m_j} - \sum_{j=0}^{N-1} \sum_{k=j+1}^{N-1} \frac{\mathbb{G}m_j m_k}{|\mathbf{r}_j - \mathbf{r}_k|}, \tag{4.3}$$

where $\mathbf{p}_j = m_j \dot{\mathbf{r}}_j$ is the momentum conjugate to $\mathbf{r}_j$.

4.1.1 Barycentric coordinates

The most widely used inertial frame is the **center-of-mass** or **barycentric** frame, in which the center of mass of the system is at rest at the origin

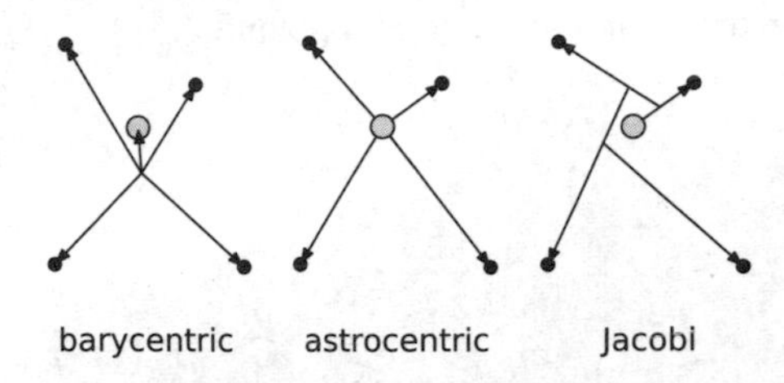

Figure 4.1: Schematic diagrams of the three coordinate systems defined in §4.1. The shaded circle represents the host star.

(Figure 4.1). This condition can be written as

$$\sum_{j=0}^{N-1} m_j \mathbf{r}_j = \mathbf{0}, \quad \sum_{j=0}^{N-1} \mathbf{p}_j = \sum_{j=0}^{N-1} m_j \dot{\mathbf{r}}_j = \mathbf{0}. \tag{4.4}$$

It is straightforward to show from (4.1) that if the conditions (4.4) hold at any time, then they hold at all times.

Despite their simplicity, barycentric coordinates are not ideal for many problems in celestial mechanics. One drawback is inefficiency: there are $3N$ second-order differential equations (4.1), or $6N$ first-order equations if we write the equations for $\dot{\mathbf{r}}_j$ and $\dot{\mathbf{p}}_j$ separately, but because of the conserved quantities (4.4), six of the $6N$ equations are redundant so the work done to solve them numerically is wasted. A second drawback is that for many purposes, it is more convenient to have the host star located at the origin.

4.1.2 Astrocentric coordinates

The non-inertial, non-rotating frame centered on the host star is called the **astrocentric frame**, or the **heliocentric frame** for the solar system (Figure 4.1). Let $\mathbf{r}_j^\star = \mathbf{r}_j - \mathbf{r}_0$ be the position of planet j relative to the host star. Then from equations (4.1) it is straightforward to show that

$$\ddot{\mathbf{r}}_j^\star = -\frac{\mathbb{G}(m_0 + m_j)}{|\mathbf{r}_j^\star|^3}\mathbf{r}_j^\star + \sum_{k=1,k\neq j}^{N-1} \mathbb{G}m_k \left[\frac{(\mathbf{r}_k^\star - \mathbf{r}_j^\star)}{|\mathbf{r}_k^\star - \mathbf{r}_j^\star|^3} - \frac{\mathbf{r}_k^\star}{|\mathbf{r}_k^\star|^3}\right], \; j = 1, \ldots, N-1. \tag{4.5}$$

We can rewrite this equation in terms of a gravitational potential:

$$\ddot{\mathbf{r}}_j^\star = -\frac{\partial \Phi_j(\mathbf{r}_j^\star, t)}{\partial \mathbf{r}_j^\star}, \tag{4.6}$$

where

$$\Phi_j(\mathbf{r}, t) = -\frac{\mathbb{G}(m_0 + m_j)}{|\mathbf{r}|} - \sum_{k=1, k\neq j}^{N-1} \frac{\mathbb{G}m_k}{|\mathbf{r} - \mathbf{r}_k^\star(t)|} + \sum_{k=1, k\neq j}^{N-1} \frac{\mathbb{G}m_k \mathbf{r}_k^\star(t) \cdot \mathbf{r}}{|\mathbf{r}_k^\star(t)|^3}.$$

The last term is called the **indirect potential**.

To derive a Hamiltonian for the N-body system in astrocentric coordinates, we must (i) define an additional coordinate $\mathbf{r}_0^\star$ to provide a complete set of N coordinate 3-vectors $\mathbf{r}_j^\star$; and (ii) find N momentum 3-vectors $\mathbf{p}_j^\star$ conjugate to these coordinates. To do so we use the generating function

$$\begin{aligned} S_2(\mathbf{p}_0^\star, \ldots, \mathbf{p}_{N-1}^\star, \mathbf{r}_0, \ldots, \mathbf{r}_{N-1}) \\ = \mathbf{p}_0^\star \cdot \Big[\mathbf{r}_0 + \beta \sum_{k=1}^{N-1} m_k(\mathbf{r}_k - \mathbf{r}_0)\Big] + \sum_{k=1}^{N-1} \mathbf{p}_k^\star \cdot (\mathbf{r}_k - \mathbf{r}_0), \end{aligned} \tag{4.7}$$

where β is a constant parameter. From equations (D.63),

$$\mathbf{r}_0^\star = \frac{\partial S_2}{\partial \mathbf{p}_0^\star} = \mathbf{r}_0 + \beta \sum_{k=1}^{N-1} m_k(\mathbf{r}_k - \mathbf{r}_0), \tag{4.8a}$$

$$\mathbf{p}_0 = \frac{\partial S_2}{\partial \mathbf{r}_0} = \mathbf{p}_0^\star - \sum_{k=1}^{N-1} (\beta m_k \mathbf{p}_0^\star + \mathbf{p}_k^\star), \tag{4.8b}$$

$$\mathbf{r}_j^\star = \frac{\partial S_2}{\partial \mathbf{p}_j^\star} = \mathbf{r}_j - \mathbf{r}_0, \tag{4.8c}$$

$$\mathbf{p}_j = \frac{\partial S_2}{\partial \mathbf{r}_j} = \beta m_j \mathbf{p}_0^\star + \mathbf{p}_j^\star, \tag{4.8d}$$

for $j = 1, \ldots, N-1$. The inverse transformations are

$$\mathbf{p}_0^\star = \sum_{k=0}^{N-1} \mathbf{p}_k, \tag{4.9a}$$

$$\mathbf{r}_0 = \mathbf{r}_0^\star - \beta \sum_{k=1}^{N-1} m_k \mathbf{r}_k^\star, \tag{4.9b}$$

$$\mathbf{p}_j^\star = \mathbf{p}_j - \beta m_j \sum_{k=0}^{N-1} \mathbf{p}_k, \tag{4.9c}$$

$$\mathbf{r}_j = \mathbf{r}_j^\star + \mathbf{r}_0^\star - \beta \sum_{k=1}^{N-1} m_k \mathbf{r}_k^\star. \tag{4.9d}$$

Equation (4.9a) is proved by substituting the right side of (4.8d) into (4.8b); it shows that $\mathbf{p}_0^\star$ is the total momentum. Then (4.9c) is proved by substituting (4.9a) into (4.8d). A similar process is used to derive (4.9b) and (4.9d).

The Hamiltonian is derived by replacing $\mathbf{r}_j$ and $\mathbf{p}_j$ in the Hamiltonian (4.3) by $\mathbf{r}_j^\star$ and $\mathbf{p}_j^\star$:

$$\begin{aligned} &H(\mathbf{r}_0^\star, \dots, \mathbf{r}_{N-1}^\star, \mathbf{p}_0^\star, \dots, \mathbf{p}_{N-1}^\star) \\ &= \tfrac{1}{2}\mathbf{p}_0^{\star 2}\left[\frac{1 - 2\beta(M - m_0)}{m_0} + \beta^2 \frac{M}{m_0}(M - m_0)\right] + \sum_{j=1}^{N-1}\left(\frac{\mathbf{p}_j^{\star 2}}{2m_j} - \frac{\mathbb{G} m_0 m_j}{|\mathbf{r}_j^\star|}\right) \\ &\quad - \sum_{j=1}^{N-1}\sum_{k=j+1}^{N-1} \frac{\mathbb{G} m_j m_k}{|\mathbf{r}_j^\star - \mathbf{r}_k^\star|} + \frac{\beta M - 1}{m_0}\mathbf{p}_0^\star \cdot \sum_{j=1}^{N-1} \mathbf{p}_j^\star + \frac{1}{2m_0}\sum_{j=1}^{N-1}\sum_{k=1}^{N-1} \mathbf{p}_j^\star \cdot \mathbf{p}_k^\star, \end{aligned} \tag{4.10}$$

where $M \equiv \sum_{k=0}^{N-1} m_k$ is the total mass of the system. We can extract the terms from the final sum in which $k = j$ and add them to the terms proportional to $\mathbf{p}_j^{\star 2}$. Introducing $\mu_j \equiv m_0 m_j/(m_0 + m_j)$ for the reduced mass of the two-body system consisting of the host star and planet j (cf. eq. 1.10), we have

$$\begin{aligned} &H(\mathbf{r}_0^\star, \dots, \mathbf{r}_{N-1}^\star, \mathbf{p}_0^\star, \dots, \mathbf{p}_{N-1}^\star) \\ &= \tfrac{1}{2}\mathbf{p}_0^{\star 2}\left[\frac{1 - 2\beta(M - m_0)}{m_0} + \beta^2 \frac{M}{m_0}(M - m_0)\right] \\ &\quad + \sum_{j=1}^{N-1}\left[\frac{\mathbf{p}_j^{\star 2}}{2\mu_j} - \frac{\mathbb{G}\mu_j(m_0 + m_j)}{|\mathbf{r}_j^\star|}\right] \\ &\quad - \sum_{j=1}^{N-1}\sum_{k=j+1}^{N-1} \frac{\mathbb{G} m_j m_k}{|\mathbf{r}_j^\star - \mathbf{r}_k^\star|} + \frac{\beta M - 1}{m_0}\mathbf{p}_0^\star \cdot \sum_{j=1}^{N-1} \mathbf{p}_j^\star + \frac{1}{m_0}\sum_{j=1}^{N-1}\sum_{k=j+1}^{N-1} \mathbf{p}_j^\star \cdot \mathbf{p}_k^\star. \end{aligned} \tag{4.11}$$

There are two natural choices for the parameter β. In **convention A** or **Poincaré coordinates**, $\beta = 0$ (Poincaré 1896). Then $\mathbf{r}_0^\star = \mathbf{r}_0$ is the position of the host star in the barycentric frame and equation (4.11) becomes

$$H_A = \frac{\mathbf{p}_0^{\star 2}}{2m_0} + \sum_{j=1}^{N-1}\left[\frac{\mathbf{p}_j^{\star 2}}{2\mu_j} - \frac{\mathbb{G}\mu_j(m_0+m_j)}{|\mathbf{r}_j^\star|}\right] \tag{4.12}$$
$$- \sum_{j=1}^{N-1}\sum_{k=j+1}^{N-1}\frac{\mathbb{G}m_j m_k}{|\mathbf{r}_j^\star - \mathbf{r}_k^\star|} - \frac{\mathbf{p}_0^\star}{m_0}\cdot\sum_{j=1}^{N-1}\mathbf{p}_j^\star + \frac{1}{m_0}\sum_{j=1}^{N-1}\sum_{k=j+1}^{N-1}\mathbf{p}_j^\star\cdot\mathbf{p}_k^\star.$$

If the original inertial coordinates are barycentric then $\mathbf{p}_0^\star = \mathbf{0}$; however, we cannot remove the terms proportional to $\mathbf{p}_0^\star\cdot\mathbf{p}_j^\star$ from the Hamiltonian, since otherwise Hamilton's equations would erroneously imply that $\dot{\mathbf{r}}_0^\star = \mathbf{0}$. The first line of equation (4.12) is simply the kinetic energy of the host star plus the sum of the two-body Hamiltonians for all the planets. The second line is the perturbing Hamiltonian. The first term is the sum of the gravitational potential energies of the pairwise interactions between the planets. The cross term involving $\mathbf{p}_j^\star\cdot\mathbf{p}_k^\star$ is an inconvenient but inevitable complication of the change to a non-inertial frame.

An interesting feature of convention A, easy to see from equations (4.8) with $\beta = 0$, is that the coordinates of the planets $\mathbf{r}_j^\star$ are astrocentric but the momenta $\mathbf{p}_j^\star$ are barycentric.

In some cases we may want to express the Kepler terms in the Hamiltonian (4.12) in terms of Delaunay variables, as described in Box 4.1.

In **convention B** or **democratic coordinates**, $\beta = M^{-1}$ (Duncan et al. 1998). Then $\mathbf{r}_0^\star$ is the position of the center of mass of the system, rather than the position of the host star. The Hamiltonian from equation (4.10) is

$$H_B = \frac{\mathbf{p}_0^{\star 2}}{2M} + \sum_{j=1}^{N-1}\left(\frac{\mathbf{p}_j^{\star 2}}{2m_j} - \frac{\mathbb{G}m_0 m_j}{|\mathbf{r}_j^\star|}\right) - \sum_{j=1}^{N-1}\sum_{k=j+1}^{N-1}\frac{\mathbb{G}m_j m_k}{|\mathbf{r}_j^\star - \mathbf{r}_k^\star|} + \frac{1}{2m_0}\left|\sum_{j=1}^{N-1}\mathbf{p}_j^\star\right|^2, \tag{4.13}$$

while equation (4.11) yields

$$H_B = \frac{\mathbf{p}_0^{\star 2}}{2M} + \sum_{j=1}^{N-1}\left[\frac{\mathbf{p}_j^{\star 2}}{2\mu_j} - \frac{\mathbb{G}\mu_j(m_0+m_j)}{|\mathbf{r}_j^\star|}\right]$$

Box 4.1: Momenta and actions in N-body systems

In descriptions of the two-body problem or of the motion of a test particle in a given potential, it is usually simplest to define the momentum to have dimensions of (length)/(time) (see footnote 10 in §1.4). Similarly, angular momentum and actions such as the Delaunay variables Λ, L, and L_z (eq. 1.84) have dimensions of (length)^2/(time). In systems of N bodies with different masses, it is better to follow the usual convention in physics, in which momentum has dimensions of (mass)×(length)/(time) and actions have dimensions of (mass)×(length)^2/(time).

Thus, suppose that a Kepler Hamiltonian has the form (cf. eq. 1.80)

$$H = \frac{\mathbf{p}^2}{2m_a} - \frac{\mathbb{G}m_a m_b}{|\mathbf{q}|}. \tag{a}$$

The corresponding Delaunay variables are

$$\Lambda = m_a(\mathbb{G}m_b a)^{1/2}, \quad L = \Lambda(1-e^2)^{1/2}, \quad L_z = L\cos I, \tag{b}$$

where a, e and I are the semimajor axis, eccentricity and inclination. The Hamiltonian can be written

$$H = -\frac{\mathbb{G}m_a m_b}{2a} = -\frac{\mathbb{G}^2 m_a^3 m_b^2}{2\Lambda^2}. \tag{c}$$

For example, in astrocentric coordinates the Kepler Hamiltonian for planet j is contained in the square brackets in equation (4.12) or (4.14). Then $m_a = \mu_j$, $m_b = m_0 + m_j$, and the first Delaunay variable and the Hamiltonian are

$$\Lambda_j = \frac{m_0 m_j}{m_0 + m_j}[\mathbb{G}(m_0 + m_j)a_j]^{1/2}, \quad H_{\mathrm{K},j} = -\frac{\mathbb{G}^2 m_0^3 m_j^3}{2(m_0+m_j)\Lambda_j^2}, \tag{d}$$

where the semimajor axis is measured relative to the host star. On the other hand, if the Hamiltonian is written in the form (4.13), then $m_a = m_j$, $m_b = m_0$, and

$$\Lambda_j = m_j(\mathbb{G}m_0 a_j)^{1/2}, \quad H_{\mathrm{K},j} = -\frac{\mathbb{G}^2 m_0^2 m_j^3}{2\Lambda_j^2}. \tag{e}$$

In Jacobi coordinates, the Kepler Hamiltonian is contained in the square brackets in equation (4.39). Then $m_a = \tilde{\mu}_j$, $m_b = M_j$, and

$$\Lambda_j = \frac{m_j M_{j-1}}{M_j}(\mathbb{G}M_j a_j)^{1/2}, \quad H_{\mathrm{K},j} = -\frac{\mathbb{G}^2 m_j^3 M_{j-1}^3}{2M_j\Lambda_j^2}. \tag{f}$$

Here M_j is the sum of the host-star mass and the masses of all the planets interior to and including planet j, and the semimajor axis is measured relative to the center of mass of the host star and all of the planets interior to j.

$$-\sum_{j=1}^{N-1}\sum_{k=j+1}^{N-1}\frac{\mathbb{G}m_j m_k}{|\mathbf{r}_j^\star-\mathbf{r}_k^\star|}+\frac{1}{m_0}\sum_{j=1}^{N-1}\sum_{k=j+1}^{N-1}\mathbf{p}_j^\star\cdot\mathbf{p}_k^\star. \tag{4.14}$$

In either Poincaré or democratic coordinates, the equations of motion derived from the astrocentric Hamiltonian are the same as equations (4.5), supplemented by an equation of motion for $\mathbf{r}_0^\star$ and the constraint $\mathbf{p}_0^\star$ = const.

The total angular momentum of the system can also be expressed in astrocentric coordinates. We write the generating function (4.7) in the form

$$S_2(\mathbf{p}_0^\star,\ldots,\mathbf{p}_{N-1}^\star,\mathbf{r}_0,\ldots,\mathbf{r}_{N-1})=\sum_{j,k=0}^{N-1}A_{jk}\mathbf{p}_j^\star\cdot\mathbf{r}_k, \tag{4.15}$$

where $\mathbf{A}$ is a constant $N\times N$ matrix. Then

$$\mathbf{r}_i^\star=\frac{\partial S_2}{\partial\mathbf{p}_i^\star}=\sum_{k=0}^{N-1}A_{ik}\mathbf{r}_k,\quad \mathbf{p}_i=\frac{\partial S_2}{\partial\mathbf{r}_i}=\sum_{n=0}^{N-1}A_{ni}\mathbf{p}_n^\star. \tag{4.16}$$

The first equation can be inverted to give

$$\mathbf{r}_i=\sum_{k=0}^{N-1}A_{ik}^{-1}\mathbf{r}_k^\star. \tag{4.17}$$

The total angular momentum is

$$\mathbf{L}=\sum_{i=0}^{N-1}\mathbf{r}_i\times\mathbf{p}_i=\sum_{i,k,n=0}^{N-1}A_{ik}^{-1}A_{ni}\mathbf{r}_k^\star\times\mathbf{p}_n^\star=\sum_{i=0}^{N-1}\mathbf{r}_i^\star\times\mathbf{p}_i^\star. \tag{4.18}$$

The last equality follows from the identity $\sum_i A_{ni}A_{ik}^{-1}=\delta_{ik}$, where δ_{ik} is the Kronecker delta (eq. C.1). Equation (4.18) shows that the total angular momentum has the same form in inertial and astrocentric coordinates.

Astrocentric coordinates are poor choices for the description of the trajectories of bodies orbiting at large distances from a planetary system. The reason can be illustrated by a simplified system consisting of a star, one planet and a distant test particle. The motion of the star around the barycenter introduces a fictitious force in the astrocentric frame that does not decline with distance. Because this force oscillates with the period of the planet, any integrator following the test particle has to use a timestep that is much less than this period, no matter how long the period of the test particle may be. Numerical integrations of distant objects in astrocentric coordinates are therefore extremely inefficient.

As this illustration suggests, astrocentric coordinates work well for inner planets, and barycentric coordinates work well for outer planets. The coordinate system described in the next section incorporates some of the advantages of both systems.

4.1.3 Jacobi coordinates

The limitations of barycentric and astrocentric coordinates are partly removed by **Jacobi coordinates**. In Jacobi coordinates the position and velocity of body m_k are given relative to the barycenter of bodies $m_0, \ldots, m_{k-1}$. Thus the transformation to Jacobi coordinates depends on the ordering of the labels assigned to the planets; it is said to be "undemocratic." If the planets are on well separated orbits, the best approach is to arrange the labels in order of increasing semimajor axis. Thus m_0 is the host star, m_1 is the innermost planet, and m_{N-1} is the outermost. More generally the ordering should be from the subsystems with the shortest orbital period to those with the longest period; for example, in the Earth–Moon–Sun system we should choose m_0 to be the Earth, m_1 the Moon, and m_2 the Sun.

The Jacobi coordinates and momenta are labeled $(\tilde{\mathbf{r}}_j, \tilde{\mathbf{p}}_j)$ with the index j running from 0 to $N-1$ in an N-body system (star plus $N-1$ planets). We define Jacobi coordinates in terms of barycentric coordinates $\mathbf{r}_j$ by

$$\tilde{\mathbf{r}}_0 \equiv \frac{1}{M_{N-1}} \sum_{k=0}^{N-1} m_k \mathbf{r}_k, \quad \tilde{\mathbf{r}}_j \equiv \mathbf{r}_j - \frac{1}{M_{j-1}} \sum_{k=0}^{j-1} m_k \mathbf{r}_k, \quad j = 1, \ldots, N-1, \tag{4.19}$$

where $M_j \equiv \sum_{k=0}^{j} m_k$. The coordinate $\tilde{\mathbf{r}}_0$ is the position of the barycenter of the system. For $k = 1, \ldots, N-1$, the Jacobi coordinate $\tilde{\mathbf{r}}_k$ is the difference between the position of planet j and the center of mass of all the bodies with smaller labels (see Figure 4.1). Thus the Jacobi coordinates are astrocentric for the innermost body and barycentric for a test particle orbiting outside the planetary system. We can rewrite equations (4.19) as

$$\tilde{\mathbf{r}}_j = \sum_{k=0}^{N-1} Q_{jk}\mathbf{r}_k, \tag{4.20}$$

where

$$Q_{jk} = \begin{cases} m_k/M_{N-1}, & j = 0; \\ -m_k/M_{j-1}, & k < j; \\ 1, & k = j > 0; \\ 0, & k > j > 0. \end{cases} \tag{4.21}$$

We also need the inverse transformation from Jacobi to barycentric coordinates. To derive this, note that equation (4.19) for $\tilde{\mathbf{r}}_{N-1}$ can be rewritten as $M_{N-2}\tilde{\mathbf{r}}_{N-1} = M_{N-2}\mathbf{r}_{N-1} - \sum_{k=0}^{N-2} m_k\mathbf{r}_k = M_{N-2}\mathbf{r}_{N-1} - M_{N-1}\tilde{\mathbf{r}}_0 + m_{N-1}\mathbf{r}_{N-1} = M_{N-1}\mathbf{r}_{N-1} - M_{N-1}\tilde{\mathbf{r}}_0$; and this can be solved to give

$$\mathbf{r}_{N-1} = \tilde{\mathbf{r}}_0 + \frac{M_{N-2}}{M_{N-1}}\tilde{\mathbf{r}}_{N-1}. \tag{4.22}$$

This gives $\mathbf{r}_{N-1}$ in terms of Jacobi coordinates. Similarly we can solve for $\mathbf{r}_{N-2}$ in terms of $\mathbf{r}_{N-1}$ and Jacobi coordinates, and eliminate $\mathbf{r}_{N-1}$ using equation (4.22) to obtain

$$\mathbf{r}_{N-2} = \tilde{\mathbf{r}}_0 + \frac{M_{N-3}}{M_{N-2}}\tilde{\mathbf{r}}_{N-2} - \frac{m_{N-1}}{M_{N-1}}\tilde{\mathbf{r}}_{N-1}. \tag{4.23}$$

This process can be continued, and we find

$$\mathbf{r}_0 = \tilde{\mathbf{r}}_0 - \sum_{k=1}^{N-1}\frac{m_k}{M_k}\tilde{\mathbf{r}}_k, \quad \mathbf{r}_j = \tilde{\mathbf{r}}_0 + \frac{M_{j-1}}{M_j}\tilde{\mathbf{r}}_j - \sum_{k=j+1}^{N-1}\frac{m_k}{M_k}\tilde{\mathbf{r}}_k, \quad j = 1, \ldots, N-1; \tag{4.24}$$

here it is understood that the summation is zero for $j = N - 1$, where the lower limit of the index k is larger than the upper limit. This transformation can also be written

$$\mathbf{r}_j = \sum_{k=0}^{N-1} Q_{jk}^{-1}\tilde{\mathbf{r}}_k, \tag{4.25}$$

where

$$Q^{-1}_{jk} = \begin{cases} 1, & k = 0; \\ -m_k/M_k, & k > j; \\ M_{k-1}/M_k, & k = j > 0; \\ 0, & j > k > 0. \end{cases} \tag{4.26}$$

We must now find the momenta conjugate to the Jacobi coordinates. To do so we choose the generating function (cf. eq. 4.15)

$$S_2(\tilde{\mathbf{p}}_0, \ldots, \tilde{\mathbf{p}}_{N-1}, \mathbf{r}_0, \ldots, \mathbf{r}_{N-1}) = \sum_{j,k=0}^{N-1} Q_{jk}\tilde{\mathbf{p}}_j \cdot \mathbf{r}_k. \tag{4.27}$$

Then

$$\tilde{\mathbf{r}}_j = \frac{\partial S_2}{\partial \tilde{\mathbf{p}}_j} = \sum_{k=0}^{N-1} Q_{jk}\mathbf{r}_k, \quad \mathbf{p}_j = \frac{\partial S_2}{\partial \mathbf{r}_j} = \sum_{k=0}^{N-1} Q_{kj}\tilde{\mathbf{p}}_k. \tag{4.28}$$

The first of these is identical to (4.20). The second is a matrix equation that can be inverted to give

$$\tilde{\mathbf{p}}_k = \sum_{j=0}^{N-1} Q^{-1}_{jk}\mathbf{p}_j. \tag{4.29}$$

Then using equations (4.26), wc have

$$\tilde{\mathbf{p}}_0 = \sum_{j=0}^{N-1} \mathbf{p}_j, \quad \tilde{\mathbf{p}}_k = \frac{M_{k-1}}{M_k}\mathbf{p}_k - \frac{m_k}{M_k}\sum_{j=0}^{k-1} \mathbf{p}_j, \quad k = 1, \ldots, N-1. \tag{4.30}$$

The inverse transformation is given by the second of equations (4.28),

$$\mathbf{p}_0 = \frac{m_0}{M_{N-1}}\tilde{\mathbf{p}}_0 - \sum_{k=1}^{N-1} \frac{m_0}{M_{k-1}}\tilde{\mathbf{p}}_k, \quad \mathbf{p}_j = \frac{m_j}{M_{N-1}}\tilde{\mathbf{p}}_0 + \tilde{\mathbf{p}}_j - \sum_{k=j+1}^{N-1} \frac{m_j}{M_{k-1}}\tilde{\mathbf{p}}_k, \tag{4.31}$$

for $j = 1, \ldots, N-1$. As in equations (4.24) it is understood that the summation is zero for $j = N-1$, where the lower limit of the dummy index is larger than the upper limit.

The arguments leading to equation (4.18) apply both to astrocentric and Jacobi coordinates, so the total angular momentum is

$$\mathbf{L} = \sum_{k=0}^{N-1} \tilde{\mathbf{r}}_k \times \tilde{\mathbf{p}}_k. \tag{4.32}$$

The total energy can also be expressed with a simple formula. Using the second of equations (4.28), the kinetic energy can be written

$$T = \sum_{j=0}^{N-1} \frac{\mathbf{p}_j^2}{2m_j} = \sum_{j,k,n=0}^{N-1} \frac{Q_{kj}Q_{nj}}{2m_j} \tilde{\mathbf{p}}_k \cdot \tilde{\mathbf{p}}_n. \tag{4.33}$$

Using equations (4.21) it is straightforward to show that

$$\sum_{j=0}^{N-1} \frac{Q_{kj}Q_{nj}}{m_j} = \begin{cases} 1/M_{N-1}, & k = n = 0; \\ M_k/(m_k M_{k-1}), & k = n > 0; \\ 0, & k \neq n. \end{cases} \tag{4.34}$$

Therefore the kinetic energy can be written

$$T = \sum_{k=0}^{N-1} \frac{\tilde{\mathbf{p}}_k^2}{2\tilde{\mu}_k}, \tag{4.35}$$

where

$$\tilde{\mu}_k = \begin{cases} M_{N-1}, & k = 0; \\ \dfrac{m_k M_{k-1}}{M_k}, & k = 1, \ldots, N-1. \end{cases} \tag{4.36}$$

Thus the kinetic energy in Jacobi coordinates has the same form as in barycentric coordinates, except that the masses m_k are replaced by $\tilde{\mu}_k$: this is far simpler than the expression for the kinetic energy in astrocentric coordinates, equations (4.10) or (4.11).

The Hamiltonian in Jacobi coordinates is

$$H = \frac{\tilde{\mathbf{p}}_0^2}{2\tilde{\mu}_0} + \sum_{j=1}^{N-1} \frac{\tilde{\mathbf{p}}_j^2}{2\tilde{\mu}_j} - \sum_{j=0}^{N-1} \sum_{k=j+1}^{N-1} \frac{\mathbb{G} m_j m_k}{|\mathbf{r}_j - \mathbf{r}_k|}, \tag{4.37}$$

where $\mathbf{r}_k$ is defined in terms of $\tilde{\mathbf{r}}_k$ by equations (4.24). The Hamiltonian is independent of $\tilde{\mathbf{r}}_0$ so the total momentum $\tilde{\mathbf{p}}_0$ is conserved, and the time derivatives of the Jacobi coordinates are related to their conjugate momenta in the usual way,

$$\frac{\mathrm{d}\tilde{\mathbf{r}}_k}{\mathrm{d}t} = \frac{\partial H}{\partial \tilde{\mathbf{p}}_k} = \frac{\tilde{\mathbf{p}}_k}{\tilde{\mu}_k}. \tag{4.38}$$

The Hamiltonian can also be written in the form

$$H = \frac{\tilde{\mathbf{p}}_0^2}{2\tilde{\mu}_0} + \sum_{j=1}^{N-1}\left[\frac{\tilde{\mathbf{p}}_j^2}{2\tilde{\mu}_j} - \frac{\mathbb{G}M_j\tilde{\mu}_j}{|\tilde{\mathbf{r}}_j|}\right]$$
$$- \sum_{j=1}^{N-1}\sum_{k=j+1}^{N-1}\frac{\mathbb{G}m_j m_k}{|\mathbf{r}_j - \mathbf{r}_k|} + \sum_{k=1}^{N-1}\mathbb{G}m_k\left(\frac{M_{k-1}}{|\tilde{\mathbf{r}}_k|} - \frac{m_0}{|\mathbf{r}_k - \mathbf{r}_0|}\right). \quad (4.39)$$

The terms on the first line consist of the kinetic energy of the system as a whole—which we can drop since it is constant—and the usual two-body Hamiltonian for the motion of a body of mass $\tilde{\mu}_k$ around a fixed point located at the barycenter of the first k masses ($k - 1$ planets and the host star). The terms on the second line are smaller—by a factor of roughly the ratio of the planetary masses to the stellar mass—and depend only on the Jacobi coordinates, since $\mathbf{r}_j$ can be converted to $\tilde{\mathbf{r}}_j$ using equations (4.24). Therefore Jacobi coordinates can be used in the Wisdom–Holman integrators described in §2.5.3.

Jacobi coordinates are well suited for systems of planets on orbits that never cross. If the orbits cross, the equations of motion based on Jacobi coordinates remain valid, but the planets are no longer ordered by their distances from the host, so numerical integrators based on these coordinates may be inefficient.

4.2 Hamiltonian perturbation theory

The Hamiltonian for a system of $N - 1$ planets orbiting a common host star can be written in the form

$$H(\mathbf{q}_1, \cdots, \mathbf{q}_{N-1}, \mathbf{p}_1, \cdots, \mathbf{p}_{N-1})$$
$$= \sum_{i=1}^{N-1} H_{\mathrm{K},i}(\mathbf{q}_i, \mathbf{p}_i) + \sum_{i=1}^{N-1}\sum_{j=i+1}^{N-1} H_{ij}(\mathbf{q}_i, \mathbf{p}_i, \mathbf{q}_j, \mathbf{p}_j). \quad (4.40)$$

Here $H_{\mathrm{K},i}$ is the Kepler Hamiltonian for planet i, and H_{ij} represents the interaction between planets i and j. In general $|H_{ij}| \ll |H_{\mathrm{K},i}|$ because planetary masses are much smaller than stellar masses. The goal of perturbation theory in celestial mechanics is to exploit the small masses of planets to find approximate descriptions of the evolution of trajectories in multi-planet systems—in fact, it was for this purpose that perturbation theory was originally invented.

The Hamiltonian in astrocentric coordinates has the form (4.40) in either convention A or B (eqs. 4.12 or 4.14), provided we treat the total momentum $\mathbf{p}_0^\star$ as a constant—which it is, since the Hamiltonian is independent of the conjugate coordinate $\mathbf{r}_0^\star$—and drop the constant term proportional to $\mathbf{p}_0^{\star 2}$. Similarly the Hamiltonian in Jacobi coordinates (4.37) has this form, provided we treat $\tilde{\mathbf{p}}_0$ as a constant and drop the term proportional to $\frac{1}{2}\tilde{\mathbf{p}}_0^2$.

The Delaunay variables for the Kepler Hamiltonian $H_{\mathrm{K},i}$ may be written $(\boldsymbol{\theta}_i, \mathbf{J}_i)$, where the vector of coordinates or angles is $\boldsymbol{\theta}_i \equiv (\ell_i, \omega_i, \Omega_i)$ and the vector of momenta or actions is $\mathbf{J}_i \equiv (\Lambda_i, L_i, L_{zi})$, as defined in Box 4.1. We can rewrite the Hamiltonian (4.40) in these variables as

$$H(\boldsymbol{\theta}_1,\cdots,\boldsymbol{\theta}_{N-1},\mathbf{J}_1,\cdots,\mathbf{J}_{N-1}) = \sum_{i=1}^{N-1} H_{\mathrm{K},i}(\mathbf{J}_i) + \sum_{i=1}^{N-1}\sum_{j=i+1}^{N-1} H_{ij}(\boldsymbol{\theta}_i,\mathbf{J}_i,\boldsymbol{\theta}_j,\mathbf{J}_j), \tag{4.41}$$

where the Kepler Hamiltonians $H_{\mathrm{K},i}$ are given in Box 4.1 for astrocentric and Jacobi coordinates. The remaining terms H_{ij} are called the **disturbing function.**[1]

Calculating the disturbing function is much harder in Delaunay variables than in Cartesian coordinates (for example, it depends on all six Delaunay variables but only three Cartesian coordinates), but the advantage of working in angle-action variables like the Delaunay variables is so great that this price is often worth paying. Wisdom–Holman integrators (§2.5.3) are based on splitting the Hamiltonian into two parts: a Kepler Hamiltonian that is simple to integrate in Delaunay variables, and a disturbing function that is simple to integrate in Cartesian coordinates.

[1] Unfortunately, for many authors this name refers to $-H_{ij}$, not H_{ij}.

4.2.1 First-order perturbation theory

Equation (4.41) is an example of a Hamiltonian of the form

$$H(\boldsymbol{\theta}, \mathbf{J}) = H_0(\mathbf{J}) + \epsilon H_1(\boldsymbol{\theta}, \mathbf{J}), \tag{4.42}$$

where $\boldsymbol{\theta}$ and $\mathbf{J}$ are vectors of angle-action variables (of dimension $3N - 3$ in eq. 4.41). Here $\epsilon \ll 1$ is an ordering parameter that is used to keep track of the size of the changes due to the perturbing Hamiltonian.

Because $\boldsymbol{\theta}$ is a vector of angles, H_1 must be periodic with period 2π in each of its components θ_j. Therefore it can be expanded as a Fourier series (Appendix B.4),

$$H(\boldsymbol{\theta}, \mathbf{J}) = H_0(\mathbf{J}) + \epsilon \sum_{\mathbf{m}} h_{\mathbf{m}}(\mathbf{J}) \exp(\mathrm{i}\mathbf{m} \cdot \boldsymbol{\theta}), \tag{4.43}$$

where $\mathbf{m}$ is a vector of integers having the same dimension as $\boldsymbol{\theta}$. Since both H_0 and H_1 are real, when we take the complex conjugate of this equation and replace the dummy index $\mathbf{m}$ by $-\mathbf{m}$ we have

$$H(\boldsymbol{\theta}, \mathbf{J}) = H_0(\mathbf{J}) + \epsilon \sum_{\mathbf{m}} h^*_{-\mathbf{m}}(\mathbf{J}) \exp(\mathrm{i}\mathbf{m} \cdot \boldsymbol{\theta}). \tag{4.44}$$

The two equations must be the same for all $\boldsymbol{\theta}$, which requires that $h^*_{-\mathbf{m}}(\mathbf{J}) = h_{\mathbf{m}}(\mathbf{J})$. If we write $h_{\mathbf{m}}(\mathbf{J}) = H_{\mathbf{m}}(\mathbf{J}) \exp[-\mathrm{i}\phi_{\mathbf{m}}(\mathbf{J})]$ where $H_{\mathbf{m}}$ and $\phi_{\mathbf{m}}$ are real, then this requirement implies that

$$H_{-\mathbf{m}} = H_{\mathbf{m}} \quad \text{and} \quad \phi_{-\mathbf{m}} = -\phi_{\mathbf{m}}. \tag{4.45}$$

Equation (4.43) can now be rewritten as

$$H(\boldsymbol{\theta}, \mathbf{J}) = H_0(\mathbf{J}) + \epsilon \sum_{\mathbf{m}} H_{\mathbf{m}}(\mathbf{J})[\cos(\mathbf{m}\cdot\boldsymbol{\theta} - \phi_{\mathbf{m}}) + \mathrm{i}\sin(\mathbf{m}\cdot\boldsymbol{\theta} - \phi_{\mathbf{m}})]. \tag{4.46}$$

Because of the conditions (4.45) the sine terms with indices $\mathbf{m}$ and $-\mathbf{m}$ cancel and the sine term with $\mathbf{m} = \mathbf{0}$ vanishes, so this result simplifies to

$$H(\boldsymbol{\theta}, \mathbf{J}) = H_0(\mathbf{J}) + \epsilon \sum_{\mathbf{m}} H_{\mathbf{m}}(\mathbf{J}) \cos(\mathbf{m} \cdot \boldsymbol{\theta} - \phi_{\mathbf{m}}). \tag{4.47}$$

Hamilton's equations are

$$\frac{\mathrm{d}\mathbf{J}}{\mathrm{d}t} = -\frac{\partial H}{\partial \boldsymbol{\theta}} = -\mathrm{i}\epsilon \sum_{\mathbf{m}} \mathbf{m}\, h_{\mathbf{m}}(\mathbf{J}) \exp(\mathrm{i}\mathbf{m}\cdot\boldsymbol{\theta}), \tag{4.48a}$$

$$\frac{\mathrm{d}\boldsymbol{\theta}}{\mathrm{d}t} = \frac{\partial H}{\partial \mathbf{J}} = \frac{\partial H_0}{\partial \mathbf{J}}(\mathbf{J}) + \epsilon \sum_{\mathbf{m}} \frac{\partial h_{\mathbf{m}}}{\partial \mathbf{J}}(\mathbf{J}) \exp(\mathrm{i}\mathbf{m}\cdot\boldsymbol{\theta}). \tag{4.48b}$$

If $\epsilon = 0$ the solution is

$$\mathbf{J} = \mathbf{J}_{\mathrm{in}}, \quad \boldsymbol{\theta} = \boldsymbol{\theta}_{\mathrm{in}} + \boldsymbol{\Omega}(\mathbf{J}_{\mathrm{in}})t, \tag{4.49}$$

where $\mathbf{J}_{\mathrm{in}}$ and $\boldsymbol{\theta}_{\mathrm{in}}$ are the values of the actions and angles at the initial time $t = 0$, and $\boldsymbol{\Omega}(\mathbf{J}) \equiv \partial H_0/\partial \mathbf{J}$. For example, in Delaunay variables the actions are $\mathbf{J} = (\Lambda, L, L_z)$, and in astrocentric coordinates the Kepler Hamiltonian is given by equation (d) of Box 4.1, $H_{\mathrm{K}}(\mathbf{J}) = -\,\mathbb{G}^2 m_0^3 m^3/[2(m_0 + m)\Lambda^2]$, so

$$\boldsymbol{\Omega}(\mathbf{J}) = \left(\frac{\mathbb{G}^2 m_0^3 m^3}{(m_0 + m)\Lambda^3}, 0, 0\right) = \left(\frac{[\,\mathbb{G}(m_0 + m)]^{1/2}}{a^{3/2}}, 0, 0\right) = (n, 0, 0), \tag{4.50}$$

where n is the usual mean motion.

Now assume that ϵ is small but nonzero. We write the trajectory in angle-action variables as a power series in ϵ,

$$\mathbf{J}(t) = \mathbf{J}^{(0)}(t) + \epsilon \mathbf{J}^{(1)}(t) + \epsilon^2 \mathbf{J}^{(2)}(t) + \ldots,$$
$$\boldsymbol{\theta}(t) = \boldsymbol{\theta}^{(0)}(t) + \epsilon \boldsymbol{\theta}^{(1)}(t) + \epsilon^2 \boldsymbol{\theta}^{(2)}(t) + \ldots. \tag{4.51}$$

We substitute these expressions in Hamilton's equations (4.48), expand factors like $h_{\mathbf{m}}(\mathbf{J}^{(0)} + \epsilon \mathbf{J}^{(1)} + \cdots)$ as Taylor series in ϵ, and collect all of the terms with the same power of ϵ.

The terms independent of ϵ have the solutions (4.49),

$$\mathbf{J}^{(0)}(t) = \mathbf{J}_{\mathrm{in}}, \quad \boldsymbol{\theta}^{(0)}(t) = \boldsymbol{\theta}_{\mathrm{in}} + \boldsymbol{\Omega}(\mathbf{J}_{\mathrm{in}})t. \tag{4.52}$$

The terms proportional to ϵ are

$$\frac{\mathrm{d}J_j^{(1)}}{\mathrm{d}t} = -\mathrm{i} \sum_{\mathbf{m}} m_j\, h_{\mathbf{m}} \exp\left(\mathrm{i}\mathbf{m}\cdot\boldsymbol{\theta}^{(0)}\right), \tag{4.53a}$$

$$\frac{\mathrm{d}\theta_j^{(1)}}{\mathrm{d}t} = \sum_k \frac{\partial^2 H_0}{\partial J_j \partial J_k} J_k^{(1)} + \sum_{\mathbf{m}} \frac{\partial h_{\mathbf{m}}}{\partial J_i} \exp\big(\mathrm{i}\mathbf{m}\cdot\boldsymbol{\theta}^{(0)}\big). \tag{4.53b}$$

In these equations all functions of the actions on the right side, such as $h_{\mathbf{m}}$ and $\partial^2 H_0/\partial J_j \partial J_k$, are evaluated at the zero-order actions $\mathbf{J}^{(0)}$.

At this point we divide the terms in the Fourier series into two sets. Let S denote the set of all integer triples $\mathbf{m}$ such that $\mathbf{m}\cdot\boldsymbol{\Omega}(\mathbf{J}^{(0)}) = 0$. When $\mathbf{m} \in S$, the corresponding component of the Fourier series is called a **secular** term, otherwise it is a **short-period** term. In Delaunay variables the Hamiltonian H_0 depends only on $J_1 = \Lambda$, so the secular terms are those with $m_1 = 0$.

We now substitute equations (4.52) into the right side of equation (4.53a) and integrate:

$$\begin{aligned} J_j^{(1)}(t) = {} & -\mathrm{i}t \sum_{\mathbf{m}\in S} m_j h_{\mathbf{m}}(\mathbf{J}_{\rm in}) \exp(\mathrm{i}\mathbf{m}\cdot\boldsymbol{\theta}_{\rm in}) \\ & - \sum_{\mathbf{m}\notin S} \frac{m_j\, h_{\mathbf{m}}(\mathbf{J}_{\rm in})}{\mathbf{m}\cdot\boldsymbol{\Omega}(\mathbf{J}_{\rm in})} \exp(\mathrm{i}\mathbf{m}\cdot\boldsymbol{\theta}_{\rm in})\big\{\exp[\mathrm{i}\mathbf{m}\cdot\boldsymbol{\Omega}(\mathbf{J}_{\rm in})t] - 1\big\}. \end{aligned} \tag{4.54}$$

This expression includes an integration constant chosen such that $\mathbf{J}^{(1)}(t = 0) = \mathbf{0}$, consistent with the initial conditions. The secular terms on the first line lead to changes in the actions that are linear in time; however, the secular terms have $m_1 = 0$ so there is no secular change in the action $J_1 = \Lambda$, which implies that there is no secular change in the semimajor axis.

The solution for the angles $\boldsymbol{\theta}^{(1)}(t)$ is similar, although we must substitute the solution (4.54) for $\mathbf{J}^{(1)}(t)$ into the right side of the differential equation (4.53b) before integrating. This procedure can be repeated to determine the angles and actions to order ϵ^2, ϵ^3, and so on, but the labor grows rapidly at higher orders, and the insight gained from the solution diminishes equally rapidly. More efficient approaches are described in the following subsections, but in practice perturbation theory is rarely applied to Hamiltonian systems beyond order ϵ^2.

In principle, equation (4.54) and its analog for the angles provide a complete description of the dominant effects of mutual gravitational interactions on planetary orbits. However, the practical application of these results encounters two obstacles. The first of these is the presence of terms proportional to t in the equation for $\mathbf{J}(t)$. These imply that perturbations in the actions grow without limit, so long as there are any secular terms with $\mathbf{m} \neq \mathbf{0}$—for Delaunay variables these would be all terms with $m_1 = 0$ and m_2 or m_3 nonzero, so changes in the eccentricity and inclination can grow linearly, but changes in the semimajor axis cannot. This result raises the uncomfortable possibility that the eccentricities and inclinations of the planets in the solar system may grow until the orbits cross and planets collide.[2] Fortunately this is not the case: as described in §5.2, the perturbations in eccentricity and inclination are oscillatory but with a frequency that is smaller than the mean motion by of order ϵ, so the oscillations look linear in a perturbation series[3] that is valid only to $\mathrm{O}(\epsilon)$.

The second problem is the presence of the factor $\mathbf{m}\cdot\boldsymbol{\Omega}$ in the denominator in equations (4.54). Since $\mathbf{m}$ is the set of all integer N-tuples in a $2N$ dimensional phase space, $\mathbf{m}\cdot\boldsymbol{\Omega}$ can be arbitrarily close to zero for some value of $\mathbf{m}$, which means that the series does not converge. This is a physical phenomenon, not just a mathematical inconvenience: in general, nonlinear Hamiltonian systems with more than one degree of freedom are not integrable, so chaotic trajectories are found in the neighborhood of every point in phase space, and this complexity is not captured by perturbation theory (see Appendix D.8). The results of perturbation theory should be thought of instead as asymptotic series: over a fixed time interval they provide a better and better description of the motion as the parameter ϵ becomes smaller and smaller. In practical terms, perturbation theory is useful for describing the motion of planets and satellites over intervals of hundreds or thousands of orbits but not for the lifetime of the universe.

4.2.2 The Poincaré–von Zeipel method

The perturbation theory developed in the preceding subsection is straightforward but cumbersome. An alternative and simpler approach is to find a

---

[2] This was a fundamental scientific issue in the eighteenth century. See §4.5 for a brief history.

[3] A simple example of this phenomenon is given by the differential equation $\dot{z} = \mathrm{i}\epsilon z$ with initial condition $z = 1$ at $t = 0$. The solution is $z(t) = \exp(\mathrm{i}\epsilon t)$, which is oscillatory, but a power-series expansion of the solution gives linear growth, $z(t) = 1 + \mathrm{i}\epsilon t + \mathrm{O}(\epsilon^2)$.

canonical transformation to new variables that obey a simpler Hamiltonian (Poincaré 1892–1897; von Zeipel 1916).

Let $\mathbf{J}'$ and $\boldsymbol{\theta}'$ be new variables derived from $\mathbf{J}$ and $\boldsymbol{\theta}$ by the generating function

$$S_2(\mathbf{J}',\boldsymbol{\theta}) = \mathbf{J}'\cdot\boldsymbol{\theta} + \epsilon\sum_{\mathbf{m}} s_{\mathbf{m}}(\mathbf{J}')\exp(\mathrm{i}\mathbf{m}\cdot\boldsymbol{\theta}). \tag{4.55}$$

Then from equations (D.63),

$$\mathbf{J} = \frac{\partial S}{\partial\boldsymbol{\theta}} = \mathbf{J}' + \mathrm{i}\epsilon\sum_{\mathbf{m}} \mathbf{m}\, s_{\mathbf{m}}(\mathbf{J}')\exp(\mathrm{i}\mathbf{m}\cdot\boldsymbol{\theta}), \tag{4.56a}$$

$$\boldsymbol{\theta}' = \frac{\partial S}{\partial\mathbf{J}'} = \boldsymbol{\theta} + \epsilon\sum_{\mathbf{m}} \frac{\partial s_{\mathbf{m}}}{\partial\mathbf{J}'}(\mathbf{J}')\exp(\mathrm{i}\mathbf{m}\cdot\boldsymbol{\theta}). \tag{4.56b}$$

These equations can be rewritten to express the old variables in terms of the new ones. To order ϵ,

$$\mathbf{J} = \mathbf{J}' + \mathrm{i}\epsilon\sum_{\mathbf{m}} \mathbf{m}\, s_{\mathbf{m}}(\mathbf{J}')\exp(\mathrm{i}\mathbf{m}\cdot\boldsymbol{\theta}') + \mathrm{O}(\epsilon^2), \tag{4.57a}$$

$$\boldsymbol{\theta} = \boldsymbol{\theta}' - \epsilon\sum_{\mathbf{m}} \frac{\partial s_{\mathbf{m}}}{\partial\mathbf{J}'}(\mathbf{J}')\exp(\mathrm{i}\mathbf{m}\cdot\boldsymbol{\theta}') + \mathrm{O}(\epsilon^2). \tag{4.57b}$$

Since the generating function is independent of time, the value of the Hamiltonian is the same in the old and new variables, $H'(\mathbf{J}',\boldsymbol{\theta}') = H(\mathbf{J},\boldsymbol{\theta})$ with $H(\mathbf{J},\boldsymbol{\theta})$ given by equation (4.43). Eliminating the old variables using equations (4.57), expanding the result to $\mathrm{O}(\epsilon)$ in a Taylor series, and replacing the derivatives $\partial H_0(\mathbf{J})/\partial\mathbf{J}$ by $\boldsymbol{\Omega}(\mathbf{J})$, we find

$$\begin{aligned} H'(\mathbf{J}',\boldsymbol{\theta}') = H_0(\mathbf{J}') &+ \mathrm{i}\epsilon\sum_{\mathbf{m}} \mathbf{m}\cdot\boldsymbol{\Omega}(\mathbf{J}')s_{\mathbf{m}}(\mathbf{J}')\exp(\mathrm{i}\mathbf{m}\cdot\boldsymbol{\theta}') \\ &+ \epsilon\sum_{\mathbf{m}} h_{\mathbf{m}}(\mathbf{J}')\exp(\mathrm{i}\mathbf{m}\cdot\boldsymbol{\theta}') + \mathrm{O}(\epsilon^2). \end{aligned} \tag{4.58}$$

We now set

$$s_{\mathbf{m}}(\mathbf{J}') = \begin{cases} 0, & \mathbf{m}\in S, \\ \mathrm{i}\dfrac{h_{\mathbf{m}}(\mathbf{J}')}{\mathbf{m}\cdot\boldsymbol{\Omega}(\mathbf{J}')}, & \mathbf{m}\notin S, \end{cases} \tag{4.59}$$

where as usual S is the set of integer multiplets $\mathbf{m}$ for which $\mathbf{m} \cdot \mathbf{\Omega} = 0$. With this substitution

$$H'(\mathbf{J}') = H_0(\mathbf{J}') + \epsilon \sum_{\mathbf{m} \in S} h_{\mathbf{m}}(\mathbf{J}') \exp(\mathrm{i}\mathbf{m} \cdot \boldsymbol{\theta}') + \mathrm{O}(\epsilon^2). \tag{4.60}$$

Thus at first order in ϵ the new Hamiltonian is equal to the old Hamiltonian with all of the short-period terms removed.

The Poincaré–von Zeipel method is better than the approach in §4.2.1 for two reasons. First, the effects of the perturbation are encapsulated in two scalar functions, the Hamiltonian and the generating function, rather than in $2N$ actions and angles; second, the secular Hamiltonian—the sum over terms with $\mathbf{m} \in S$ in equation (4.60)—is retained intact, so it can be analyzed more accurately to eliminate the misleading linear growth in the perturbations in the actions or angles (see §5.2).

Despite these advantages, the Poincaré–von Zeipel method is still awkward when extended to second or higher order in the perturbation strength ϵ, mostly because of the need to work with a mixture of the old and new variables.

4.2.3 Lie operator perturbation theory

The basic idea of the Poincaré–von Zeipel method is to find a new set of canonical variables close to the original ones in which the Hamiltonian has a simple form. Lie operator perturbation theory is based on the same idea but avoids the use of mixed-variable generating functions.

Let $\mathbf{z}$ be a set of canonical variables and $H(\mathbf{z})$ the Hamiltonian in these variables. For brevity we assume that the Hamiltonian is time-independent; this assumption is not restrictive, because any time-dependent Hamiltonian can be converted to a time-independent one in an extended phase space as described in Box 2.1.

Now let $\mathbf{z}'(\mathbf{z}, \epsilon)$ be a new family of canonical variables, depending on the parameter ϵ. We assume that $\epsilon \ll 1$ and that $\mathbf{z}'(\mathbf{z}, 0) = \mathbf{z}$.

Write $\mathbf{z}'_1 = \mathbf{z}'(\mathbf{z}, \epsilon)$ and $\mathbf{z}'_2 = \mathbf{z}'(\mathbf{z}, \epsilon + \mathrm{d}\epsilon)$. The transformation from $\mathbf{z}'_1$ to $\mathbf{z}'_2$ is canonical and so can be written in terms of a generating function

$S_2(\mathbf{q}_1', \mathbf{p}_2') = \mathbf{q}_1' \cdot \mathbf{p}_2' + w(\mathbf{q}_1', \mathbf{p}_2', \epsilon)\mathrm{d}\epsilon$. From equations (D.63) we have

$$\mathbf{q}_2' - \mathbf{q}_1' = \frac{\partial}{\partial \mathbf{p}_2'} w(\mathbf{q}_1', \mathbf{p}_2', \epsilon)\mathrm{d}\epsilon, \quad \mathbf{p}_2' - \mathbf{p}_1' = -\frac{\partial}{\partial \mathbf{q}_1'} w(\mathbf{q}_1', \mathbf{p}_2', \epsilon)\mathrm{d}\epsilon. \quad (4.61)$$

Now let ϵ shrink toward zero. Then the difference between $\mathbf{z}_1'$ and $\mathbf{z}_2'$ also shrinks to zero, so we lose no accuracy if we replace $\mathbf{p}_2'$ by $\mathbf{p}_1'$ on the right side. Thus

$$\frac{\mathbf{q}_2' - \mathbf{q}_1'}{\mathrm{d}\epsilon} = \frac{\partial}{\partial \mathbf{p}_1'} w(\mathbf{q}_1', \mathbf{p}_1'), \qquad \frac{\mathbf{p}_2' - \mathbf{p}_1'}{\mathrm{d}\epsilon} = -\frac{\partial}{\partial \mathbf{q}_1'} w(\mathbf{q}_1', \mathbf{p}_1'), \qquad (4.62)$$

which can be written more compactly as

$$\frac{\partial \mathbf{z}'}{\partial \epsilon} = \{\mathbf{z}', w\} = \mathsf{L}_w \mathbf{z}', \qquad (4.63)$$

in which $\{\cdot,\cdot\}$ is the Poisson bracket (Appendix D.3), and we have introduced the Lie operator L_w defined in Appendix D.4. Although this result was derived using the coordinates $\mathbf{z}_1'$, the Poisson bracket is the same in any canonical coordinates (Appendix D.6). The function $w(\mathbf{z}', \epsilon)$ is called the **local generating function**.

The Lie operator L_w maps the space of scalar phase-space functions onto itself. For each ϵ we define a second operator of this kind, T_ϵ, as follows: if $f(\mathbf{z})$ is any function, then

$$(\mathsf{T}_\epsilon f)(\mathbf{z}) = f[\mathbf{z}'(\mathbf{z}, \epsilon)]; \qquad (4.64)$$

in words, $\mathsf{T}_\epsilon f$ evaluates f at the transformed point $\mathbf{z}'$. The derivative of equation (4.64) is

$$\begin{aligned} \frac{\partial}{\partial \epsilon}(\mathsf{T}_\epsilon f)(\mathbf{z}) &= \sum_i \frac{\partial f}{\partial z_i'} \frac{\partial z_i'}{\partial \epsilon} = \sum_i \frac{\partial f}{\partial z_i'} \{z_i', w\} = \{f, w\}_{\mathbf{z}'} \\ &= (\mathsf{L}_w f)(\mathbf{z}') = (\mathsf{T}_\epsilon \mathsf{L}_w f)(\mathbf{z}). \end{aligned} \qquad (4.65)$$

This result can be written more compactly as

$$\frac{\partial \mathsf{T}_\epsilon}{\partial \epsilon} = \mathsf{T}_\epsilon \mathsf{L}_w. \qquad (4.66)$$

We also need the inverse operator $\mathbf{T}_\epsilon^{-1}$, defined by $\mathbf{T}_\epsilon \mathbf{T}_\epsilon^{-1} = \mathbf{T}_\epsilon^{-1}\mathbf{T}_\epsilon = \mathbf{I}$ where $\mathbf{I}$ is the identity operator. Differentiating $\mathbf{T}_\epsilon \mathbf{T}_\epsilon^{-1} = \mathbf{I}$ with respect to ϵ and using equation (4.66) gives

$$\mathbf{T}_\epsilon \mathbf{L}_w \mathbf{T}_\epsilon^{-1} + \mathbf{T}_\epsilon \frac{\partial \mathbf{T}_\epsilon^{-1}}{\partial \epsilon} = \mathbf{0}, \tag{4.67}$$

and left-multiplying by $\mathbf{T}_\epsilon^{-1}$ gives

$$\frac{\partial \mathbf{T}_\epsilon^{-1}}{\partial \epsilon} = -\mathbf{L}_w \mathbf{T}_\epsilon^{-1}. \tag{4.68}$$

We now apply these results to a system with Hamiltonian $h(\mathbf{z}) = h_0(\mathbf{z}) + \epsilon h_1(\mathbf{z})$. Usually the Hamiltonian $h_0(\mathbf{z})$ is integrable and the variables $\mathbf{z}$ are angle-action variables for $h_0(\mathbf{z})$, but these assumptions are not necessary for the derivation below. We denote the Hamiltonian in the new canonical variables $\mathbf{z}'(\mathbf{z}, \epsilon)$ by $H(\mathbf{z}')$. Since the canonical transformation has no explicit time dependence, $H(\mathbf{z}') = h_0(\mathbf{z}) + \epsilon h_1(\mathbf{z})$ or, using equation (4.64),

$$\mathbf{T}_\epsilon H = h_0 + \epsilon h_1. \tag{4.69}$$

Differentiating this expression with respect to ϵ and using equation (4.66),

$$\mathbf{T}_\epsilon \mathbf{L}_w H + \mathbf{T}_\epsilon \frac{\partial H}{\partial \epsilon} = h_1. \tag{4.70}$$

Then left-multiplying by $\mathbf{T}_\epsilon^{-1}$ gives

$$\mathbf{L}_w H + \frac{\partial H}{\partial \epsilon} = \mathbf{T}_\epsilon^{-1} h_1. \tag{4.71}$$

We now expand the functions w and H and the operators $\mathbf{T}_\epsilon$ and $\mathbf{T}_\epsilon^{-1}$ in power series,

$$w = \sum_{n=0}^{\infty} w_{n+1}\epsilon^n, \quad H = \sum_{n=0}^{\infty} H_n \epsilon^n, \quad \mathbf{T}_\epsilon = \sum_{n=0}^{\infty} \epsilon^n \mathbf{T}_n, \quad \mathbf{T}_\epsilon^{-1} = \sum_{n=0}^{\infty} \epsilon^n \mathbf{T}_n^{-1}. \tag{4.72}$$

When $\epsilon = 0$, $\mathbf{z}'(\mathbf{z}, 0) = \mathbf{z}$, so $H_0 = h_0$ and $\mathsf{T}_0 = \mathsf{T}_0^{-1} = \mathsf{I}$. Notice however that in this notation T_n^{-1} is *not* the inverse of T_n for $n > 0$.

Since L_w is linear in w, we can write

$$\mathsf{L}_w = \sum_{n=0}^{\infty} \epsilon^n \mathsf{L}_{n+1}, \quad \text{where} \quad \mathsf{L}_n f = \mathsf{L}_{w_n} f = \{f, w_n\}. \tag{4.73}$$

Substituting these expansions into equations (4.66) and (4.68) and collecting the terms proportional to ϵ^{n-1} with $n > 0$,

$$\mathsf{T}_n = \frac{1}{n} \sum_{m=0}^{n-1} \mathsf{T}_m \mathsf{L}_{n-m}, \quad \mathsf{T}_n^{-1} = -\frac{1}{n} \sum_{m=0}^{n-1} \mathsf{L}_{n-m} \mathsf{T}_m^{-1}. \tag{4.74}$$

Since $\mathsf{T}_0 = \mathsf{T}_0^{-1} = \mathsf{I}$, these relations can be used recursively to determine any T_n or T_n^{-1} in terms of the operators $\mathsf{L}_1, \ldots, \mathsf{L}_n$. Up to $n = 2$,

$$\begin{aligned} \mathsf{T}_0 &= \mathsf{I}, & \mathsf{T}_1 &= \mathsf{L}_1, & \mathsf{T}_2 &= \tfrac{1}{2}\mathsf{L}_2 + \tfrac{1}{2}\mathsf{L}_1^2, \\ \mathsf{T}_0^{-1} &= \mathsf{I}, & \mathsf{T}_1^{-1} &= -\mathsf{L}_1, & \mathsf{T}_2^{-1} &= -\tfrac{1}{2}\mathsf{L}_2 + \tfrac{1}{2}\mathsf{L}_1^2. \end{aligned} \tag{4.75}$$

A similar expansion of equation (4.71) yields for the terms proportional to ϵ^n

$$\sum_{m=0}^{n} \mathsf{L}_{n-m+1} H_m + (n+1) H_{n+1} - \mathsf{T}_n^{-1} h_1 = 0. \tag{4.76}$$

We evaluate these equations for $n = 0, 1, 2$, using equations (4.75) to simplify the results:

$$\begin{gathered} \mathsf{L}_1 H_0 + H_1 - h_1 = 0, \quad \mathsf{L}_2 H_0 + 2H_2 + \mathsf{L}_1 (h_1 + H_1) = 0, \\ \mathsf{L}_3 H_0 + 3H_3 + \mathsf{L}_2 H_1 + \mathsf{L}_1 H_2 + \tfrac{1}{2}(\mathsf{L}_2 - \mathsf{L}_1^2) h_1 = 0. \end{gathered} \tag{4.77}$$

Similar expressions are straightforward to derive for higher orders. Equations (4.77) can be rewritten as

$$\begin{gathered} \{w_1, H_0\} = H_1 - h_1, \quad \{w_2, H_0\} = 2H_2 + \mathsf{L}_1 (h_1 + H_1), \\ \{w_3, H_0\} = 3H_3 + \mathsf{L}_2 H_1 + \mathsf{L}_1 H_2 + \tfrac{1}{2}(\mathsf{L}_2 - \mathsf{L}_1^2) h_1. \end{gathered} \tag{4.78}$$

We are now free to choose the terms in the new Hamiltonian, H_n, and solve the differential equations (4.78) for the corresponding generating function w_n. Typically H_n is chosen so that w_n has only oscillating terms, thereby ensuring that the new and old variables differ only by oscillating terms.

For example, suppose that the Hamiltonian has the form (4.43). In the present notation the canonical variables $\mathbf{z}$ are the angles and actions $(\boldsymbol{\theta}, \mathbf{J})$, the unperturbed Hamiltonian is $h_0(\mathbf{J})$, and the perturbation is $h_1(\mathbf{J}, \boldsymbol{\theta}) = \sum_{\mathbf{m}} h_{\mathbf{m}}(\mathbf{J}) \exp(\mathrm{i}\mathbf{m} \cdot \boldsymbol{\theta})$. Now consider the first of equations (4.78), corresponding to first-order perturbation theory. We choose H_1 so the average of the right side over angles vanishes. This requires $H_1 = \sum_{\mathbf{m} \in S} h_{\mathbf{m}}(\mathbf{J})$, where S is the set of all integer multiplets $\mathbf{m}$ for which $\mathbf{m} \cdot \boldsymbol{\Omega}(\mathbf{J}) = \mathbf{m} \cdot \partial h_0 / \partial \mathbf{J} = 0$ (the secular terms). Then the first of equations (4.78) becomes

$$\{w_1, H_0\} = \boldsymbol{\Omega}(\mathbf{J}) \cdot \frac{\partial w_1}{\partial \boldsymbol{\theta}} = H_1 - h_1 = -\sum_{\mathbf{m} \notin S} h_{\mathbf{m}}(\mathbf{J}) \exp(\mathrm{i}\mathbf{m} \cdot \boldsymbol{\theta}), \tag{4.79}$$

which is easily solved for w_1:

$$w_1(\mathbf{J}, \boldsymbol{\theta}) = \mathrm{i} \sum_{\mathbf{m} \notin S} \frac{h_{\mathbf{m}}(\mathbf{J}) \exp(\mathrm{i}\mathbf{m} \cdot \boldsymbol{\theta})}{\mathbf{m} \cdot \boldsymbol{\Omega}(\mathbf{J})}. \tag{4.80}$$

Once w_1 is determined we can find the relation between the old and new phase-space variables, $\mathbf{z}$ and $\mathbf{z}'(\epsilon, \mathbf{z})$. Let $Z_k(\mathbf{z})$ be the function that gives the kth coordinate of $\mathbf{z}$, $z_k = Z_k(\mathbf{z})$. Then equation (4.64) implies that

$$z'_k = (\mathsf{T}_\epsilon Z_k)\mathbf{z}. \tag{4.81}$$

From equations (4.75),

$$z'_k = z_k + \epsilon\{z_k, w_1\} + \mathrm{O}(\epsilon^2). \tag{4.82}$$

Thus to $\mathrm{O}(\epsilon)$,

$$\mathbf{J}' = \mathbf{J} + \epsilon\{\mathbf{J}, w_1\} = \mathbf{J} - \epsilon \frac{\partial w_1}{\partial \boldsymbol{\theta}} = \mathbf{J} + \epsilon \sum_{\mathbf{m} \notin S} \frac{\mathbf{m} h_{\mathbf{m}}(\mathbf{J})}{\mathbf{m} \cdot \boldsymbol{\Omega}} \exp(\mathrm{i}\mathbf{m} \cdot \boldsymbol{\theta}), \tag{4.83}$$

$$\boldsymbol{\theta}' = \boldsymbol{\theta} + \epsilon\{\boldsymbol{\theta}, w_1\} = \boldsymbol{\theta} + \epsilon\frac{\partial w_1}{\partial \mathbf{J}} = \boldsymbol{\theta} + \mathrm{i}\epsilon \sum_{\mathbf{m}\notin S} \frac{\partial}{\partial \mathbf{J}}\left[\frac{h_{\mathbf{m}}(\mathbf{J})}{\mathbf{m}\cdot\boldsymbol{\Omega}(\mathbf{J})}\right]\exp(\mathrm{i}\mathbf{m}\cdot\boldsymbol{\theta}).$$

Notice that at order ϵ the local generating function w_1 is identical to the mixed-variable generating function used in the Poincaré–von Zeipel approach (eq. 4.59), as is the relation between the primed and unprimed angles and actions (eqs. 4.56).

In second-order perturbation theory, we use the second of equations (4.78), which can be written

$$\{w_2, H_0\} = \boldsymbol{\Omega}(\mathbf{J})\cdot\frac{\partial w_2}{\partial \boldsymbol{\theta}} = 2H_2 + \{h_1, w_1\} + \{H_1, w_1\}. \tag{4.84}$$

We choose H_2 so the average of the right side over angles vanishes. Denoting this average by $\langle\cdot\rangle$ we have

$$H_2 = -\tfrac{1}{2}\langle\{h_1, w_1\} + \{H_1, w_1\}\rangle. \tag{4.85}$$

Since H_1 depends only on the actions,

$$\langle\{H_1, w_1\}\rangle = -\sum_i \frac{\partial H_1}{\partial J_i}\left\langle\frac{\partial w_1}{\partial \theta_i}\right\rangle = 0; \tag{4.86}$$

thus the second Poisson bracket in equation (4.85) vanishes. Since h_1 is real, it can be replaced by its complex conjugate h_1^* and then

$$\begin{aligned} H_2 &= -\tfrac{1}{2}\langle\{h_1, w_1\}\rangle \\ &= -\tfrac{1}{2}\mathrm{i}\sum_{\mathbf{m}}\sum_{\mathbf{m}'\notin S}\left\langle\left\{h_{\mathbf{m}}^*(\mathbf{J})\exp(-\mathrm{i}\mathbf{m}\cdot\boldsymbol{\theta}), \frac{h_{\mathbf{m}'}\exp(\mathrm{i}\mathbf{m}'\cdot\boldsymbol{\theta})}{\mathbf{m}'\cdot\boldsymbol{\Omega}}\right\}\right\rangle \\ &= -\tfrac{1}{2}\sum_{\mathbf{m}\notin S}\mathbf{m}\cdot\frac{\partial}{\partial \mathbf{J}}\frac{|h_{\mathbf{m}}(\mathbf{J})|^2}{\mathbf{m}\cdot\boldsymbol{\Omega}(\mathbf{J})}. \end{aligned} \tag{4.87}$$

This result can be substituted into equation (4.84), which can then be solved for $w_2(\boldsymbol{\theta}, \mathbf{J})$; then w_2 can be used to find the relation between the old and new angle-action variables through equation (4.82).

Thus we have shown that the solution to Hamilton's equations for the Hamiltonian $h_0(\mathbf{J}) + \epsilon h_1(\boldsymbol{\theta}, \mathbf{J})$ is given by the solution for the Hamiltonian

$$\begin{aligned} H(\boldsymbol{\theta}', \mathbf{J}') &= H_0(\mathbf{J}') + \epsilon H_1(\mathbf{J}') + \epsilon^2 H_2(\mathbf{J}') + \mathrm{O}(\epsilon^3) \\ &= h_0(\mathbf{J}') + \sum_{\mathbf{m}\in S} h_{\mathbf{m}}(\mathbf{J}') - \tfrac{1}{2} \sum_{\mathbf{m}\notin S} \mathbf{m} \cdot \frac{\partial}{\partial \mathbf{J}'} \frac{|h_{\mathbf{m}}(\mathbf{J}')|^2}{\mathbf{m} \cdot \boldsymbol{\Omega}(\mathbf{J}')} + \mathrm{O}(\epsilon^3), \end{aligned} \tag{4.88}$$

in which $(\boldsymbol{\theta}', \mathbf{J}')$ and $(\mathbf{J}, \boldsymbol{\theta})$ differ by terms that are $\mathrm{O}(\epsilon)$ and average to zero over the angles.

This process is straightforward to continue to higher orders in ϵ. Of course, all of the perturbation theories described in this section continue to suffer from one of the problems described at the end of §4.2.1: the presence of small divisors. Because of these, the sum over $\mathbf{m}$ only converges like an asymptotic series, that is, the results of perturbation theory at any order are valid over a fixed time interval as $\epsilon \to 0$, but not over an arbitrarily large time interval for any fixed value of ϵ.

For more thorough treatments of Lie operator perturbation theory, see Deprit (1969), Cary (1981), Lichtenberg & Lieberman (1992), or Sussman & Wisdom (2001).

4.3 The disturbing function

In most planetary systems the planets travel on nearly circular orbits close to a common plane. In this case the interaction Hamiltonian or disturbing function $H_{ij}(\boldsymbol{\theta}_i, \mathbf{J}_i, \boldsymbol{\theta}_j, \mathbf{J}_j)$ between planets i and j (eq. 4.41) can be evaluated as an expansion in powers of the eccentricities and inclinations of the two planets. Thousands of pages have been written about these expansions, but the analysis is straightforward in principle, and using computer algebra it is straightforward in practice as well.

The interaction Hamiltonian contains both the gravitational interaction potential and the additional terms that arise in non-inertial coordinate systems such as astrocentric or Jacobi coordinates (§4.1). We focus first on the interaction potential,

$$\Phi_{ij}(\mathbf{r}_i, \mathbf{r}_j) = -\frac{\mathbb{G} m_i m_j}{\Delta_{ij}}, \quad \text{where} \quad \Delta_{ij} \equiv |\mathbf{r}_i - \mathbf{r}_j|. \tag{4.89}$$

We shall work with the orbital elements a, e, I, λ, ϖ, Ω—semimajor axis, eccentricity, inclination, mean longitude, longitude of periapsis, and longitude of the ascending node—which are related to the Delaunay variables by

$$\Lambda = (\mathbb{G}Ma)^{1/2}, \qquad L = \Lambda(1-e^2)^{1/2}, \qquad L_z = L\cos I,$$
$$\ell = \lambda - \varpi, \qquad \omega = \varpi - \Omega; \tag{4.90}$$

the variable Ω is common to both sets. We set the dummy variables $i = 1$ and $j = 2$.

Some restrictions on the form of Δ_{12}^{-1} can be derived from symmetry arguments. If we replace λ_1 by $\lambda_1 + 2\pi$ the position of the planet is unchanged, so Δ_{12}^{-1} is unchanged. Thus Δ_{12}^{-1} must be a periodic function of λ_1 with period 2π. Similarly it must be a periodic function of ϖ_1 and Ω_1 and of the same variables with label 2. Thus it can be written as a Fourier series (Appendix B.4):

$$\frac{1}{\Delta_{12}} = \sum_{\substack{j_1 k_1 m_1 \\ j_2 k_2 m_2}} H_{j_1 k_1 m_1 j_2 k_2 m_2}(a_1, e_1, I_1, a_2, e_2, I_2) \times \cos(j_1\lambda_1 + k_1\varpi_1 + m_1\Omega_1 - j_2\lambda_2 - k_2\varpi_2 - m_2\Omega_2). \tag{4.91}$$

A general Fourier series contains both sine and cosine terms, but the sine terms in this expansion all vanish.[4]

When the eccentricity e_1 is zero, Δ_{12}^{-1} cannot depend on the periapsis direction ϖ_1. Similarly, when the inclination I_1 is zero, Δ_{12}^{-1} cannot depend

[4] Proof: Consider two coordinate systems defined by the Cartesian axes $(\hat{\mathbf{x}}, \hat{\mathbf{y}}, \hat{\mathbf{z}})$ and $(\hat{\mathbf{x}}', \hat{\mathbf{y}}', \hat{\mathbf{z}}')$. The primed axes are obtained by a rotation of $180°$ around the x-axis; thus $\hat{\mathbf{x}}' = \hat{\mathbf{x}}$, $\hat{\mathbf{y}}' = -\hat{\mathbf{y}}$, $\hat{\mathbf{z}}' = -\hat{\mathbf{z}}$. If the elements of a point on an orbit in the unprimed coordinates are a, e, I, λ, ϖ, Ω, then the elements of the same point in the primed coordinates are a, e, I', λ', ϖ', Ω' with $\lambda' = \lambda - 2\Omega$, $\varpi' = \varpi - 2\Omega$, $\Omega' = \pi - \Omega$ and $I' = \pi - I$ (all quantities modulo 2π). If we now time-reverse the orbit, the elements become (eq. 2.37) a, e, I'', λ'', ϖ'', Ω'' with $\lambda'' = 2\Omega' - \lambda'$, $\varpi'' = 2\Omega' - \varpi'$, $\Omega'' = \pi + \Omega'$ and $I'' = \pi - I'$. Then $\lambda'' = -\lambda$, $\varpi'' = -\varpi$, and $\Omega'' = -\Omega$. The gravitational potential between two mass points must be independent of which coordinate system we use and whether or not the orbits are time reversed. Therefore the potential has to be the same in the unprimed and doubly primed coordinates, which implies that it must be an even function of $j_1\lambda_1 + k_1\varpi_1 + m_1\Omega_1 - j_2\lambda_2 - k_2\varpi_2 - m_2\Omega_2$, so no sine terms are present.

on the nodal longitude Ω_1. More generally, H_1 has to be an analytic function of the position and this restricts the possible forms of H_1 to satisfy the **d'Alembert property**,[5] which states the following: if the eccentricities and inclinations e_1, e_2, I_1, I_2 are small, then

$$H_{j_1k_1m_1j_2k_2m_2}(a_1, e_1, I_1, a_2, e_2, I_2) = \mathrm{O}(e_1^{|k_1|}e_2^{|k_2|}I_1^{|m_1|}I_2^{|m_2|}). \quad (4.92)$$

Now suppose that we rotate the origin of the azimuthal coordinate system by some angle ψ. Then all of the angular variables increase by ψ, so the argument of the cosine increases by $(j_1 + k_1 + m_1 - j_2 - k_2 - m_2)\psi$. Since the interaction potential must be independent of the choice of the origin, we require that

$$j_1 + k_1 + m_1 - j_2 - k_2 - m_2 = 0. \quad (4.93)$$

To obtain an explicit expression for Δ_{12}^{-1} we assume that the eccentricities e_1 and e_2 and the inclinations I_1 and I_2 are of order $\epsilon \ll 1$. To make this dependence explicit, we replace e_1 and I_1 by ϵe_1 and ϵI_1, with the same replacement for e_2 and I_2, then set ϵ to unity once we have obtained a consistent ordering. We refer to terms in the disturbing function proportional to ϵ^k as having **degree** k.

We seek an expression for Δ_{12}^{-1} that is correct to $\mathrm{O}(\epsilon^k)$. We write

$$\frac{1}{\Delta_{12}} = \frac{1}{|\mathbf{r}_1 - \mathbf{r}_2|} = \frac{1}{(r_1^2 + r_2^2 - 2\,\mathbf{r}_1 \cdot \mathbf{r}_2)^{1/2}}, \quad (4.94)$$

then use equations (1.70) to evaluate $\mathbf{r}_1 \cdot \mathbf{r}_2$ in terms of r_1 and the orbital elements f_1, ω_1, Ω_1, and I_1 and their analogs for body 2. We replace r_1, f_1, r_2 and f_2 by their expressions (1.151) and (1.155) in terms of a, e, and ℓ, keeping terms up to $\mathrm{O}(\epsilon^k)$. We also expand $\cos I$ and $\sin I$, keeping terms up to $\mathrm{O}(\epsilon^k)$. We then substitute these expansions into equation (4.94) and expand the result to $\mathrm{O}(\epsilon^k)$. For example, if $k = 1$,

$$\frac{1}{\Delta_{12}} = \frac{1}{[a_1^2+a_2^2-2a_1a_2\cos(\lambda_1-\lambda_2)]^{1/2}} + \frac{\epsilon}{[a_1^2+a_2^2-2a_1a_2\cos(\lambda_1-\lambda_2)]^{3/2}}$$

[5] Similarly, if a smooth function $f(x, y)$ of the Cartesian coordinates x and y is rewritten in polar coordinates (r, ϕ), then its expansion in powers of r will have the form $c_0 + c_1 r\cos(\phi - \phi_1) + c_2 r^2 \cos 2(\phi - \phi_2) + \cdots$; that is, the coefficient of $\cos m(\phi - \phi_n)$ will be $\mathrm{O}(r^m)$.

$$\times \left[a_1^2 e_1 \cos(\lambda_1 - \varpi_1) + a_2^2 e_2 \cos(\lambda_2 - \varpi_2) + \tfrac{1}{2} a_1 a_2 e_1 \cos(2\lambda_1 - \lambda_2 - \varpi_1) + \tfrac{1}{2} a_1 a_2 e_2 \cos(2\lambda_2 - \lambda_1 - \varpi_2) - \tfrac{3}{2} a_1 a_2 e_1 \cos(\lambda_2 - \varpi_1) - \tfrac{3}{2} a_1 a_2 e_2 \cos(\lambda_1 - \varpi_2)\right] + \mathrm{O}(\epsilon^2). \tag{4.95}$$

This expression and its more complicated cousins for higher k contain denominators of the form $[a_1^2 + a_2^2 - 2a_1 a_2 \cos(\lambda_1 - \lambda_2)]^{-s}$ where $s = \frac{1}{2}, \frac{3}{2}, \frac{5}{2}, \ldots$. We replace these denominators by a Fourier series (Appendix B.4), using the expansion (eq. 4.105)

$$\left(1 + \alpha^2 - 2\alpha \cos\phi\right)^{-s} = \tfrac{1}{2} \sum_{m=-\infty}^{\infty} b_s^m(\alpha) \cos m\phi, \tag{4.96}$$

where $b_s^m(\alpha)$ is a **Laplace coefficient**. The Laplace coefficients play a central role in the evaluation of the disturbing function, and their properties are described in detail in §4.4. In most cases the body labels 1 and 2 are chosen such that $\alpha = a_1/a_2$ and $\alpha < 1$, but the Laplace coefficients are well defined for any positive value of α.

Once this replacement is made, products of trigonometric functions are converted to sums of trigonometric functions using the identity $\cos a \cos b = \frac{1}{2}\cos(a + b) + \frac{1}{2}\cos(a - b)$. We then adjust the dummy index m such that (i) all terms with the same argument of the cosine function are collected together; (ii) the variable λ_1 always appears in the argument in the combination $m\lambda_1$; (iii) the variable λ_2 always appears in the combination $(m+j)\lambda_2$ where $j \geq 0$. We have

$$\frac{1}{\Delta_{12}} = \frac{1}{2a_2} \sum_{m=-\infty}^{\infty} b_{1/2}^m \cos(m\lambda_1 - m\lambda_2) + \frac{\epsilon}{a_2} \sum_{m=-\infty}^{\infty} \Big\{\left[\tfrac{1}{4}\alpha b_{3/2}^{m+2} + \tfrac{1}{2}\alpha^2 b_{3/2}^{m+1} - \tfrac{3}{4}\alpha b_{3/2}^m\right] e_1 \cos[m\lambda_1 - (m+1)\lambda_2 + \varpi_1] + \left[\tfrac{1}{4}\alpha b_{3/2}^{m-1} + \tfrac{1}{2} b_{3/2}^m - \tfrac{3}{4}\alpha b_{3/2}^{m+1}\right] e_2 \cos[m\lambda_1 - (m+1)\lambda_2 + \varpi_2]\Big\} + \epsilon^2 \sum_{m=-\infty}^{\infty} X_m + \mathrm{O}(\epsilon^3). \tag{4.97}$$

In this expression the argument of all the Laplace coefficients is $\alpha = a_1/a_2$, and X_m represents the terms of degree $k = 2$.

Notice that there are no inclination-dependent terms at order ϵ; these only appear at $\mathrm{O}(\epsilon^2)$.

As described in §4.4 there are many relations between different Laplace coefficients, so the factors in front of the cosines can be written in many different forms. The most compact form for equation (4.97) is

$$\begin{aligned}\frac{1}{\Delta_{12}} = {} & \frac{1}{2a_2}\sum_{m=-\infty}^{\infty} b_{1/2}^{m}\cos(m\lambda_1 - m\lambda_2) + \frac{\epsilon}{a_2}\sum_{m=-\infty}^{\infty} \\ & \big\{-(m+1+\tfrac{1}{2}\alpha D)b_{1/2}^{m+1}e_1\cos[m\lambda_1-(m+1)\lambda_2+\varpi_1] \\ & \quad +(m+\tfrac{1}{2}+\tfrac{1}{2}\alpha D)b_{1/2}^{m}e_2\cos[m\lambda_1-(m+1)\lambda_2+\varpi_2]\big\} \\ & +\epsilon^2\sum_{m=-\infty}^{\infty} X_m + \mathrm{O}(\epsilon^3),\end{aligned} \tag{4.98}$$

where $Db_s^m \equiv \mathrm{d}b_s^m(\alpha)/\mathrm{d}\alpha$.

The second-degree terms are given by

$$\begin{aligned} a_2X_m = {} & \big[(e_1^2+e_2^2)(\tfrac{1}{8}\alpha^2D^2+\tfrac{1}{4}\alpha D-\tfrac{1}{2}m^2)b_{1/2}^{m} \\ & \quad -\tfrac{1}{16}(I_1^2+I_2^2)\alpha(b_{3/2}^{m-1}+b_{3/2}^{m+1})\big]\cos(m\lambda_1-m\lambda_2) \\ & +\tfrac{1}{8}e_1^2\left[6+11m+4m^2+(6+4m)\alpha D+\alpha^2D^2\right]b_{1/2}^{m+2} \\ & \quad \times\cos[m\lambda_1-(m+2)\lambda_2+2\varpi_1] \\ & +\tfrac{1}{8}e_2^2\left[4+9m+4m^2+(6+4m)\alpha D+\alpha^2D^2\right]b_{1/2}^{m} \\ & \quad \times\cos[m\lambda_1-(m+2)\lambda_2+2\varpi_2] \\ & -\tfrac{1}{4}e_1e_2\left[6+10m+4m^2+(6+4m)\alpha D+\alpha^2D^2\right]b_{1/2}^{m+1} \\ & \quad \times\cos[m\lambda_1-(m+2)\lambda_2+\varpi_1+\varpi_2] \\ & +\tfrac{1}{4}e_1e_2\left(2+6m+4m^2-2\alpha D-\alpha^2D^2\right)b_{1/2}^{m+1} \\ & \quad \times\cos(m\lambda_1-m\lambda_2+\varpi_1-\varpi_2) \\ & -\tfrac{1}{4}I_1I_2\alpha b_{3/2}^{m+1}\cos[m\lambda_1-(m+2)\lambda_2+\Omega_1+\Omega_2] \\ & +\tfrac{1}{4}I_1I_2\alpha b_{3/2}^{m+1}\cos(m\lambda_1-m\lambda_2+\Omega_1-\Omega_2)\end{aligned}$$

$$+ \tfrac{1}{8} I_1^2 \alpha b_{3/2}^{m+1} \cos[m\lambda_1 - (m+2)\lambda_2 + 2\Omega_1]$$
$$+ \tfrac{1}{8} I_2^2 \alpha b_{3/2}^{m+1} \cos[m\lambda_1 - (m+2)\lambda_2 + 2\Omega_2]. \tag{4.99}$$

Since Δ_{12} is a symmetric function of the labels 1 and 2, the expansions (4.98) and (4.99) are also symmetric; that is, they are invariant under the **exchange operator**, which swaps the labels 1 and 2. More precisely, we set $a_1 = a_2'$, $a_2 = a_1'$, $\alpha = a_1/a_2 = a_2'/a_1' = 1/\alpha'$, $\lambda_1 = \lambda_2'$, and so forth. In some cases a single term of the disturbing function is invariant under the action of the exchange operator, that is, it looks the same in the primed and unprimed variables. Thus, for example, the term $I \equiv (2a_2)^{-1} b_{1/2}^m(\alpha) \cos m(\lambda_1 - \lambda_2)$ in the first line of equation (4.98) is transformed to $I' \equiv (2a_1')^{-1} b_{1/2}^m(1/\alpha') \cos m(\lambda_2' - \lambda_1')$ and using (4.108) it is straightforward to show that $I' = (2a_2')^{-1} b_{1/2}^m(\alpha') \cos m(\lambda_1' - \lambda_2')$, the same as I except in primed variables. In other cases a pair of terms are transformed into one other by the exchange operator. For example, the term

$$J = -\frac{1}{a_2}[(m+1) b_{1/2}^{m+1}(\alpha) + \tfrac{1}{2}\alpha D b_{1/2}^{m+1}(\alpha)] \\ \times e_1 \cos[m\lambda_1 - (m+1)\lambda_2 + \varpi_1] \tag{4.100}$$

is transformed to

$$J' = -\frac{1}{a_1'}[(m+1) b_{1/2}^{m+1}(1/\alpha') + \tfrac{1}{2}\alpha'^{-1} D b_{1/2}^{m+1}(1/\alpha')] \\ \times e_2' \cos[m\lambda_2' - (m+1)\lambda_1' + \varpi_2']. \tag{4.101}$$

Setting $m = -m' - 1$ and using the symmetry relations (4.106), (4.108) and (4.137), we find

$$J' = \frac{1}{a_2'}[(m' + \tfrac{1}{2}) b_{1/2}^{m'}(\alpha') + \tfrac{1}{2}\alpha' D b_{1/2}^{m'}(\alpha')] \\ \times e_2' \cos[m'\lambda_1' - (m'+1)\lambda_2' + \varpi_2'], \tag{4.102}$$

which is the same as the term on the third line of equation (4.98), except in primed variables.

These relations provide valuable checks of the accuracy of algebraic expansions of the disturbing function.[6]

In astrocentric coordinates the indirect potential contributes additional terms to the disturbing function. The indirect potential exerted by body 2 on body 1 is (eq. 4.6)

$$\Phi_1^{\text{ind}} = \frac{\mathbb{G}m_2\mathbf{r}_2\cdot\mathbf{r}_1}{|\mathbf{r}_2|^3} = \frac{\mathbb{G}m_2 a_1}{a_2^2}\frac{a_2^2\,\mathbf{r}_2\cdot\mathbf{r}_1}{a_1|\mathbf{r}_2|^3}. \tag{4.103}$$

The expansion to degree 2 is

$$\begin{aligned}\frac{a_2^2\,\mathbf{r}_2\cdot\mathbf{r}_1}{a_1|\mathbf{r}_2|^3} = {} & \cos(\lambda_1-\lambda_2)+\epsilon\big[\tfrac{1}{2}e_1\cos(2\lambda_1-\lambda_2-\varpi_1)-\tfrac{3}{2}e_1\cos(\lambda_2-\varpi_1)\\ & +2e_2\cos(\lambda_1-2\lambda_2+\varpi_2)\big]\\ & +\epsilon^2\big[-\big(\tfrac{1}{2}e_1^2+\tfrac{1}{2}e_2^2+\tfrac{1}{4}I_1^2+\tfrac{1}{4}I_2^2\big)\cos(\lambda_1-\lambda_2)\\ & +\tfrac{3}{8}e_1^2\cos(3\lambda_1-\lambda_2-2\varpi_1)+\tfrac{1}{8}e_1^2\cos(\lambda_1+\lambda_2-2\varpi_1)\\ & +\tfrac{1}{8}e_2^2\cos(\lambda_1+\lambda_2-2\varpi_2)+\tfrac{27}{8}e_2^2\cos(\lambda_1-3\lambda_2+2\varpi_2)\\ & +e_1e_2\cos(2\lambda_1-2\lambda_2-\varpi_1+\varpi_2)-3e_1e_2\cos(2\lambda_2-\varpi_1-\varpi_2)\\ & +\tfrac{1}{2}I_1I_2\cos(\lambda_1-\lambda_2-\Omega_1+\Omega_2)-\tfrac{1}{2}I_1I_2\cos(\lambda_1+\lambda_2-\Omega_1-\Omega_2)\\ & +\tfrac{1}{4}I_1^2\cos(\lambda_1+\lambda_2-2\Omega_1)+\tfrac{1}{4}I_2^2\cos(\lambda_1+\lambda_2-2\Omega_2)\big]+\mathrm{O}(\epsilon^3).\end{aligned} \tag{4.104}$$

The indirect potential exerted by body 1 on body 2 is found by exchanging the labels 1 and 2. Notice that the indirect potential is *not* a symmetric function of the labels and is not invariant under the exchange operator.

Expansions to higher degree in the eccentricities and inclinations are available in a variety of references. Most of these, with the exception of Murray & Dermott (1999), give the expansion in terms of the mutual inclination and the mutual ascending node—in effect, they assume that the coordinate system is chosen such that one of the inclinations is zero.

[6] Expansions of the disturbing function would be shorter and simpler if the Laplace coefficients were re-defined to be symmetric functions of a_1 and a_2. We have kept the traditional definition to facilitate comparison to expressions in the literature, and because symmetrizing the Laplace coefficients would eliminate the ability of the exchange operator to detect algebraic errors.

Murray & Dermott (1999) give the expansion to fourth degree in the eccentricities and inclinations. Among older references, Peirce (1849) gives the expansion to sixth degree, Le Verrier (1855) to seventh degree, and Boquet (1889) to eighth degree. Le Verrier made only one nontrivial error, discovered a century later by computer algebra (Murray 1985).

There are two disadvantages to this approach. First, in many cases we are only interested in a single cosine term (for example, when the dynamics is dominated by a resonance), but to find the factor in front of the cosine requires the complete expansion of the disturbing function to the degree of that term. Second, we would like to find the disturbing function using computer algebra, and the approach we have described requires sophisticated algebra including Taylor series expansions, derivatives, and the like, as well as judgments on how to simplify the results. Alternative approaches are described by Murray & Dermott (1999). A review of early work on computer determination of the disturbing function is given by Henrard (1989).

4.4 Laplace coefficients

The **Laplace coefficients** $b_s^m(\alpha)$ are defined by the generating function

$$\left(1 - 2\alpha\cos\phi + \alpha^2\right)^{-s} = \tfrac{1}{2}\sum_{m=-\infty}^{\infty} b_s^m(\alpha)\cos m\phi \tag{4.105}$$

and the assumption

$$b_s^{-m}(\alpha) = b_s^m(\alpha). \tag{4.106}$$

Equation (4.105) is equivalent to

$$(1+\alpha^2 - \alpha z - \alpha z^{-1})^{-s} = (1-\alpha z)^{-s}(1-\alpha z^{-1})^{-s} = \tfrac{1}{2}\sum_{m=-\infty}^{\infty} b_s^m(\alpha) z^m. \tag{4.107}$$

For our purposes α is always real and positive, m is an integer, and $s = \frac{1}{2}, \frac{3}{2}, \frac{5}{2}, \ldots$, although most of the results in this section do not require these restrictions. It is easy to show that

$$b_s^m(\alpha^{-1}) = \alpha^{2s} b_s^m(\alpha). \tag{4.108}$$

Now multiply equation (4.105) by z^{-k} where k is an integer, set $z = \exp(\mathrm{i}\phi)$, and integrate over ϕ from 0 to 2π. Since $\int_0^{2\pi} \mathrm{d}\phi \, \exp[\mathrm{i}(m-k)\phi] = 0$ if $k \neq m$ and 2π if $k = m$, we have

$$b_s^m(\alpha) = \frac{1}{\pi}\int_0^{2\pi} \frac{\mathrm{d}\phi \, \exp(-\mathrm{i}m\phi)}{(1-2\alpha\cos\phi+\alpha^2)^s} = \frac{2}{\pi}\int_0^{\pi} \frac{\mathrm{d}\phi \, \cos m\phi}{(1-2\alpha\cos\phi+\alpha^2)^s}. \tag{4.109}$$

In terms of standard special functions

$$b_s^m(\alpha) = 2\frac{\Gamma(s+|m|)}{\Gamma(s)\Gamma(|m|+1)}\alpha^{|m|}F(s, s+|m|; |m|+1; \alpha^2), \quad 0 \le \alpha < 1, \tag{4.110}$$

and the Laplace coefficient with argument $\alpha > 1$ is found from equation (4.108). Here $\Gamma(z)$ is the gamma function (Appendix C.3), and $F(a, b; c; z)$ is the hypergeometric function defined by the series

$$F(a, b; c; z) = \frac{\Gamma(c)}{\Gamma(a)\Gamma(b)} \sum_{n=0}^{\infty} \frac{\Gamma(a+n)\Gamma(b+n)}{\Gamma(c+n)\Gamma(n+1)} z^n, \tag{4.111}$$

which converges for $|z| < 1$; this function is also written as ${}_2F_1(a, b; c; z)$ in some references. Thus

$$b_s^m(\alpha) = \frac{2}{\Gamma^2(s)}\alpha^{|m|} \sum_{n=0}^{\infty} \frac{\Gamma(s+n)\Gamma(|m|+s+n)}{\Gamma(|m|+1+n)\Gamma(n+1)}\alpha^{2n}, \quad 0 \le \alpha < 1. \tag{4.112}$$

The most important special cases are

$$\begin{aligned} b_{1/2}^0(\alpha) &= \frac{4}{\pi}K(\alpha) = \frac{4}{\pi}R_F(0, 1-\alpha^2, 1), \\ b_{1/2}^1(\alpha) &= \frac{4}{\pi\alpha}\big[K(\alpha) - E(\alpha)\big] = \frac{4\alpha}{3\pi}R_D(0, 1-\alpha^2, 1). \end{aligned} \tag{4.113}$$

Here $K(\cdot)$, $E(\cdot)$, $R_F(\cdot)$ and $R_D(\cdot)$ are elliptic integrals (Appendix C.4).

Any Laplace coefficient can be evaluated by numerical quadrature of equation (4.109). A second approach is to use equation (4.110) to express the Laplace coefficient in terms of a hypergeometric function, since many programming languages have procedures for evaluating hypergeometric functions. A third approach, which was preferred when calculations were done by hand, is to use recursion relations as described in the next subsection.

4.4.1 Recursion relations

In principle, all of the Laplace coefficients with half-integral values of s can be found from the coefficients in equations (4.113) using a set of recursion relations. To derive the first of these, differentiate equation (4.107) with respect to z to obtain

$$\frac{\alpha s(1 - z^{-2})}{(1 + \alpha^2 - \alpha z - \alpha z^{-1})^{s+1}} = \frac{1}{2} \sum_{m=-\infty}^{\infty} b_s^m(\alpha) m z^{m-1}. \tag{4.114}$$

Now multiply the result by $z(1 + \alpha^2 - \alpha z - \alpha z^{-1})$:

$$\frac{\alpha s(z - z^{-1})}{(1 + \alpha^2 - \alpha z - \alpha z^{-1})^s} = \frac{1}{2} \sum_{m=-\infty}^{\infty} m b_s^m(\alpha) \left[z^m(1 + \alpha^2) - \alpha z^{m+1} - \alpha z^{m-1} \right]. \tag{4.115}$$

We replace the denominator on the left side using equation (4.107) and increment or decrement the dummy index m such that the variable z enters as z^m in all terms:

$$\sum_{m=-\infty}^{\infty} s\left[b_s^{m-1}(\alpha) - b_s^{m+1}(\alpha)\right] z^m \tag{4.116}$$
$$= \sum_{m=-\infty}^{\infty} \left[m b_s^m(\alpha)(\alpha + \alpha^{-1}) - (m-1) b_s^{m-1}(\alpha) - (m+1) b_s^{m+1}(\alpha) \right] z^m.$$

Since the coefficient of z^m must be the same on each side of the equation,

$$b_s^{m+1}(\alpha) = \frac{m}{m - s + 1}\left(\alpha + \frac{1}{\alpha}\right) b_s^m(\alpha) - \frac{m + s - 1}{m - s + 1} b_s^{m-1}(\alpha). \tag{4.117}$$

Similarly, we can solve for $b_s^{m-1}(\alpha)$ to find

$$b_s^{m-1}(\alpha) = \frac{m}{m + s - 1}\left(\alpha + \frac{1}{\alpha}\right) b_s^m(\alpha) - \frac{m - s + 1}{m + s - 1} b_s^{m+1}(\alpha). \tag{4.118}$$

This result can also be derived from equation (4.117) by replacing m with $-m$ and using the symmetry relation (4.106). With these two equations we can find the Laplace coefficients for all values of the superscript m at given s from any two coefficients with adjacent values of m.

To derive a second recursion relation, replace the denominator in the left side of equation (4.114) using equation (4.105) in the form

$$(1+\alpha^2-\alpha z-\alpha z^{-1})^{-s-1}=\tfrac{1}{2}\sum_{m=-\infty}^{\infty} b_{s+1}^m(\alpha)z^m. \tag{4.119}$$

Then increment or decrement the index m such that the variable z enters as z^{m-1} in all terms. The requirement that the equation is satisfied for each m yields

$$\alpha s b_{s+1}^{m-1}(\alpha)-\alpha s b_{s+1}^{m+1}(\alpha)=m b_s^m(\alpha). \tag{4.120}$$

Now use (4.117) to replace b_{s+1}^{m+1} by b_{s+1}^m and b_{s+1}^{m-1} (this derivation is not valid for $m=0$ but the formula below is still correct):

$$\frac{2s\alpha}{m-s}b_{s+1}^{m-1}(\alpha)-\frac{s}{m-s}\left(1+\alpha^2\right)b_{s+1}^m(\alpha)=b_s^m(\alpha). \tag{4.121}$$

Replacing m by $1-m$ and using the symmetry relation (4.106) gives

$$\frac{s}{m+s-1}\left(1+\alpha^2\right)b_{s+1}^{m-1}(\alpha)-\frac{2s\alpha}{m+s-1}b_{s+1}^m(\alpha)=b_s^{m-1}(\alpha). \tag{4.122}$$

We can eliminate $b_{s+1}^{m-1}(\alpha)$ from (4.121) and (4.122) to obtain

$$b_{s+1}^m(\alpha)=\frac{(s-m)(1+\alpha^2)b_s^m(\alpha)+2(m+s-1)\alpha b_s^{m-1}(\alpha)}{s(1-\alpha^2)^2}. \tag{4.123}$$

If we replace m by $-m$ and use the symmetry relation (4.106), we find

$$b_{s+1}^m(\alpha)=\frac{2(s-1-m)\alpha b_s^{m+1}(\alpha)+(s+m)(1+\alpha^2)b_s^m(\alpha)}{s(1-\alpha^2)^2}. \tag{4.124}$$

Thus if we know the Laplace coefficients with a given subscript s, we can compute the coefficients with subscript $s+1$.

The recursion relations (4.117), (4.118), (4.123) and (4.124) allow the determination of $b_s^m(\alpha)$ for all m and $s=\frac{1}{2},\frac{3}{2},\frac{5}{2},\ldots$ given the expressions for $b_{1/2}^0(\alpha)$ and $b_{1/2}^1(\alpha)$ in equation (4.113).

4.4.2 Limiting cases

When the argument $\alpha \ll 1$, the Laplace coefficient can be approximated using the first few terms of the Taylor series (4.112). For $m \geq 0$

$$b_s^m(\alpha) = \frac{2\Gamma(m+s)\alpha^m}{\Gamma(m+1)\Gamma(s)}\left[1+\frac{s(m+s)}{m+1}\alpha^2+\frac{s(s+1)(m+s)(m+s+1)}{2(m+1)(m+2)}\alpha^4\right.$$
$$\left.+\frac{s(s+1)(s+2)(m+s)(m+s+1)(m+s+2)}{3!(m+1)(m+2)(m+3)}\alpha^6+\mathrm{O}(\alpha^8)\right], \quad (4.125)$$

and this result can be extended to $m < 0$ using (4.106). The asymptotic behavior as $\alpha \to \infty$ is found from (4.125) using equation (4.108).

The Laplace coefficients diverge as $\alpha \to 1$. Inspection of equation (4.109) shows that the main contribution to this divergence occurs when $\cos\phi$ is nearly unity, thus when ϕ is nearly zero. In this case $\cos\phi = 1 - \frac{1}{2}\phi^2 + \mathrm{O}(\phi^4)$, so

$$b_s^m(\alpha) \underset{\alpha\to 1}{\longrightarrow} \frac{2}{\pi}\int_0^\pi \frac{\mathrm{d}\phi\cos m\phi}{[(1-\alpha)^2+\alpha\phi^2]^s}. \quad (4.126)$$

The factor α multiplying ϕ^2 can be replaced by unity and the integration variable ϕ can be replaced by $u \equiv \phi/|1-\alpha|$. The upper limit of the integration is at $u = \pi/|1-\alpha|$, which approaches infinity as $\alpha \to 1$. Thus

$$b_s^m(\alpha) \underset{\alpha\to 1}{\longrightarrow} \frac{2}{\pi|1-\alpha|^{2s-1}}\int_0^\infty \frac{\mathrm{d}u\cos(m|1-\alpha|u)}{(1+u^2)^s} \quad (4.127)$$
$$= \frac{2^{3/2-s}}{\pi^{1/2}\Gamma(s)}\left|\frac{m}{1-\alpha}\right|^{s-1/2} K_{s-1/2}(|m(1-\alpha)|), \quad s > 0,$$

where $K_\nu(\cdot)$ is a modified Bessel function (eq. C.30). If m is fixed and nonzero and $\alpha \to 1$ the Laplace coefficient is obtained from the limiting form of the modified Bessel function, equation (C.31):

$$b_s^m(\alpha) \underset{\alpha\to 1}{\longrightarrow} \begin{cases} \dfrac{\Gamma(s-\frac{1}{2})}{\pi^{1/2}\Gamma(s)|1-\alpha|^{2s-1}}, & s > \frac{1}{2}, \\ \frac{2}{\pi}\{-\gamma-\log[\frac{1}{2}|m(1-\alpha)|]\}, & s = \frac{1}{2}, \end{cases} \quad (4.128)$$

where $\gamma = 0.577\,216\cdots$ is Euler's constant.

To determine the behavior of the Laplace coefficients for large m, we use the relation (C.15a), $\Gamma(z+a)/\Gamma(z+b) \to z^{a-b}$ as $|z| \to \infty$. For $z = |m|$, $a = s + n$, $b = 1 + n$, equation (4.112) yields

$$b_s^m(\alpha) \xrightarrow[m\to\infty]{} \frac{2}{\Gamma^2(s)} \alpha^{|m|} |m|^{s-1} \sum_{n=0}^{\infty} \frac{\Gamma(s+n)}{\Gamma(n+1)} \alpha^{2n}, \quad 0 \le \alpha < 1. \tag{4.129}$$

The binomial series for $(1-\alpha^2)^{-s}$ is

$$(1-\alpha^2)^{-s} = \sum_{n=0}^{\infty} \frac{(-1)^n \Gamma(1-s)}{\Gamma(1-s-n)\Gamma(n+1)} \alpha^{2n}. \tag{4.130}$$

Using the relation (C.14b), this expression can be rewritten as

$$(1-\alpha^2)^{-s} = \frac{1}{\Gamma(s)} \sum_{n=0}^{\infty} \frac{\Gamma(s+n)}{\Gamma(n+1)} \alpha^{2n}. \tag{4.131}$$

Thus equation (4.129) simplifies to

$$b_s^m(\alpha) \xrightarrow[m\to\infty]{} \frac{2\alpha^{|m|}|m|^{s-1}}{\Gamma(s)(1-\alpha^2)^s}, \quad 0 \le \alpha < 1, \tag{4.132}$$

showing that the Laplace coefficients decay as $|m|^{s-1}\exp(|m|\log\alpha)$ when $|m| \to \infty$. The behavior for $\alpha > 1$ is found using equation (4.108).

4.4.3 Derivatives

Expansions of the disturbing function such as (4.98) and (4.99) are often expressed concisely using derivatives of the Laplace coefficients with respect to the argument α. To find these we differentiate equation (4.107):

$$\frac{s(z+z^{-1}-2\alpha)}{(1+\alpha^2-\alpha z-\alpha z^{-1})^{s+1}} = \tfrac{1}{2} \sum_{-\infty}^{\infty} \frac{\mathrm{d}b_s^m(\alpha)}{\mathrm{d}\alpha} z^m. \tag{4.133}$$

Using equation (4.119) and requiring that the coefficient of z^m be the same on both sides of the equation, we find

$$Db_s^m(\alpha) \equiv \frac{\mathrm{d}b_s^m(\alpha)}{\mathrm{d}\alpha} = s\big[b_{s+1}^{m-1}(\alpha) - 2\alpha b_{s+1}^m(\alpha) + b_{s+1}^{m+1}(\alpha)\big]. \tag{4.134}$$

We then use equation (4.120) to eliminate b_{s+1}^{m+1}, (4.123) to eliminate b_{s+1}^m and (4.124) with $m \to m-1$ to eliminate b_{s+1}^{m-1}. We find

$$Db_s^m(\alpha) = \frac{2(s+m-1)\alpha b_s^{m-1}(\alpha) - [m+\alpha^2(m-2s)]b_s^m(\alpha)}{\alpha(1-\alpha^2)}. \quad (4.135)$$

Replacing m by $-m$ and using the symmetry relation (4.106), we obtain a formula involving b_s^m and b_s^{m+1},

$$Db_s^m(\alpha) = \frac{[m+\alpha^2(m+2s)]b_s^m(\alpha) - 2(m-s+1)\alpha b_s^{m+1}(\alpha)}{\alpha(1-\alpha^2)}. \quad (4.136)$$

The analogs to the symmetry relations (4.106) and (4.108) are[7]

$$Db_s^{-m}(\alpha) = Db_s^m(\alpha), \quad Db_s^m(\alpha^{-1}) = -\alpha^{2+2s}Db_s^m(\alpha) - 2s\alpha^{1+2s}b_s^m(\alpha). \quad (4.137)$$

4.5 The stability of the solar system

This is one of the oldest and most famous problems in theoretical physics and one of the simplest to state. According to perturbation theory (§4.2), each planet excites small oscillatory variations in the orbits of the other planets. Although the fractional variations in the orbits are small (typically less than 10^{-3}–10^{-4}), the age of the system is large (10^8–10^{10} orbital periods). Over these vary large times, do the variations in the orbits remain strictly oscillatory or do they gradually grow, leading eventually to the catastrophic disruption of the solar system?

This question has fascinated physicists and mathematicians since the time of Newton. Newton apparently believed that the perturbations *did* grow, stating in his book *Opticks* in 1730 that the "irregularities" in the solar system arising "from the mutual actions of comets[8] and planets upon one

[7] The notation $Db_s^m(1/\alpha)$ is somewhat ambiguous. More precisely, let $f(x) \equiv \mathrm{d}b_s^m(x)/\mathrm{d}x$; then $Db_s^m(1/\alpha) = f(\alpha^{-1})$.

[8] Newton did not know that the masses of comets are many orders of magnitude smaller than the masses of the planets.

another" would gradually grow until the solar system "wants a reformation," that is, until God intervenes to restore order (e.g., Iliffe 2017). This theistic view was mocked by Leibniz, Newton's rival, who believed that the perfection of God required the perfection of the solar system, and complained in 1715 that "according to [Newton's] doctrine, God Almighty wants to wind up his watch from time to time . . . he had not, it seems, sufficient foresight to make it a perpetual motion."

4.5.1 Analytic results

The controversy between Newton and Leibniz was influenced by observations of Jupiter and Saturn dating back to Johannes Kepler in 1625, which seemed to show that their semimajor axes were changing linearly in time. Adding to the confusion, the development of perturbation theory during the eighteenth century showed that terms linear in time are present in the secular perturbations to the actions (see §5.1).

These issues were only resolved a half-century after Newton's death, when Joseph–Louis Lagrange (1736–1813) and Pierre–Simon Laplace (1749–1827) showed that the secular perturbations are not linear in time but oscillatory with long periods (see §5.2), and the amplitudes of the oscillations are small enough that they do not compromise the stability of the solar system. Then in 1785, Laplace showed that the apparent drift in the semimajor axes of Jupiter and Saturn arises because of a near-resonance between the two planets: their mean motions are related by $2n_{\rm Jupiter} \simeq 5n_{\rm Saturn}$.[9] This near-resonance, sometimes called the Great Inequality, leads to oscillations in the mean motions with a period of about 900 years, as shown in Figure 4.2, and this variation appeared nearly linear over the 150 years between Kepler and Laplace. See Laskar (2013) and Wilson (1985) for historical reviews.

These investigations led to three fundamental results. First, they showed that Newton's law of gravitation is universal, in the sense that it determines

[9] The d'Alembert property (4.92) implies that terms in the disturbing function with this frequency are third degree in the eccentricities and inclinations of Jupiter and Saturn, which is why they had escaped earlier researchers.

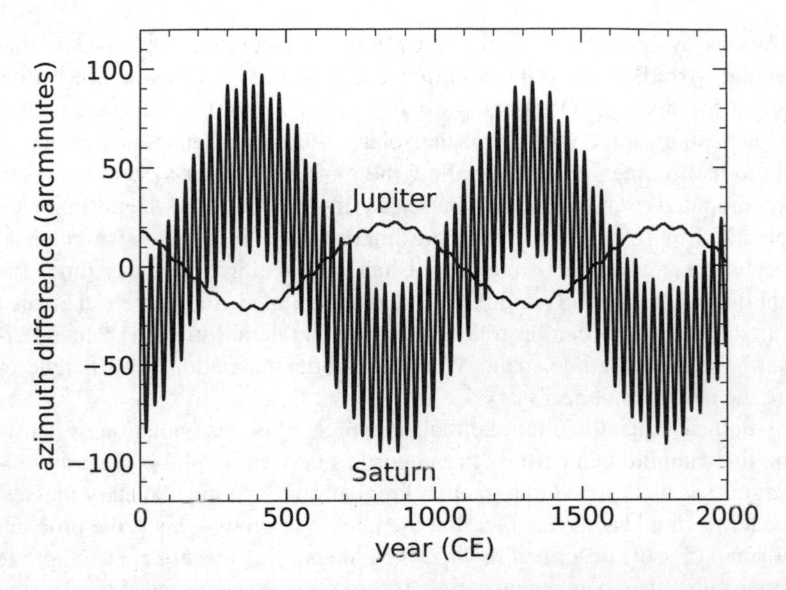

Figure 4.2: Oscillations of the longitude of Jupiter and Saturn. The plot shows $\psi \equiv \phi - \omega t - \phi_0$ where ϕ is the longitude of Jupiter or Saturn, and ϕ_0 and ω are chosen such that $\langle\psi\rangle = \langle\dot{\psi}\rangle = 0$. The time interval shown is from 0 CE to 2 000 CE. The longitudes have been smoothed to eliminate much larger short-period oscillations, by convolving ψ with a Gaussian with standard deviation 10 yr. The slow oscillations visible in the diagram, with a period of about 900 yr, are called the Great Inequality.

not just the forces between the planets and the Sun but also the forces between the planets. Second, they provided tools that allowed the motions of solar-system bodies to be computed and predicted to any desired level of accuracy over historical timescales (apart from the small corrections due to general relativity described in Appendix J). Finally, they demonstrated that the solar system was stable, at least according to perturbation theory carried to first order in the planetary masses using a disturbing function that is second degree in the inclinations and eccentricities.

The belief that the solar system is stable was buttressed by the demonstrations by Laplace in 1776 and by Siméon Denis Poisson in 1809 that secular perturbations to the semimajor axes vanish at first and second order in the masses. Of course, a belief is not a proof and these analyses do not establish the stability of the solar system over timescales comparable to its lifetime, nor do they allow the motions of solar-system bodies to be computed over geological timescales of $\sim 100\,\mathrm{Myr}$. As Henri Poincaré remarked in 1897, "Those who are interested in the progress of celestial mechanics... must feel some astonishment at seeing how many times the stability of the Solar System has been demonstrated. Lagrange established it first, Poisson has demonstrated it again, other demonstrations came afterward, others will come again. Were the old demonstrations insufficient, or are the new ones unnecessary?"

Poincaré quantified the skepticism expressed in this quotation by showing that Hamiltonian perturbation theory as derived in §4.2 is generally divergent: as it is carried to higher and higher orders in the planetary masses, the terms first become smaller and then begin to grow—this is the problem of small divisors described in §4.2.1. Mathematically, these are asymptotic series rather than convergent series. If the series are truncated at first or second order, they can accurately predict the motion of the planets over long times—certainly over the hundreds or thousands of years needed for most purposes—but they cannot be used to establish rigorously whether the solar system is stable forever, or even over timescales comparable to the lifetime of the Sun.

Motion in the potential of a point mass like the Sun is integrable, that is, there exist angle-action variables like the Delaunay variables (eq. 1.84) and the Hamiltonian depends only on the actions, $H = H_0(\mathbf{J})$. The trajectories of an integrable Hamiltonian system are restricted to an N-dimensional torus in a $2N$-dimensional phase space, and they are quasi-periodic—that is, the Fourier transform of the trajectory $\mathbf{r}(t)$ is a line spectrum consisting of integer combinations of N fundamental frequencies $\Omega_i = \partial H/\partial J_i$ (in a point-mass or Kepler potential the actions can be chosen such that two of the three frequencies are zero). The fundamental question that determines the stability of a non-integrable Hamiltonian system like the solar system is whether or not a small perturbation to an integrable system leads to trajectories that are still permanently restricted to tori in phase space. This question is not addressed by the perturbation theory of §4.2, which *always* produces trajectories that lie on tori. A fundamental insight into this question was provided by the Kolmogorov–Arnold–Moser or KAM theorems, developed in a series of papers in the 1950s and 1960s by Andrei Kolmogorov, Vladimir Arnold and Jürgen Moser. These theorems show that in Hamiltonian systems that are subjected to a sufficiently small perturbation some of the tori survive, and that these occupy a nonzero volume of phase space. The orbits whose tori are destroyed become chaotic. The chaotic and quasi-periodic trajectories are mixed together in phase space, in the sense that any finite phase-space neighborhood of any point, no matter how small, contains both types of trajectory: both the chaotic and quasi-periodic orbits are said to be dense in phase space.

It is difficult to apply the KAM theorems directly to the solar system, for several reasons. First, the mathematical proofs require planetary masses, eccentricities and inclinations that are far smaller than the actual values in the solar system, although it is probably safe to assume that the theorems are valid within a much larger parameter space than their current proofs require. Second, because the quasi-periodic orbits are mixed with chaotic orbits, an arbitrarily small external perturbation can convert one kind of trajectory to the other. Finally, for systems with more than $N = 2$ degrees of freedom, like the solar system, the N-dimensional tori do not divide up the phase space; in other words the phase space occupied by chaotic orbits is not only dense but also connected. Thus any chaotic trajectory can, and eventually will, visit every neighborhood of the phase space, a phenomenon known as **Arnold diffusion**. In principle, an arbitrarily small perturbation to any orbit can cause it to eventually pass arbitrarily close to any point in phase space.

Although we know that Arnold diffusion exists, there is no reliable way to calculate the rate of diffusion in any realistic model of a planetary system. The only rigorous results are **Nekhoroshev estimates**, which state that the chaotic orbits will remain close to their quasi-periodic neighbors for a time that is an exponential function of the strength of the perturbation (Yalinewich & Petrovich 2020).

4.5.2 Numerical results

To investigate the stability of the actual solar system over the Sun's lifetime,[10] we must use numerical integrations. The first long integrations of the solar system followed only Jupiter, Saturn, Uranus, Neptune, and Pluto, since this subsystem is not expected to be strongly influenced by the inner planets, which have much smaller masses. Early milestones were the outer solar system integrations lasting 0.12 Myr (Cohen & Hubbard 1965), 1 Myr (Cohen et al. 1973), 5 Myr (Kinoshita & Nakai 1984), 100 Myr (Roy et al. 1988), 200 Myr (Applegate et al. 1986), 845 Myr (Sussman & Wisdom 1988), and 5.5 Gyr (Kinoshita & Nakai 1996). All eight planets and Pluto were followed for 4 400 yr by Newhall et al. (1983), 3 Myr by Quinn et al. (1991), 100 Myr by Sussman & Wisdom (1992) and ~ 10 Gyr by Ito & Tanikawa (2002). A careful comparison of integration methods is given by Hayes (2008).

In a parallel line of investigation, Laskar (1986, 1989) used computer algebra to develop secular perturbation theory to second order in the planetary masses, using a disturbing function that was sixth degree in the eccentricities and inclinations—for comparison recall that the Lagrange–Laplace theory is only first order in the masses and second degree in the eccentricities and inclinations. The resulting differential equations contained some 150 000 polynomial terms; nevertheless they were straightforward to integrate numerically and much faster than N-body integrations.

These integrations reveal that in most cases the solar system is stable over timescales of a few Gyr. All of the planets survive, and mostly they remain in orbits very similar to their present ones. On 10^5-year timescales, the Lagrange–Laplace theory gives a reasonably good description of the variations of the eccentricities and inclinations (see Figure 5.1). However, integrations over timescales of 100 Myr and longer reveal that all the planetary orbits are chaotic, with a Liapunov time $t_\mathrm{L} \sim 10^7$ years. Thus, on timescales $\gg t_\mathrm{L}$, small changes in the orbits grow exponentially, as $\exp(t/t_\mathrm{L})$. This means that small changes now in the orbits of the planets—for example

[10] By "lifetime" we mean about 10 Gyr: the solar system was born 4.57 Gyr ago and the Sun will survive in its present form for about 7.6 Gyr into the future (see Appendix A and Box 1.2).

from the difference in the gravitational attraction of Jupiter on Earth caused when you raise your coffee cup to drink—will be amplified by a factor of $\sim \exp(7.6\,\mathrm{Gyr}/t_\mathrm{L}) \sim 10^{330}$ before the death of the Sun.

The nature of the chaotic behavior is different in the inner and outer solar system. In the outer solar system (Jupiter, Saturn, Uranus, Neptune), the chaotic behavior is restricted to narrow regions in phase space associated with high-order mean-motion resonances (see the discussion at the start of Chapter 6). As a result, the exponential divergence is limited to the orbital phase and the other elements (such as semimajor axis or eccentricity) remain close to their initial values. In contrast, in the inner solar system (Mercury, Venus, Earth, Mars) chaos arises from overlap of secular resonances and directly affects the eccentricities and inclinations of the planets.

This difference is illustrated in Figure 4.3, which shows the mean eccentricity of the planets over intervals of 10 Myr, long enough that the oscillations in Lagrange–Laplace theory average to nearly zero. If Lagrange–Laplace theory were correct—or more generally if the eccentricity oscillations were quasi-periodic with periods much less than 10 Myr—the curves should be flat, and this is approximately correct for the outer planets, where the chaos is due to high-order mean-motion resonances. For the inner planets, in contrast, the eccentricity undergoes a chaotic diffusion or random walk, which is strongest for Mercury but still significant for Venus, Earth and Mars.

Of course, this figure only shows one possible trajectory for the planets, as the evolution is chaotic and therefore extremely sensitive to the initial conditions. To illustrate this sensitivity, Figure 4.4 shows an expanded view of the future evolution of Mercury's eccentricity. There are five curves, each resulting from a change in the initial position of Mercury of a few centimeters. After ~ 1 Gyr these tiny differences have grown to be comparable to the size of Mercury's orbit. The conclusion is that there is no practical way to predict Mercury's eccentricity over Gyr timescales.

Integrations of a large ensemble of solar systems with slightly different initial conditions by Laskar & Gastineau (2009) show that there is about a 1% probability that Mercury will experience some catastrophic event—a collision with the Sun, Venus, or even Earth—some time in the next 5 Gyr. Of course, long after such an event there would be no obvious sign that

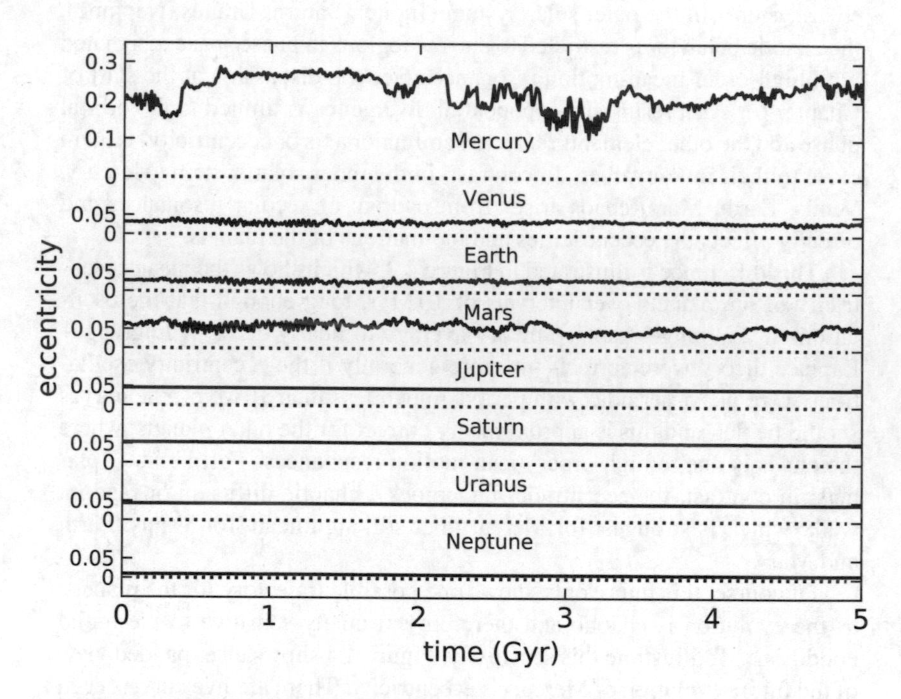

Figure 4.3: The mean eccentricity of each planet in successive intervals of 10 Myr, over the next 5 Gyr. The behavior of the outer four planets is sufficiently regular that the curves appear as straight horizontal lines, while the inner four planets exhibit chaotic diffusion. Data from Brown & Rein (2020).

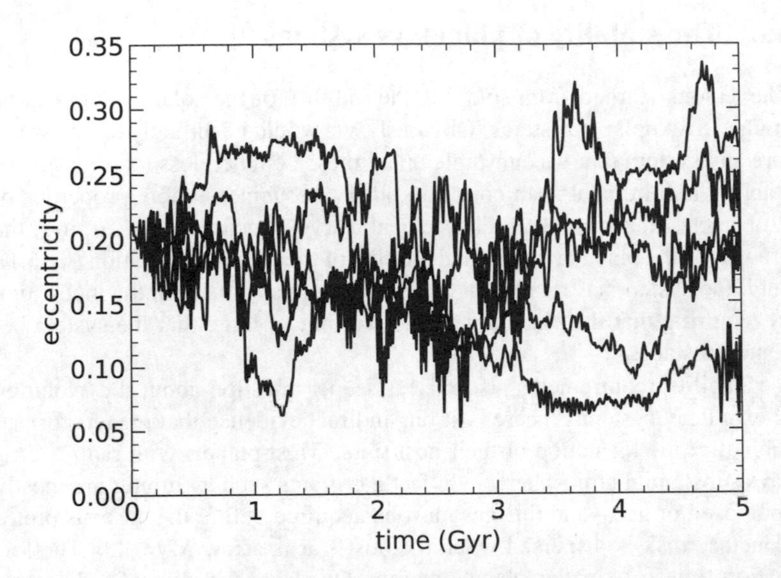

Figure 4.4: The mean eccentricity of Mercury in successive intervals of 10 Myr, over the next 5 Gyr. The plot shows five integrations starting from slightly different initial conditions. Data from Brown & Rein (2020).

Mercury had ever been present. Thus it is plausible that the solar system had more planets early in its history, and that one or more of these has been lost.

We can conclude that the question "is the solar system stable?" does not have a simple "yes" or "no" answer: in the future the solar system is probably stable, at least up to the time when the Sun swallows the inner planets about 7–8 Gyr from now; in the past it may well have been unstable but we will never know for sure.

4.6 The stability of planetary systems

The lessons learned from studying the stability of the solar system can be applied to exoplanet systems. Obviously, we would be unlikely to find in nature any system that was unstable on a timescale much less than its age, so stability requirements can constrain otherwise unmeasurable properties of exoplanets. For example, radial-velocity surveys measure only $m \sin I$, the product of the planetary mass and the sine of the orbital inclination (§1.6.1), but if the system has more than one planet we may constrain the inclination by requiring that the planetary masses are not so large that the system becomes unstable.

Stability requirements also raise a deeper question about the evolution of exoplanet systems. There is strong indirect evidence that planets formed soon after the formation of their host stars. Most planets with radii $\gtrsim 2R_\oplus$ have substantial atmospheres—in fact gas giants such as Jupiter are mostly composed of gas—and this gas must be acquired before the gaseous protoplanetary disk is dispersed when the host star is a few Myr old. The formation time for smaller planets may be longer, but studies of radioactive isotope systems in the oldest Earth rocks and in meteorites suggest that the formation of the Earth was largely complete when the solar system was only 1% of its current age (Dalrymple 2001).

These considerations prompt the simple question: did the solar system, and by extension other planetary systems, look the same after 50–100 Myr as they do after 5–10 Gyr? In other words, do planetary systems generally form in states that are stable over much longer timescales? Or do instabilities lead to continued evolution of planetary systems, with the number of planets slowly whittled down by collisions and ejections resulting from instabilities that emerge over longer and longer timescales?

To address this question effectively, we would like to have theoretical tools that allow us to determine whether a given planetary system is stable, without having to integrate the planetary orbits for billions of years. When developing these tools, we should bear in mind two discouraging lessons from studies of the stability of the solar system. First, small changes in the initial conditions or system parameters can lead to large changes in behavior. Second, chaos does not necessarily imply instability: in the solar system the

Liapunov time of $\sim 10^7$ yr is 500 times smaller than the age, and in some planetary systems (such as Kepler-36) the Liapunov time for stable orbits with initial conditions consistent with the observations can be as short as ~ 10 yr, even though the orbits are stable for at least 10^7 yr (Deck et al. 2012).

The stability of two-planet systems is described in §3.5. A characteristic feature of two-planet systems is that instability—if it occurs at all—occurs quickly. In other words the instability boundary in a two-planet system is sharp, dividing the phase space into stable trajectories that persist forever, and unstable ones that evolve on a timescale of only a few orbital periods. In contrast, instabilities in systems of three or more planets develop over timescales that can vary by many orders of magnitude, depending on the masses and initial orbits.

N-planet systems have a large number of free parameters—six orbital elements and a mass for each of the N planets. We focus here on the relatively simple case in which the planets all have the same mass and are initially on circular, coplanar orbits with constant logarithmic spacing, that is, the ratio of the semimajor axes of adjacent planets is fixed. The lifetime of the system is defined to be the time elapsed until the first close encounter between two planets, since these usually lead rapidly to ejection of a planet or a collision between two planets.

We expect that systems with smaller planets can survive with smaller separations between the orbits. To parametrize this dependence, we define the characteristic radius associated with a pair of adjacent planets to be

$$r_\mu = \tfrac{1}{2}(a_n + a_{n+1})\left(\frac{m_n + m_{n+1}}{M_*}\right)^\mu, \tag{4.138}$$

where a_n and m_n are the semimajor axes and masses of the two planets, and M_* is the mass of the host star.

Different values of the exponent μ are relevant in different contexts. The mutual Hill radius defined in (3.112) is $r_\mathrm{H} = 3^{-1/3} r_{1/3}$. A system of two small planets on nearby circular, coplanar orbits is Hill stable if $|a_2 - a_1| > 2 \cdot 3^{1/6} r_{1/3}$ (eq. 3.135), and chaotic if $|a_2 - a_1| \lesssim 1.4 r_{2/7}$ (eq. 3.140). For systems of three or more planets, both numerical experiments and analytic models of the long-term dynamics suggest that the relevant characteristic radius is defined by $\mu = \frac{1}{4}$ (Quillen 2011; Petit et al. 2020).

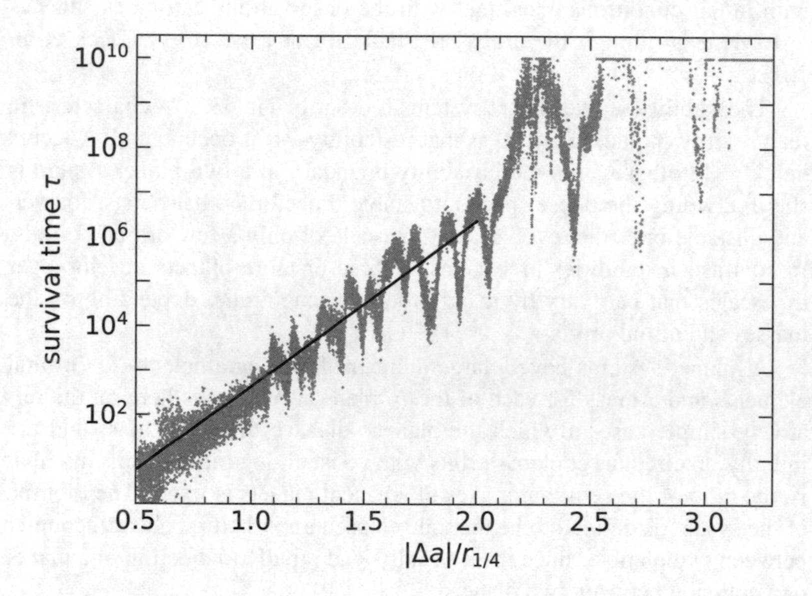

Figure 4.5: The lifetimes of systems of five Earth-mass planets on initially circular, coplanar orbits around a solar-mass star. The initial semimajor axes have equal logarithmic spacing. The horizontal axis is the initial separation in units of the characteristic radius (4.138) with $\mu = \frac{1}{4}$ and the vertical axis is the lifetime τ in units of the initial orbital period P_0 of the innermost planet. Integrations of surviving systems were terminated at $\tau = 10^{10} P_0$. The straight line shows the fit (4.139). Data from Obertas et al. (2017).

Figure 4.5 plots the lifetimes of $\sim 2 \times 10^4$ systems of five equally spaced Earth-mass planets. The orbits are followed for up to $10^{10} P_0$, where P_0 is the initial orbital period of the innermost planet, and the lifetime τ of each system is plotted as a function of the semimajor axis difference in units of the characteristic radius $r_{1/4}$.

On average, the lifetime grows exponentially with separation; the fit shown in the figure is

$$\log_{10} \frac{\tau}{P_0} = -1.10 + 3.74 \frac{|\Delta a|}{r_{1/4}}. \tag{4.139}$$

The scatter around the mean at a given separation is substantial, about 0.4 in $\log_{10} \tau/P_0$, measured as the distance between the median and the quartiles. The scatter is also asymmetric, mostly because the lifetimes drop periodically by up to two orders of magnitude at the mean-motion resonances between the planets, where small perturbations are amplified by the resonance (see Chapter 6). At separations larger than about $|\Delta a| = 2r_{1/4}$, the lifetimes grow substantially above the predictions of the fit (4.139), with many systems surviving for at least 10^{10} orbits.

Most of these features are generic for systems of three or more planets on nearly circular orbits. In particular, if the planetary masses and separations are fixed, the instability time is not a strong function of the number of planets so long as $N \geq 3$.

It is likely that this behavior arises from the overlap of weak but numerous resonances that contribute terms to the disturbing function that are either $\mathrm{O}(m^2)$ where m is the planet mass or $\mathrm{O}(e^k)$ where e is the eccentricity and $k > 1$ (Quillen 2011; Petit et al. 2020); however, our analytic understanding of the origin of long-term instabilities in multi-planet systems is still incomplete.

These considerations illuminate the answer to the question raised at the beginning of this section: have long-term instabilities led to significant evolution of planetary systems after their initial formation was complete? Figure 4.6 shows the distribution of separations in units of the characteristic radius for 543 pairs of adjacent planets discovered by the Kepler mission (Weiss et al. 2018). The density of separations shows a sharp rise at $|\Delta a|/r_{1/4} \simeq$ 2–3, not far from the separations required for stability over the lifetime of a typical planetary system. This finding suggests that many planets with smaller separations may have been lost due to dynamical instabilities long after the original planet-formation process was complete.

There are several caveats to any argument of this kind. First, since Kepler only measures radii, not masses, the masses have been determined using

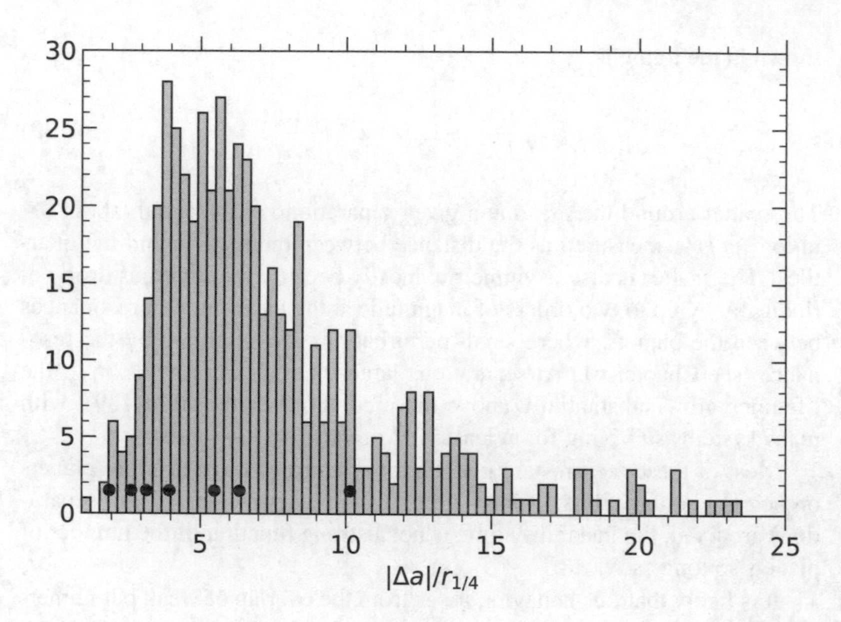

Figure 4.6: The distribution of separations between adjacent planets in units of the characteristic radius, for a sample of 349 multi-planet systems discovered by Kepler. Black dots mark the separations of adjacent planets in the solar system. Data from Weiss et al. (2018).

an empirical relation between mass and radius (Weiss et al. 2018); however, since the masses only enter the characteristic radius as $m^{1/4}$, the results are not very sensitive to errors in the mass. Second, we have few direct measurements of the eccentricities of the Kepler planets, and eccentric orbits are usually less stable than the circular ones used to construct Figure 4.5. And finally, not all planets in a given system were discovered by Kepler, and accounting for missing planets would squeeze the distribution toward smaller separations. In particular, many of the planets with the largest separations, $|\Delta a|/r_{1/4} \gtrsim 10$, are probably adjacent to undiscovered planets.

Chapter 5

Secular dynamics

5.1 Introduction

To introduce this chapter, let us imagine a test particle orbiting a mass M with semimajor axis a and mean motion $n = (\mathbb{G}M/a^3)^{1/2}$. The test particle is perturbed by a distant, nearly stationary mass that exerts a small force per unit mass of order $\epsilon\,\mathbb{G}M/a^2$ with $0 < \epsilon \ll 1$. This perturbing force induces small periodic variations in the position and velocity of the test particle, with a fractional amplitude of order ϵ and frequency of order n.

In addition, the distant mass can produce changes in the orbit that are much slower and larger than these periodic variations. For example, if the orbit is circular and the perturber lies outside the orbital plane, it exerts a steady torque that changes the direction of the angular momentum of the test particle, just as the torque from the Earth's gravity causes a spinning top to precess. If the orbit is eccentric the perturber can exert a steady torque even if it lies in the orbital plane, since the test particle spends most of its time near apoapsis and the orientation of the line of apsides is fixed for a Kepler orbit. The rate of change of the eccentricity or inclination is only of order ϵn, but the changes accumulate over a time of order $(\epsilon n)^{-1}$, at which point the total change in the eccentricity or inclination has grown to be of order unity.

As indicated by these arguments, the effects of a perturber with fractional strength $\epsilon \ll 1$ on the test-particle orbit can usefully be divided into two broad classes: **short-period perturbations**, which have oscillation frequencies of order n and fractional amplitudes of order ϵ; and **secular perturbations**, which have oscillation frequencies of order ϵn and amplitudes of order unity.

To provide a concrete example, we now solve the equations of motion for an abstract dynamical system that illustrates the distinction between short-period and secular dynamics. The system has two degrees of freedom, with coordinates (q_1, q_2), momenta (p_1, p_2) and Hamiltonian

$$H(q_1, q_2, p_1, p_2) = H_0(q_1, p_1) + \epsilon H_1(q_1, q_2, p_1, p_2), \tag{5.1}$$

where ϵ is a small positive number and

$$H_0(q_1, p_1) = \tfrac{1}{2}p_1^2 + \tfrac{1}{2}q_1^2, \quad H_1(q_1, q_2, p_1, p_2) = \tfrac{1}{2}p_2^2 + \tfrac{1}{2}q_2^2 + q_1 q_2. \tag{5.2}$$

The key features of this system are that (i) the unperturbed Hamiltonian H_0 describes a harmonic oscillator with unit frequency; (ii) the coordinate-momentum pair (q_2, p_2) does not appear in the unperturbed Hamiltonian; and (iii) the perturbation Hamiltonian H_1 contains a term $q_1 q_2$ that couples the two coordinate-momentum pairs. The (q_1, p_1) pair are called "fast" variables since their frequency in the unperturbed system is $\omega_\mathrm{f} = 1$, and (q_2, p_2) are called "slow" variables since their frequency in the unperturbed system is $\omega_\mathrm{s} = 0$.

Hamilton's equations are

$$\dot{p}_1 = -q_1 - \epsilon q_2, \quad \dot{q}_1 = p_1, \quad \dot{p}_2 = -\epsilon(q_1 + q_2), \quad \dot{q}_2 = \epsilon p_2; \tag{5.3}$$

or we can eliminate the momenta to obtain

$$\ddot{q}_1 = -q_1 - \epsilon q_2, \quad \ddot{q}_2 = -\epsilon^2(q_1 + q_2). \tag{5.4}$$

The solution to these equations of motion can be written

$$\begin{aligned} q_1(t) &= a_\mathrm{f} \cos\omega_\mathrm{f} t + b_\mathrm{f} \sin\omega_\mathrm{f} t + a_\mathrm{s} \cos\omega_\mathrm{s} t + b_\mathrm{s} \sin\omega_\mathrm{s} t, \\ q_2(t) &= f_\mathrm{f} a_\mathrm{f} \cos\omega_\mathrm{f} t + f_\mathrm{f} b_\mathrm{f} \sin\omega_\mathrm{f} t + f_\mathrm{s} a_\mathrm{s} \cos\omega_\mathrm{s} t + f_\mathrm{s} b_\mathrm{s} \sin\omega_\mathrm{s} t, \end{aligned} \tag{5.5}$$

where

$$f_{\mathrm{f}} \equiv \frac{\omega_{\mathrm{f}}^2 - 1}{\epsilon}, \quad f_{\mathrm{s}} \equiv \frac{\omega_{\mathrm{s}}^2 - 1}{\epsilon}, \tag{5.6}$$

and the constants a_{f}, b_{f}, a_{s}, b_{s} are determined by the initial conditions. The frequencies are

$$\begin{aligned} \omega_{\mathrm{f}} &= \frac{1}{\sqrt{2}}\Big[1 + \epsilon^2 + \big(1 - 2\epsilon^2 + 4\epsilon^3 + \epsilon^4\big)^{1/2}\Big]^{1/2} \\ &= 1 + \tfrac{1}{2}\epsilon^3 + \mathrm{O}(\epsilon^5), \\ \omega_{\mathrm{s}} &= \frac{1}{\sqrt{2}}\Big[1 + \epsilon^2 - \big(1 - 2\epsilon^2 + 4\epsilon^3 + \epsilon^4\big)^{1/2}\Big]^{1/2} \\ &= \epsilon - \tfrac{1}{2}\epsilon^2 - \tfrac{1}{8}\epsilon^3 - \tfrac{9}{16}\epsilon^4 + \mathrm{O}(\epsilon^5). \end{aligned} \tag{5.7}$$

The oscillations with frequency ω_{f} are short-period perturbations since ω_{f} is close to unity, while those characterized by ω_{s} are secular perturbations since ω_{s} is of order ϵ.

To illustrate the behavior of the system, assume that the initial conditions are $q_1 = q_2 = 1$, $p_1 = p_2 = 0$ at time $t = 0$. The unperturbed motion ($\epsilon = 0$) corresponding to these initial conditions is $q_1(t) = \cos t$, $q_2(t) = 1$. It is straightforward to show that when ϵ is nonzero,

$$\begin{aligned} q_1(t) &= \tfrac{1}{2}[(1 + 2\epsilon - \epsilon^2)\Delta + 1]\cos\omega_{\mathrm{f}}t - \tfrac{1}{2}[(1 + 2\epsilon - \epsilon^2)\Delta - 1]\cos\omega_{\mathrm{s}}t, \\ q_2(t) &= \tfrac{1}{2}[(3\epsilon^2 - 1)\Delta + 1]\cos\omega_{\mathrm{f}}t - \tfrac{1}{2}[(3\epsilon^2 - 1)\Delta - 1]\cos\omega_{\mathrm{s}}t, \end{aligned} \tag{5.8}$$

where $\Delta \equiv (1 - 2\epsilon^2 + 4\epsilon^3 + \epsilon^4)^{-1/2}$. An expansion of the amplitudes of the cosine terms in powers of ϵ gives

$$\begin{aligned} q_1(t) &= [1 + \epsilon + \mathrm{O}(\epsilon^4)]\cos\omega_{\mathrm{f}}t - [\epsilon + \mathrm{O}(\epsilon^4)]\cos\omega_{\mathrm{s}}t, \\ q_2(t) &= [\epsilon^2 + \mathrm{O}(\epsilon^3)]\cos\omega_{\mathrm{f}}t + [1 + \mathrm{O}(\epsilon^2)]\cos\omega_{\mathrm{s}}t. \end{aligned} \tag{5.9}$$

These equations illustrate the rich, and occasionally confusing, behavior of secular perturbations. The perturbations modify the fast frequency from 1 to $1+\mathrm{O}(\epsilon^3)$ and modify the slow frequency from 0 to $\epsilon+\mathrm{O}(\epsilon^2)$. The amplitude of the oscillations in the fast variable $q_1(t)$ are also modified by $\mathrm{O}(\epsilon)$.

The effect on the slow variable $q_2(t)$ is more dramatic: the perturbation causes it to oscillate on the secular timescale ω_s^{-1} with approximately unit amplitude. As the perturbation becomes smaller and smaller, the frequency of this slow oscillation declines but its amplitude does not. In other words, in this system an arbitrarily small perturbation can cause a large change in the phase-space position, provided that we wait long enough.

Notice that a straightforward expansion of equations (5.7) and (5.9) in powers of ϵ gives

$$\begin{aligned} q_1(t) &= \cos t + \epsilon(\cos t - 1) + \tfrac{1}{2}\epsilon^3(t^2 - t\sin t) + \mathrm{O}(\epsilon^4), \\ q_2(t) &= 1 + \epsilon^2(\cos t - 1 - \tfrac{1}{2}t^2) + \mathrm{O}(\epsilon^3). \end{aligned} \tag{5.10}$$

This result is formally correct but seriously misleading, as it suggests that the perturbations in both the slow and fast variables contain components that grow as polynomials in time rather than oscillating slowly. The problem is that the expansion (5.10) breaks down after a time $t \sim \epsilon^{-1}$; mathematically, the expansion is convergent, but not uniformly convergent. The techniques for handling differential equations such as (5.4) that have disparate timescales depending on the perturbation parameter are known as "multiple-scale analysis" (Kevorkian & Cole 1996; Bender & Orszag 1999).

The most important of these techniques in celestial mechanics is **orbit averaging**, an example of the averaging principle described in Appendix D.9: since the changes in the slow variables occur on much longer timescales than the oscillation period of the fast variables, we time-average the perturbing Hamiltonian H_1 over one period of the fast variables, assuming that the trajectories of the fast variables are determined by the unperturbed Hamiltonian H_0 and that the slow variables are frozen in time. The resulting **orbit-averaged Hamiltonian** is then used to describe the secular evolution of the system.

For the Hamiltonian (5.2) the orbit average of H_1 is just

$$\langle H_1 \rangle = \tfrac{1}{2}p_2^2 + \tfrac{1}{2}q_2^2 + \langle q_1 \rangle q_2. \tag{5.11}$$

For the example described above, the unperturbed trajectory is $q_1(t) = \cos t$, so $\langle q_1 \rangle = 0$ and the total Hamiltonian is

$$H_0 + \epsilon\langle H_1 \rangle = \tfrac{1}{2}p_1^2 + \tfrac{1}{2}q_1^2 + \epsilon(\tfrac{1}{2}p_2^2 + \tfrac{1}{2}q_2^2), \tag{5.12}$$

consisting of two uncoupled harmonic oscillators, one with frequency 1 and one with frequency ϵ. The solution satisfying the initial conditions $q_1 = q_2 = 1$, $p_1 = p_2 = 0$ at time $t = 0$ is

$$q_1(t) = \cos t, \quad q_2(t) = \cos \epsilon t. \tag{5.13}$$

Comparing this result to equations (5.7), we find that the orbit-averaged approximation gives the correct fast and slow frequencies ω_f and ω_s to leading order in ϵ. Similarly, equations (5.9) show that the amplitudes of the fast oscillations in $q_1(t)$ and the slow oscillations in $q_2(t)$ are correct to leading order. Similar results hold in most applications: orbit-averaging correctly finds the amplitude and frequency of the oscillations, both short-period and secular, but only to leading order in the strength of the perturbation.

The application of orbit averaging to near-Kepler systems is simplest using angle-action variables such as the Delaunay variables (eq. 1.84). The Hamiltonian for a test particle orbiting a mass M and perturbed by a time-independent Hamiltonian ϵH_1 is

$$H = -\frac{(\mathbb{G}M)^2}{2\Lambda^2} + \epsilon H_1(\ell, \omega, \Omega, \Lambda, L, L_z). \tag{5.14}$$

We now orbit average H_1. In the unperturbed system, the actions Λ, L and L_z are all fixed, as are the two angles ω and Ω. Since the mean anomaly ℓ increases uniformly with time in the unperturbed system, the average of H_1 over one unperturbed orbit is simply the average over ℓ with all the other Delaunay variables fixed. Thus the orbit-averaged Hamiltonian becomes

$$\langle H\rangle = -\frac{(\mathbb{G}M)^2}{2\Lambda^2} + \epsilon\langle H_1\rangle, \quad \text{where} \quad \langle H_1\rangle = \frac{1}{2\pi}\int_0^{2\pi} \mathrm{d}\ell\, H_1. \tag{5.15}$$

Since we have averaged over ℓ, $\langle H_1\rangle$ is independent of ℓ; therefore $\langle H\rangle$ is independent of ℓ as well, so $\mathrm{d}\Lambda/\mathrm{d}t = -\partial\langle H\rangle/\partial\ell = 0$. Since $\Lambda = (\mathbb{G}Ma)^{1/2}$ we conclude that *the semimajor axis is conserved in the orbit-averaged dynamical system.* More precisely, just as in the example system with Hamiltonian (5.2), the semimajor axis undergoes both fast and slow oscillations with fractional amplitude $\mathrm{O}(\epsilon)$, while the slow variables—L and L_z or the eccentricity and inclination—undergo fast oscillations with amplitude $O(\epsilon)$ and slow oscillations with amplitude $O(1)$.

Physically, orbit-averaging the potential or force due to a particle of mass m on a Kepler orbit corresponds to replacing the particle by an eccentric wire with the same shape as its orbit—essentially the wire is a long time exposure of the orbit. If the orbit is eccentric, then the mass density of the wire is nonuniform. In particular, if the orbital radius as a function of azimuth is $r(\phi)$, then the density (mass per unit length) $\lambda(\phi)$ of the wire must satisfy $\lambda r \mathrm{d}\phi = m\mathrm{d}t/P$, where $\mathrm{d}t$ is the time the orbit spends in the azimuthal interval $\mathrm{d}\phi$. Then Kepler's laws imply that

$$\lambda(\phi) = \frac{mr(\phi)}{2\pi a^2(1-e^2)^{1/2}}, \tag{5.16}$$

where a and e are the semimajor axis and eccentricity of the orbit. This insight, and the equations describing the forces from the wire, are due to Carl Friedrich Gauss (1777–1855). For a recent treatment see Touma et al. (2009).

The most important application of these results is to multi-planet systems such as the solar system. Here the perturbation Hamiltonian represents the gravitational interactions among the planets, and the small parameter ϵ can be thought of as the typical ratio of the masses of the planets to the mass of the host star. Thus in multi-planet systems we expect that the fast and slow oscillations of the planetary semimajor axes will be much smaller than the oscillations in eccentricity and inclination.[1]

The oscillations of the semimajor axes can be larger if there is a near-resonance between two or more planets. For example, Uranus and Neptune are not far from a 2:1 resonance ($P_{\mathrm{Neptune}}/P_{\mathrm{Uranus}} - 2 = -0.0385$), and Jupiter and Saturn are not far from a 5:2 resonance ($P_{\mathrm{Saturn}}/P_{\mathrm{Jupiter}} - \frac{5}{2} = -0.0167$), and the effects of these near-resonances dominate the long-term variations in the semimajor axes of these planets (Figure 4.2).

These arguments are consistent with numerical integrations of the orbits of planets in the solar system. The root-mean-square fractional variations

[1] We have shown that the slow oscillations in the semimajor axes are no larger than $O(\epsilon)$, but according to a theorem of Poisson (1809) they are even smaller, $O(\epsilon^2)$.

of the semimajor axes are $\lesssim 4 \times 10^{-3}$ for the giant planets Jupiter, Saturn, Uranus and Neptune, and $\lesssim 4 \times 10^{-5}$ for the terrestrial planets Mercury, Venus, Earth and Mars, while the variations in eccentricity and inclination are much larger (see Figures 4.3 and 5.1).

5.2 Lagrange–Laplace theory

Lagrange–Laplace theory is an approximate description of the secular dynamics of a system of planets on nearly circular, nearly coplanar orbits.[2] The description is approximate because it uses orbit averaging and thus, as described in the preceding section, the results are only accurate to lowest order in the planetary masses. Moreover the secular Hamiltonian is truncated by keeping only terms up to $\mathrm{O}(e^2, I^2)$ in the eccentricities and inclinations of the planets. Despite these drastic simplifications, Lagrange–Laplace theory provided the most accurate description of the long-term behavior of the solar system for over a century. In particular it showed that variations in the orbital elements that were polynomial functions of time in less sophisticated analyses (cf. eq. 5.10) were actually oscillatory with long periods, and hence did not threaten the long-term stability of the solar system. Lagrange–Laplace theory has been superseded by more accurate secular theories, but these are usually too complicated to be investigated without the aid of computers. In the most important of these theories, due to Laskar (1986, 1989), the differential equations are accurate to second order in the planetary masses and contain polynomials of up to degree five in the eccentricities and inclinations [corresponding to a Hamiltonian that retains all terms up to $\mathrm{O}(e^6, I^6)$], but these equations contain over 150 000 terms.

The disturbing function of a system of planets on nearly circular, nearly coplanar orbits was derived in §4.3 as a series of terms involving cosines of the form $\cos(k_i\lambda_i - k_j\lambda_j + \theta_i - \theta_j)$. Here λ_i is the mean longitude of planet i, k_i is an integer, and θ_i is some combination of the longitude of periapsis ϖ_i and the longitude of the node Ω_i. As we described in §5.1, to find the secular

[2] The history of the development of this theory is described by Laskar (2013) and in greater detail by Wilson (1985).

behavior of this system to first order in the masses, we simply average the disturbing function over the mean longitudes. This average vanishes unless $k_i = k_j = 0$, in which case the average is $\cos(\theta_i - \theta_j)$. From equations (4.98) and (4.99) it is straightforward to show that the orbit-averaged value of the inverse distance between bodies 1 and 2 is

$$\left\langle \frac{1}{\Delta_{12}} \right\rangle = \frac{1}{2a_2} b^0_{1/2} + \frac{\epsilon^2}{a_2}\Big[(e_1^2 + e_2^2)(\tfrac{1}{8}\alpha^2 D^2 + \tfrac{1}{4}\alpha D) b^0_{1/2} - \tfrac{1}{8}(I_1^2 + I_2^2)\alpha b^1_{3/2} \\ + \tfrac{1}{4} e_1 e_2 (2 - 2\alpha D - \alpha^2 D^2) b^1_{1/2} \cos(\varpi_1 - \varpi_2) \\ + \tfrac{1}{4} I_1 I_2 \alpha b^1_{3/2} \cos(\Omega_1 - \Omega_2)\Big]. \tag{5.17}$$

Here the argument of the Laplace coefficients is $\alpha = a_1/a_2$ and ϵ is an ordering parameter, which from now on we set to unity. The orbit-averaged contribution from the indirect potential (4.104) vanishes. The initial term proportional to $b^0_{1/2}$ depends only on the semimajor axes so the resulting term in Hamilton's equations does not affect the inclinations or eccentricities and can be dropped. Equation (5.17) can be simplified further using the relations

$$(2\alpha D + \alpha^2 D^2) b^0_{1/2} = \alpha b^1_{3/2}, \quad (2 - 2\alpha D - \alpha^2 D^2) b^1_{1/2} = -\alpha b^2_{3/2}, \tag{5.18}$$

so we have

$$\left\langle \frac{1}{\Delta_{12}} \right\rangle = \frac{1}{a_2}\big\{\tfrac{1}{8}\alpha b^1_{3/2}(\alpha)[e_1^2 + e_2^2 - I_1^2 - I_2^2 + 2 I_1 I_2 \cos(\Omega_1 - \Omega_2)] \\ - \tfrac{1}{4}\alpha b^2_{3/2}(\alpha) e_1 e_2 \cos(\varpi_1 - \varpi_2)\big\}. \tag{5.19}$$

Since Δ_{12} is symmetric in the arguments 1 and 2, the right side must be a symmetric function as well, and this can be checked using equation (4.108).

It is straightforward to generalize this result to a system of N planets. The potential that governs the motion of planet i is the **Lagrange–Laplace disturbing function**

$$\Phi_i^{\mathrm{LL}} = -\sum_{\substack{j=1 \\ j \neq i}}^{N} \left\langle \frac{\mathbb{G} m_j}{\Delta_{ij}} \right\rangle$$

$$= -\sum_{\substack{j=1\\j\neq i}}^{N} \frac{\mathbb{G}m_j}{a_j}\big\{\tfrac{1}{8}\alpha_{ij}b_{3/2}^1(\alpha_{ij})[e_i^2+e_j^2-I_i^2-I_j^2+2I_iI_j\cos(\Omega_i-\Omega_j)] \\ -\tfrac{1}{4}\alpha_{ij}b_{3/2}^2(\alpha_{ij})e_ie_j\cos(\varpi_i-\varpi_j)\big\}, \tag{5.20}$$

with $\alpha_{ij} = a_i/a_j$. The Lagrange-Laplace disturbing function is limited to terms that are secular (no dependence on the mean longitudes) and quadratic in the eccentricities or inclinations.

We may now analyze the dynamics of an N-planet system governed by the Lagrange–Laplace disturbing function. In astrocentric coordinates (eq. 4.6) the Hamiltonian for planet i is $H_i = H_{\mathrm{K},i} + \Phi_i^{\mathrm{LL}}$, where $H_{\mathrm{K},i} = -\frac{1}{2}\mathbb{G}^2(m_0+m_i)^2/\Lambda_i^2$ is the Kepler Hamiltonian for planet i, m_0 is the mass of the host star and $\Lambda_i = [\,\mathbb{G}(m_0+m_i)a_i]^{1/2} = n_ia_i^2$, where n_i is the mean motion. Because the Lagrange–Laplace disturbing function is averaged over the mean longitudes it is independent of them, so Hamilton's equations give $\mathrm{d}\Lambda_i/\mathrm{d}t = -\partial H_i/\partial\lambda_i = 0$. Thus the semimajor axes of the planets are conserved. The evolution of the mean longitude λ_i is dominated by the Kepler Hamiltonian and is not of interest here. We define dimensionless variables (a simplification of eqs. 1.71 in the limit $I \ll 1$):

$$k_i \equiv e_i\cos\varpi_i, \quad h_i \equiv e_i\sin\varpi_i, \quad q_i = I_i\cos\Omega_i, \quad p_i = I_i\sin\Omega_i. \tag{5.21}$$

Then the remaining Hamilton's equations can be written (eq. 1.193):

$$\frac{\mathrm{d}k_i}{\mathrm{d}t} = \frac{1}{\Lambda_i}\frac{\partial\Phi_i^{\mathrm{LL}}}{\partial h_i}, \qquad \frac{\mathrm{d}h_i}{\mathrm{d}t} = -\frac{1}{\Lambda_i}\frac{\partial\Phi_i^{\mathrm{LL}}}{\partial k_i},$$
$$\frac{\mathrm{d}q_i}{\mathrm{d}t} = \frac{1}{\Lambda_i}\frac{\partial\Phi_i^{\mathrm{LL}}}{\partial p_i}, \qquad \frac{\mathrm{d}p_i}{\mathrm{d}t} = -\frac{1}{\Lambda_i}\frac{\partial\Phi_i^{\mathrm{LL}}}{\partial q_i}. \tag{5.22}$$

Apart from the constant factors of Λ_i, these resemble Hamilton's equations if k_i and q_i are interpreted as coordinates and h_i and p_i as momenta. In terms of these variables, equation (5.20) becomes

$$\Phi_i^{\mathrm{LL}} = -\sum_{\substack{j=1\\j\neq i}}^{N} \frac{\mathbb{G}m_j}{a_j}\big[\tfrac{1}{8}\alpha_{ij}b_{3/2}^1(\alpha_{ij})(k_i^2+h_i^2+k_j^2+h_j^2 \\ -q_i^2-p_i^2-q_j^2-p_j^2+2q_iq_j+2p_ip_j)-\tfrac{1}{4}\alpha_{ij}b_{3/2}^2(\alpha_{ij})(k_ik_j+h_ih_j)\big]. \tag{5.23}$$

The terms k_j^2, h_j^2, q_j^2 and p_j^2 can be dropped since they do not involve k_i, h_i, q_i or p_i and hence do not contribute to the equations of motion.

The equations of motion (5.22) become

$$\frac{\mathrm{d}k_i}{\mathrm{d}t} = -\frac{d_i}{\Lambda_i}h_i + \frac{1}{\Lambda_i}\sum_{j=1}^{N} b_{ij}h_j, \qquad \frac{\mathrm{d}h_i}{\mathrm{d}t} = \frac{d_i}{\Lambda_i}k_i - \frac{1}{\Lambda_i}\sum_{j=1}^{N} b_{ij}k_j,$$
$$\frac{\mathrm{d}q_i}{\mathrm{d}t} = \frac{d_i}{\Lambda_i}p_i - \frac{1}{\Lambda_i}\sum_{j=1}^{N} a_{ij}p_j, \qquad \frac{\mathrm{d}p_i}{\mathrm{d}t} = -\frac{d_i}{\Lambda_i}q_i + \frac{1}{\Lambda_i}\sum_{j=1}^{N} a_{ij}q_j, \tag{5.24}$$

where

$$a_{ij} = (1-\delta_{ij})\frac{\mathbb{G}m_j}{4a_j}\alpha_{ij}b_{3/2}^1(\alpha_{ij}), \quad b_{ij} = (1-\delta_{ij})\frac{\mathbb{G}m_j}{4a_j}\alpha_{ij}b_{3/2}^2(\alpha_{ij}),$$
$$d_i = \sum_{\substack{k=1\\k\neq i}}^{N}\frac{\mathbb{G}m_k}{4a_k}\alpha_{ik}b_{3/2}^1(\alpha_{ik}). \tag{5.25}$$

Here δ_{ij} is the Kronecker delta (Appendix C.1).

These can be simplified by introducing rescaled variables

$$K_i \equiv (m_i\Lambda_i)^{1/2}k_i, \qquad H_i \equiv (m_i\Lambda_i)^{1/2}h_i,$$
$$Q_i \equiv (m_i\Lambda_i)^{1/2}q_i, \qquad P_i \equiv (m_i\Lambda_i)^{1/2}p_i, \tag{5.26}$$

and rescaled matrices with components

$$A_{ij} = \frac{m_i^{1/2}a_{ij}}{m_j^{1/2}\Lambda_i^{1/2}\Lambda_j^{1/2}} = (1-\delta_{ij})\frac{\mathbb{G}(m_im_j)^{1/2}}{4a_j\Lambda_i^{1/2}\Lambda_j^{1/2}}\alpha_{ij}b_{3/2}^1(\alpha_{ij}),$$
$$B_{ij} = \frac{m_i^{1/2}b_{ij}}{m_j^{1/2}\Lambda_i^{1/2}\Lambda_j^{1/2}} = (1-\delta_{ij})\frac{\mathbb{G}(m_im_j)^{1/2}}{4a_j\Lambda_i^{1/2}\Lambda_j^{1/2}}\alpha_{ij}b_{3/2}^2(\alpha_{ij}),$$
$$D_{ij} = \delta_{ij}\frac{d_i}{\Lambda_i} = \frac{\delta_{ij}}{\Lambda_i}\sum_{\substack{k=1\\k\neq i}}^{N}\frac{\mathbb{G}m_k}{4a_k}\alpha_{ik}b_{3/2}^1(\alpha_{ik}). \tag{5.27}$$

The rescaling has been chosen such that the matrices $\mathbf{A}$ and $\mathbf{B}$ are symmetric in the indices i and j; this claim can be checked using equation (4.108).

The matrix $\mathbf{D}$ is automatically symmetric since it is diagonal. Then we can write in vector notation

$$\frac{\mathrm{d}\mathbf{K}}{\mathrm{d}t} = (\mathbf{B}-\mathbf{D})\mathbf{H}, \qquad \frac{\mathrm{d}\mathbf{H}}{\mathrm{d}t} = -(\mathbf{B}-\mathbf{D})\mathbf{K},$$

$$\frac{\mathrm{d}\mathbf{Q}}{\mathrm{d}t} = -(\mathbf{A}-\mathbf{D})\mathbf{P}, \qquad \frac{\mathrm{d}\mathbf{P}}{\mathrm{d}t} = (\mathbf{A}-\mathbf{D})\mathbf{Q}. \tag{5.28}$$

Finally we introduce complex vectors $\mathbf{Z}_e \equiv \mathbf{K} + \mathrm{i}\mathbf{H}$, $\mathbf{Z}_I \equiv \mathbf{Q} + \mathrm{i}\mathbf{P}$. These obey the equations

$$\frac{\mathrm{d}\mathbf{Z}_e}{\mathrm{d}t} = -\mathrm{i}(\mathbf{B}-\mathbf{D})\mathbf{Z}_e, \qquad \frac{\mathrm{d}\mathbf{Z}_I}{\mathrm{d}t} = \mathrm{i}(\mathbf{A}-\mathbf{D})\mathbf{Z}_I. \tag{5.29}$$

These equations show that the evolution of $\mathbf{Z}_e$ and $\mathbf{Z}_I$ is decoupled: in the Lagrange–Laplace approximation, the evolution of the eccentricities and longitudes of periapsis is independent of the evolution of the inclinations and the longitudes of the nodes. The solutions of these equations are sums of terms of the form $\mathbf{R}_e \exp(\mathrm{i}gt)$ or $\mathbf{R}_I \exp(\mathrm{i}ft)$, where

$$-(\mathbf{B}-\mathbf{D})\mathbf{R}_e = g\mathbf{R}_e, \qquad (\mathbf{A}-\mathbf{D})\mathbf{R}_I = f\mathbf{R}_I. \tag{5.30}$$

Thus g and f are eigenvalues of $-(\mathbf{B}-\mathbf{D})$ and $(\mathbf{A}-\mathbf{D})$ respectively, and $\mathbf{R}_e$ and $\mathbf{R}_I$ are the corresponding eigenvectors. The eigenvalues or **secular frequencies** are given by the solutions of

$$\det[(\mathbf{B}-\mathbf{D}) + g\mathbf{I}] = 0, \quad \det[-(\mathbf{A}-\mathbf{D}) + f\mathbf{I}] = 0, \tag{5.31}$$

where "det" is shorthand for the determinant and $\mathbf{I}$ is the $N \times N$ unit matrix. Since $\mathbf{A}$, $\mathbf{B}$, and $\mathbf{D}$ are all real, symmetric matrices and the eigenvalues of such matrices are real, g and f must be real; thus the solutions of the Lagrange–Laplace equations are all stable.

Let $\mathbf{R}_\star \equiv [(m_1\Lambda_1)^{1/2}, \ldots, (m_N\Lambda_N)^{1/2}]$. Then it is straightforward to show that $(\mathbf{A}-\mathbf{D})\mathbf{R}_\star = \mathbf{0}$. In other words, $\mathbf{R}_\star$ is an eigenvector of $\mathbf{A}-\mathbf{D}$ with eigenvalue 0, so one of the solutions for the evolution of the inclinations and nodes has frequency $f = 0$. Physically, this solution arises because the planetary system is neutrally stable to an overall tilt.

Table 5.1: Lagrange–Laplace secular frequencies for the solar system

	Lagrange–Laplace (arcsec yr^{-1})	Laskar et al. (2004b) (arcsec yr^{-1})	Period (yr)
g_1	5.462	5.59	2.318×10^5
g_2	7.347	7.452	1.739×10^5
g_3	17.332	17.368	7.462×10^4
g_4	18.006	17.916	7.234×10^4
g_5	3.733	4.257	3.044×10^5
g_6	22.512	28.245	4.588×10^4
g_7	2.707	3.088	4.197×10^5
g_8	0.635	0.673	1.926×10^6
f_1	−5.201	−5.59	2.318×10^5
f_2	−6.571	−7.05	1.838×10^5
f_3	−18.747	−18.850	6.875×10^4
f_4	−17.637	−17.755	7.299×10^4
f_5	0	0	—
f_6	−25.989	−26.348	4.919×10^4
f_7	−2.908	−2.993	4.331×10^5
f_8	−0.679	−0.692	1.874×10^6

Lagrange–Laplace frequencies are determined from equations (5.31). Laskar et al. (2004b) frequencies are determined from numerical integration of the equations of motion over tens of Myr. Periods are $2\pi/|g_i|$, $2\pi/|f_i|$, as derived from Laskar et al.

If there are N planets, then $\mathbf{A}$, $\mathbf{B}$ and $\mathbf{D}$ are $N \times N$ matrices and there are N eigenvalues g_n and N eigenvectors $\mathbf{R}_e^n$ that describe the evolution of the eccentricities. The same number of eigenvalues and eigenvectors f_n, $\mathbf{R}_I^n$ describe the inclinations. The general solution of the Lagrange–Laplace equations of motion is thus

$$\mathbf{Z}_e = \mathbf{K} + \mathrm{i}\mathbf{H} = \sum_{n=1}^{N} \alpha_n \mathrm{e}^{\mathrm{i}g_n t} \mathbf{R}_e^n, \quad \mathbf{Z}_I = \mathbf{Q} + \mathrm{i}\mathbf{P} = \sum_{n=1}^{N} \beta_n \mathrm{e}^{\mathrm{i}f_n t} \mathbf{R}_I^n, \tag{5.32}$$

where α_n and β_n are constants. Given the initial conditions $\mathbf{Z}_e(t = t_0) = \mathbf{Z}_{e0}$, we can determine the constants α_n as follows: evaluating the first of equations (5.32) at $t = t_0$ yields

$$\mathbf{Z}_{e0} = \sum_{n=1}^{N} \alpha_n \mathrm{e}^{\mathrm{i}g_n t_0} \mathbf{R}_e^n. \tag{5.33}$$

The eigenvectors of a real, symmetric matrix with distinct eigenvalues are orthogonal, that is, $\mathbf{R}_e^m \cdot \mathbf{R}_e^n = \sum_{i=1}^N R_{ei}^m R_{ei}^n = 0$ if $g_m \neq g_n$. Assuming all the eigenvalues are distinct, as is usually the case, we can multiply (5.33) by $\mathbf{R}_e^m$ to obtain

$$\alpha_m = \frac{\mathbf{R}_e^m \cdot \mathbf{Z}_{e0}}{\mathbf{R}_e^m \cdot \mathbf{R}_e^m} \mathrm{e}^{-\mathrm{i}g_m t_0}. \tag{5.34}$$

A similar derivation yields an expression for β_m.

Geometrically, equations (5.32) can be interpreted as representing each eccentricity vector and angular-momentum vector as the sum of N vectors of fixed length, each rotating uniformly at a different frequency.

The secular frequencies for the solar system, as derived from equations (5.31), are given in Table 5.1. Also shown are the best current estimates for these frequencies, obtained from the power spectrum of the positions of the planets over an interval of 20 Myr from the present (Laskar et al. 2004b).[3] The Lagrange–Laplace estimates are accurate to within 20% and usually much better than this; the largest errors arise because of the 5:2

[3] Over longer time intervals some of the secular frequencies vary substantially due to the chaotic nature of the planetary orbits (§4.5.2); the largest variations, in g_3, g_4 and f_2, reach 0.2 arcsec yr^{-1} over 100 Myr.

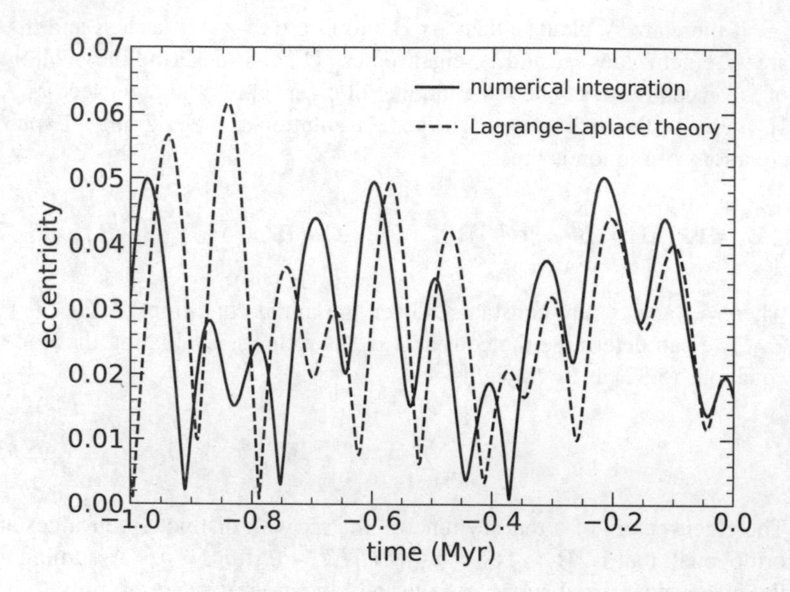

Figure 5.1: The history of the Earth's eccentricity over the last Myr. The solid line shows the result from a numerical integration of the orbits of the planets in the solar system, and the dashed line shows the result from Lagrange–Laplace theory.

near-resonance between Jupiter and Saturn, which contributes to the secular frequencies at second order in the planetary masses.

Figure 5.1 shows the behavior of the Earth's eccentricity over the past million years (studies of this kind are usually run backward, rather than forward, because the past history of the Earth's orbit is reflected in the geological record). The figure compares the results from a numerical integration of the orbits of the Earth and other planets, which has negligible uncertainties over this timescale, and from Lagrange–Laplace theory. The latter theory is reasonably accurate over the past 20 000 yr or so; beyond that time, it qualitatively reproduces the amplitude and period of the oscillations but is not quantitatively accurate.

An important special case is the motion of a test particle—a comet, asteroid, or even a very small planet—in the gravitational field of a set of N massive planets. If we denote the orbital elements of the test particle by variables without subscripts, equations (5.24) and (5.25) become

$$\frac{\mathrm{d}k}{\mathrm{d}t} + gh = \frac{1}{\Lambda}\sum_{j=1}^{N} b_j h_j, \qquad \frac{\mathrm{d}h}{\mathrm{d}t} - gk = -\frac{1}{\Lambda}\sum_{j=1}^{N} b_j k_j,$$

$$\frac{\mathrm{d}q}{\mathrm{d}t} - gp = -\frac{1}{\Lambda}\sum_{j=1}^{N} a_j p_j, \qquad \frac{\mathrm{d}p}{\mathrm{d}t} + gq = \frac{1}{\Lambda}\sum_{j=1}^{N} a_j q_j, \tag{5.35}$$

where

$$a_i = \frac{\mathbb{G}m_j}{4a_j}\alpha_j b^1_{3/2}(\alpha_j), \quad b_j = \frac{\mathbb{G}m_j}{4a_j}\alpha_j b^2_{3/2}(\alpha_j),$$

$$g = \frac{1}{\Lambda}\sum_{k=1}^{N}\frac{\mathbb{G}m_k}{4a_k}\alpha_k b^1_{3/2}(\alpha_k). \tag{5.36}$$

Here $\alpha_j = a/a_j$ and $\Lambda = (\mathbb{G}m_0 a)^{1/2}$. By setting $z_e \equiv k + \mathrm{i}h$ and $z_I \equiv q + \mathrm{i}p$ equations (5.35) can be simplified to

$$\frac{\mathrm{d}z_e}{\mathrm{d}t} - \mathrm{i}gz_e = -\frac{\mathrm{i}}{\Lambda}\sum_{j=1}^{N} b_j(k_j + \mathrm{i}h_j), \qquad \frac{\mathrm{d}z_I}{\mathrm{d}t} + \mathrm{i}gz_I = \frac{\mathrm{i}}{\Lambda}\sum_{j=1}^{N} a_j(q_j + \mathrm{i}p_j). \tag{5.37}$$

The functions on the right side of equations (5.37) are linear combinations of terms oscillating with the frequencies $\{g_n, f_n\}$. Thus we can write

$$\frac{\mathrm{d}z_e}{\mathrm{d}t} - \mathrm{i}gz_c = \sum_{n=1}^{N} \nu_n \mathrm{e}^{\mathrm{i}g_n t}, \qquad \frac{\mathrm{d}z_I}{\mathrm{d}t} + \mathrm{i}gz_I = \sum_{n=1}^{N} \mu_n \mathrm{e}^{\mathrm{i}f_n t}, \tag{5.38}$$

where the complex constants ν_n and μ_n are determined by the initial conditions for the massive planets. The solutions are

$$z_e(t) = \nu\mathrm{e}^{\mathrm{i}gt} + \mathrm{i}\sum_{n=1}^{N}\frac{\nu_n}{g - g_n}\mathrm{e}^{\mathrm{i}g_n t}, \quad z_I(t) = \mu\mathrm{e}^{-\mathrm{i}gt} - \mathrm{i}\sum_{n=1}^{N}\frac{\mu_n}{g + f_n}\mathrm{e}^{\mathrm{i}f_n t}, \tag{5.39}$$

where the complex constants ν and μ are determined by the initial conditions for the test particle. The eccentricity and inclination associated with the natural precession frequencies g and $-g$ are given by $|\nu|$ and $|\mu|$, and these are sometimes called the **free** or **proper eccentricity** and **free inclination**. Similarly, the absolute values of the summations in equations (5.39) are called the **forced eccentricity** and **forced inclination**. The forced eccentricity and inclination diverge at **secular resonances**, where $g = g_n$ or $g = -f_n$ respectively; in the vicinity of these resonances, the Lagrange–Laplace theory does not give an accurate description of the secular dynamics. See §6.6 for a more complete description of secular resonance.

The free and forced eccentricity have a simple geometrical interpretation. If we write $z_e = k + \mathrm{i}h$ and treat k and h as Cartesian coordinates, then the vector from the origin to (k, h) has length e and makes an angle ϖ with the k-axis. This vector is the sum of two others: the tip of the first has length equal to the proper eccentricity and rotates uniformly at angular speed g, while the second has time-varying coordinates $[-\mathrm{Im}\,\sum_n \nu_n \exp(\mathrm{i}g_n t)/(g - g_n), \mathrm{Re}\,\sum_n \nu_n \exp(\mathrm{i}g_n t)/(g - g_n)]$ and represents the forced eccentricity. The interpretation of the proper and forced inclinations is similar.

The free eccentricity and inclination are constant in time, apart from short-period perturbations, and contain more information on the dynamical history of a particle than the forced eccentricity and inclination, which are determined by the semimajor axis of the particle and the orbits of the massive bodies in the system. Asteroids exhibit clustering in phase space and the clustering is stronger when plotted using the proper eccentricity and inclination rather than the total or osculating eccentricity and inclination (Figure 5.2). These clusters, or **families**, are probably the fragments from collisions that have led to the breakup of, or large craters in, parent asteroids.

5.3 The Milankovich equations

We have shown in §5.1 how secular dynamics can be described by the orbit-averaged Hamiltonian $\langle H \rangle$ (eq. 5.15) and how the semimajor axis is conserved in this description. The evolution of the mean longitude λ is

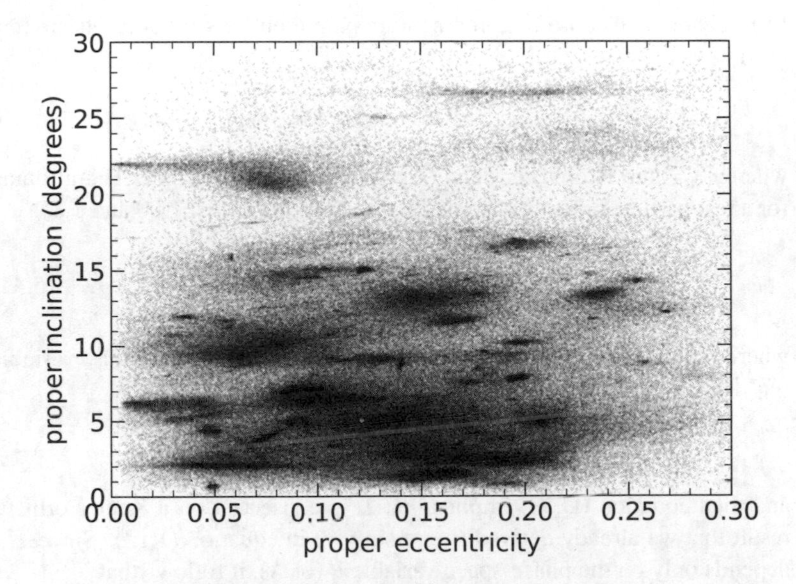

Figure 5.2: The distribution of proper inclination versus proper eccentricity for 500 000 asteroids. The clumps are asteroid families. For example the Vesta family, consisting of fragments from a collision with Vesta, is centered at $e = 0.10$, $I = 6.7°$. Data from https://newton.spacedys.com/astdys/index.php?pc=0.

dominated by the unperturbed mean motion and is not of interest in secular theory. The behavior of the remaining four orbital elements e, I, ϖ, Ω is described by Lagrange's equations (1.187). Unfortunately these equations are complicated and lack any natural structure; moreover they are ill-defined when the eccentricity e or inclination I is zero, or when $e = 1$. These disadvantages can be remedied in a vector-based formalism for secular theory, which we now derive.

The angular momentum per unit mass $\mathbf{L} = \mathbf{r} \times \mathbf{v} = \epsilon_{ijk}\hat{\mathbf{n}}_i r_j v_k$, where ϵ_{ijk} is the permutation symbol (Appendix C.1); throughout this section the summation convention described in Appendix B is in force. Since $\mathbf{r}$ and $\mathbf{v}$

form a canonical coordinate-momentum pair, their Poisson brackets are (eq. D.31)

$$\{r_i, r_j\} = 0, \quad \{v_i, v_j\} = 0, \quad \{r_i, v_j\} = -\{v_i, r_j\} = \delta_{ij}, \tag{5.40}$$

where δ_{ij} is the Kronecker delta (Appendix C.1). The Kepler Hamiltonian for a test particle orbiting a mass M at the origin is (eqs. 1.80 and 1.85)

$$H_{\mathrm{K}}(\mathbf{r}, \mathbf{v}) = \tfrac{1}{2}v^2 - \frac{\mathbb{G}M}{r} = -\frac{\mathbb{G}M}{2a} = -\frac{(\mathbb{G}M)^2}{2\Lambda^2}, \tag{5.41}$$

where a is the semimajor axis and $\Lambda = (\mathbb{G}Ma)^{1/2}$ is a Delaunay variable (eq. 1.84). It is straightforward to show that

$$\{H_{\mathrm{K}}, L_i\} = 0, \tag{5.42}$$

and then equation (D.38) implies that $\mathbf{L}$ is conserved on a Kepler orbit, a result that we already derived (more simply) in equation (1.13). Since H_{K} depends only on the phase-space variable a (or Λ), it follows that

$$\{a, L_i\} = 0 \quad \text{or} \quad \{\Lambda, L_i\} = 0. \tag{5.43}$$

Using equations (5.40) and the identity (C.2), it is straightforward to show that the Poisson brackets of the components of angular momentum are

$$\{L_i, L_j\} = \epsilon_{ijk} L_k. \tag{5.44}$$

Since $|L| = [\mathbb{G}Ma(1-e^2)]^{1/2} = \Lambda(1-e^2)^{1/2}$ (eq. 1.28), it proves useful to define a dimensionless angular momentum

$$\mathbf{j} \equiv \frac{\mathbf{L}}{\Lambda} = (1-e^2)^{1/2}\hat{\mathbf{L}}, \tag{5.45}$$

whose magnitude varies between 0 and 1. Using equations (D.32c) and (D.32d), it is straightforward to show that equations (5.43) and (5.44) can be rewritten as

$$\{a, j_i\} = 0, \quad \{j_i, j_j\} = \frac{1}{\Lambda}\epsilon_{ijk} j_k. \tag{5.46}$$

The eccentricity vector defined in Box 1.1 is

$$\mathbf{e} = \frac{\mathbf{v} \times (\mathbf{r} \times \mathbf{v})}{\mathbb{G}M} - \hat{\mathbf{r}} = \mathbf{r}\frac{v^2}{\mathbb{G}M} - \mathbf{v}\frac{\mathbf{r} \cdot \mathbf{v}}{\mathbb{G}M} - \frac{\mathbf{r}}{r}. \tag{5.47}$$

In terms of the usual orbital elements (§1.3.2), the Cartesian components of these vectors can be written

$$\begin{aligned}\mathbf{e} &= e(\cos\Omega\cos\omega - \cos I\sin\Omega\sin\omega, \sin\Omega\cos\omega + \cos I\cos\Omega\sin\omega, \\ &\qquad \sin I\sin\omega), \\ \mathbf{j} &= (1-e^2)^{1/2}(\sin I\sin\Omega, -\sin I\cos\Omega, \cos I).\end{aligned} \tag{5.48}$$

Similarly to equation (5.42), we can show that

$$\{H_{\mathrm{K}}, e_i\} = 0, \tag{5.49}$$

which confirms that the eccentricity vector is conserved on a Kepler orbit. Since H_{K} depends only on a (or Λ), it follows that

$$\{a, e_i\} = 0 \quad \text{or} \quad \{\Lambda, e_i\} = 0. \tag{5.50}$$

It is straightforward, though tedious, to show that the Poisson brackets of the components of the eccentricity vector are

$$\{e_i, e_j\} = \frac{1}{\Lambda}\epsilon_{ijk}j_k. \tag{5.51}$$

Similarly we can show that

$$\{j_i, e_j\} = \{e_i, j_j\} = \frac{1}{\Lambda}\epsilon_{ijk}e_k. \tag{5.52}$$

The elegant relations (5.46), (5.51) and (5.52) arise because the symmetry group of the Kepler problem is the group of rotations in 4-dimensional space, SO(4).

The orbit-averaged Hamiltonian $\langle H \rangle$ is a function of the size and shape of the orbit, and possibly the time. These can be specified by the orbital elements a, e, I, ϖ, Ω (the mean longitude λ does not appear because of the orbit averaging). Alternatively, we can specify the orbit by the semi-major axis a and the two vectors $\mathbf{j}$ and $\mathbf{e}$. Thus we can write the orbit-averaged Hamiltonian as $\langle H\rangle(a,\mathbf{j},\mathbf{e},t)$. Note that these arguments contain seven phase-space variables (a and the three components of each of the two vectors), but only five are independent, because they are related by the constraints

$$\mathbf{j}\cdot\mathbf{e}=0, \quad j^2+e^2=1. \tag{5.53}$$

The time evolution of j_i under the influence of $\langle H\rangle$ is given by equation (D.38),

$$\frac{\mathrm{d}j_i}{\mathrm{d}t}=\{j_i,\langle H\rangle\}. \tag{5.54}$$

Then from the chain rule

$$\frac{\mathrm{d}j_i}{\mathrm{d}t}=\{j_i,j_k\}\frac{\partial\langle H\rangle}{\partial j_k}+\{j_i,e_k\}\frac{\partial\langle H\rangle}{\partial e_k}+\{j_i,a\}\frac{\partial\langle H\rangle}{\partial a}. \tag{5.55}$$

Using the evaluations of the Poisson brackets in equations (5.46) and (5.52), the result simplifies to

$$\frac{\mathrm{d}j_i}{\mathrm{d}t}=\frac{1}{\Lambda}\epsilon_{ikm}j_m\frac{\partial\langle H\rangle}{\partial j_k}+\frac{1}{\Lambda}\epsilon_{ikm}e_m\frac{\partial\langle H\rangle}{\partial e_k}. \tag{5.56}$$

This can be rewritten in vector notation as

$$\frac{\mathrm{d}\mathbf{j}}{\mathrm{d}t}=-\frac{1}{\Lambda}\left(\mathbf{j}\times\frac{\partial}{\partial\mathbf{j}}\langle H\rangle+\mathbf{e}\times\frac{\partial}{\partial\mathbf{e}}\langle H\rangle\right), \tag{5.57}$$

where $\partial f/\partial\mathbf{j}$ is the vector having components $(\partial f/\partial j_1,\partial f/\partial j_2,\partial f/\partial j_3)$ for any function $f(j_1,j_2,j_3)$. Similarly, the time evolution of the eccentricity vector is given by

$$\frac{\mathrm{d}\mathbf{e}}{\mathrm{d}t}=-\frac{1}{\Lambda}\left(\mathbf{e}\times\frac{\partial}{\partial\mathbf{j}}\langle H\rangle+\mathbf{j}\times\frac{\partial}{\partial\mathbf{e}}\langle H\rangle\right). \tag{5.58}$$

Equations (5.57) and (5.58) are the **Milankovich equations.**[4]

[4] Milutin Milankovich (1879–1958) was a Serbian applied mathematician. He is responsible for the concept that long-term changes in Earth's climate are quasi-periodic and driven mainly by secular variations in the Earth's orbit and the precession of its spin axis. These **Milankovich cycles** have periods between 2×10^4 yr and 1×10^5 yr. The equations first appear in Milankovich (1939).

It is straightforward to show that the Milankovich equations conserve $\mathbf{j} \cdot \mathbf{e}$ and $j^2 + e^2$. Thus if the constraints (5.53) are satisfied by the initial conditions, they continue to be satisfied for all time. Because of this property the formula for a given Hamiltonian in terms of $\mathbf{j}$ and $\mathbf{e}$ is not unique—for example, $\langle H \rangle = j^2$ could also be written $\langle H \rangle = -e^2$ or $\langle H \rangle = j^2 + \mathbf{e} \cdot \mathbf{j}$—but the trajectories determined by the Milankovich equations are the same for all of these.

A more compact form of these equations is obtained by defining new variables

$$\hat{\mathbf{b}}_+ \equiv \mathbf{j} + \mathbf{e}, \quad \hat{\mathbf{b}}_- \equiv \mathbf{j} - \mathbf{e}, \tag{5.59}$$

which imply that

$$\mathbf{j} = \tfrac{1}{2}(\hat{\mathbf{b}}_+ + \hat{\mathbf{b}}_-), \quad \mathbf{e} = \tfrac{1}{2}(\hat{\mathbf{b}}_+ - \hat{\mathbf{b}}_-). \tag{5.60}$$

Using the constraints (5.53), it is straightforward to show that $|\hat{\mathbf{b}}_\pm|^2 = (\mathbf{j} \pm \mathbf{e}) \cdot (\mathbf{j} \pm \mathbf{e}) = 1$; thus $\hat{\mathbf{b}}_+$ and $\hat{\mathbf{b}}_-$ are unit vectors that can be represented by points on the unit sphere. In terms of these the Milankovich equations become

$$\frac{d\hat{\mathbf{b}}_+}{dt} = -\frac{2}{\Lambda}\hat{\mathbf{b}}_+ \times \frac{\partial}{\partial \hat{\mathbf{b}}_+}\langle H \rangle, \quad \frac{d\hat{\mathbf{b}}_-}{dt} = -\frac{2}{\Lambda}\hat{\mathbf{b}}_- \times \frac{\partial}{\partial \hat{\mathbf{b}}_-}\langle H \rangle. \tag{5.61}$$

Despite the simplicity of these equations, we shall usually work with equations (5.57) and (5.58), since the vectors $\mathbf{j}$ and $\mathbf{e}$ have a direct physical interpretation.

We have written the Milankovich equations for an orbit-averaged Hamiltonian, which is independent of the mean anomaly. A more general form of these equations, valid for Hamiltonians that are not necessarily orbit-averaged, is described by Allan & Ward (1963).

5.3.1 The Laplace surface

Most of the inner satellites orbiting the giant planets in the solar system likely formed from a thin disk of gas and solid material surrounding the planet. The orientation of this disk is determined by the shape of the gravitational potential in which it orbits. If the disk were oriented in some arbitrary direction, the angular-momentum vector $\mathbf{j}$ of each fluid element in the disk would precess at a different rate, depending on its semimajor axis, so the disk would rapidly dissolve. Therefore the only possible orientations for the disk are those for which the angular-momentum vectors do not precess.[5]

In the most common case, the disk orientation required for zero precession rate is determined by the competition between the torques from the quadrupole potential of the planet's equatorial bulge and the quadrupole potential from the Sun. Consider a flattened planet with mass M_p, quadrupole moment J_2 and radius R_p. Assume that the higher multipole moments J_3, J_4, and so forth are negligible. Its non-Kepler potential is given by equation (1.135),

$$H_\mathrm{p} = \frac{\mathbb{G}M_\mathrm{p}J_2R_\mathrm{p}^2}{2r^3}(3\cos^2\theta - 1) = \frac{\mathbb{G}M_\mathrm{p}J_2R_\mathrm{p}^2}{2r^5}(2z^2 - x^2 - y^2); \qquad (5.62)$$

here the first expression is in spherical coordinates and the second in Cartesian coordinates, with the equators of both coordinate systems aligned with the equator of the planet. Using equations (1.70), the potential can be rewritten as

$$H_\mathrm{p} = \frac{\mathbb{G}M_\mathrm{p}J_2R_\mathrm{p}^2}{2r^3}\big[3\sin^2 I\sin^2(f+\omega) - 1\big]. \qquad (5.63)$$

We orbit average using equations (1.66c)–(1.66e) to obtain

$$\begin{aligned}\langle H_\mathrm{p}\rangle &= \frac{\mathbb{G}M_\mathrm{p}J_2R_\mathrm{p}^2}{4a^3(1-e^2)^{3/2}}\big(3\sin^2 I - 2\big) = \frac{(\mathbb{G}M_\mathrm{p})^4J_2R_\mathrm{p}^2}{4L^5\Lambda^3}\big(L^2 - 3L_z^2\big)\\ &= \frac{\mathbb{G}M_\mathrm{p}J_2R_\mathrm{p}^2}{4a^3}\frac{j^2 - 3j_z^2}{j^5},\end{aligned} \qquad (5.64)$$

where the dimensionless angular momentum $\mathbf{j}$ is defined in equation (5.45). We can write this result in a coordinate-free form by defining a unit vector $\hat{\mathbf{n}}_\mathrm{p}$ that is normal to the planet's equator, which so far we have assumed to

---

[5] Strictly, they must precess at the same rate as the planet's spin axis precesses due to the torque from its host star (see §7.1), but this rate is usually slow enough that the spin axis can be assumed to be fixed.

be the z-axis:

$$\langle H_\mathrm{p}\rangle = \frac{\mathbb{G}M_\mathrm{p}J_2R_\mathrm{p}^2}{4a^3}\frac{j^2 - 3(\mathbf{j}\cdot\hat{\mathbf{n}}_\mathrm{p})^2}{j^5}. \tag{5.65}$$

The Milankovich equation (5.58) is satisfied if $\mathbf{e} = \mathbf{0}$; thus a circular orbit will remain circular. Since we expect the orbits in a gas disk to be circular anyway, we set $\mathbf{e} = \mathbf{0}$ henceforth.[6] Then the Milankovich equation (5.57) becomes

$$\frac{\mathrm{d}\mathbf{j}}{\mathrm{d}t} = \frac{3(\mathbb{G}M_\mathrm{p})^{1/2}J_2R_\mathrm{p}^2}{2a^{7/2}}(\mathbf{j}\cdot\hat{\mathbf{n}}_\mathrm{p})\,\mathbf{j}\times\hat{\mathbf{n}}_\mathrm{p}. \tag{5.66}$$

The simplest solutions to this equation have $\mathbf{j}\times\hat{\mathbf{n}}_\mathrm{p} = \mathbf{0}$ or $\mathbf{j}\cdot\hat{\mathbf{n}}_\mathrm{p} = 0$. In the first of these, $\mathbf{j}$ is parallel or antiparallel to the planet's spin axis so the disk lies in the equatorial plane of the planet; in the second, $\mathbf{j}$ is perpendicular to $\hat{\mathbf{n}}_\mathrm{p}$ so the disk crosses the pole of the planet. In practice, equatorial disks are favored over polar disks, in part because gas disks dissipate energy so they tend to evolve toward a state of minimum energy at a given semimajor axis; and equation (5.65) shows that the Hamiltonian is minimized at fixed a and j when the disk is equatorial.

At large distances from the planet, the orientation of the disk is governed by the tidal field from the host star. Consider a planet that orbits a host star with mass M_*, on an orbit with semimajor axis a_p and eccentricity e_p. In coordinates centered on the planet the quadrupole tidal potential from the star is given by equation (3.71):

$$H_* = \frac{\mathbb{G}M_*r^2}{2r_*^3} - \frac{3\mathbb{G}M_*(\mathbf{r}\cdot\mathbf{r}_*)^2}{2r_*^5}, \tag{5.67}$$

with $\mathbf{r}_*$ the position of the star.

Let $(\hat{\mathbf{x}},\hat{\mathbf{y}},\hat{\mathbf{z}})$ be an orthogonal triad of unit vectors, with $\hat{\mathbf{x}}$ pointing to the periapsis of the orbit of a disk particle around the planet, $\hat{\mathbf{z}}$ parallel to its orbital angular momentum, and $\hat{\mathbf{y}} \equiv \hat{\mathbf{z}}\times\hat{\mathbf{x}}$, with similar definitions for $(\hat{\mathbf{x}}_*,\hat{\mathbf{y}}_*,\hat{\mathbf{z}}_*)$. The position of the disk particle is $\mathbf{r} = r(\hat{\mathbf{x}}\cos f + \hat{\mathbf{y}}\sin f)$, where f is the true anomaly, and similarly $\mathbf{r}_* = r_*(\hat{\mathbf{x}}_*\cos f_* + \hat{\mathbf{y}}_*\sin f_*)$.

[6] The behavior of orbits with nonzero eccentricity is discussed in Problem 5.4.

We now average H_* over the orbit of the test particle. Using equations (1.65c)–(1.65f), we have

$$\langle H_* \rangle = \frac{\mathbb{G}M_* a^2}{2r_*^5}\left[(1+\tfrac{3}{2}e^2)r_*^2 - 3(\hat{\mathbf{x}}\cdot\mathbf{r}_*)^2(\tfrac{1}{2}+2e^2) - \tfrac{3}{2}(\hat{\mathbf{y}}\cdot\mathbf{r}_*)^2(1-e^2)\right]. \tag{5.68}$$

Now eliminate $\hat{\mathbf{y}}$ using the relation $(\hat{\mathbf{y}}\cdot\mathbf{r}_*)^2 = r_*^2 - (\hat{\mathbf{x}}\cdot\mathbf{r}_*)^2 - (\hat{\mathbf{z}}\cdot\mathbf{r}_*)^2$ (eq. B.6a), and replace $\hat{\mathbf{x}}$ and $\hat{\mathbf{z}}$ with the eccentricity and dimensionless angular-momentum vectors using the relations $\mathbf{e} = e\hat{\mathbf{x}}$, $\mathbf{j} = (1-e^2)^{1/2}\hat{\mathbf{z}}$:

$$\langle H_* \rangle = \frac{\mathbb{G}M_* a^2}{4r_*^5}\left[(6e^2-1)r_*^2 - 15(\mathbf{e}\cdot\mathbf{r}_*)^2 + 3(\mathbf{j}\cdot\mathbf{r}_*)^2\right]. \tag{5.69}$$

We now carry out a similar average over the orbit of the host star around the planet, using equations (1.66b)–(1.66e). We denote the second average by double angle brackets:

$$\langle\langle H_* \rangle\rangle = \frac{\mathbb{G}M_* a^2}{8a_*^3(1-e_*^2)^{3/2}}\left[15(\mathbf{e}\cdot\hat{\mathbf{n}}_*)^2 - 6e^2 + 1 - 3(\mathbf{j}\cdot\hat{\mathbf{n}}_*)^2\right], \tag{5.70}$$

where $\hat{\mathbf{n}}_* = \hat{\mathbf{z}}_*$ is normal to the orbit of the host star. If we adopt a coordinate system in which the host-star orbit is in the equatorial plane, the Hamiltonian can be rewritten in terms of orbital elements or Delaunay variables using equations (5.48):

$$\begin{aligned}\langle\langle H_* \rangle\rangle &= \frac{\mathbb{G}M_* a^2}{8a_*^3(1-e_*^2)^{3/2}}(3\sin^2 I - 2 - 3e^2 - 3e^2\sin^2 I + 15e^2\sin^2 I\sin^2\omega)\\ &= \frac{\mathbb{G}M_* a^2}{8a_*^3(1-e_*^2)^{3/2}}\left[\frac{6L^2}{\Lambda^2} - 5 - \frac{3L_z^2}{\Lambda^2} + 15\left(1 - \frac{L_z^2}{L^2} - \frac{L^2}{\Lambda^2} + \frac{L_z^2}{\Lambda^2}\right)\sin^2\omega\right].\end{aligned} \tag{5.71}$$

Using the Hamiltonian (5.70), the Milankovich equations read

$$\frac{\mathrm{d}\mathbf{j}}{\mathrm{d}t} = \frac{3\,\mathbb{G}^{1/2}M_* a^{3/2}}{4M_\mathrm{p}^{1/2}a_*^3(1-e_*^2)^{3/2}}\left[(\mathbf{j}\cdot\hat{\mathbf{n}}_*)\mathbf{j}\times\hat{\mathbf{n}}_* - 5(\mathbf{e}\cdot\hat{\mathbf{n}}_*)\mathbf{e}\times\hat{\mathbf{n}}_*\right], \tag{5.72}$$

$$\frac{\mathrm{d}\mathbf{e}}{\mathrm{d}t} = \frac{3\,\mathbb{G}^{1/2} M_* a^{3/2}}{4M_\mathrm{p}^{1/2} a_*^3 (1-e_*^2)^{3/2}} \big[(\mathbf{j}\cdot\hat{\mathbf{n}}_*)\mathbf{e}\times\hat{\mathbf{n}}_* - 5(\mathbf{e}\cdot\hat{\mathbf{n}}_*)\mathbf{j}\times\hat{\mathbf{n}}_* + 2\mathbf{j}\times\mathbf{e}\big].$$

Once again we assume that the disk orbits are circular, $\mathbf{e} = \mathbf{0}$, so the second of these equations is automatically satisfied. The simplest solutions have $\mathbf{j}\times\hat{\mathbf{n}}_* = \mathbf{0}$ or $\mathbf{j}\cdot\hat{\mathbf{n}}_* = 0$, so the disk lies either in the orbital plane or perpendicular to the orbital plane, with the former configuration favored because the energy (5.70) is smaller at fixed semimajor axis when $\mathbf{j}$ is parallel or antiparallel to $\hat{\mathbf{n}}_*$.

The **obliquity** of a planet is the angle between its spin angular momentum and orbital angular momentum. The obliquities of planets in the solar system range from less than a few degrees for Mercury and Jupiter to 98° for Uranus and 177° for Venus (see Appendix A). If the obliquity of a planet is nonzero, then the preferred orientation of the disk must transition from the equatorial plane of the planet at small distances to the orbital plane of the planet at large distances. To determine the shape of this transition, we add the Milankovich equations (5.66) for the effects of the planetary quadrupole and (5.72) for the effects of the tide from the host star and look for solutions with zero precession. For circular disks the equation $\mathrm{d}\mathbf{j}/\mathrm{d}t = \mathbf{0}$ is satisfied if

$$\frac{J_2 R_\mathrm{p}^2}{a^5}(\mathbf{j}\cdot\hat{\mathbf{n}}_\mathrm{p})\mathbf{j}\times\hat{\mathbf{n}}_\mathrm{p} + \frac{M_*}{2M_\mathrm{p} a_*^3(1-e_*^2)^{3/2}}(\mathbf{j}\cdot\hat{\mathbf{n}}_*)\mathbf{j}\times\hat{\mathbf{n}}_* = \mathbf{0}. \tag{5.73}$$

We define the **Laplace radius** r_L by

$$r_\mathrm{L}^5 \equiv J_2 R_\mathrm{p}^2 a_*^3 (1-e_*^2)^{3/2}\frac{M_\mathrm{p}}{M_*}; \tag{5.74}$$

then

$$2r_\mathrm{L}^5(\mathbf{j}\cdot\hat{\mathbf{n}}_\mathrm{p})\mathbf{j}\times\hat{\mathbf{n}}_\mathrm{p} + a^5(\mathbf{j}\cdot\hat{\mathbf{n}}_*)\mathbf{j}\times\hat{\mathbf{n}}_* = \mathbf{0}. \tag{5.75}$$

In the simplest solutions, $\hat{\mathbf{n}}_\mathrm{p}$, $\hat{\mathbf{n}}_*$ and $\mathbf{j}$ all lie in a common plane. The obliquity ϵ is the angle between $\hat{\mathbf{n}}_\mathrm{p}$ and $\hat{\mathbf{n}}_*$; by convention the obliquity lies between 0 and 180°. Let ϕ be the angle between $\mathbf{j}$ and $\hat{\mathbf{n}}_\mathrm{p}$, measured in the same direction as the obliquity in the common plane. Then the constraint (5.75) becomes

$$2r_\mathrm{L}^5 \sin 2\phi + a^5 \sin 2(\phi - \epsilon) = 0. \tag{5.76}$$

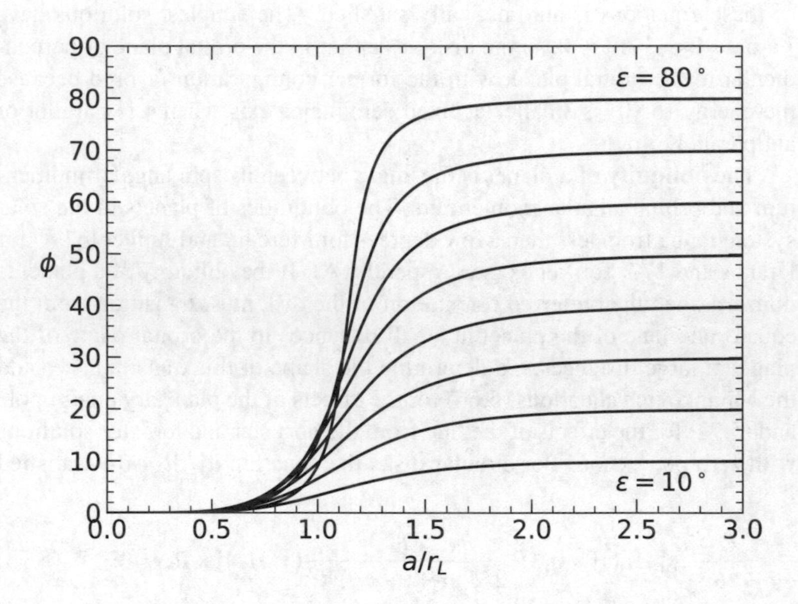

Figure 5.3: The shape of the Laplace surface as a function of the obliquity ϵ, obtained by solving equation (5.76). The horizontal axis is $a/r_{\rm L}$, where a is the semi-major axis and the Laplace radius $r_{\rm L}$ is defined in equation (5.74). The vertical axis is ϕ, where $\phi = 0$ when the surface coincides with the equatorial plane of the planet, and $\phi = \epsilon$ when the surface coincides with the orbital plane of the planet. Solutions for planets with obliquity $> 90^\circ$ can be obtained by the transformation $\epsilon \to 180^\circ - \epsilon$, $\phi \to 180^\circ - \phi$.

The surface mapped out by this constraint is the **Laplace surface**. When $a \ll r_\mathrm{L}$ the constraint requires $\sin 2\phi = 0$, so $\phi = 0, \frac{1}{2}\pi$, or π. The energy arguments following equation (5.66) favor the solutions $\phi = 0, \pi$, in which the Laplace surface coincides with the equatorial plane of the planet. Similarly, for $a \gg r_\mathrm{L}$ we must have $\sin 2(\phi - \epsilon) = 0$, and energy arguments favor $\phi = \epsilon$ or $\epsilon + \pi$ so the Laplace surface lies in the plane of the orbit. The shape at intermediate radii is shown in Figure 5.3. See Tremaine et al. (2009) for a comprehensive description of the Laplace surface.

The shape of the Laplace surface motivates the division of satellites into **inner satellites**, which have semimajor axes $a \lesssim r_\mathrm{L}$ and orbit close to the planet's equator; and **outer satellites** with $a \gtrsim r_\mathrm{L}$, which orbit close to the ecliptic. The Moon is an outer satellite but almost all of the major satellites of the giant planets except for Saturn's satellite Iapetus are inner satellites.

Of course, the gas disks from which the satellites of the giant planets formed disappeared long ago. Nevertheless, we still expect that the present orbits of these satellites will lie close to the Laplace surface. Whether or not this is true, the angular-momentum vector of a satellite will precess around the direction defined by the normal to the Laplace surface at its semimajor axis. This is why the JPL Solar System Dynamics database, https://ssd.jpl.nasa.gov/, quotes the inclinations and nodes of many satellites relative to the local Laplace surface.

5.3.2 Stellar flybys

The solar system, like most planetary systems, is far from the other stars in the solar neighborhood: the nearest star, Proxima Centauri, is almost 10^4 times farther than the outermost planet, Neptune. Nevertheless, many stars, perhaps including the Sun, were born in dense star clusters that later dissolved, and/or migrated through the Galactic disk from locations closer to the Galactic center with much higher stellar densities. Therefore we need to examine the possible effects of encounters with a passing star on a planetary system.

The velocities of stars in the solar neighborhood relative to the Sun are sufficiently high ($V \sim 50\,\mathrm{km\,s^{-1}}$) and their distances sufficiently large that they pass us on straight-line orbits, unaffected by the Sun's gravity. If the impact parameter or distance of closest approach to the Sun is b, then the duration of an encounter, when the gravitational force from the star is strongest, is roughly

$$\tau \equiv \frac{b}{V} = 94.8\,\mathrm{yr}\frac{b}{10^3\,\mathrm{au}}\frac{50\,\mathrm{km\,s^{-1}}}{V}; \tag{5.77}$$

thus all but the closest encounters last much longer than a typical planetary orbital period, so their effects can be studied using secular dynamics.

We first calculate the rate of stellar encounters with the Sun in its present environment. Let $f(\mathbf{v})\mathrm{d}\mathbf{v}$ be the number of stars per unit volume with velocities in the range $\mathbf{v}$ to $\mathbf{v} + \mathrm{d}\mathbf{v}$. The number of encounters with these stars per unit time with impact parameter less than b is $f(\mathbf{v})\mathrm{d}\mathbf{v}$ times the volume of an annulus with radius b and length equal to the relative speed $|\mathbf{v} - \mathbf{v}_\odot|$. The total rate of encounters with impact parameter less than b is then

$$R(b) = \int \mathrm{d}\mathbf{v}\, f(\mathbf{v})\pi b^2|\mathbf{v} - \mathbf{v}_\odot|. \tag{5.78}$$

To evaluate the integral we make two simplifying assumptions. First, we assume that the velocity distribution is Maxwellian with dispersion σ in one dimension,

$$f(\mathbf{v}) = \frac{n}{(2\pi\sigma^2)^{3/2}}\exp\left(-\frac{v^2}{2\sigma^2}\right), \tag{5.79}$$

where n is the number of stars per unit volume. A more accurate representation of the velocity distribution in the solar neighborhood is a triaxial Gaussian distribution, equation (9.17), but this refinement is not needed for the estimates we make here. The root-mean-square velocity of the stars is approximately $50\,\mathrm{km\,s^{-1}}$ (Dehnen & Binney 1998) so the dispersion in one dimension can be taken to be $\sigma = (50\,\mathrm{km\,s^{-1}})/\sqrt{3} \simeq 30\,\mathrm{km\,s^{-1}}$. Second, we assume that the Sun is at rest with respect to the mean velocity of nearby stars. The actual value of the Sun's velocity in this frame is only $18\,\mathrm{km\,s^{-1}}$ (Schönrich et al. 2010) compared to the root-mean-square velocity of $50\,\mathrm{km\,s^{-1}}$, so this approximation is not too bad. Then

$$R(b) = \frac{b^2 n}{2^{3/2}\pi^{1/2}\sigma^3}\int \mathrm{d}\mathbf{v}\,|\mathbf{v}|\exp\left(-\frac{v^2}{2\sigma^2}\right)$$

$$= \frac{2^{1/2}\pi^{1/2}b^2 n}{\sigma^3}\int_0^\infty dv\, v^3 \exp\left(-\frac{v^2}{2\sigma^2}\right) = 2^{3/2}\pi^{1/2}b^2 n\sigma$$

$$= \frac{0.36}{\text{Gyr}}\left(\frac{b}{1\,000\,\text{au}}\right)^2 \frac{n}{0.10\,\text{pc}^{-3}}\frac{\sigma}{30\,\text{km s}^{-1}}. \tag{5.80}$$

We may conclude that the closest encounter the Sun has had in its lifetime $t_{ss} = 4.57$ Gyr is $b \sim 800$ au, corresponding to $R(b)t_{ss} = 1$. Of course, there are large uncertainties in this result, both because the closest encounter is a random event[7] and because the Sun's environment at the time of its birth was probably much denser than the present solar neighborhood.

We now examine the effect of an encounter on a planetary orbit. The trajectory of the star relative to the Sun can be written $\mathbf{r}_*(t) = \mathbf{b} + \mathbf{V}t$, where $\mathbf{V}$ is the velocity of the star relative to the Sun, $t = 0$ is the time of closest approach and $\mathbf{b}$ is the impact parameter vector, the position of the star at the periapsis or point of closest approach. This definition implies that $\mathbf{b}$ is perpendicular to $\mathbf{V}$ and $|\mathbf{b}| = b$, the impact parameter. We use Cartesian coordinates in which the origin is at the Sun, the trajectory of the star is in the x-y plane and the x-axis points to the periapsis of the passing star. In these coordinates $\mathbf{r}_*(t) = (b, Vt, 0)$ and the quadrupole tidal field from the star at position $\mathbf{r} = (x, y, z)$ is given by equation (3.71),

$$H_* = \frac{\mathbb{G}M_* r^2}{2(b^2 + V^2t^2)^{3/2}} - \frac{3\,\mathbb{G}M_*(xb + yVt)^2}{2(b^2 + V^2t^2)^{5/2}}, \tag{5.81}$$

where M_* is the mass of the passing star. We want to average over the planetary orbit. The averaging can be written symbolically as

$$\langle H_*\rangle = \frac{\mathbb{G}M_*\langle r^2\rangle}{2(b^2 + V^2t^2)^{3/2}} - \frac{3\,\mathbb{G}M_*(\langle x^2\rangle b^2 + 2\langle xy\rangle bVt + \langle y^2\rangle V^2t^2)}{2(b^2 + V^2t^2)^{5/2}}. \tag{5.82}$$

As usual with orbit averaging the Hamiltonian $\langle H_*\rangle$ is independent of the mean longitude, so the semimajor axis of the planet is conserved during the encounter.

[7] A more precise statement is that the probability that the closest encounter has an impact parameter greater than b is $\exp[-R(b)t_{ss}]$.

Since the changes $\Delta\mathbf{j}$ and $\Delta\mathbf{e}$ in the angular-momentum and eccentricity vectors from a single encounter are expected to be small, we can solve the Milankovich equations (5.57) and (5.58) by evaluating the right sides on the original planetary orbit. In this case the only time-varying component of the right sides is $\langle H\rangle$, so we can write

$$\Delta\mathbf{j} = -\frac{1}{\Lambda}\left(\mathbf{j}\times\frac{\partial}{\partial\mathbf{j}} + \mathbf{e}\times\frac{\partial}{\partial\mathbf{e}}\right)\int_{-\infty}^{\infty}\mathrm{d}t\,\langle H_*\rangle,$$
$$\Delta\mathbf{e} = -\frac{1}{\Lambda}\left(\mathbf{e}\times\frac{\partial}{\partial\mathbf{j}} + \mathbf{j}\times\frac{\partial}{\partial\mathbf{e}}\right)\int_{-\infty}^{\infty}\mathrm{d}t\,\langle H_*\rangle, \tag{5.83}$$

where $\Lambda = (\mathbb{G}M_\odot a)^{1/2}$. Since the term involving $\langle xy\rangle$ in equation (5.82) is odd in t, its integral will vanish and it can be dropped from this analysis. We also use the integrals

$$\int_{-\infty}^{\infty}\frac{\mathrm{d}t}{(b^2+V^2t^2)^{3/2}} = \frac{2}{b^2V}, \quad \int_{-\infty}^{\infty}\frac{\mathrm{d}t}{(b^2+V^2t^2)^{5/2}} = \frac{4}{3b^4V},$$
$$\int_{-\infty}^{\infty}\frac{\mathrm{d}t\,t^2}{(b^2+V^2t^2)^{5/2}} = \frac{2}{3b^2V^3}. \tag{5.84}$$

Then

$$\int_{-\infty}^{\infty}\mathrm{d}t\,\langle H_*\rangle = \frac{\mathbb{G}M_*}{b^2V}\big(\langle r^2\rangle - 2\langle x^2\rangle - \langle y^2\rangle\big). \tag{5.85}$$

The average $\langle r^2\rangle = a^2(1+\frac{3}{2}e^2)$ by equation (1.65c) and the averages $\langle x^2\rangle$ and $\langle y^2\rangle$ are evaluated in equation (P.3). We find

$$\int_{-\infty}^{\infty}\mathrm{d}t\,\langle H_*\rangle = \frac{\mathbb{G}M_*a^2}{b^2V}\big[3e^2-\tfrac{1}{2}-5(\mathbf{e}\cdot\hat{\mathbf{b}})^2+(\mathbf{j}\cdot\hat{\mathbf{b}})^2-\tfrac{5}{2}(\mathbf{e}\cdot\hat{\mathbf{V}})^2+\tfrac{1}{2}(\mathbf{j}\cdot\hat{\mathbf{V}})^2\big]. \tag{5.86}$$

Here we have used the unit vector $\hat{\mathbf{b}}$ pointing toward the point of closest approach $\mathbf{b}$ (the x-axis) and the unit vector $\hat{\mathbf{V}}$ along the direction of the velocity $\mathbf{V}$ (the y-axis).

Substituting this result into the Milankovich equations (5.83), we have (Heggie & Rasio 1996)

$$\Delta\mathbf{j} = \frac{\mathbb{G}^{1/2}M_*a^{3/2}}{M_\odot^{1/2}b^2V}\big[-2(\mathbf{j}\cdot\hat{\mathbf{b}})\mathbf{j}\times\hat{\mathbf{b}} - (\mathbf{j}\cdot\hat{\mathbf{V}})\mathbf{j}\times\hat{\mathbf{V}}$$

$$+ 10(\mathbf{e}\cdot\hat{\mathbf{b}})\mathbf{e}\times\hat{\mathbf{b}} + 5(\mathbf{e}\cdot\hat{\mathbf{V}})\mathbf{e}\times\hat{\mathbf{V}}\big],$$

$$\Delta\mathbf{e} = \frac{\mathbb{G}^{1/2}M_* a^{3/2}}{M_\odot^{1/2}b^2V}\big[-2(\mathbf{j}\cdot\hat{\mathbf{b}})\mathbf{e}\times\hat{\mathbf{b}} - (\mathbf{j}\cdot\hat{\mathbf{V}})\mathbf{e}\times\hat{\mathbf{V}} - 6\mathbf{j}\times\mathbf{e} + 10(\mathbf{e}\cdot\hat{\mathbf{b}})\mathbf{j}\times\hat{\mathbf{b}} + 5(\mathbf{e}\cdot\hat{\mathbf{V}})\mathbf{j}\times\hat{\mathbf{V}}\big]. \quad (5.87)$$

The right side of each equation contains a geometric factor in square brackets that is of order unity, multiplied by a pre-factor that gives the typical magnitude of the changes. The geometric factor in the second equation shows that a circular orbit ($\mathbf{e} = \mathbf{0}$) remains circular ($\Delta\mathbf{e} = \mathbf{0}$); physically, this is because the torque exerted by any gravitational potential on any axisymmetric mass distribution must be perpendicular to its symmetry axis. The pre-factor is

$$\frac{\mathbb{G}^{1/2}M_* a^{3/2}}{M_\odot^{1/2}b^2V} = 6.66\times10^{-6}\frac{M_*}{M_\odot}\left(\frac{a}{5\,\mathrm{au}}\right)^{3/2}\left(\frac{1\,000\,\mathrm{au}}{b}\right)^2\frac{50\,\mathrm{km\,s^{-1}}}{V}. \quad (5.88)$$

Thus the changes in the orbits of the solar-system planets due to passing stars are negligible, even for an encounter with the smallest plausible impact parameter predicted by equation (5.80). One immediate consequence of this conclusion is that the direction of the ecliptic pole is fixed so long as the stellar environment of the Sun is similar to the present one.[8]

This result is based on three main approximations: that the duration of the encounter is much larger than the orbital period of the planet (eq. 5.77), that the stars pass the planetary system on straight-line orbits and that the smallest impact parameter over the lifetime of the system is much larger than the semimajor axis of the planet (eq. 5.80). In denser environments such as star clusters, the conditions are quite different—in particular the star density can be 10^6 times larger than in the solar neighborhood—so some or all of these assumptions may fail and different tools are needed to determine the fate of a planetary system (Spurzem et al. 2009).

[8] An additional effect is precession of the ecliptic pole due to the tidal field of the Galaxy, but this is also negligible. See Problem 9.6.

5.4 ZLK oscillations

In some planetary systems a distant body such as a companion star can excite large oscillations in the eccentricity and inclination of the planetary orbits. A remarkable feature of these **von Zeipel–Lidov–Kozai (ZLK) oscillations**[9] is that the amplitude of the oscillations is independent of the mass of the perturber.

We borrow the analysis of equations (5.67)–(5.72), except that now we are investigating a planet orbiting a host star in the tidal field of a companion star, rather than a satellite orbiting a planet in the tidal field of its host star. The planet is represented as a zero-mass test particle with semimajor axis a, eccentricity vector $\mathbf{e}$ and dimensionless angular-momentum vector $\mathbf{j}$; as usual these are related by $|\mathbf{e}|^2 + |\mathbf{j}|^2 = e^2 + j^2 = 1$. The host star has mass M_h and belongs to a binary system in which the companion star has mass M_c. The relative orbit of the two stars has semimajor axis a_c and eccentricity e_c. In this notation, the disturbing function (5.70) has the form

$$\langle\!\langle H_\mathrm{quad} \rangle\!\rangle = \frac{\mathbb{G} M_\mathrm{c} a^2}{8 a_\mathrm{c}^3 (1 - e_\mathrm{c}^2)^{3/2}} \left[15(\mathbf{e} \cdot \hat{\mathbf{n}}_\mathrm{c})^2 - 6e^2 + 1 - 3(\mathbf{j} \cdot \hat{\mathbf{n}}_\mathrm{c})^2 \right], \quad (5.89)$$

and the Milankovich equations (5.72) read

$$\frac{\mathrm{d}\mathbf{j}}{\mathrm{d}t} = \frac{3\,\mathbb{G}^{1/2} M_\mathrm{c} a^{3/2}}{4 M_\mathrm{h}^{1/2} a_\mathrm{c}^3 (1 - e_\mathrm{c}^2)^{3/2}} \left[(\mathbf{j} \cdot \hat{\mathbf{n}}_\mathrm{c}) \mathbf{j} \times \hat{\mathbf{n}}_\mathrm{c} - 5 (\mathbf{e} \cdot \hat{\mathbf{n}}_\mathrm{c}) \mathbf{e} \times \hat{\mathbf{n}}_\mathrm{c} \right], \quad (5.90)$$

$$\frac{\mathrm{d}\mathbf{e}}{\mathrm{d}t} = \frac{3\,\mathbb{G}^{1/2} M_\mathrm{c} a^{3/2}}{4 M_\mathrm{h}^{1/2} a_\mathrm{c}^3 (1 - e_\mathrm{c}^2)^{3/2}} \left[(\mathbf{j} \cdot \hat{\mathbf{n}}_\mathrm{c}) \mathbf{e} \times \hat{\mathbf{n}}_\mathrm{c} - 5 (\mathbf{e} \cdot \hat{\mathbf{n}}_\mathrm{c}) \mathbf{j} \times \hat{\mathbf{n}}_\mathrm{c} + 2\,\mathbf{j} \times \mathbf{e} \right].$$

[9] Named after three astronomers: Hugo von Zeipel (1873–1959) from Sweden, Mikhail Lidov (1926–1993) from Russia and Yoshihide Kozai (1928–2018) from Japan. The history of ZLK oscillations is complicated (Ito & Ohtsuka 2020). They were discovered and investigated in detail by von Zeipel (1910), but this paper was mostly forgotten. Lidov discovered the phenomenon independently, probably in the course of investigating the trajectory of Luna 3, the first spacecraft to image the far side of the Moon. He presented his results at a conference in Moscow in 1961 (Lidov 1961) that was attended by Kozai, who extended Lidov's results substantially and disseminated them outside the Soviet Union (Kozai 1962).

Let us assume that the planet is on a circular orbit, $\mathbf{e} = \mathbf{0}$. Then the second Milankovich equation is trivially satisfied and the first reads

$$\frac{\mathrm{d}\mathbf{j}}{\mathrm{d}t} = \frac{3\,\mathbb{G}^{1/2} M_\mathrm{c} a^{3/2}}{4 M_\mathrm{h}^{1/2} a_\mathrm{c}^3 (1 - e_\mathrm{c}^2)^{3/2}} (\mathbf{j} \cdot \hat{\mathbf{n}}_\mathrm{c}) \mathbf{j} \times \hat{\mathbf{n}}_\mathrm{c}. \tag{5.91}$$

It is straightforward to show from this equation that $\mathbf{j} \cdot \mathrm{d}\mathbf{j}/\mathrm{d}t = 0$ so $j^2 = 1 - e^2$ is conserved, confirming that a circular planet orbit remains circular. Similarly $\hat{\mathbf{n}}_\mathrm{c} \cdot \mathrm{d}\mathbf{j}/\mathrm{d}t = 0$, and since the binary orbit normal $\hat{\mathbf{n}}_\mathrm{c}$ is fixed, $\mathbf{j} \cdot \hat{\mathbf{n}}_\mathrm{c}$ is conserved. Since the orbit is circular, $\mathbf{j} \cdot \hat{\mathbf{n}}_\mathrm{c} = \cos I$, where I is the fixed inclination of the planetary orbit relative to the binary-star orbit. Then equation (5.91) describes uniform precession of the angular-momentum vector $\mathbf{j}$ around the orbital axis of the binary, at a constant rate

$$\boldsymbol{\omega}_\mathrm{c} = -\frac{3\,\mathbb{G}^{1/2} M_\mathrm{c} a^{3/2}}{4 M_\mathrm{h}^{1/2} a_\mathrm{c}^3 (1 - e_\mathrm{c}^2)^{3/2}} \cos I\, \hat{\mathbf{n}}_\mathrm{c}. \tag{5.92}$$

Note that $\boldsymbol{\omega}_\mathrm{c} \cdot \hat{\mathbf{n}}_\mathrm{c} < 0$, so the precession is retrograde in the frame in which the positive z-axis is parallel to $\hat{\mathbf{n}}_\mathrm{c}$.

We now ask whether these precessing circular orbits are stable. We write $\mathbf{e} = \epsilon\, \mathbf{e}_1$ and expand the Milankovich equations (5.90) to first order in ϵ. The first of these is unchanged at this order, so the angular-momentum vector of the planet continues to precess uniformly at the rate $\boldsymbol{\omega}_\mathrm{c}$. The second equation becomes

$$\frac{\mathrm{d}\mathbf{e}_1}{\mathrm{d}t} = \frac{3\,\mathbb{G}^{1/2} M_\mathrm{c} a^{3/2}}{4 M_\mathrm{h}^{1/2} a_\mathrm{c}^3 (1 - e_\mathrm{c}^2)^{3/2}} \big[(\mathbf{j}\cdot\hat{\mathbf{n}}_\mathrm{c}) \mathbf{e}_1 \times \hat{\mathbf{n}}_\mathrm{c} - 5 (\mathbf{e}_1 \cdot \hat{\mathbf{n}}_\mathrm{c}) \mathbf{j} \times \hat{\mathbf{n}}_\mathrm{c} + 2 \mathbf{j} \times \mathbf{e}_1 \big]. \tag{5.93}$$

To analyze the solutions of this equation it is helpful to transform to a frame that rotates with the unperturbed precession of the angular-momentum vector, so the unperturbed solution is stationary. In this frame the rate of change of the eccentricity vector is (eq. D.16) $(\mathrm{d}\mathbf{e}_1/\mathrm{d}t)_\mathrm{rot} = \mathrm{d}\mathbf{e}_1/\mathrm{d}t - \boldsymbol{\omega}_\mathrm{c} \times \mathbf{e}_1$. Thus the linearized Milankovich equation is simplified to

$$\left(\frac{\mathrm{d}\mathbf{e}_1}{\mathrm{d}t}\right)_\mathrm{rot} = \frac{3\,\mathbb{G}^{1/2} M_\mathrm{c} a^{3/2}}{4 M_\mathrm{h}^{1/2} a_\mathrm{c}^3 (1 - e_\mathrm{c}^2)^{3/2}} \big[2 \mathbf{j} \times \mathbf{e}_1 - 5 (\mathbf{e}_1 \cdot \hat{\mathbf{n}}_\mathrm{c}) \mathbf{j} \times \hat{\mathbf{n}}_\mathrm{c} \big]$$

$$\equiv \mathbf{A}\mathbf{e}_1 . \tag{5.94}$$

In Cartesian coordinates with $\hat{\mathbf{n}}_c = \hat{\mathbf{z}}$ we have

$$\mathbf{A} = \frac{3\,\mathbb{G}^{1/2} M_c a^{3/2}}{4 M_h^{1/2} a_c^3 (1 - e_c^2)^{3/2}} \begin{bmatrix} 0 & -2j_z & -3j_y \\ 2j_z & 0 & 3j_x \\ -2j_y & 2j_x & 0 \end{bmatrix} \tag{5.95}$$

where $\mathbf{j} = (j_x, j_y, j_z)$ is now a constant. We may assume that $\mathbf{e}_1 \propto \exp(\lambda t)$ and find the solution of the resulting equation for λ, which can be written $\det(\mathbf{A} - \lambda\mathbf{I}) = 0$. We find that either $\lambda = 0$ or

$$\lambda = \pm \frac{3\,\mathbb{G}^{1/2} M_c a^{3/2}}{2^{3/2} M_h^{1/2} a_c^3 (1 - e_c^2)^{3/2}} (3 - 5\cos^2 I)^{1/2}, \tag{5.96}$$

where $\cos I = \mathbf{j} \cdot \hat{\mathbf{n}}_c = j_z$ since the unperturbed planet orbit is circular. The motion is unstable if any solution for λ has a positive real part, which occurs if $|\cos I| < (\frac{3}{5})^{1/2}$. Thus stability requires either $0 \le I \le I_{\text{ZLK}}$ or $\pi - I_{\text{ZLK}} \le I \le \pi$, where $I_{\text{ZLK}} = \cos^{-1}(\frac{3}{5})^{1/2} = 39.23°$ is the critical **ZLK angle**. If these conditions are not satisfied the circular orbit is unstable and the planet undergoes **ZLK oscillations**.[10]

To explore the nature of ZLK oscillations it is more convenient to use Delaunay variables (eq. 1.84). The orbit-averaged Hamiltonian arising from the quadrupole tidal field of the companion star is given by equation (5.71). The Hamiltonian is independent of the mean anomaly ℓ because it is orbit-averaged. Therefore the conjugate momentum Λ and the semimajor axis a are conserved, as usual in secular dynamics. Moreover the Hamiltonian is independent of the longitude of the node Ω. Therefore the conjugate

[10] This analysis shows that ZLK oscillations arise in the quadrupole field from a distant companion star, i.e., a mass that is *exterior* to the body whose orbit we are following. A quadrupole field can also arise from mass *interior* to the body we are following, such as a quadrupole moment of the host star. In this case there are no ZLK oscillations, because the orbit-averaged Hamiltonian (5.64) is independent of the argument of periapsis ω when expressed in Delaunay variables, or independent of the direction of the eccentricity vector when expressed in terms of $\mathbf{e}$ and $\mathbf{j}$. This is an accidental property of the potential due to an interior quadrupole moment. See Tremaine & Yavetz (2014) for more detail.

momentum L_z, the z-component of angular momentum, is also conserved (as we discuss below, this is an accidental but helpful property of the potential due to an exterior quadrupole moment, which does not extend to more general potentials). Since both $L_z = [\mathbb{G}M_\mathrm{h}a(1-e^2)]^{1/2}\cos I$ and a are conserved, we can write $L_z = (\mathbb{G}M_\mathrm{h}a)^{1/2}\cos I_0$, where I_0 is an integral of motion equal to the inclination of the circular orbit with the given z-component of angular momentum. Then the conservation of the Hamiltonian (5.71) implies that the **ZLK function**

$$C_\mathrm{ZLK}(e,\omega) \equiv 5\left(1 - \frac{\cos^2 I_0}{1-e^2}\right)e^2\sin^2\omega - 2e^2 \tag{5.97}$$

is conserved along a trajectory. In words, the existence of the two conserved momenta Λ and L_z has reduced the dynamics from three degrees of freedom to one degree of freedom, corresponding to two phase-space dimensions. Since the Hamiltonian is conserved, the trajectory in these two dimensions must lie along contours of constant Hamiltonian, which coincide with contours of constant ZLK function.[11]

Not all values of e correspond to physical trajectories. In particular the definition of I_0 implies that $\cos^2 I = \cos^2 I_0/(1-e^2)$, and since $\cos^2 I < 1$ we must have $e \le \sin I_0$.

The contours of the ZLK function are shown in Figure 5.4. These are plotted using (e,ω) as polar coordinates, which is appropriate since ω is ill-defined when $e=0$. There is an extremum of $C_\mathrm{ZLK}(e,\omega)$ at eccentricity $e = 0$, and the onset of ZLK oscillations corresponds to a change in the nature of this extremum, from a maximum when $|\cos I_0| > \cos I_\mathrm{ZLK} = (\frac{3}{5})^{1/2}$ to a saddle point when $|\cos I_0| < \cos I_\mathrm{ZLK}$. In the former case a nearly circular orbit remains nearly circular, while in the latter case an initially circular orbit oscillates between eccentricity $e = 0$ and a maximum value e_max, which is reached when $\omega = \pm\frac{1}{2}\pi$. The value of e_max is found by solving the equation $C_\mathrm{ZLK}(e,\pm\frac{1}{2}\pi) = 0$:

$$e_\mathrm{max} = \left(1 - \tfrac{5}{3}\cos^2 I_0\right)^{1/2}. \tag{5.98}$$

[11] The corresponding equations of motion can be solved in terms of elliptic integrals (Kinoshita & Nakai 2007) but we shall not use these solutions.

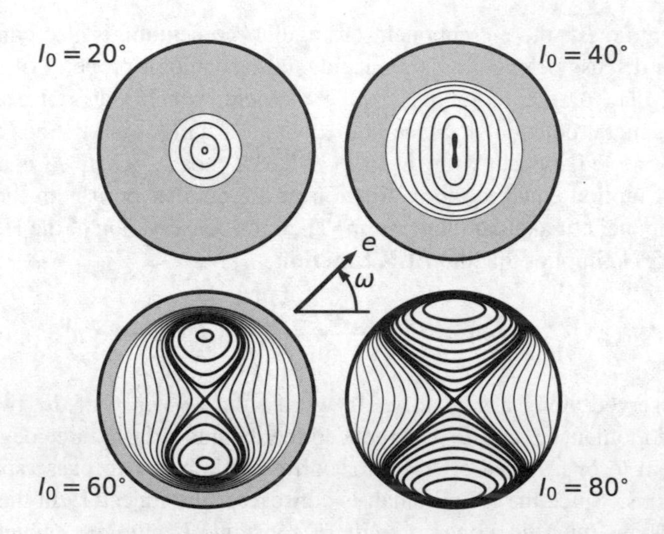

Figure 5.4: Contours of the ZLK function $C_{\mathrm{ZLK}}(e, \omega)$ (eq. 5.97), with the eccentricity e and argument of periapsis ω plotted as polar coordinates. Shaded regions have $e > \sin I_0$ and are unphysical. Plots are shown for four values of I_0, the inclination of a circular orbit. Prograde circular orbits are unstable for $I_0 > I_{\mathrm{ZLK}} = 39.23°$. Heavy contours have $C_{\mathrm{ZLK}}(e, \omega) \geq 0$. Compare Figure 9.5.

Thus $e_{\max} \to 1$ as $I_0 \to \pm\frac{1}{2}\pi$, showing that a planet on an initially circular, nearly polar orbit executes ZLK oscillations that are so large that it may collide with its host star.

For $|\cos I_0| > \cos I_{\mathrm{ZLK}}$, all orbits circulate in the sense that ω rotates between 0 and 2π and $\dot{\omega}$ is never zero. For $|\cos I_0| < \cos I_{\mathrm{ZLK}}$, orbits with $C_{\mathrm{ZLK}} < 0$ circulate but orbits with $C_{\mathrm{ZLK}} \geq 0$ librate in the sense that ω oscillates around $\pm\frac{1}{2}\pi$. The boundary between the circulating and librating orbits is a **separatrix** passing through the origin. In the presence of other perturbations the separatrix becomes chaotic, so circular orbits are chaotic whenever $|\cos I_0| < \cos I_{\mathrm{ZLK}}$.

The trajectories in Figure 5.4 are independent of the strength of the tidal field from the companion star. Shrinking the tidal field leaves the amplitude of the eccentricity oscillation the same but increases the period of the oscillations.

The time dependence of the oscillations in e and ω can be found using the Hamiltonian (5.71) and Hamilton's equations in Delaunay variables:

$$\frac{\mathrm{d}\omega}{\mathrm{d}t} = \frac{3\,\mathbb{G}M_\mathrm{c}}{4na_\mathrm{c}^3(1-e_\mathrm{c}^2)^{3/2}}\left[\frac{5\cos^2 I_0 \sin^2\omega}{(1-e^2)^{3/2}} + (1-e^2)^{1/2}(2-5\sin^2\omega)\right]$$
$$\frac{\mathrm{d}e}{\mathrm{d}t} = \frac{15\,\mathbb{G}M_\mathrm{c}}{4na_\mathrm{c}^3(1-e_\mathrm{c}^2)^{3/2}} e(1-e^2)^{1/2}\left(1-\frac{\cos^2 I_0}{1-e^2}\right)\sin\omega\cos\omega, \qquad (5.99)$$

where $n = (\mathbb{G}M_\mathrm{h}/a^3)^{1/2}$ is the mean motion of the planet around its host. There is a stable equilibrium state—a maximum of $C_\mathrm{ZLK}(e,\omega)$—at $\omega = \pm\frac{1}{2}\pi$ and $e = e_0 \equiv [1-(\frac{5}{3})^{1/2}|\cos I_0|]^{1/2}$. The libration period around this equilibrium is of order the square of the orbital period of the binary divided by the orbital period of the planet (see Problem 5.6).

ZLK oscillations are important in a remarkable variety of astrophysical contexts: the irregular satellites of the giant planets, the excitation of exoplanet eccentricities by companion stars, high-eccentricity planetary migration (§5.4.2), the formation of close binary stars, blue-straggler stars, Type Ia supernovae, black-hole mergers, comets (§9.4), and so forth.

5.4.1 Beyond the quadrupole approximation

The analysis so far assumes that that the *only* non-Kepler forces arise from the quadrupole tidal field of the companion star. Any additional sources of apsidal precession, even small ones, can dramatically alter the nature of the ZLK oscillations.

Relativistic precession According to equation (J.15) of Appendix J the orbit-averaged relativistic correction to the Kepler Hamiltonian is

$$\langle H\rangle_\mathrm{gr} = -\frac{3\,\mathbb{G}^2M_\mathrm{h}^2}{c^2a^2(1-e^2)^{1/2}} + \frac{15\,\mathbb{G}^2M_\mathrm{h}^2}{8c^2a^2} + \mathrm{O}(c^{-4}). \qquad (5.100)$$

We drop the second term on the right since it has no effect on the secular dynamics, and drop all terms that are $\mathrm{O}(c^{-4})$. The relativistic Hamiltonian becomes

$$\langle H\rangle_{\mathrm{gr}} = -\frac{3\,\mathbb{G}^2 M_{\mathrm{h}}^2}{c^2 a^2 j}, \tag{5.101}$$

where as usual $j = (1-e^2)^{1/2}$. The addition of this Hamiltonian has no effect on the Milankovich equation (5.57) for the evolution of the dimensionless angular momentum $\mathbf{j}$. The linearized Milankovich equation for the evolution of the eccentricity vector $\mathbf{e}$, equation (5.94), becomes

$$\left(\frac{\mathrm{d}\mathbf{e}_1}{\mathrm{d}t}\right)_{\mathrm{rot}} = (\mathbf{A} + \mathbf{A}_{\mathrm{gr}})\mathbf{e}_1, \tag{5.102}$$

where we have assumed that $j = |\mathbf{j}| \simeq 1$, $\mathbf{A}$ is given by equation (5.95), and

$$\mathbf{A}_{\mathrm{gr}} = \frac{3\,\mathbb{G}^{3/2} M_{\mathrm{h}}^{3/2}}{c^2 a^{5/2}} \begin{bmatrix} 0 & -j_z & j_y \\ j_z & 0 & -j_x \\ -j_y & j_x & 0 \end{bmatrix}. \tag{5.103}$$

Then $\mathbf{e}_1 \propto \exp(\lambda t)$ and equation (5.96) is modified to

$$\lambda = \pm \frac{3\,\mathbb{G}^{1/2} M_{\mathrm{c}} a^{3/2}}{2^{3/2} M_{\mathrm{h}}^{1/2} a_{\mathrm{c}}^3 (1-e_{\mathrm{c}}^2)^{3/2}} (1+2\epsilon_{\mathrm{gr}})^{1/2} \left(3 - 4\epsilon_{\mathrm{gr}} - 5\cos^2 I\right)^{1/2}, \tag{5.104}$$

where

$$\begin{aligned} \epsilon_{\mathrm{gr}} &\equiv \frac{\mathbb{G} M_{\mathrm{h}}^2 a_{\mathrm{c}}^3 (1-e_{\mathrm{c}}^2)^{3/2}}{c^2 a^4 M_{\mathrm{c}}} \\ &= 0.009\,871 \left(\frac{M_{\mathrm{h}}}{M_\odot}\right)^2 \left(\frac{M_\odot}{M_{\mathrm{c}}}\right) \left(\frac{a_{\mathrm{c}}}{100\,\mathrm{au}}\right)^3 \left(\frac{1\,\mathrm{au}}{a}\right)^4 (1-e_{\mathrm{c}}^2)^{3/2} \end{aligned} \tag{5.105}$$

parametrizes the relative rates of apsidal precession from general relativity and the tidal field from the companion star. The circular orbit is unstable, and ZLK oscillations set in, when $\cos^2 I < \frac{3}{5} - \frac{4}{5}\epsilon_{\mathrm{gr}}$. When $\epsilon_{\mathrm{gr}} > \frac{3}{4}$ the circular orbit is stable at all inclinations. Thus relativistic precession can completely suppress ZLK oscillations, one of the few cases in which general relativity plays an important role in celestial mechanics.

Octopole potential The relatively simple characteristics of ZLK oscillations described so far arise because the z-component of angular momentum is conserved, a property that reduces the dynamics to a single degree of freedom. This conservation law arises because both the quadrupole tidal potential described by the Hamiltonian in equation (5.71) and the relativistic Hamiltonian (5.100) are independent of the longitude of the ascending node Ω. This property does not extend to more general tidal potentials. To see this, we examine the next order in the multipole expansion of the tidal potential in equation (3.68). This is the octopole potential, given by the terms of order λ^{-4} in that equation:

$$H_{\rm oct} = \frac{3\,\mathbb{G}M_{\rm c}r^2(\mathbf{r}\cdot\mathbf{r}_{\rm c})}{2r_{\rm c}^5} - \frac{5\,\mathbb{G}M_{\rm c}(\mathbf{r}\cdot\mathbf{r}_{\rm c})^3}{2r_{\rm c}^7}. \tag{5.106}$$

Following steps similar to those in equations (5.67)–(5.70), we can average this potential over the orbits of both the planet and the companion star to obtain (Brown 1936; Breiter & Vokrouhlický 2015)

$$\langle\langle H_{\rm oct}\rangle\rangle = \frac{15\,\mathbb{G}M_{\rm c}a^3}{64a_{\rm c}^4(1-e_{\rm c}^2)^{5/2}}\{\mathbf{e}\cdot\mathbf{e}_{\rm c}[8e^2-1+5(\mathbf{j}\cdot\hat{\mathbf{j}}_{\rm c})^2-35(\mathbf{e}\cdot\hat{\mathbf{j}}_{\rm c})^2] + 10(\mathbf{e}\cdot\hat{\mathbf{j}}_{\rm c})(\mathbf{j}\cdot\mathbf{e}_{\rm c})(\mathbf{j}\cdot\hat{\mathbf{j}}_{\rm c})\}. \tag{5.107}$$

The Hamiltonian can be converted to the usual orbital elements using equations (5.48). Notice that $\langle\langle H_{\rm oct}\rangle\rangle = 0$ if the companion-star orbit is circular ($\mathbf{e}_{\rm c} = \mathbf{0}$); because of this feature, the influence of the octopole tidal potential on ZLK oscillations is sometimes called the **eccentric ZLK effect** (Ford et al. 2000; Naoz 2016; Shevchenko 2017).

Motion in the orbit-averaged quadrupole plus octopole Hamiltonian, the sum of equations (5.89) and (5.107), conserves semimajor axis because the Hamiltonians are independent of mean longitude as a result of the orbit-averaging. However, the motion does not conserve the z-component of angular momentum L_z. As a result, the addition of even a weak octopole component to the quadrupole Hamiltonian leads to much richer dynamical behavior and enables the trajectory to explore a much wider volume of phase space. In particular, in the quadrupole approximation a distant planet must be on a nearly polar orbit ($L_z \simeq 0$) if it is to make a close approach to its host star as a result of ZLK oscillations, but once the octopole contribution is included planets on a much wider range of orbits may eventually interact with their hosts.

Nonlinear dynamics All of the analysis above is based on first-order perturbation theory. If the ratio of the mean motion of the companion to the mean motion of the planet is $m \equiv n_{\rm c}/n$, then second-order perturbations change the apsidal precession rate of the planet by a fraction that is $\mathrm{O}(m)$, as in equation (3.79) for the case of nearly circular and coplanar orbits. The strongest second-order perturbations arise from terms that were eliminated when we averaged over the companion orbit having period $P_{\rm c} = 2\pi/n_{\rm c}$. These are much more important than terms that are eliminated by averaging over the planet orbit, with period $2\pi/n \ll P_{\rm c}$. For this reason, calculations that include the strongest second-order perturbations are sometimes said to use the **single-averaging approximation** as opposed to the **double-averaging** approximation used elsewhere in this section.

The effects of these second-order perturbations can be incorporated by adding a secular Hamiltonian proportional to the square of the strength of the field from the companion, derived from the Poincaré–von Zeipel method or Lie operator perturbation theory (eq. 4.85). Following Breiter & Vokrouhlický (2015) we have

$$\begin{aligned} H_2 = \frac{9\,\mathbb{G}M_{\rm c}^2 a^{7/2}}{16 M_{\rm h}^{1/2}(M_{\rm h}+M_{\rm c})^{1/2} a_{\rm c}^{9/2}(1-e_{\rm c}^2)^3}\Big(&A(e_{\rm c})(\mathbf{j}\cdot\hat{\mathbf{j}}_{\rm c})\big[24e^2 - 15(\mathbf{e}\cdot\hat{\mathbf{j}}_{\rm c})^2 \\ &- (\mathbf{j}\cdot\hat{\mathbf{j}}_{\rm c})^2 + 1\big] + B(e_{\rm c})\big\{(\mathbf{j}\cdot\hat{\mathbf{j}}_{\rm c})\big[1 - 2(\mathbf{j}\cdot\hat{\mathbf{e}}_{\rm c})^2 - (\mathbf{j}\cdot\hat{\mathbf{j}}_{\rm c})^2 + 4e^2 \\ &- 10(\mathbf{e}\cdot\hat{\mathbf{e}}_{\rm c})^2 - 15(\mathbf{e}\cdot\hat{\mathbf{j}}_{\rm c})^2\big] - 20(\mathbf{e}\cdot\hat{\mathbf{e}}_{\rm c})(\mathbf{j}\cdot\hat{\mathbf{e}}_{\rm c})(\mathbf{e}\cdot\hat{\mathbf{j}}_{\rm c})\big\}\Big). \end{aligned} \qquad (5.108)$$

Here

$$A(e_{\rm c}) \equiv -\frac{3+2e_{\rm c}^2}{12}, \quad B(e_{\rm c}) \equiv \frac{4(1-e_{\rm c}^2)^{3/2} - 4 + 6e_{\rm c}^2 + 3e_{\rm c}^4}{12 e_{\rm c}^2}. \qquad (5.109)$$

The function $B(e_{\rm c})$ is defined as $e_{\rm c} \to 0$ by its limit, $B(e_{\rm c}) = \frac{3}{8}e_{\rm c}^2 + \mathrm{O}(e_{\rm c}^4)$. In some systems the magnitude of the nonlinear Hamiltonian H_2 can exceed the magnitude of the octopole Hamiltonian $\langle\langle H_{\mathrm{oct}} \rangle\rangle$. An approximate criterion for this is that $n_\mathrm{c}/n \gtrsim a/a_\mathrm{c}$ (Luo et al. 2016), which by Kepler's law is equivalent to $1 + M_\mathrm{c}/M_\mathrm{h} \gtrsim a_\mathrm{c}/a$. Thus the double-averaged octopole approximation is usually appropriate in hierarchical systems ($a_\mathrm{c} \gg a$) consisting of a planet orbiting one member of a binary-star system with $M_\mathrm{c} \sim M_\mathrm{h}$, but not necessarily in systems consisting of a satellite orbiting a planet, where M_h is the planet mass and $M_\mathrm{c} \gg M_\mathrm{h}$ is the mass of the host star.

In the Earth–Moon–Sun system we have $M_\mathrm{c} = M_\odot \gg M_\mathrm{h} = M_\oplus \gg M_☽$, and as a first approximation we can assume that the Sun is on a circular orbit, $e_\mathrm{c} = 0$. In the frame with polar axis normal to the ecliptic, the Hamiltonian (5.108) becomes

$$H_2 = -\frac{9\,\mathbb{G}M_\oplus}{64a}\frac{n_\odot^3}{n^3}(25e^2 + I^2) + \mathrm{O}(e^4, I^4, e^2 I^2); \tag{5.110}$$

here $n_\odot = (\mathbb{G}M_\odot/a_\mathrm{c}^3)^{1/2}$ and $n = (\mathbb{G}M_\oplus/a^3)^{1/2}$ are the mean motions of the Sun and Moon around the Earth. Using the Lagrange equations (1.188), we find that this Hamiltonian contributes to the apsidal and nodal precession

$$\frac{\mathrm{d}\varpi}{\mathrm{d}t} = \tfrac{225}{32}nm^3, \quad \frac{\mathrm{d}\Omega}{\mathrm{d}t} = \tfrac{9}{32}nm^3, \tag{5.111}$$

where $m \equiv n_\odot/n$. These are the second terms in the series (3.79).

5.4.2 High-eccentricity migration

As described in §3.6, hot Jupiters—giant planets with orbital periods less than 10 days—likely formed at much larger distances from their host stars and migrated to their current orbits. One possibility is that they migrated early in the history of their planetary system, within the first few Myr after the formation of their host star, through gravitational torques exerted by the gaseous protoplanetary disk from which they formed (§3.6). An alternative hypothesis is that they migrated long after planet formation was complete, through **high-eccentricity migration**.

Hot Jupiters formed by high-eccentricity migration begin their lives in nearly circular orbits with orbital periods in the range $\gtrsim 300\,\mathrm{d}$ where most giant planets are now found (Figure 3.15). High-eccentricity migration then requires two steps. First, some process excites the planetary orbit to high eccentricity e—close enough to $e = 1$ that the periapsis distance $a(1 - e)$ is only a few stellar radii. Second, tidal friction drains energy from the planetary orbit until the planet settles on a circular orbit with a period of only a few days (§8.5.1).

Planetary orbits can be excited to high eccentricities either by a close encounter between two planets or through ZLK oscillations induced by a distant companion, either a planet or a star, on an inclined orbit.

One important feature of high-eccentricity migration is that in the final state the planet's orbital angular-momentum vector is generally not aligned with the host star's spin angular momentum, even if the two vectors were aligned when the planet was originally formed. This misalignment can be probed observationally if the hot Jupiter transits the host star, by accurate radial-velocity measurements of the host star during the transit (the **Rossiter–McLaughlin effect**; see for example Winn & Fabrycky 2015). These observations show that the spin and orbital angular momenta are misaligned in about one-third of all systems containing hot Jupiters, often by more than 90°—supporting, but not proving, the hypothesis that high-eccentricity migration is responsible for many hot Jupiters.

The classic example of high-eccentricity migration is the system HD 80606. This is a solar-mass star hosting a single planet with mass 3.9 Jupiter masses, orbital period 111.4 d, and semimajor axis $a = 0.45$ au. The planet is detected through both radial-velocity variations and transits. The orbit has a remarkably large eccentricity, $e = 0.934$. Its periapsis $a(1 - e)$ is only 0.030 au $= 6.4R_\odot$, small enough that tidal friction from its interaction with the host is sapping energy from the orbit. Thus the planet in HD 80606 likely will eventually settle on a circular orbit with a period of < 10 d—that is, it will become a hot Jupiter (Wu & Murray 2003).

One difficulty with models of high-eccentricity migration is that they only produce planets with orbital periods of a few days, whereas the observations show a relatively flat distribution of orbital periods out to ~ 100 d (Figure 3.15). Some other mechanism is needed to produce warm Jupiters with periods between 10 d and 100 d, and this mechanism must be roughly as efficient as high-eccentricity migration—an unlikely coincidence.

Chapter 6

Resonances

Informally, a resonance occurs when two or more of the fundamental frequencies governing the dynamics of one or more planets are in a simple integer ratio. In **mean-motion resonances**, the frequencies are the mean motions of the two planets. In **spin-orbit resonances**, one frequency is the spin angular speed of the planet and the other is its mean motion (§7.2). In **secular resonances**, the two frequencies are the secular frequencies governing the slow precession of the apsides and nodes (§6.6). There are other varieties of resonance as well, for example between the precession frequency of a planet's spin angular momentum and one of the secular frequencies (§7.1.2), or mean-motion resonances involving three or more bodies called **resonant chains**.

A resonance is typically labeled by $(p+q):p$, where p and q are integers and $(p+q)/p$ is the ratio of the resonant frequencies. Thus, for example, in a 2:1 mean-motion resonance the mean motion of the inner planet is twice the mean motion of the outer planet. The co-orbital satellites described in §3.2 may be said to be in a 1:1 mean-motion resonance. A satellite like the Moon that is in synchronous rotation, with equal spin and orbital periods, is in a 1:1 spin-orbit resonance.

There are two preliminary questions to address before embarking on a study of resonant dynamics in planetary systems.

First, since every real number can be approximated arbitrarily closely by the ratio of two integers, is not every pair of planets in resonance? Technically, of course, the answer is "yes," but the strength of most of these resonances is very small. To see why, consider the typical case of two planets on low-eccentricity, low-inclination orbits. Formally this means that e_1, e_2, I_1, I_2 are $\mathrm{O}(\epsilon)$, where ϵ is a small parameter. The disturbing function between these planets can be expanded in a Fourier series as shown in equation (4.91), in which each term has the form $H_\mathbf{j} \cos \Phi_\mathbf{j}$ where $\mathbf{j} \equiv \{j_1, k_1, m_1, j_2, k_2, m_2\}$ is a vector of integers, $H_\mathbf{j}$ is a function of the actions of the two planets, and $\Phi_\mathbf{j} \equiv j_1\lambda_1 + k_1\varpi_1 + m_1\Omega_1 - j_2\lambda_2 - k_2\varpi_2 - m_2\Omega_2$. The terms that govern the dynamics of a $(p+q):p$ mean-motion resonance are those in which $\Phi_\mathbf{j}$ varies slowly when the mean motions $\dot\lambda_1$ and $\dot\lambda_2$ are in the ratio $(p+q)/p$, which requires that $(p+q)/q = j_2/j_1$. Without loss of generality, we can assume that $p+q$ and q have no common factor, so this condition implies that $j_1 = rp$ and $j_2 = r(p+q)$ where r is a nonzero integer. Then equation (4.93) implies that $k_1 + m_1 - k_2 - m_2 = rq$. Since $\sum |a_i| \geq \sum_i a_i$ for any sequence $\{a_i\}$, we must have $|k_1| + |k_2| + |m_1| + |m_2| \geq |rq|$ and then equation (4.92) implies that the amplitude $H_{j_1 k_1 m_1 j_2 k_2 m_2}(a_1, e_1, I_1, a_2, e_2, I_2)$ associated with the resonance is $\mathrm{O}(\epsilon^{|rq|})$. Since $|r| \geq 1$ the largest amplitudes, corresponding to the strongest resonances, have $|r| = 1$ and these are $\mathrm{O}(\epsilon^{|q|})$. Therefore if the eccentricities and inclinations are small, resonances of the form $(p \pm 1) : p$ are much stronger than those of the form $(p \pm 2) : p$, these are much stronger than $(p \pm 3) : p$ resonances, and so on. In practice this means that only a small set of resonances are likely to be strong enough to have an important effect on the dynamics.

Second, if planets or satellites form independently and their masses are small, should not the fraction that are in resonance be small, and if so then why is the dynamics of resonances important? In fact, the fraction of solar-system objects found in some kind of resonance is remarkably large. Neptune and Pluto are in a 3:2 mean-motion resonance (§6.4); Mercury is in a 3:2 spin-orbit resonance (§7.2); Saturn's satellites Mimas and Tethys are in a 2:1 mean-motion resonance, as are its satellites Enceladus and Dione; Saturn's satellites Titan and Hyperion are in a 4:3 mean-motion resonance; and Jupiter's satellites Io, Europa and Ganymede are in a three-body resonant chain (the **Laplace resonance**) in which their mean motions are related by

$n_{\rm Io} - 3n_{\rm Europa} + 2n_{\rm Ganymede} = 0$. Members of the Hilda group of asteroids are in a 3:2 resonance with Jupiter, and many trans-Neptunian objects are in resonances with Neptune (§9.6). Most inner satellites and the Moon are in synchronous rotation, corresponding to a 1:1 spin-orbit resonance. Saturn's satellites Janus and Epimetheus are in a 1:1 mean-motion resonance (§3.2), and there are thousands of asteroids in 1:1 mean-motion resonances with Jupiter (the Trojan asteroids, see §3.1.1).

The most likely explanation for the large number of resonant configurations is that slow changes in the properties of the system have led to slow changes in the mean motions, spins and other dynamical frequencies. These changes can cause the ratios of mean motions or other frequencies to drift through resonance. As we show later in this chapter, if the drift is slow enough and the resonance is strong enough, the system can be trapped in resonance and then will remain in resonance even as the planetary system continues to evolve.

Mean-motion resonances are also found in exoplanetary systems. One of the most exotic examples is Kepler-223, which contains four planets that are found in two three-body resonant chains and one four-body chain, with the mean motions satisfying[1] $n_b - 2n_c + n_d = 0$, $n_c - 3n_d + 2n_e = 0$ and $3n_b - 4n_c - 3n_d + 4n_e = 0$ (Mills et al. 2016). Another example is the planetary system in TRAPPIST-1, a nearby (12 pc), low-mass ($0.08M_\odot$) star containing seven transiting planets with orbital periods ranging from 1.51 d to 18.77 d, named TRAPPIST-1b to TRAPPIST-1h. The planets are found in a complex, interlocking set of five three-body mean-motion resonances such as $2n_b - 5n_c + 3n_d = 0$ and $n_c - 3n_d + 2n_e = 0$ (Luger et al. 2017).

A broader view of the occurrence of resonances in exoplanet systems is provided by Figure 6.1, which plots the period ratios of all pairs of planets in a given system. The figure shows two histograms, one for a sample of 130 multi-planet systems discovered by radial-velocity surveys, and one for 439 systems discovered by the Kepler transit survey. The plots show tantalizing hints of a concentration of planets near the strongest resonances—in particular there is a strong, narrow peak at the 3:2 resonance in the Kepler

[1] Typically planets in a given system are labeled b, c, d, and so forth, from the inside out. Label "a" is reserved for the host star.

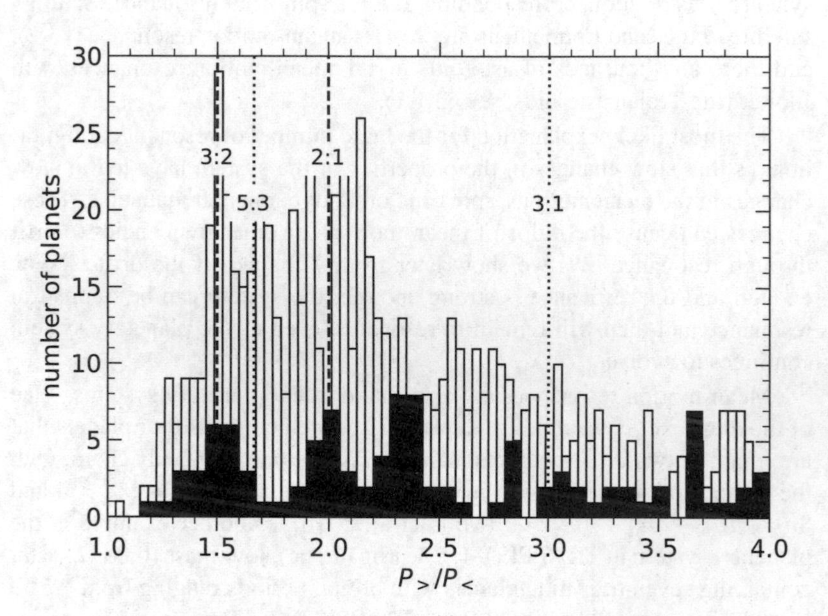

Figure 6.1: Distribution of period ratios in multi-planet systems. The open histogram shows systems discovered by the Kepler mission and the filled histogram shows systems discovered by radial-velocity surveys. A system with n planets contributes $\frac{1}{2}n(n-1)$ data points, found by taking the ratio of the larger to the smaller period for all possible pairs of planets. Some of the $(p+1):p$ and $(p+2):p$ resonance locations are marked by dashed and dotted lines respectively. Data from the NASA Exoplanet Archive, https://exoplanetarchive.ipac.caltech.edu/index.html.

sample and peaks at the 3:2 and 2:1 resonances in the radial-velocity sample. In contrast, the Kepler sample exhibits a significant *dip* in the number of planets near the 2:1 resonance.

6.1 The pendulum

The pendulum is the fundamental model for resonance, and we begin our study of resonances with a review of its properties.

The pendulum Hamiltonian for a particle of mass m can be written

$$H(q,p) = \frac{p^2}{2m} - m\omega^2 \cos q. \tag{6.1}$$

For a simple pendulum of length L in a gravitational field g we have $\omega^2 = g/L$, but the pendulum Hamiltonian is more general than this specific system. We assume that $m > 0$; if not, we simply replace q by $q' = q - \pi$ and m by $m' = -m$.

Hamilton's equations read

$$\dot{q} = \frac{\partial H}{\partial p} = \frac{p}{m}, \quad \dot{p} = -\frac{\partial H}{\partial q} = -m\omega^2 \sin q, \tag{6.2}$$

and by eliminating p we have

$$\ddot{q} = -\omega^2 \sin q. \tag{6.3}$$

The equilibrium solutions have $p = 0$ and $q = n\pi$, where n is an integer. The stability of the equilibria can be determined by writing $q = n\pi + q_1$, where $q_1 \ll 1$. Then equation (6.3) becomes $\ddot{q}_1 = -\omega^2(-1)^n q_1 + \mathrm{O}(q_1^3)$; if we drop the higher order terms, the solutions are the sum of terms $\propto \exp(\pm \mathrm{i}\omega t)$ if n is even, and $\propto \exp(\pm\omega t)$ if n is odd. Thus the equilibria with even n are stable and those with odd n are unstable; the former are minima of the Hamiltonian and the latter are saddle points (Figure 6.2).

The Hamiltonian is conserved along a trajectory, so if $m\epsilon$ is the constant value of the Hamiltonian we can rewrite equations (6.1) and (6.2) as

$$p = m\dot{q} = \pm m[2(\epsilon + \omega^2 \cos q)]^{1/2}. \tag{6.4}$$

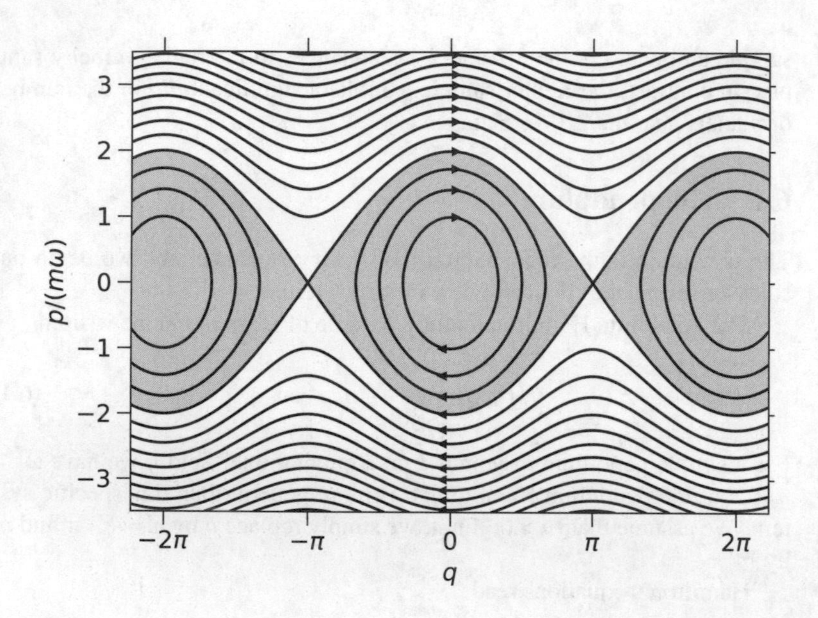

Figure 6.2: Phase plane of the pendulum Hamiltonian (6.1). Librating regions with $H < m\omega^2$ are shaded in gray, and circulating regions with $H > m\omega^2$ are unshaded. The directions of motion are indicated by arrows.

From equation (6.1) the smallest possible value of the Hamiltonian is $-m\omega^2$ so $\epsilon \geq -\omega^2$. If $-\omega^2 < \epsilon < \omega^2$, then $\dot{q} = 0$ at $q = 2n\pi \pm q_{\max}$ where $q_{\max} = \cos^{-1}(-\epsilon/\omega^2)$, and motion in the regions where $\cos q < \cos q_{\max}$ is forbidden; for example, if $n = 0$ then q oscillates between $-q_{\max}$ and $q_{\max}$. In this case the pendulum is said to **librate**. The period of the oscillation is[2]

$$P = 2\int_{-q_{\max}}^{q_{\max}} \frac{\mathrm{d}q}{|\dot{q}|} = 2\int_{-q_{\max}}^{q_{\max}} \frac{\mathrm{d}q}{[2(\epsilon + \omega^2 \cos q)]^{1/2}} = \frac{4}{\omega}K(k), \quad 0 \leq k < 1, \tag{6.5}$$

[2] The last equality is proved by changing the integration variable to ϕ, where ϕ is defined by equation (6.9), and then using the first of equations (C.16).

where $K(\cdot)$ is an elliptic integral (Appendix C.4) and

$$k = \left(\frac{\epsilon + \omega^2}{2\omega^2}\right)^{1/2}. \tag{6.6}$$

The argument k is related to the amplitude $q_{\max}$ by $k = \sin\frac{1}{2}q_{\max}$. In the limit $\epsilon \to -\omega^2$ $(k \to 0)$, the pendulum equation of motion becomes that of a harmonic oscillator and $P \to 2\pi/\omega$. In contrast as $\epsilon \to \omega^2$ $(k \to 1)$, the period approaches infinity.

The action for librating orbits is

$$\begin{aligned} J &= \frac{1}{2\pi}\oint \mathrm{d}q\, p = \frac{2m}{\pi}\int_0^{q_{\max}} \mathrm{d}q\,[2(\epsilon + \omega^2\cos q)]^{1/2} \\ &= \frac{8m\omega}{\pi}[E(k) - (1-k^2)K(k)], \quad 0 \le k < 1, \end{aligned} \tag{6.7}$$

where $E(\cdot)$ is also an elliptic integral. We choose the zero point of the angle variable θ that is conjugate to this action to be at $q = 0, p > 0$. Then θ increases from 0 to $\frac{1}{2}\pi$ as q increases from 0 to $q_{\max}$ with $p > 0$, increases from $\frac{1}{2}\pi$ to $\frac{3}{2}\pi$ as q varies from $q_{\max}$ to $-q_{\max}$ with $p < 0$, and finally increases to 2π as q varies from $-q_{\max}$ to 0 with $p > 0$. An explicit formula for the range 0 to $\frac{1}{2}\pi$ is (Problem 6.1)

$$\theta = \frac{\pi F(\phi, k)}{2K(k)}, \quad 0 \le k < 1. \tag{6.8}$$

Here $0 \le \phi \le \frac{1}{2}\pi$ and ϕ is related to q by

$$k\sin\phi = \sin\tfrac{1}{2}q; \tag{6.9}$$

the function $F(\phi, k)$ is an incomplete elliptic integral (eq. C.17). The relation between q and θ in other quadrants is easy to determine using symmetry arguments.

If $\epsilon > \omega^2$ then equation (6.4) shows that $\dot{q}$ can never vanish, and the pendulum is said to **circulate**. In this case the period is defined as the time needed for q to change by $\pm 2\pi$,[3]

$$P = \int_0^{2\pi} \frac{\mathrm{d}q}{|\dot{q}|} = \int_0^{2\pi} \frac{\mathrm{d}q}{[2(\epsilon + \omega^2 \cos q)]^{1/2}} = \frac{2}{k\omega} K(k^{-1}), \quad k > 1. \tag{6.10}$$

In the limit $\epsilon \to \infty$, the period approaches $\pi(2/\epsilon)^{1/2}$, as expected since the influence of the potential $-\omega^2 \cos q$ becomes negligible so the trajectory has constant velocity $\dot{q} = \pm(2\epsilon)^{1/2}$. As $\epsilon \to \omega^2$ the period approaches infinity.

The action for circulating orbits is found by treating the coordinate q as an angle that varies between 0 and 2π:

$$\begin{aligned} J &= \frac{1}{2\pi} \int_0^{2\pi} \mathrm{d}q\, p = \frac{m}{\pi} \int_0^{\pi} \mathrm{d}q\, [2(\epsilon + \omega^2 \cos q)]^{1/2} \\ &= \frac{4m\omega k}{\pi} E(k^{-1}), \quad k > 1. \end{aligned} \tag{6.11}$$

The zero point of the conjugate angle θ is chosen to correspond to $q = 0$. Then θ increases from 0 to 2π as q increases from 0 to 2π when $p > 0$, or as q decreases from 0 to -2π when $p < 0$. An explicit formula for the range 0 to π when $p > 0$ is (Problem 6.1)

$$\theta = \frac{\pi F(\frac{1}{2}q, k^{-1})}{K(k^{-1})}, \quad k > 1. \tag{6.12}$$

The trajectory with $\epsilon = \omega^2$ has $k = 1$, separating librating from circulating trajectories, and is called the **separatrix**. The period of the separatrix orbit is infinite. Notice that the action of the separatrix orbit is $8m\omega/\pi$ according to equation (6.7) and $4m\omega/\pi$ according to equation (6.11), a consequence of the different geometry used to compute the action for librating and circulating orbits. The trajectory of a particle on the separatrix is given in Problem 6.2.

The **width** of the resonance is defined as the difference between the largest and smallest momentum in librating orbits and is given by

$$w = p_{\max} - p_{\min} = 4m\omega. \tag{6.13}$$

[3] The last equality is proved by changing the integration variable to $\phi = \frac{1}{2}q$ and using the first of equations (C.16).

6.1.1 The torqued pendulum

The simplest model for possible resonance trapping in Hamiltonian systems is the torqued pendulum. Its equation of motion is

$$\ddot{q} = -\omega^2 \sin q + N(t), \tag{6.14}$$

which describes a pendulum of length L in a gravitational field $g = \omega^2 L$ subjected to a torque $mL^2 N(t)$. For brevity let us call $N(t)$ the torque.

This equation of motion can be derived from the Hamiltonian

$$H_1(q, p, t) = \frac{p^2}{2m} - m\omega^2 \cos q - mN(t)q, \tag{6.15}$$

or

$$H_2(q, p', t) = \frac{[p' + c(t)]^2}{2m} - m\omega^2 \cos q, \tag{6.16}$$

where $m\dot{q} = p = p' + c(t)$ and $\dot{c}(t) = mN(t)$.

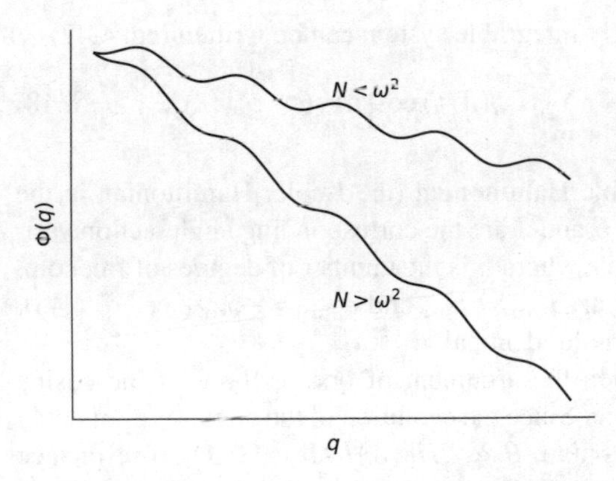

Figure 6.3: The potential $\Phi(q)$ (eq. 6.17) for a pendulum subjected to a constant torque $N > 0$. The lower curve shows $\Phi(q)$ for the case of a large torque, $N > \omega^2$, and the upper curve shows the case $N < \omega^2$.

If the torque is time-independent, the Hamiltonian of equation (6.15) is also time-independent and describes motion of a particle in the potential $m\Phi(q)$, where

$$\Phi(q) = -\omega^2 \cos q - Nq. \tag{6.17}$$

This potential is plotted in Figure 6.3. Notice that there are equilibrium points ($\mathrm{d}\Phi/\mathrm{d}q = 0$) if and only if $|N| < \omega^2$. Thus if the torque is sufficiently large, $|N| > \omega^2$, the pendulum can only circulate. In particular, if $N > 0$ and the initial motion is retrograde ($\dot{q} < 0$), the pendulum will travel to the left in Figure 6.3, climbing the potential and circulating more and more slowly until eventually it reverses direction and starts traveling to the right, circulating more and more rapidly in the prograde direction ($\dot{q} > 0$).

If however $|N| < \omega^2$, the pendulum can either circulate or librate. A librating pendulum is restricted to one of the potential wells in Figure 6.3, and the angle q varies only over a limited range. In this case we say that the pendulum is **trapped** in resonance. In this simple model an initially circulating pendulum can never librate; we say that the pendulum can never be **captured** in resonance. In general resonance capture requires that either the torque or the characteristic frequency ω of the pendulum varies with time; the mechanics of resonance capture in this case are described in §6.3.

6.1.2 Resonances in Hamiltonian systems

The Hamiltonian for a nearly integrable system can be written (eq. 4.47)

$$H(\boldsymbol{\theta}, \mathbf{J}, t) = H_0(\mathbf{J}) + \epsilon \sum_{\mathbf{m}} H_{\mathbf{m}}(\mathbf{J}, t) \cos[\mathbf{m} \cdot \boldsymbol{\theta} - \phi_{\mathbf{m}}(t)]. \tag{6.18}$$

Here $H_0(\mathbf{J})$ is the integrable Hamiltonian (the Kepler Hamiltonian in the case of planetary systems), $\boldsymbol{\theta}$ and $\mathbf{J}$ are the corresponding angle-action variables, $\mathbf{m}$ is an integer n-tuple where n is the number of degrees of freedom, and $\phi_{\mathbf{m}}$ is a phase. Note that $H_{-\mathbf{m}} = H_{\mathbf{m}}$ and $\phi_{-\mathbf{m}} = -\phi_{\mathbf{m}}$ (see eq. 4.45), so the terms with index $\pm\mathbf{m}$ are identical.

A resonance arises when the argument of one or more of the cosine terms varies slowly with time. Since the evolution of the angles is dominated by the unperturbed Hamiltonian, $\dot{\boldsymbol{\theta}} \simeq \partial H_0(\mathbf{J})/\partial \mathbf{J} = \boldsymbol{\Omega}(\mathbf{J})$ so resonance requires $\mathbf{m} \cdot \boldsymbol{\Omega}(\mathbf{J}) - \dot{\phi}_{\mathbf{m}} \simeq 0$. This condition can only be satisfied if $\dot{\phi}_{\mathbf{m}}$ is approximately constant, so we write $\phi_{\mathbf{m}} = \phi_{0\mathbf{m}} + \omega_{\mathbf{m}} t$ where $\omega_{\mathbf{m}}$ is a constant. Then exact resonance occurs when

$$\mathbf{m} \cdot \boldsymbol{\Omega} = \omega_{\mathbf{m}}. \tag{6.19}$$

In the simplest and most common case, a single term in the potential dominates the resonant response, so we are interested in the dynamics governed by the Hamiltonian

$$H(\boldsymbol{\theta}, \mathbf{J}, t) = H_0(\mathbf{J}) + \epsilon H_{\mathbf{m}}(\mathbf{J}) \cos(\mathbf{m} \cdot \boldsymbol{\theta} - \phi_0 - \omega t); \tag{6.20}$$

here we have assumed for simplicity that the amplitude $H_{\mathbf{m}}(\mathbf{J})$ is time-independent and dropped the subscripts "$\mathbf{m}$" on ϕ_0 and ω. To study this Hamiltonian we perform a canonical transformation to new angle-action variables. The new variables consist of one angle-action pair $(\phi_{\mathrm{s}}, J_{\mathrm{s}})$ ("s" for "slow") in which we will isolate the effects of the resonance, and $n - 1$ angle-action pairs $(\phi_{\mathrm{f},i}, J_{\mathrm{f},i})$, $i = 1, \ldots, n-1$ ("f" for "fast"). We use the notation $\boldsymbol{\phi}_{\mathrm{f}}$ and $\mathbf{J}_{\mathrm{f}}$ for the $(n - 1)$-dimensional vectors $(\phi_{\mathrm{f},1}, \ldots, \phi_{\mathrm{f},n-1})$ and $(J_{f1}, \ldots, J_{\mathrm{f},n-1})$. The transformation is defined by a generating function that depends on the old angles, the new actions, and time:

$$S_2(\boldsymbol{\theta}, J_{\mathrm{s}}, \mathbf{J}_{\mathrm{f}}, t) = J_{\mathrm{s}}(\mathbf{m} \cdot \boldsymbol{\theta} - \phi_0 - \omega t) + J_{f1}\theta_1 + \cdots + J_{\mathrm{f},n-1}\theta_{n-1}. \tag{6.21}$$

Then from equations (D.63),

$$\begin{aligned} J_n &= \frac{\partial S_2}{\partial \theta_n} = m_n J_{\mathrm{s}}, & \phi_{\mathrm{s}} &= \frac{\partial S_2}{\partial J_{\mathrm{s}}} = \mathbf{m} \cdot \boldsymbol{\theta} - \phi_0 - \omega t, \\ J_i &= \frac{\partial S_2}{\partial \theta_i} = m_i J_{\mathrm{s}} + J_{\mathrm{f},i}, & \phi_{\mathrm{f},i} &= \frac{\partial S_2}{\partial J_{\mathrm{f},i}} = \theta_i, \quad i = 1, \ldots, n - 1. \end{aligned} \tag{6.22}$$

The motivation for the terms "fast" and "slow" is that near the resonance, ϕ_{s} varies much more slowly than any of the other new angles.

The Hamiltonian in the new variables is

$$\begin{aligned} H_{\mathrm{res}}(\phi_{\mathrm{s}}, \boldsymbol{\phi}_{\mathrm{f}}, J_{\mathrm{s}}, \mathbf{J}_{\mathrm{f}}) &= H(\boldsymbol{\theta}, \mathbf{J}, t) + \frac{\partial S_2}{\partial t} = H(\boldsymbol{\theta}, \mathbf{J}, t) - \omega J_{\mathrm{s}} \\ &= H_0[\mathbf{J}(J_{\mathrm{s}}, \mathbf{J}_{\mathrm{f}})] + \epsilon H_{\mathbf{m}}[\mathbf{J}(J_{\mathrm{s}}, \mathbf{J}_{\mathrm{f}})] \cos \phi_{\mathrm{s}} - \omega J_{\mathrm{s}}. \end{aligned} \tag{6.23}$$

Here we have written the relation (6.22) between the old and new actions as $\mathbf{J}(J_{\mathrm{s}}, \mathbf{J}_{\mathrm{f}})$.

The Hamiltonian H_{res} is autonomous (time-independent) and is also independent of the fast angles $\phi_{f1}, \ldots, \phi_{\mathrm{f},n-1}$. Thus both the Hamiltonian and the $n-1$ fast actions are conserved, and the dynamics have been reduced from those of a time-dependent Hamiltonian with n degrees of freedom to an autonomous Hamiltonian with only one degree of freedom.

Since the perturbation is weak ($\epsilon \ll 1$) the variations in J_s are expected to be small, even near resonance. Therefore we write $J_\mathrm{s} = J_{s,\mathrm{res}} + \Delta J_\mathrm{s}$, where $J_{s,\mathrm{res}}$ is the slow action corresponding to exact resonance (eq. 6.19) at the fixed values of the fast actions. We expand H_0 in a Taylor series around $J_{s,\mathrm{res}}$, again at the fixed values of the fast actions. Equations (6.22) show that $(\partial J_i/\partial J_\mathrm{s})_{\mathbf{J}_\mathrm{f}} = m_i$ so

$$\left(\frac{\partial H_0}{\partial J_\mathrm{s}}\right)_{\mathbf{J}_\mathrm{f}} = \sum_{i=1}^{n} m_i \frac{\partial H_0}{\partial J_i} = \sum_{i=1}^{n} m_i \Omega_i, \qquad \left(\frac{\partial^2 H_0}{\partial J_\mathrm{s}^2}\right)_{\mathbf{J}_\mathrm{f}} = \sum_{i,j=1}^{n} m_i m_j \frac{\partial^2 H_0}{\partial J_i \partial J_j}, \tag{6.24}$$

and the Taylor series becomes

$$H_0[\mathbf{J}(J_\mathrm{s},\mathbf{J}_\mathrm{f})] = H_0[\mathbf{J}(J_{s,\mathrm{res}},\mathbf{J}_\mathrm{f})] + \sum_{i=1}^{n} m_i \frac{\partial H_0}{\partial J_i}\Delta J_\mathrm{s} + \tfrac{1}{2}\sum_{i,j=1}^{n} m_i m_j \frac{\partial^2 H_0}{\partial J_i \partial J_j}(\Delta J_\mathrm{s})^2 + \mathrm{O}[(\Delta J_\mathrm{s})^3]. \tag{6.25}$$

We write the final term in the resonant Hamiltonian (6.23) as $-\omega(J_{s,\mathrm{res}} + \Delta J_\mathrm{s})$. Using the resonance condition (6.19), we find that the terms linear in ΔJ_s cancel in the resonant Hamiltonian. The constant terms play no role in the dynamics and can be dropped, so the resonant Hamiltonian simplifies to

$$H_\mathrm{res} = \tfrac{1}{2}\sum_{i,j=1}^{n} m_i m_j \frac{\partial^2 H_0}{\partial J_i \partial J_j}(\Delta J_\mathrm{s})^2 + \epsilon H_\mathbf{m}[\mathbf{J}(J_{s,\mathrm{res}} + \Delta J_\mathrm{s}, \mathbf{J}_\mathrm{f})]\cos\phi_\mathrm{s}. \tag{6.26}$$

We can drop the dependence of the small term $\epsilon H_\mathbf{m}$ on ΔJ_s to obtain finally

$$H_\mathrm{res} = \tfrac{1}{2}\alpha(\Delta J_\mathrm{s})^2 + \beta\cos\phi_\mathrm{s}, \tag{6.27}$$

where

$$\alpha \equiv \sum_{i,j=1}^{n} m_i m_j \frac{\partial^2 H_0}{\partial J_i \partial J_j}, \qquad \beta \equiv \epsilon H_\mathbf{m}, \tag{6.28}$$

both evaluated at $\mathbf{J}(J_{s,\mathrm{res}}, \mathbf{J}_{\mathrm{f}})$. This is the pendulum Hamiltonian of equation (6.1), apart from the changes in notation $m^{-1} \to \alpha$ and $-m\omega^2 \to \beta$. In the pendulum Hamiltonian the inverse mass m^{-1} is always positive and $-m\omega^2$ is always negative. This is not so for α and β, which can have either sign. However, we can always change the signs of α and β by the transformations described in Box 6.1. There is therefore no loss in generality if we assume that $\alpha > 0$ and $\beta < 0$, as they are for the pendulum.

Box 6.1: Symmetry relations for resonant Hamiltonians

A general resonant Hamiltonian can be written in the form

$$H(I, \phi) = A(I) + B(I)\cos\phi, \tag{a}$$

where (I, ϕ) is an angle-action pair for the unperturbed Hamiltonian $A(I)$. The Hamiltonians (6.27), (6.37) and (6.58) are all of this form.

Changes in the definition of the angle variable ϕ lead to modifications in the Hamiltonian (a). If we let $\phi = \phi' + \pi$, then (I, ϕ') satisfy Hamilton's equations for the Hamiltonian

$$H'(I, \phi') = A(I) - B(I)\cos\phi'. \tag{b}$$

Similarly, if we let $\phi = -\phi''$, then (I, ϕ'') satisfy Hamilton's equations for the Hamiltonian

$$H''(I, \phi'') = -A(I) - B(I)\cos\phi''. \tag{c}$$

Finally, if $\phi = \pi - \phi'''$, then (I, ϕ''') satisfy Hamilton's equations for the Hamiltonian

$$H'''(I, \phi''') = -A(I) + B(I)\cos\phi'''. \tag{d}$$

These arguments show that the pendulum Hamiltonian describes the generic behavior of nearly integrable Hamiltonian systems at resonances. In particular the width of the resonance, defined as in equation (6.13), is obtained by replacing m by $|\alpha|^{-1}$ and $m\omega^2$ by $|\beta|$:

$$w = J_{s,\max} - \Delta J_{s,\min} = 4\left|\frac{\beta}{\alpha}\right|^{1/2}; \tag{6.29}$$

thus the width of the resonance in action space varies as the square root of the strength of the perturbation. The libration period for small amplitudes is given by equation (6.5) in the limit $k \to 0$,

$$P = \frac{2\pi}{|\alpha\beta|^{1/2}}, \tag{6.30}$$

so the libration period varies as the inverse square root of the strength of the perturbation. Weaker resonances are narrower and have larger libration periods.

6.2 Resonance for circular orbits

The pendulum model of the preceding section describes resonances in a wide variety of Hamiltonian systems, but not all of them. The most important exception in celestial mechanics occurs for nearly circular orbits.

To discuss resonant dynamics in this case we shall assume that the motion is restricted to the equatorial plane, so there are only two degrees of freedom. We use the angle-action variables defined in equation (1.88),

$$\begin{aligned} J_1 &= \Lambda = (\mathbb{G}Ma)^{1/2}, \quad & J_2 &= \Lambda - L = (\mathbb{G}Ma)^{1/2}[1-(1-e^2)^{1/2}], \\ \theta_1 &= \lambda = \ell + \varpi, & \theta_2 &= -\varpi, \end{aligned} \tag{6.31}$$

where as usual a, e, ϖ and λ are the semimajor axis, eccentricity, longitude of periapsis and mean longitude. We examine the behavior of a test particle[4] subjected to perturbations from a planet of mass M_p with orbital elements a_p, e_p, ϖ_p and λ_p. To first order in the eccentricities e and e_p, the disturbing function is given by equations (4.98) and (4.104) as

$$\begin{aligned} H_1 = \epsilon \sum_{m=-\infty}^{\infty} \big\{ & A_m \cos m(\lambda - \lambda_\mathrm{p}) + B_m e \cos[m\lambda - (m+1)\lambda_\mathrm{p} + \varpi] \\ & + C_m e_\mathrm{p} \cos[m\lambda - (m+1)\lambda_\mathrm{p} + \varpi_\mathrm{p}] \big\}. \end{aligned} \tag{6.32}$$

Here A_m, B_m and C_m are functions of the semimajor axes, which we can treat as constants since the variations in semimajor axes are small—even in

[4] The case of a resonance between two bodies of comparable mass is discussed in Problem 6.8.

resonance—if the planet mass is small. The terms proportional to A_m are only in resonance if the mean motions $n = \dot{\lambda}$ and $n_\mathrm{p} = \dot{\lambda}_\mathrm{p}$ are equal; this case requires that the orbits nearly coincide and is better treated using the analysis of §3.2 for co-orbital satellites. The terms proportional to C_m involve the eccentricity e_p of the perturbing planet and can be treated using the pendulum Hamiltonian, as described in the preceding section. Thus from now on, we focus on the terms proportional to B_m. For small eccentricity we can write $J_2 \simeq \frac{1}{2}(\mathbb{G}Ma)^{1/2}e^2$, so the factor $B_m e$ can be rewritten without additional loss of accuracy as $(2J_2)^{1/2}H_m$, where $H_m = B_m(\mathbb{G}Ma)^{-1/4}$. Isolating the resonance belonging to a single m, we arrive at the Hamiltonian (cf. eq. 6.20)

$$H(\boldsymbol{\theta}, \mathbf{J}, t) = H_0(\mathbf{J}) + \epsilon(2J_2)^{1/2}H_m \cos(m\theta_1 - \theta_2 - \omega t - \phi_0), \tag{6.33}$$

where $\omega = (m+1)n_\mathrm{p}$ and ϕ_0 is a constant.

We specialize to the case where H_0 is the Kepler Hamiltonian, $H_0 = -\frac{1}{2}(\mathbb{G}M)^2/J_1^2$. In the absence of the perturbation $\dot{\theta}_1 = n = \partial H_0/\partial J_1 = (\mathbb{G}M/a^3)^{1/2}$ and $\dot{\theta}_2 = 0$, so exact resonance occurs when $n = (1+1/m)n_\mathrm{p}$. For $m > 0$ this condition implies that the test particle orbits interior to the perturbing planet, while for $m < 0$ the test particle is exterior to the planet.

We now follow the route from equation (6.20) to (6.27) to derive the resonant Hamiltonian. We define the canonical transformation to fast and slow angle-action variables as

$$S_2(\theta_1, \theta_2, J_\mathrm{s}, J_\mathrm{f}, t) = J_\mathrm{s}(m\theta_1 - \theta_2 - \omega t - \phi_0) + J_\mathrm{f}\theta_1. \tag{6.34}$$

Then from equations (D.63),

$$\begin{aligned} J_1 &= \frac{\partial S_2}{\partial \theta_1} = mJ_\mathrm{s} + J_\mathrm{f}, & \phi_\mathrm{s} &= \frac{\partial S_2}{\partial J_\mathrm{s}} = m\theta_1 - \theta_2 - \omega t - \phi_0, \\ J_2 &= \frac{\partial S_2}{\partial \theta_2} = -J_\mathrm{s}, & \phi_\mathrm{f} &= \frac{\partial S_2}{\partial J_\mathrm{f}} = \theta_1. \end{aligned} \tag{6.35}$$

The resonant Hamiltonian in the new variables is

$$H_\mathrm{res} = H(\boldsymbol{\theta}, \mathbf{J}, t) + \frac{\partial S_2}{\partial t}$$

$$= -\frac{(\mathbb{G}M)^2}{2(mJ_\mathrm{s} + J_\mathrm{f})^2} + \epsilon(-2J_\mathrm{s})^{1/2} H_m \cos\phi_\mathrm{s} - \omega J_\mathrm{s}. \tag{6.36}$$

The Hamiltonian is independent of the fast angle ϕ_f, so the fast action J_f is constant.

In the preceding section we assumed that the fractional variations in the slow action J_s were small, so the coefficient of $\cos\phi_\mathrm{s}$ could be set to a constant. This assumption is no longer correct in the case we are examining here, since the eccentricity and therefore $J_\mathrm{s} \propto e^2$ is initially small but can be excited by the resonance. Instead we can assume that the coefficient H_m is constant.

We now expand the first term as a Taylor series around $J_\mathrm{s} = 0$, using the result $(1 + x)^{-2} = 1 - 2x + 3x^2 + \mathrm{O}(x^3)$. After dropping an unimportant constant term, we have

$$H_\mathrm{res} = \alpha J_\mathrm{s} + \beta J_\mathrm{s}^2 + \gamma(-2J_\mathrm{s})^{1/2} \cos\phi_\mathrm{s}, \tag{6.37}$$

where

$$\alpha \equiv \frac{(\mathbb{G}M)^2 m}{J_\mathrm{f}^3} - \omega, \quad \beta \equiv -\frac{3(\mathbb{G}M)^2 m^2}{2J_\mathrm{f}^4}, \quad \gamma \equiv \epsilon H_m. \tag{6.38}$$

The factor α has a simple physical interpretation. If the eccentricity is small then J_s is small, so $J_\mathrm{f} \simeq J_1 = \Lambda = (\mathbb{G}Ma)^{1/2}$. Moreover $\omega = (m + 1)n_\mathrm{p}$. Then $\alpha = mn - (m + 1)n_p$, which parametrizes the distance of the test particle away from exact resonance with the planet.

To study the behavior of orbits in this Hamiltonian, we first rescale it to a standard form. To do this, we begin with Hamilton's equations

$$\frac{\mathrm{d}\phi_\mathrm{s}}{\mathrm{d}t} = \frac{\partial H_\mathrm{res}}{\partial J_\mathrm{s}}, \quad \frac{\mathrm{d}J_\mathrm{s}}{\mathrm{d}t} = -\frac{\partial H_\mathrm{res}}{\partial \phi_\mathrm{s}}. \tag{6.39}$$

Now let $J_\mathrm{s} = cR$, $H_\mathrm{res}(J_\mathrm{s}, \phi_\mathrm{s}) = bH'_\mathrm{res}(R, \phi_\mathrm{s})$, $t = t_0\tau$, where b, c and t_0 are constants. Then R is a momentum conjugate to the coordinate ϕ_s and H'_res is a Hamiltonian relative to the rescaled time τ if

$$\frac{\mathrm{d}\phi_\mathrm{s}}{\mathrm{d}\tau} = \frac{\partial H'_\mathrm{res}}{\partial R}, \quad \frac{\mathrm{d}R}{\mathrm{d}\tau} = -\frac{\partial H'_\mathrm{res}}{\partial \phi_\mathrm{s}}. \tag{6.40}$$

Equations (6.39) and (6.40) can be satisfied simultaneously if and only if $t_0 = c/b$.

We now set

$$b = \frac{|\gamma|^{4/3}}{2^{4/3}|\beta|^{1/3}}, \quad c = -\left|\frac{\gamma}{2\beta}\right|^{2/3}, \quad t_0 = \left|\frac{4}{\beta\gamma^2}\right|^{1/3}. \tag{6.41}$$

The rescaled Hamiltonian is

$$H'_{\rm res} = -\Delta\,\mathrm{sgn}(\beta)R + \mathrm{sgn}(\beta)R^2 + 2\,\mathrm{sgn}(\gamma)(2R)^{1/2}\cos\phi_{\rm s}, \tag{6.42}$$

where

$$\Delta \equiv \mathrm{sgn}(\beta)\frac{2^{2/3}\alpha}{|\beta\gamma^2|^{1/3}} \tag{6.43}$$

and $\mathrm{sgn}(x)$ is +1 (−1) if x is positive (negative). We can eliminate the dependence of the Hamiltonian on the signs of β and γ by defining a new angle, following the arguments given in Box 6.1. Let

$$r = \begin{cases} \mathrm{sgn}(\beta)\phi_{\rm s} & \text{if } \beta\gamma < 0, \\ \mathrm{sgn}(\beta)\phi_{\rm s} + \pi & \text{if } \beta\gamma > 0. \end{cases} \tag{6.44}$$

Then (r, R) is a canonical coordinate-momentum pair, with Hamiltonian

$$\Gamma = -\Delta R + R^2 - 2(2R)^{1/2}\cos r. \tag{6.45}$$

This is the **Henrard–Lemaitre Hamiltonian** (Henrard & Lemaitre 1983), sometimes called the **second fundamental model for resonance**—the first one being the pendulum.

For some purposes it is more useful to write the Hamiltonian in terms of a new canonical momentum x and conjugate coordinate y defined by the generating function $S_2(r, x) = \frac{1}{2}x^2\tan r$. Then from equations (D.63),

$$R = \frac{\partial S_2}{\partial r} = \frac{x^2}{2\cos^2 r}, \quad y = x\tan r, \tag{6.46}$$

and these equations yield

$$x = (2R)^{1/2}\cos r, \quad y = (2R)^{1/2}\sin r. \tag{6.47}$$

In these coordinates

$$\Gamma = -\tfrac{1}{2}\Delta(x^2 + y^2) + \tfrac{1}{4}(x^2 + y^2)^2 - 2x. \tag{6.48}$$

By rescaling the Hamiltonian to a standard form, we can make order-of-magnitude estimates of the response of the orbit to the resonance. The strength of the perturbing Hamiltonian is proportional to γ and γ is of order M_p/M, the ratio of the planet mass to the stellar mass. The scalings (6.41) imply that the characteristic eccentricity excited by the resonance scales as $J_\mathrm{s}^{1/2} \propto c^{1/2} \propto \gamma^{1/3} \propto (M_\mathrm{p}/M)^{1/3}$. The small exponent of $\frac{1}{3}$ means that even a small planet can excite a relatively large eccentricity at resonance. The characteristic timescale t_0 for resonant orbits is given by $nt_0 \propto \gamma^{-2/3} \propto (M_\mathrm{p}/M)^{-2/3}$.

The contours of the Hamiltonian Γ are plotted in the Cartesian coordinates (x, y) for several values of the parameter Δ in Figure 6.4. The plots show that for $\Delta < 3$ there is only one extremum of the Hamiltonian, representing an equilibrium solution. For $\Delta > 3$ there are three extrema, of which the leftmost is a saddle point and an unstable equilibrium, while the other two are stable. Two separatrices emerge from the saddle point, and these divide the phase space into three zones: an **internal zone** inside the smaller separatrix, a **resonance zone** between the two separatrices, and an **external zone** outside the larger separatrix.

As defined in the preceding section, an orbit is librating if it has a slow angle ϕ_s that oscillates between fixed values, so $\dot\phi_\mathrm{s}$ periodically changes sign, while circulating orbits have $\dot\phi_\mathrm{s}$ of fixed sign. In the Henrard–Lemaitre Hamiltonian, it is not useful to distinguish librating orbits from circulating ones. All of the panels of Figure 6.4 contain both librating and circulating orbits, depending on whether the two intersections of the corresponding level curve of the Hamiltonian with the x-axis have the same or opposite signs, but this distinction has no particular dynamical significance. The division into internal, external and resonant zones replaces the division into librating and circulating zones.

Since y and x are a canonical coordinate-momentum pair, the area enclosed by a contour of the Hamiltonian in Figure 6.4 is the same in all canonical variables. Moreover the area is an adiabatic invariant (Appendix D.10),

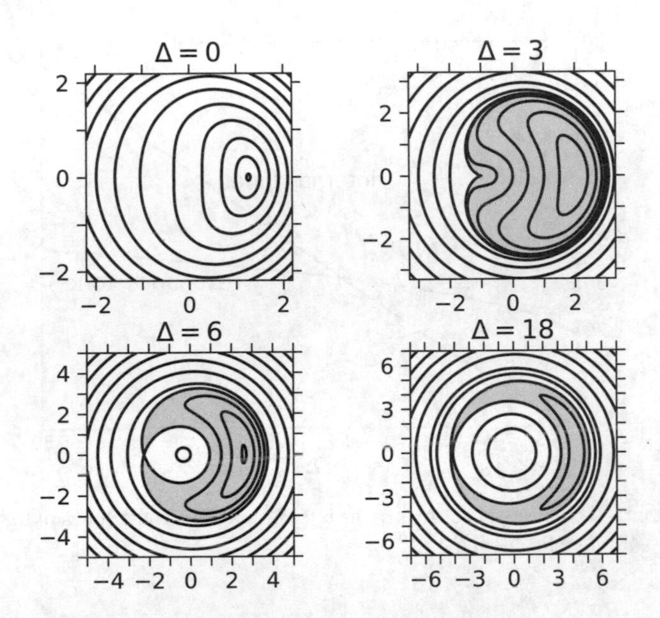

Figure 6.4: Contours of the Hamiltonian (6.48). The gray shading in plots with $\Delta \geq 3$ marks the resonance zone, which surrounds the internal zone and is surrounded by the external zone. The coordinates are x and y.

that is, the area enclosed by a trajectory is invariant under slow changes in the parameters of the Hamiltonian. Figure 6.5 shows the areas of the internal and resonance zones as a function of the parameter Δ. This plot will be used in §6.3 to describe evolution through resonance.

The behavior described in this section is governed by a resonant term in the Hamiltonian proportional to $e\cos[m\lambda-(m+1)\lambda_{\rm p}+\varpi]$ (eq. 6.32). The characteristic feature of this term is that the strength of the perturbation is proportional to the first power of the eccentricity or the square root of the corresponding action. These are sometimes called **eccentricity resonances**. There are also **inclination resonances**, in which the resonant Hamiltonian is proportional to $I^2\cos[m\lambda-(m+2)\lambda_{\rm p}+2\Omega]$ (eq. 4.99). Here the strength

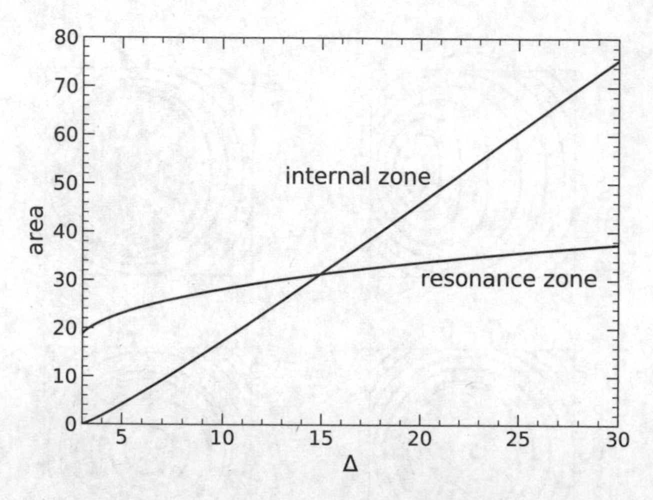

Figure 6.5: The phase-space areas of the internal and resonance zones of the Hamiltonian (6.48). See Problem 6.4.

is proportional to the square of the inclination or the first power of the corresponding action $L-L_z$ (eq. 1.88). Inclination resonances can be investigated using methods similar to those used in this section (Borderies & Goldreich 1984).

More generally, any resonant Hamiltonian contains a cosine function with argument $m\lambda-(m+k)\lambda_{\rm p}$ plus some combination of angles such as ϖ and Ω that are stationary in the unperturbed Kepler orbit. Exact resonance occurs when $mn-(m+k)n_{\rm p}=0$, where n and $n_{\rm p}$ are the mean motions of the test particle and the perturbing body. Resonances with $k=0$, called co-orbital resonances, are discussed in §3.2. Resonances with $m=k=0$ are called **secular resonances** and all others are **mean-motion resonances**. Such resonances are sometimes labeled by the ratio $n/n_{\rm p}$ (e.g., a 2:1 resonance). Resonances with $k/m>0$ have $n>n_{\rm p}$, so the test particle has smaller semimajor axis than the perturber; these are **interior resonances**, while resonances with $n<n_{\rm p}$ are **exterior resonances**.

6.2.1 The resonance-overlap criterion for nearly circular orbits

The analysis so far in this section has been based on the assumption that a single near-resonant term in the disturbing function (6.32) dominates the dynamics. This assumption is valid if the width of the resonance is much less than the distance in action space to adjacent resonances; if not, the resonances overlap and the motion is likely to be chaotic (Appendix F.1).

For the Hamiltonian (6.33), exact resonance occurs when the mean motion is $n = (1 + 1/m)n_\mathrm{p}$, where n_p is the mean motion of the perturber and m is the index characterizing the resonance. Thus as $|m| \to \infty$, the separation of the resonances in mean motion or semimajor axis becomes smaller. Moreover the test particle and perturber become closer, so the strength of the perturbations and thus the resonance width become larger. Therefore resonance overlap is inevitable if $|m|$ is large enough. We now make this argument quantitative.

From equations (6.33) or (6.88) the strength of the resonance is parametrized by

$$H_m = \frac{\mathbb{G}^{3/4} M_\mathrm{p}}{a_\mathrm{p} a^{1/4} M^{1/4}} \left[\tfrac{1}{2}\alpha\delta_{m,-2} + (m + 1 + \tfrac{1}{2}\alpha D) b_{1/2}^{m+1}\right]. \tag{6.49}$$

Here M_p and a_p are the mass and semimajor axis of the perturber, M is the mass of the central star, $\alpha = a/a_\mathrm{p}$, and we have assumed that $m \neq 0$. The term involving the Laplace coefficient $b_{1/2}^{m+1}(\alpha)$ comes from equation (4.98) and the term involving the Kronecker delta $\delta_{m,-2}$ comes from the indirect potential. As usual we assume that $M_\mathrm{p}/M \ll 1$, which implies that resonance overlap occurs only close to the planet, when $|m| \gg 1$. Thus we can drop the contribution from the indirect potential and assume that $a \simeq a_\mathrm{p}$ and $\alpha \simeq 1$. More precisely, the resonance condition $n = (1 + 1/m)n_\mathrm{p}$ together with Kepler's law implies

$$a - a_\mathrm{p} = a_\mathrm{p}(1 + 1/m)^{-2/3} - a_\mathrm{p} \simeq -\frac{2a_\mathrm{p}}{3m} \quad \text{or} \quad \alpha \simeq 1 - \frac{2}{3m}. \tag{6.50}$$

For $|m| \gg 1$, equations (4.127) and (C.34) imply that

$$b_{1/2}^{m+1}(\alpha) \simeq \frac{2}{\pi} K_0[|m(1 - \alpha)|],$$

$$Db_{1/2}^{m+1}(\alpha) \simeq \frac{2m}{\pi}\,\mathrm{sgn}[m(1-\alpha)]K_1[|m(1-\alpha)|], \tag{6.51}$$

where $K_0(\cdot)$ and $K_1(\cdot)$ are modified Bessel functions, described in Appendix C.5. Equation (6.50) implies that at resonance the argument of the Bessel functions is $\frac{2}{3}$, so

$$H_m = \frac{2.5195\, m\, \mathbb{G}^{3/4} M_\mathrm{p}}{\pi a_\mathrm{p}^{5/4} M^{1/4}}\big[1 + \mathrm{O}(m^{-1})\big], \tag{6.52}$$

where the factor $2.5195 = 2K_0(\frac{2}{3}) + K_1(\frac{2}{3})$ (cf. eq. 3.131).

The resonance-overlap criterion is only approximate, so we parametrize the resonance "width" to be used in this criterion as δR, where R is the rescaled dimensionless momentum and δR is of order unity. Then the width in terms of the slow action is $\delta J_\mathrm{s} = |c|\delta R$, where c is defined by equation (6.41). The fast action $J_\mathrm{f} = J_1 - mJ_\mathrm{s}$ (eq. 6.35) is conserved, so the width in terms of J_1 is $\delta J_1 = |mc|\delta R$. Since $J_1 = (\mathbb{G}Ma)^{1/2}$, the width in terms of semimajor axis is $\delta a \simeq 2|mc|a^{1/2}\delta R/(\mathbb{G}M)^{1/2} \simeq 2|mc|\delta R/(n_\mathrm{p}a_\mathrm{p})$. The distance between resonances differing in index by $\Delta m \ll m$ is obtained by differentiating equation (6.50) to obtain $\Delta a \simeq 2a_\mathrm{p}\Delta m/(3m^2)$. The distance between adjacent resonances corresponds to $|\Delta m| = 1$, thus $|\Delta a| = 2a_\mathrm{p}/(3m^2)$. Resonances overlap if $\delta a > |\Delta a|$ or

$$|m|^3 > \frac{n_\mathrm{p}a_\mathrm{p}^2}{3|c|\delta R}. \tag{6.53}$$

For small eccentricities and $|m| \gg 1$, equations (6.38), (6.41) and (6.52) imply that

$$|c| = \frac{2.5195^{2/3}\,\mathbb{G}^{1/2} M_\mathrm{p}^{2/3} a_\mathrm{p}^{1/2}}{(3\pi)^{2/3}|m|^{2/3} M^{1/6}}; \tag{6.54}$$

in deriving this we have replaced J_f by $J_1 = (\mathbb{G}Ma)^{1/2}$, which is valid so long as the eccentricity is sufficiently small. Combining this result with (6.53), we find that resonance overlap occurs when

$$|m| > 0.910\left(\frac{M}{M_\mathrm{p}}\right)^{2/7}(\delta R)^{-3/7}, \tag{6.55}$$

or in terms of the semimajor axis difference, when

$$\frac{|a - a_{\rm p}|}{a_{\rm p}} < 0.732\left(\frac{M_{\rm p}}{M}\right)^{2/7}(\delta R)^{3/7}. \tag{6.56}$$

This result demonstrates that the width of the chaotic zone around a planet of mass $M_{\rm p}$ should scale as $M_{\rm p}^{2/7}$ for nearly circular orbits. We have not attempted to determine the appropriate value of the width δR since the resonance-overlap criterion is approximate anyway, but it should be of order unity. Analytic calculations of the width, supplemented by numerical orbit integrations, imply that the boundary of the chaotic zone is (Wisdom 1980; Gladman 1993; Deck et al. 2013; Morrison & Malhotra 2015)

$$\frac{|a - a_{\rm p}|}{a_{\rm p}} \lesssim f\left(\frac{M_{\rm p}}{M}\right)^{2/7}, \tag{6.57}$$

where $f \simeq 1.2$–1.5. The boundary is fuzzy because it contains a mixture of chaotic and regular orbits. A more general discussion of the stability and orbits in two-planet systems, including a heuristic derivation of equation (6.57), is given in §3.5.

6.3 Resonance capture

A dynamical system that is described by angle-action variables is said to be **trapped** in resonance if some linear combination of the angle variables librates, or if the system is found in a resonance zone as described in the preceding section. The system is said to be **captured** in resonance if the combination of angle variables changes from circulation to libration, or if the system enters the resonance zone. In the absence of dissipation, external forces, or changes in the parameters of the dynamical system, permanent capture in resonance can never occur: if there is a trajectory leading to capture, then there must also be a time-reversed trajectory leading to escape from the resonance, and eventually the system will take this path to exit the resonance.

Permanent capture *can* occur if the parameters of the system vary slowly with time due to processes such as mass loss or tidal dissipation. To explore

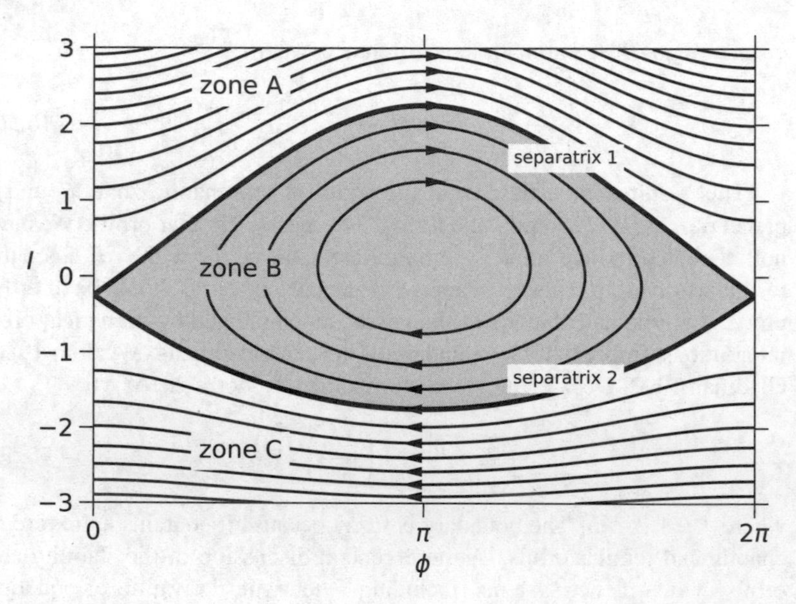

Figure 6.6: The level surfaces of the Hamiltonian (6.58) near a separatrix. The libration or resonance zone (zone B) is colored gray. The directions of motion of the orbits are plotted on the assumption that the stable equilibrium is a minimum of the Hamiltonian.

this process, we can investigate the behavior of trajectories governed by such resonant Hamiltonians as (6.27) or (6.45) as their parameters are varied. To do so, we follow Henrard (1982) and work with a more general Hamiltonian,

$$H(I, \phi; \lambda) = A(I, \lambda) + B(I, \lambda)\cos\phi, \tag{6.58}$$

where (I, ϕ) is an angle-action pair in the unperturbed Hamiltonian $A(I, \lambda)$, and $\lambda(\tau)$ represents some parameter of the system that varies slowly with time τ. Hamilton's equations of motion are

$$\frac{d\phi}{d\tau} = \frac{\partial H}{\partial I} = \frac{\partial A}{\partial I} + \frac{\partial B}{\partial I}\cos\phi, \qquad \frac{dI}{d\tau} = -\frac{\partial H}{\partial \phi} = B\sin\phi. \tag{6.59}$$

When λ is frozen there are equilibria at $(I^\star, \phi^\star)$, where

$$\cos\phi^\star = \pm 1, \qquad \left.\frac{\partial(A \pm B)}{\partial I}\right|_{I^\star} = 0. \tag{6.60}$$

The contours of the Hamiltonian that pass through the unstable equilibria are the separatrices and the value of the Hamiltonian on the separatrices is denoted by $E^\star(\lambda)$.

To be definite, we shall assume that one of the two equilibria is stable and the other is unstable, and that (i) the unstable equilibria are at $\phi^\star = 0$ and 2π and the stable equilibrium is at $\phi^\star = \pi$; (ii) the stable equilibrium is a minimum of the Hamiltonian. Conditions (i) and (ii) can always be satisfied by transformations of the angle variable as described in Box 6.1. The overall geometry is sketched in Figure 6.6. Note that with these assumptions the orbits in the top half of the figure travel to the right, and those in the bottom half travel to the left. The upper and lower separatrices have actions $I_{1,2}(\phi, \lambda)$. We shall say that the Hamiltonian governs the motion of a "particle," although of course the results are valid for any Hamiltonian system.

The action for the Hamiltonian (6.58) is proportional to the area in phase space enclosed by the trajectory at fixed λ (eq. D.72),

$$J(E, \lambda) = \frac{1}{2\pi} \oint \mathrm{d}\phi\, I(E, \phi; \lambda), \tag{6.61}$$

where $I(E, \phi; \lambda)$ is obtained by solving $E = H(I, \phi; \lambda)$ for the action I in the unperturbed Hamiltonian. The actions for librating and circulating orbits in the pendulum Hamiltonian are given by equations (6.7) and (6.11). The action is an adiabatic invariant (Appendix D.10), which means that if the parameter λ varies slowly compared to the orbital period then $J(E, \lambda)$ is approximately conserved, and this requirement determines how the energy E changes with λ.

To understand resonance capture, we need to know what happens as an orbit approaches one of the separatrices in Figure 6.6 as a result of slow changes in the Hamiltonian. In particular, we would like to know whether an orbit that is initially circulating to the right ($\dot{\phi} > 0$) evolves into a librating orbit (**resonance capture**) or an orbit circulating to the left (**resonance crossing**).

Unfortunately the adiabatic invariant is not conserved near the separatrices, for two reasons. The first (trivial) reason is that there is a discontinuity in $J(E, \lambda)$ across the separatrix, because the geometry of librating and circulating orbits is different. We already encountered this phenomenon in the pendulum Hamiltonian, as discussed after equation (6.12). A second, more fundamental problem is that the period of the separatrix orbit is infinite, so the assumption that λ changes slowly compared to the orbital period is not valid.

Thus we need a more careful treatment of the dynamics near a separatrix. The following is a shortened heuristic version of the rigorous analysis by Henrard (1982). We define the **relative energy** to be the difference between the energy of an orbit and the energy of a nearby separatrix,

$$K(I, \phi; \lambda) = H(I, \phi; \lambda) - E^{\star}(\lambda). \tag{6.62}$$

This expression can be rewritten as

$$K(I, \phi; \lambda) = H(I, \phi; \lambda) - H(I, \phi; \lambda)\big|_{I=I_{1,2}(\phi,\lambda)}, \tag{6.63}$$

depending on whether the orbit is close to the upper or lower separatrix in Figure 6.6. The rate of change of the relative energy is

$$\begin{aligned}\frac{\mathrm{d}K}{\mathrm{d}\tau} &= \dot{\lambda}\frac{\partial}{\partial\lambda}H(I, \phi; \lambda) - \dot{\lambda}\left[\frac{\partial}{\partial\lambda}H(I, \phi; \lambda)\right]_{I=I_{1,2}(\phi,\lambda)} \\ &\quad - \dot{\lambda}\frac{\partial}{\partial I}\left[H(I, \phi; \lambda)\right]_{I=I_{1,2}(\phi,\lambda)}\frac{\partial}{\partial\lambda}I_{1,2}(\phi, \lambda),\end{aligned} \tag{6.64}$$

where $\dot{\lambda} \equiv \mathrm{d}\lambda/\mathrm{d}\tau$. The contributions from the first two terms cancel if the orbit is sufficiently close to the separatrix. The final term can be simplified by observing that $\dot{\phi} = \mathrm{d}\phi/\mathrm{d}t = \partial H/\partial I$, so we can divide both sides of the equation by $\dot{\phi}$ to obtain

$$\frac{\mathrm{d}K}{\mathrm{d}\phi} = -\dot{\lambda}\frac{\partial}{\partial\lambda}I_{1,2}(\phi, \lambda). \tag{6.65}$$

We can now compute the total change in the relative energy as an orbit traverses the separatrix through an angle of 2π as[5]

$$\Delta E_{1,2} = \mp\dot{\lambda} \int_0^{2\pi} \mathrm{d}\phi \, \frac{\partial}{\partial \lambda} I_{1,2}(\phi, \lambda), \tag{6.66}$$

where the – sign applies if the orbit travels from $\phi = 0$ to $\phi = 2\pi$ on separatrix 1, and the + sign applies if it travels in the reverse direction on separatrix 2. The quantity $\Delta E_{1,2}$ is sometimes called the **energy balance**.

Now look again at Figure 6.6, where the phase space is divided into two circulating zones A and C and a librating or resonance zone B. By definition, the relative energy K is zero on each separatrix, and since we have assumed that the stable equilibrium is a minimum of the Hamiltonian we must have $K > 0$ in zones A and C and $K < 0$ in zone B. A particle in zone A that is close to the separatrix circulates with $\mathrm{d}\phi/\mathrm{d}\tau > 0$. At each orbit, as ϕ grows from 0 to 2π, it gains relative energy ΔE_1. Similarly, a particle in zone C circulates with $\mathrm{d}\phi/\mathrm{d}\tau < 0$. At each orbit it gains relative energy ΔE_2. A particle in zone B near the separatrices alternately gains energy ΔE_1 when it orbits near separatrix 1 and ΔE_2 when it orbits near separatrix 2.

Consider a trajectory that starts in zone A, with $K > 0$. If $\Delta E_1 > 0$, the trajectory recedes from the separatrix as K increases by ΔE_1 at each orbit, so it never crosses the separatrix. Thus the only case of interest is $\Delta E_1 < 0$. In this case the trajectory approaches closer and closer to separatrix 1 until eventually K becomes negative, and it transitions to separatrix 2. Now if $\Delta E_2 < 0$ the trajectory continues to alternate between separatrix 1 and separatrix 2, losing energy on both, so it librates in zone B with smaller and smaller amplitude until it approaches the stable minimum-energy equilibrium.

In contrast, if $\Delta E_1 < 0$, $\Delta E_2 > 0$ and $\Delta E_1 + \Delta E_2 > 0$, then the energy lost on the last passage near separatrix 1 will be more than regained on the first passage near separatrix 2. Therefore at the end of this passage, the trajectory has $K > 0$ and transitions directly from zone A to zone C,

[5] Since the time required to traverse the separatrix is infinite, this statement is strictly true only for orbits that are close to but not on the separatrix, in a sense defined precisely by Henrard.

gaining energy ΔE_2 at every orbit and receding from the separatrix. Thus the particle has crossed the resonance without being captured.

Finally we consider the case where $\Delta E_1 < 0$, $\Delta E_2 > 0$ and $\Delta E_1 + \Delta E_2 < 0$. Since $|\Delta E_1|$ is of order $\dot{\lambda}$ and therefore small, we can assume that the relative energy is uniformly distributed between 0 and $|\Delta E_1|$ as the particle begins its last passage along separatrix 1. At the end of this passage, the energy will be uniformly distributed between $\Delta E_1 < 0$ and 0. At the end of the subsequent first passage along separatrix 2, the energy will be uniformly distributed between $\Delta E_1 + \Delta E_2 < 0$ and $\Delta E_2 > 0$. If this energy is negative the particle will continue to librate with smaller and smaller amplitude, losing energy on separatrix 1 and gaining a smaller amount on separatrix 2. However, if the energy after the first passage along separatrix 2 is positive, the particle will continue to circulate near separatrix 2, gaining energy at each passage until it escapes into zone C. Thus the probability of capture in the resonance zone B is

$$p_B = \frac{\Delta E_1 + \Delta E_2}{\Delta E_1}, \tag{6.67}$$

while the probability of crossing into zone C is $p_C = 1 - p_B$.

Similarly we can describe how a particle librating in zone B escapes from resonance. The libration amplitude will grow if $\Delta E_1 + \Delta E_2 > 0$, until the particle eventually approaches the separatrix. Then if $\Delta E_1 > 0$ and $\Delta E_2 < 0$, the particle will escape from resonance into zone A. If $\Delta E_1 < 0$ and $\Delta E_2 > 0$, the particle will escape from resonance into zone C. And if $\Delta E_1 > 0$ and $\Delta E_2 > 0$, the particle will escape into zone A with probability

$$p_A = \frac{\Delta E_1}{\Delta E_1 + \Delta E_2}, \tag{6.68}$$

and it will escape into zone C with probability $p_C = 1 - p_A$.

This analysis provides a complete description of how particles are captured in and escape from resonance when the parameters of the governing Hamiltonian vary slowly with time.

6.3.1 Resonance capture in the pendulum Hamiltonian

Here we analyze resonance capture in the torqued pendulum (§6.1.1). We write the governing Hamiltonian (6.16) as

$$H(I,\phi;\lambda) = \tfrac{1}{2}\alpha(\lambda)[I + c(\lambda)]^2 + \beta(\lambda)\cos\phi. \tag{6.69}$$

For simplicity we assume that $\alpha > 0$ and $\beta > 0$; if necessary this can be arranged using the transformations in Box 6.1. When λ is frozen there are unstable equilibria at $\phi^\star = 0, 2\pi$ and $I^\star = -c(\lambda)$. The energy at the unstable equilibria is $E^\star = \beta(\lambda)$, and the separatrices are located at

$$I_{1,2}(\phi,\lambda) = -c(\lambda) \pm [2\beta(\lambda)/\alpha(\lambda)]^{1/2}(1-\cos\phi)^{1/2}. \tag{6.70}$$

The corresponding energy balances (6.66) are

$$\Delta E_{1,2} = \mp\dot{\lambda}\int_0^{2\pi} \mathrm{d}\phi \frac{\partial}{\partial\lambda} I_{1,2}(\phi,\lambda). \tag{6.71}$$

Substituting from equation (6.70) and using the result $\dot{\lambda}(\partial f/\partial\lambda) = \mathrm{d}f/\mathrm{d}\tau = \dot{f}$ for arbitrary functions f and the integral $\int_0^{2\pi} \mathrm{d}\phi\,(1-\cos\phi)^{1/2} = 2^{5/2}$, we have

$$\Delta E_{1,2} = \pm 2\pi\dot{c} - \frac{4\dot{\beta}}{(\alpha\beta)^{1/2}} + \frac{4\beta^{1/2}\dot{\alpha}}{\alpha^{3/2}}. \tag{6.72}$$

The system discussed in §6.1.1 provides a simple example of the application of these results. From the Hamiltonian (6.16) we have $\alpha = m^{-1}$ and $\beta = m\omega^2$ (after setting $q = \phi + \pi$ to ensure that $\beta > 0$). The mass m and frequency ω are constants, so $\dot{\alpha} = \dot{\beta} = 0$. Thus $\Delta E_1 = -\Delta E_2 = 2\pi\dot{c}$, so $\Delta E_1 + \Delta E_2 = 0$. This result implies that (i) the probability of capture for an initially circulating orbit is zero according to equation (6.67); and (ii) a librating orbit has no net gain of relative energy, so it does not escape from resonance. We reached these conclusions already in §6.1.1, but the analysis here makes it easy to generalize the results to pendulums with slowly varying mass or length.

6.3.2 Resonance capture for nearly circular orbits

To understand resonance capture in this case, we investigate the Henrard–Lemaitre Hamiltonian (6.45),

$$\Gamma(R, r; \Delta) = -\Delta(t)R + R^2 - 2(2R)^{1/2}\cos r, \tag{6.73}$$

in which $\Delta(t)$ is slowly varying. The analysis below follows Henrard & Lemaitre (1983).

First consider the case in which $\Delta(\tau)$ decreases with time from a large positive value. Initially, any low-eccentricity trajectory should be in the internal zone, since this is large when Δ is large and positive—in this limit the equation for the inner separatrix is $R = \frac{1}{2}\Delta + \mathrm{O}(\Delta^{1/4})$. As Δ shrinks, the trajectory conserves its adiabatic invariant, which is equal to $(2\pi)^{-1}$ times the phase-space area enclosed by the trajectory. However, as Δ shrinks the area of the internal zone decreases, becoming zero when $\Delta = 3$ (see Figure 6.5). Thus the orbit crosses the separatrix before $\Delta = 3$, at the point where its initial area equals the area of the internal zone. When this occurs, the orbit cannot enter the resonance zone, since the area of the resonance zone is also decreasing as Δ shrinks (Figure 6.5). Therefore the orbit must jump to the external zone; its enclosed area after the jump is given by the sum of the areas of the internal and resonance zones, and this area is conserved as Δ continues to decrease. Thus the orbit cannot be captured in resonance, but its eccentricity is excited as it crosses the resonance. If the eccentricities long before and long after resonance crossing are e_i and e_f, then $e_f^2/e_i^2 = (A_\mathrm{r} + A_\mathrm{i})/A_\mathrm{i}$, where A_i and A_r are the areas in Figure 6.5 at the point where the orbit transitions from the internal zone to the external zone.

Now suppose that $\Delta(\tau)$ increases with time. In the distant past, Δ was negative and all trajectories were external. As Δ slowly increases, the area enclosed by the trajectories is conserved. When $\Delta = 3$, the resonance zone suddenly appears and all trajectories with area less than $A = 6\pi$ (see Problem 6.4) or $R_0 < A/(2\pi) = 3$ enter the resonance zone. The remaining trajectories stay in the external zone, but since the area of the internal plus resonance zones grows without bound as Δ increases they must eventually cross the separatrix at the outer edge of the resonance zone. When they cross, they can be captured into the resonance or cross the resonance into the

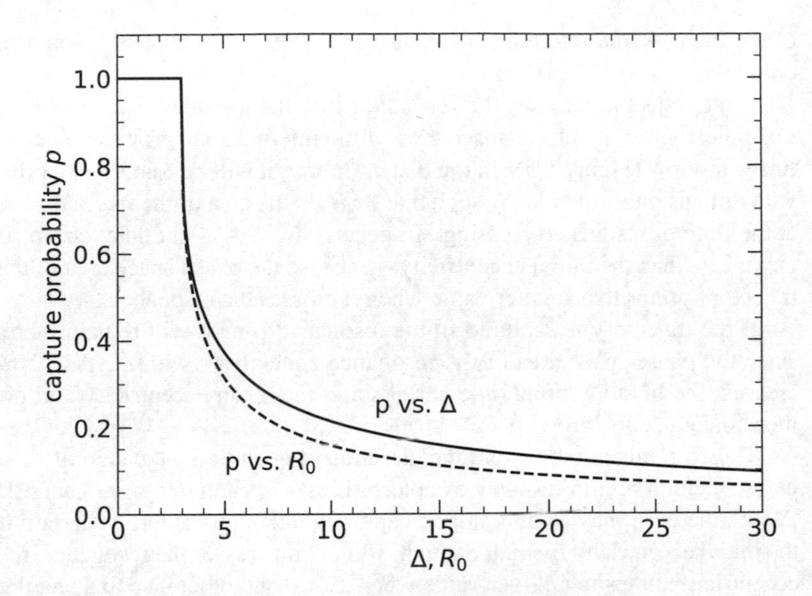

Figure 6.7: The probability of capture in the Hamiltonian (6.73) as a function of the parameter Δ at resonance crossing (solid line) or the initial action R_0 (dashed line). Capture is certain if $\Delta < 3$ or $R_0 < 3$.

internal zone. The probability of capture can be computed using equations (6.66) and (6.67), with separatrices 1 and 2 replaced by the outer and inner separatrices that bound the resonance zone. The calculation can be done either numerically (Henrard & Lemaitre 1983) or analytically (Borderies & Goldreich 1984). Figure 6.7 shows the capture probability in two ways: first as a function of the value of Δ at which resonance capture or crossing occurs, and second as a function of R_0, the dimensionless action of the orbit in the distant past. The two are related by adiabatic invariance, which ensures that the phase-space area enclosed by the orbit remains constant except when the orbit encounters a separatrix at resonance crossing. Thus $2\pi R_0$, the area enclosed by the orbit in the distant past, must equal the sum of the areas of the internal and resonance zones at the time of resonance crossing.

If the trajectory crosses the resonance into the internal zone, it will remain there since the phase-space area of the internal zone grows as Δ continues to grow (Figure 6.5). In the distant future, it will be found in an orbit with dimensionless action R_f such that $2\pi R_\mathrm{f}$ is the area of the internal zone at the time of resonance crossing. In general $R_\mathrm{f} < R_0$—the final eccentricity is less than the initial eccentricity—because the phase-space area of the trajectory jumps to a smaller value when it crosses the resonance.

If the trajectory is captured in the resonance zone it will remain there, since the phase-space area of the resonance zone grows with Δ. As Δ increases, the libration amplitude shrinks and the mean eccentricity and dimensionless action grow.

These arguments show that the direction of evolution—the sign of $\dot{\Delta}$—plays a central role in resonance capture. If $\dot{\Delta} < 0$, capture never occurs. If $\dot{\Delta} > 0$, the orbit may or may not be captured, although capture is certain if the initial eccentricity is small enough. If the orbit crosses the resonance, the eccentricity jumps to a larger value when $\dot{\Delta} < 0$ and otherwise to a smaller one.

From equation (6.43), $\dot{\Delta}$ depends on the parameters α, β, γ defined in equations (6.38). Of these, the fractional rate of change of α is largest since it is nearly zero near resonance. Moreover β is always negative. Thus the sign of $\dot{\Delta}$ is the opposite of the sign of $\dot{\alpha}$. From equations (6.31), (6.35) and (6.38), we have

$$\alpha = mn[1 + m - m(1 - e^2)^{1/2}]^{-3} - (m + 1)n_\mathrm{p}, \tag{6.74}$$

where n_p and n are the mean motions of the perturber and the test particle. For small eccentricities,

$$\alpha \simeq mn - (m + 1)n_\mathrm{p}. \tag{6.75}$$

For $m > 0$ resonance occurs when $n > n_\mathrm{p}$, so the test particle orbits interior to the perturber and $\dot{\alpha} < 0$ if $\dot{n} < 0$ or $\dot{n}_\mathrm{p} > 0$, which means that the ratio of semimajor axes n/n_p is approaching unity from above. On the other hand for $m < -1$ the test particle orbits exterior to the perturber, and $\dot{\alpha} <$

0 if n/n_p approaches unity from below. Slow changes of the semimajor axes of the test particle and/or the perturber that cause n/n_p to approach unity are called **convergent migration**, and when n/n_p recedes from unity we have **divergent migration**. We conclude that resonance capture from a low-eccentricity orbit requires convergent migration, and if the migration is convergent then capture is certain if the eccentricity is small enough.

This analysis is based on the assumption that the evolution of the Hamiltonian is slow, but how slow is slow enough? In particular we have shown that capture in resonance is certain if the initial eccentricity is zero, $\dot{\Delta} > 0$, and the evolution of $\Delta(t)$ is slow enough. We may then ask, what is the maximum value of $\dot{\Delta}$ such that capture is certain from an initially circular orbit? Since the Henrard–Lemaitre Hamiltonian (6.73) has no free parameters once $\Delta(t)$ is specified, we expect that this maximum is some number f of order unity, that is, capture is certain if $0 < \mathrm{d}\Delta/\mathrm{d}\tau < f$. Numerical integrations of Hamilton's equations—best done using equations (6.47) and (6.48)—show that $f = 8.2567$ (Friedland 2001; Quillen 2006). We can express this result in physical units using equation (6.43) for Δ, equations (6.38) for β and γ, equation (6.41) for t_0, and equation (6.75) for α. A zero-eccentricity orbit is always captured at resonance if

$$0 < \frac{\mathrm{d}}{\mathrm{d}t}[(m+1)n_\mathrm{p} - mn] < \frac{3^{2/3}|m|^{4/3} f \epsilon^{4/3} |H_m|^{4/3}}{4a^{4/3}}. \tag{6.76}$$

6.4 The Neptune–Pluto resonance

Pluto's perihelion and aphelion distances, 29.65 au and 49.30 au, span the semimajor axis of Neptune at 30.07 au. Therefore it is possible that Pluto could collide with Neptune. In one of the first long-term numerical integrations of the orbits of the planets, Cohen & Hubbard (1965) discovered that such collisions are avoided because Pluto is locked into a 3:2 mean motion resonance with Neptune.

The geometry of the resonance is shown in Figure 6.8, which plots the orbits of the outer planets and Pluto in a frame rotating with the mean motion of Neptune. The positions are projected onto the plane of the ecliptic,

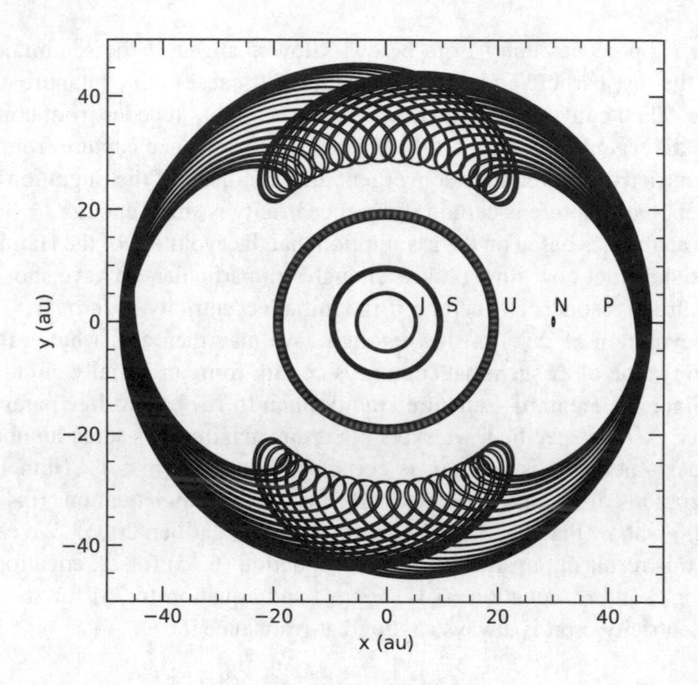

Figure 6.8: The orbits of Jupiter, Saturn, Uranus, Neptune and Pluto for 20 000 yr into the future, in a frame rotating with the mean motion of Neptune. The orbits have been projected onto the ecliptic plane. The Sun is at the origin.

represented by coordinates x and y, and the orientation of the x-axis is chosen such that Neptune lies near $x = 30\,\mathrm{au}$, $y = 0$. Over the 20 000 yr timespan shown in the figure, the perihelion of Pluto librates around mean positions 90° ahead of and 90° behind Neptune. Thus although the radial ranges of the orbits of Neptune and Pluto overlap, the two bodies never come close. Much longer integrations confirm that this resonance protects Pluto from close encounters with Neptune for the lifetime of the solar system (Malhotra & Williams 1997).

The resonance involves a slowly varying angle

$$\phi_\mathrm{s} = 3\lambda_\mathrm{P} - 2\lambda_\mathrm{N} - \varpi_\mathrm{P}, \tag{6.77}$$

where λ_P and ϖ_P are the mean longitude and longitude of perihelion of Pluto and λ_N is the mean longitude of Neptune. This angle librates around 180° with a period of about 20 000 yr and an amplitude of about 85°. When $\phi_\mathrm{s} = 180°$ and Pluto is at perihelion ($\lambda_\mathrm{P} = \varpi_\mathrm{P}$), then $2(\lambda_\mathrm{P} - \lambda_\mathrm{N}) = 180°$ so the two bodies are separated by approximately ±90° in azimuth, as seen in Figure 6.8.

We may treat Pluto as a test particle, since its mass is only about 10^{-4} Neptune masses. We assume for heuristic purposes that Pluto's eccentricity and inclination are small—this is not a very good assumption, since $e = 0.250$ and $I = 17.09°$, but it allows us to treat the dynamics analytically. The relevant term in the disturbing function is obtained from equation (4.98). We replace body 1 by Pluto and body 2 by Neptune; we drop the terms in the first line, since these are resonant only for bodies co-orbiting with Neptune; and we drop the terms in the third line, since these are proportional to the eccentricity of Neptune which is small ($e_\mathrm{N} = 0.0086$). We also set the ordering parameter $\epsilon = 1$. Then the term containing the resonant angle (6.77) corresponds to $m = -3$, and we have

$$H = -\frac{\mathbb{G}m_\mathrm{N}}{\Delta} = -\frac{\mathbb{G}m_\mathrm{N}}{a_\mathrm{N}} e_\mathrm{P}(2 - \tfrac{1}{2}\alpha D)b^2_{1/2}(\alpha)\cos\phi_\mathrm{s}, \tag{6.78}$$

where $\alpha = a_\mathrm{P}/a_\mathrm{N}$, the ratio of semimajor axes, and we have used the relation $b^{-2}_{1/2}(\alpha) = b^2_{1/2}(\alpha)$ (eq. 4.106). At the 3:2 resonance $\alpha = (\frac{3}{2})^{2/3} = 1.3104$, $b^2_{1/2} = 0.4712$ and $Db^2_{1/2} = \mathrm{d}b^2_{1/2}/\mathrm{d}\alpha = -1.4549$, so

$$H = -1.8957\frac{\mathbb{G}m_\mathrm{N}}{a_\mathrm{N}} e_\mathrm{P}\cos\phi_\mathrm{s}. \tag{6.79}$$

Analogous expressions for other resonances are given in Table 6.1.

In terms of the angle-action variables (6.31), the slow angle $\phi_\mathrm{s} = m_1\theta_1 + m_2\theta_2 - 2n_\mathrm{N}t - 2\lambda_{\mathrm{N},0}$ where $m_1 = 3$, $m_2 = 1$, n_N is the mean motion of Neptune and $\lambda_{\mathrm{N},0}$ is its mean longitude at $t = 0$. The slow and fast actions are given by equations (6.22),

$$J_{\rm s} = J_2 = (\mathbb{G}M_\odot a_{\rm P})^{1/2}[1-(1-e_{\rm P}^2)^{1/2}],$$
$$J_{\rm f} = J_1 - 3J_2 = (\mathbb{G}M_\odot a_{\rm P})^{1/2}[3(1-e_{\rm P}^2)^{1/2}-2]. \tag{6.80}$$

The Kepler Hamiltonian is $H_0 = -\frac{1}{2}(\mathbb{G}M_\odot)^2/J_1^2 = -\frac{1}{2}(\mathbb{G}M_\odot)^2/(J_{\rm f} + 3J_{\rm s})^2$ and the resonant Hamiltonian can be approximated by the pendulum Hamiltonian (6.27),

$$H_{\rm res} = \tfrac{1}{2}\alpha(\Delta J_{\rm s})^2 + \beta\cos\phi_{\rm s}, \tag{6.81}$$

with

$$\alpha = -\frac{27(\mathbb{G}M_\odot)^2}{J_{1,\rm res}^4}, \quad \beta = -1.8957\frac{\mathbb{G}m_{\rm N}}{a_{\rm N}}e_{\rm P}. \tag{6.82}$$

Here $J_{1,\rm res}$ is the action corresponding to exact resonance, given by $3n_{\rm P} = 3(\mathbb{G}M_\odot)^2/J_{1,\rm res}^3 = 2n_{\rm N}$. Since $\alpha < 0$ and $\beta < 0$, the stable equilibria are at $\phi_{\rm s} = \pm 180°$. The libration period is given by equation (6.5),

$$P = \frac{4}{\omega}K(k); \tag{6.83}$$

here $\omega^2 = |\alpha\beta|$ and $k = \sin\frac{1}{2}\phi_{\rm s,max}$, where $\phi_{\rm s,max}$ is the libration amplitude. Inserting values for Neptune's mass and semimajor axis, $m_{\rm N} = 5.151 \times 10^{-5}M_\odot$ and $a_{\rm N} = 30.0699$ au, and Pluto's current eccentricity $e_{\rm P} = 0.2502$, we find $\omega = 2.37\times10^{-11}\,{\rm s}^{-1}$. For a libration amplitude of 85° we have $k = 0.676$ and $P = 9.74\times10^3$ yr. This analytic calculation of the libration period yields about half of the correct value of 1.99×10^4 yr; the main reason for this disagreement is that Pluto's eccentricity is large enough that the expansion to $\mathrm{O}(e)$ in equation (6.79) is inadequate to represent the disturbing function (see Problem 6.5).

The width of the resonance is the difference between the largest and smallest value of the slow action in librating orbits and is given by (cf. eq. 6.29)

$$w = 4\left|\frac{\beta}{\alpha}\right|^{1/2} = 1.389\,n_{\rm N}a_{\rm N}^2\left(\frac{m_{\rm N}}{M_\odot}\right)^{1/2}e_{\rm P}^{1/2}. \tag{6.84}$$

Inserting values for the parameters, we have $w/(n_\mathrm{N}a_\mathrm{N}^2) = 0.0050$, but because of the error due to the use of the low-eccentricity approximation for the disturbing function, the correct value is closer to 0.002. The width is a rough measure of the fractional volume of phase space occupied by the resonance. It is highly improbable that Pluto formed by accident in this small volume, so we must explain why Pluto is found in the 3:2 resonance.

The most plausible answer (Malhotra 1993) is that Pluto was captured into the resonance during the late stages of planet formation, when Neptune's semimajor axis grew through planetesimal-driven migration (see the discussion at the end of §9.3). We can explore this process with a simple model for the evolution of Neptune and Pluto. We assume that (i) Neptune is born on a circular orbit with a semimajor axis a_N0, 6–10 au inside its current position at $a_\mathrm{N} = 30.07$ au; (ii) Neptune's orbit is always circular and expands slowly and smoothly from a_N0 to a_N; (iii) Pluto is born on a nearly circular orbit with a semimajor axis a_P0 that is outside the 3:2 resonance when Neptune is at its initial semimajor axis.

Since Pluto's orbit is nearly circular, capture in resonance is governed by the Henrard–Lemaitre Hamiltonian as described in §6.3.2. The migration of Neptune relative to Pluto is convergent, that is, the ratio of mean motions $n_\mathrm{P}/n_\mathrm{N}$ is approaching unity. As argued following equation (6.75), capture in resonance is then certain if Pluto's initial eccentricity is small enough and the evolution is slow enough.

After capture, Neptune continues to migrate outward, and Pluto's semimajor axis must increase as well to maintain the resonant ratio of mean motions. The fast action given by equation (6.80) is conserved during the evolution; thus as Pluto's semimajor axis grows, its eccentricity must grow as well. Conservation of the fast action, combined with our assumption that the initial eccentricity was nearly zero, enables us to compute Pluto's initial semimajor axis a_P0 from its current semimajor axis and eccentricity. We find $a_\mathrm{P0} = 32.3$ au. Capture occurred when Neptune migrated through semimajor axis $(\frac{2}{3})^{2/3}a_\mathrm{P0} = 24.6$ au.

These analytic arguments can be extended by numerical simulations. Figure 6.9, which follows Malhotra (1993), shows an N-body simulation of the evolution of Pluto's orbit in a solar system containing Jupiter, Saturn, Uranus and Neptune. The planets are assumed to form on circular, coplanar

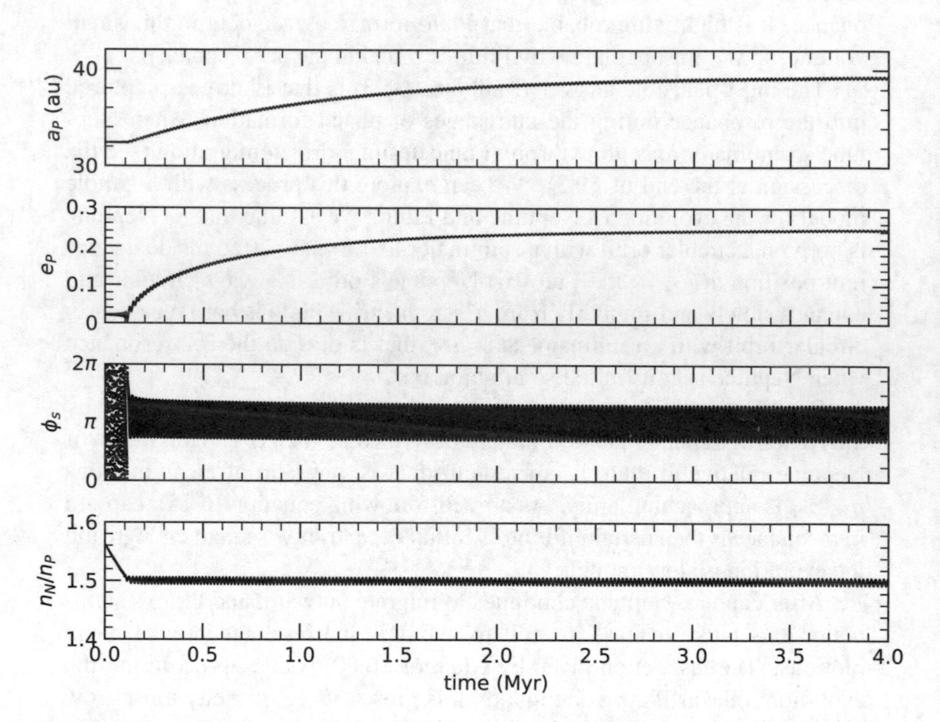

Figure 6.9: Evolution of Pluto's semimajor axis, eccentricity and resonant angle ϕ_s (eq. 6.77) when the giant planets migrate outward according to equation (6.85). The bottom panel shows the ratio of Neptune's mean motion to Pluto's.

orbits and then to migrate outward in semimajor axis following the simple model

$$a(t) = a_0 - \Delta a \exp(-t/\tau), \tag{6.85}$$

where a_0 is the planet's current semimajor axis and Δa, the migration range, is set to 0, 1, 3 and 6 au for Jupiter, Saturn, Uranus and Neptune, respectively. The migration timescale τ is set to 1 Myr. Pluto's initial semimajor axis is $a_{\rm P0} = 32.3$ au. The figure shows that Pluto is captured into the 3:2 resonance with Neptune at $t = 0.11$ Myr, after which it migrates outward such that its mean motion remains in resonance, while growing in eccentricity to conserve the fast action. The libration amplitude reflects the small but nonzero initial eccentricity of Pluto in this simulation ($e = 0.02$).

The discussion so far has focused on the case in which all of the giant planets and Pluto are coplanar. In fact Pluto's inclination of 17.1° is far larger than the inclination of any of the planets in the solar system, and it is natural to ask whether this inclination could also be generated by the resonance with Neptune. Much longer integrations than the one shown in Figure 6.8 show that Pluto exhibits a rich set of additional resonances, the most important of which involves Pluto's nodal longitude $\Omega_{\rm P}$, with slow angle $6\lambda_{\rm P} - 4\lambda_{\rm N} - 2\Omega_{\rm P}$. Pluto's inclination can be excited through this resonance, although for most initial conditions the excitation is somewhat less than the observed inclination.

Additional support for the hypothesis that Pluto was captured into the 3:2 resonance when Neptune migrated outward comes from the observation that many objects ("Plutinos") are also found in the 3:2 resonance (see §9.6). Like Pluto, these were presumably swept into the resonance when Neptune migrated, and their current eccentricities reflect their initial semimajor axes through conservation of the fast action. Additional objects are found in other resonances with Neptune; see Figure 9.10.

We must also ask why Neptune stopped migrating at its current semimajor axis of 30.07 au. The most likely explanation is that 30 au marked the outer boundary of the massive planetesimal disk that drove the migration.

Table 6.1: Resonance coefficients

m	$n_1 : n_2$	$\alpha = a_1/a_2$	F_m	G_m
1	$2:1$	0.629 96	1.190 49	−0.428 39
2	$3:2$	0.763 14	2.025 22	−2.484 01
3	$4:3$	0.825 48	2.840 43	−3.283 26
4	$5:4$	0.861 77	3.649 62	−4.083 71
−2	$1:2$	1.587 40	−0.269 87	0.749 96
−3	$2:3$	1.310 37	−1.895 65	1.545 53
−4	$3:4$	1.211 41	−2.710 27	2.344 72
−5	$4:5$	1.160 40	−3.519 23	3.145 15

The principal coefficients of the disturbing function at $(m+1):m$ resonances. Here n_2 and $n_1 = (1+1/m)n_2$ are the mean motions of the perturber and the perturbed particle, respectively. The constants F_m and G_m are defined in equations (6.89). They are related by equation (P.40).

6.5 Transit timing variations

A transiting planet in an isolated single-planet system obscures its host star at perfectly regular intervals. **Transit timing variations** or TTVs are periodic oscillations in the transit time induced by gravitational forces from another planet or a companion star. TTVs provide a powerful tool for measuring or constraining the masses and orbital properties of both the transiting planet and the perturber, whether or not it also transits (Agol et al. 2005; Holman & Murray 2005; Agol & Fabrycky 2018).

The TTV amplitude is enhanced if the planets are near a mean-motion resonance, so many of the systems with well measured TTVs are near resonance. The goal of this section is to describe the relation between TTVs and orbit dynamics in such systems (Lithwick et al. 2012).

We assume that the two planets are in nearly circular, coplanar orbits around a star of mass M. The planets are labeled 1 and 2, and initially we examine the transits of planet 1. The transiting planet is near resonance in the sense that its mean motion satisfies the relation

$$n_1 \simeq (1 + 1/m)n_2, \tag{6.86}$$

where m is an integer; the perturber is external to the transiting planet ($a_2 > a_1$ and $n_2 < n_1$) if $m > 0$ and internal if $m < -1$. The dominant terms in the perturbing Hamiltonian are given by equation (4.98) for the direct terms and (4.104) for the indirect ones:

$$\begin{aligned} H_2 = \frac{\mathbb{G}m_2}{a_2}\big\{\big[\tfrac{1}{2}\alpha\delta_{m,-2} + (m+1+\tfrac{1}{2}\alpha D)b_{1/2}^{m+1}\big]e_1\cos[m\lambda_1 - (m+1)\lambda_2 + \varpi_1] \\ + \big[2\alpha\delta_{m1} - (m+\tfrac{1}{2}+\tfrac{1}{2}\alpha D)b_{1/2}^{m}\big]e_2\cos[m\lambda_1 - (m+1)\lambda_2 + \varpi_2]\big\}, \end{aligned} \tag{6.87}$$

where λ_1 and λ_2 are the mean longitudes, ϖ_1 and ϖ_2 are the longitudes of periapsis, m_2 is the mass of the perturber, and δ_{ij} is the Kronecker delta (Appendix C.1). The Laplace coefficients $b_{1/2}^m$ (eq. 4.109) are functions of $\alpha = a_1/a_2$, and $D \equiv \mathrm{d}/\mathrm{d}\alpha$. Since the orbits are near resonance, $\alpha \simeq \alpha_{\rm res}(m) = (1+1/m)^{-2/3}$ and we can write

$$\begin{aligned} H_2 = \frac{\mathbb{G}m_2}{a_2}\big\{F_m e_1\cos[m\lambda_1 - (m+1)\lambda_2 + \varpi_1] \\ + G_m e_2\cos[m\lambda_1 - (m+1)\lambda_2 + \varpi_2]\big\}, \end{aligned} \tag{6.88}$$

where

$$\begin{aligned} F_m &\equiv \big[\tfrac{1}{2}\alpha\delta_{m,-2} + (m+1+\tfrac{1}{2}\alpha D)b_{1/2}^{m+1}\big]_{\alpha_{\rm res}}, \\ G_m &\equiv \big[2\alpha\delta_{m1} - (m+\tfrac{1}{2}+\tfrac{1}{2}\alpha D)b_{1/2}^{m}\big]_{\alpha_{\rm res}}. \end{aligned} \tag{6.89}$$

Values of these constants for the principal low-order resonances are given in Table 6.1.

The perturbing Hamiltonian (6.88) can be rewritten in terms of $k_i = e_i\cos\varpi_i$, $h_i = e_i\sin\varpi_i$, as

$$H_2 = \frac{\mathbb{G}m_2}{a_2}\left[F_m(k_1\cos\psi - h_1\sin\psi) + G_m(k_2\cos\psi - h_2\sin\psi)\right], \tag{6.90}$$

where $\psi \equiv m\lambda_1 - (m+1)\lambda_2$.

Since the eccentricities are small and we restrict ourselves to the case of coplanar motion, Hamilton's equations can be written in the form (1.193):

$$\frac{\mathrm{d}\lambda_1}{\mathrm{d}t} = n_1 + \frac{2}{n_1 a_1}\frac{\partial H_2}{\partial a_1}, \qquad \frac{\mathrm{d}a_1}{\mathrm{d}t} = -\frac{2}{n_1 a_1}\frac{\partial H_2}{\partial \lambda_1},$$
$$\frac{\mathrm{d}k_1}{\mathrm{d}t} = \frac{1}{n_1 a_1^2}\frac{\partial H_2}{\partial h_1}, \qquad \frac{\mathrm{d}h_1}{\mathrm{d}t} = -\frac{1}{n_1 a_1^2}\frac{\partial H_2}{\partial k_1}. \tag{6.91}$$

For the moment we drop the term involving $\partial H_2/\partial a_1$ (this approximation will be justified below). Then we have

$$\frac{\mathrm{d}a_1}{\mathrm{d}t} = \frac{2m\,\mathbb{G}m_2}{n_1 a_1 a_2}\left[F_m(k_1 \sin\psi + h_1\cos\psi) + G_m(k_2\sin\psi + h_2\cos\psi)\right],$$
$$\frac{\mathrm{d}\lambda_1}{\mathrm{d}t} = n_1, \quad \frac{\mathrm{d}k_1}{\mathrm{d}t} = -\frac{\mathbb{G}m_2 F_m}{n_1 a_1^2 a_2}\sin\psi, \quad \frac{\mathrm{d}h_1}{\mathrm{d}t} = -\frac{\mathbb{G}m_2 F_m}{n_1 a_1^2 a_2}\cos\psi. \tag{6.92}$$

We solve these equations using first-order perturbation theory, evaluating the right sides along the unperturbed trajectories on which the semimajor axes a_i, the mean motions $n_i = (\mathbb{G}M/a_i^3)^{1/2}$, and the eccentricity parameters k_i and h_i are constant, with values $\overline{a}_i$, $\overline{n}_i$, $\overline{k}_i$ and $\overline{h}_i$. The angle ψ along the unperturbed trajectory is denoted $\overline{\psi}$, and this changes at a rate

$$\frac{\mathrm{d}\overline{\psi}}{\mathrm{d}t} = m\overline{n}_1 - (m+1)\overline{n}_2 \equiv \Delta_{\mathrm{res}}. \tag{6.93}$$

The solutions of the last two equations are

$$k_1 = \overline{k}_1 + \delta k_1, \quad h_1 = \overline{h}_1 + \delta h_1, \tag{6.94}$$

where

$$\delta k_1 = \frac{\mathbb{G}m_2 F_m}{\overline{n}_1 \overline{a}_1^2 \overline{a}_2 \Delta_{\mathrm{res}}}\cos\overline{\psi}, \quad \delta h_1 = -\frac{\mathbb{G}m_2 F_m}{\overline{n}_1 \overline{a}_1^2 \overline{a}_2 \Delta_{\mathrm{res}}}\sin\overline{\psi}. \tag{6.95}$$

The constants $\overline{k}_1$ and $\overline{h}_1$ are sometimes referred to as components of the **free** or **proper eccentricity** since these can take on any value, independent of the actions of the perturber. The terms δk_1 and δh_1 are components of the **forced eccentricity**, which is proportional to the perturber mass m_2.

We next solve the equation for $\mathrm{d}a/\mathrm{d}t$ in (6.92). Again we work to first order in the perturber mass, which means that the eccentricity parameters k_i and h_i on the right side can be replaced by their unperturbed values $\overline{k}_i$ and $\overline{h}_i$. Thus $a_1 = \overline{a}_1 + \delta a_1$, where

$$\delta a_1 = \frac{2m\,\mathbb{G}m_2}{\overline{n}_1\overline{a}_1\overline{a}_2\Delta_{\rm res}}\big[F_m(\overline{h}_1\sin\overline{\psi} - \overline{k}_1\cos\overline{\psi}) + G_m(\overline{h}_2\sin\overline{\psi} - \overline{k}_2\cos\overline{\psi})\big]. \tag{6.96}$$

Finally, the equation for $\mathrm{d}\lambda_1/\mathrm{d}t = n_1 = (\mathbb{G}M/a_1^3)^{1/2}$ can be solved by replacing a_1 by $\overline{a}_1 + \delta a_1$, expanding the result to first order in the perturber mass, and then integrating. We find $\lambda = \lambda_{1,0} + \overline{n}_1 t + \delta\lambda_1$, where

$$\delta\lambda_1 = \frac{3m\,\mathbb{G}m_2}{\overline{a}_1^2\overline{a}_2\Delta_{\rm res}^2}\big[F_m(\overline{k}_1\sin\overline{\psi} + \overline{h}_1\cos\overline{\psi}) + G_m(\overline{k}_2\sin\overline{\psi} + \overline{h}_2\cos\overline{\psi})\big]. \tag{6.97}$$

This result shows that the perturbation to λ_1 is proportional to $\overline{e}_{1,2}m_2/\Delta_{\rm res}^2$ where $\overline{e}_i = (\overline{k}_i^2 + \overline{h}_i^2)^{1/2}$ is the free eccentricity; and it justifies our neglect of the term proportional to $\partial H_2/\partial a$ in the first of equations (6.91), since this term yields a perturbation proportional to $\overline{e}_{1,2}m_2/\Delta_{\rm res}$ which is much smaller for near-resonant planets, by of order $\Delta_{\rm res}/\overline{n}_1$.

The azimuth of planet 1 is $\phi_1 = f_1 + \varpi_1$, where f_1 is its true anomaly. Using the relation (1.151) between the true anomaly and the mean anomaly $\ell_1 = \lambda_1 - \varpi_1$, we have

$$\phi_1 = \lambda_1 + 2e_1\sin(\lambda_1 - \varpi_1) + \mathrm{O}(e_1^2) = \lambda_1 + 2k_1\sin\lambda_1 - 2h_1\cos\lambda_1 + \mathrm{O}(e_1^2). \tag{6.98}$$

The perturbation to the azimuth is $\delta\phi_1$ and to first order in the perturber mass m_2,

$$\begin{aligned}\delta\phi_1 = \delta\lambda_1[1 + 2\overline{k}_1\cos(\lambda_{1,0} + \overline{n}_1 t) + 2\overline{h}_1\sin(\lambda_{1,0} + \overline{n}_1 t)]\\ + 2\delta k_1\sin(\lambda_{1,0} + \overline{n}_1 t) - 2\delta h_1\cos(\lambda_{1,0} + \overline{n}_1 t).\end{aligned} \tag{6.99}$$

Since the eccentricity is small, the second and third terms in the square bracket are small compared to unity and can be dropped. The term involving $\delta\lambda_1$ is of order $\overline{e}_{1,2}m_2/\Delta_{\rm res}^2$ while the terms involving δk_1 and δh_1 are of order $m_2/\Delta_{\rm res}$; either set can dominate depending on whether the free eccentricities $\overline{e}_i$ are larger or smaller than the fractional distance from resonance $|\Delta_{\rm res}/n_0|$. Thus we arrive at

$$\delta\phi_1 = \delta\lambda_1 + 2\delta k_1 \sin(\lambda_{1,0} + \overline{n}_1 t) - 2\delta h_1 \cos(\lambda_{1,0} + \overline{n}_1 t). \tag{6.100}$$

The TTV is approximately $\delta t_1 = -\delta\phi_1/\overline{n}_1$. The transit occurs when the planet crosses a fixed value of ϕ, which we may take to be $\phi = 0$. Since $\phi = 0$ at transit, the mean longitude is also nearly zero, so when evaluating $\delta\phi_1$ to leading order we can set $\lambda_{1,0} + n_1 t = 0$. Using equations (6.95) and (6.97), we then have

$$\begin{aligned}\delta t_1 = \frac{1}{\overline{n}_1}(2\delta h_1 - \delta\lambda_1) = &-\frac{3m\,\mathbb{G}m_2}{\overline{n}_1\overline{a}_1^2\overline{a}_2\Delta_{\rm res}^2}\big[F_m(\overline{k}_1 \sin\overline{\psi} + \overline{h}_1 \cos\overline{\psi}) \\ &+ G_m(\overline{k}_2 \sin\overline{\psi} + \overline{h}_2 \cos\overline{\psi})\big] - \frac{2\,\mathbb{G}m_2}{\overline{n}_1^2\overline{a}_1^2\overline{a}_2\Delta_{\rm res}} F_m \sin\overline{\psi}.\end{aligned} \tag{6.101}$$

Using the resonance relation $a_1/a_2 = m^{2/3}/(m+1)^{2/3}$ and Kepler's law $\overline{n}_1^2\overline{a}_1^3 = \mathbb{G}M$, this result can be written as

$$\delta t_1 = \alpha_1 \cos\overline{\psi} + \beta_1 \sin\overline{\psi}, \tag{6.102}$$

where

$$\alpha_1 = -\frac{3m^{5/3}\overline{n}_1}{(m+1)^{2/3}\Delta_{\rm res}^2}\frac{m_2}{M}(F_m\overline{h}_1 + G_m\overline{h}_2), \tag{6.103}$$

$$\beta_1 = -\frac{3m^{5/3}\overline{n}_1}{(m+1)^{2/3}\Delta_{\rm res}^2}\frac{m_2}{M}(F_m\overline{k}_1 + G_m\overline{k}_2) - \frac{2m^{2/3}}{(m+1)^{2/3}\Delta_{\rm res}}\frac{m_2}{M}F_m.$$

The perturbations to the orbit of planet 2 are described by the Hamiltonian $H_1 = (m_1/m_2)H_2$, where H_2 is given by equation (6.90). Repeating the analysis from that point, we find that if planet 2 transits and is mainly influenced by planet 1, its TTV will be

$$\delta t_2 = \alpha_2 \cos\overline{\psi} + \beta_2 \sin\overline{\psi}, \tag{6.104}$$

where

$$\alpha_2 = \frac{3m\overline{n}_1}{\Delta_{\rm res}^2}\frac{m_1}{M}(F_m\overline{h}_1 + G_m\overline{h}_2), \tag{6.105}$$
$$\beta_2 = \frac{3m\overline{n}_1}{\Delta_{\rm res}^2}\frac{m_1}{M}(F_m\overline{k}_1 + G_m\overline{k}_2) - \frac{2}{\Delta_{\rm res}}\frac{m_1}{M}G_m.$$

The TTV signal is characterized by its amplitude $A_{\rm TTV}$ and period P_s, commonly called the **superperiod**:

$$P_s = \frac{2\pi}{|\Delta_{\rm res}|}, \quad A_{\rm TTV,1} = (\alpha_1^2 + \beta_1^2)^{1/2}, \quad A_{\rm TTV,2} = (\alpha_2^2 + \beta_2^2)^{1/2}. \tag{6.106}$$

We may now ask what we can determine about the system if both of the near-resonant planets transit. From the transits we know the mean motions n_1 and n_2, and the index m is the integer such that equation (6.86) is approximately satisfied. Once we know these three quantities, we can determine $\Delta_{\rm res} = mn_1 - (m+1)n_2$ and the expected superperiod (6.106). If the superperiod derived in this way agrees with the periods of the TTVs from each planet, then we have confirmed that their resonant interactions dominate the TTV signal.

Measurement of the amplitudes and phases of the two TTVs then yields the parameters α_1, β_1, α_2 and β_2. These can be used to determine the four unknown quantities in equations (6.103) and (6.105), the two mass ratios m_1/M and m_2/M, and the combinations $F_m\overline{k}_1 + G_m\overline{k}_2$ and $F_m\overline{h}_1 + G_m\overline{h}_2$.

An important special case is when the free eccentricities are negligible, as we expect if the planets formed in an environment with dissipation from a gas disk. Then $\alpha_1 = \alpha_2 = 0$ and the amplitudes β_1 and β_2 directly yield the mass ratios m_1/M and m_2/M. The converse is not correct—the observation that $\alpha_1 = \alpha_2 = 0$ does not imply that the free eccentricities vanish—but if this condition is satisfied in most double TTV systems, it is probably safe to assume that most of them have zero free eccentricity.

As an example, the system Kepler-18 contains three planets (Cochran et al. 2011; Lithwick et al. 2012). The outer two planets, c and d, are close to a $2:1$ resonance and exhibit TTVs. The periods are $P_c = 7.641\,596$ d and $P_d = 14.8589$ d, so the expected superperiod is $P_{\rm s} = 267.6$ d, consistent with the observed periods of the TTVs in both planets to within ~ 1%. The amplitude ratios $|\alpha_1/\beta_1| \lesssim 0.15$ and $|\alpha_2/\beta_2| \lesssim 0.28$, so we may assume that the free eccentricities are much less than the forced eccentricities. The amplitudes are $A_{\mathrm{TTV},1} = 0.0037 \pm 0.0003\,\mathrm{d}$ and $A_{\mathrm{TTV},2} = 0.0028 \pm 0.0003\,\mathrm{d}$. The derived planet masses are $m_c = 19 \pm 1 M_\oplus$ and $m_d = 25 \pm 3 M_\oplus$, assuming a stellar mass $M = 0.97 M_\odot$.

6.6 Secular resonance

The dynamics of secular resonances can be illustrated using Lagrange–Laplace theory (§5.2). We consider a system of N planets and one zero-mass test particle representing an asteroid. The trajectory of the test particle is described by equations (5.35) and (5.36). The complex eccentricity $z_e = k + \mathrm{i}h = e\exp(\mathrm{i}\varpi)$ and the complex inclination $z_I = q + \mathrm{i}p = I\exp(\mathrm{i}\Omega)$ are the sum of a free part that circulates at frequency g for z_e and $-g$ for z_I, and a forced part that is the sum of N terms, each circulating at one of the eigenfrequencies of the N-planet Lagrange–Laplace system. The forced eccentricity or inclination diverges at the secular resonances, where the frequency g or $-g$ is equal to one of the eigenfrequencies (see eq. 5.39).

Figure 6.10 shows the precession frequency as a function of semimajor axis in the solar system, along with the eigenfrequencies from the third column of Table 5.1. Within the asteroid belt, the two most important resonances occur when the precession frequency equals $g_6 = 28.245\,\mathrm{arcsec\ yr^{-1}}$ and $f_6 = -26.348\,\mathrm{arcsec\ yr^{-1}}$. These are sometimes referred to as the ν_6 and ν_{16} resonances (the convention that associates these labels with the eigenfrequencies is $\nu_i = g_i$, $\nu_{10+i} = f_i$, $i = 1, \ldots, 8$).

More accurate treatments that include terms of higher degree in the eccentricities and inclinations and higher order in the masses show that the secular resonances occupy surfaces in the 3-dimensional space with coordinates a, e, and I (Williams & Faulkner 1981; Knežević et al. 1991), and this structure is reflected in the distribution of asteroids, probably because the resonance excites eccentricities that are large enough to allow the asteroid to collide with Mars or have a close encounter with Jupiter.

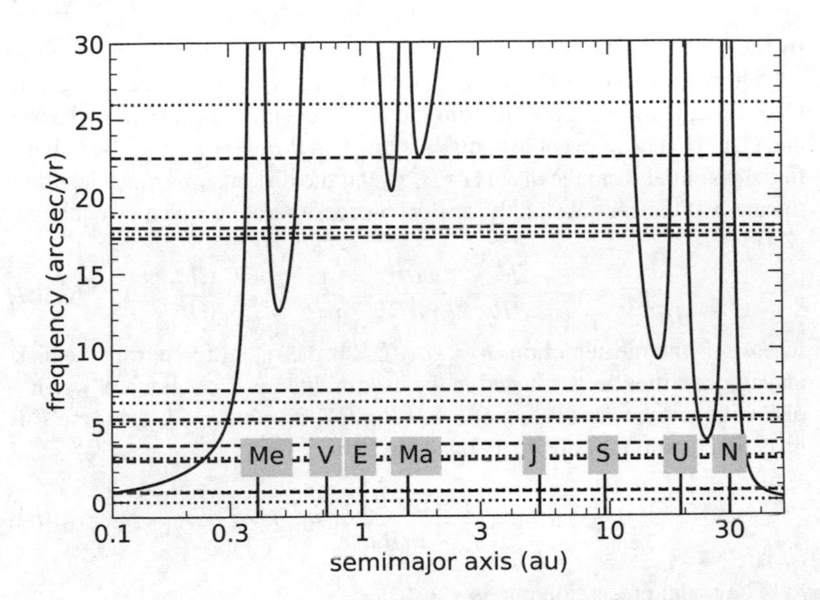

Figure 6.10: The apsidal and nodal precession rates of a test particle as a function of semimajor axis in the solar system. The horizontal lines mark the eigenfrequencies g_j and $-f_j$ from the third column of Table 5.1, where $g_j > 0$ are the apsidal frequencies (dashed lines) and $f_j < 0$ are the nodal frequencies (dotted lines). Secular eccentricity resonances occur when the free precession frequency $g = g_j$, and secular inclination resonances occur when $-g = f_j$.

6.6.1 Resonance sweeping

Planets or planetesimals embedded in a protoplanetary disk experience apsidal and nodal precession due to the gravitational field from the disk. Although the mass of the disk is much smaller than the mass of the host star, the Kepler potential from the star induces no precession, so even a low-mass disk can dominate both free precession rates: in effect, when a disk is present the precession frequencies of the periapsis and the node in equations (5.37)—$\dot{\varpi} = g$ and $\dot{\Omega} = -g$—are determined mainly by the disk rather than by the other planets.

The precession rates for a nearly circular orbit can be derived either from Gauss's equations (1.200) or from the epicycle equations (1.161), (1.168), and (1.174). The relevant formulas are given in Problem 1.13. If the gravitational potential from the disk is $\phi(R, z)$ and the disk mass is much less than the mass of the central star, the apsidal precession rate is given by equations (P.4),

$$\dot{\varpi} = -\frac{1}{nR}\frac{\partial\phi}{\partial R} - \frac{1}{2n}\frac{\partial^2\phi}{\partial R^2} = \frac{1}{nR}F + \frac{1}{2n}\frac{\mathrm{d}F}{\mathrm{d}R}, \tag{6.107}$$

where n is the mean motion, $F = -\partial\phi/\partial R$ is the radial force from the disk, and all quantities are evaluated in the disk midplane $z = 0$. For a power-law disk with surface density $\Sigma(R) = \Sigma_0(R_0/R)^{k+1}$, $-1 < k < 1$, the force F is given in equation (e) of Box 6.2 and we have

$$\dot{\varpi} = -(1-k)\frac{\mathbb{G}\Sigma_0 C(k)}{2nR_0}(R_0/R)^{k+2}. \tag{6.108}$$

The nodal precession rate is

$$\dot{\Omega} = \frac{1}{2nR}\frac{\partial\phi}{\partial R} - \frac{1}{2n}\frac{\partial^2\phi}{\partial z^2}, \tag{6.109}$$

with all quantities evaluated in the midplane. If the disk is thin the second term will be much larger than the first, and Poisson's equation (B.44) can be approximated by its 1-dimensional form $\partial^2\phi/\partial z^2 = 4\pi\mathbb{G}\rho(R)$, where $\rho(R)$ is the density of the disk in the midplane. Thus

$$\dot{\Omega} = -\frac{2\pi\mathbb{G}\rho(R)}{n}. \tag{6.110}$$

The simplest model of the Sun's protoplanetary disk is constructed as follows. The mass of solid material or heavy elements belonging to each planet in the solar system is spread out into an annulus around the Sun reaching halfway to the adjacent planets. Then the mass in each annulus is augmented by adding gas until the annulus has the same composition as the Sun, and a smooth surface-density profile is fit to the result. This recipe gives the

Box 6.2: The gravitational potential of a disk

To find the apsidal precession rate due to a disk, we must first determine the gravitational potential and force arising from a disk-like mass distribution. There are many ways to do this with varying degrees of generality (Binney & Tremaine 2008) and we focus on the simplest. Consider **Kuzmin's potential**

$$\Phi_{\mathrm{Kuz}}(R, z) = -\frac{\mathbb{G}M}{[R^2 + (a + |z|)^2]^{1/2}}, \quad a > 0, \tag{a}$$

with R and z the usual cylindrical coordinates. For $z < 0$, $\Phi_{\mathrm{Kuz}}(R, z)$ equals the potential of a point mass M located at $(R, z) = (a, 0)$. Similarly when $z > 0$, $\Phi_{\mathrm{Kuz}}(R, z)$ equals the potential of a mass M at $(R, z) = (-a, 0)$. Hence $\nabla^2\Phi_{\mathrm{Kuz}}(R, z)$ vanishes everywhere except on the plane $z = 0$, and by Poisson's equation (B.44) $\Phi_{\mathrm{Kuz}}(R, z)$ must therefore be generated by a mass distribution confined to this plane. By applying Gauss's theorem to a volume containing a small portion of the plane, we find that Φ_{Kuz} is generated by a surface density

$$\Sigma_{\mathrm{Kuz}}(R, z) = \frac{aM}{2\pi(R^2 + a^2)^{3/2}}, \tag{b}$$

known as **Kuzmin's disk**. The corresponding radial force in the disk plane is

$$F_{\mathrm{Kuz}} = -\frac{\partial\Phi_{\mathrm{Kuz}}}{\partial R}\bigg|_{z=0} = -\frac{\mathbb{G}MR}{(R^2 + a^2)^{3/2}}. \tag{c}$$

More general models can be constructed by combining a continuous distribution of Kuzmin disks of different masses and scale lengths:

$$\Sigma(R) = \frac{1}{2\pi}\int_0^\infty \frac{\mathrm{d}a\, aM(a)}{(R^2 + a^2)^{3/2}}, \quad \Phi(R, z) = -\mathbb{G}\int_0^\infty \frac{\mathrm{d}a\, M(a)}{[R^2 + (a + |z|)^2]^{1/2}}, \tag{d}$$

where $M(a)$ is now an arbitrary function. For example, we can construct disks with a power-law surface density distribution by setting $M(a) \propto a^{-k}$ with $-1 < k < 1$. We find the surface-density and force pair[a]

$$\Sigma(R) = \Sigma_0(R_0/R)^{k+1}, \quad F(R) = -\mathbb{G}\Sigma_0 C(k)(R_0/R)^{k+1}, \tag{e}$$

where

$$C(k) \equiv 2\pi\frac{\Gamma(\frac{1}{2} - \frac{1}{2}k)\Gamma(1 + \frac{1}{2}k)}{\Gamma(1 - \frac{1}{2}k)\Gamma(\frac{1}{2} + \frac{1}{2}k)}, \tag{f}$$

and $\Gamma(z)$ is the gamma function described in Appendix C.3.

[a] The potential is only well defined over the smaller range $0 < k < 1$.

minimum-mass solar nebula (Weidenschilling 1977; Hayashi 1981),

$$\Sigma(R) = 1.7 \times 10^3 \text{ g cm}^{-2} \left(\frac{1 \text{ au}}{R}\right)^{1.5}. \tag{6.111}$$

Given this surface density, the thickness and therefore the midplane density can be estimated by assuming that the gas is in thermal equilibrium with the solar radiation (Hayashi 1981):

$$\rho(R) = 1.4 \times 10^{-9} \text{ g cm}^{-3} \left(\frac{1 \text{ au}}{R}\right)^{2.75}. \tag{6.112}$$

The corresponding precession rates are

$$\dot{\varpi} = -8.52 \times 10^2 \text{ arcsec yr}^{-1} \left(\frac{1 \text{ au}}{R}\right),$$
$$\dot{\Omega} = -1.92 \times 10^4 \text{ arcsec yr}^{-1} \left(\frac{1 \text{ au}}{R}\right)^{1.25}. \tag{6.113}$$

Notice that both the apsidal and nodal precession rates due to the disk are negative.[6] Thus as the disk disperses, the free precession rates for ϖ and Ω both approach zero from below, then stabilize at positive and negative values respectively once they are dominated by the gravitational fields of the planets rather than the disk. During this evolution, the secular resonances can sweep through a large fraction of the planetary system.

To examine how this process of **resonance sweeping** affects the eccentricities and inclinations of small bodies, we write the first of equations (5.38) in a simplified form involving only one resonance:

$$\frac{\mathrm{d}z_e}{\mathrm{d}t} = \mathrm{i}g(t)z_e + \nu_n \exp(\mathrm{i}g_n t), \tag{6.114}$$

where ν_n is a complex constant and g_n is one of the Lagrange–Laplace eigenfrequencies. We assume that g_n is constant, although in practice the

[6] Although if the planet opens a gap in the disk, as described in §3.6, the apsidal precession rate due to the disk mass will be positive.

eigenfrequencies are likely to evolve as the disk disperses. This equation can be solved to give

$$z_e(t) = \nu_n \exp[\mathrm{i}\phi(t)] \int_{-\infty}^{t} \mathrm{d}t' \exp[\mathrm{i}g_n t' - \mathrm{i}\phi(t')], \tag{6.115}$$

where

$$\phi(t) = \int^{t} \mathrm{d}t'\, g(t'); \tag{6.116}$$

in this equation we have assumed that the eccentricity $|z_e(t)|$ was zero in the distant past.

Resonance occurs at a time $t = t_0$ when $g(t_0) = g_n$, and we can approximate $\phi(t)$ in the vicinity of the resonance by $\phi(t) = \phi_0 + g_n t + \frac{1}{2}\dot{g}(t-t_0)^2$, where $\dot{g} = \mathrm{d}g/\mathrm{d}t|_{t_0}$. Then equation (6.116) becomes

$$z_e(t) = \nu_n \exp[\mathrm{i}g_n t + \tfrac{1}{2}\mathrm{i}\dot{g}(t-t_0)^2] \int_{-\infty}^{t-t_0} \mathrm{d}\tau \exp(-\mathrm{i}\dot{g}\tau^2). \tag{6.117}$$

Long after resonance crossing, the eccentricity is

$$|z_e(t \to \infty)| = |\nu_n| \left| \int_{-\infty}^{\infty} \mathrm{d}\tau \exp(-\mathrm{i}\dot{g}\tau^2) \right| = \left(\frac{\pi}{2|\dot{g}|} \right)^{1/2} |\nu_n|. \tag{6.118}$$

Thus the secular resonance excites the eccentricities and/or inclinations of small bodies and planets as it sweeps through the system. Slower dispersal (smaller $|\dot{g}|$) excites larger eccentricities and inclinations.

Resonance sweeping has a wide variety of possible effects, still only partly explored. By exciting the eccentricities and inclinations of small bodies, it may inhibit planet formation and even clear a wide gap in a planetesimal disk if high-velocity collisions grind the planetesimals into dust. It may trap planets in secular resonances in which their nodes or apsidal lines are locked together. Although the semimajor axes of the planets and planetesimals are conserved by secular resonances, subsequent damping of the eccentricities may lead to radial migration of the planetary orbits.

Chapter 7

Planetary spins

7.1 Precession of planetary spins

Tidal torques exerted on the equatorial bulge of a planet cause the planet's spin to precess. The most common sources of these torques are the planet's host star or a massive satellite. For example, the Earth's spin precesses with a period of about 25 700 yr due to torques from the Moon and Sun. Traditionally this effect is called the **precession of the equinoxes**: the equinoxes are the points on the celestial sphere where the plane perpendicular to the Earth's orbital angular momentum (the ecliptic) intersects the plane perpendicular to its spin (the equator).

We first consider a rotating, flattened planet with no satellites that orbits a host star of mass M_*. We assume that the host star is sufficiently far away that it can be approximated as a point mass. At large distances, the non-spherical component of the gravitational field from the planet can be written (eq. 1.135)

$$\Phi(\mathbf{r}) = \frac{\mathbb{G}M_\mathrm{p}J_2R_\mathrm{p}^2}{2r^5}[3(\hat{\mathbf{S}}\cdot\mathbf{r})^2 - r^2], \tag{7.1}$$

where M_p, R_p and J_2 are the mass, radius and quadrupole moment of the planet, and $\hat{\mathbf{S}}$ is a unit vector parallel to its spin angular momentum $\mathbf{S}$, which is assumed to coincide with its symmetry axis (for reasons given in Box 7.1). If the host star is located at position $\mathbf{r}$ relative to the center of the planet, the torque exerted by the planet on the star is

$$\mathbf{N} = -M_*\mathbf{r} \times \nabla\Phi = \frac{3\,\mathbb{G}M_*M_\mathrm{p}J_2R_\mathrm{p}^2}{r^5}(\hat{\mathbf{S}}\cdot\mathbf{r})\hat{\mathbf{S}}\times\mathbf{r}. \tag{7.2}$$

The total angular momentum residing in the motions of the centers of mass of the planet and star is the sum of two components (eq. 1.9). The first is $(M_* + M_\mathrm{p})\mathbf{r}_\mathrm{cm} \times \dot{\mathbf{r}}_\mathrm{cm}$, where $\mathbf{r}_\mathrm{cm}$ is the position of the barycenter of the system in an inertial frame; this angular momentum is fixed if there are no external forces, so we can ignore it from now on. The second[1] is $\mathbf{L} = \mu\mathbf{r}\times\dot{\mathbf{r}}$, where $\mu = M_*M_\mathrm{p}/(M_* + M_\mathrm{p})$ is the reduced mass of the star and planet. If the star and planet are in a Kepler orbit with semimajor axis a and eccentricity e, then $|\mathbf{L}| = \mu[\,\mathbb{G}(M_* + M_\mathrm{p})a(1-e^2)]^{1/2}$ (eq. 1.28).

Now erect a Cartesian coordinate system in which the positive z-axis is parallel to $\mathbf{L}$ and the positive x-axis points toward periapsis. Thus $\hat{\mathbf{z}} = \mathbf{L}/|\mathbf{L}| = \hat{\mathbf{L}}$ and the position of the star can be written $\mathbf{r} = \hat{\mathbf{x}}\cos f + \hat{\mathbf{y}}\sin f$, where f is the true anomaly of the star relative to the planet. Then orbit-averaging equation (7.2) using equations (1.66c)–(1.66e) gives

$$\langle\mathbf{N}\rangle = -\frac{3\,\mathbb{G}M_*M_\mathrm{p}J_2R_\mathrm{p}^2}{2a^3(1-e^2)^{3/2}}(\hat{\mathbf{S}}\cdot\hat{\mathbf{L}})\hat{\mathbf{S}}\times\hat{\mathbf{L}}. \tag{7.3}$$

The torque exerted on the planet is $-\langle\mathbf{N}\rangle$, so the orbit-averaged equations of motion for the spin and orbital angular momenta are

$$\frac{\mathrm{d}\mathbf{L}}{\mathrm{d}t} = \langle\mathbf{N}\rangle, \quad \frac{\mathrm{d}\mathbf{S}}{\mathrm{d}t} = -\langle\mathbf{N}\rangle. \tag{7.4}$$

Equations (7.4) imply that $\mathrm{d}(\mathbf{L}+\mathbf{S})/\mathrm{d}t = \mathbf{0}$ so the total angular momentum $\mathbf{L} + \mathbf{S}$ is conserved, as expected. Moreover $\mathbf{L}\cdot\langle\mathbf{N}\rangle = 0$ so $\mathbf{L}\cdot\mathrm{d}\mathbf{L}/\mathrm{d}t = \frac{1}{2}\mathrm{d}L^2/\mathrm{d}t = 0$, where $L = |\mathbf{L}|$ is the scalar orbital angular momentum.

[1] There are two differences in notation from earlier chapters: (i) the symbol $\mathbf{L}$ denotes angular momentum rather than angular momentum per unit mass; (ii) the relative position vector $\mathbf{r}$ points from the planet to the star rather than from the star to the planet.

We conclude that the torque changes the *direction* of the orbital angular-momentum vector but not its magnitude. Similarly, the torque changes the direction of the spin angular-momentum vector but not its magnitude. Moreover it is straightforward to show from equations (7.3) and (7.4) that $\mathrm{d}(\mathbf{L}\cdot\mathbf{S})/\mathrm{d}t = 0$; since the magnitudes of **L** and **S** are conserved we conclude that the angle between **L** and **S** (the **obliquity**, denoted by ϵ) also remains constant.

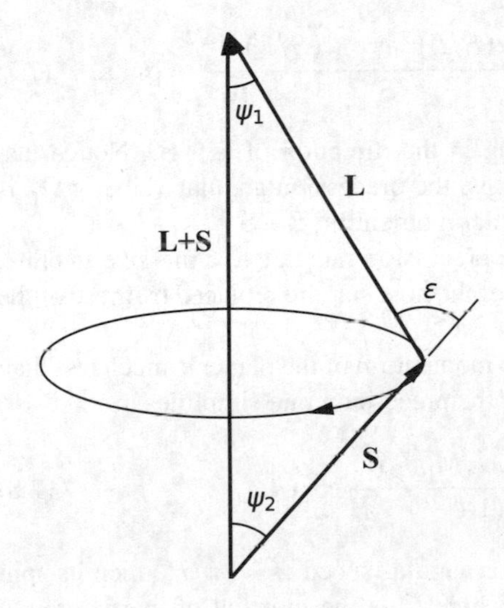

Figure 7.1: Precession of a planet's spin and orbital angular momenta, **S** and **L**, due to the torque between the host star and the planet's equatorial bulge. The total angular momentum **L** + **S** is constant. The angle ϵ is the obliquity.

The geometry is illustrated in Figure 7.1. The spin and orbital angular momenta precess around the fixed vector **L** + **S**; the lengths $L = |\mathbf{L}|$ and $S = |\mathbf{S}|$ remain fixed, as do the angles ψ_1, ψ_2 and ϵ. From equations (7.3) and (7.4) we have

$$\begin{aligned}\frac{\mathrm{d}\mathbf{S}}{\mathrm{d}t} &= \frac{3\,\mathbb{G}M_*M_\mathrm{p}J_2R_\mathrm{p}^2}{2a^3(1-e^2)^{3/2}LS}(\hat{\mathbf{S}}\cdot\hat{\mathbf{L}})\mathbf{S}\times\mathbf{L} \\ &= \frac{3\,\mathbb{G}M_*M_\mathrm{p}J_2R_\mathrm{p}^2}{2a^3(1-e^2)^{3/2}LS}(\hat{\mathbf{S}}\cdot\hat{\mathbf{L}})\mathbf{S}\times(\mathbf{L}+\mathbf{S})\end{aligned}$$

$$= \boldsymbol{\Omega}_{\rm p} \times \mathbf{S}, \tag{7.5}$$

where

$$\boldsymbol{\Omega}_{\rm p} = -\frac{3\,\mathbb{G}M_* M_{\rm p} J_2 R_{\rm p}^2}{2a^3(1-e^2)^{3/2} LS} \cos\epsilon\,(\mathbf{L}+\mathbf{S}) \tag{7.6}$$

is the angular speed or precession rate of the spin vector (cf. eq. D.16). Using the law of cosines, this can be rewritten as

$$\boldsymbol{\Omega}_{\rm p} = -\frac{3\,\mathbb{G}M_* M_{\rm p} J_2 R_{\rm p}^2 [1 + 2(S/L)\cos\epsilon + (S/L)^2]^{1/2}}{2a^3(1-e^2)^{3/2} S} \cos\epsilon\,\hat{\mathbf{a}}, \tag{7.7}$$

where $\hat{\mathbf{a}}$ is a unit vector pointing in the direction of $\mathbf{L} + \mathbf{S}$. Notice that the precession is *retrograde*, that is, the precession angular velocity $\boldsymbol{\Omega}_{\rm p}$ is antiparallel to the conserved angular momentum $\mathbf{L} + \mathbf{S}$.

Equation (7.7) also yields the precession rate due to a massive satellite, if the parameters M_*, L, a and e of the host star are replaced by those of the satellite.

In most cases the spin angular momentum of the planet is much less than its orbital angular momentum, so the precession rate simplifies to

$$\boldsymbol{\Omega}_{\rm p} = -\frac{3\,\mathbb{G}M_* M_{\rm p} J_2 R_{\rm p}^2}{2a^3(1-e^2)^{3/2} S} \cos\epsilon\,\hat{\mathbf{L}}. \tag{7.8}$$

If the planet has spin period $P_{\rm s}$ or angular speed $\omega = 2\pi/P_{\rm s}$, then its spin angular momentum is $S = C\omega$ where C is the moment of inertia around the polar axis (eqs. 1.133 and D.91). The dimensionless ratio $C/(M_{\rm p}R_{\rm p}^2)$ is sometimes called the **moment of inertia factor**; this factor is $\frac{2}{5}$ for a homogeneous planet and smaller in the usual case where the density increases toward the planet's center (moment of inertia factors for the solar-system planets are given in Appendix A). Then

$$\boldsymbol{\Omega}_{\rm p} = -\frac{3\,\mathbb{G}M_* J_2 P_{\rm s}}{4\pi a^3(1-e^2)^{3/2}} \frac{M_{\rm p}R_{\rm p}^2}{C} \cos\epsilon\,\hat{\mathbf{L}} \tag{7.9a}$$

$$\simeq -\frac{3\pi J_2 P_{\rm s}}{P^2(1-e^2)^{3/2}} \frac{M_{\rm p}R_{\rm p}^2}{C} \cos\epsilon\,\hat{\mathbf{L}}; \tag{7.9b}$$

in the second of these equations we have replaced $\mathbb{G}M_*/a^3$ by $4\pi^2/P^2$ where P is the orbital period of the planet, an approximation that is valid when $M_\mathrm{p} \ll M_*$ (eq. 1.43). An alternative expression replaces J_2 by the difference between the equatorial and polar moments of inertia, using equation (1.134):

$$\mathbf{\Omega}_\mathrm{p} = -\frac{3\,\mathbb{G}M_* P_\mathrm{s}}{4\pi a^3(1-e^2)^{3/2}}\frac{C-A}{C}\cos\epsilon\,\hat{\mathbf{L}}. \tag{7.10}$$

Thus a measurement of the precession rate of the spin axis of the planet yields the combination of moments of inertia $(C-A)/C$, called the **dynamical ellipticity**. Similarly, measurements of the apsidal and nodal precession rates of satellites of the planet measure its quadrupole moment or $C - A$, so the two measurements can be combined to determine both moments of inertia of an axisymmetric planet.

As an example, we calculate the rate of precession of the Earth's spin. First consider the precession due to the Sun. The Earth's spin angular momentum is far smaller than its orbital angular momentum, so we can use expression (7.9a) for the precession rate. From Appendix A, we have $\mathbb{G}M_* = \mathbb{G}M_\odot = 1.327\times10^{20}\,\mathrm{m^3\,s^{-2}}$, $a = 1.496\times10^{11}$ m, $e = 0.0167$, $J_2 = 0.001\,082\,6$, $P_\mathrm{s} = 86\,164.1$ s and $\epsilon = 23.44°$. These yield $|\mathbf{\Omega}_\mathrm{p}| = 8.1028\times10^{-13}\,\mathrm{s^{-1}}M_\oplus R_\oplus^2/C$.

Calculating the precession due to the Moon is more complicated. In contrast to the Sun, the Moon's orbital angular momentum is only about five times the Earth's spin angular momentum. Thus, if the Earth and Moon were an isolated system we would use equation (7.7) to compute the precession rate. However, the torques on the lunar orbit due to the Sun are far larger than the torques due to the Earth's equatorial bulge. As a result, the Moon's angular-momentum vector precesses rapidly around the normal to the ecliptic, with a period of only 18.6 yr. To compute the torque that the Moon exerts on the Earth's equatorial bulge, we must average equation (7.3) over this period. To do so, we write $\hat{\mathbf{S}}$ and $\hat{\mathbf{L}}$ in Cartesian coordinates (x, y, z), with the z-axis normal to the ecliptic. Then the averages of the components of $\hat{\mathbf{L}}$ are $\langle\hat{L}_x\rangle = \langle\hat{L}_y\rangle = \langle\hat{L}_x\hat{L}_y\rangle = 0$, $\langle\hat{L}_x^2\rangle = \langle\hat{L}_y^2\rangle = \frac{1}{2}\sin^2 I$ and $\hat{L}_z = \cos I$, where I is the inclination of the lunar orbit to the ecliptic. From these results it is straightforward to show that $\langle(\hat{\mathbf{S}}\cdot\hat{\mathbf{L}})\hat{\mathbf{S}}\times\hat{\mathbf{L}}\rangle = (1-\frac{3}{2}\sin^2 I)(\hat{\mathbf{S}}\cdot\hat{\mathbf{z}})\hat{\mathbf{S}}\times\hat{\mathbf{z}}$, so equation (7.3) is modified to

$$\langle\!\langle \mathbf{N} \rangle\!\rangle = -\frac{3\,\mathbb{G}M_{☽}M_\oplus J_2 R_\oplus^2}{2a^3(1-e^2)^{3/2}}(1-\tfrac{3}{2}\sin^2 I)(\hat{\mathbf{S}}\cdot\hat{\mathbf{z}})\hat{\mathbf{S}}\times\hat{\mathbf{z}}, \tag{7.11}$$

where $M_{☽}$ denotes the lunar mass and the orbital elements now refer to the lunar orbit. The corresponding precession rate is

$$\boldsymbol{\Omega}_\mathrm{p} = -\frac{3\,\mathbb{G}M_{☽}J_2 P_\mathrm{s}}{4\pi a^3(1-e^2)^{3/2}}\frac{M_\oplus R_\oplus^2}{C}(1-\tfrac{3}{2}\sin^2 I)\cos\epsilon\,\hat{\mathbf{z}}. \tag{7.12}$$

From Appendix A, we have $\mathbb{G}M_{☽} = 4.903\times10^{12}\,\mathrm{m}^3\,\mathrm{s}^{-2}$, $a = 384\,400$ km, $e = 0.0549$, $I = 5.145°$, and we find $|\boldsymbol{\Omega}_\mathrm{p}| = 1.7503\times10^{-12}\,\mathrm{s}^{-1}M_\oplus R_\oplus^2/C$, about twice as large as the contribution from the Sun.

We add the contributions of the Moon and Sun to obtain the total precession rate, $|\boldsymbol{\Omega}_\mathrm{p}| = 2.5606\times10^{-12}\,\mathrm{s}^{-1}M_\oplus R_\oplus^2/C$. In practice this estimate is combined with the observed precession rate,[2] $|\boldsymbol{\Omega}_\mathrm{p}| = 7.7405\times10^{-12}\,\mathrm{s}^{-1}$, to determine the moment of inertia C and thus constrain the properties of the Earth's interior. We find $C/(M_\oplus R_\oplus^2) = 0.3308$, close to the accepted value of 0.3307 (Appendix A).

This analysis illustrates that the precession of a planet can be strongly influenced by gravitational interactions with its satellites, and we now turn to a more complete description of these interactions.

7.1.1 Precession and satellites

If a satellite orbit is fixed in inertial space, its inclination relative to the equator of its host planet varies as the planetary spin precesses. For example,

[2] In the literature the precession rate is variously defined in two different frames: (i) an inertial frame whose polar axis is perpendicular to the plane of the Earth's orbit on the date J2000.0, and (ii) a non-inertial frame whose polar axis is always perpendicular to the plane of the Earth's orbit, which varies slowly with time due to secular perturbations from the other planets. We work in the inertial frame, where $|\boldsymbol{\Omega}_\mathrm{p}| = 7.7405\times10^{-12}\,\mathrm{s}^{-1}$, while in the non-inertial frame the rate is $7.7257\times10^{-12}\,\mathrm{s}^{-1}$. These rates are known to many more significant digits than quoted here; in more traditional units they correspond to 50.384 815 07 and 50.287 961 95 arcseconds per Julian year (arcsec yr^{-1}), respectively (Capitaine et al. 2003).

if the obliquity is ϵ and the initial inclination is zero, then the inclination varies periodically between 0 and 2ϵ. Nevertheless, both satellites of Mars as well as the inner satellites of Jupiter, Saturn, Uranus and Neptune all orbit near the equator of their host planet.[3] Thus we are obliged to explain why these satellites have remained near the equator of their host planet as it precesses (Goldreich 1965).

We restrict our attention to an axisymmetric planet orbited by a single satellite of negligible mass. To start the analysis, we introduce two reference frames. The first is an inertial set of Cartesian coordinates (X, Y, Z) in which the X-Y plane coincides with the orbital plane of the planet. Thus both the orbital angular momentum of the star around the planet $\mathbf{L}$ and the angular velocity of precession $\mathbf{\Omega}_\mathrm{p}$ lie along the Z-axis. The second is a set of Cartesian coordinates (x, y, z) in which the x-y plane coincides with the planet's equator, the spin angular momentum $\mathbf{S}$ is parallel to $\hat{\mathbf{z}}$, and the x-axis lies in the X-Y or orbital plane. The second system is non-inertial because of the precession of the planet's spin.

The transformation from (X, Y, Z) to (x, y, z) is described by the Euler angles of Appendix B.6, with $\beta = \epsilon$, the obliquity of the planet; $\gamma = 0$ because the x-axis lies in the X-Y plane; and $\dot{\alpha} = \Omega_\mathrm{p}$ where $\mathbf{\Omega}_\mathrm{p} = \Omega_\mathrm{p}\hat{\mathbf{Z}}$. The planet precesses at fixed obliquity ϵ, so $\dot{\beta} = \dot{\gamma} = 0$ and equation (B.65a) implies that the transformation from (X, Y, Z) to (x, y, z) is a rotation with angular speed $\mathbf{\Omega}_\mathrm{p}$. The non-inertial nature of the (x, y, z) frame gives rise to a disturbing function (eq. D.24)

$$H_{\mathrm{non-in}} = -\mathbf{\Omega}_\mathrm{p} \cdot \mathbf{L}_\mathrm{s} = -\Lambda_\mathrm{s}\mathbf{\Omega}_\mathrm{p} \cdot \mathbf{j}_\mathrm{s}. \tag{7.13}$$

Here $\mathbf{L}_\mathrm{s}$ is the orbital angular momentum of the satellite, $\Lambda_\mathrm{s} = (\mathbb{G}M_\mathrm{p}a_\mathrm{s})^{1/2}$ where a_s is the semimajor axis of the satellite, $\mathbf{j}_\mathrm{s} = \mathbf{L}_\mathrm{s}/\Lambda_\mathrm{s}$ is its dimensionless angular momentum, and $j_\mathrm{s} = |\mathbf{j}_\mathrm{s}| = (1 - e_\mathrm{s}^2)^{1/2}$ where e_s is the eccentricity of the satellite. The disturbing function also has a component that

[3] The ring systems of the four outer planets also lie close to the equator, but for a different reason: planetary rings are dissipative, and circular orbits in the equatorial plane are a minimum-energy state for a given semimajor axis. See discussion after equation (5.66).

arises from the quadrupole moment of the planet, given by equation (5.64):

$$\langle H_\mathrm{p}\rangle = \frac{\mathbb{G}M_\mathrm{p}J_2R_\mathrm{p}^2}{4a_\mathrm{s}^3}\frac{j_\mathrm{s}^2 - 3(\mathbf{j}_\mathrm{s}\cdot\hat{\mathbf{S}})^2}{j_\mathrm{s}^5}, \tag{7.14}$$

where as usual the angle bracket $\langle\cdot\rangle$ denotes an average over the orbit of the satellite.

The Milankovich equations (5.57) and (5.58) for the evolution of the angular-momentum and eccentricity vectors $\mathbf{j}_\mathrm{s}$ and $\mathbf{e}_\mathrm{s}$ under the influence of the disturbing function $H_\text{non-in} + \langle H_\mathrm{p}\rangle$ are

$$\begin{aligned}
\frac{\mathrm{d}\mathbf{j}_\mathrm{s}}{\mathrm{d}t} &= -\boldsymbol{\Omega}_\mathrm{p}\times\mathbf{j}_\mathrm{s} + \frac{3(\mathbb{G}M_\mathrm{p})^{1/2}J_2R_\mathrm{p}^2}{2a_\mathrm{s}^{7/2}j_\mathrm{s}^5}(\mathbf{j}_\mathrm{s}\cdot\hat{\mathbf{S}})\mathbf{j}_\mathrm{s}\times\hat{\mathbf{S}},\\
\frac{\mathrm{d}\mathbf{e}_\mathrm{s}}{\mathrm{d}t} &= -\boldsymbol{\Omega}_\mathrm{p}\times\mathbf{e}_\mathrm{s} + \frac{3(\mathbb{G}M_\mathrm{p})^{1/2}J_2R_\mathrm{p}^2}{2a_\mathrm{s}^{7/2}j_\mathrm{s}^7}\Big[j_\mathrm{s}^2(\mathbf{j}_\mathrm{s}\cdot\hat{\mathbf{S}})\mathbf{e}_\mathrm{s}\times\hat{\mathbf{S}}\\
&\qquad + \tfrac{1}{2}j_\mathrm{s}^2\mathbf{e}_\mathrm{s}\times\mathbf{j}_\mathrm{s} - \tfrac{5}{2}(\mathbf{j}_\mathrm{s}\cdot\hat{\mathbf{S}})^2\mathbf{e}_\mathrm{s}\times\mathbf{j}_\mathrm{s}\Big].
\end{aligned} \tag{7.15}$$

To investigate the behavior of trajectories governed by these equations of motion, we specialize to the case of a circular orbit ($\mathbf{e}_\mathrm{s} = \mathbf{0}$, $|\mathbf{j}_\mathrm{s}| = j_\mathrm{s} = 1$), and initially small inclination ($\mathbf{j}_\mathrm{s}\cdot\hat{\mathbf{S}} \simeq 1$). Then the second of equations (7.15) is trivially satisfied and the first simplifies to

$$\frac{\mathrm{d}\mathbf{j}_\mathrm{s}}{\mathrm{d}t} = (\boldsymbol{\omega}_\mathrm{Q} - \boldsymbol{\Omega}_\mathrm{p})\times\mathbf{j}_\mathrm{s}, \tag{7.16}$$

where the precession due to the planet's quadrupole moment is

$$\boldsymbol{\omega}_\mathrm{Q} \equiv -\frac{3(\mathbb{G}M_\mathrm{p})^{1/2}J_2R_\mathrm{p}^2}{2a_\mathrm{s}^{7/2}}\hat{\mathbf{S}}. \tag{7.17}$$

When the planetary spin precession $\boldsymbol{\Omega}_\mathrm{p} = \mathbf{0}$, this equation describes the usual retrograde precession of the line of nodes due to the planet's quadrupole moment (cf. eq. 1.180b). When the spin precession is nonzero, the precession rate shifts in direction. However, so long as $|\boldsymbol{\Omega}_\mathrm{p}| \ll |\boldsymbol{\omega}_\mathrm{Q}|$ the shift is small, and the normal to the satellite orbit will continue to precess around an axis close to the spin axis. Thus, the condition that the satellite's orbital inclination remains small is

$$|\boldsymbol{\Omega}_{\rm p}| \ll \frac{(\mathbb{G}M_{\rm p})^{1/2}J_2R_{\rm p}^2}{a_{\rm s}^{7/2}}. \tag{7.18}$$

In words, if the nodal precession rate of the satellite is much faster than the spin precession rate of the planet, the satellite orbit remains locked in the equator even as the orientation of the equator changes due to the precession of the planet. For example, in the case of Mars the nodal precession periods of the satellites, 2.26 yr for Phobos and 54.5 yr for Deimos,[4] are much shorter than the spin precession period of the planet, $2\pi/|\boldsymbol{\Omega}_{\rm p}| = 1.7\,\text{Myr}$ (Table 7.1).

This discussion neglects the effects of perturbations caused by the host star, described by the disturbing function (5.70), and mutual perturbations between satellites. In some cases these may need to be included, but the physical principles remain the same.

One curious feature of the present analysis deserves to be mentioned. In most cases the orbital elements in a perturbed Kepler potential are osculating (Box 1.4), that is, if the perturbation were turned off instantaneously the resulting Kepler orbit would have the same elements. In this analysis, however, the elements are not osculating (Goldreich 1965). The reason is that the generalized momentum in the non-inertial frame is $\mathbf{p} = \dot{\mathbf{r}} + \boldsymbol{\Omega}_{\rm p} \times \mathbf{r}$ (eq. D.22), which is not equal to the velocity. A more physical reason is that the disturbing function depends on both position and velocity, whereas in most problems in celestial mechanics it depends only on position.

When one or more massive satellites are locked to the planet's equatorial plane in this way, the derivation of the spin precession rate of the planet must treat these satellites as if they were part of the planet. Thus the planet's quadrupole moment J_2 must be augmented by the orbit-averaged mass distribution of the satellites (eq. 1.132), and the spin angular momentum $\mathbb{S}$ must be augmented by the orbital angular momentum of the satellites. If the satellites have masses $m_i \ll M_{\rm p}$, semimajor axes a_i, inclinations I_i relative

[4] See https://ssd.jpl.nasa.gov/?sat_elem.

Table 7.1: Spin precession periods of solar-system planets

	spin period $P_{\rm s}$(d)	obliquity ϵ	quadrupole moment $J_2(J_2')$	moment of inertia $C/M_{\rm p}R_{\rm p}^2$	precession period (yr)
Venus	243.0	177.36°	4.404×10^{-6}	0.34	2.9×10^4
Earth	0.9973	23.44°	0.001 08	0.3307	2.57×10^4
Mars	1.026	25.19°	0.001 96	0.364	1.70×10^5
Jupiter	0.4135	3.13°	0.0147(0.0450)	0.276	5.1×10^5
Saturn	0.4440	26.73°	0.0163(0.0650)	0.22	1.8×10^6
Uranus	0.7183	97.77°	0.003 51(0.0191)	0.22	2.1×10^8
Neptune	0.6713	28.32°	0.003 41(0.0193)	0.26	1.5×10^8

Mercury is not included because it has near-zero obliquity. The modified quadrupole moment J_2' is defined by equation (7.19a). The satellite Callisto and all satellites interior to it precess with Jupiter's spin; similarly for Hyperion (Saturn), Oberon (Uranus) and Triton (Neptune). Data from Appendix A and https://ssd.jpl.nasa.gov.

to the planet's equator, and zero eccentricity, then the modified quadrupole moment and spin angular momentum are

$$J_2' = J_2 + \tfrac{1}{2}\sum_i \frac{m_i}{M_{\rm p}}\left(\frac{a_i}{R_{\rm p}}\right)^2 \left(1-\tfrac{3}{2}\sin^2 I_i\right), \tag{7.19a}$$

$$\mathbf{S}' = \hat{\mathbf{S}}\big[C\omega + \sum_i m_i(\mathbb{G}M_{\rm p}a_i)^{1/2}\cos I_i\big], \tag{7.19b}$$

where the sum is over all the satellites whose orbits are locked to the planet's equator. For the giant planets in the solar system, the correction to the quadrupole moment varies from a factor of 3 to a factor of 6, while the correction to the spin angular momentum is at most a few percent. The precession periods of solar-system planets are shown in Table 7.1.

7.1.2 The chaotic obliquity of Mars

The precession frequency of Mars's pole is given by equation (7.10),

$$\mathbf{\Omega}_{\mathrm{p}} = -\alpha(\hat{\mathbf{L}} \cdot \hat{\mathbf{S}})\hat{\mathbf{L}}; \tag{7.20}$$

and using the properties of Mars and its orbit from Appendix A, we find $\alpha = 8.397\,\text{arcsec yr}^{-1}$. At Mars's current obliquity $\epsilon = 25.19°$, the corresponding precession frequency is $|\mathbf{\Omega}_{\mathrm{M}}| = \alpha(\hat{\mathbf{L}} \cdot \hat{\mathbf{S}}) = \alpha \cos \epsilon = 7.599\,\text{arcsec yr}^{-1}$.

According to Lagrange–Laplace theory (§5.2), the normal $\hat{\mathbf{L}}$ to Mars's orbit also precesses, with a more complicated motion that is the sum of eight components with different amplitudes and frequencies. The precession frequencies f_i (Table 5.1) are negative—the motion of the tip of $\hat{\mathbf{L}}$ is clockwise as viewed from the north ecliptic pole—as is the precession of Mars's spin axis $\hat{\mathbf{S}}$. Moreover, one of the eight Lagrange–Laplace precession frequencies, $|f_2| = 7.05\,\text{arcsec yr}^{-1}$, is not far from Mars's spin precession frequency. Thus we expect that secular oscillations in the orientation of Mars's orbit may drive oscillations in its spin direction.

To investigate this interaction, we first rewrite equation (7.5) as

$$\frac{\mathrm{d}\hat{\mathbf{S}}}{\mathrm{d}t} = -\alpha(\hat{\mathbf{L}} \cdot \hat{\mathbf{S}})\hat{\mathbf{L}} \times \hat{\mathbf{S}}. \tag{7.21}$$

The orientation of Mars's orbit can be written in terms of the ascending node Ω and the inclination I as

$$\hat{\mathbf{L}} = (\sin I \sin \Omega, -\sin I \cos \Omega, \cos I). \tag{7.22}$$

The most accurate way to determine the time history of the orientation $\hat{\mathbf{L}}$ is through N-body integrations of the solar system. We sacrifice some accuracy for the sake of simplicity and insight by expanding $\sin I \cos \Omega$ and $\sin I \sin \Omega$ as trigonometric series in the time. These series can be written compactly as

$$\sin I \exp(\mathrm{i}\Omega) = \sum_{j=1}^{N} A_j \exp[\mathrm{i}(\nu_j t + \phi_j)]. \tag{7.23}$$

The simplest version of this approach is Lagrange–Laplace theory, in which $N = 8$ and the amplitudes A_j, phases ϕ_j and frequencies ν_j are determined by the analytic approach of §5.2. This approach neglects additional harmonics that arise from higher powers of the eccentricities and inclinations and nonlinear contributions to the secular dynamics. A more accurate approach (Laskar 1988) is to fit the amplitudes, frequencies and phases to numerical integrations. For our purposes we keep only the strongest few terms with frequencies close to the current spin precession frequency:

$$\begin{aligned}
\nu_1 &= -7.053\,11\,\text{arcsec yr}^{-1}, & A_1 &= 0.001\,318, & \phi_1 &= 144.957^\circ, \\
\nu_2 &= -6.963\,11\,\text{arcsec yr}^{-1}, & A_2 &= 0.001\,0367, & \phi_2 &= 311.799^\circ, \\
\nu_3 &= -7.002\,51\,\text{arcsec yr}^{-1}, & A_3 &= 0.000\,7314, & \phi_3 &= 118.035^\circ, \\
\nu_4 &= -7.148\,32\,\text{arcsec yr}^{-1}, & A_4 &= 0.000\,6883, & \phi_4 &= 327.622^\circ, \\
\nu_5 &= -6.860\,59\,\text{arcsec yr}^{-1}, & A_5 &= 0.000\,6112, & \phi_5 &= 298.460^\circ. \quad (7.24)
\end{aligned}$$

The origin of time is J2000.0, and the reference frame is the ecliptic coordinate system at J2000.0.

We now integrate the equation of motion (7.21) backward in time, using equations (7.22), (7.23) and (7.24) to determine the orbit orientation $\hat{\mathbf{L}}$ at any time. Figure 7.2 shows three histories of the obliquity $\epsilon = \cos^{-1}\hat{\mathbf{L}}\cdot\hat{\mathbf{S}}$. Each history is based on the same equation of motion but uses initial conditions for the spin axis that differ by random amounts of about 1 arcsecond. The differences between the integrations grow rapidly, and by 100 Myr the trajectories are completely different. This behavior is a signature of chaos (Laskar & Robutel 1993; Touma & Wisdom 1993). The corresponding Liapunov time is about 5 Myr.

The figure shows that the obliquity of Mars varies between about 25° and 40°. The actual variations are even larger—from near-zero to 65° over this interval—since the restriction of the series (7.23) to near-resonant frequencies suppresses terms with shorter periods that also lead to obliquity oscillations, although they play no role in the chaotic behavior (Laskar et al. 2004a).

The chaos arises because of the overlap of resonances associated with the terms in equation (7.21). If a single term dominates, such that the orbital

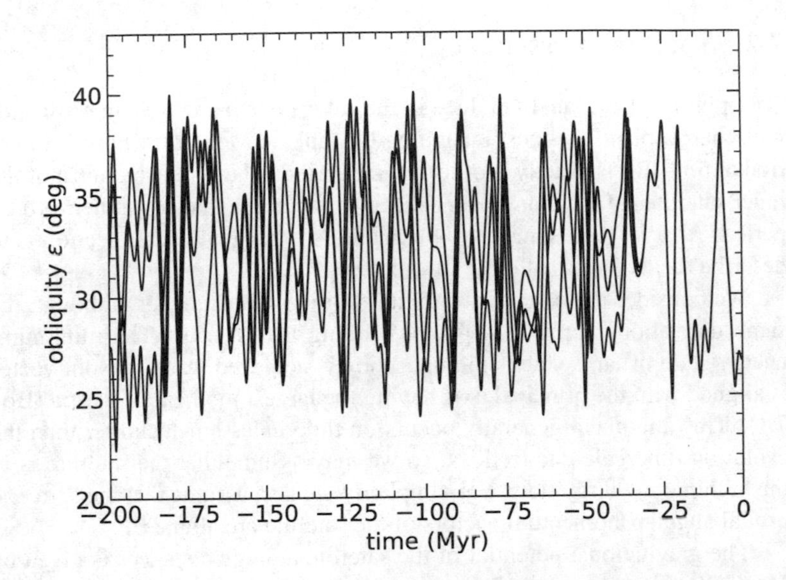

Figure 7.2: The obliquity of Mars over the past 200 Myr. The plot shows three integrations with differences in the orientation of the initial spin vector of about one part in 10^5. Oscillations with periods $\lesssim 0.2\,\mathrm{Myr}$ have been suppressed.

angular momentum of Mars precesses uniformly, the motion is integrable as described in §7.4. However, the terms in equation (7.24) have similar amplitudes and are sufficiently closely spaced in frequency ν_i that the resonances associated with them overlap. Remarkably, the chaos persists even if we simplify the model further by discarding all but the strongest two terms.

In contrast to Mars, Earth's obliquity is quite stable, varying only by about $\pm 1^\circ$. The reason is that Earth's spin precession is dominated by torques from the Moon, not the Sun, so its precession frequency is higher than the frequencies of the secular variations in its orbit.

7.2 Spin-orbit resonance

The spins of planets and satellites in the solar system are often in resonance with their orbits. The most intriguing example is Mercury, whose spin period of 58.646 d is exactly $\frac{2}{3}$ of its orbital period. The Moon and most of the inner satellites of the giant planets have spin periods that equal their orbital periods; that is, their spins are **synchronous**. The goal of this section is to describe the dynamics of these resonances.

We consider the case of a satellite orbiting a host planet, although the same derivations apply to a planet orbiting a host star. The minimum-energy state of an isolated spinning body is achieved when its spin vector is aligned with the principal axis having the largest moment of inertia (Box 7.1). This alignment generally occurs on timescales much shorter than the evolution timescale due to tides, so we may assume that the satellite is in this minimum-energy state. For simplicity we also assume that the spin and orbital angular-momentum vectors of the satellite are aligned.

The gravitational potential of the satellite at large distances is given by MacCullagh's formula (1.129),

$$\Phi = \frac{3\mathbb{G}}{2r^5}(Ax'^2 + By'^2 + Cz'^2) - \frac{\mathbb{G}}{2r^3}(A + B + C), \tag{7.25}$$

where (x', y', z') are the coordinates along the principal axes, $A \leq B \leq C$ are the corresponding moments of inertia (eqs. D.87), and $r = (x'^2 + y'^2 + z'^2)^{1/2}$ is the distance from the center of mass of the satellite. Since any spherical potential has no effect on the spin dynamics, we can simplify this expression by subtracting $\mathbb{G}(A - \frac{1}{2}B - \frac{1}{2}C)/r^3$ to obtain

$$\Phi = \frac{3\mathbb{G}}{2r^5}[(B - A)y'^2 + (C - A)z'^2]. \tag{7.26}$$

Our assumption that the spin vector is aligned with the principal axis having the largest moment of inertia implies that the spin vector is parallel to the z'-axis.

We now introduce a non-rotating reference frame centered on the satellite, with coordinates (x, y, z), which is oriented so the host planet orbits

Box 7.1: Principal-axis rotation

Spinning bodies such as planets and satellites experience internal stresses that can dissipate their rotational energy.

In the principal-axis frame (x', y', z') the inertia tensor of a body is diagonal, with components $A = I_{x'x'}$, $B = I_{y'y'}$, $C = I_{z'z'}$ (eqs. D.87). Without loss of generality we may assume that $A \le B \le C$. The rotational energy is given by equation (D.96),

$$T = \tfrac{1}{2}A\omega_{x'}^2 + \tfrac{1}{2}B\omega_{y'}^2 + \tfrac{1}{2}C\omega_{z'}^2. \tag{a}$$

In the principal-axis frame the spin angular momentum $\mathbf{S} = S_{x'}\hat{\mathbf{x}}' + S_{y'}\hat{\mathbf{y}}' + S_{z'}\hat{\mathbf{z}}' = A\omega_{x'}\hat{\mathbf{x}}' + B\omega_{y'}\hat{\mathbf{y}}' + C\omega_{z'}\hat{\mathbf{z}}'$. Thus the kinetic energy can be written

$$T = \frac{S_{x'}^2}{2A} + \frac{S_{y'}^2}{2B} + \frac{S_{z'}^2}{2C}. \tag{b}$$

If the body is isolated the spin angular momentum $\mathbf{S}$ is conserved. Therefore the rotational energy is minimized if $S_{x'} = S_{y'} = 0$ and $S_{z'} = \pm|\mathbf{S}|$; in other words the spin angular momentum in the minimum-energy state is parallel to the principal axis with the largest moment of inertia. Usually this is the shortest principal axis—the one with the smallest distance from the center of mass to the surface.

If the spin angular momentum is not aligned with one of the principal axes, rotation induces time-varying internal stresses that dissipate energy, so the orientation of the principal axes evolves toward the minimum-energy state. In principle, the body could persist indefinitely in a state with $\mathbf{S}$ parallel to a different principal axis. However, rotation around the principal axis with the intermediate moment of inertia is dynamically unstable, even without dissipation (Problem 7.1). Rotation around the principal axis with the smallest moment of inertia is an energy maximum, so any small misalignment of the angular momentum from the principal axis will initiate dissipation that forces the misalignment to grow.

We therefore expect that most solid astronomical bodies should rotate around the principal axis with the largest moment of inertia. Some small solar-system bodies such as comets and asteroids do not spin around any of the principal axes—they are sometimes said to be "tumbling"—presumably because of recent impacts, outgassing, or the YORP effect (§7.5).

the satellite in the (x, y) plane and the positive z-axis points in the direction of the orbital angular momentum. Then the position of the planet relative to the satellite is $\mathbf{r} = r\cos(f + \varpi)\hat{\mathbf{x}} + r\sin(f + \varpi)\hat{\mathbf{y}}$, where f and ϖ are the true anomaly and longitude of periapsis of the planet, and $r = a(1-e^2)/(1+e\cos f)$, where a and e are the semimajor axis and eccentricity of the orbit (eq. 1.29). Since the spin and orbital axes are aligned by assumption, we have $z = z'$ and

$$x' = x\cos\theta + y\sin\theta, \quad y' = -x\sin\theta + y\cos\theta, \tag{7.27}$$

where $\theta(t)$ is the angle between the $\hat{\mathbf{x}}$ and $\hat{\mathbf{x}}'$ axes. When evaluated at the position of the planet, equation (7.26) becomes

$$\Phi = \frac{3\,\mathbb{G}(B-A)}{2r^3}\sin^2(f + \varpi - \theta). \tag{7.28}$$

The torque exerted on the planet by the satellite is $\mathbf{N} = -M(\partial\Phi/\partial f)\hat{\mathbf{z}}$, where M is the planet mass. This torque affects both the orbit of the planet around the satellite and the spin of the satellite, but in most cases the orbital angular momentum is much larger than the satellite's spin angular momentum, so changes in the orbit can be neglected compared to changes in the spin. The spin angular momentum $\mathbf{S} = C\dot{\theta}\hat{\mathbf{z}}$ changes at a rate $\dot{\mathbf{S}} = -\mathbf{N}$, so

$$\ddot{\theta} = \frac{3\,\mathbb{G}M}{2r^3}\frac{B-A}{C}\sin 2(f + \varpi - \theta). \tag{7.29}$$

In the absence of resonances, the angles f and θ circulate independently and the right side of this equation averages to zero. More interesting behavior occurs when there is a spin-orbit resonance, which occurs when the mean motion in the orbit n and the spin rate of the satellite $\dot{\theta}$ are related by $\dot{\theta} \simeq pn$, where n is the mean motion and p is an integer (or half-integer, as we shall see). To investigate this behavior we write $\psi \equiv 2(\theta - p\ell - \varpi)$ where ℓ is the mean anomaly, eliminate θ from equation (7.29), and then average over time, or equivalently over ℓ. The result can be written as

$$\ddot{\psi} = -\frac{3\,\mathbb{G}M}{a^3}\frac{B-A}{C}Y(p, e)\sin\psi, \tag{7.30}$$

where

$$Y(p,e) \equiv \langle (a/r)^3 \cos 2(f - p\ell) \rangle. \tag{7.31}$$

Here $\langle \cdot \rangle$ denotes the orbit average; note that $\langle (a/r)^3 \sin 2(f - p\ell) \rangle = 0$, since f is an odd function of ℓ while r is even. The average in equation (7.31) can only be nonzero if $2p\ell$ varies by an integer multiple of 2π when ℓ varies by 2π, so p must be an integer or half-integer. The values of $Y(p,e)$ can be determined numerically or as power series in the orbital eccentricity e:

$$\begin{aligned}
Y(2,e) &= \tfrac{17}{2}e^2 - \tfrac{115}{6}e^4 + \mathrm{O}(e^6), & Y(\tfrac{3}{2},e) &= \tfrac{7}{2}e - \tfrac{123}{16}e^3 + \mathrm{O}(e^5),\\
Y(1,e) &= 1 - \tfrac{5}{2}e^2 + \tfrac{13}{16}e^4 + \mathrm{O}(e^6), & Y(\tfrac{1}{2},e) &= -\tfrac{1}{2}e + \tfrac{1}{16}e^3 + \mathrm{O}(e^5),\\
Y(0,e) &= 0, & Y(-\tfrac{1}{2},0) &= \tfrac{1}{48}e^3 + \mathrm{O}(e^5),\\
Y(-1,e) &= \tfrac{1}{24}e^4 + \mathrm{O}(e^6), & Y(-\tfrac{3}{2},e) &= \tfrac{81}{1280}e^5 + \mathrm{O}(e^7).
\end{aligned} \tag{7.32}$$

Equation (7.30) is the equation of motion for a pendulum (§6.1). The variable ψ can either circulate or librate; if ψ librates the satellite is said to be in a **spin-orbit resonance**. In a spin-orbit resonance the average of $\dot{\psi}$ vanishes, so the time average of $\dot{\theta} = pn$. Since $B - A > 0$ the resonant angle librates around $\psi = 0$ if $Y(p,e) > 0$ and otherwise around π. Equilibrium at $\psi = 0$ implies that when $\ell = 0$ we have $\theta = \varpi$; in words, the long axis of the satellite points toward the host planet at periapsis. The resonance is stable even in the presence of additional torques on the satellite, so long as the torques are not too large (see discussion following eq. 6.17).

There is an infinite number of spin-orbit resonances corresponding to different half-integer values of p, but most of these are very weak. If the planetary orbit has a small eccentricity, the strongest resonances are those in which $Y(p,e)$ depends on the lowest power of the eccentricity: according to equations (7.32) these have $p = 1$ [$Y(p,e) = \mathrm{O}(1)$] and $p = \frac{1}{2}$ or $p = \frac{3}{2}$ [$Y(p,e) = \mathrm{O}(e)$]. For circular orbits the only resonance is $p = 1$, the synchronous resonance.

The nature of the synchronous resonance is different from other spin-orbit resonances. As we shall see in Chapter 8, tidal friction naturally tends to decrease the spin rate of a satellite if this rate exceeds the mean motion and increase the spin rate if it is less than the mean motion. Thus satellites subjected to tidal friction generally evolve toward the synchronous state, even if there is no resonance (e.g., if the satellite is axisymmetric, so $B-A = 0$).

The Moon and most of the inner satellites in the solar system are synchronous, as are Pluto and its satellite Charon, while Mercury is found in the $p = \frac{3}{2}$ resonance. The current libration amplitudes of these bodies are all near zero, having been damped long ago by internal stresses.

The existence of stable resonances does not address the question of *why* satellites and planets are found in these resonances—or why, for example, Mercury is found in the $p = \frac{3}{2}$ resonance and not a resonance with larger p or the synchronous state. In general, satellites and planets found in spin-orbit resonance have timescales for tidal despinning that are short compared to the age of the solar system, which strongly suggests that resonance capture is a consequence of tidal evolution. However, for the simplest model of despinning, in which tides produce a constant torque with the opposite sign to $\dot{\theta} - n$, we showed in §6.1.1 that capture in resonance cannot occur.

7.2.1 The chaotic rotation of Hyperion

We derived equation (7.30) by averaging equation (7.29) over time, keeping the slow variable $\psi(t)$ fixed. We can derive an equation of motion for $\psi(t)$ without averaging by treating the functions $Y(p, e)$ as coefficients in a Fourier series for $(a/r)^3 \exp[2\mathrm{i}(f - p\ell)]$ (Appendix B.4). We find

$$\ddot{\psi} = -\frac{3\mathbb{G}M}{a^3}\frac{B-A}{C}\sum_{k=-\infty}^{\infty} Y(p + \tfrac{1}{2}k, e)\sin(\psi - k\ell). \tag{7.33}$$

To simplify the notation we use units in which $\mathbb{G}M/a^3 = 1$. Then the mean motion $n = 1$, and we can choose the origin of time so $\ell = t$. Equation (7.33) is derived from a Hamiltonian

$$H(Q, P, t) = \tfrac{1}{2}P^2 - 3\frac{B-A}{C}\sum_{k=-\infty}^{\infty} Y(p + \tfrac{1}{2}k, e)\cos(Q - k\ell), \tag{7.34}$$

where $Q = \psi$ is the coordinate and $P = \dot{\psi}$ is the conjugate momentum.

Assume for the moment that a single term in the sum dominates the Hamiltonian. We change to a new coordinate-momentum pair (Q', P') using the generating function $S_2(Q, P') = P'Q + kQ - kP't$. Then equations (D.63) imply that $P = \partial S_2/\partial Q = P' + k$, $Q' = \partial S_2/\partial P' = Q - kt$, and the new Hamiltonian is

$$H'(Q', P') = H + \frac{\partial S_2}{\partial t} = \tfrac{1}{2}P'^2 + \tfrac{1}{2}k^2 - 3\frac{B-A}{C}Y(p+\tfrac{1}{2}k, e)\cos Q'. \quad (7.35)$$

The constant $\frac{1}{2}k^2$ plays no role and can be dropped. Then the Hamiltonian is the pendulum Hamiltonian of §6.1, with $m = 1$ and $\omega^2 = 3Y(p+\frac{1}{2}k, e)(B-A)/C$. The equilibria are at $P' = 0$, corresponding to $P = k$, and the width of the resonance is given by equation (6.13),

$$w_k = 4\left[3\frac{B-A}{C}Y(p+\tfrac{1}{2}k, e)\right]^{1/2}. \quad (7.36)$$

According to the resonance-overlap criterion (Appendix F.1), a region of phase space near a resonance is chaotic when the sum of the half-widths of two adjacent resonances exceeds their separation. The separation of adjacent resonances ($|\Delta k| = 1$) is $\Delta P = 1$, and for small eccentricities, the largest pair of adjacent coefficients in equations (7.32) are $Y(1, e) = 1 + \mathrm{O}(e^2)$ and $Y(\frac{3}{2}, e) = \frac{7}{2}e + \mathrm{O}(e^3)$. Therefore we expect to find a region of chaotic spin when (Wisdom et al. 1984)

$$\left(3\frac{B-A}{C}\right)^{1/2}[2 + (14e)^{1/2}] \gtrsim f, \quad (7.37)$$

where $f \simeq 0.6$ from numerical investigations of the standard map (Appendix F.1). The chaotic region surrounds the $p = 1$ and $p = \frac{3}{2}$ resonances, corresponding to spins $\dot{\theta}$ between 1 and 1.5 times the mean motion.

The most interesting case in the solar system is Saturn's satellite Hyperion, which has $e = 0.123$ and $(B - A)/C = 0.29$, so the left side of (7.37) is 3.2, well above what is needed for chaos. Numerical integrations confirm that Hyperion's spin is chaotic. Moreover the integrations show that rotation in which the spin and orbital angular momenta are aligned is unstable. Thus Hyperion is tumbling chaotically, with a remarkably short Liapunov time of only ~ 40 days (Wisdom et al. 1984).

7.3 Andoyer variables

The behavior of a rotating rigid body is usually described by Euler's equations (Appendix D.11.2). However, there are several reasons why a Hamiltonian description of the motion is more useful. First, the powerful tool of Hamiltonian perturbation theory is not available when the differential equations do not have a Hamiltonian structure. Second, numerical studies of Hamiltonian systems can employ symplectic integration algorithms (§2.2.2), which are more accurate than general-purpose integration algorithms over long times. Finally, we shall find that the internal symmetries of the dynamics are more explicit in the Hamiltonian formulation than in Euler's equations.

Let (x, y, z) be the coordinates in a reference frame, usually but not always inertial. Similarly, let (x', y', z') be coordinates in the principal-axis frame of the body, in which the inertia tensor is diagonal with components $A = I_{x'x'}$, $B = I_{y'y'}$, $C = I_{z'z'}$ (eq. D.87). Without loss of generality we may assume that $A \leq B \leq C$.

The Euler angles (α, β, γ) describe rotation from the reference frame to the principal-axis frame. The Hamiltonian for rotation of a rigid body is given in equation (D.99) in terms of these angles and their conjugate momenta $(p_\alpha, p_\beta, p_\gamma)$, but the expressions are relatively cumbersome. The dynamics can be simplified by a canonical transformation to **Andoyer variables**.

The Andoyer variables are derived by supplementing (α, β, γ) with two additional sets of Euler angles. If the body is in free rotation (i.e., there are no external torques), its spin angular momentum $\mathbf{S}$ is conserved, so the plane perpendicular to $\mathbf{S}$ is fixed. The presence of this fixed plane prompts us to define a new set of Cartesian coordinates (x_s, y_s, z_s) in which the x_s-axis lies in the x-y plane of the original reference coordinate system and the positive z_s-axis is parallel to $\mathbf{S}$. The Euler angles of the spin frame relative to the reference frame are denoted $(h, i, 0)$, and the Euler angles of the principal-axis frame relative to the spin frame are denoted (g, J, ℓ). The geometry is illustrated in Figure 7.3.

The components of the spin angular momentum in the principal-axis or

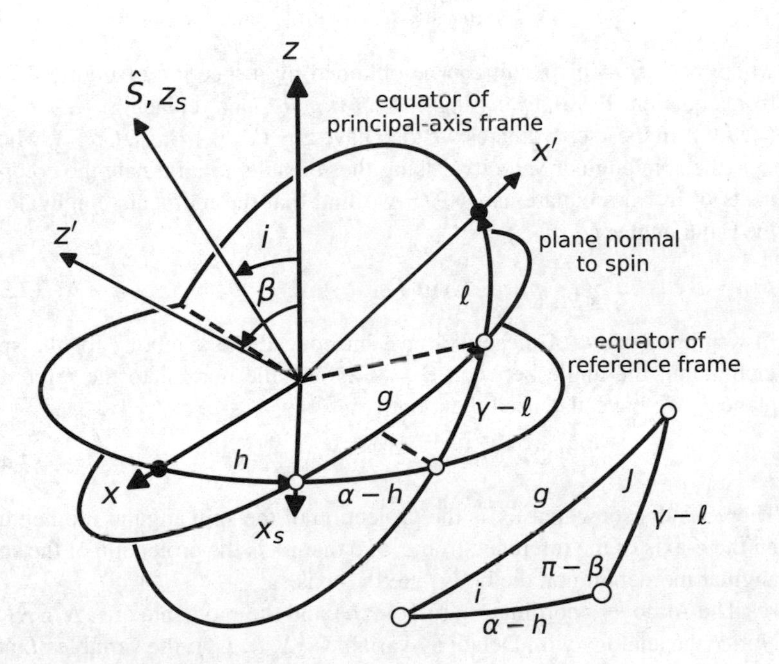

Figure 7.3: Andoyer coordinates for a rotating body. The figure shows three Cartesian frames: the reference frame (x, y, z), the principal-axis frame (x', y', z') fixed to the rotating body, and the spin frame (x_s, y_s, z_s) in which the positive z_s-axis is parallel to the body's spin angular momentum and the x_s-axis lies in the reference plane. The Euler angles (α, β, γ) specify the rotation from the reference frame to the principal-axis frame, the Euler angles $(h, i, 0)$ specify the rotation from the reference frame to the spin frame, and the Euler angles (g, J, ℓ) specify the rotation from the spin frame to the principal-axis frame. The spherical triangle in the lower right is an enlarged view of the triangle whose vertices are marked by open circles in the main figure. The Andoyer coordinates are (ℓ, g, h).

(x', y', z') frame are

$$\mathbf{S} = S(\sin \ell \sin J, \cos \ell \sin J, \cos J), \tag{7.38}$$

where $S = |\mathbf{S}|$—this result can be obtained by inspection of Figure 7.3 or from equation (B.60), by setting $x = y = 0$, $z = S$ and replacing (α, β, γ) by (g, J, ℓ). In these coordinates we also have $\mathbf{S} = (A\omega_{x'}, B\omega_{y'}, C\omega_{z'})$, where $\boldsymbol{\omega}$ is the spin angular velocity. Using these results to eliminate the components of $\boldsymbol{\omega}$ from equations (D.97), we find that the momenta conjugate to the Euler angles (α, β, γ) are

$$p_\alpha = S \cos i \equiv S_z, \quad p_\beta = -S \sin J \sin(\gamma - \ell), \quad p_\gamma = S \cos J \equiv \Lambda. \tag{7.39}$$

In writing the first of these we have introduced the symbol i for the spin inclination, the angle between $\mathbf{S} = S\hat{\mathbf{z}}_s$ and the normal to the reference plane $\hat{\mathbf{z}}$. We have also used the identity[5]

$$\cos i = \cos \beta \cos J + \sin \beta \sin J \cos(\gamma - \ell). \tag{7.40}$$

These results show that p_α is the projection of the spin angular momentum on the z-axis of the reference frame, and that p_γ is the projection of the spin angular momentum on the body-fixed z'-axis.

The Andoyer coordinates are (ℓ, g, h) and the momenta are (Λ, S, S_z). Notice the analogy with Delaunay variables (Λ, L, L_z): the variables L and L_z are the orbital angular momentum and its z-component in Delaunay variables, while S and S_z are the spin angular momentum and its z-component in Andoyer variables.

We now show that Andoyer coordinates and momenta are canonical variables (Deprit 1967). We examine the triangle whose vertices are marked with open circles in Figure 7.3. Using the spherical cosine law, the first of equations (B.49) with $A = \pi - \beta$, we find

$$\cos g = \cos(\alpha - h) \cos(\gamma - \ell) - \sin(\alpha - h) \sin(\gamma - \ell) \cos \beta. \tag{7.41}$$

[5] This can be derived in two ways. (i) Apply the first of equations (B.56) to the triangle marked with open circles in Figure 7.3, using $A = i$. (ii) Evaluate $\hat{\mathbf{z}}$ and $\hat{\mathbf{z}}_s$ in the principal-axis frame, using equation (B.60) and the Euler angles (α, β, γ) and (g, J, ℓ) respectively; then $\cos i = \hat{\mathbf{z}} \cdot \hat{\mathbf{z}}_s$.

Taking the differential of this expression, we have

$$\begin{aligned}\sin g\,\mathrm{d}g = &\left[\sin(\alpha-h)\cos(\gamma-\ell)+\cos(\alpha-h)\sin(\gamma-\ell)\cos\beta\right]\mathrm{d}(\alpha-h)\\ &+\left[\cos(\alpha-h)\sin(\gamma-\ell)+\sin(\alpha-h)\cos(\gamma-\ell)\cos\beta\right]\mathrm{d}(\gamma-\ell)\\ &-\sin(\alpha-h)\sin(\gamma-\ell)\sin\beta\,\mathrm{d}\beta.\end{aligned} \tag{7.42}$$

This result can be simplified by eliminating $\sin\beta$ using the spherical sine law, equation (B.55), in the form $\sin\beta = \sin g\sin J/\sin(\alpha-h)$; and by eliminating $\cos\beta$ using equation (7.41):

$$\begin{aligned}\mathrm{d}g = &\frac{\cos(\gamma-\ell)-\cos(\alpha-h)\cos g}{\sin g\sin(\alpha-h)}\mathrm{d}(\alpha-h)\\ &+\frac{\cos(\alpha-h)-\cos(\gamma-\ell)\cos g}{\sin g\sin(\gamma-\ell)}\mathrm{d}(\gamma-\ell)-\sin(\gamma-\ell)\sin J\,\mathrm{d}\beta.\end{aligned} \tag{7.43}$$

The spherical cosine law can be applied again to simplify each of the first two terms, giving

$$\mathrm{d}g = \cos i\,\mathrm{d}(\alpha-h)+\cos J\,\mathrm{d}(\gamma-\ell)-\sin(\gamma-\ell)\sin J\,\mathrm{d}\beta. \tag{7.44}$$

Using equations (7.39) this result can be rewritten as

$$\Lambda\mathrm{d}\ell + S\mathrm{d}g + S_z\mathrm{d}h = p_\alpha\mathrm{d}\alpha + p_\beta\mathrm{d}\beta + p_\gamma\mathrm{d}\gamma. \tag{7.45}$$

This equation relates the new momenta (Λ, S, S_z) and the differentials of the new coordinates (ℓ, g, h) to the old momenta $(p_\alpha, p_\beta, p_\gamma)$ and the differentials of the old coordinates (α, β, γ). The proof that the transformation is then canonical is given after equation (D.52).

The Hamiltonian for a rotating body H_{rot} is given in terms of the old variables in equation (D.99). To convert this to Andoyer variables, we set

$$\begin{aligned}p_\gamma &= S\cos J, \quad p_\beta = -S\sin J\sin(\gamma-\ell),\\ p_\alpha &= S\cos\beta\cos J + S\sin\beta\sin J\cos(\gamma-\ell),\end{aligned} \tag{7.46}$$

using equations (7.39) and (7.40). The Hamiltonian simplifies to

$$H_{\mathrm{rot}} = S^2\sin^2 J\left(\frac{\sin^2\ell}{2A}+\frac{\cos^2\ell}{2B}\right)+\frac{S^2\cos^2 J}{2C}+\Phi. \tag{7.47}$$

Since $\cos^2 J = \Lambda^2/S^2$ we obtain finally

$$H_{\rm rot} = (S^2 - \Lambda^2)\left(\frac{\sin^2 \ell}{2A} + \frac{\cos^2 \ell}{2B}\right) + \frac{\Lambda^2}{2C} + \Phi. \tag{7.48}$$

In the old variables the orientation of the body, and thus the potential Φ, is determined entirely by the Euler angles (α, β, γ), independent of the momenta $(p_\alpha, p_\beta, p_\gamma)$. In Andoyer variables the orientation is determined by both the coordinates and the momenta: the orientation of the spin frame (x_s, y_s, z_s) relative to the reference frame (x, y, z) is determined by h and $i = \cos^{-1} S_z/S$, while the orientation of the principal-axis frame relative to the spin frame is determined by g, $J = \cos^{-1} \Lambda/S$ and ℓ.

If the body is rotating freely and the reference frame is inertial, then $\Phi = 0$ and the Hamiltonian is independent of both g and h, so their conjugate momenta S and S_z are both conserved. The Hamiltonian is also independent of S_z, so h is conserved. All of these are consequences of the conservation of the spin angular momentum $\mathbf{S}$. The only remaining variables are the coordinate-momentum pair (ℓ, Λ), so the problem has one degree of freedom. The trajectories can be explored through the contours of constant $H_{\rm rot}$ (Deprit 1967) or by solving the equations of motion using elliptic functions (Kinoshita 1972). If the body is axisymmetric ($B = A$) then the Hamiltonian is independent of ℓ, so the conjugate momentum Λ is also conserved.

In the spin frame the spin vector has coordinates $S(0, 0, 1)$. Its coordinates in the principal-axis frame are given by equation (7.38). Thus in the principal-axis frame, the spin vector precesses or **nutates** around the symmetry axis at a rate $|\dot{\ell}| = \Lambda(A^{-1} - C^{-1})$. When the angle J between the symmetry axis and the spin angular momentum is small, and the dynamical ellipticity $(C - A)/C \ll 1$, the precession rate is smaller than the spin rate by a factor equal to the dynamical ellipticity.

Note that when the satellite is rotating around its principal axis ($\hat{\mathbf{z}}'$ parallel to $\hat{\mathbf{z}}_s$), the angle J is zero, $\Lambda = S$, and the coordinates g and ℓ are not well defined: only their sum $g + \ell$ is physically meaningful. A similar singularity occurs when the spin is normal to the reference plane ($\hat{\mathbf{z}}$ parallel to $\hat{\mathbf{z}}_s$), so g and h are not well defined. In such cases we can transform to modified Andoyer variables with the same approach that was used to modify the Delaunay variables in equations (1.87)–(1.92). For example the generating function $S_2(\mathbf{q}, \mathbf{P}) = (\ell + g + h)P_1 + (g + h)P_2 - hP_3$ yields new coordinate-momentum pairs

$$\begin{aligned} \theta_1 &= \ell + g + h, & \theta_2 &= g + h, & \theta_3 &= -h, \\ P_1 &= \Lambda, & P_2 &= S - \Lambda = S(1 - \cos J), & P_3 &= S - S_z = S(1 - \cos i). \end{aligned} \tag{7.49}$$

7.4 Colombo's top and Cassini states

Earlier sections in this chapter described the spin dynamics of a satellite traveling on a fixed orbit around its host planet. In practice, the orientation of satellite orbits often precesses due to forces from the planet's equatorial bulge, other satellites, or the host star. These variations can in turn strongly affect the evolution of the satellite's spin. The study of these effects dates back more than three centuries (Box 7.2).

The simplest example is a satellite such as the Moon, whose orbit precesses due to torques from the planet's host star.[6] For simplicity we assume that the satellite is axisymmetric and that its orbit around the planet is circular. To describe the spin evolution we use Andoyer variables with the reference (x, y, z) coordinate system chosen such that the satellite orbits the planet in the x-y plane. Then $\hat{\mathbf{z}}$ is parallel to the satellite's orbital angular momentum. As usual, the (x', y', z') coordinate system is the principal-axis frame, with the largest moment of inertia C around the z'-axis.

Because the satellite's orbit precesses, the (x, y, z) frame is not inertial: it precesses around the Z-axis of an inertial frame (X, Y, Z) with angular speed $\boldsymbol{\Omega}_\mathrm{s} = \Omega_\mathrm{s}\hat{\mathbf{Z}}$. We assume that the dominant torque on the satellite orbit arises from the host star, so the (X, Y) plane is the plane of the planet's orbit around the Sun, which we assume to be fixed. We may assume that the positive x-axis lies in the (X, Y) plane, at the ascending node of the satellite's orbit, and that the inclination of the satellite orbit to this plane is I.

[6] The derivation here follows Henrard & Murigande (1987) but beware of differences in notation

Box 7.2: Cassini's laws

In 1693 Giovanni Cassini stated three empirical laws that characterize the motion of the Moon. In modern language, these are:

(i) The Moon's spin period equals the period of its orbit around the Earth.

(ii) The angle between the Moon's spin axis and the normal to its orbit around the Earth is constant (about 1.54°), even as the orbital plane precesses due to the gravitational influence of the Sun.

(iii) The Moon's spin axis, the normal to the Moon's orbit around the Earth, and the normal to the Earth's orbit around the Sun all lie in the same plane.

The first law is a consequence of tidal friction, which has transferred angular momentum from the Moon's spin to the Earth–Moon orbit (§8.4.2). The second and third laws arise because the Moon is in a Cassini state, an extremum of its rotational energy in the sense described in the paragraphs following equation (7.67).

Cassini's third law implies that the precession period of the Moon's spin axis equals the precession period of the axis of its orbit around the Earth, about 18.60 years.

The non-inertial nature of the (x, y, z) frame gives rise to a disturbing function (cf. eq. 7.13)

$$H_{\text{non-in}} = -\boldsymbol{\Omega}_{\text{s}} \cdot \mathbf{S}, \tag{7.50}$$

where $\mathbf{S}$ is the satellite's spin angular momentum. In the (x, y, z) coordinates the precession frequency and the spin angular momentum are

$$\begin{aligned} \boldsymbol{\Omega}_{\text{s}} &= \Omega_{\text{s}}\hat{\mathbf{Z}} = \Omega_{\text{s}}(0, \sin I, \cos I), \\ \mathbf{S} &= S(\sin i \sin h, -\sin i \cos h, \cos i) \\ &= [(S^2 - S_z^2)^{1/2} \sin h, -(S^2 - S_z^2)^{1/2} \cos h, S_z], \end{aligned} \tag{7.51}$$

where i and h are defined in Figure 7.3. Then

$$H_{\text{non-in}} = \Omega_{\text{s}}\big[(S^2 - S_z^2)^{1/2} \sin I \cos h - S_z \cos I\big]. \tag{7.52}$$

Note that Ω_{s} is usually negative because the nodes of satellite orbits usually regress.

The next step is to determine the contribution to the Hamiltonian from the host planet's tidal field. This is given by equation (7.26),

$$M\Phi = \frac{3\mathbb{G}M}{2r^5}(C - A)z'^2, \tag{7.53}$$

where M is the planet mass, z' and r are evaluated at the position of the planet, and we have set $A = B$ since the satellite is axisymmetric. As usual, let (α, β, γ) be the Euler angles describing the rotation from the (x, y, z) reference frame to the (x', y', z') principal-axis frame; then from (B.60) $z' = x \sin\alpha \sin\beta - y \cos\alpha \sin\beta + z \cos\beta$. At the position of the planet $z = 0$, $x = r\cos f$ and $y = r \sin f$, where f is the true anomaly. Averaging over the circular orbit of the satellite around the planet using equations (1.66c)–(1.66e), we have

$$M\langle\Phi\rangle = \frac{3\mathbb{G}M}{4a^3}(C - A)\sin^2\beta, \tag{7.54}$$

where a is the satellite's semimajor axis and we assume that the orbit eccentricity $e = 0$. To express the Euler angle β in terms of Andoyer variables, we apply the first of the trigonometric identities (B.56) to the spherical triangle at the lower right of Figure 7.3, setting $A = \pi - \beta$, $B = J$, $C = i$, $a = g$ to find

$$\cos\beta = \cos i \cos J - \sin i \sin J \cos g. \tag{7.55}$$

To simplify equation (7.54) we can replace $\sin^2\beta$ by $-\cos^2\beta$—the constant term generates no torques and can be dropped—to obtain finally

$$M\langle\Phi\rangle = -\frac{3\mathbb{G}M}{4a^3}(C - A)(\cos i \cos J - \sin i \sin J \cos g)^2, \tag{7.56}$$

in which i and J are related to the Andoyer variables by $S_z = S\cos i$ and $\Lambda = S\cos J$.

The total Hamiltonian governing the spin dynamics is the sum of (i) the Hamiltonian (7.48) representing the rotational energy, with $A = B$ since the satellite is axisymmetric; (ii) the Hamiltonian (7.52) representing the fictitious forces that arise because the reference frame is non-inertial, and (iii) the Hamiltonian (7.56) representing the orbit-averaged tidal forces from the planet. Thus

$$H = H_{\rm rot} + H_{\rm non-in} + M\langle\Phi\rangle \tag{7.57}$$

$$= \frac{S^2 - \Lambda^2}{2A} + \frac{\Lambda^2}{2C} + \Omega_s[(S^2 - S_z^2)^{1/2} \sin I \cos h - S_z \cos I]$$
$$- \frac{3\,\mathbb{G}M}{4a^3 S^4}(C - A)\big[S_z\Lambda - (S^2 - S_z^2)^{1/2}(S^2 - \Lambda^2)^{1/2} \cos g\big]^2.$$

For the same reasons as in §7.2, we assume that the spin angular momentum **S** is nearly parallel to the principal axis with the largest moment of inertia. In this case the angle J is nearly zero, so Λ is nearly equal to S. Note that (i) with these assumptions, the angle i is equal to the obliquity ϵ of the satellite—the angle between the spin and orbital angular momenta—and we will change our notation accordingly; and (ii) the equations of motion depend on the *derivatives* of the Hamiltonian, so by setting $\Lambda = S$ (as we shall now do) we abandon our ability to follow the evolution of the conjugate variables g and ℓ, which determine the rotational phase of the satellite.

Replacing Λ by S, the Hamiltonian (7.57) simplifies to

$$H = \frac{S^2}{2C} + \Omega_s\big[(S^2 - S_z^2)^{1/2} \sin I \cos h - S_z \cos I\big]$$
$$- \frac{3\,\mathbb{G}M}{4a^3}(C - A)\frac{S_z^2}{S^2}. \tag{7.58}$$

Since H is independent of the coordinates ℓ and g, the momenta Λ and S are conserved. Therefore the term $\frac{1}{2}S^2/C$ is a constant and can be dropped.

The remaining Hamiltonian can be rescaled to a standard form, as described in equations (6.39) and (6.40). Let $R = S_z/S = \cos i = \cos \epsilon$, $r = h$ and $H(S_z, h) = bH_{\rm C}(R, h)$, and define a rescaled time $\tau = t/t_0$. Then R is a momentum conjugate to the coordinate h and $H_{\rm C}$ is a Hamiltonian related to the rescaled time τ if and only if $t_0 = S/b$ (recall that the spin S is constant). We choose

$$t_0 = \frac{2a^3 S}{3\,\mathbb{G}M(C - A)} = \frac{2\omega}{3n^2}\frac{C}{C - A}; \tag{7.59}$$

in the last equation we have introduced the satellite's mean motion $n = (\mathbb{G}M/a^3)^{1/2}$ and its spin angular speed $\omega = S/C$. The inverse of the characteristic time t_0 is roughly the precession frequency of the satellite's spin (cf. eq. 7.10).

The rescaled Hamiltonian can be written

$$H_{\rm C}(R,r) = -\tfrac{1}{2}(R-p)^2 - q(1-R^2)^{1/2}\cos r; \tag{7.60}$$

we have added an unimportant constant $-\frac{1}{2}p^2$ to complete the square and defined

$$p \equiv -\Omega_{\rm s}t_0\cos I, \quad q \equiv -\Omega_{\rm s}t_0\sin I. \tag{7.61}$$

Usually p and q are positive since $\Omega_{\rm s}$ is negative and $0 < I \ll 1$. If not, they can be made positive using the transformations in Box 6.1. This is the Hamiltonian for the dynamical system known as **Colombo's top** (Colombo 1966; Henrard & Murigande 1987).[7]

The equations of motion resulting from the Hamiltonian (7.60) are

$$\frac{\mathrm{d}R}{\mathrm{d}\tau} = -q(1-R^2)^{1/2}\sin r, \quad \frac{\mathrm{d}r}{\mathrm{d}\tau} = p - R + \frac{qR}{(1-R^2)^{1/2}}\cos r. \tag{7.62}$$

The equilibrium solutions of these equations are the **Cassini states**, found at

$$r = 0, \pi; \quad R - p = \pm\frac{qR}{(1-R^2)^{1/2}}. \tag{7.63}$$

Using equations (7.61) and the relation $R = \cos\epsilon$, the equilibrium condition can be rewritten as

$$\sin\epsilon\cos\epsilon + \Omega_{\rm s}t_0\sin(\epsilon \pm I) = 0. \tag{7.64}$$

Squaring the second of equations (7.63) and rearranging the terms gives us a quartic equation for R,

$$f(R) \equiv (R-p)^2(1-R^2) - q^2R^2 = 0. \tag{7.65}$$

A quartic always has 0, 2, or 4 real roots, but these are only physical if $|R| \le 1$. Since $f(\pm 1) = -q^2 < 0$ and $f(0) = p^2 > 0$, there must be at least two roots between -1 and $+1$. There cannot be three roots between -1 and

[7] Giuseppe ("Bepi") Colombo (1920–1984) was an Italian mathematician and engineer. Among other contributions, he redesigned the trajectory of the Mariner 10 spacecraft to accomplish three flybys of Mercury rather than just one.

+1 since $f(1)$ and $f(-1)$ have the same sign. Therefore there are either two or four Cassini states, depending on the values of p and q. There are four states when[8]

$$|p|^{2/3} + |q|^{2/3} \leq 1, \tag{7.66}$$

otherwise there are only two.

The natural manifold for the phase space of Colombo's top is the unit sphere traced out by the normalized spin vector $\hat{\mathbf{S}} = \mathbf{S}/S$. In the (x, y, z) coordinate system, $\hat{\mathbf{S}} = [(1 - R^2)^{1/2} \sin r, -(1 - R^2)^{1/2} \cos r, R]$ (eq. 7.51) so the Hamiltonian (7.60) can be written

$$H_{\mathrm{C}} = -\tfrac{1}{2}(\hat{S}_z - p)^2 + q\hat{S}_y. \tag{7.67}$$

A smooth function on a sphere must have at least one maximum and one minimum. Thus when there are two Cassini states, one is a maximum of the Hamiltonian, one is a minimum, and both are therefore stable. For a smooth function on a sphere the number of maxima and minima minus the number of saddle points equals 2, the Euler characteristic of a sphere. When there are four Cassini states this constraint can only be satisfied if two are maxima of the Hamiltonian, one is a minimum, and one is saddle point; or if two are minima, one a maximum, and one a saddle point. In either case there are three stable and one unstable Cassini states.

Since the Cassini states have $r = 0, \pi$, they have $\hat{S}_x = 0$. With this restriction the phase space for Colombo's top is the circle $\hat{S}_y^2 + \hat{S}_z^2 = 1$, and the Hamiltonian (7.67) is a parabola. The Cassini states occur at values of H_{C} for which the parabola is tangent to the circle. The geometry is

[8] Proof: The number of roots between $R = -1$ and $+1$ is a function of the parameters (p, q). The transition from four roots to two can occur either because (i) the roots move outside the interval $|R| \leq 1$ or (ii) the total number of roots drops from four to two. Transition (i) is not allowed because $f(\pm 1) = -q^2$ is nonzero, so there can never be a root at $R = \pm 1$. Transition (ii) occurs when there is a double root, which requires that both $f(R) = 0$ and $f'(R) = 0$ are satisfied simultaneously. Eliminating q^2 from these two equations yields $R = \mathrm{sgn}(p)|p|^{1/3}$, and substituting this result back into $f(R) = 0$ yields $|p|^{2/3} + |q|^{2/3} = 1$. This is the locus of parameters (p, q) on which there is a double root and hence the locus on which the transition between two and four roots must occur. Examination of special cases (e.g., $|p|, |q| \ll 1$, $|p|, |q| \gg 1$) shows that there are four roots when $|p|$ and $|q|$ are small and two when they are large.

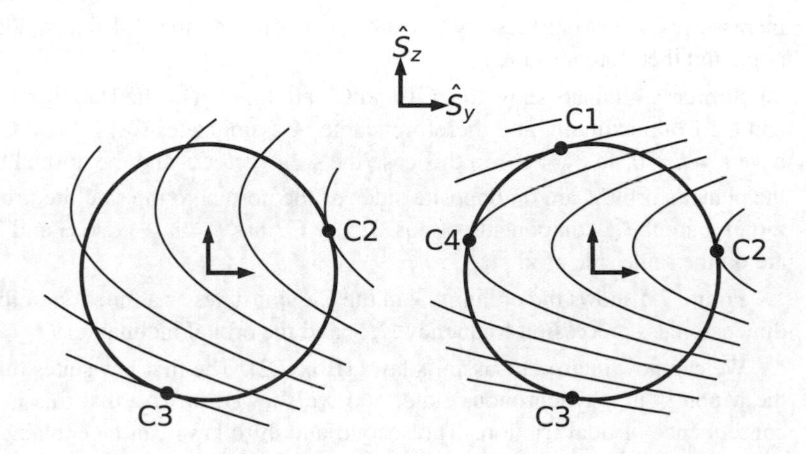

Figure 7.4: The state of Colombo's top is defined by the components $(\hat{S}_x, \hat{S}_y, \hat{S}_z)$ of the unit spin vector in the non-inertial frame, where $\hat{\mathbf{z}}$ is parallel to the satellite's orbital angular momentum and $\hat{\mathbf{x}}$ points to the ascending node of the satellite orbit on the orbit of its host planet. The figure shows the $\hat{S}_x = 0$ plane for $p = q = 0.5$ (left panel) and $p = q = 0.2$ (right panel). The unit spin vector is restricted to the circle $\hat{S}_y^2 + \hat{S}_z^2 = 1$, and the parabolas opening to the right are the contours of the Hamiltonian (7.67). The Cassini states, labeled C1, C2, C3, C4, are located where the contours of the Hamiltonian are tangent to the unit circle.

illustrated in Figure 7.4 for the usual case when $p, q > 0$. The left panel shows the case $p = q = 0.5$, for which there are two Cassini states, and the right panel shows the case $p = q = 0.2$, for which there are four.

First consider Cassini state 4, labeled C4 in the right panel. The Hamiltonian is constant along the parabolic contours and increases to the right. If we move a small distance away from C4 along the unit circle, keeping $\hat{S}_x = 0$, then the value of the Hamiltonian decreases. In contrast, if we move away from $\hat{S}_x = 0$, keeping $\hat{S}_y$ and $\hat{S}_z$ constant, the value of the Hamiltonian remains the same; however, this movement takes us outside the unit sphere. To return to the unit sphere, we must move to the right, so the Hamiltonian increases. Comparing these two results, we conclude that C4 is a saddle point and therefore unstable.

Similar arguments show that C1 and C3 are minima of the Hamiltonian and C2 is a maximum, and therefore stable. Cassini states C1, C3 and C4 have $r = h = 0$, so $\hat{S}_y < 0$—in this case the spin vector $\hat{\mathbf{S}}$ and the normal to the planet's orbit $\hat{\mathbf{Z}}$ are on opposite sides of the normal to the satellite orbit $\hat{\mathbf{z}}$ (compare the y-components of eqs. 7.51). C2 has $r = h = \pi$, so $\hat{\mathbf{S}}$ and $\hat{\mathbf{Z}}$ are on the same side of $\hat{\mathbf{z}}$.

Figure 7.5 shows the obliquity ϵ in the Cassini states as a function of the dimensionless precession frequency $\Omega_s t_0$ and the orbital inclination I.

We can now interpret Cassini's laws (Box 7.2). The first law states that the Moon is in a synchronous state, and we showed in §7.2 that this is a consequence of tidal friction. The second and third laws can be explained if the Moon is in a Cassini state. The second law states that the angle $i = \epsilon$ between the Moon's spin axis and the reference plane (the plane of its orbit around the Earth) is constant; this is obviously true for a Cassini state, as any equilibrium of Colombo's top has constant spin in the non-inertial frame defined by the satellite's orbit. The third law states that the Moon's spin axis, the normal to its orbit around the Earth, and the normal to the Earth's orbit around the Sun all lie in the same plane. In our notation the normal to the Earth's orbit is parallel to $\boldsymbol{\Omega}_s$ and the spin axis is parallel to the angular momentum $\mathbf{S}$. In the (x, y, z) coordinate frame the normal to the Moon's orbit is $\hat{\mathbf{z}} = (0, 0, 1)$. In this frame, equation (7.51) tells us that $\boldsymbol{\Omega}_s = \Omega_s(0, \sin I, \cos I)$ and that $\mathbf{S} = S(0, \mp \sin i, \cos i)$ in a Cassini state with $r = h = 0, \pi$. Thus $\boldsymbol{\Omega}_s$, $\hat{\mathbf{z}}$, and $\mathbf{S}$ all lie in the y-z plane, so Cassini's third law is satisfied.

In a Cassini state the spin angular momentum of the satellite is not aligned with its orbital angular momentum. Because of this misalignment, the material in the satellite is subject to time-dependent internal stresses leading to energy dissipation, even if the satellite is in synchronous rotation. The source of this energy is the satellite's orbit.

Planets can also be found in Cassini states (e.g., Correia 2015). In this case the precession of the planet's orbital plane is forced by other planets with nonzero mutual inclinations (for example, Mercury is in state C1). If

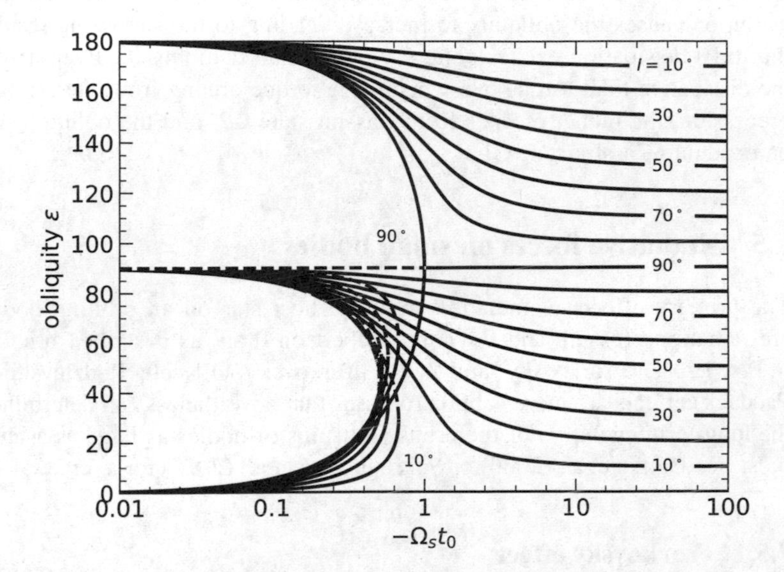

Figure 7.5: The obliquity ϵ of Cassini states as a function of the orbital precession rate Ω_s and the characteristic time t_0 (eq. 7.59). Results are shown for $\Omega_s < 0$, as is usual in planetary systems; the results for $\Omega_s > 0$ are found by replacing ϵ with $180^\circ - \epsilon$. Results are shown for $I = 10^\circ, 20^\circ, \ldots, 90^\circ$ where I is the inclination of the precessing satellite orbit relative to the orbit of its host planet; the results for $90^\circ < I < 180^\circ$ are also found by replacing ϵ by $180^\circ - \epsilon$. The curves with $\epsilon > 90^\circ$ are Cassini state C3 and the curves with $\epsilon < 90^\circ$ that extend to the right edge of the figure are Cassini state C2. The remaining curves in the region $\epsilon < 90^\circ$ and $-\Omega_s t_0 \lesssim 0.63$ are Cassini states C1 (solid) and C4 (dashed). State C4 is unstable.

a hot Jupiter is in a Cassini state, the energy dissipation rate may be large enough to cause significant decay of the planet's orbit. If the orbital precession is due to exterior planets, then as the orbit decays $\Omega_s t_0$ shrinks, and the Cassini state moves to the left in Figure 7.5. In Cassini state C2, this evolution causes the obliquity to increase, leading to the surprising result that tidal dissipation excites obliquity rather than damping it. Eventually the dissipative tidal torque overwhelms the torque arising from the orbital precession, the planet escapes from Cassini state C2, and the obliquity is damped into Cassini state C1.

7.5 Radiative forces on small bodies

The primary effects of radiation from the host star on an orbiting body are radiation pressure and Poynting–Robertson drag, as described briefly in Box 1.5. The Yarkovsky[9] and YORP (Yarkovsky–O'Keefe–Radzievskii–Paddack) effects are more subtle processes that nevertheless can determine the long-term evolution of the orbits and spins of bodies as large as asteroids. See Bottke et al. (2006) or Vokrouhlický et al. (2015) for a review.

7.5.1 Yarkovsky effect

We begin by examining the emission and absorption of radiation by an asteroid. We assume that the asteroid orbit is circular, that the asteroid is spherical, and that its spin and orbital axes are aligned (of course, none of these are good approximations for real asteroids). On average, the asteroid is in thermal equilibrium: the energy absorbed from the Sun, mostly at visible wavelengths, is re-radiated by the asteroid, mostly at infrared wavelengths. In the description of Poynting–Robertson drag in Box 1.5, we assumed that the energy was re-radiated isotropically in the rest frame of the body, in which case it exerts no net force on the body. This is equivalent to the assumption that the temperature and emissivity of the body's surface are

[9] Ivan Osipovich Yarkovsky (1844—1902) was a Polish–Russian civil engineer who first described the effect in 1901 (Beekman 2005).

uniform across its surface, and this is a good approximation for the small dust grains for which Poynting–Robertson drag is significant.

On much larger bodies such as asteroids, however, the surface temperature is not uniform. Of course, the night side pointing away from the Sun is cooler than the day side. More importantly, because of the asteroid's spin the temperature is not the same on the leading and trailing faces of the asteroid: just as on Earth, the surface is hotter during the asteroid's afternoon than in the morning, and hotter after sunset than before sunrise. This time lag is sometimes called "thermal inertia," and a precise definition of this term is given in equation (H.8). If the asteroid spin is prograde, this asymmetry implies that the momentum carried away by infrared photons is larger on the trailing face than on the leading face. As a result, the thermal emission adds angular momentum to the orbit. Similarly, if the spin is retrograde the emission drains orbital angular momentum. This **Yarkovsky effect** is negligible if the body is either very small (because the body reaches thermal equilibrium on a timescale that is much shorter than the spin period) or very large (because the area to mass ratio is small).[10]

The Yarkovsky force in this system is derived in Appendix H. The most important component, along the azimuthal direction in the orbital plane, is given by equation (H.19b):

$$F_{\mathrm{Y}} = \mathrm{sgn}(\omega_{\mathrm{syn}}) \frac{\alpha L_{\odot} R^2}{9ca^2} \frac{\Lambda_*}{1 + 2\Lambda_* + 2\Lambda_*^2}, \tag{7.68}$$

where $F_{\mathrm{Y}} > 0$ if the force points in the direction of orbital motion, and

$$\Lambda_* \equiv \frac{(|\omega_{\mathrm{syn}}| C_V \kappa)^{1/2}}{2^{5/2} \epsilon \sigma T_*^3}, \qquad \epsilon \sigma T_*^4 = \frac{3^{1/2} \alpha L_{\odot}}{8\pi^2 a^2}. \tag{7.69}$$

Here the surface properties of the asteroid are its radius R, its absorption coefficient α, and its emissivity ϵ; its bulk properties are its density ρ, its heat capacity per unit volume C_V, and its thermal conductivity κ; and its dynamical properties are the semimajor axis a and the synodic spin frequency

[10] The mechanism described is sometimes called the diurnal or daily Yarkovsky effect. There is also a seasonal Yarkovsky effect if the obliquity of the asteroid is nonzero.

$\omega_{\rm syn} = \omega_{\rm in} - n$, where $\omega_{\rm in}$ is the spin frequency in an inertial frame and n is the mean motion. The characteristic temperature of the asteroid surface is T_*, $\sigma = 5.670\,374\ldots \times 10^{-8}$ W m^{-2} K^{-4} is the Stefan–Boltzmann constant, and $L_\odot$ is the solar luminosity.

The azimuthal force changes the orbital angular momentum at a rate $\dot{L} = aF_{\rm Y}$; for an asteroid of mass $m = \frac{4}{3}\pi\rho R^3$ on a circular orbit, the semimajor axis a is related to the angular momentum by $L = m(GM_\odot a)^{1/2}$ so

$$\frac{\dot{a}}{a} = \text{sgn}(\omega_{\rm syn})\frac{\alpha L_\odot}{6\pi\rho Rc(\mathbb{G}M_\odot)^{1/2}a^{3/2}}\begin{cases}\Lambda_*, & \Lambda_* \ll 1,\\ \frac{1}{2}\Lambda_*^{-1}, & \Lambda_* \gg 1.\end{cases} \tag{7.70}$$

The thermal conductivity κ of common materials varies by more than four orders of magnitude. Iron-rich objects have $\kappa \sim$ 10–100 W m^{-1}K^{-1}, rocky and icy objects have $\kappa \sim$ 1 W m^{-1}K^{-1}, and regoliths[11] have much smaller values, $\kappa \sim 10^{-3}$ W m^{-1}K^{-1}. The volumetric heat capacity varies much less; typically $C_V \simeq 2 \times 10^6$ J m^{-3} K^{-1}. From equations (7.69) and (7.70),

$$T_* = 165\,\text{K}\left(\frac{\alpha}{\epsilon}\right)^{1/4}\left(\frac{3\,\text{au}}{a}\right)^{1/2}, \tag{7.71}$$

$$\Lambda_* = \frac{8.14}{\epsilon}\left(\frac{200\,\text{K}}{T_*}\right)^3\left(\frac{8\,\text{hr}}{P_{\rm syn}}\frac{C_V}{2\times 10^6\ \text{J m}^{-3}\,\text{K}^{-1}}\frac{\kappa}{\text{W m}^{-1}\text{K}^{-1}}\right)^{1/2},$$

$$\frac{\dot{a}}{a} = \frac{\text{sgn}(\omega_{\rm syn})\alpha}{4.86\,\text{Gyr}}\frac{3\,\text{g cm}^{-3}}{\rho}\frac{1\,\text{km}}{R}\left(\frac{3\,\text{au}}{a}\right)^{3/2}\begin{cases}\Lambda_*, & \Lambda_* \ll 1,\\ \frac{1}{2}\Lambda_*^{-1}, & \Lambda_* \gg 1.\end{cases}$$

This analysis is based on the plane-parallel approximation to the heat-conduction equation (eq. H.1) and therefore requires that the thermal diffusion length or **skin depth** $|k|^{-1}$ (eq. H.5) is less than the size of the body; in other words we need $|k|R \gtrsim 1$ or

$$R \gtrsim \left(\frac{\kappa}{|\omega_{\rm syn}|C_V}\right)^{1/2} = 0.048\,\text{m}\left(\frac{P_{\rm syn}}{8\,\text{hr}}\frac{\kappa}{\text{W m}^{-1}\text{K}^{-1}}\frac{2\times 10^6\,\text{J m}^{-3}\,\text{K}^{-1}}{C_V}\right)^{1/2}. \tag{7.72}$$

[11] Regolith is a surface layer of unconsolidated deposits, such as dust, gravel, ash, liquids, and so forth. Regolith covers most of the Earth, Moon, Mars, and asteroids. See also §8.6.2.

For bodies smaller than this limit, the Yarkovsky force is reduced below the values given by equation (7.68)—see Vokrouhlický (1998) for the generalization of the calculations in this section.

For a black body ($\alpha = \epsilon = 1$), the ratio of the Yarkovsky force to the Poynting–Robertson force, equation (c) of Box 1.5, is

$$\frac{F_{\rm Y}}{F_{\rm PR}} = -\operatorname{sgn}(\omega_{\rm syn})\frac{4c}{9v}\frac{\Lambda_*}{1+2\Lambda_*+2\Lambda_*^2}, \tag{7.73}$$

where v is the orbital speed. Since $v \ll c$ in planetary systems, the Yarkovsky force is much larger if Λ_* is of order unity, but the Poynting–Robertson drag dominates for very small bodies with $\Lambda_* \ll 1$, and very large ones with $\Lambda_* \gg 1$.

The Yarkovsky force can be measured directly from changes in the orbits of artificial satellites and near-Earth asteroids (Bottke et al. 2006; Vokrouhlický et al. 2015; Greenberg et al. 2020). According to equations (7.71), over the lifetime of the solar system it can cause significant changes in the semimajor axes (say, $\gtrsim 0.1\,$au) for asteroids as large as a few kilometers in radius. The most important consequence of these changes is to deliver asteroids at a steady rate into mean-motion and secular resonances with Jupiter, from where they can be excited onto Earth-crossing orbits (see §9.7). Without the Yarkovsky effect the flux of meteorites striking the Earth would be drastically reduced.

7.5.2 YORP effect

The absorption, scattering, and re-emission of sunlight from the surface of an asteroid of irregular shape can exert a torque relative to the center of mass of the asteroid. This is the **YORP effect**, distinct from the Yarkovsky effect because it affects the asteroid's spin rather than its orbit and does not rely on the thermal inertia of the asteroid. The acronym "YORP" was coined by Rubincam (2000) and recognizes early contributions by Yarkovsky, O'Keefe (1976), Radzievskii (1954), and Paddack (1969).

A simple, though artificial, example of the YORP effect would be a propeller or windmill with its axis facing the Sun: sunlight reflected from

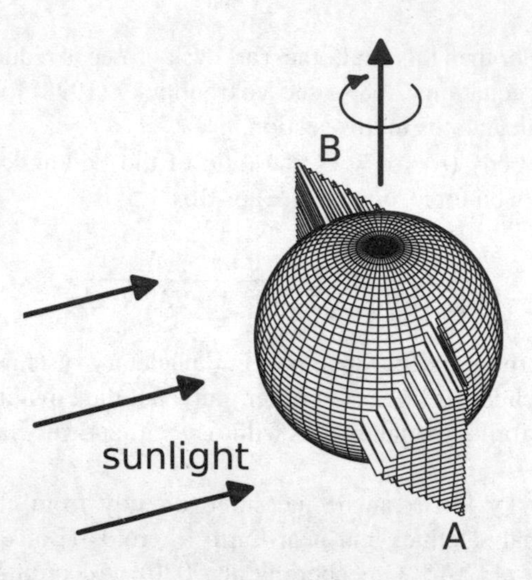

Figure 7.6: Spin-up of an asymmetrical asteroid due to the YORP effect. The asteroid consists of a sphere with 45° wedges A and B attached to its equator. The asteroid's spin angular velocity is normal to its orbital plane. The incoming solar radiation exerts the same force on each wedge, but the force exerted by the re-radiated thermal emission is in a different direction—in the orbital plane for wedge B but at ±45° to the orbital plane for wedge A (assuming that the emission from the asteroid surface satisfies Lambert's law). The resulting net torque causes the asteroid to rotate faster and faster in the clockwise direction, as viewed from above.

or absorbed by the blades causes the windmill to spin.[12] A somewhat more

[12] There is a close conceptual relation between the YORP effect and the Crookes radiometer, an airtight, partially evacuated glass bulb containing a set of four lightweight vanes mounted vertically on a low-friction spindle. Each vane is painted black on one side and white on the other. When the bulb is exposed to light, the spindle begins to turn, with the dark sides retreating from the radiation source and the white sides approaching it. The torque causing this spin is due to differences in gas pressure between the hotter black side and the cooler white side of each vane, a mechanism that is not relevant to asteroids. If the bulb were completely evacuated, the spindle would still turn due to the YORP effect, though at a much slower rate and in the opposite direction.

realistic model is described in Figure 7.6.

To quantify this effect, once again we assume that the asteroid is in a circular orbit with semimajor axis a and that its spin axis is normal to the orbital plane. The flux of incoming energy per unit surface area is $L_\odot/(4\pi a^2)$. A small area $\mathrm{d}A$ of the asteroid surface with normal $\hat{\mathbf{n}}$ intercepts this energy at a rate $\mathrm{d}A\, W(-\hat{\mathbf{X}}\cdot\hat{\mathbf{n}})L_\odot/(4\pi a^2)$, where $\hat{\mathbf{X}}$ is the unit vector pointing from the Sun to the asteroid and $W(x) = x$ if $x > 0$ (day side) and 0 if $x < 0$ (night side).[13] This corresponds to a momentum flux onto the surface or force per unit area

$$\frac{\mathrm{d}\mathbf{F}_{\rm in}}{\mathrm{d}A} = \frac{L_\odot}{4\pi a^2 c} W(-\hat{\mathbf{X}}\cdot\hat{\mathbf{n}})\hat{\mathbf{X}}. \tag{7.74}$$

Now assume for simplicity that the asteroid is a black body with negligible thermal inertia. Then the same energy flux must be re-radiated from the surface, and if the surface satisfies Lambert's law (Appendix H) the corresponding force per unit area is given by equations (H.3) and (H.17):

$$\frac{\mathrm{d}\mathbf{F}_{\rm out}}{\mathrm{d}A} = -\frac{L_\odot}{6\pi a^2 c} W(-\hat{\mathbf{X}}\cdot\hat{\mathbf{n}})\hat{\mathbf{n}}. \tag{7.75}$$

The torque on the asteroid is

$$\mathbf{N} = \int \mathrm{d}A\, \mathbf{r}\times\left(\frac{\mathrm{d}\mathbf{F}_{\rm in}}{\mathrm{d}A} + \frac{\mathrm{d}\mathbf{F}_{\rm out}}{\mathrm{d}A}\right) = \frac{L_\odot}{4\pi a^2 c}\int \mathrm{d}A\, W(-\hat{\mathbf{X}}\cdot\hat{\mathbf{n}})\mathbf{r}\times(\hat{\mathbf{X}} - \tfrac{2}{3}\hat{\mathbf{n}}), \tag{7.76}$$

where $\mathbf{r}$ is the vector from the center of mass of the asteroid to the surface area element $\mathrm{d}A$.

We now average the torque over the spin period of the asteroid. Let (x, y, z) be Cartesian coordinates that are fixed on the asteroid, with origin at the center of mass, $\hat{\mathbf{z}}$ parallel to the spin angular momentum of the asteroid, and $\hat{\mathbf{x}}$ coinciding with the unit vector $\hat{\mathbf{X}}$ from the Sun at $t = 0$. We have $\hat{\mathbf{X}} = \hat{\mathbf{x}}\cos\phi - \hat{\mathbf{y}}\sin\phi$, with $\phi = \omega_{\rm syn}t$ and $\omega_{\rm syn}$ the synodic frequency. If the integral (7.76) is done in the body-fixed coordinates (x, y, z), then $\hat{\mathbf{n}}$ and $\mathbf{r}$

[13] This calculation neglects the possibility of self-shadowing of one part of the asteroid by another, in which the line of sight from the Sun intersects the asteroid surface more than twice. Self-shadowing is best accounted for by numerical calculations.

are fixed as the asteroid spins, and the average over ϕ gives[14]

$$\begin{aligned}\langle W(-\hat{\mathbf{X}}\cdot\hat{\mathbf{n}})(\hat{\mathbf{X}}-\tfrac{2}{3}\hat{\mathbf{n}})\rangle_\phi \\ &= \langle W(-\hat{\mathbf{x}}\cdot\hat{\mathbf{n}}\cos\phi+\hat{\mathbf{y}}\cdot\hat{\mathbf{n}}\sin\phi)(\hat{\mathbf{x}}\cos\phi-\hat{\mathbf{y}}\sin\phi-\tfrac{2}{3}\hat{\mathbf{n}})\rangle_\phi \\ &= -\tfrac{1}{4}[\hat{\mathbf{n}}-\hat{\mathbf{z}}(\hat{\mathbf{n}}\cdot\hat{\mathbf{z}})]-\frac{2}{3\pi}[1-(\hat{\mathbf{n}}\cdot\hat{\mathbf{z}})^2]^{1/2}\hat{\mathbf{n}}. \end{aligned} \tag{7.77}$$

Therefore equation (7.76) becomes

$$\langle\mathbf{N}\rangle_\phi = -\frac{L_\odot}{16\pi a^2 c}\int \mathrm{d}A\,\mathbf{r}\times\Big\{\hat{\mathbf{n}}-\hat{\mathbf{z}}(\hat{\mathbf{n}}\cdot\hat{\mathbf{z}})+\frac{8}{3\pi}[1-(\hat{\mathbf{n}}\cdot\hat{\mathbf{z}})^2]^{1/2}\hat{\mathbf{n}}\Big\}. \tag{7.78}$$

The first two terms in the braces arise from the incoming radiation and the last term arises from the outgoing radiation. The first term can be rewritten as an integral over the body of the asteroid using the identity (B.37), $\int \mathrm{d}A\,\mathbf{r}\times\hat{\mathbf{n}} = -\int \mathrm{d}A\,\hat{\mathbf{n}}\times\mathbf{r} = -\int \mathrm{d}\mathbf{r}\,\nabla\times\mathbf{r} = \mathbf{0}$. Similarly, the component of the second term along axis i can be rewritten using the identity (B.32), $-\int \mathrm{d}A\,(\mathbf{r}\times\hat{\mathbf{z}})_i(\hat{\mathbf{n}}\cdot\hat{\mathbf{z}}) = -\int \mathrm{d}A\,\hat{\mathbf{n}}\cdot[\hat{\mathbf{z}}(\mathbf{r}\times\hat{\mathbf{z}})_i] = -\int \mathrm{d}\mathbf{r}\,\nabla\cdot[\hat{\mathbf{z}}(\mathbf{r}\times\hat{\mathbf{z}})_i] = 0$. We conclude that the average torque from the incoming solar radiation, arising from $\mathrm{d}\mathbf{F}_{\mathrm{in}}/\mathrm{d}A$, always vanishes. Thus any average torque arises solely from the outgoing thermal radiation, and equation (7.78) simplifies to

$$\langle\mathbf{N}\rangle_\phi = -\frac{L_\odot}{6\pi^2 a^2 c}\int \mathrm{d}A\,[1-(\hat{\mathbf{n}}\cdot\hat{\mathbf{z}})^2]^{1/2}(\mathbf{r}\times\hat{\mathbf{n}}). \tag{7.79}$$

It is straightforward to show that this integral vanishes if the asteroid is spherical, or even an ellipsoid rotating around one of its principal axes.

If the torque is along the spin axis, $\langle\mathbf{N}\rangle_\phi = N\hat{\mathbf{z}}$, we have

$$N = -\frac{L_\odot}{6\pi^2 a^2 c}\int \mathrm{d}A\,(1-n_z^2)^{1/2}(xn_y-yn_x). \tag{7.80}$$

The spin angular momentum of the asteroid is $\mathbf{S} = S\hat{\mathbf{z}}$, where $S = C\omega_{\mathrm{in}}$, ω_{in} is the spin frequency in an inertial frame, and $C = \rho\int \mathrm{d}\mathbf{r}\,(x^2+y^2)$ is the

[14] The integral involved in the average can be done by writing $-\hat{\mathbf{x}}\cdot\hat{\mathbf{n}}\cos\phi+\hat{\mathbf{y}}\cdot\hat{\mathbf{n}}\sin\phi = A\sin(\phi-B)$ where $A\sin B = \hat{\mathbf{x}}\cdot\hat{\mathbf{n}}$ and $A\cos B = \hat{\mathbf{y}}\cdot\hat{\mathbf{n}}$, then integrating from $\phi = B$ to $\phi = B+\pi$ if $A > 0$ or from $\phi = B-\pi$ to B if $A < 0$. The result is then simplified using the identities (B.6) with $\mathbf{A} = \hat{\mathbf{n}}$.

moment of inertia (eq. 1.133). Then $\dot{S} = N$ or

$$\dot{\omega}_{\rm in} = \frac{L_\odot f_{\rm YORP}}{6\pi^2 \rho a^2 c R^2} \quad \text{where} \quad f_{\rm YORP} \equiv -R^2 \frac{\int {\rm d}A\,(1-n_z^2)^{1/2}(x n_y - y n_x)}{\int {\rm d}\mathbf{r}\,(x^2+y^2)}. \tag{7.81}$$

Here R is the mean radius of the asteroid and $f_{\rm YORP}$ is the **YORP coefficient**, a dimensionless measure of the irregularity in the asteroid surface that contributes to the YORP effect. The size of the YORP coefficient is difficult to estimate, but the handful of near-Earth asteroids in which the YORP effect has been measured suggest $|f_{\rm YORP}| \sim 0.01$.

The characteristic timescale for spin-up or spin-down due to the YORP effect is given by

$$\frac{|\dot{\omega}_{\rm in}|}{\omega_{\rm in}} = \frac{1}{19.3\,\text{Myr}} \frac{P}{8\,\text{hr}} \frac{3\,\text{g}\,\text{cm}^{-3}}{\rho} \frac{f_{\rm YORP}}{0.01} \left(\frac{3\,\text{au}}{a} \frac{1\,\text{km}}{R}\right)^2. \tag{7.82}$$

Thus the YORP effect can significantly alter the spins of asteroids as large as $R \sim 10\,\text{km}$ over the lifetime of the solar system.

Since the YORP torque is independent of the asteroid's angular velocity (so long as the thermal conductivity of the asteroid is unimportant), it can spin up asteroids to angular speeds sufficiently large that the centrifugal force exceeds the self-gravitational force and tensile strength of the asteroid, causing it to break apart and perhaps forming a binary asteroid or a pair of unbound asteroids on similar orbits (Walsh 2018). A complication in such scenarios is that the YORP torque also tends to increase the obliquity, and once the obliquity exceeds 90° the torque will be reversed and the asteroid will begin to spin down. Changes in the spin angular momentum due to the YORP effect must also compete with changes due to collisions with other asteroids.

The spin evolution due to the YORP effect also determines the magnitude and direction of semimajor axis evolution due to the Yarkovsky effect.

Chapter 8

Tides

The analysis in earlier chapters of this book was focused on time-reversible Hamiltonian systems. However, this analysis does not provide a complete picture of celestial mechanics. Time-dependent tidal forces inside a satellite, planet, or star create internal strains that dissipate energy. This process of **tidal friction** ultimately converts spin or orbital energy into heat and thereby leads to irreversible changes in the dynamics of a planetary system. Our understanding of the physics of tidal friction is still incomplete, but its consequences can be explored using simple parametrized models (Darwin 1899; Cartwright 1999; Souchay et al. 2013).

The tidal force per unit mass exerted at the surface of a body of mass M and radius R by a body of mass m at distance $d \gg R$ is approximately $F_{\mathrm{tide}} = \mathbb{G}mR/d^3$ (eq. 3.71). The force at the surface due to M's self-gravity is $F_{\mathrm{self}} = \mathbb{G}M/R^2$. Thus the fractional strength of the tides is characterized by the dimensionless number

$$\epsilon_{\mathrm{tide}} = \frac{F_{\mathrm{tide}}}{F_{\mathrm{self}}} = \frac{m}{M}\left(\frac{R}{d}\right)^3. \tag{8.1}$$

On the Earth, $\epsilon_{\mathrm{tide}} = 5.6 \times 10^{-8}$ from the Moon and $\epsilon_{\mathrm{tide}} = 2.6 \times 10^{-8}$ from the Sun. Exoplanets in short-period orbits experience much stronger tides; for example, WASP-12b has $\epsilon_{\mathrm{tide}} = 0.060$. When ϵ_{tide} approaches unity the body may be tidally disrupted, a process described in §8.6.

8.1 The minimum-energy state

For an initial orientation to the effects of tidal friction, we examine a system of two bodies—planet and satellite, star and planet, or binary star—in a circular orbit. The masses are M_1 and M_2, the semimajor axis is a, and the spin angular momenta of the two bodies are parallel to the orbital angular momentum (i.e., the obliquities are zero), with magnitudes $S_1 = C_1\omega_1$ and $S_2 = C_2\omega_2$. Here ω_i and C_i are the angular speed and moment of inertia for each body.

In the center-of-mass frame, the angular momentum of the system is

$$L_{\rm tot} = L_{\rm orb}(a) + C_1\omega_1 + C_2\omega_2, \tag{8.2}$$

where the orbital angular momentum is (eqs. 1.9 and 1.28)

$$L_{\rm orb}(a) = \mu[\,\mathbb{G}(M_1 + M_2)a]^{1/2} \tag{8.3}$$

with the reduced mass

$$\mu = \frac{M_1 M_2}{M_1 + M_2}. \tag{8.4}$$

The orbital plus spin energy of the system is (eqs. 1.9, 1.32 and D.92)

$$E_{\rm tot} = -\frac{\mathbb{G}\mu(M_1 + M_2)}{2a} + \tfrac{1}{2}C_1\omega_1^2 + \tfrac{1}{2}C_2\omega_2^2 = -\frac{\mathbb{G}M_1M_2}{2a} + \tfrac{1}{2}C_1\omega_1^2 + \tfrac{1}{2}C_2\omega_2^2. \tag{8.5}$$

Tidal friction dissipates energy but conserves angular momentum, so tides drive the system toward the minimum-energy state consistent with its initial angular momentum. To find this equilibrium state, we rewrite the energy as

$$E_{\rm tot} = -\frac{\mathbb{G}M_1M_2}{2a} + \tfrac{1}{2}C_1\omega_1^2 + \frac{[L_{\rm tot} - L_{\rm orb}(a) - C_1\omega_1]^2}{2C_2}. \tag{8.6}$$

Suppose that the semimajor axis and the spin frequencies in the minimum-energy state are $a_{\rm eq}$, $\omega_{1,\rm eq}$ and $\omega_{2,\rm eq}$. Set $a = a_{\rm eq} + \Delta a$, $\omega_1 = \omega_{1,\rm eq} + \Delta\omega_1$ and expand the result in the small quantities Δa and $\Delta\omega_1$. To first order,

$$E_{\rm tot} = \left[\frac{\mathbb{G}M_1M_2}{2a_{\rm eq}^2} - \frac{\mathbb{G}^{1/2}M_1M_2\,\omega_{2,\rm eq}}{2(M_1 + M_2)^{1/2}a_{\rm eq}^{1/2}}\right]\Delta a$$

$$+ C_1(\omega_{1,\text{eq}} - \omega_{2,\text{eq}})\Delta\omega_1 + \text{const}, \tag{8.7}$$

where $\omega_{2,\text{eq}}$ is obtained by evaluating equation (8.2) using a_eq and $\omega_{1,\text{eq}}$. If the parameters $(a_\text{eq}, \omega_{1,\text{eq}}, \omega_{2,\text{eq}})$ are an extremum of the energy, then the first-order variation in energy must vanish, which requires

$$\omega_{1,\text{eq}} = \omega_{2,\text{eq}} = \left[\frac{\mathbb{G}(M_1 + M_2)}{a_\text{eq}^3}\right]^{1/2} = n_\text{eq}; \tag{8.8}$$

in the last equality we have introduced the mean motion $n(a) = [\mathbb{G}(M_1 + M_2)/a^3]^{1/2}$. These conditions mean that an equilibrium system is **synchronous**: the angular speeds of both spins and the orbit are the same. A local example of such an equilibrium is Pluto and its satellite Charon.

Substituting (8.8) into (8.2), we arrive at an implicit equation for the equilibrium semimajor axis a_eq,

$$L_\text{tot} = \mu[\mathbb{G}(M_1 + M_2)a_\text{eq}]^{1/2} + (C_1 + C_2)\left[\frac{\mathbb{G}(M_1 + M_2)}{a_\text{eq}^3}\right]^{1/2}. \tag{8.9}$$

This equation has no roots if L_tot is smaller than

$$L_\text{min} \equiv 4\left[\frac{\mathbb{G}^2 M_1^3 M_2^3 (C_1 + C_2)}{27(M_1 + M_2)}\right]^{1/4}, \tag{8.10}$$

while if $L_\text{tot} > L_\text{min}$ there are two roots. One of the two roots for the semimajor axis is smaller than a_0 and the other is larger than a_0, where

$$a_0^2 = \frac{3(C_1 + C_2)}{\mu}. \tag{8.11}$$

Any equilibrium state is an extremum of the energy—a maximum, minimum, or saddle point—but the equilibrium is only stable in the presence of dissipation if it is a minimum of the energy. To determine whether an equilibrium is stable, we must therefore expand equation (8.6) to second order in the small quantities Δa and $\Delta\omega_1$. Since the first-order terms vanish at equilibrium, we have

$$E_\text{tot} - E_\text{eq} = (\Delta a)^2 \frac{\mu n_\text{eq}^2}{8C_2}(\mu a_\text{eq}^2 - 3C_2) + (\Delta a \Delta\omega_1)\frac{C_1}{2C_2}\mu n_\text{eq} a_\text{eq}$$

$$+ (\Delta\omega_1)^2 \frac{C_1(C_1 + C_2)}{2C_2}. \tag{8.12}$$

The right side of this expression is the quadratic form $px^2 + rxy + qy^2$. If p is negative, the expression is negative whenever $y = 0$ and x is nonzero, so $x = y = 0$ cannot be a minimum. If p is non-negative, the expression can be rewritten as $(p^{1/2}x + \frac{1}{2}rp^{-1/2}y)^2 + (q - \frac{1}{4}r^2/p)y^2$. Since the square is non-negative, the complete expression is non-negative for all x and y if and only if $pq \geq \frac{1}{4}r^2$. Therefore the equilibrium is a minimum-energy state if and only if

$$\mu a_{\rm eq}^2 \geq 3C_2 \quad \text{and} \quad \mu a_{\rm eq}^2 \geq 3(C_1 + C_2). \tag{8.13}$$

The second condition is always stronger, and corresponds to $a_{\rm eq} > a_0$ where a_0 is defined in equation (8.11). We conclude that for $L_{\rm tot} > L_{\rm min}$, there is a unique minimum-energy state having a semimajor axis that is the larger of the two solutions of equation (8.9). A generalization of these arguments to non-circular orbits and nonzero obliquities is given by Hut (1980).

So far we have only used two basic laws of physics: the total angular momentum of the binary system is conserved, and the total kinetic and potential energy must decrease when tides convert this energy to heat. As we have seen, these are sufficient to describe completely the stable equilibrium state of the binary—if there is one. The description of the evolutionary path by which the system reaches this equilibrium is much more complicated and uncertain, and this description will occupy most of the rest of this chapter. However, one basic feature of almost every dynamical model of tides is that when the spin of a body exceeds the angular speed of its companion in its orbit, then tides transfer angular momentum from the spin to the orbit, and vice versa.

The implications of this feature are easiest to describe in the case where the binary orbit is circular and the spin and orbital angular momenta are aligned. Let n and ω be the mean motion and the spin rate of the body we are examining. Then if $\omega > n$, tides drain angular momentum from the spin, so ω shrinks, eventually approaching the synchronous state if we ignore changes to the orbit. Similarly if $\omega < n$, tides add angular momentum to the spin so ω grows.

The situation for the orbit is reversed: if $\omega > n$, tides add angular momentum to the orbit, but the mean motion of a circular orbit is a decreasing function of its angular momentum, $n \propto L^{-3}$, so the orbit expands and the mean motion shrinks, diverging from the spin rate—another example of the donkey principle described in §3.2. Similarly if $\omega < n$, tides drain angular momentum from the orbit, which shrinks, and the mean motion grows, again diverging from the spin rate.

We can now describe what happens when $L_{\rm tot} < L_{\rm min}$ so there is no equilibrium state for the binary. For simplicity, suppose that body 2 is a point mass, so $C_2 = 0$, and suppose that the mean motion is initially larger than the angular speed of body 1. Then tidal friction transfers angular momentum from the orbit to body 1, so ω_1 grows. In so doing the orbit loses angular momentum, its semimajor axis shrinks, and n grows. If n grows faster than ω_1, the synchronous state can never be achieved—the orbit will continue to shrink until the two bodies collide. Similarly, if n is initially smaller than ω_1, the orbit will expand to infinity without ever reaching the synchronous state.[1]

As an example, we can apply these arguments to the Earth–Moon system. To keep the example simple, we neglect the obliquity of the Earth, the eccentricity of the lunar orbit, and the influence of the Sun, although all of these play an important role in the evolution of the lunar orbit. At present the semimajor axis is a = 384 400 km and 83% of the total angular momentum is in the orbit, with almost all the rest in the Earth's spin. The total angular momentum is $L_{\rm tot} = 2.33 L_{\rm min}$ and the critical semimajor axis is a_0 = 57 670 km. The mean motion is much smaller than the angular speed of the Earth—the month is much longer than the day—so angular momentum is being transferred by tides from the Earth's spin to the Moon's orbit. As a result the Moon's orbit is expanding and the Earth's spin is slowing down, and both the month and the day are getting longer. If and when the system reaches the synchronous equilibrium, the semimajor axis will be 554 200 km = $86.9 R_\oplus$ and the spin and orbital periods will be 47.2 d. Descriptions of the past and future evolution of the Earth–Moon system are

[1] This behavior is often called the **Darwin instability**, after G. H. Darwin (1845–1912), the son of Charles Darwin and one of the first physicists to investigate tides.

given by Goldreich (1966) and Touma & Wisdom (1994).

In contrast to the Earth–Moon system, most of the exoplanets that are close enough to their host stars to have significant tidal interactions (orbital periods less than a few days) have angular momenta less than the critical value $L_{\rm min}$ and mean motions larger than the stellar spin velocity, and thus their orbits will shrink until the planet is consumed by its host star (Matsumura et al. 2010).

8.2 The equilibrium tide

In this section we describe the response of a body to the tidal field from an orbiting companion. To keep the analysis general, we refer to the two objects as the "responding body" (subscript "r") and the "forcing body" (subscript "f"). In planetary systems the forcing body or the responding body may be a planetary satellite, a planet or a star. In almost all circumstances we can treat the forcing body as a point mass, and we shall do so here. For a complete picture of the tidal interactions in a two-body system, we must treat first one body then the other as the forcing body.

We work in a frame centered on the responding body. The potential due to a forcing body of mass $m_{\rm f}$ at position $\mathbf{r}_{\rm f}$ is given in terms of spherical harmonics by equation (C.55),

$$\Phi(\mathbf{r}) = -\frac{\mathbb{G}m_{\rm f}}{|\mathbf{r}_{\rm f} - \mathbf{r}|} = -\sum_{l=0}^{\infty} \frac{4\pi\,\mathbb{G}m_{\rm f}}{2l+1} \frac{r^l}{r_{\rm f}^{l+1}} \sum_{m=-l}^{l} Y_{lm}^*(\theta_{\rm f}, \phi_{\rm f}) Y_{lm}(\theta, \phi). \quad (8.14)$$

Here (r, θ, ϕ) and $(r_{\rm f}, \theta_{\rm f}, \phi_{\rm f})$ are the spherical coordinates of the points $\mathbf{r}$ and $\mathbf{r}_{\rm f}$ respectively, and we have assumed that $r_{\rm f} > r$ since the two bodies cannot overlap. The monopole or $l = 0$ term is independent of $\mathbf{r}$ and therefore generates no force. The dipole or $l = 1$ term is canceled by the fictitious force due to the acceleration of the reference frame (see the discussion following eq. 3.70). Thus the forcing potential can be written

$$\Phi_{\rm f}(\mathbf{r}) = -\sum_{l=2}^{\infty} \frac{4\pi\,\mathbb{G}m_{\rm f}}{2l+1} \frac{r^l}{r_{\rm f}^{l+1}} \sum_{m=-l}^{l} Y_{lm}^*(\theta_{\rm f}, \phi_{\rm f}) Y_{lm}(\theta, \phi). \quad (8.15)$$

In most cases $r_\mathrm{f} \gg r$, so we can drop all terms with $l > 2$. Then if we write the time dependence explicitly, we have

$$\Phi_\mathrm{f}(\mathbf{r},t) = -\frac{4\pi \mathbb{G} m_\mathrm{f}}{5} \frac{r^2}{r_\mathrm{f}^3(t)} \sum_{m=-2}^{2} Y_{2m}^*[\theta_\mathrm{f}(t), \phi_\mathrm{f}(t)] Y_{2m}(\theta,\phi). \tag{8.16}$$

An equivalent expression is (cf. eq. 3.71)

$$\Phi_\mathrm{f}(\mathbf{r},t) = \frac{\mathbb{G} m_\mathrm{f} r^2}{2 r_\mathrm{f}^3(t)} - \frac{3\,\mathbb{G} m_\mathrm{f} [\mathbf{r}_\mathrm{f}(t)\cdot\mathbf{r}]^2}{2 r_\mathrm{f}^5(t)}. \tag{8.17}$$

The **equilibrium tide** is the distortion of the responding body in response to a static tidal potential. In practice, "static" means either that (i) the orbital period of the forcing body is much longer than the characteristic response time of the responding body's interior (and atmosphere and oceans, if present), or (ii) the two bodies are static in a rotating frame of reference, as in the case of a planet and satellite in the synchronous state.

For example, suppose that the responding body is rigid and surrounded by a low-density atmosphere. The surface of the atmosphere subjected to a static tide is an equipotential surface (see footnote to Box 1.3), given by the equation

$$-\frac{\mathbb{G} m_\mathrm{r}}{r} + \Phi_\mathrm{f}(r,\theta,\phi) = \text{const}, \tag{8.18}$$

where m_r is the mass of the responding body. Writing $r(\theta,\phi) = R_\mathrm{r} + \Delta R(\theta,\phi)$ (where R_r is the mean radius of the responding body), expanding to first order in ΔR, and assuming that the tidal potential is small, we find

$$\Delta R(\theta,\phi) \simeq -\frac{\Phi_\mathrm{f}(R_\mathrm{r},\theta,\phi)}{g}, \quad \text{where} \quad g = \frac{\mathbb{G} m_\mathrm{r}}{R_\mathrm{r}^2} \tag{8.19}$$

is the gravitational acceleration at the surface of the unperturbed responding body. Using equation (8.17) we have

$$\Delta R = \frac{m_\mathrm{f}}{m_\mathrm{r}} \frac{R_\mathrm{r}^4}{r_\mathrm{f}^3} \left[\tfrac{3}{2}(\hat{\mathbf{r}}_\mathrm{f}\cdot\hat{\mathbf{r}})^2 - \tfrac{1}{2}\right], \tag{8.20}$$

where $\hat{\mathbf{r}}$ and $\hat{\mathbf{r}}_\mathrm{f}$ are unit vectors pointing from the center of the planet to the location on the surface and to the forcing body. Low tide occurs when $\mathbf{r}_\mathrm{f}$ is perpendicular to $\mathbf{r}$, and high tide occurs when they are parallel or antiparallel, which of course is why there are two tides per day on Earth.

8.2.1 Love numbers

In practice, the responding body is not rigid and so it is distorted by any imposed tidal potential. The tidal potential can be decomposed into a sum of spherical harmonics, as in equation (8.15), and each term of the sum can be written in the form $\Phi_\mathrm{f}(r,\theta,\phi) = f^\mathrm{f}_{lm}(r/R_\mathrm{r})^l Y_{lm}(\theta,\phi)$, where f^f_{lm} is a function of time depending on the trajectory of the forcing body. If the unperturbed responding body is spherical and non-rotating, and the tidal field is small enough for its response to be linear, then a tidal potential of this form creates a density response that also has angular dependence proportional to $Y_{lm}(\theta,\phi)$. This density response generates a potential response due to the self-gravity of the responding body, and this response $\Phi_\mathrm{r}(r,\theta,\phi)$ will also be proportional to $Y_{lm}(\theta,\phi)$. The potential response must satisfy Laplace's equation outside the radius of the responding body R_r; thus its radial dependence for $r > R_\mathrm{r}$ is determined by its angular dependence, $\Phi_\mathrm{r}(\mathbf{r}) = f^\mathrm{r}_{lm}(R_\mathrm{r}/r)^{l+1} Y_{lm}(\theta,\phi)$. The response is a linear function of the forcing, so $f^\mathrm{r}_{lm} \propto f^\mathrm{f}_{lm}$ or[2]

$$\frac{\Phi_\mathrm{r}(R_\mathrm{r},\theta,\phi)}{\Phi_\mathrm{f}(R_\mathrm{r},\theta,\phi)} = \frac{f^\mathrm{r}_{lm}}{f^\mathrm{f}_{lm}} \equiv k_l. \tag{8.21}$$

where the **gravitational Love number** k_l is a constant[3] that depends only on the internal properties of the responding body; because of the assumed

[2] In equation (8.21) we have assumed that Φ_r is well defined at the surface of the responding body, even though the density may be discontinuous there. This assumption is justified by Poisson's equation (B.44), which implies that a discontinuous density gives rise to a potential that is continuous and has a continuous first derivative.

[3] The Love number measures the same physical property as the apsidal-motion constant in the theory of binary stars. The apsidal-motion constant is a factor of two smaller than the Love number, but unfortunately both are denoted by k_2.

spherical symmetry of the planet, the Love number is independent of m at fixed l.

Other properties of the response can also be characterized by dimensionless numbers. The **displacement Love number** h_l is defined by the ratio of the height of the tidal distortion of the surface to the height of the tide in a rigid body with a low-density atmosphere (eq. 8.19),

$$-\frac{g\Delta R(\theta,\phi)}{\Phi_{\mathrm{f}}(R_{\mathrm{r}},\theta,\phi)} \equiv h_l. \tag{8.22}$$

Small solar-system bodies can be approximated as incompressible elastic solids. The response of such bodies to a tidal field is derived in Appendix I. In particular the Love numbers for the dominant $l = 2$ tidal fields are given by equations (I.30) and (I.31),

$$k_2 = \frac{3}{2[1 + 19\mu/(2g\rho R_{\mathrm{r}})]}, \quad h_2 = \frac{5}{2[1 + 19\mu/(2g\rho R_{\mathrm{r}})]}, \tag{8.23}$$

where ρ and μ are the density and rigidity of the solid, assumed to be constant.[4] For completely rigid planets ($\mu \to \infty$) the Love numbers vanish, $k_2 = h_2 = 0$. For large planets of uniform density, in which self-gravity is much more important than rigidity, $k_2 = \frac{3}{2}$ and $h_2 = \frac{5}{2}$. In gaseous planets, the density increases toward the center and for these the Love numbers are smaller, typically $0.1 \lesssim k_2 \lesssim 0.6$. The Love numbers of stars are even smaller; for the Sun $k_2 \simeq 0.030$ (Claret 2019).

The Earth's Love numbers are $k_2 = 0.295$ and $h_2 = 0.608$. Its tides arise from forcing by the Moon and Sun. The difference in height between high and low tide is given by equation (8.20) as $\frac{3}{2}h_2(m_{\mathrm{f}}/m_{\mathrm{r}})(R_{\mathrm{r}}^4/r_{\mathrm{f}}^3)$, or 0.33 m for lunar tides and 0.15 m for solar tides. These numbers refer to solid Earth tides. The tides in the Earth's oceans are larger—the difference between high and low tide can exceed 16 m—in part because the oceans

[4] The rigidities of typical materials are ~ 3 GPa for ice, ~ 20 GPa for rock, and ~ 80 GPa for steel. One gigapascal = 10^9 Pa = 10^9 kg m^{-1} s^{-2}.

have no rigidity.[5] The ocean tides are also much more complicated than the solid tides, because the natural oscillation periods of ocean basins and bays can be comparable to the tidal forcing frequency of about half a day.

Bodies much smaller than Earth have Love numbers much less than unity because of the factor $\mu/(g\rho R_\mathrm{r})$ in equations (8.23) reflecting the difference between fluid and elastic response; for example the Moon has $k_2 = 0.024$ and $h_2 = 0.042$. The gravitational Love numbers of the four inner planets plus Jupiter and Saturn have been measured by tracking orbiting spacecraft (see Appendix A).

8.3 Tidal friction

To explore the effects of dissipation due to gravitational tides, we first examine the simple case of a damped harmonic oscillator subjected to a sinusoidally varying external force. The equation of motion is

$$\ddot{x} + \omega_0^2 x = -\nu\dot{x} + F\cos\omega_\mathrm{f}t, \tag{8.24}$$

where $\nu > 0$ is the damping rate; $\omega_0 > 0$ is the oscillation frequency of the undamped, unforced oscillator; and ω_f is the forcing frequency. After any initial transients have died away, the solution is $x = A\cos(\omega_\mathrm{f}t + \delta)$ where δ is the phase offset or lag angle. We allow A to have either sign so we can assume that $|\delta| \le \frac{1}{2}\pi$. Substituting this solution into the equation of motion, we find

$$A = \frac{F\,\mathrm{sgn}(\omega_0^2 - \omega_\mathrm{f}^2)}{[(\omega_0^2 - \omega_\mathrm{f}^2)^2 + \nu^2\omega_\mathrm{f}^2]^{1/2}}, \quad \tan\delta = -\frac{\nu\omega_\mathrm{f}}{\omega_0^2 - \omega_\mathrm{f}^2}. \tag{8.25}$$

With our conventions the phase offset can be positive or negative, and the relation between the signs of F, A, and δ is

$$\mathrm{sgn}(AF) = -\,\mathrm{sgn}(\omega_\mathrm{f}\delta) = \mathrm{sgn}(\omega_0^2 - \omega_\mathrm{f}^2). \tag{8.26}$$

[5] In 1863 William Thompson, later Lord Kelvin (1824–1907), pointed out that if the Earth had no rigidity the solid tides would have the same amplitude as the ocean tides, so the shorelines would move up and down with the oceans and there would be no relative ocean tides. This was one of the first important constraints on the structure of the Earth's interior.

We are mostly interested in systems with weak damping, in which $|\delta| \ll 1$.

The rate at which the external force does work on the oscillator is the product of force and velocity,

$$\begin{aligned}\dot{W} &= (F\cos\omega_{\rm f}t)\dot{x} = -FA\omega_{\rm f}\cos\omega_{\rm f}t\sin(\omega_{\rm f}t+\delta) \\ &= -\tfrac{1}{2}FA\omega_{\rm f}[\sin(2\omega_{\rm f}t+\delta)+\sin\delta].\end{aligned} \tag{8.27}$$

This result can be integrated to give

$$W(t) = FA\left[\tfrac{1}{4}\cos(2\omega_{\rm f}t+\delta) - \tfrac{1}{2}\omega_{\rm f}t\sin\delta\right] + \text{const.} \tag{8.28}$$

The total work done over one oscillation period, $2\pi/|\omega_{\rm f}|$, is

$$\Delta W = W(t+2\pi/\omega_{\rm f}) - W(t) = -\pi FA\,\text{sgn}(\omega_{\rm f})\sin\delta, \tag{8.29}$$

which is always positive since work must be done to replace the energy lost through dissipation. The maximum energy stored in the oscillation is the difference between the maximum and minimum values of $W(t)$; neglecting the linear term in equation (8.28) since $|\delta| \ll 1$, this difference is

$$W^* = \tfrac{1}{2}|FA|. \tag{8.30}$$

The **quality factor** is defined as

$$Q \equiv \frac{2\pi W^*}{\Delta W}, \tag{8.31}$$

and equations (8.29) and (8.30) imply that

$$Q = -\frac{\text{sgn}(FA\omega_{\rm f})}{\sin\delta} = \frac{1}{|\sin\delta|} \simeq \frac{1}{|\delta|}. \tag{8.32}$$

The behavior of harmonic oscillators is relevant to tidal friction because the response of a planet or satellite to weak tidal forces can be decomposed into normal modes, each of which acts like a damped harmonic oscillator. Each normal mode has an eigenfrequency ω_0 and a quality factor Q. The equilibrium tide described in §8.2 is based on the assumption that the forcing frequency is much less than the eigenfrequency, $|\omega_{\rm f}| \ll \omega_0$, so equations (8.25) simplify to

$$A = \frac{F}{\omega_0^2}, \quad \tan\delta = -\frac{\nu\omega_{\rm f}}{\omega_0^2}. \tag{8.33}$$

Since $\nu > 0$, the sign of the phase shift is the opposite of the sign of the forcing frequency; thus the replacement rule for adding friction from the equilibrium tide is to change

$$\omega_{\rm f} t \Rightarrow \omega_{\rm f} t + \delta, \quad \delta = -\frac{\mathrm{sgn}(\omega_{\rm f})}{Q}. \tag{8.34}$$

In particular, the dominant $l = 2$ component of the tidal potential can be written as a sum of terms of the form (cf. eq. 8.16)

$$\Phi_{\rm f}(\mathbf{r}, t) = a_m r^2 Y_{2m}(\theta, \phi) \exp(-\mathrm{i}\omega_{\rm f} t), \tag{8.35}$$

where (r, θ, ϕ) are spherical coordinates centered on the responding body that is subjected to the tides. In the absence of dissipation, the tide distorts the responding body and this distortion generates an additional potential (eq. 8.21)

$$\Phi_{\rm r}(\mathbf{r}, t) = \frac{a_m k_2 R_{\rm r}^5}{r^3} Y_{2m}(\theta, \phi) \exp(-\mathrm{i}\omega_{\rm f} t), \tag{8.36}$$

where k_2 is the Love number. Dissipation is modeled by adding a phase offset,[6] changing this result to

$$\Phi_{\rm r}(\mathbf{r}, t) = \frac{a_m k_2 R_{\rm r}^5}{r^3} Y_{2m}(\theta, \phi) \exp[-\mathrm{i}(\omega_{\rm f} t + \delta)]. \tag{8.37}$$

In effect, we have replaced the Love number k_2 by a complex Love number $k_2 \exp(-\mathrm{i}\delta)$.

---

[6] This approach accounts for the phase offset due to dissipation but not the change in the amplitude (see the first of eqs. 8.25). Changes in the amplitude of the damped harmonic oscillator are second order in the damping rate, whereas changes in the lag angle are first order, so neglecting the changes in amplitude is justified so long as the dissipation is weak.

We can also write the exponential as

$$\exp[-i\omega_f(t-\tau)], \quad \text{where} \quad \tau = -\frac{\delta}{\omega_f} = \frac{\nu}{\omega_0^2} = \frac{1}{|\omega_f|Q} \tag{8.38}$$

is the time offset. The last expression follows from the second of equations (8.33) when $|\delta| \ll 1$. Thus the time offset is always positive (i.e., there is a time lag in the peaks and troughs of the forced oscillation) for the equilibrium tide. The replacement rule for adding friction can be written

$$\omega_f t \Rightarrow \omega_f(t-\tau), \quad \tau = \frac{1}{|\omega_f|Q}. \tag{8.39}$$

The assumption of constant time lag is widely used because it leads to relatively simple expressions for the evolution of the orbital elements (see Box 8.1).

A third parametrization of the phase offset is obtained by observing that $Y_{2m}(\theta,\phi)$ varies with azimuth as $\exp(im\phi)$, so we may write

$$\begin{aligned} Y_{2m}(\theta,\phi)\exp[-i(\omega_f t+\delta)] &= Y_{2m}(\theta,0)\exp[im\phi - i(\omega_f t+\delta)] \\ &= Y_{2m}(\theta,0)\exp[im(\phi-\Delta\phi) - i\omega_f t], \quad \text{where} \quad \Delta\phi = \frac{\delta}{m}. \end{aligned} \tag{8.40}$$

Thus we have replaced the phase offset δ by an azimuthal offset $\Delta\phi$. It is this angular offset that leads to torques that slowly change the spin or orbit of a body subjected to tidal forces. The replacement rule is

$$m\phi - \omega_f t \Rightarrow m(\phi-\Delta\phi) - \omega_f t, \quad \Delta\phi = -\frac{\text{sgn}(\omega_f)}{mQ} = -\frac{\omega_f\tau}{m}. \tag{8.41}$$

These arguments suggest two distinct empirical approaches to characterizing tidal friction. In models with **constant time offset**, the response of the perturbed body is displaced in time by a fixed offset τ so $Q \propto 1/|\omega_f|$. In models with **constant angle offset**, the response is displaced in angle by a fixed offset $\Delta\phi$ so Q is independent of the forcing frequency ω_f. Constant angle offset has the unphysical property that the offset jumps abruptly from a nonzero positive value to a nonzero negative one as the forcing frequency decreases smoothly from positive values to negative ones (Efroimsky & Makarov 2013). Constant time offset also has the advantage that the resulting formulas for spin and orbit evolution are generally simpler (see Box 8.1). However, neither model is physically justified: Q is likely to have a complicated dependence on the forcing frequency ω_{f} and can also depend on the amplitude, and the characterization of tidal friction by a lag angle, or a time offset, or a quality factor is just a convenient parametric representation of a nonlinear dynamical system.

The source of the dissipation or lag depends on the nature of the responding body. In solid planets or planetary cores, the dissipation is presumably due to viscoelasticity but the nature of the appropriate viscoelastic model is not well understood. In gaseous planets the dissipation may arise from turbulent viscosity in convective regions or from the excitation of internal waves that eventually damp by viscosity or nonlinear effects.

Even in the simplest case of a weak, static tidal force our current observational and theoretical constraints on Q in satellites, planets, and stars are limited (Lainey 2016). The most accurate estimates are for the Earth–Moon system: the Earth's oceans have $Q \simeq 10$ and the solid Earth has $Q \simeq 300$, while the Moon has $Q \simeq 40$. Mars has $Q \simeq 100$ as measured from secular changes in the orbit of its satellite Phobos (Problem 8.3).

Estimates of Q for other solar-system planets have traditionally been derived from indirect arguments (Goldreich & Soter 1966). For example, Mercury's Q must be small enough, $\lesssim 200$, that tidal torques from the Sun have substantially slowed its spin angular velocity, or else it would not have been captured in the 3:2 spin-orbit resonance (§7.2). Since tidal friction from the planet causes satellite orbits to expand, the existence of the innermost satellites of Jupiter, Saturn, and Uranus sets a lower limit of $Q \gtrsim 10^5$ in the host planet, assuming that the satellites are as old as the solar system.

Astrometric measurements of orbit evolution are available for several of the satellites of the outer planets, and these yield $Q \simeq 5 \times 10^4$ for Jupiter and $Q \simeq 2\,500$ for Saturn (Lainey 2016), 2–3 orders of magnitude smaller than expected from the indirect arguments in the preceding paragraph. The reason for this discrepancy is not yet understood, although a possible explanation is that the satellite orbits are locked in resonance with the normal modes of their host planet (§8.5.2).

A complication in all these estimates is that the rate of spin or orbital evolution depends on both the quality factor Q and the amplitude of the response to tidal forces, which is proportional to the Love number k_2. Thus arguments based on tidal evolution constrain only Q/k_2. This is not a big concern for the larger planets, where k_2 is of order unity, but is much more important for small bodies and stars where $k_2 \ll 1$.

Tidal friction in stars can change the orbits of planets with small semimajor axes and can even cause the planets to spiral into the star. Fortunately, fairly reliable estimates of Q in stars can be obtained from binary stars found in clusters. All of the stars in a cluster have nearly the same age (say, T), and within a cluster it is found that binary stars with orbital periods less than some critical value $P(T)$ have nearly circular orbits, while binaries with periods exceeding $P(T)$ have a wide range of eccentricities (Meibom & Mathieu 2005). The interpretation of this finding is that binaries are born on eccentric orbits, and those with periods $< P(T)$ have been circularized by tidal friction, so $P(T)$ can be used to estimate Q. The results are consistent with $Q/k_2 \sim 10^6$ for solar-type stars. More massive stars are expected to have quality factors that are larger by 2–3 orders of magnitude as the outer parts of the star are in radiative rather than convective equilibrium, so the effective viscosity is much lower (see Ogilvie 2014 for a review).

8.4 Spin and orbit evolution

We assume that the forcing body orbits the responding body with semimajor axis a. Its mean motion $n = [\mathbb{G}(m_\mathrm{r}+m_\mathrm{f})/a^3]^{1/2}$, where m_r and m_f are the masses of the responding and forcing bodies. The responding body's spin angular velocity is ω_r. As usual we may treat the forcing body as a point mass, since interactions between the quadrupole moments of the two bodies are negligible compared to the interaction between the quadrupole moment of one and the monopole moment of the other.

In the absence of dissipation, the potential generated by the tidal distortion of the responding body can be determined from equations (8.16), (8.35) and (8.36):

$$\Phi_\mathrm{r} = -\frac{4\pi\,\mathbb{G}m_\mathrm{f}k_{2,\mathrm{r}}R_\mathrm{r}^5}{5r_\mathrm{f}^3r^3}\sum_{m=-2}^{2}Y_{2m}^*(\theta_\mathrm{f},\phi_\mathrm{f})Y_{2m}(\theta,\phi), \tag{8.42}$$

where $k_{2,\mathrm{r}}$ is the gravitational Love number of the responding body. Using equation (8.17), this relation can also be written as

$$\Phi_\mathrm{r} = \frac{\mathbb{G}m_\mathrm{f}k_{2,\mathrm{r}}R_\mathrm{r}^5}{2r_\mathrm{f}^5r^5}\big[r_\mathrm{f}^2r^2 - 3(\mathbf{r}_\mathrm{f}\cdot\mathbf{r})^2\big]. \tag{8.43}$$

8.4.1 Semimajor axis migration

We may assume without loss of generality that the orbit lies in the plane $\theta_\mathrm{f} = \frac{1}{2}\pi$; then from the definitions of the spherical harmonics in equations (C.56) we have

$$\Phi_\mathrm{r} = -\frac{3\,\mathbb{G}m_\mathrm{f}k_{2,\mathrm{r}}R_\mathrm{r}^5}{4r_\mathrm{f}^3r^3}\sin^2\theta\cos 2(\phi-\phi_\mathrm{f}) + \frac{\mathbb{G}m_\mathrm{f}k_{2,\mathrm{r}}R_\mathrm{r}^5}{4r_\mathrm{f}^3r^3}(3\cos^2\theta - 1). \tag{8.44}$$

For the following derivation, we also assume that the forcing body travels on a circular orbit and that the spin axis of the responding body is aligned with the orbital axis. In this case we can replace $r_\mathrm{f}(t)$ by a, the semimajor axis, and we can drop the second term in the potential (8.44) since it is time-independent and does not contribute to tidal friction. According to equation (8.41), we can model tidal dissipation by adding an azimuthal phase shift to the time-dependent tidal potential,

$$\Phi_\mathrm{r} = -\frac{3\,\mathbb{G}m_\mathrm{f}k_{2,\mathrm{r}}R_\mathrm{r}^5}{4a^3r^3}\sin^2\theta\cos 2(\phi-\phi_\mathrm{f}-\Delta\phi). \tag{8.45}$$

To relate the phase shift $\Delta\phi$ to the quality factor Q_r we work in the frame in which the responding body is stationary, which rotates at angular speed ω_r. In this frame the azimuth of the forcing body changes at a rate

Box 8.1: Tides with constant time lag

The spin and orbit evolution are simplified if the quality factor is inversely proportional to frequency, $Q \propto 1/|\omega_\mathrm{f}|$, so the time lag τ is independent of frequency.

When the forcing body is at $\mathbf{r}_\mathrm{f}$ the tidal potential at position $\mathbf{r}$ is given by equation (8.43); we write this as $\Phi_\mathrm{r}(\mathbf{r}, \mathbf{r}_\mathrm{f})$. We model tidal friction by evaluating the potential at time t using the position of the forcing body relative to the responding body at time $t - \tau$. To find this, we want to know the velocity of the forcing body in the frame in which the responding body is non-rotating. If the latter's spin angular velocity is $\boldsymbol{\omega}_\mathrm{r}$, the velocity of the former in this frame is $\mathbf{v}_\mathrm{rot} = \mathbf{v} - \boldsymbol{\omega}_\mathrm{r} \times \mathbf{r}_\mathrm{f}$, where $\mathbf{v}$ is the velocity in a non-rotating frame (eq. D.17). Then if the time lag is small, $\mathbf{r}_\mathrm{f}(t - \tau) \simeq \mathbf{r}_\mathrm{f}(t) - \tau \mathbf{v}_\mathrm{rot}(t)$, so the potential is $\Phi_\mathrm{r}[\mathbf{r}, \mathbf{r}_\mathrm{f}(t-\tau)] \simeq \Phi_\mathrm{r}[\mathbf{r}, \mathbf{r}_\mathrm{f}(t)] - \tau(\mathbf{v} - \boldsymbol{\omega}_\mathrm{r} \times \mathbf{r}_\mathrm{f}) \cdot \partial \Phi_\mathrm{r}[\mathbf{r}, \mathbf{r}_\mathrm{f}(t)]/\partial \mathbf{r}_\mathrm{f}$. The components of the force on the forcing body are

$$F_{\mathrm{tide},i} = -\left.\frac{\partial \Phi_\mathrm{r}(\mathbf{r}, \mathbf{r}_\mathrm{f})}{\partial r_i}\right|_{\mathbf{r}_\mathrm{f}=\mathbf{r}} + \tau[\mathbf{v} - (\boldsymbol{\omega}_\mathrm{r} \times \mathbf{r}_\mathrm{f})]_j \left.\frac{\partial^2 \Phi_\mathrm{r}(\mathbf{r}, \mathbf{r}_\mathrm{f})}{\partial r_i \partial r_{\mathrm{f},j}}\right|_{\mathbf{r}_\mathrm{f}=\mathbf{r}}. \tag{a}$$

For the potential (8.43) the term $-\partial \Phi_\mathrm{r}/\partial r_i$ contributes only a radial force, which does not lead to any evolution of the orbit. The remaining term yields

$$\mathbf{F}_\mathrm{tide} = -\frac{3\,\mathbb{G} m_\mathrm{f} k_{2,\mathrm{r}} R_\mathrm{r}^5 \tau}{r_\mathrm{f}^8}[\mathbf{v} + 2(\mathbf{v} \cdot \hat{\mathbf{r}}_\mathrm{f})\hat{\mathbf{r}}_\mathrm{f} - \boldsymbol{\omega}_\mathrm{r} \times \mathbf{r}_\mathrm{f}]. \tag{b}$$

It is now straightforward to evaluate the evolution of the orbital elements. As an example, we find the rate of change of the semimajor axis a. The Kepler energy per unit mass is $E = -\frac{1}{2}\,\mathbb{G}(m_\mathrm{r} + m_\mathrm{f})/a$ (eq. 1.32) and multiplying this by the reduced mass $m_\mathrm{r} m_\mathrm{f}/(m_\mathrm{r} + m_\mathrm{f})$ we find the total energy in the center-of-mass frame $E = -\frac{1}{2}\,\mathbb{G} m_\mathrm{f} m_\mathrm{r}/a$. The rate of change of this energy is $\mathrm{d}E/\mathrm{d}t = \mathbf{F}_\mathrm{r} \cdot \mathbf{v}_\mathrm{r} + \mathbf{F}_\mathrm{f} \cdot \mathbf{v}_\mathrm{f}$, where $\mathbf{v}_i$ is the velocity of body i in the center-of-mass frame and $\mathbf{F}_i$ is the force on body i. Now $\mathbf{v}_\mathrm{f} = m_\mathrm{r}\mathbf{v}/(m_\mathrm{r} + m_\mathrm{f})$, $\mathbf{v}_\mathrm{r} = -m_\mathrm{f}\mathbf{v}/(m_\mathrm{r} + m_\mathrm{f})$ (eq. 1.6), $\mathbf{F}_\mathrm{f} = \mathbf{F}_\mathrm{tide}$, and $\mathbf{F}_\mathrm{r} = -\mathbf{F}_\mathrm{tide}$ by Newton's third law. Therefore $\mathrm{d}E/\mathrm{d}t = \mathbf{F}_\mathrm{tide} \cdot \mathbf{v}$. We substitute equation (b) and then average over the orbit using equation (1.64). If the spin vector $\boldsymbol{\omega}_\mathrm{r}$ is normal to the orbit plane, then

$$\begin{aligned}\frac{\mathrm{d}a}{\mathrm{d}t} = -\frac{6\,\mathbb{G} m_\mathrm{f} k_{2,\mathrm{r}} R_\mathrm{r}^5 \tau}{a^7(1-e^2)^{15/2}} \frac{m_\mathrm{r} + m_\mathrm{f}}{m_\mathrm{r}} \Big[& 1 + \tfrac{31}{2}e^2 + \tfrac{255}{8}e^4 + \tfrac{185}{16}e^6 + \tfrac{25}{64}e^8 \\ & - \frac{\omega_\mathrm{r}}{n}(1-e^2)^{3/2}(1 + \tfrac{15}{2}e^2 + \tfrac{45}{8}e^4 + \tfrac{5}{16}e^6)\Big]\end{aligned} \tag{c}$$

This result agrees with equation (8.50) in the limit $e \to 0$, if we make the substitution $Q = 1/(|\omega_\mathrm{f}|\tau)$ (eq. 8.39) with $\omega_\mathrm{f} = 2(n - \omega_\mathrm{r})$.

Expressions for other orbital elements are given in the text and in Alexander (1973), Hut (1981) and Matsumura et al. (2010).

$\dot{\phi}_\mathrm{f} = n - \omega_\mathrm{r}$, so $\cos 2(\phi - \phi_\mathrm{f}) = \cos(2\phi - \omega_\mathrm{f} t + \text{const})$ with[7] $\omega_\mathrm{f} = 2(n - \omega_\mathrm{r})$. Then according to equation (8.41),

$$\Delta\phi \simeq \frac{\operatorname{sgn}(\omega_\mathrm{r} - n)}{2Q_\mathrm{r}}. \tag{8.46}$$

If the semimajor axis is larger than the synchronous radius r_sync where the mean motion and the spin rate are equal, then $\omega_\mathrm{r} > n$ and $\Delta\phi > 0$, which means that the tidal bulge on the responding body is carried ahead of the forcing body; while if the forcing body is interior to the synchronous radius the bulge lags behind it.

The torque exerted on the forcing body is $N = -m_\mathrm{f}\partial\Phi_\mathrm{r}/\partial\phi$ evaluated at $r = a$, $\theta = \frac{1}{2}\pi$ and $\phi = \phi_\mathrm{f}$:

$$N = \frac{3\,\mathbb{G} m_\mathrm{f}^2 k_{2,\mathrm{r}} R_\mathrm{r}^5}{2a^6} \sin 2\Delta\phi \simeq \frac{3\,\mathbb{G} m_\mathrm{f}^2 k_{2,\mathrm{r}} R_\mathrm{r}^5}{2Q_\mathrm{r} a^6} \operatorname{sgn}(\omega_\mathrm{r} - n). \tag{8.47}$$

The orbital angular momentum is (eqs. 1.9 and 1.28)

$$L = \frac{m_\mathrm{f} m_\mathrm{r}}{m_\mathrm{f} + m_\mathrm{r}} [\mathbb{G}(m_\mathrm{f} + m_\mathrm{r})a]^{1/2}. \tag{8.48}$$

The torque causes the orbital angular momentum to change at a rate $\dot{L} = N$, which causes the semimajor axis to migrate at a rate

$$\dot{a} = \frac{3\,\mathbb{G}^{1/2}(m_\mathrm{f} + m_\mathrm{r})^{1/2} m_\mathrm{f} k_{2,\mathrm{r}} R_\mathrm{r}^5}{m_\mathrm{r} Q_\mathrm{r} a^{11/2}} \operatorname{sgn}(\omega_\mathrm{r} - n) \tag{8.49}$$

or

$$\frac{\dot{a}}{a} = \frac{3k_{2,\mathrm{r}} n}{Q_\mathrm{r}} \frac{m_\mathrm{f}}{m_\mathrm{r}} \left(\frac{R_\mathrm{r}}{a}\right)^5 \operatorname{sgn}(\omega_\mathrm{r} - n). \tag{8.50}$$

This result deserves some comments. (i) Note the strong dependence on semimajor axis, $\dot{a} \propto a^{-11/2}$; only close binaries are subject to significant tidal friction. (ii) The migration rate in (8.50) is proportional to the

[7] Note that ω_f is the forcing frequency, not the spin rate of the forcing body, which is always treated as a point mass.

mass of the forcing body; the orbits of more massive forcing bodies evolve more quickly. (iii) Bodies outside the synchronous radius spiral out, while those inside the synchronous radius spiral in; in both cases the orbits evolve *away* from the synchronous radius. (iv) The discontinuous jump in the semimajor axis evolution rate at the synchronous radius, $\dot{a} \propto \mathrm{sgn}(\omega_\mathrm{r} - n)$, is unphysical; more realistic models such as those with constant time offset rather than constant phase offset have $1/Q_\mathrm{r} \to 0$ as the forcing frequency $\omega_\mathrm{f} = 2(n - \omega_\mathrm{r}) \to 0$ and thus avoid this discontinuity. (v) There is an additional contribution to the migration from tides raised by the responding body on the forcing body. In general the response of the larger of the two bodies dominates the evolution, because small bodies subject to significant tidal friction rapidly achieve synchronous rotation and thereafter make no contribution to the semimajor axis evolution (see discussion following eq. 8.54).

The Moon lies far outside the synchronous radius for the Earth, $r_\mathrm{sync} = 42\,164\,\mathrm{km}$ (Problem 1.1), so its orbit is expanding from tidal friction. Equation (8.50) for the rate of semimajor axis evolution due to the Earth yields

$$\dot{a} = 4.50\,\mathrm{cm\,yr^{-1}}\,\frac{k_{2\oplus}}{0.30}\frac{10}{Q_\oplus}. \tag{8.51}$$

The actual value is $3.83\,\mathrm{cm\,yr^{-1}}$.

Most of the other solar-system satellites also lie outside their host planets' synchronous radii. The exceptions are the Martian satellite Phobos and a handful of small satellites orbiting Jupiter, Uranus and Neptune. The orbit of Phobos is shrinking at $3.85\,\mathrm{cm\,yr^{-1}}$. Phobos will crash into the surface of Mars in about 40 Myr (see Problem 8.3); apparently we are observing it in the last few percent of its lifetime.

Semimajor axis migration due to tidal friction is also important for several satellites of the outer planets (Peale 1999). These lie outside the synchronous radius, so they gain angular momentum and migrate outward. Not coincidentally, these satellites are usually found in mean-motion resonances, since migration can lead to resonance capture (§6.3) if it is convergent (see discussion following eq. 6.75). Convergent migration of two satellites requires that the ratio of their mean motions approaches unity, and this is expected for outward migration if the satellite masses are not too different, since the migration rate is a strongly decreasing function of semi-major axis. The list of resonant satellite pairs includes Mimas and Tethys, Enceladus and Dione, and Titan and Hyperion, all satellites of Saturn.

Planets also migrate due to tidal friction from their host star. This process is negligible for any of the planets in the solar system but can be important for exoplanets, which are often found much closer to their host stars. Most tidally evolving exoplanets lie inside the synchronous radius and so migrate inward. The best current example is the exoplanet WASP-12b, which has orbital period $P = 1.091\,\mathrm{d}$ and mass $m = 1.5M_\mathrm{J}$ (M_J = Jupiter mass). Assuming that the planet rotates synchronously with its orbit—the justification for this assumption is described after equation (8.54)—the torques on the orbit arise solely from the tidal response of the star to forcing by the planet. Applying equation (8.50) we find

$$\frac{\dot{a}}{a} = -\frac{1}{3.9\times 10^7\,\mathrm{yr}}\frac{10^6 k_{2,*}}{Q_*}\frac{m}{1.5M_\mathrm{J}}\left(\frac{1.4M_\odot}{M_*}\right)^{8/3}\left(\frac{R_*}{1.65R_\odot}\right)^5. \tag{8.52}$$

Here we have used estimates of the host star's mass and radius, and $k_{2,*}$ and Q_* are the star's Love number and quality factor. The decay rate measured from transit timing is given by $a/\dot{a} = (4.9 \pm 0.4) \times 10^6\,\mathrm{yr}$, which implies $Q_*/k_{2*} \sim 1 \times 10^5$ (Yee et al. 2020), consistent with plausible models for tidal friction in stars of this type. Indirect evidence that tidal friction has played a role in modifying the distribution of short-period giant exoplanets includes the scarcity of gas giants with periods less than a day, and the anomalously rapid rotation of some host stars of these planets.

8.4.2 Spinup and spindown

In the preceding subsection we examined the effects of the torque N (eq. 8.47) exerted by the responding body on the forcing body (recall that we are approximating the forcing body as a point mass). The forcing body exerts an equal and opposite torque on the responding body, which changes its spin angular momentum. This equals $C_\mathrm{r}\omega_\mathrm{r}$, where ω_r is the spin angular velocity and C_r is the moment of inertia of the responding body around the polar axis (eq. 1.133). Thus

$$\dot{\omega}_{\rm r} = -\frac{3\,\mathbb{G}m_{\rm f}^2 k_{2,{\rm r}} R_{\rm r}^3}{2Q_{\rm r} m_{\rm r} a^6} \frac{m_{\rm r} R_{\rm r}^2}{C_{\rm r}} \,{\rm sgn}(\omega_{\rm r} - n). \tag{8.53}$$

For example, the Earth's rotation is slowing due to the tidal torque from the Moon.

In general each of the two members of a binary system exerts a torque on the other. If we label the bodies by 1 and 2, then equation (8.53) implies that if neither body rotates synchronously with the orbit,

$$\frac{|\dot{\omega}_1|}{|\dot{\omega}_2|} = \frac{k_{2,1}}{k_{2,2}} \frac{Q_2}{Q_1} \frac{m_1 R_1^2}{C_1} \frac{C_2}{m_2 R_2^2} \frac{m_2^3}{m_1^3} \frac{R_1^3}{R_2^3}. \tag{8.54}$$

If the two bodies have similar Love numbers, quality factors and dimensionless moments of inertia C/mR^2, and similar mean densities, then $|\dot{\omega}_1/\dot{\omega}_2| \sim R_2^6/R_1^6$; thus the smaller body in the binary spins down much faster. It is therefore usually safe to assume that the smaller body—the satellite in a planet-satellite system or the planet in a star-planet system—rotates synchronously whenever there has been significant tidal evolution. This is consistent with the observation that many of the satellites in the solar system are in synchronous rotation or **tidally locked**, with $\omega = n$; these include the Moon, the two Martian satellites Phobos and Deimos, all four of the Galilean satellites of Jupiter, and the five large satellites of Uranus.

A tidal model with constant time lag (Box 8.1) allows equation (8.53) to be generalized to arbitrary eccentricity (Matsumura et al. 2010):

$$\dot{\omega}_{\rm r} = \frac{3\,\mathbb{G}m_{\rm f}^2 k_{2,{\rm r}} R_{\rm r}^3}{m_{\rm r} a^6 (1-e^2)^6} \frac{m_{\rm r} R_{\rm r}^2}{C_{\rm r}} n\tau \Big[1 + \tfrac{15}{2}e^2 + \tfrac{45}{8}e^4 + \tfrac{5}{16}e^6 - \frac{\omega_{\rm r}}{n}(1-e^2)^{3/2}(1 + 3e^2 + \tfrac{3}{8}e^4)\Big]. \tag{8.55}$$

This result agrees with equation (8.53) in the limit $e \to 0$, if we make the substitution $Q_{\rm r} = 1/(|\omega_{\rm f}|\tau)$ (eq. 8.39) with $\omega_{\rm f} = 2(n - \omega_{\rm r})$.

A body on an eccentric orbit is said to be **pseudo-synchronous** if the spin rate does not evolve under the influence of tides from the forcing body.

Equation (8.55) implies that the pseudo-synchronous spin rate is

$$\omega_{\mathrm{ps}} = n\frac{1 + \frac{15}{2}e^2 + \frac{45}{8}e^4 + \frac{5}{16}e^6}{(1 - e^2)^{3/2}(1 + 3e^2 + \frac{3}{8}e^4)}. \tag{8.56}$$

Since tidal forces are a strongly decreasing function of distance, the pseudo-synchronous spin rate is always close to (within 20% of) the angular speed of the orbit at periapsis, $n(1 + e)^{1/2}(1 - e)^{-3/2}$.

These arguments have traditionally been applied to planet-satellite systems in which the satellite is the forcing body and the planet is the responding body. However, they are also relevant to star-planet systems in which the star is the forcing body and the planet is the responding body. In terms of typical parameters, the deceleration of the planet's spin can be written

$$\frac{\dot{\omega}_{\mathrm{p}}}{\omega_{\mathrm{p}}} = -\frac{\mathrm{sgn}(\omega_{\mathrm{p}} - n)}{7.62 \times 10^7\,\mathrm{yr}}\frac{10^6 k_{2,\mathrm{p}}}{Q_{\mathrm{p}}}\frac{0.4 m_{\mathrm{p}} R_{\mathrm{p}}^2}{C_{\mathrm{p}}}\frac{M_{\mathrm{J}}}{m_{\mathrm{p}}}\left(\frac{R_{\mathrm{p}}}{R_{\mathrm{J}}}\right)^3\left(\frac{10\,\mathrm{d}}{P_{\mathrm{orb}}}\right)^4\frac{P_{\mathrm{spin}}}{1\,\mathrm{d}}, \tag{8.57}$$

where P_{orb} and P_{spin} are the orbital and the spin period of the planet and M_{J} and R_{J} are the mass and radius of Jupiter. This result suggests that most giant planets in orbits with periods less than ~ 30 d should be synchronous.

8.4.3 Eccentricity damping

We now ask whether and how tidal friction damps the eccentricity of the orbit. To begin we must evaluate the tidal response potential (8.44) for an eccentric orbit. To simplify the calculations, we assume that the eccentricity is small and work only to first order in e. We also assume that the orbit lies in the equatorial plane $\theta_{\mathrm{f}} = \frac{1}{2}\pi$. Then $r_{\mathrm{f}}^{-3} \simeq a^{-3}(1 + 3e\cos\ell_{\mathrm{f}}) + \mathrm{O}(e^2)$ and $\phi_{\mathrm{f}} = f_{\mathrm{f}} + \varpi_{\mathrm{f}} = \ell_{\mathrm{f}} + \varpi_{\mathrm{f}} + 2e\sin\ell_{\mathrm{f}} + \mathrm{O}(e^2)$ (from eqs. 1.155 and 1.151), where f_{f}, ℓ_{f} and ϖ_{f} are the true anomaly, mean anomaly and longitude of periapsis of the forcing body. Thus

$$\Phi_{\mathrm{r}} = -\frac{3\mathbb{G}m_{\mathrm{f}}k_{2,\mathrm{r}}R_{\mathrm{r}}^5}{4a^3r^3}\sin^2\theta\big[\cos(2\phi - 2\ell_{\mathrm{f}} - 2\varpi_{\mathrm{f}}) - \tfrac{1}{2}e\cos(2\phi - \ell_{\mathrm{f}} - 2\varpi_{\mathrm{f}}) + \tfrac{7}{2}e\cos(2\phi - 3\ell_{\mathrm{f}} - 2\varpi_{\mathrm{f}})\big]$$

$$+\frac{\mathbb{G}m_{\mathrm{f}}k_{2,\mathrm{r}}R_{\mathrm{r}}^5}{4a^3r^3}(3\cos^2\theta-1)(1+3e\cos\ell_{\mathrm{f}})+\mathrm{O}(e^2). \tag{8.58}$$

In the frame rotating with the responding body at angular speed ω_{r}, the mean anomaly and longitude of periapsis change at rates $\dot{\ell}_{\mathrm{f}}=n$ and $\dot{\varpi}_{\mathrm{f}}=-\omega_{\mathrm{r}}$, where n is the mean motion. Therefore the first three time-varying cosine terms have the form $\cos(2\phi-\omega_{\mathrm{f}}t+\text{const})$ with the forcing frequency ω_{f} equal to $2n-2\omega_{\mathrm{r}}$, $n-2\omega_{\mathrm{r}}$, and $3n-2\omega_{\mathrm{r}}$ respectively. The final time-varying term, $\cos\ell_{\mathrm{f}}$, corresponds to a forcing frequency $\omega_{\mathrm{f}}=n$. The effects of tidal friction can be modeled by adding an offset to each of these terms:

$$\begin{aligned}\Phi_{\mathrm{r}}=&-\frac{3\,\mathbb{G}m_{\mathrm{f}}k_{2,\mathrm{r}}R_{\mathrm{r}}^5}{4a^3r^3}\sin^2\theta\big[\cos(2\phi-2\ell_{\mathrm{f}}-2\varpi_{\mathrm{f}}-2\Delta\phi_1)\\&-\tfrac{1}{2}e\cos(2\phi-\ell_{\mathrm{f}}-2\varpi_{\mathrm{f}}-2\Delta\phi_2)+\tfrac{7}{2}e\cos(2\phi-3\ell_{\mathrm{f}}-2\varpi_{\mathrm{f}}-2\Delta\phi_3)\big]\\&+\frac{\mathbb{G}m_{\mathrm{f}}k_{2,\mathrm{r}}R_{\mathrm{r}}^5}{4a^3r^3}(3\cos^2\theta-1)[1+3e\cos(\ell_{\mathrm{f}}+\delta_4)]+\mathrm{O}(e^2).\end{aligned} \tag{8.59}$$

These offsets are related to the responding body's quality factor by equations (8.41) and (8.34),

$$\Delta\phi_1=\frac{\mathrm{sgn}(\omega_{\mathrm{r}}-n)}{2Q_{\mathrm{r}}},\qquad \Delta\phi_2=\frac{\mathrm{sgn}(2\omega_{\mathrm{r}}-n)}{2Q_{\mathrm{r}}},$$
$$\Delta\phi_3=\frac{\mathrm{sgn}(2\omega_{\mathrm{r}}-3n)}{2Q_{\mathrm{r}}},\qquad \delta_4=-\frac{1}{Q_{\mathrm{r}}}. \tag{8.60}$$

Note that the quality factor in each of these expressions may be different, since Q_{r} likely depends on the forcing frequency.

We now determine the radial and azimuthal forces per unit mass on the forcing body,

$$R=-\frac{\partial\Phi_{\mathrm{r}}}{\partial r},\quad T=-\frac{1}{r}\frac{\partial\Phi_{\mathrm{r}}}{\partial\phi}. \tag{8.61}$$

The forces must be evaluated at the position of the forcing body, so after differentiating we set $r=r_{\mathrm{f}}=a(1-e\cos\ell_{\mathrm{f}})+\mathrm{O}(e_{\mathrm{f}}^2)$, $\theta=\theta_{\mathrm{f}}=\frac{1}{2}\pi$ and

$\phi = \phi_\mathrm{f} = \ell_\mathrm{f} + \varpi_\mathrm{f} + 2e\sin\ell_\mathrm{f} + \mathrm{O}(e_\mathrm{f}^2)$. To first order in the eccentricity,[8]

$$\begin{aligned} R = &-\frac{9\,\mathbb{G}m_\mathrm{f}k_{2,\mathrm{r}}R_\mathrm{r}^5}{4a^7}\frac{m_\mathrm{r}+m_\mathrm{f}}{m_\mathrm{r}}\big[\cos 2\Delta\phi_1 + 4e\cos(\ell_\mathrm{f}-2\Delta\phi_1)\\ &-\tfrac{1}{2}e\cos(\ell_\mathrm{f}-2\Delta\phi_2)+\tfrac{7}{2}e\cos(\ell_\mathrm{f}+2\Delta\phi_3)\\ &+\tfrac{1}{3}+\tfrac{4}{3}e\cos\ell_\mathrm{f}+e\cos(\ell_\mathrm{f}+\delta_4)\big],\\ T = &\frac{3\,\mathbb{G}m_\mathrm{f}k_{2,\mathrm{r}}R_\mathrm{r}^5}{2a^7}\frac{m_\mathrm{r}+m_\mathrm{f}}{m_\mathrm{r}}\big[\sin 2\Delta\phi_1 - 4e\sin(\ell_\mathrm{f}-2\Delta\phi_1)\\ &+\tfrac{1}{2}e\sin(\ell_\mathrm{f}-2\Delta\phi_2)+\tfrac{7}{2}e\sin(\ell_\mathrm{f}+2\Delta\phi_3)\big]. \end{aligned} \tag{8.62}$$

From equations (1.200), the rate of change of the eccentricity due to these forces is

$$\frac{\mathrm{d}e}{\mathrm{d}t} = R\,\frac{(1-e^2)^{1/2}\sin f_\mathrm{f}}{na} + T\,\frac{(1-e^2)^{1/2}(\cos u_\mathrm{f}+\cos f_\mathrm{f})}{na}, \tag{8.63}$$

where f_f and u_f are the true and eccentric anomaly, and $n = [\mathbb{G}(m_\mathrm{r} + m_\mathrm{f})/a^3]^{1/2}$ is the mean motion. To first order in the eccentricity we can write $f_\mathrm{f} = \ell_\mathrm{f} + 2e\sin\ell_\mathrm{f} + \mathrm{O}(e^2)$ (eq. 1.151), $u_\mathrm{f} = \ell_\mathrm{f} + e\sin\ell_\mathrm{f} + \mathrm{O}(e^2)$ (eq. 1.150) and $(1-e^2)^{1/2} = 1 + \mathrm{O}(e^2)$, so

$$\frac{\mathrm{d}e}{\mathrm{d}t} = R\,\frac{\sin\ell_\mathrm{f} + e\sin 2\ell_\mathrm{f}}{na} + T\,\frac{4\cos\ell_\mathrm{f} - 3e + 3e\cos 2\ell_\mathrm{f}}{2na} + \mathrm{O}(e^2). \tag{8.64}$$

Combining equations (8.62) and (8.64) and orbit-averaging, we find the rate of change of eccentricity:

$$\begin{aligned} \frac{1}{e}\frac{\mathrm{d}e}{\mathrm{d}t} = &-\frac{3G^{1/2}m_\mathrm{f}(m_\mathrm{r}+m_\mathrm{f})^{1/2}k_{2,\mathrm{r}}R_\mathrm{r}^5}{16m_\mathrm{r}a^{13/2}}\\ &\times\big(4\sin 2\Delta\phi_1 + \sin 2\Delta\phi_2 - 49\sin 2\Delta\phi_3 - 6\sin\delta_4\big). \end{aligned} \tag{8.65}$$

Thus tides can cause the eccentricity to either decay or grow, depending on the spin and the quality factors associated with the four components of the tide.

---

[8] The origin of the factor $(m_\mathrm{r} + m_\mathrm{f})/m_\mathrm{r}$ is described in the footnote following equation (1.181).

The situation is simpler if the responding body is synchronously rotating, $\omega_\mathrm{r} = n$. In this case $\Delta\phi_1 = 0$ by symmetry. Moreover the forcing frequencies associated with the offsets $\Delta\phi_2$ and $\Delta\phi_3$ are $-n$ and $+n$ (see discussion following eq. 8.58), so $\Delta\phi_2 = -\Delta\phi_3$ and $\Delta\phi_2 > 0$ by equations (8.60). Therefore equation (8.65) is modified to

$$\frac{1}{e}\frac{\mathrm{d}e}{\mathrm{d}t} = -\frac{3G^{1/2}m_\mathrm{f}(m_\mathrm{r}+m_\mathrm{f})^{1/2}k_{2,\mathrm{r}}R_\mathrm{r}^5}{8m_\mathrm{r}a^{13/2}}\big(25\sin 2\Delta\phi_2 - 3\sin\delta_4\big). \tag{8.66}$$

Equations (8.60) imply that $\Delta\phi_2 > 0$ and $\delta_4 < 0$, so tides raised in a synchronously rotating body always damp the eccentricity. If we assume that the phase lags are small (quality factor $Q_\mathrm{r} \gg 1$) and that Q_r is the same for both phase lags in equations (8.60), then

$$\frac{1}{e}\frac{\mathrm{d}e}{\mathrm{d}t} = -\frac{21G^{1/2}m_\mathrm{f}(m_\mathrm{r}+m_\mathrm{f})^{1/2}k_{2,\mathrm{r}}R_\mathrm{r}^5}{2m_\mathrm{r}Q_\mathrm{r}a^{13/2}}. \tag{8.67}$$

We conclude that in a system consisting of a planet and a synchronously rotating satellite, tides raised on the planet can damp or excite the eccentricity, but tides raised on the satellite always damp the eccentricity. If we label the two bodies in a binary system by 1 and 2, and the two bodies have similar Love numbers and phase lags, then the ratio of the rate of change of eccentricity due to dissipation in body 1 to the rate due to dissipation in body 2 is

$$\left|\frac{\dot{e}_1}{\dot{e}_2}\right| \sim \frac{m_2^2}{m_1^2}\frac{R_1^5}{R_2^5}. \tag{8.68}$$

If the two bodies have similar mean densities, then this ratio is $\sim R_2/R_1$. Thus dissipation in the smaller body tends to dominate the eccentricity evolution, so if the smaller body is synchronously rotating the eccentricity will damp. This conclusion is consistent with the observation that all of the solar-system satellites for which the eccentricity evolution time is shorter than the age of the solar system have very small eccentricities.

In a model with constant time lag (quality factor inversely proportional to frequency), the rate of eccentricity damping can be evaluated for arbitrary eccentricity (Matsumura et al. 2010):

$$\frac{1}{e}\frac{\mathrm{d}e}{\mathrm{d}t} = -\frac{27\,\mathbb{G}m_{\mathrm{f}}k_{2,\mathrm{r}}R_{\mathrm{r}}^5\tau}{a^8(1-e^2)^{13/2}}\frac{m_{\mathrm{r}}+m_{\mathrm{f}}}{m_{\mathrm{r}}}\Big[1+\tfrac{15}{2}e^2+\tfrac{15}{8}e^4+\tfrac{5}{64}e^6 - \frac{11\omega_{\mathrm{r}}}{18n}(1-e^2)^{3/2}(1+\tfrac{3}{2}e^2+\tfrac{1}{8}e^4)\Big]. \tag{8.69}$$

This result agrees with equation (8.65) if we assume that $|\Delta\phi_i|, |\delta| \ll 1$ and use equations (8.39) and (8.60) and the forcing frequencies given above equation (8.59) to set $2\Delta\phi_1 = 2(\omega_{\mathrm{r}} - n)\tau$, $2\Delta\phi_2 = (2\omega_{\mathrm{r}} - n)\tau$, $2\Delta\phi_3 = (2\omega_{\mathrm{r}} - 3n)\tau$ and $\delta_4 = -n\tau$.

The energy dissipated by tidal friction heats satellites. The most extreme case in the solar system is Jupiter's satellite Io, which has a molten interior and widespread vulcanism powered by tidal friction (Peale 2003).

The eccentricities of the orbits of planets can also be altered by tides from the host star. The rates can be determined from equations (8.65) and (8.66). From the arguments in the paragraph containing equation (8.68) we expect that the eccentricity evolution is dominated by tides raised in the planet. As described at the end of §8.4.2, planets that are subject to significant tidal evolution are expected to be synchronously rotating, so tidal friction damps the eccentricity, consistent with the observation that most giant exoplanets with orbital periods less than a few days are on nearly circular orbits (see Problem 8.5).

8.5 Non-equilibrium tides

The theory of equilibrium tides developed earlier in this chapter is based on two ingredients: a description of the equilibrium response of a star or planet to a static tidal potential, and a parametrized framework that modifies this response to describe tidal friction (the parameter being the lag angle, time lag, or quality factor). The equilibrium tide should be an adequate description of tides in which the forcing frequency is much smaller than the natural oscillation frequencies of the body subjected to the tide. However, the lag angle cannot be computed from first principles, and moreover depends on the frequency and amplitude of the forcing. Therefore it is difficult to make clean quantitative predictions from the theory, and even more difficult to test them in the solar system. Moreover, there are situations in which the equilibrium tide cannot capture the important physics involved in the tidal interaction.

The next step beyond equilibrium tides is the theory of **dynamical tides**, which accounts for the dynamical response of the star or planet to a time-varying tidal field.

The forcing frequencies of tidal potentials are generally lower than the characteristic frequency $(\mathbb{G}M/R^3)^{1/2}$ of a star or planet of mass M and radius R; hence the only modes that couple well to tidal forcing are those with low eigenfrequencies. For gaseous bodies such as stars or giant planets, these include the **g modes** ("g" for "gravity") found in stably stratified regions and the **inertial modes** found in rotating bodies. The restoring force for g modes is buoyancy and that for inertial modes is the Coriolis force.

Linear calculations allow us to determine how each of these modes is excited by a tidal potential. A more complicated question is how the modes are damped once they have been excited. The ordinary viscosity in a star or planet due to two-body collisions between ions, atoms, or molecules is far too small to damp these waves on any timescale of interest. Instead the waves are damped by turbulent viscosity in convectively unstable regions, or radiative viscosity in stably stratified regions, and possibly also by nonlinear effects such as wave breaking (Ogilvie 2014).

A mature theory of dynamical tides would allow us to calculate the quality factor associated with a given forcing frequency, at least to order of magnitude. Unfortunately this theory is not yet available for most stars and planets.

8.5.1 Planets on high-eccentricity orbits

We described in §5.4.2 how hot Jupiters may be formed through high-eccentricity migration, in which a planet on a nearly radial orbit—typically a semimajor axis of several au and a periapsis of only a few times the stellar radius—loses energy and orbital eccentricity through tidal friction, eventually settling onto a circular orbit close to the star.

For high-eccentricity orbits, the physical description of tidal friction is quite different from that of the equilibrium tide. During periapsis passage, the tidal force from the star excites nonradial oscillations in the planet. The energy deposited in these oscillations comes from the orbit. The nature of the dissipation—that is, the value of the quality factor—in the planet is unimportant so long as the damping timescale is long compared to the duration of periapsis passage and short compared to the orbital period.

The approximate energy loss per periapsis passage is straightforward to estimate. The tidal or forcing potential from the star at the surface of the planet is $\Phi_{\rm f}(R_{\rm p},\theta,\phi) \sim \mathbb{G}M_* R_{\rm p}^2/r^3$, where M_* is the stellar mass, $R_{\rm p}$ is the planetary radius, and r is the distance between the planet and the star (eq. 8.17). The corresponding potential due to the response of the planet is $\Phi_{\rm r}(r,\theta,\phi) \sim k_{2,\rm p}\Phi_{\rm f}(R_{\rm p},\theta,\phi)(R_{\rm p}/r)^3 \sim k_{2,\rm p}\,\mathbb{G}M_* R_{\rm p}^5/r^6$, where $k_{2,\rm p}$ is the planet's gravitational Love number (eq. 8.21). The force on the star is $\mathbf{F} \sim -M_*\nabla\Phi_{\rm r}$ so $|\mathbf{F}| \sim M_*\Phi_{\rm r}/r$. During periapsis passage, the force acts over a distance $\sim r$, leading to an energy change $\Delta E \sim |\mathbf{F}|r \sim k_{2,\rm p}\,\mathbb{G}M_*^2 R_{\rm p}^5/q^6$ where q is the periapsis distance. A more accurate formula is

$$\Delta E = f(\eta)\frac{\mathbb{G}M_*^2 R_{\rm p}^5}{q^6}, \quad \eta \equiv \left(\frac{M_{\rm p}}{M_* + M_{\rm p}}\right)^{1/2}\left(\frac{q}{R_{\rm p}}\right)^{3/2}; \tag{8.70}$$

here η is the ratio of the characteristic frequency $(\mathbb{G}M_{\rm p}/R_{\rm p}^3)^{1/2}$ of the normal modes in the planet to the circular angular speed $[\mathbb{G}(M_* + M_{\rm p})/q^3]^{1/2}$ at periapsis passage, and the dimensionless function $f(\eta)$ measures how well the normal modes of the planet couple to external tides as a function of frequency. In a stably stratified planet, $f(\eta)$ declines as an inverse power of η when $\eta \gg 1$ because of the dense spectrum of low-frequency g modes.

The planet also excites tides in the star, but the energy change due to these is smaller by a factor $(k_{2,*}/k_{2,\rm p})(M_{\rm p}/M_*)^2(R_*/R_{\rm p})^5$. In terms of the mean densities, this ratio is $(k_{2,*}/k_{2,\rm p})(\rho_{\rm p}/\rho_*)^2(R_{\rm p}/R_*)$; we are mostly interested in giant planets, which have mean densities similar to those of stars, larger Love numbers, and smaller radii, so the dominant energy change is due to tides induced in the planet.

The process is more complicated if the excitations are not fully damped in one orbital period. In this case energy can flow back and forth between the oscillations and the orbit, depending on the phase of the oscillations at the time of the next periapsis passage. As a result the semimajor axis of the planet orbit may evolve chaotically. A further complication is that the orbital energy that must be dissipated often exceeds the internal binding energy of the planet, so unless this energy is radiated away efficiently the planet may be disrupted.

8.5.2 Resonance locking

The internal structures of giant planets can evolve significantly over their lifetimes as they lose energy and cool. As they evolve, the frequencies of their normal modes evolve as well, and in some cases these frequencies may cross a resonance with the mean motion of a satellite. By analogy with the arguments in §6.3 the satellite may then be captured in resonance, and its orbit will thereafter evolve at a rate determined by the evolution of the planet's normal-mode frequency rather than by the equilibrium tide (Fuller et al. 2016). For distant satellites the rate of orbit evolution will be much faster after resonance locking, and the rate of tidal dissipation will be correspondingly higher.

Resonance locking may explain the rates of evolution for several of the satellites of the outer planets, which appear to be 2–3 orders of magnitude faster than predicted by the equilibrium tide (Lainey 2016). In particular if more than one satellite is locked the outward migration rate $\dot{a}/a$ will be similar for all the locked satellites, in contrast to models based on the equilibrium tide, which imply that more distant satellites evolve much more slowly.

Resonance locking may also determine the rate of semimajor axis evolution for some hot Jupiters on very short-period orbits (Ma & Fuller 2021).

8.6 Tidal disruption

A satellite can be disrupted by tidal forces if it comes too close to its host planet. The dynamics and the outcome of this process depend on many factors: Is the satellite held together by self-gravity or tensile strength? Is it homogeneous or centrally concentrated? Does it approach the planet on an inspiraling circular orbit or a highly eccentric one? And so forth. This complex process is best studied by examining a variety of simplified model problems.

We consider a satellite of mass m_s orbiting a planet of mass m_p, and we assume for simplicity that $m_s \ll m_p$. If the satellite has radius R and is separated from the planet by a distance r_p, the tidal force at its surface is $F_{\rm tide} \sim \mathbb{G} m_p R/r_p^3$ (the gradient of the quadrupole potential 3.71). The force from the self-gravity of the satellite is $F_{\rm self} \sim \mathbb{G} m_s/R^2$. If the satellite is held together by its own gravity, we expect that tidal disruption will occur when $F_{\rm tide} \gtrsim F_{\rm self}$, or

$$r_p \lesssim R\left(\frac{m_p}{m_s}\right)^{1/3}. \tag{8.71}$$

We can also express this result as

$$\gamma \equiv \frac{m_p}{2\pi\rho r_p^3} = \frac{2\overline{\rho}_p}{3\rho} \gtrsim 1, \tag{8.72}$$

where ρ is the mean density of the satellite and $\overline{\rho}_p = m_p/(\frac{4}{3}\pi r_p^3)$ is the mean density of the planet within the radius of the satellite's orbit. The problems that we examine below confirm these crude arguments and provide exact criteria for tidal disruption in idealized situations.

8.6.1 The Roche limit

We first examine a satellite composed of incompressible fluid, on a circular orbit around the planet. The satellite is assumed to be in synchronous rotation, so the fluid is stationary in a frame rotating with the orbit. The effective potential $\Phi_{\rm eff}$ in the rotating frame is given by the sum of three components, the tidal potential $\Phi_{\rm tide}$, the potential due to self-gravity $\Phi_{\rm self}$, and the centrifugal potential $\Phi_{\rm cent}$ (eq. D.21). Since the satellite is much less massive than the planet, tidal disruption will occur at a distance much larger than the size of the satellite, so the tidal potential inside the satellite is dominated by the quadrupole terms and hence is a quadratic function of the Cartesian coordinates relative to the center of the satellite. Similarly, the potential due to self-gravity satisfies Poisson's equation (B.44), $\nabla^2\Phi_{\rm self} = 4\pi\mathbb{G}\rho$, and ρ is constant because the satellite's material is incompressible, so $\Phi_{\rm self}$ is also a quadratic function of the coordinates; and the centrifugal potential is quadratic as well. Therefore the effective potential is quadratic.[9] The surface of the satellite must be an equipotential (see footnote in Box 1.3) and therefore is an ellipsoid, called the **Roche ellipsoid**.

The properties of this ellipsoid as a function of the orbital radius $r_{\rm p}$ were computed by Édouard Roche (1820–1883). When $r_{\rm p}$ is sufficiently large the tidal and centrifugal forces are negligible compared to the self-gravity of the satellite, and the ellipsoid is a sphere. As the orbital radius shrinks the ellipsoid deviates more and more from a sphere. There is no solution if $r_{\rm p}$ is less than the **Roche limit**

$$r^{e=0}_{\rm Roche} = 1.523\left(\frac{m_{\rm p}}{\rho}\right)^{1/3}, \tag{8.73}$$

corresponding to $\gamma > 0.0450$. At the Roche limit the axes of the ellipsoid are in the ratio $1.5947 : 0.8151 : 0.7693$, with the longest axis pointing toward the planet and the shortest axis perpendicular to the orbital plane (Chandrasekhar 1963, 1969).

If an incompressible fluid satellite is in a parabolic orbit, it will be disrupted at periapsis passage if the periapsis distance is less than (Sridhar & Tremaine 1992)

$$r^{e=1}_{\rm Roche} = 1.05\left(\frac{m_{\rm p}}{\rho}\right)^{1/3}, \tag{8.74}$$

corresponding to $\gamma > 0.137$ if $r_{\rm p}$ is taken to be the periapsis distance in equation (8.72).

---

[9] This argument is not rigorous, because the potential includes a solution of Laplace's equation that must be added to match the boundary conditions at the surface of the satellite and at infinity. It is a fortunate accident that this solution vanishes inside the satellite if its surface is ellipsoidal.

8.6.2 Tidal disruption of regolith

A more realistic model for small satellites is a rigid spherical body of density ρ and radius R surrounded by a thin ocean, atmosphere, or regolith.[10] For simplicity we assume that the satellite is in a circular orbit around its host planet and is synchronously rotating. We also assume—less realistically—that the density of the regolith is negligible, so it does not contribute to the gravitational field. We work in rotating coordinates fixed on the satellite, with the positive x-axis pointing toward the planet and the z-axis normal to the plane of the satellite orbit.

The gravitational potential outside the satellite is $\Phi(\mathbf{r}) = \Phi_{\text{self}}(\mathbf{r}) + \Phi_{\text{cent}}(\mathbf{r}) + \Phi_{\text{tide}}(\mathbf{r})$. The first component is the potential due to the rigid central body, $\Phi_{\text{self}} = -\mathbb{G}m_{\text{s}}/r = -\frac{4}{3}\pi\,\mathbb{G}\rho R^3/r$ where $r = (x^2 + y^2 + z^2)^{1/2}$. The centrifugal potential $\Phi_{\text{cent}}(\mathbf{r}) = -\frac{1}{2}n^2(x^2 + y^2)$ (eq. D.21), where $n = (\mathbb{G}m_{\text{p}}/r_{\text{p}}^3)^{1/2}$ is the mean motion, which is equal to the spin angular velocity of the satellite since it is synchronous by assumption. The tidal potential is given by equation (8.17),

$$\Phi_{\text{tide}}(\mathbf{r}) = \frac{\mathbb{G}m_{\text{p}}}{2r_{\text{p}}^3}(y^2 + z^2 - 2x^2). \tag{8.75}$$

The radial force exerted on an element of regolith on the surface of the asteroid, at position $\mathbf{r} = (R, \theta, \phi)$ in spherical coordinates, is

$$\begin{aligned} F_r &= -\hat{\mathbf{r}}\cdot\nabla(\Phi_{\text{self}} + \Phi_{\text{cent}} + \Phi_{\text{tide}}) \\ &= -\tfrac{4}{3}\pi\,\mathbb{G}\rho R + \frac{\mathbb{G}m_{\text{p}}R}{r_{\text{p}}^3}(3\sin^2\theta\cos^2\phi - \cos^2\theta). \end{aligned} \tag{8.76}$$

The radial force is negative (inward) at all positions on the surface if

$$r_{\text{p}} > \left(\frac{9}{4\pi}\right)^{1/3}\left(\frac{m_{\text{p}}}{\rho}\right)^{1/3} = 0.8947\left(\frac{m_{\text{p}}}{\rho}\right)^{1/3} \quad \text{or } R < r_{\text{p}}\left(\frac{m_{\text{s}}}{3m_{\text{p}}}\right)^{1/3}, \tag{8.77}$$

[10] Regolith is a surface layer of unconsolidated deposits such as dust, gravel, ash, or liquid. Regolith covers most of the Earth, Moon, Mars and asteroids. Because regolith has little or no ability to withstand tensile or shear stresses, it can be approximated for our purposes as a fluid.

corresponding to $\gamma = 0.222$ in equation (8.72). If this condition is violated, regolith near $x = \pm R$, $y = 0$, $z = 0$ (the points on the satellite surface closest to and farthest from the planet) will be levitated from the surface by the tidal force and lost from the satellite. Moreover the forces parallel to the satellite surface arising from the tidal potential push regolith toward these points, thereby tending to deplete the entire regolith.

The conditions (8.77) are the same as the condition that the satellite's physical radius R is less than its Hill radius r_H (eq. 3.24).

8.6.3 Tidal disruption of rigid bodies

The disruption of large astronomical bodies is determined by the competition between tidal forces and self-gravity. Small bodies are held together by the tensile strength of the material of which they are composed, so these can survive in regions where massive bodies of the same density cannot.

For an approximate analysis, consider a satellite of density ρ, radius R and mass $m_s = \frac{4}{3}\pi\rho R^3$ in a circular orbit of radius r_p around a planet of mass m_p. We now imagine replacing the satellite by two points of mass $\frac{1}{2}m_s$. In the coordinate system used in the preceding subsection, the two masses are located at $x = \pm\frac{1}{2}R$, $y = z = 0$ (i.e., symmetrically placed along the line from the planet to the satellite and separated by a distance R). The two points are held together by a string. The force per unit mass on each body is $\mathbf{F} = -\nabla(\Phi_{\text{cent}} + \Phi_{\text{tide}}) = (3n^2 x, 0, 0) = \pm\frac{3}{2}(n^2 R, 0, 0)$, so the tension on the string is $\sim n^2 m_s R$. The force between the two halves of a single spherical satellite should be similar, so the tensile stress at the midplane of the satellite should be approximately force/area $= n^2 m_s R/(\pi R^2) \sim \rho n^2 R^2$. If the tensile strength of the material is T, the tides will disrupt the body unless

$$T \gtrsim \rho n^2 R^2 = \frac{\mathbb{G} m_p \rho R^2}{r_p^3}, \tag{8.78}$$

where m_p is the mass of the host planet and r_p is the satellite's orbital radius. The more careful calculation in Appendix I shows that the satellite survives if

$$T > \frac{5}{19}\frac{\mathbb{G}m_{\rm p}\rho R^2}{r_{\rm p}^3}, \qquad \gamma = \frac{m_{\rm p}}{2\pi\rho r_{\rm p}^3} \le \frac{19}{60};$$
$$> \frac{25}{19}\frac{\mathbb{G}m_{\rm p}\rho R^2}{r_{\rm p}^3}\left(1 - \frac{38\pi\rho r_{\rm p}^3}{75m_{\rm p}}\right), \qquad \gamma = \frac{m_{\rm p}}{2\pi\rho r_{\rm p}^3} > \frac{19}{60}. \tag{8.79}$$

In the first case, the body fractures first at the edge, in the second case at the center. The threshold between these two modes of fracture occurs at a distance from the planet $r_{\rm p} = (30/19\pi)^{1/3}(m_{\rm p}/\rho)^{1/3} = 0.795(m_{\rm p}/\rho)^{1/3}$ corresponding to $\gamma = 0.317$.

A summary of the characteristic disruption distances is in Table 8.1.

Typical tensile strengths are ~ 1 MPa for ice, ~ 10 MPa for rock and ~ 300 MPa for steel (1 MPa = 1 megapascal = 10^6 Pa = 10^6 kg m^{-1} s^{-2}). According to the first of equations (8.79), at the Roche limit (8.73) a body with tensile strength T can survive without the assistance of its own self-gravity if its radius is smaller than

$$R = 473\,\text{km}\,\frac{3\,\text{g cm}^{-3}}{\rho}\left(\frac{T}{10\,\text{MPa}}\right)^{1/2}. \tag{8.80}$$

These considerations also determine the shapes of satellites. Objects much smaller than the radius defined by equation (8.80), like rocks, can have arbitrary shapes, while larger objects must be nearly spherical because the stresses due to gravity exceed the ability of solid material to maintain any other shape.[11] The transition from small, irregular bodies to large, spherical ones can be seen in the asteroids and the satellites of the giant planets of the solar system. The largest asteroid, Ceres (radius R = 470 km), and Saturn's satellite Mimas (R = 198 km) are nearly spherical, but smaller asteroids and satellites are not (e.g., Saturn's satellite Hyperion with R = 135 km). The transition radius is larger for rocky bodies like Ceres than for icy bodies like Mimas, because rock is stronger than ice.[12]

[11] According to the International Astronomical Union, a spherical shape is one of the defining characteristics of a "planet." For further discussion see the end of §9.2.

[12] Weisskopf (1975) gives an elegant order-of-magnitude derivation of this transition radius in terms of fundamental constants.

Table 8.1: Tidal disruption radii

	γ
Roche limit for circular orbit	0.045
Roche limit for parabolic orbit	0.137
regolith escapes	0.222
cracks propagate from center	0.317

The quantity γ is defined by equation (8.72).

Most small bodies in the solar system, in particular asteroids and comets, have strengths much smaller than expected for solid rock or ice, probably because they are **rubble piles**: a collection of unconsolidated smaller bodies held together by gravity, with significant interior voids (Walsh 2018).

Striking evidence that most asteroids are rubble piles comes from the distribution of asteroid rotation periods as a function of size, shown in Figure 8.1. The figure shows a sharply defined minimum rotation period, $P \simeq$ 2.5 hours, independent of the asteroid size over the range 0.3 km–10 km. The maximum rotation rate for a spherical body covered with regolith is attained when the outward centrifugal force at the equator equals the gravitational force, corresponding to $\omega_{\mathrm{max}}^2 R = \mathbb{G}m/R^2$ or $\omega_{\mathrm{max}}^2 = \frac{4}{3}\pi\,\mathbb{G}\rho$ for a body with uniform density ρ. In terms of the rotation period $P = 2\pi/\omega$,

$$P_{\mathrm{min}} = \left(\frac{3\pi}{\mathbb{G}\rho}\right)^{1/2} = 2.334\,\mathrm{h}\left(\frac{2\,\mathrm{g\,cm^{-3}}}{\rho}\right)^{1/2}, \tag{8.81}$$

close to the observed limit for a plausible estimate of the mean density of rubble-pile asteroids. If asteroids had significant tensile strength they could rotate much faster, and the absence of any such rapidly rotating asteroids suggests that their tensile strength is $\lesssim 0.001\,\mathrm{MPa}$.

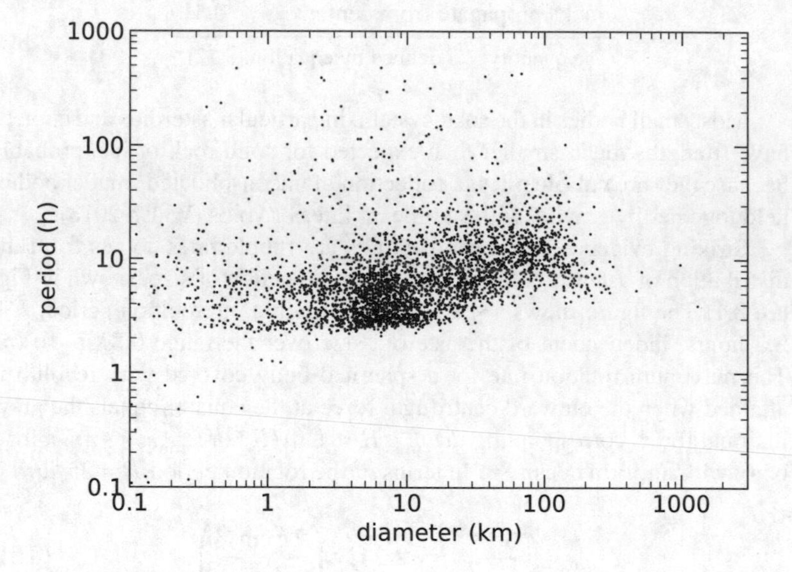

Figure 8.1: The distribution of spin periods, in hours, of $\simeq$ 3 700 asteroids as a function of diameter. Data from Warner et al. (2019), restricted to asteroids with rotation period quality code U3 (highest).

Chapter 9

Planet-crossing orbits

Almost all plausible models of planet formation begin with the condensation of small solid particles ("dust") from the gaseous disk surrounding a newly formed star. The dust particles stick together to form small solid bodies ("planetesimals"), and the planetesimals collide and coalesce, eventually forming planets and the cores of gas-giant planets. This process is remarkably complicated, involving growth over 40 orders of magnitude in mass and a wide variety of complicated and often messy physical processes. Despite this complexity, and despite the many gaps in our understanding of planet formation, a clear prediction of these models is that both planetesimals and planets should be formed on nearly circular, nearly coplanar orbits—and indeed the mean eccentricity and inclination of the planets in the solar system are only 0.06 and 2.3°.

A characteristic feature of systems that contain multiple planets on low-eccentricity orbits is that the planets cannot collide or even suffer close encounters. Adjacent low-mass planets with semimajor axes a_1 and $a_2 > a_1$ and eccentricities e_1 and e_2 cannot collide if $a_1e_1 + a_2e_2 < a_2 - a_1$, since the apoapsis of the inner planet is smaller than the periapsis of the outer, so the orbits cannot cross. It is the absence of close encounters that allows the perturbation techniques developed earlier in this book, which are based on expansions in the small parameters $e_1a_1/(a_2 - a_1)$ and $e_2a_2/(a_2 - a_1)$, to offer accurate descriptions of planetary motion.

Nevertheless there are both theoretical and observational motivations to understand the behavior of particles on planet-crossing orbits. From the theoretical side, many processes can excite eccentricity after a planet is formed; these include long-term instabilities (§4.5), ZLK oscillations (§5.4) and resonance capture (§6.4). From the observational side, many exoplanets have high eccentricities—HD 80606b has $e = 0.933$—and the mean eccentricity of exoplanets discovered through radial-velocity surveys is 0.22, much larger than for solar-system planets. Debris disks detected by thermal emission from dust surround many stars (Wyatt 2008), and the small particles in these disks are probably produced in ongoing collisions between larger bodies on crossing orbits. The mean eccentricity of the asteroids is 0.14; many asteroids cross the orbits of Earth and Mars; and Earth-crossing asteroids pose a potential hazard to civilization. A more extreme solar-system example is the population of comets, icy kilometer-sized bodies that are believed to be residual planetesimals formed at distances ~ 20–40 au from the Sun. Many comets have eccentricities of 0.9999 or higher, and the vast majority of known comets cross the orbits of one or more planets.

9.1 Local structure of a planetesimal disk

So far we have studied the dynamics of individual objects in planetary systems. The numbers of planetesimals, asteroids and comets are so large that a continuum description can be more useful. As an illustration of this approach, we discuss the macroscopic properties of a large population of test particles orbiting in a disk. We shall call this a "planetesimal disk," but the results are applicable to other systems.

In disks that contain large numbers of particles, the mean eccentricity and inclination $\langle e \rangle$ and $\langle I \rangle$ are related to the dispersion in velocities of the particles in a small volume of the disk (see eqs. 9.19 below). Thus it is useful to think of the mean eccentricity and inclination at a given semimajor axis as a kind of "temperature" of the disk: disks composed of particles on circular, coplanar orbits are "cold" and as the mean eccentricity and inclination grow the disk becomes "hotter."

Planetesimal disks are expected to have mean eccentricities and inclinations $\langle e\rangle, \langle I\rangle \ll 1$ and thus are "cool." The large-scale properties of such disks are determined by the distribution of eccentricity and inclination and the number density of particles as a function of semimajor axis a. In this section we focus instead on the distribution of particles in phase space on small scales, on which the gradients with semimajor axis can be ignored.

We use cylindrical coordinates (r, ϕ, z) with the equatorial plane $z = 0$ coinciding with the midplane of the disk, and we assume that the disk is axisymmetric. The number density of planetesimals is $n(r, z)$, and the surface number density is $\Sigma_N(r) = \int_{-\infty}^{\infty} \mathrm{d}z\, n(r, z)$.

We can write the number of planetesimals in a small volume element of phase space as

$$\mathrm{d}N = f(\mathbf{r}, \mathbf{v})\, \mathrm{d}\mathbf{r}\mathrm{d}\mathbf{v}, \tag{9.1}$$

where $f(\mathbf{r}, \mathbf{v})$ is the **distribution function**. We can also specify the phase-space position of a particle by its Delaunay variables (1.84) and write

$$\mathrm{d}N = f'(\Lambda, L, L_z, \ell, \omega, \omega)\, \mathrm{d}\Lambda \mathrm{d}L \mathrm{d}L_z \mathrm{d}\ell \mathrm{d}\omega \mathrm{d}\Omega. \tag{9.2}$$

Because phase-space volumes are conserved in a canonical transformation (Appendix D.6), $\mathrm{d}\mathbf{r}\mathrm{d}\mathbf{v} = \mathrm{d}\Lambda \mathrm{d}L \mathrm{d}L_z \mathrm{d}\ell \mathrm{d}\omega \mathrm{d}\Omega$ and so

$$f(\mathbf{r}, \mathbf{v}) = f'(\Lambda, L, L_z, \ell, \omega, \omega). \tag{9.3}$$

In words, the distribution function is the same in all canonical variables. Given this result, for notational simplicity we can adopt the convention that the distribution function is a function of position in phase space rather than a function of the coordinates. Thus $f(\mathbf{r}, \mathbf{v})$ and $f(\Lambda, L, L_z, \ell, \omega, \Omega)$ are taken to have the same value if $(\mathbf{r}, \mathbf{v})$ and $(\Lambda, L, L_z, \ell, \omega, \Omega)$ denote the same phase-space position in different coordinate systems.

Since the distribution function is time-independent, it must be independent of the mean anomaly ℓ. If the disk is axisymmetric, then the distribution function must also be independent of the ascending node Ω. In principle the distribution function could depend on the argument of periapsis ω, but the rate of apsidal precession due to effects such as the self-gravity of the disk is generally large enough that any dependence on ω is washed out. Thus the distribution function can be written $f(\Lambda, L, L_z)$.

The Delaunay variables can be expanded as Taylor series in the eccentricity and inclination:

$$\Lambda - L \simeq \tfrac{1}{2}\Lambda e^2 + \mathrm{O}(e^4), \quad L - L_z \simeq \tfrac{1}{2}\Lambda I^2 + \mathrm{O}(I^4, e^2 I^2). \tag{9.4}$$

In a disk with small eccentricities and inclinations, the variations of the distribution function with $\Lambda - L$ and $L - L_z$ are much faster than the variation with Λ. Therefore to describe the local dynamics of the disk, we can ignore the variation of the distribution function with Λ and treat it as a function of just the two variables $\Lambda - L$ and $L - L_z$. In contrast to the case of a gas in a box, where the distribution function must be a Maxwellian function of the velocity because of the Boltzmann H-theorem, there is no simple physical argument that predicts the distribution function in planetesimal disks. However, the following empirical form is widely used and generally consistent with numerical experiments on the evolution of planetesimal disks:

$$\begin{aligned} f(L, L_z) &= A \exp\left[-\beta(\Lambda - L) - \beta\gamma^2(L - L_z)\right] \\ &\simeq A \exp\left[-\tfrac{1}{2}\Lambda\beta(e^2 + \gamma^2 I^2)\right], \end{aligned} \tag{9.5}$$

where A, β and γ are constants. Our assumption that the eccentricities and inclinations are small implies that $\Lambda\beta \gg 1$.

The number of particles in a small range of semimajor axis $\mathrm{d}a$ is

$$\begin{aligned} \mathrm{d}N &= \frac{\mathrm{d}\Lambda}{\mathrm{d}a}\mathrm{d}a \int_0^\Lambda \mathrm{d}L \int_{-L}^{L} \mathrm{d}L_z \int_0^{2\pi} \mathrm{d}\ell \int_0^{2\pi} \mathrm{d}\omega \int_0^{2\pi} \mathrm{d}\Omega\, f(L, L_z) \\ &= (2\pi)^3 \frac{\mathrm{d}\Lambda}{\mathrm{d}a}\mathrm{d}a \int_{-\infty}^{\Lambda} \mathrm{d}L \int_{-\infty}^{L} \mathrm{d}L_z\, f(L, L_z) \\ &= 4\pi^3 \frac{A}{\beta^2\gamma^2}\left(\frac{\mathbb{G}M_*}{a}\right)^{1/2} \mathrm{d}a, \end{aligned} \tag{9.6}$$

where M_* is the mass of the host star. Here we have extended the lower limits of the integrals over L and L_z to minus infinity, which introduces almost no error since $\Lambda\beta \gg 1$. The surface number density $\Sigma_N = \mathrm{d}N/(2\pi a\,\mathrm{d}a)$, so

$$\Sigma_N = 2\pi^2 \frac{A}{\beta^2\gamma^2}\left(\frac{\mathbb{G}M_*}{a^3}\right)^{1/2}. \tag{9.7}$$

The probability that a planetesimal has eccentricity and inclination in the ranges $e \to e + \mathrm{d}e$, $I \to I + \mathrm{d}I$ is

$$\begin{aligned} p(e,I)\,\mathrm{d}e\mathrm{d}I &\propto \mathrm{d}L\mathrm{d}L_z \exp\big[-\beta(\Lambda - L) - \beta\gamma^2(L - L_z)\big] \\ &\propto \mathrm{d}(\Lambda - L)\mathrm{d}(L - L_z)\exp\big[-\beta(\Lambda - L) - \beta\gamma^2(L - L_z)\big] \\ &\propto \mathrm{d}(\tfrac{1}{2}e^2)\mathrm{d}(\tfrac{1}{2}I^2)\exp[-\tfrac{1}{2}\Lambda\beta(e^2 + \gamma^2 I^2)]. \end{aligned} \tag{9.8}$$

Since $p(e,I)$ is a probability distribution we must have $\int \mathrm{d}e\mathrm{d}I\, p(e,I) = 1$, so

$$p(e,I) = p_{\mathrm{e}}(e)p_I(I), \tag{9.9}$$

where

$$p_{\mathrm{e}}(e) = \Lambda\beta e \exp\big(-\tfrac{1}{2}\Lambda\beta e^2\big), \quad p_I(I) = \Lambda\beta\gamma^2 I \exp\big(-\tfrac{1}{2}\Lambda\beta\gamma^2 I^2\big). \tag{9.10}$$

The probability distribution is separable, that is, $p(e,I)$ is the product of independent probability distributions in e and I. Each of these is a **Rayleigh distribution** of the form

$$p_{\mathrm{R}}(x) = \lambda x \exp\big(-\tfrac{1}{2}\lambda x^2\big). \tag{9.11}$$

It is straightforward to verify that the mean of a Rayleigh distribution is $\langle x \rangle = \int_0^\infty \mathrm{d}x\, x p_{\mathrm{R}}(x) = (\tfrac{1}{2}\pi/\lambda)^{1/2}$. Therefore

$$\beta = \frac{\pi}{2\Lambda\langle e\rangle^2}, \quad \gamma = \frac{\langle e\rangle}{\langle I\rangle}; \tag{9.12}$$

thus β measures the inverse "temperature" of the disk, while γ measures the anisotropy between eccentricities and inclinations. Now equations (9.10) can be rewritten as

$$p_{\mathrm{e}}(e) = \frac{\pi}{2\langle e\rangle^2} e \exp\left(-\frac{\pi e^2}{4\langle e\rangle^2}\right), \quad p_I(I) = \frac{\pi}{2\langle I\rangle^2} I \exp\left(-\frac{\pi I^2}{4\langle I\rangle^2}\right). \tag{9.13}$$

Notice that $p_{\mathrm{e}}(e) \propto e$ as e approaches zero, with a similar result for $p_I(I)$; loosely speaking, this means that the probability of finding a planetesimal on an exactly circular or equatorial orbit is zero. This property arises from the geometry of phase space and our assumption that the distribution function is approximately constant near $e = I = 0$.[1]

We can also write the distribution function (9.5) in terms of the planetesimal velocities. To do so, we introduce a reference particle traveling on a circular orbit of radius r in the disk midplane. The reference particle has speed $v_c = (\mathbb{G}M_*/r)^{1/2}$. We work in a frame rotating with the reference particle and erect Cartesian coordinates with origin at the reference particle, positive x-axis pointing radially outward, positive y-axis pointing in the direction of rotation, and z-axis perpendicular to the midplane. Then an orbit passing through the origin has $r = a(1 - e\cos u)$ (eq. 1.46) and velocity $v_x = (\mathbb{G}M_*/a)^{1/2} e\sin u/(1 - e\cos u)$ (eq. 1.54). If we keep only terms up to first order in the eccentricity then these equations imply that $v_x \simeq v_c e\sin u$. Similarly the velocity $v_y = v_\phi - v_c = (\mathbb{G}M_*/a)^{1/2}(1-e^2)^{1/2}/(1-e\cos u) - v_c$ (eq. 1.55). To evaluate this to first order in the eccentricity, we write $(\mathbb{G}M_*/a)^{1/2} = (\mathbb{G}M_*/r)^{1/2}(1 - e\cos u)^{1/2} \simeq v_c(1 - \frac{1}{2}e\cos u)$; then $v_y \simeq \frac{1}{2}v_c e\cos u$. Combining these results we have

$$v_x^2 + 4v_y^2 = v_c^2 e^2(\sin^2 u + \cos^2 u) = v_c^2 e^2. \tag{9.14}$$

Similarly, from equations (1.70) we find that to $\mathrm{O}(e, I)$, $z \simeq rI\sin(f + \omega)$ and $v_z \simeq rI\cos(f + \omega)\dot{f} \simeq v_c I\cos(f + \omega)$ so

$$\frac{v_c^2}{r^2}z^2 + v_z^2 = v_c^2 I^2[\sin^2(f + \omega) + \cos^2(f + \omega)] = v_c^2 I^2. \tag{9.15}$$

Therefore the distribution function (9.5) becomes

$$f(\mathbf{r}, \mathbf{v}) = A\exp\left[-\frac{\Lambda\beta}{2v_c^2}(v_x^2 + 4v_y^2) - \frac{\Lambda\beta\gamma^2}{2v_c^2}v_z^2 - \frac{\Lambda\beta\gamma^2}{2r^2}z^2\right]. \tag{9.16}$$

Using equations (9.12), this result can be rewritten as

$$f(\mathbf{r}, \mathbf{v}) = A\exp\left[-\frac{\pi}{4v_c^2\langle e\rangle^2}(v_x^2 + 4v_y^2) - \frac{\pi}{4v_c^2\langle I\rangle^2}v_z^2 - \frac{\pi}{4r^2\langle I\rangle^2}z^2\right]. \tag{9.17}$$

[1] An analogy is the case of a circle of unit radius that is uniformly filled with N points. The number of points with distance from the center between r and $r + \mathrm{d}r$ is $n(r)\mathrm{d}r$, where $n(r) = 2Nr$. Thus $n(r) \propto r$ as $r \to 0$.

This is a triaxial Gaussian distribution in velocity, called a **Schwarzschild distribution** because Karl Schwarzschild (1873–1916) used it to describe the distribution of velocities of stars in the solar neighborhood. The normalization constant A is simply related to the number density of planetesimals in the disk midplane, n_0: since $n_0 = \int \mathrm{d}\mathbf{v}\, f(z=0,\mathbf{v}) = 4Av_\mathrm{c}^3\langle e\rangle^2\langle I\rangle$,

$$A = \frac{n_0}{4v_\mathrm{c}^3\langle e\rangle^2\langle I\rangle}. \tag{9.18}$$

This analysis shows that an *exponential* distribution in the actions or Delaunay variables corresponds to a *Rayleigh* distribution in the eccentricities and inclinations, and a *Gaussian* distribution in the velocities (relative to the local circular velocity). The velocity dispersions or root-mean-square velocities in the three orthogonal directions are

$$\langle v_x^2\rangle^{1/2} = \left(\frac{2}{\pi}\right)^{1/2} v_\mathrm{c}\langle e\rangle, \quad \langle v_y^2\rangle^{1/2} = \left(\frac{1}{2\pi}\right)^{1/2} v_\mathrm{c}\langle e\rangle, \quad \langle v_z^2\rangle^{1/2} = \left(\frac{2}{\pi}\right)^{1/2} v_\mathrm{c}\langle I\rangle. \tag{9.19}$$

Thus the velocity dispersion in the azimuthal direction is always half the dispersion in the radial direction,[2] while the ratio of dispersions in the radial and vertical directions depends on the arbitrary ratio $\gamma = \langle e\rangle/\langle I\rangle$. Neither observations nor theory point to a unique value of γ. Dynamical models of planetesimal disks that evolve through gravitational scattering have $\gamma \simeq 2$ (Ida et al. 1993); the asteroid belt has $\gamma \simeq 0.8$, although this value is largely determined by the requirement that the orbits be stable in the presence of Jupiter; and the trans-Neptunian belt has $\gamma \simeq 0.7$, although this sample is subject to strong observational selection effects depending on the eccentricity and inclination.

The distribution function (9.17) also implies that the number density of planetesimals as a function of height above the disk midplane is

$$n(z) = n_0 \exp\left(-\tfrac{1}{2}z^2/z_0^2\right), \quad z_0 \equiv (2/\pi)^{1/2} r\langle I\rangle = 0.798\, r\langle I\rangle; \tag{9.20}$$

[2] This result contrasts with the root-mean-square azimuthal velocity of a single particle relative to its guiding center, which is *twice* as large as its root-mean-square radial velocity (see text following eq. 1.171). The difference arises because the analysis here refers to the azimuthal velocity relative to the circular speed at the same radius, rather than the azimuthal velocity relative to the guiding center.

here z_0 is the root-mean-square height of the planetesimals relative to the midplane, and the central number density n_0 is related to the surface number density Σ_N by $\Sigma_N = \int_{-\infty}^{\infty} \mathrm{d}z\, n(z) = (2\pi)^{1/2} n_0 z_0 = 2n_0 r\langle I \rangle$.

9.2 Disk-planet interactions

A planet embedded in a planetesimal disk interacts with the disk mainly through two mechanisms: (i) collisions between the planetesimals and the planet, which lead to growth of the planet mass and depletion of the planetesimal population in the vicinity of the planet; and (ii) gravitational scattering or "stirring" of the disk by the planet, which excites the eccentricities and inclinations of the planetesimals near the planet. Both processes are central to planet formation and to the evolution of the disk.

We study these in a disk surrounding a star of mass M_* that contains a planet of mass M_p and radius R, on a circular orbit of semimajor axis a_p in the midplane of the disk. The planet's orbital speed is the circular speed $v_\mathrm{c} = (\mathbb{G}M_*/a_\mathrm{p})^{1/2}$, and the escape speed from its surface is $v_\mathrm{esc} = (2\,\mathbb{G}M_\mathrm{p}/R)^{1/2}$ (eq. 1.21).

The response of the disk depends strongly on its velocity dispersion (the root-mean-square velocity of the planetesimals relative to the local circular speed), which we denote by σ. We only use this symbol in approximate arguments, for which a sufficiently good approximation is that $\sigma \simeq v_\mathrm{c}\langle e \rangle$, where $\langle e \rangle$ is the mean eccentricity of the particles in the vicinity of the planet (cf. eq. 9.19). A second characteristic velocity that governs the local dynamics is $s \equiv v_\mathrm{c} r_\mathrm{H}/a_\mathrm{p}$; here $r_\mathrm{H} = a_\mathrm{p}[M_\mathrm{p}/(3M_*)]^{1/3}$ is the Hill radius (eq. 3.24). Since the disk is differentially rotating, s is a measure of the shear in the disk across one Hill radius. A **shear-dominated disk** is one in which $\sigma \lesssim s$; in this case the relative velocity between planet and planetesimal before an encounter is dominated by the contribution from differential rotation; the interactions with the planet are strongest if the impact parameter $b \lesssim r_\mathrm{H}$; and the interaction decays rapidly when $b \gg r_\mathrm{H}$ (see Figure 3.13). In contrast, in a **dispersion-dominated disk**, with $\sigma \gtrsim s$, the relative velocity before an encounter is dominated by the non-circular motion of the planetesimal, and all encounters with impact parameter $b \lesssim a_\mathrm{p}\sigma/v_\mathrm{c}$ can lead to strong interactions with the planet.

In general, the effect of encounters with a planet in a shear-dominated disk is to "heat" or "stir" the disk until it becomes dispersion-dominated in the vicinity of the planet. When the impact parameter $b \lesssim r_{\rm H}$, this heating requires only a few encounters with the planet, that is, a time interval of a few times $r_{\rm H}/v_{\rm c}$. As a consequence, most planetesimal disks containing planets are dispersion-dominated near each planet, and we restrict ourselves to dispersion-dominated disks from now on. For more general discussions of the collision rate see Greenzweig & Lissauer (1992) and Dones & Tremaine (1993), and for more general discussions of gravitational stirring see Stewart & Ida (2000) and Rafikov (2003).

9.2.1 Collisions

First we examine the rate of collisions between the planetesimals and the planet. Suppose that a planetesimal makes a close approach to the planet on a trajectory with relative velocity v. The collision cross section is (eq. 1.41)[3]

$$\pi b_{\rm coll}^2 = \pi R^2 + \frac{2\pi \mathbb{G} M_{\rm p} R}{v^2}. \tag{9.21}$$

Let $f(\mathbf{v})$ be the distribution function, so $f(\mathbf{v})\mathrm{d}\mathbf{v}$ is the number of planetesimals per unit volume with velocities in the range $\mathbf{v} \to \mathbf{v}+\mathrm{d}\mathbf{v}$, measured in the rotating frame centered on the planet. Then the number of collisions per unit time with planetesimals in this velocity range is just $f(\mathbf{v})\mathrm{d}\mathbf{v}$ times the volume of a cylinder with cross section $\pi b_{\rm coll}^2$ and length $v = |\mathbf{v}|$. The total collision rate is thus

$$\Gamma_{\rm coll} = \int \mathrm{d}\mathbf{v}\, f(\mathbf{v}) \left(\pi R^2 v + \frac{2\pi \mathbb{G} M_{\rm p} R}{v} \right). \tag{9.22}$$

We use the Schwarzschild distribution function defined by equations (9.17) and (9.18), which we evaluate at $z = 0$ since the planet orbits in the disk

[3] This expression ignores the differential effects of the gravitational field of the host star on the orbits of the planetesimal and the planet during the encounter between them, an approximation that is generally valid in dispersion-dominated disks.

midplane. Then

$$\Gamma_{\text{coll}} = \frac{n_0}{4v_c^3\langle e\rangle^2\langle I\rangle}\int d\mathbf{v}\left(\pi R^2 v + \frac{2\pi\mathbb{G}M_pR}{v}\right)$$
$$\times \exp\left[-\frac{\pi}{4v_c^2\langle e\rangle^2}(v_x^2 + 4v_y^2) - \frac{\pi}{4v_c^2\langle I\rangle^2}v_z^2\right], \tag{9.23}$$

where n_0 is the number density in the midplane, and $\langle e\rangle$ and $\langle I\rangle$ are the mean eccentricity and inclination of the planetesimals in the disk. By replacing the integration variable $\mathbf{v}$ with $\mathbf{u} = \mathbf{v}/(v_c\langle e\rangle)$, the expression can be rewritten as

$$\Gamma_{\text{coll}} = n_0 v_c R^2\langle e\rangle\left[\Psi_a(\gamma) + \Theta\Psi_b(\gamma)\right], \tag{9.24}$$

where $\gamma = \langle e\rangle/\langle I\rangle$ and

$$\Psi_a(\gamma) \equiv \tfrac{1}{4}\pi\gamma\int d\mathbf{u}\,u\exp\left[-\tfrac{1}{4}\pi(u_x^2 + 4u_y^2 + \gamma^2u_z^2)\right],$$
$$\Psi_b(\gamma) \equiv \tfrac{1}{4}\pi\gamma\int d\mathbf{u}\,u^{-1}\exp\left[-\tfrac{1}{4}\pi(u_x^2 + 4u_y^2 + \gamma^2u_z^2)\right]. \tag{9.25}$$

Here $d\mathbf{u} = du_x du_y du_z$, $u = (u_x^2 + u_y^2 + u_z^2)^{1/2}$,

$$\Theta \equiv \frac{2\,\mathbb{G}M_p}{Rv_c^2\langle e\rangle^2} = \frac{2}{\langle e\rangle^2}\frac{M_p}{M_*}\frac{a_p}{R} = \frac{v_{\text{esc}}^2}{v_c^2\langle e\rangle^2} \tag{9.26}$$

is the **Safronov number**,[4] and v_{esc} is the escape speed from the planet's surface (eq. 1.21).

The first term in equation (9.24) is the collision rate that would occur if the planet's mass (but not radius) were zero; the second term accounts for the enhancement in the collision rate due to the deflection of orbits by the planet's mass (gravitational focusing).

The functions $\Psi_a(\gamma)$ and $\Psi_b(\gamma)$ are plotted in Figure 9.1. Typical values are $\Psi_a(1.5) = 2.959$, $\Psi_b(1.5) = 4.388$.

---

[4] There is no common standard for the definition of the Safronov number; for example, some authors replace $v_c\langle e\rangle$ by the radial dispersion $\langle v_x^2\rangle^{1/2}$ (eq. 9.19), others remove a factor of two, and others drop the factor $\langle e\rangle^2$. See equation (9.46) for an alternative definition.

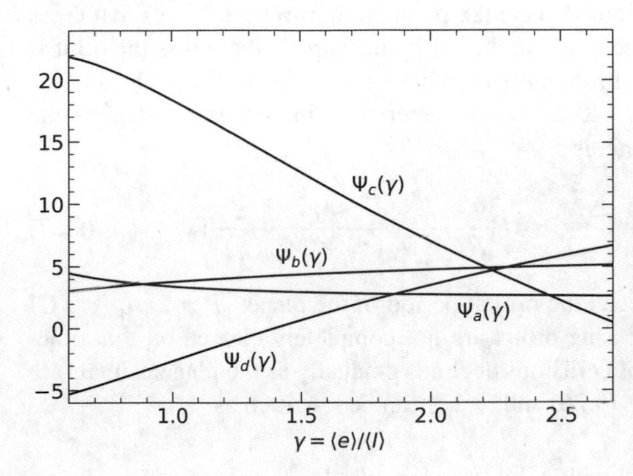

Figure 9.1: The functions $\Psi_a(\gamma)$, $\Psi_b(\gamma)$, $\Psi_c(\gamma)$ and $\Psi_d(\gamma)$ defined in equations (9.25) and (9.39). The argument γ is the ratio of mean eccentricity to mean inclination.

Table 9.1: Safronov numbers, collision times and stirring times for solar-system planets

	a_p (au)	$M_\odot/M_\mathrm{p}$	Θ	t_coll (Myr)	t_stir (Myr)
Mercury	0.387	6.024×10^6	0.79	5.64	4.06
Venus	0.723	4.085×10^5	8.75	1.27	0.069
Earth	1.000	3.329×10^5	14.09	2.26	0.077
Mars	1.524	3.099×10^6	4.33	103	9.09
Jupiter	5.203	1 047.6	2 079	0.041	0.0011
Saturn	9.537	3 498.8	1 353	0.741	0.0012
Uranus	19.19	22 905	981	65.7	0.057
Neptune	30.07	19 416	1 871	177	0.085

The Safronov number Θ, collision time t_coll and eccentricity stirring time t_stir are defined in equations (9.26), (9.27) and (9.42). The numerical values in the table are for mean eccentricity $\langle e\rangle = 0.1$ and $\gamma = \langle e\rangle/\langle I\rangle = 1.5$. The Coulomb logarithm $\log\Lambda$ appearing in t_stir is given by equation (9.36).

Eventually collisions deplete the population of planetesimals that cross the planet's orbit. The total number of planetesimals that cross the orbit is $\Delta N = 4\pi a_{\rm p}^2\Sigma_N\langle e\rangle$ (Problem 9.1), where $\Sigma_N = 2n_0 a_{\rm p}\langle I\rangle$ is the surface number density (eq. 9.20). The characteristic time needed to deplete the disk, the **collision time**, is

$$t_{\rm coll} \equiv \frac{\Delta N}{\Gamma_{\rm coll}} = 4P\frac{a_{\rm p}^2}{R^2}\frac{\langle e\rangle}{\gamma[\Psi_a(\gamma)+\Theta\Psi_b(\gamma)]}; \tag{9.27}$$

here we have introduced the orbital period of the planet, $P = 2\pi a_{\rm p}/v_{\rm c}$. Of course the planet-crossing orbits are not completely cleared on this timescale, since the rate of collisions declines gradually as the planetesimals are consumed. Equations (9.26) and (9.27) can be rewritten as

$$\Theta = 18.91\frac{\rho_{\rm p}}{3\,{\rm g\,cm^{-3}}}\frac{M_\odot}{M_*}\frac{a_{\rm p}}{1\,{\rm au}}\left(\frac{R}{10^4\,{\rm km}}\frac{0.1}{\langle e\rangle}\right)^2,$$

$$t_{\rm coll} = 20.2\,{\rm Myr}\left(\frac{a_{\rm p}}{1\,{\rm au}}\right)^{7/2}\left(\frac{M_\odot}{M_*}\right)^{1/2}\left(\frac{10^4\,{\rm km}}{R}\right)^2\frac{\langle e\rangle}{0.1}(1+1.48\,\Theta)^{-1}; \tag{9.28}$$

here we have written the results in terms of the planet's density $\rho_{\rm p}$ and radius R, related to the mass through $M_{\rm p} = \frac{4}{3}\pi\rho_{\rm p}R^3$, and in the second equation we have set $\gamma = \langle e\rangle/\langle I\rangle = 1.5$.

The Safronov numbers and collision times for planets in the solar system are given in Table 9.1, for representative values $\langle e\rangle = 0.1$, $\gamma = \langle e\rangle/\langle I\rangle = 1.5$.

9.2.2 Gravitational stirring

Planetesimals that pass close to the planet, but not close enough to collide, are deflected from their trajectories by the planet's gravitational field. These deflections gradually accumulate, leading to growth in the mean eccentricity and inclination ("stirring") of the population of planetesimals in the region of the disk neighboring the planet. As described in the following section, in extreme cases planetesimals can even be ejected from the planetary system by this process.

The calculation of the rate of eccentricity and inclination growth due to gravitational stirring has some subtleties. In the rotating frame in which we are working the planet is stationary, and in a dispersion-dominated disk the effects of the Coriolis, centrifugal and tidal forces during a close encounter with the planet are generally negligible. Therefore the kinetic energy of the planetesimal in the rotating frame is unchanged during the encounter, just as it would be for a test particle that scattered off a stationary gravitational potential in an inertial frame. How then can the mean eccentricity and inclination increase, since they are related to the kinetic energy through equations (9.19)? To resolve this apparent paradox we must recognize that scattering by a stationary body tends to isotropize the velocity distribution function, that is, to produce a distribution function in which $\langle v_x^2 \rangle = \langle v_y^2 \rangle = \langle v_z^2 \rangle$. However, an isotropic distribution function is inconsistent with equations (9.19). Therefore, the scattering process must be accompanied by a redistribution of the semimajor axes of the scattered planetesimals that returns the distribution function to the shape (9.19). This redistribution liberates energy that is transferred to the non-circular motion of the planetesimals, causing the mean eccentricities and inclinations to grow.[5]

The squared eccentricity and inclination of an orbit are given by equations (9.14) and (9.15):

$$e^2 = \frac{1}{v_\mathrm{c}^2}[(\hat{\mathbf{x}} \cdot \mathbf{v})^2 + 4(\hat{\mathbf{y}} \cdot \mathbf{v})^2], \quad I^2 = \frac{(\hat{\mathbf{z}} \cdot \mathbf{r})^2}{a_\mathrm{p}^2} + \frac{(\hat{\mathbf{z}} \cdot \mathbf{v})^2}{v_\mathrm{c}^2}. \tag{9.29}$$

Recall that the origin of the coordinate system coincides with the planet, the positive x-axis points radially outward and the positive y-axis points in the direction of rotation, and the velocity $\mathbf{v}$ is measured relative to the planet.

During a close encounter with the planet, the change in position $\mathbf{r}$ of the planetesimal is negligible and the post-encounter velocity $\mathbf{v}'$ is given in terms of the deflection angle θ by equation (1.37). Thus the change in the squared eccentricity and inclination during an encounter are

$$\Delta e^2 = \frac{\cos^2\theta - 1}{v_\mathrm{c}^2}\left[(\hat{\mathbf{x}} \cdot \mathbf{v})^2 + 4(\hat{\mathbf{y}} \cdot \mathbf{v})^2\right] \quad \frac{2v\sin\theta\cos\theta}{v_\mathrm{c}^2}[(\hat{\mathbf{x}} \cdot \mathbf{v})(\hat{\mathbf{x}} \cdot \hat{\mathbf{b}})$$

[5] A similar process occurs in shear-dominated disks, as described in §3.4.2.

$$+4(\hat{\mathbf{y}}\cdot\mathbf{v})(\hat{\mathbf{y}}\cdot\hat{\mathbf{b}})\big]+\frac{v^2\sin^2\theta}{v_c^2}\big[(\hat{\mathbf{x}}\cdot\hat{\mathbf{b}})^2+4(\hat{\mathbf{y}}\cdot\hat{\mathbf{b}})^2\big], \tag{9.30}$$

$$\Delta I^2=\frac{\cos^2\theta-1}{v_c^2}(\hat{\mathbf{z}}\cdot\mathbf{v})^2-\frac{2v\sin\theta\cos\theta}{v_c^2}(\hat{\mathbf{z}}\cdot\mathbf{v})(\hat{\mathbf{z}}\cdot\hat{\mathbf{b}})+\frac{v^2\sin^2\theta}{v_c^2}(\hat{\mathbf{z}}\cdot\hat{\mathbf{b}})^2.$$

Now let $\hat{\mathbf{u}}$ and $\hat{\mathbf{w}}$ be unit vectors perpendicular to $\mathbf{v}$, so $\hat{\mathbf{u}}$, $\hat{\mathbf{v}}$ and $\hat{\mathbf{w}}$ form an orthonormal triad. Since $\hat{\mathbf{b}}$ is also perpendicular to $\mathbf{v}$, we may write $\hat{\mathbf{b}}=\hat{\mathbf{u}}\cos\psi+\hat{\mathbf{w}}\sin\psi$. The angle ψ—the azimuthal angle relative to the positive v-axis at which the particle makes its closest approach to the planet—should be uniformly distributed between 0 and 2π for close encounters, so we can average over it. Denoting this average by $\langle\cdot\rangle_\psi$, we have $\langle\hat{\mathbf{x}}\cdot\hat{\mathbf{b}}\rangle_\psi=\langle\hat{\mathbf{y}}\cdot\hat{\mathbf{b}}\rangle_\psi=0$ and $\langle(\hat{\mathbf{x}}\cdot\hat{\mathbf{b}})^2\rangle_\psi=\frac{1}{2}(\hat{\mathbf{x}}\cdot\hat{\mathbf{u}})^2+\frac{1}{2}(\hat{\mathbf{x}}\cdot\hat{\mathbf{w}})^2$. Using the identity (B.6a), the last of these expressions must equal $\frac{1}{2}-\frac{1}{2}(\hat{\mathbf{x}}\cdot\hat{\mathbf{v}})^2$, with a similar result for $\langle(\hat{\mathbf{y}}\cdot\hat{\mathbf{b}})^2\rangle_\psi$ and $\langle(\hat{\mathbf{z}}\cdot\hat{\mathbf{b}})^2\rangle_\psi$. Then equations (9.30) can be rewritten to give the average change in eccentricity and inclination,

$$\langle\Delta e^2\rangle_\psi=\frac{\sin^2\theta}{2v_c^2}\big[5v^2-3(\hat{\mathbf{x}}\cdot\mathbf{v})^2-12(\hat{\mathbf{y}}\cdot\mathbf{v})^2\big],$$

$$\langle\Delta I^2\rangle_\psi=\frac{\sin^2\theta}{2v_c^2}\big[v^2-3(\hat{\mathbf{z}}\cdot\mathbf{v})^2\big]. \tag{9.31}$$

The number of encounters with the planet per unit time by planetesimals in the velocity range $\mathbf{v}\to\mathbf{v}+\mathrm{d}\mathbf{v}$ having impact parameters in the range $b\to b+\mathrm{d}b$ is $2\pi bvf(\mathbf{v})\,\mathrm{d}b\,\mathrm{d}\mathbf{v}$. Then if $\sum_i e_i^2$ is the sum of the squared eccentricities of all the planetesimals in the disk, the total heating or stirring rate is

$$\frac{\mathrm{d}\sum_i e_i^2}{\mathrm{d}t}=\frac{\pi}{v_c^2}\int\mathrm{d}\mathbf{v}\,vf(\mathbf{v})\int\mathrm{d}b\,b\sin^2\theta\big[5v^2-3(\hat{\mathbf{x}}\cdot\mathbf{v})^2-12(\hat{\mathbf{y}}\cdot\mathbf{v})^2\big], \tag{9.32}$$

with a similar expression for the inclinations.

From equation (1.39) the deflection angle is related to the impact parameter through

$$\sin^2\theta=\frac{4\tan^2\frac{1}{2}\theta}{(1+\tan^2\frac{1}{2}\theta)^2}=\frac{4(\mathbb{G}M_\mathrm{p})^2b^2v^4}{[(\mathbb{G}M_\mathrm{p})^2+b^2v^4]^2}. \tag{9.33}$$

The integral over the impact parameter b is carried out between lower and upper limits $b_{\rm min}$ and $b_{\rm max}$. The minimum impact parameter is given by the trajectory that just skims the planet's surface (eq. 9.21), since trajectories with smaller impact parameters collide with the planet. For simplicity we neglect the possibility of collisions in this calculation, so we set $b_{\rm min} = 0$; then using equation (9.33), the integral becomes

$$\int_0^{b_{\rm max}} db\, b \sin^2\theta = \frac{2(\mathbb{G}M_{\rm p})^2}{v^4}\left[\log(1+\Lambda^2) - \frac{\Lambda^2}{1+\Lambda^2}\right] \text{ with } \Lambda \equiv \frac{b_{\rm max}v^2}{\mathbb{G}M_{\rm p}}. \tag{9.34}$$

We shall see shortly that $\Lambda \gg 1$ in typical cases, so we can approximate the quantity in square brackets by $2\log\Lambda$ with a fractional error of order $(\log\Lambda)^{-1}$. Then equation (9.32) becomes

$$\frac{d\sum_i e_i^2}{dt} = \frac{4\pi(\mathbb{G}M_{\rm p})^2}{v_{\rm c}^2}\int \frac{d\mathbf{v}}{v^3} f(\mathbf{v}) \log\Lambda\left[5v^2 - 3(\hat{\mathbf{x}}\cdot\mathbf{v})^2 - 12(\hat{\mathbf{y}}\cdot\mathbf{v})^2\right]. \tag{9.35}$$

The quantity $\log\Lambda$ is called the **Coulomb logarithm**; a similar logarithm appears in calculations of the transport properties of electrostatic plasmas and stellar systems (e.g., Binney & Tremaine 2008).

The choice of the appropriate value for the maximum impact parameter $b_{\rm max}$ needs some care. We cannot set $b_{\rm max} \to \infty$ since then $\log\Lambda$ diverges. However, the divergence is slow ($\propto \log b_{\rm max}$) so an approximate estimate of $b_{\rm max}$ is sufficient. Our calculation so far has been based on the assumption that the density of planetesimals is homogeneous near the planet and therefore fails once the impact parameter is comparable to the disk thickness. Therefore we should choose $b_{\rm max}$ to be roughly the root-mean-square disk thickness $z_0 = 0.798\, a_{\rm p}\langle I\rangle$ (eq. 9.20); the density at this impact parameter is $\exp(-\frac{1}{2}) = 0.606$ of the central density in the $\hat{\mathbf{z}}$ direction and close to unity in the $\hat{\mathbf{x}}$ and $\hat{\mathbf{y}}$ directions. Given the approximations we have made so far, there is little further loss of accuracy if we make some further simplifications to the definition of Λ: we assume that $\langle e\rangle = \langle I\rangle$; we replace the variable v by $v_{\rm c}\langle e\rangle$, and we drop all dimensionless factors of order unity.

Then $b_{\rm max} \simeq a_{\rm p}\langle e\rangle$, $v^2 = v_{\rm c}^2\langle e\rangle^2 = (\mathbb{G}M_*/a_{\rm p})\langle e\rangle^2$ and

$$\Lambda = \frac{M_*}{M_{\rm p}}\langle e\rangle^3. \tag{9.36}$$

For example, a disk around a solar-mass star with mean eccentricity $\langle e\rangle = 0.1$ that contains an Earth-mass planet has $\Lambda \simeq 330$, confirming our earlier assertion that $\Lambda \gg 1$ in typical cases. In this case the fractional error due to the approximations we have made in evaluating Λ is $\sim (\log\Lambda)^{-1} \sim 0.2$.

Substituting the Schwarzschild distribution function, equations (9.17) and (9.18), into equation (9.35) we find that the rate of eccentricity growth is

$$\begin{aligned}\frac{\mathrm{d}\sum_i e_i^2}{\mathrm{d}t} = &\frac{\pi n_0(\mathbb{G}M_{\rm p})^2}{v_{\rm c}^5\langle e\rangle^2\langle I\rangle}\log\Lambda\int\frac{\mathrm{d}v_x\mathrm{d}v_y\mathrm{d}v_z}{v^3} \\ &\times\exp\left[-\frac{\pi}{4v_{\rm c}^2\langle e\rangle^2}(v_x^2+4v_y^2)-\frac{\pi}{4v_{\rm c}^2\langle I\rangle^2}v_z^2\right]\left(5v^2-3v_x^2-12v_y^2\right).\end{aligned} \tag{9.37}$$

As in equations (9.23) and (9.24), this expression and the analogous expression for the inclinations can be rewritten as

$$\begin{aligned}\frac{\mathrm{d}\sum_i e_i^2}{\mathrm{d}t} &= \frac{n_0(\mathbb{G}M_{\rm p})^2\log\Lambda}{v_{\rm c}^3\langle e\rangle}\Psi_c(\gamma), \\ \frac{\mathrm{d}\sum_i I_i^2}{\mathrm{d}t} &= \frac{n_0(\mathbb{G}M_{\rm p})^2\log\Lambda}{v_{\rm c}^3\langle e\rangle}\Psi_d(\gamma),\end{aligned} \tag{9.38}$$

where $\gamma = \langle e\rangle/\langle I\rangle$ and

$$\begin{aligned}\Psi_c(\gamma) &\equiv \pi\gamma\int\frac{\mathrm{d}\mathbf{u}}{u^3}\exp\left[-\tfrac{1}{4}\pi(u_x^2+4u_y^2+\gamma^2u_z^2)\right]\left(2u_x^2-7u_y^2+5u_z^2\right), \\ \Psi_d(\gamma) &\equiv \pi\gamma\int\frac{\mathrm{d}\mathbf{u}}{u^3}\exp\left[-\tfrac{1}{4}\pi(u_x^2+4u_y^2+\gamma^2u_z^2)\right]\left(u_x^2+u_y^2-2u_z^2\right).\end{aligned} \tag{9.39}$$

The functions $\Psi_c(\gamma)$ and $\Psi_d(\gamma)$ are plotted in Figure 9.1. Typical values are $\Psi_c(1.5) = 12.586$, $\Psi_d(1.5) = 0.694$. When $\gamma < 1.390$ the coefficient $\Psi_d(\gamma)$ is negative, so gravitational scattering causes the mean inclination to shrink as the vertical velocities of the planetesimals are converted to horizontal velocities.

This analysis neglects the excitation of the eccentricities of planetesimals that do not cross the planet's orbit, but these make a contribution to $\mathrm{d}\sum_i e_i^2/\mathrm{d}t$ that is smaller by a factor of order $1/\log\Lambda$ (e.g., Stewart & Ida 2000).

Equations (9.38) imply that the eccentricities grow faster than the inclinations if $\Psi_c(\gamma) > \Psi_d(\gamma)$ and vice versa. If the distribution function in the disk has the Schwarzschild form, as we have assumed throughout this section, then

$$\frac{\sum_i e_i^2}{\sum_i I_i^2} = \frac{\langle e\rangle^2}{\langle I\rangle^2} = \gamma^2. \tag{9.40}$$

Taking the time derivative of this equation and using equations (9.38),

$$\frac{\mathrm{d}\gamma^2}{\mathrm{d}t} = \frac{\mathrm{d}}{\mathrm{d}t}\log\Big(\sum_i I_i^2\Big)\left[\frac{\Psi_c(\gamma)}{\Psi_d(\gamma)} - \gamma^2\right]. \tag{9.41}$$

If gravitational stirring by planets or large planetesimals is the primary mechanism that excites the eccentricities and inclinations, the evolution will approach a self-similar form in which $\Psi_c(\gamma) = \gamma^2\Psi_d(\gamma)$, which in turn requires $\gamma = 1.832$.

Eventually gravitational stirring by the planet excites the eccentricities of a significant fraction of the planetesimals that cross the planet's orbit, which depletes the surface density of the disk near the planets because the planetesimals visit a wider range of radii. The sum of the squared eccentricities of all the planetesimals that cross the planet's orbit is $\sum_i e_i^2 = 24a_\mathrm{p}^2\Sigma_N\langle e\rangle^3$, where $\Sigma_N = 2n_0a_\mathrm{p}\langle I\rangle$ is the surface number density (Problem 9.1). Then the characteristic depletion time is

$$t_\mathrm{stir} \equiv \frac{\sum_i e_i^2}{\mathrm{d}\sum_i e_i^2/\mathrm{d}t} = \frac{24P}{\pi}\left(\frac{M_*}{M_\mathrm{p}}\right)^2 \frac{\langle e\rangle^5}{\gamma\Psi_c(\gamma)\log\Lambda}. \tag{9.42}$$

The depletion times for planets in the solar system are given in Table 9.1 for representative values $\langle e\rangle = 0.1$, $\gamma = \langle e\rangle/\langle I\rangle = 1.5$. Note that the results are quite sensitive to the assumed value of $\langle e\rangle$ since $t_\mathrm{stir} \propto \langle e\rangle^5$.

Gravitational stirring also excites the inclinations of the planetesimals. This process does not directly affect the surface density of the planetesimal disk but does reduce the number density in the midplane.

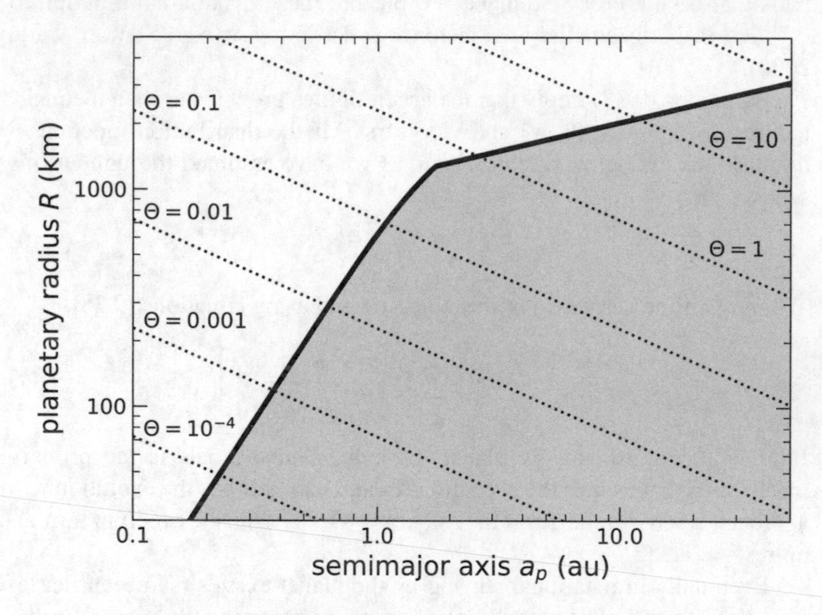

Figure 9.2: In the unshaded regions a body of radius R (vertical axis) and semimajor axis $a_{\rm p}$ (horizontal axis) can clear the neighborhood of its orbit from planetesimals. The broken heavy line is the solution of the equation $\min(t_{\rm coll}, t_{\rm stir}) = t_{\rm ss}$ where $t_{\rm ss} = 4.57\,$Gyr is the age of the solar system, $t_{\rm coll}$ is the collision time (9.27) and $t_{\rm stir}$ is the stirring time (9.42). The mean eccentricity in the disk is assumed to be $\langle e\rangle = 0.1$, the mean density of the body is $3\,\mathrm{g\,cm^{-3}}$, the ratio of mean eccentricity to mean inclination is $\gamma = \langle e\rangle/\langle I\rangle = 1.5$, and the mass of the host star is $1M_\odot$. The figure also shows contours of constant Safronov number (9.26) as dotted lines.

We can now ask whether collisions or gravitational stirring dominates the depletion of a planetesimal disk in the vicinity of a planet. Collisions dominate if $t_{\rm coll} \lesssim t_{\rm stir}$. Using the definition of the Safronov number Θ (eq.

9.26), this condition requires

$$\Theta^2 \lesssim \frac{24}{\pi} \frac{\Psi_a(\gamma) + \Theta\Psi_b(\gamma)}{\Psi_c(\gamma)\log\Lambda}. \tag{9.43}$$

If we replace the inequality by an equality, we have a quadratic equation for Θ with two roots. One is always negative and therefore of no interest since Θ must be positive. If we call the positive root $\Theta_{\rm crit}$, we may conclude that collisions are more important than stirring when $\Theta \lesssim \Theta_{\rm crit}$. For example, if $\gamma = 1.5$ and $\log\Lambda = 5$, we find $\Theta_{\rm crit} = 0.922$. For the range $1 \leq \gamma \leq 2$ and $1 \leq \log\Lambda \leq 10$, $\Theta_{\rm crit}$ varies between 0.46 and 5.7. Given these variations, and the approximations we have made, an appropriate heuristic conclusion is that collisions dominate the evolution if $\Theta \ll 1$, while gravitational stirring dominates if $\Theta \gg 1$.

In 2006 the International Astronomical Union (IAU) defined a "planet" to be a celestial body that (i) is in orbit around the Sun, (ii) has sufficient mass for its self-gravity to overcome rigid-body forces such that it assumes a hydrostatic equilibrium (nearly round) shape (see the discussion following eq. 8.80), and (iii) has cleared the neighborhood around its orbit. The last of these is approximately equivalent to the condition that the smaller of the collision time $t_{\rm coll}$ and the gravitational stirring time $t_{\rm coll}$ is less than the age of the solar system, 4.57 Gyr. This constraint on the planetary radius R as a function of semimajor axis $a_{\rm p}$ is plotted in Figure 9.2 for planetary density $\rho_{\rm p} = 3\,{\rm g\,cm^{-3}}$ and mean eccentricity $\langle e\rangle = 0.1$. Bodies in the shaded region below the line are not "planets" according to the IAU definition.

9.3 Evolution of high-eccentricity orbits

We have seen in the preceding section that particles can be pumped into high-eccentricity orbits by gravitational stirring. The goal of this section is to investigate the behavior of particles once their orbits achieve high eccentricity.

We consider a test particle belonging to a system containing a single planet on a circular orbit with semimajor axis $a_{\rm p}$. The test-particle orbit has an eccentricity e that is not far from unity and is planet-crossing. We shall assume that the periapsis $q = a(1 - e)$ is not far inside the planet and that the apoapsis is far outside it.

For an orbit with $a \gg a_p$, the Tisserand parameter (eq. c of Box 3.1) becomes

$$T = a_p/a + 2(a/a_p)^{1/2}(1 - e^2)^{1/2} \cos I \simeq 2^{3/2}(q/a_p)^{1/2} \cos I. \qquad (9.44)$$

Thus particles in low-inclination orbits conserve their periapsis distance so long as they remain at low inclination. Even if we allow for inclination variations, the periapsis of the test particle can never grow to be significantly larger than a_p. The reason is that if the test particle never comes close to the planet it cannot interact strongly with it (note the rapid falloff in the root-mean-square energy change in Figure 9.3 as q/a_p grows beyond unity), so such an orbit would survive for an indefinite period in the future without becoming planet-crossing; but since Newton's laws are time-reversible this means it could never have been planet-crossing in the past.[6]

The energy of the test-particle orbit is $-\frac{1}{2}\mathbb{G}M_*/a$, where M_* is the mass of the host star. To simplify the notation it is easier to follow the variable $x \equiv 1/a$, which is proportional to the energy (with a negative proportionality constant); where the chance of confusion is small we shall adopt the common practice of calling x the "energy." Let the root-mean-square change in x resulting from a single periapsis passage—from $r \gg a$ pre-periapsis to $r \gg a$ post-periapsis—be σ_x. Simple scaling arguments imply that $\sigma_x = f_x(M_p/M_*)/a_p$ where f_x is a constant of order unity that depends on the inclination, the periapsis distance, and the argument of periapsis (but not the semimajor axis so long as $a \gg a_p$, or the longitude of the node, which is random since the phase of the encounter is random).[7] Since the mean

[6] This argument bears on the formation of the curious trans-Neptunian object Sedna and a handful of objects on similar orbits, called the **detached disk**. Sedna has periapsis distance 76.4 au, eccentricity $e = 0.85$, inclination 11.9° relative to the ecliptic, and orbital period $P = 11\,500$ yr. Sedna's periapsis distance is 2.5 times the semimajor axis of Neptune, and its orbit is unaffected by perturbations from Neptune or other known planets over the lifetime of the solar system. How then did it acquire its current orbit?

[7] A more careful analysis shows that f_x is of order $\log(M_*/M_p)$ rather than unity. This factor is the Coulomb logarithm described in the preceding section (Binney & Tremaine 2008).

energy change is zero (see below), the root-mean-square change in energy after N passages is $\langle(\Delta x)^2\rangle_N^{1/2} = N^{1/2}\sigma_x = N^{1/2} f_x(M_\mathrm{p}/M_*)/a_\mathrm{p}$. We define N_relax to be the characteristic number of orbits needed for the orbit to random-walk from $x \simeq 0$ to $x \sim 1/a_\mathrm{p}$; more precisely $\langle(\Delta x)^2\rangle_N^{1/2} = 1/a_\mathrm{p}$ when $N = N_\mathrm{relax}$, so $N_\mathrm{relax} = f_x^{-2}(M_*/M_\mathrm{p})^2$.

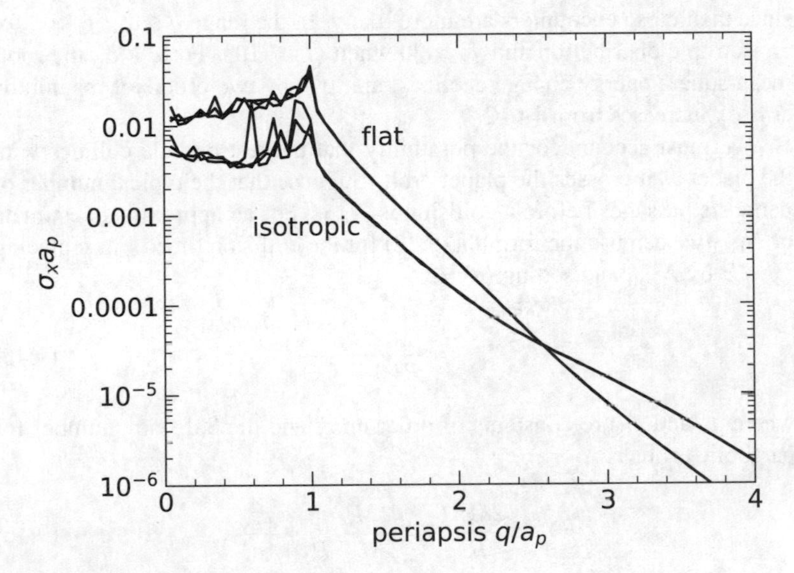

Figure 9.3: Root-mean-square change in $x = 1/a$ for an initially parabolic orbit during a single periapsis passage through a planetary system. The system contains a single planet on a circular orbit with semimajor axis a_p, having mass 0.001 times the stellar mass. Each curve is determined from 10^4 incoming orbits with random values of the longitude of node and periapsis. Three Monte–Carlo realizations are shown for each curve to indicate the uncertainties. The inclinations of the parabolic orbits are chosen from either a Rayleigh distribution with mean inclination $\langle I\rangle = 10°$ (labeled "flat") or a uniform distribution in $\cos I$ (labeled "isotropic").

Figure 9.3 shows the root-mean-square change in x per periapsis passage for incoming particles on parabolic orbits, with a planet mass M_p =

$0.001M_*$ (roughly appropriate for Jupiter and the Sun). The figure shows results for two inclination distributions: one corresponding to parabolic orbits from a spherical source (labeled "isotropic"); and one for a Rayleigh distribution (eq. 9.11) with mean inclination 10° (labeled "flat"), corresponding to parabolic orbits in nearly the same plane as the planet. The curves are relatively flat when the periapsis distance $q \lesssim a_\mathrm{p}$, with a weak peak at $q \simeq a_\mathrm{p}$ since then close encounters are more likely. In the range $q \lesssim a_\mathrm{p}$, $f_x \simeq 6$ for an isotropic distribution and $f_x \simeq 20$ when $\langle I \rangle = 10°$. For $q \gtrsim a_\mathrm{p}$ the root-mean-square energy change declines rapidly—by two orders of magnitude as q/a_p increases from 1 to 2.

We must account for the possibility that the test particle collides with the planet as it crosses the planet orbit. Suppose that the typical number of periapsis passages before a collision is N_coll. For an approximate estimate of N_coll we can use the formula (9.27) for the collision time t_coll, replacing t_coll/P by N_coll and setting $\langle e \rangle = 1$,

$$N_\mathrm{coll} = \frac{a_\mathrm{p}^2}{R^2}\frac{f_\mathrm{c}}{1 + f_\mathrm{s}\Theta_1}, \tag{9.45}$$

where f_c and f_s are constants of order unity and the Safronov number for parabolic orbits is

$$\Theta_1 \equiv \frac{2\mathbb{G}M_\mathrm{p}}{Rv_\mathrm{c}^2} = \frac{2M_\mathrm{p}}{M_*}\frac{a_\mathrm{p}}{R} = \frac{v_\mathrm{esc}^2}{v_\mathrm{c}^2}. \tag{9.46}$$

Here v_c is the circular speed at the planet's semimajor axis, and v_esc is the escape speed from the planet's surface (eq. 1.21). The constant f_c grows roughly as the inverse of the mean inclination of the planet-crossing orbits (Problem 9.5).

The ratio of the collision lifetime to the relaxation lifetime is

$$\frac{N_\mathrm{coll}}{N_\mathrm{relax}} = \frac{f_x^2 f_\mathrm{c}}{1 + f_\mathrm{s}\Theta_1}\left(\frac{a_\mathrm{p}}{R}\frac{M_\mathrm{p}}{M_*}\right)^2 = \frac{f_x^2 f_\mathrm{c}}{4}\frac{\Theta_1^2}{1 + f_\mathrm{s}\Theta_1}. \tag{9.47}$$

Thus if the parabolic Safronov number $\Theta_1 \ll 1$, $N_\mathrm{coll} \ll N_\mathrm{relax}$, and the test particle is likely to collide with the planet long before its orbit suffers

Table 9.2: Safronov numbers Θ_1 for solar-system planets

	a_p (au)	$M_\odot/M_\mathrm{p}$	radius (km)	Θ_1
Mercury	0.387	6.024×10^6	2 440	0.007 88
Venus	0.723	4.085×10^5	6 051	0.0875
Earth	1.000	3.329×10^5	6 378	0.1409
Mars	1.524	3.099×10^6	3 396	0.0433
Jupiter	5.203	1 047.6	71 492	20.79
Saturn	9.537	3 498.8	60 330	13.52
Uranus	19.19	22 905	25 559	9.81
Neptune	30.07	19 416	24 764	18.71

The parabolic Safronov number is defined in equation (9.46).

significant changes. However, if $\Theta_1 \gg 1$ collisions are unimportant and the fate of the test particle is determined by its orbital evolution, as if the planet were a point mass (Wyatt et al. 2017).

Table 9.2 gives the parabolic Safronov numbers for the solar-system planets. The terrestrial planets have Safronov numbers in the range ~ 0.01–0.1 and therefore bodies that cross their orbits are likely to collide with them long before their orbits evolve to high eccentricities. In contrast, the giant planets, from Jupiter to Neptune, have Safronov numbers $\gtrsim 10$ so collisions are unimportant compared to orbital evolution. The goal of the following discussion is to investigate how planet-crossing orbits in the outer solar system evolve and what their fate is.

Figure 9.4 shows the root-mean-square change in the energy $x = 1/a$ for a parabolic orbit passing through the solar system. As in Figure 9.3, two sets of curves are shown, one for an isotropic distribution of incoming orbits and one for a Rayleigh distribution with a mean inclination of 10°. There are spikes when the periapsis distance is close to the semimajor axis of one of the giant planets (Table 9.2), and the root-mean-square energy change declines rapidly when the periapsis distance exceeds the semimajor axis of the outermost planet, Neptune with $a_\mathrm{Nep} = 30.07$ au.

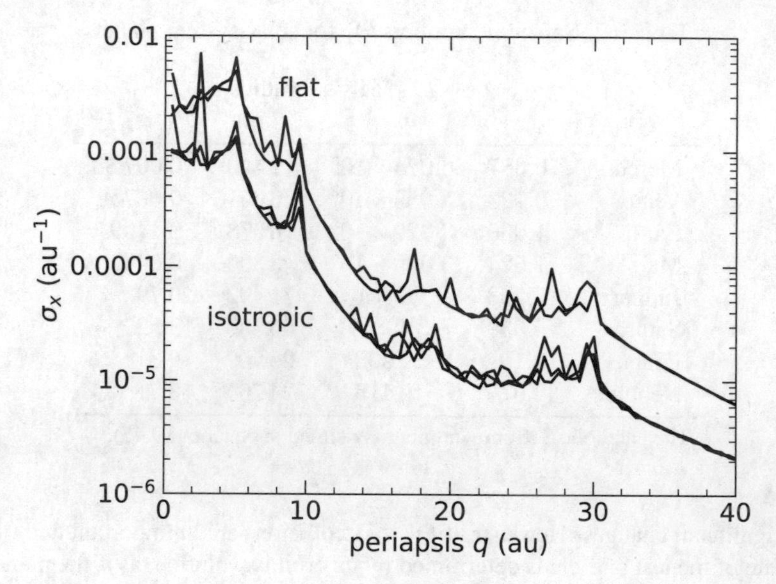

Figure 9.4: As in Figure 9.3, but for a system containing the four giant planets of the solar system.

We now have the tools to investigate the evolution of a test particle on a high-eccentricity orbit in the solar system, with initial semimajor axis $a \gg a_{\rm Nep}$ but periapsis distance $q \lesssim a_{\rm Nep}$. This analysis is based on three assumptions:

(i) We assume that the semimajor axis and orbital period of the particle are large enough that the phases of the planets can be treated as random variables at each periapsis passage (the random-phase approximation).[8] Thus we can describe the changes in $x = 1/a$ during each

[8] Here is a heuristic derivation of the accuracy of this approximation. The period of the orbit is $P = 2\pi(\mathbb{G}M_* x^3)^{-1/2}$, so the typical change in period per periapsis passage is $\Delta P \sim 3\pi(\mathbb{G}M_* x^5)^{-1/2}\sigma_x$, which represents the change in arrival time of the particle at its next periapsis. If the change in arrival time is larger than the time needed for the

periapsis passage by a probability distribution $g(\Delta x)$, where

$$\int \mathrm{d}\Delta x\, g(\Delta x) = 1, \quad \int \mathrm{d}\Delta x\, g(\Delta x)\Delta x = 0,$$
$$\int \mathrm{d}\Delta x\, g(\Delta x)(\Delta x)^2 = \sigma_x^2. \tag{9.48}$$

The first expression holds because $g(\Delta x)$ is a probability distribution, so its integral must be unity. The second holds because Newton's laws are time-reversible, so the mean has to be the same whether we trace the particle forward or backward in time and therefore must be zero.[9] The third is simply the definition of the root-mean-square energy change per orbit σ_x.

(ii) We ignore changes in the periapsis distance and inclination (see the discussion at the start of this section), which allows us to assume that σ_x is constant as the orbit evolves.

(iii) With assumptions (i) and (ii) the orbital evolution is a random walk in the energy x at constant q and I. If the energy changes are small, $\sigma_x \ll x$, then the random walk can be approximated by the diffusion equation that we now derive.

Let the probability that the particle lies in a small range of $x = 1/a$ at time t be $p(x,t)\mathrm{d}x$, and let the change in x in a single periapsis passage be denoted by Δx. In a small time interval $\mathrm{d}t$ the probability that a particle with energy x passes through periapsis is $\mathrm{d}t/P(x)$, where $P(x) = 2\pi a^{3/2}/(\mathbb{G}M_*)^{1/2} = 2\pi/(\mathbb{G}M_* x^3)^{1/2}$ is the orbital period. The particle is

phase of the planet to change by a radian or so, then the phase can be treated as a random variable; this requires $n_\mathrm{p}\Delta P \gtrsim 1$, where $n_\mathrm{p} = (\mathbb{G}M_*/a_\mathrm{p}^3)^{1/2}$ and a_p are the mean motion and semimajor axis of the planet. Thus the random-phase approximation requires $x \lesssim (3\pi)^{2/5}\sigma_x^{2/5}/a_\mathrm{p}^{3/5}$. For a particle crossing Neptune's orbit at low inclination, we have $a_\mathrm{p} = 30.1\,\mathrm{au}$ and $\sigma_x \simeq 3 \times 10^{-5}\,\mathrm{au}^{-1}$ from Figure 9.4, so we require $x \lesssim 0.005\,\mathrm{au}^{-1}$ or $a \gtrsim 200\,\mathrm{au}$.

[9] This argument is not strictly correct. A more accurate statement is that when $\sigma_x a_\mathrm{p} \ll 1$, the integral is of order $\sigma_x^2 a_\mathrm{p}$ rather than of order σ_x. In the analysis of this section we shall assume that the integral is exactly zero.

perturbed to a different energy as it passes through the planetary system, so $p(x,t)\mathrm{d}x$ is reduced by an amount $p(x,t)\mathrm{d}x\mathrm{d}t/P(x)$. On the other hand, if the particle has some other energy it can be scattered into the energy interval $(x, x+\mathrm{d}x)$, and the probability for this process is $\int \mathrm{d}\Delta x\, p(x-\Delta x,t)g(\Delta x)\mathrm{d}t\mathrm{d}x/P(x-\Delta x)$. Combining these two processes, we find

$$\frac{\partial p(x,t)}{\partial t} = \int \cdot d\Delta x\, g(\Delta x)\frac{p(x-\Delta x,t)}{P(x-\Delta x)} - \frac{p(x,t)}{P(x)}. \tag{9.49}$$

Since the energy changes are small by assumption (iii)—that is, since $p(x,t)$ and $P(x)$ vary on characteristic scales much larger than Δx—we can expand $p(x,t)/P(x)$ in a Taylor series:

$$\begin{aligned}\frac{\partial p(x,t)}{\partial t} &= \frac{p(x,t)}{P(x)}\int d\Delta x\, g(\Delta x) - \frac{\partial}{\partial x}\left[\frac{p(x,t)}{P(x)}\right]\int d\Delta x\, g(\Delta x)\Delta x \\ &\quad + \tfrac{1}{2}\frac{\partial^2}{\partial x^2}\left[\frac{p(x,t)}{P(x)}\right]\int d\Delta x\, g(\Delta x)(\Delta x)^2 - \frac{p(x,t)}{P(x)} + \mathrm{O}(\Delta x)^3.\end{aligned} \tag{9.50}$$

We drop the term that is $\mathrm{O}(\Delta x)^3$; then using equations (9.48) this result simplifies to

$$\frac{\partial p(x,t)}{\partial t} = \tfrac{1}{2}\sigma_x^2\frac{\partial^2}{\partial x^2}\left[\frac{p(x,t)}{P(x)}\right] = \frac{\sigma_x^2(\mathbb{G}M_*)^{1/2}}{4\pi}\frac{\partial^2}{\partial x^2}x^{3/2}p(x,t). \tag{9.51}$$

We seek a Green's function solution to this equation, that is, a solution in which $p(x,t=0) = \delta(x-x_0)$, implying that the particle is at $x = x_0$ at $t = 0$. The procedure for finding this solution is outlined in Yabushita (1980), and yields

$$p(x,t) = \frac{2kx_0^{1/2}}{xt}\exp\left[-\frac{4k(x_0^{1/2}+x^{1/2})}{t}\right]I_2\left[\frac{8k}{t}(x_0x)^{1/4}\right], \tag{9.52}$$

where $I_2(\cdot)$ is a modified Bessel function (Appendix C.5) and

$$k \equiv \frac{4\pi}{(\mathbb{G}M_*)^{1/2}\sigma_x^2}. \tag{9.53}$$

The probability that the particle will survive for a time t or longer is found from equation (C.35),

$$f(t) = \int_0^\infty \mathrm{d}x\, p(x,t) = 1 - (1+z)\mathrm{e}^{-z}, \quad \text{where} \quad z \equiv \frac{4kx_0^{1/2}}{t}. \tag{9.54}$$

At the initial time $t = 0$, $f(0) = 1$ as required; at large times

$$f(t) \to \frac{8k^2 x_0}{t^2} = \frac{128\pi^2 x_0}{\mathbb{G}M_* \sigma_x^4 t^2}. \tag{9.55}$$

The half-life of a particle is defined by $f(t_{1/2}) = \frac{1}{2}$ and given by

$$\begin{aligned} t_{1/2}(x_0) &= 2.3833\, kx_0^{1/2} = \frac{29.95}{(\mathbb{G}M_* a_0)^{1/2}\sigma_x^2} \\ &= 5.30 \times 10^8\,\mathrm{yr}\left(\frac{100\,\mathrm{au}}{a_0}\right)^{1/2}\left(\frac{3\times 10^{-5}\,\mathrm{au}^{-1}}{\sigma_x}\right)^2, \end{aligned} \tag{9.56}$$

in which we have used a typical root-mean-square energy change σ_x for orbits with perihelion in the Uranus–Neptune region, 20–30 au (Figure 9.4).

This simple model captures much of the dynamics that governs the long-term evolution of particles on planet-crossing orbits. In particular, it shows that the lifetime of particles is a decreasing function of semimajor axis—particles with large semimajor axes interact with the planets less often but have a smaller distance to diffuse to reach escape energy. However, the diffusion equation (9.51) is not accurate enough for quantitative calculations for several reasons: (i) The root-mean-square change per orbit σ_x is a strong function of the periapsis distance q (Figure 9.4) and enters the half-life as σ_x^{-2}, so modest changes in periapsis distance can have a dramatic influence on the lifetime. (ii) Particles may be "protected" from energy diffusion for extended periods because of resonances with the planet, as in the case of Pluto and Neptune (§6.4). (iii) ZLK oscillations (§5.4) can lead to large changes in the inclination and periapsis distance if the particle is initially on a highly inclined orbit. (iv) Close encounters cannot be treated with the diffusion approximation, nor can encounters in the region near escape energy where $x \lesssim \sigma_x$. (v) We have assumed that the mean energy change is exactly zero, when in fact its influence on the diffusion can be comparable to that of the root-mean-square energy change (see footnote 9). The insights provided by the diffusion approximation are important, but in the end there is no substitute for long-term numerical integrations to study the survival and fate of particles on planet-crossing orbits, as described in §9.5.

We have approximated the planetesimals as test particles, but in reality they have a small but nonzero mass. Therefore the ejection of planetesimals on escape orbits with positive energy causes the semimajor axis of the planet to shrink, a process known as **planetesimal-driven migration**. The migration will be substantial when the mass of the ejected planetesimals is comparable to the mass of the planet. When multiple planets are present, the amount and even the direction of planetesimal-driven migration depend on the configuration of the planets. For example, in the outer solar system Saturn, Uranus and Neptune migrate *outward*, while Jupiter migrates inward by a much smaller amount. The reason is that particles scattered by (for example) Neptune diffuse to both larger and smaller semimajor axes, and the particles scattered to smaller semimajor axes come under the influence of Jupiter and are rapidly ejected. Thus there is an absorbing boundary due to Jupiter at small semimajor axes, which is more important than the more distant absorbing boundary at the escape energy. As a result, on average the particles diffuse inward causing Neptune to migrate outward (Fernández & Ip 1984). It is this outward migration that enables Neptune to capture Pluto in the 3:2 resonance, as described in §6.4.

9.4 The Galactic tidal field

The Sun and most nearby stars are part of the disk of our home galaxy, known as the Milky Way or simply the Galaxy. The gravitational effects of the Galaxy are negligible at typical planetary distances of a few au or less but play a critical role in the evolution of solar-system bodies with much larger semimajor axes.

The distribution of stars and gas in the Galactic disk is approximately symmetric around a plane known as the **Galactic midplane**. We work with a Cartesian coordinate system (X, Y, Z) in which the $Z = 0$ plane coincides with the midplane. The $\hat{\mathbf{Z}}$-direction is often referred to as "vertical" or the "north Galactic pole." The north pole of the ecliptic is inclined to the north Galactic pole by $60.2°$.

We denote the density of the stars and gas in the Galaxy as $\rho(X, Y, Z)$. Since the variation of density within the plane is much slower than the vertical variation (i.e., the disk is thin), we can approximate the density in the solar neighborhood as $\rho(Z)$. The corresponding gravitational potential is $\Phi(Z)$, and Poisson's equation (B.44) implies that

$$\frac{\mathrm{d}^2\Phi}{\mathrm{d}Z^2} = 4\pi \mathbb{G}\rho(Z). \tag{9.57}$$

The vertical acceleration of the Sun due to the Galactic potential is $\ddot{Z}_\odot = -\mathrm{d}\Phi/\mathrm{d}Z|_{Z_\odot}$. Similarly, the vertical acceleration of a particle orbiting the Sun at position Z is $\ddot{Z} = -\mathbb{G}M_\odot(Z - Z_\odot)/r^3 \quad \mathrm{d}\Phi/\mathrm{d}Z|_Z$, where r is the distance of the particle from the Sun. We now change to heliocentric coordinates (x, y, z) aligned with (X, Y, Z), so $z = Z - Z_\odot$ and the Sun is at $x = y = z = 0$. Since the size of bound heliocentric orbits (less than a parsec, even for the most distant comets) is much less than the thickness of the Galactic disk (several hundred parsecs), we can write

$$\begin{aligned}
\ddot{z} = \ddot{Z} - \ddot{Z}_\odot &= -\frac{\mathbb{G}M_\odot}{r^3}z - \left.\frac{\mathrm{d}\Phi}{\mathrm{d}Z}\right|_Z + \left.\frac{\mathrm{d}\Phi}{\mathrm{d}Z}\right|_{Z_\odot} \simeq -\frac{\mathbb{G}M_\odot}{r^3}z - \left.\frac{\mathrm{d}^2\Phi}{\mathrm{d}Z^2}\right|_{Z_\odot} z \\
&= -\frac{\mathbb{G}M_\odot}{r^3}z - 4\pi\mathbb{G}\rho(Z_\odot)z = -\frac{\partial}{\partial z}\left[-\frac{\mathbb{G}M_\odot}{r} + \Phi_{\mathrm{G}}(z, t)\right],
\end{aligned} \tag{9.58}$$

where the Galactic tidal potential is

$$\Phi_{\mathrm{G}}(z, t) = 2\pi\mathbb{G}\rho(t)z^2, \tag{9.59}$$

with $\rho(t)$ the mass density at the Sun's location $Z_\odot(t)$. The other coordinates are governed by the Kepler equations of motion,

$$\ddot{x} = -\frac{\mathbb{G}M_\odot}{r^3}x, \quad \ddot{y} = -\frac{\mathbb{G}M_\odot}{r^3}y. \tag{9.60}$$

The current vertical position and velocity of the Sun are $Z_\odot \simeq 20\,\mathrm{pc}$ (Bennett & Bovy 2019) and $\dot{Z}_\odot \simeq 7\,\mathrm{km\,s^{-1}}$ (Schönrich et al. 2010). For plausible estimates of the potential $\Phi(z,t)$, the Sun oscillates above and below the Galactic midplane with an amplitude of about 90 pc and a half-period (time between midplane crossings) of about 40 Myr. The density $\rho(t)$ varies substantially along this orbit, mostly because of a thin, dense gas layer in the midplane (Holmberg & Flynn 2000; Guo et al. 2020). However, for the sake of simplicity we will ignore this variation and approximate the density by the average over the solar orbit, $\rho \simeq 0.1 M_\odot\,\mathrm{pc}^{-3}$.

Averaging over the orbit of the bound particle using equations (1.70) and (1.65d)–(1.65f), we have

$$\langle \Phi_\mathrm{G} \rangle = \pi\,\mathbb{G}\rho a^2 \sin^2 I (1 - e^2 + 5e^2 \sin^2 \omega). \tag{9.61}$$

Here the inclination I and argument of perihelion ω are measured relative to the x–y plane parallel to the Galactic midplane, *not* relative to the ecliptic. In terms of the Delaunay variables (1.84),

$$\langle \Phi_\mathrm{G} \rangle = \frac{\pi \rho \Lambda^2}{\mathbb{G} M_\odot^2 L^2}(L^2 - L_z^2)[L^2 + 5(\Lambda^2 - L^2)\sin^2 \omega]. \tag{9.62}$$

Kepler orbits in this potential undergo secular oscillations similar to ZLK oscillations (§5.4), as we now describe (Heisler & Tremaine 1986; Hamilton & Rafikov 2019). The Hamiltonian is independent of the mean anomaly ℓ because it is orbit-averaged. Therefore the conjugate momentum Λ and the semimajor axis a are conserved, as usual in secular dynamics. Moreover the Hamiltonian is independent of the longitude of the node Ω, so the conjugate momentum $L_z = [\,\mathbb{G} M_\odot a(1 - e^2)]^{1/2} \cos I$ is conserved. Therefore we can write $L_z = (\,\mathbb{G} M_\odot a)^{1/2} \cos I_0$, where I_0 is an integral of motion equal to the inclination of the circular orbit with the given z-component of angular momentum. Then the conservation of the orbit-averaged Hamiltonian $-\frac{1}{2}\,\mathbb{G} M_\odot / a + \langle \Phi_\mathrm{G} \rangle$ implies that the function

$$C_\mathrm{G}(e,\omega) \equiv \left(1 - \frac{\cos^2 I_0}{1 - e^2}\right)(1 - e^2 + 5e^2 \sin^2 \omega) - \sin^2 I_0 \tag{9.63}$$

is conserved along a trajectory (the term $\sin^2 I_0$ is added so that $C_G = 0$ at $e = 0$). In words, the existence of the two conserved quantities Λ and L_z has reduced the dynamics from three degrees of freedom to one degree of freedom, corresponding to two phase-space dimensions, which can be taken to be the eccentricity e and the argument of periapsis ω.

Not all values of $C_G(e, \omega)$ correspond to physical trajectories. In particular the definition of I_0 implies that $\cos^2 I = \cos^2 I_0/(1 - e^2)$, and since $\cos^2 I < 1$ we must have $e \leq \sin I_0$. If this condition is satisfied the large bracket in equation (9.63) is positive or zero, and since the small bracket in that equation is also positive $C_G(e, \omega) \geq -\sin^2 I_0$. Therefore contours with $C_G(e, \omega) < -\sin^2 I_0$ are unphysical.

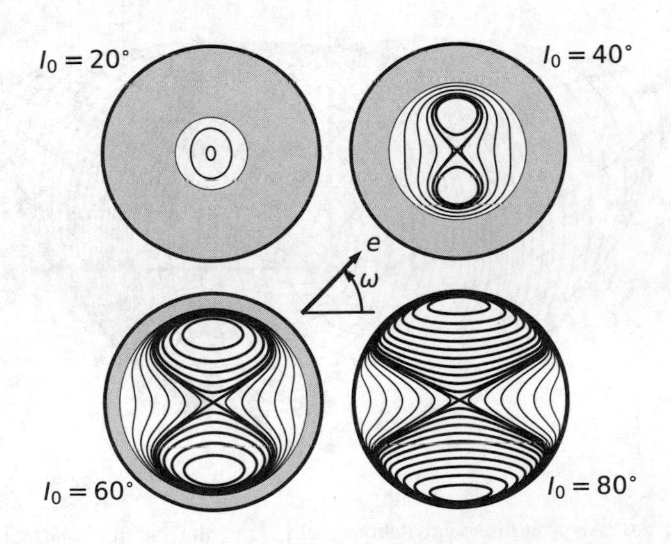

Figure 9.5: Contours of the function $C_G(e, \omega)$ (eq. 9.63), with the eccentricity e and argument of periapsis ω plotted as polar coordinates. Shaded regions have $e > \sin I_0$ and are unphysical. Plots are shown for four values of I_0, the inclination of a circular orbit. Circular orbits are unstable for $I_0 > 26.57°$. Heavy contours have $C_G(e, \omega) \geq 0$. Compare Figure 5.4.

The contours of $C_{\rm G}(e,\omega)$ are shown in Figure 9.5, which resembles Figure 5.4. It is straightforward to show that there are two stationary solutions, corresponding to maxima of $\langle\Phi_{\rm G}\rangle$, when $|L_z|/\Lambda = |\cos I_0| < (\frac{4}{5})^{1/2}$; the maxima are located at $\omega = \pm\frac{1}{2}\pi$ and $L^2/\Lambda^2 = 1 - e^2 = (\frac{5}{4})^{1/2}L_z/\Lambda$. Circular orbits are an additional stationary solution, and these are stable if and only if the contours near $e = 0$ describe a maximum or minimum rather than a saddle point; this requires that the inclination I_0 relative to the Galactic plane satisfies $|\cos I_0| > (\frac{4}{5})^{1/2}$, that is, $I_0 < 26.57°$ or $I_0 > 153.43°$ (Problem 9.7).

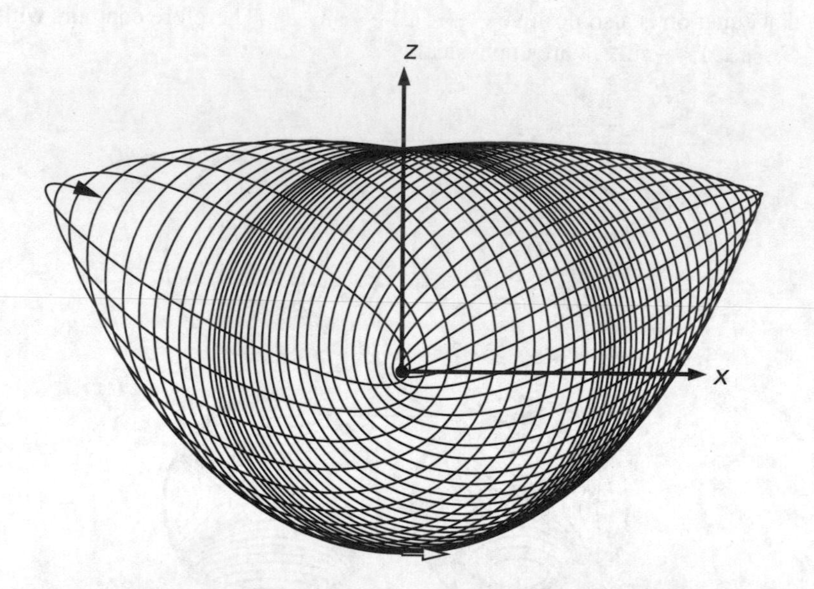

Figure 9.6: An orbit in the gravitational field of the Sun and the Galactic tide (eq. 9.59). The orbit has inclination $I = I_0 = 90°$ and initial eccentricity $e = 0.1$. The z-axis is perpendicular to the Galactic plane. The initial and final positions are marked by white and black arrows.

Figure 9.6 shows an orbit with inclination $I = 90°$ that is confined to the x–z plane. The orbit is initially nearly circular and counterclockwise, with eccentricity $e = 0.1$ and eccentricity vector pointing along the negative z-axis. Under the influence of the torque from the Galactic tidal field, the eccentricity grows until it reaches nearly unity; at this point the angle ω between the apoapsis direction and the positive x-axis is $\omega - \pi = \sin^{-1} 5^{-1/2} = 26.57°$. The continuing torque from the tide reverses the direction of motion, so the orbit is now clockwise, and the eccentricity shrinks, becomes nearly zero, then grows again until it reaches nearly unity with $\omega - \pi = 153.43°$—at which point the cycle repeats but our integration stops.

In terms of the eccentricity vector $\mathbf{e}$ (Box 1.1) and the dimensionless angular-momentum vector $\mathbf{j}$ (eq. 1.33), the orbit-averaged Galactic potential is

$$\langle \Phi_{\rm G} \rangle = \pi \mathbb{G} \rho a^2 [j^2 - (\mathbf{j} \cdot \hat{\mathbf{z}})^2 + 5(\mathbf{e} \cdot \hat{\mathbf{z}})^2]. \tag{9.64}$$

The Milankovich equations (5.57) and (5.58) read

$$\begin{aligned} \frac{\mathrm{d}\mathbf{j}}{\mathrm{d}t} &= \frac{2\pi \mathbb{G}^{1/2} \rho a^{3/2}}{M_\odot^{1/2}} [(\mathbf{j} \cdot \hat{\mathbf{z}})(\mathbf{j} \times \hat{\mathbf{z}}) - 5(\mathbf{e} \cdot \hat{\mathbf{z}})(\mathbf{e} \times \hat{\mathbf{z}})], \\ \frac{\mathrm{d}\mathbf{e}}{\mathrm{d}t} &= \frac{2\pi \mathbb{G}^{1/2} \rho a^{3/2}}{M_\odot^{1/2}} [(\mathbf{j} \cdot \hat{\mathbf{z}})(\mathbf{e} \times \hat{\mathbf{z}}) - 5(\mathbf{e} \cdot \hat{\mathbf{z}})(\mathbf{j} \times \hat{\mathbf{z}}) + \mathbf{j} \times \mathbf{e}]. \end{aligned} \tag{9.65}$$

We are particularly interested in particles that are initially on planet-crossing orbits with large semimajor axes, and for these orbits j is initially nearly zero. Thus the conserved quantity $\mathbf{j} \cdot \hat{\mathbf{z}} = L_z/\Lambda \simeq 0$, and the conserved quantity $\langle \Phi_{\rm G} \rangle = \pi \mathbb{G} \rho a^2 (1 - e^2 + 5e_z^2)$. The orbit oscillates between unit eccentricity and a minimum eccentricity $e_{\rm min}$ along contours on which $1 - e^2 + 5e_z^2 = 5\cos^2\theta$, where θ is the angle between the eccentricity vector and the z-axis (the north Galactic pole) at the initial time. It is straightforward to show that if $\cos^2\theta > \frac{1}{5}$ then $e_{\rm min} = \frac{1}{2}(5\cos^2\theta - 1)^{1/2}$ and is achieved when the eccentricity vector points along the z-axis, while if $\cos^2\theta < \frac{1}{5}$ then $e_{\rm min} = (1 - 5\cos^2\theta)^{1/2}$, achieved when the eccentricity vector is perpendicular to the z-axis.

When the angular momentum is near zero, the Milankovich equations simplify to

$$\frac{\mathrm{d}\mathbf{j}}{\mathrm{d}t} \simeq -\frac{10\pi\,\mathbb{G}^{1/2}\rho a^{3/2}}{M_\odot^{1/2}}(\mathbf{e}\cdot\hat{\mathbf{z}})(\mathbf{e}\times\hat{\mathbf{z}}) + \mathrm{O}(j^2), \quad \frac{\mathrm{d}\mathbf{e}}{\mathrm{d}t} = \mathrm{O}(j). \tag{9.66}$$

The rate of change $\mathrm{d}\mathbf{j}/\mathrm{d}t$ seen in the first of these equations simply reflects the orbit-averaged torque exerted by the Galactic tide (see Problem 9.8). The direction of the torque is determined by the direction of $\mathbf{e}\times\hat{\mathbf{z}}$ and is independent of $\mathbf{j}$ so long as $j \ll 1$; thus we can write $\mathbf{j}(t) \simeq \mathbf{j}(t_0) + (t - t_0)\mathrm{d}\mathbf{j}/\mathrm{d}t$. The perihelion $q \ll a$ corresponds to a dimensionless angular momentum $j = (2q/a)^{1/2}$ so we expect that if the initial perihelion is q_0, then it will exceed q ($q_0 \ll q \ll a$) after a **drift time** t_q given by $t_q|\mathrm{d}\mathbf{j}/\mathrm{d}t| = (2q/a)^{1/2}$. We have

$$\begin{aligned} t_q(a) &= \frac{M_\odot^{1/2} q^{1/2}}{50^{1/2}\pi\,\mathbb{G}^{1/2}\rho a^2|\sin\theta\cos\theta|} \\ &= \frac{3.44\,\mathrm{Gyr}}{|\sin\theta\cos\theta|}\left(\frac{q}{30\,\mathrm{au}}\right)^{1/2}\left(\frac{1\,000\,\mathrm{au}}{a}\right)^2\frac{0.1M_\odot\,\mathrm{pc}^{-3}}{\rho}, \end{aligned} \tag{9.67}$$

where θ is the angle between the eccentricity vector and the north Galactic pole. We shall also define the **critical semimajor axis**, where the drift time $t_q(a)$ equals the orbital period $P(a) = 2\pi a^{3/2}/(\mathbb{G}M_\odot)^{1/2}$,

$$\begin{aligned} a_{\mathrm{crit}}(q) &= \frac{0.244\,q^{1/7}}{|\sin\theta\cos\theta|^{2/7}}\left(\frac{M_\odot}{\rho}\right)^{2/7} \\ &= \frac{2.75\times10^4\,\mathrm{au}}{|\sin\theta\cos\theta|^{2/7}}\left(\frac{q}{30\,\mathrm{au}}\right)^{1/7}\left(\frac{0.1M_\odot\,\mathrm{pc}^{-3}}{\rho}\right)^{2/7}. \end{aligned} \tag{9.68}$$

The primary importance of this drift in angular momentum or perihelion distance induced by the Galactic tide is that it can lift the perihelion of a particle on a planet-crossing orbit beyond the orbits of the planets in the system, at which point the energy perturbations at perihelion passage become negligible and the semimajor axis is frozen in place. This is the mechanism that forms the Oort comet cloud described in the following section.

We have focused so far on steady tides arising from the time-averaged density distribution of gas and stars in the Galaxy. The time-varying forces from individual stars in the solar neighborhood that happen to pass close to the solar system can also perturb the semimajor axes and perihelia of particles with large semimajor axes. In the case of the Sun, the effects of these encounters are qualitatively similar to those of the overall Galactic tide at the semimajor axes $\sim 3 \times 10^4$ au where both play their most important roles (Heisler & Tremaine 1986; Duncan et al. 1987; Collins & Sari 2010). One important difference is that the secular evolution of the orbit due to the tide is periodic, whereas the evolution due to stellar encounters is stochastic. In particular, under the influence of the tide alone an orbit with near-zero angular momentum would eventually return to the same small angular momentum; however, gravitational kicks from individual stellar encounters cause the angular momentum to random-walk to larger values long before this occurs. A second difference is that the semimajor axis or energy is conserved by the Galactic tide, whereas kicks from individual stars introduce a slow random walk in energy that can lead to escape.

We shall not discuss the influence of individual stellar encounters analytically since the best approach is to analyze the effects of tides and stellar encounters simultaneously using N-body simulations, as described in the following section.

9.5 The Oort cloud

Comets are chunks of ice and rock with typical sizes of ~ 1 km (Fernández 2005). Comets are visible—and have inspired awe since prehistoric times—because of the long tails of dust and ionized gas evaporated from the comet's surface when it comes near the Sun. At distances $d \gg 1$ au evaporation is negligible and the comets are visible only by reflected sunlight from their surface; thus the brightness of a comet varies as $1/d^4$ (the incident flux varies as $1/d^2$ and the fraction of the reflected flux that we receive also varies as $1/d^2$). The result of this rapid falloff in brightness is that few comets are detected at distances larger than a few au. Moreover, a typical comet that comes close enough to the Sun to be visible loses material by evaporation sufficiently rapidly that it cannot survive for more than a few hundred orbits, much less than the age of the solar system.

Comets are believed to be planetesimals, or fragments of them: small solid bodies that grew from dust that condensed out of the gaseous disk surrounding the newly formed Sun. The planetesimals collided and grew to form the terrestrial planets and the cores of the giant planets, but not all planetesimals were incorporated in the planets, and the residual population of planetesimals is presumably the original source of comets.

These considerations imply that there must be one or more cometary "reservoirs"—regions in phase space in which vast numbers of comets have been stored since soon after the formation of the solar system. Orbits in the reservoir must be sufficiently stable to remain there for Gyr timescales, but sufficiently unstable to provide a slow, steady supply of comets that will eventually come close enough to the Sun to become visible. In other words the reservoir has to leak, but not too much.

Figure 9.7 shows the semimajor axes and inclinations (relative to the ecliptic) of known comets. The figure shows two distinct classes of comet. The first appears as a dense clump with semimajor axes $a \lesssim 10$ au and inclination $I \lesssim 30°$; these are the **Jupiter-family comets**, usually defined to be comets with orbital period $P < 20$ yr ($a < 7.37$ au). The second class has an approximately isotropic distribution of inclinations (uniform in $\cos I$, as expected for a spherical source) and a wide range of semimajor axes, from $a \simeq 10$ au to $a = 10^4$ au and beyond. These are classified as **Halley-type comets** if $20\,\text{yr} < P < 200\,\text{yr}$ and **long-period comets** if $P > 200$ yr, but this distinction is based on tradition rather than physical or kinematical differences between Halley-type and long-period comets.

The flat distribution of the Jupiter-family comets compared to the isotropic distribution of the Halley-type and long-period comets strongly suggests that they originate in different reservoirs, one flat and the other spherical. It is believed that the reservoir for the Jupiter-family comets is the trans-Neptunian belt, to be discussed in the following section, and the reservoir for the Halley-type and long-period comets is the Oort cloud, which we now describe. It is highly likely that analogs of the Oort cloud and trans-Neptunian belt exist in many exoplanetary systems.

The analyses earlier in this chapter show that the evolution of a particle

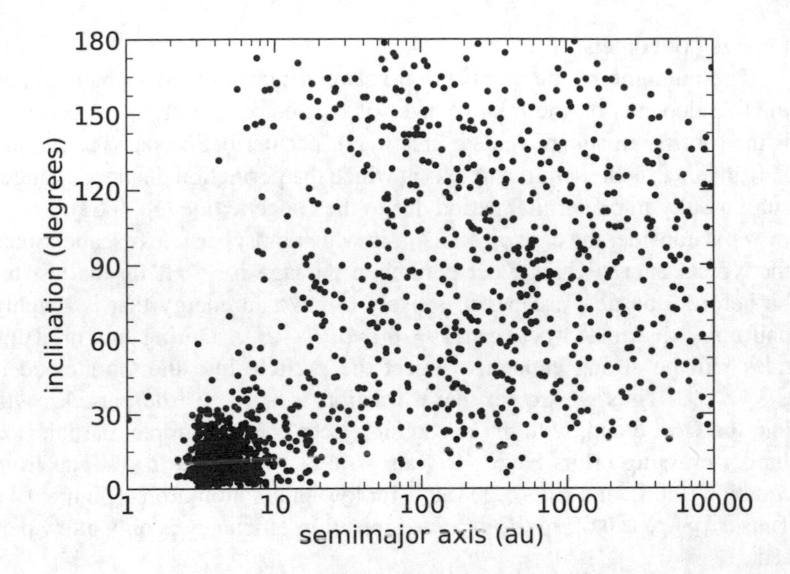

Figure 9.7: Semimajor axis versus inclination relative to the ecliptic, for comets from the JPL Small Body Database, https://ssd.jpl.nasa.gov/sbdb_query.cgi. Comets with semimajor axes $\gtrsim 10$ au are underrepresented, since comets are only detected near perihelion and astronomical survey techniques are improving on a timescale shorter than the orbital period $P = 31.6\,\mathrm{yr}(a/10\,\mathrm{au})^{3/2}$.

on a high-eccentricity planet-crossing orbit is determined by a competition between two dynamical processes: random walk of the semimajor axis or energy due to interactions with the planets near perihelion (§9.3), and drift in the perihelion distance due to Galactic tidal forces near aphelion (§9.4). If the first process dominates, the particle eventually reaches escape energy and is lost from the solar system. If the second process dominates, the perihelion expands outside Neptune's orbit and the energy perturbations at perihelion passage are quenched (Figure 9.4). In the latter case we say that the particle has been injected into the **Oort cloud**, named for the Dutch astronomer Jan Oort (1900–1992), whose 1950 paper was a watershed in the study of comets.

The outcome of the competition between planetary perturbations and the tide depends on the relative sizes of σ_x and $x_{\rm crit} = 1/a_{\rm crit}$. Here σ_x is the root-mean-square change in $x = 1/a$ per perihelion passage (Figure 9.4) and $a_{\rm crit}$ is the semimajor axis at which the perihelion distance changes substantially in one orbital period due to the Galactic tide (eq. 9.68).

First consider the case $\sigma_x \gg x_{\rm crit}$, in which most particles escape. Since the typical energy change per perihelion passage is $\sim \sigma_x$, on the last orbit before a particle escapes it is likely to have an energy that is roughly uniformly distributed between $x = 0$ and $x \sim \sigma_x$. During this orbit the tides will be strong enough to inject the particle into the Oort cloud if $x < x_{\rm crit}$. Thus we expect that a fraction $x_{\rm crit}/\sigma_x$ of the particles will join the Oort cloud, with the rest being ejected. For example, particles on Jupiter-crossing orbits have $q < 5\,$au, so $a_{\rm crit} = x_{\rm crit}^{-1} \simeq 2 \times 10^4\,$au from equation (9.68), and $\sigma_x \simeq 0.003\,\text{au}^{-1}$ for low-inclination orbits (Figure 9.4). Thus $x_{\rm crit}/\sigma_x \simeq 0.02$ so the expected injection efficiency is only a few percent.

In contrast, if $\sigma_x \ll x_{\rm crit}$, most particles are injected into the Oort cloud. We can describe how this occurs using the diffusion approximation for the evolution due to planetary perturbations. Since the diffusion half-life $t_{1/2}$ varies as $a^{-1/2}$ (eq. 9.56) while the drift time t_q varies as a^{-2} (eq. 9.67), semimajor axis diffusion dominates the evolution at small semimajor axes while the Galactic tide dominates at large semimajor axes. The crossover occurs where $t_{1/2} = t_q$, corresponding to the semimajor axis

$$a_{\rm Oort} = \frac{7.50 \times 10^3\,\text{au}}{|\sin\theta\cos\theta|^{2/3}} \left(\frac{0.1 M_\odot\,\text{pc}^{-3}}{\rho}\right)^{2/3} \left(\frac{q}{30\,\text{au}}\right)^{1/3} \left(\frac{\sigma_x}{3 \times 10^{-5}\,\text{au}^{-1}}\right)^{4/3}. \tag{9.69}$$

Numerical simulations show that comets are injected into the Oort cloud with a wide range of semimajor axes centered on $a_{\rm Oort}$, from $\sim 2 \times 10^3\,$au to $10^5\,$au. The wide variation reflects the stochastic nature of the energy diffusion process, variations in the projection factor $|\sin\theta\cos\theta|$, and variations in the root-mean-square energy change σ_x with perihelion distance.

After they are injected into the Oort cloud, orbits continue to change, although more slowly. The Galactic tide causes the eccentricity and angular-momentum vectors to evolve according to equations (9.65). Perturbations from passing stars lead to a random walk in all of the orbital elements including semimajor axis and gradually deplete the cloud as particles diffuse to escape energy. Estimates of the rate of the random walk in the orbital elements, confirmed by numerical simulations, show that the comets in the Oort cloud fill the available phase space at a given semimajor axis approximately uniformly,[10] at least for $a \gtrsim 10^4$ au. This result implies that the cloud is spherical and that the probability that a comet has eccentricity less than e is (Problem 9.9)

$$p(< e) = e^2. \tag{9.70}$$

From time to time the comets wandering through the Oort cloud reach eccentricities near unity. If $q = a(1 - e)$ is less than ~ 30 au, then the energy will once again begin to suffer random perturbations at each perihelion passage, and the comet may rapidly escape the solar system or evolve to much smaller semimajor axes.

There is direct evidence of the existence and properties of the Oort cloud, or at least its outer parts. We have argued that most comets are only visible when they come within a few au of the Sun. If the semimajor axis of the comet exceeds $a_{\text{crit}}(q = 40\,\text{au}) \simeq 3 \times 10^4$ au (eq. 9.68) then the Galactic tide can bring the comet from $q \gtrsim 40$ au, where it is immune to planetary perturbations, to $q \lesssim 1$ au, where it is easily visible, in a single orbit. Such comets are called **new comets**, since we are seeing them during their first passage through the planetary system since leaving the Oort cloud. The semimajor axis of a new comet at the time of observation equals the semimajor axis it had in the Oort cloud modified by the perturbations it received in its passage through the planetary system so far. These modifications are straightforward to remove by numerical integration of the comet orbit backward from the time of observation until it is well outside the planetary system, and its semimajor axis at this point, relative to the barycenter of the solar system, is called the **original semimajor axis** a_{orig}.

Figure 9.8 plots the inverse of the original semimajor axes of comets having well-determined orbits with near-zero energy. A handful of comets

[10] In statistical mechanics a distribution that is uniform in phase space at a given energy is called a microcanonical ensemble.

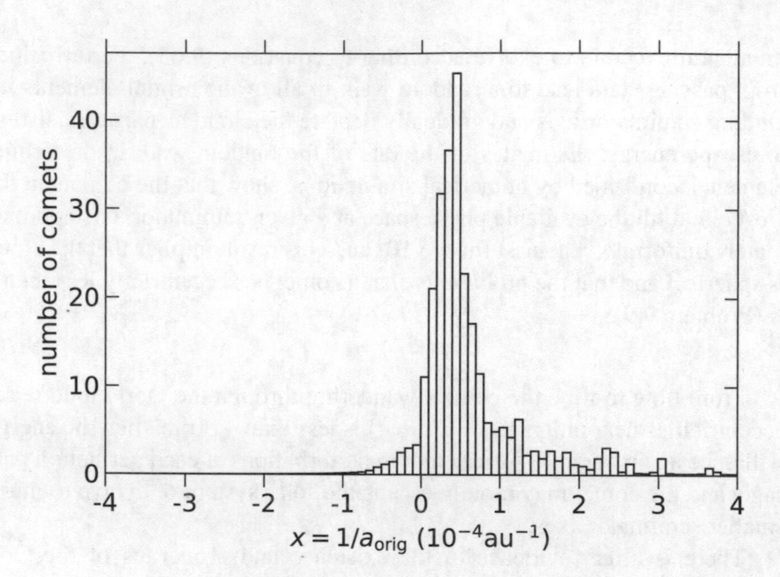

Figure 9.8: Inverse original semimajor axes of comets. The original semimajor axis is the semimajor axis that the comet had before it entered the planetary system. Only the range $|a| > 2\,500$ au is shown. Negative semimajor axes correspond to unbound orbits. The contribution of each of the 277 comets is represented by a unit Gaussian with standard deviation equal to the estimated observational error. Data from Królikowska & Dybczyński (2020).

have negative energies (unbound orbits), but all of these can be attributed to observational error or non-gravitational forces due to outgassing from the comet surface. Most of the comets are bound to the solar system and there is a strong concentration, known as the **Oort spike**, centered at $a = (2–3) \times 10^4$ au.[11] The existence of the Oort spike, and its impressive agreement with the estimate of $a_{\rm crit} = 3 \times 10^4$ au from the preceding paragraph, provide

[11] Remarkably, Oort's seminal paper (Oort 1950) describing and interpreting the spike was based on a catalog that contained only 14 comets with inverse semimajor axes between 0 and 1×10^{-4} au.

convincing evidence of the existence of the Oort cloud.

The Oort cloud is often divided by semimajor axis into an inner and outer part, with the dividing line at $a = 2 \times 10^4$ au. The inner cloud is mostly invisible to us, since the orbits of comets from the inner cloud are drastically altered by the giant planets before they become visible.

We can now estimate the total population of comets in the outer cloud. The flux of new comets with perihelion $q < 2.5$ au and diameter $> D_{\rm min} =$ 2.3 km, where current surveys are reasonably complete, is roughly 1 per year (Dones et al. 2015). At these large distances, the distribution of eccentricities is given by equation (9.70). A comet with semimajor axis a and perihelion $q \ll a$ has eccentricity $e = 1 - q/a$. Therefore the fraction of comets with perihelion less than 2.5 au is $f = 1 - [1 - (2.5\,\text{au}/a)]^2 \simeq 5\,\text{au}/a$; for a typical semimajor axis of new comets, $a = 3 \times 10^4$ au, this fraction is $f = 0.000\,17$. The orbital period at this semimajor axis is $P = 5.2$ Myr. Thus the total reservoir of potential new comets is $(1/\,\text{yr}) \times f^{-1} \times P = 3 \times 10^{10}$. For a typical comet density of $0.5\,\text{g cm}^{-3}$ the mass of a comet with diameter $D_{\rm min}$ is 3.2×10^{12} kg, so if the cloud were composed entirely of comets with diameter $D_{\rm min}$ the total mass of the outer cloud would be $0.016 M_\oplus$. The actual mass is several times larger, and quite uncertain, since most of the mass of the cloud appears to be in rare high-mass comets. Adding the uncertain contribution of the inner Oort cloud increases the mass still further, so the best we can say is that the total Oort cloud mass is probably $\sim 1 M_\oplus$, with an uncertainty of almost an order of magnitude.

These arguments are no substitute for long-term N-body integrations of planetesimal orbits in the outer solar system. These integrations typically include the four giant planets, the Galactic tide, and a Monte Carlo model of the perturbations from passing stars. The comets are represented by an ensemble of test particles, initially on nearly circular orbits near the plane of the ecliptic with a smooth distribution of semimajor axes out to 40–50 au. In the "standard model" for the formation and evolution of the Oort cloud, the planets are in their current orbits, the Sun is at its current distance from the center of the Galaxy—that is, there is no migration of either the planets or the Sun—and the gravitational effects of the dense cluster of stars in which the Sun was likely to have been born are neglected. More general models relax one or more of these assumptions.

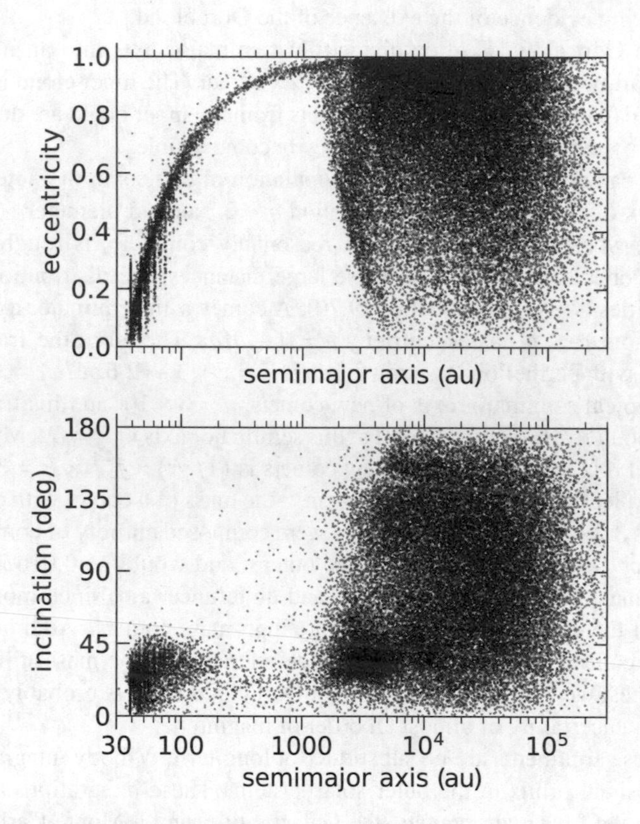

Figure 9.9: The present distribution of orbital elements of comets in a simulation of Oort cloud formation (Vokrouhlický et al. 2019). The simulation began with 10^6 test particles distributed in low-eccentricity, low-inclination orbits near the ecliptic, with semimajor axes between 25 au and 30 au. Uranus and Neptune migrated during the simulation from 17 au and 24 au to their current semimajor axes (19 au and 30 au). The plots show the ~ 50 000 particles still bound to the solar system after 4.5 Gyr. The top plot shows eccentricity as a function of semimajor axis and the bottom plot shows inclination relative to the ecliptic.

Figure 9.9 shows the distribution of comets surviving $4.5\,$Gyr after the formation of the solar system, about its current age, in a simulation from Vokrouhlický et al. (2019). Most of the parameters were those of the standard model, except that Uranus and Neptune migrated outward to their current semimajor axis over the first $\sim 0.1\,$Gyr after the formation of the solar system. The plots show eccentricity (top panel) and inclination (bottom panel) versus semimajor axis.

A striking feature is the arc extending from eccentricity $e \simeq 0$ at $a \simeq 40\,$au to $e \simeq 1$ at $a \simeq 10^3\,$au. The arc traces a line of constant perihelion distance $q = a(1-e) \simeq 35\,$au and contains comets whose semimajor axes are being perturbed at perihelion passage by the giant planets. This population of comets is called the **scattered disk**. They have median inclination of only $25°$. Many of the comets in the scattered disk with $a \lesssim 100\,$au are trapped in mean-motion resonances with the giant planets, which appear in the figure as vertical bars. About 9% of the surviving comets belong to the scattered disk.

The simulations show that the Oort cloud extends from a few thousand au to more than 10^5 au. The cloud comets have a wide range of eccentricities and inclinations; for semimajor axes $\gtrsim 5 \times 10^3$ au the cloud is approximately spherical and the squared eccentricity is approximately uniformly distributed, as expected from the arguments in the paragraph containing equation (9.70). The inner and outer clouds, separated at $a = 2 \times 10^4$ au, contain 51% and 40% of the surviving comets.

In the absence of resonances the eventual fate of almost every solar-system object on a planet-crossing orbit is ejection, or collision with a planet or the host star. This ejection process, modified by the different planetary configurations around different stars, is probably the origin of interstellar comets and asteroids such as ʻOumuamua.

9.6 The trans-Neptunian belt

The trans-Neptunian belt is a disk of bodies in the outer solar system extending roughly from Neptune's orbit at $30\,$au to $\gtrsim 50\,$au. In the decades following the discovery of Pluto in 1930, several astronomers argued that there were two possible explanations for the absence of large planets beyond Neptune: either (i) the protoplanetary disk ended at or near Neptune's orbit, or (ii) planet formation by the collision and growth of planetesimals became slower and less efficient at larger distances, where the orbital periods were longer and the density in the disk was lower. If explanation (ii) were correct then we might expect to find a disk of fossil planetesimals outside Neptune. Additional support for this hypothesis came from the observation that the low-inclination Jupiter-family comets (§9.5) must have originated in a flattened cometary reservoir somewhere outside Neptune (Fernández 1980; Duncan et al. 1988).

The first object to be found beyond Neptune was Pluto, but it took more than four decades before the larger population of trans-Neptunian objects (TNOs)[12] began to be discovered and characterized (Jewitt & Luu 1993). By now several thousand TNOs are known; most have radii ~ 100 km or smaller but the largest are ten times bigger. The total mass of the trans-Neptunian belt is poorly known but the best estimate is ~ 0.05–$0.1 M_\oplus$. See Gladman & Volk (2021) for a review.

The closest analogs of the trans-Neptunian belt around other stars are debris disks (Wyatt 2020), which are detected by thermal emission from dust produced by collisions between parent bodies that are probably similar to TNOs. These are mostly detected around stars younger than the Sun, because the population of parent bodies declines with age as they are destroyed in collisions. The debris disks detected so far all contain orders of magnitude more dust than is present in trans-Neptunian space.

By definition, a TNO has semimajor axis a greater than Neptune's semimajor axis $a_{\rm Nep} = 30.07$ au; objects in the outer solar system with smaller semimajor axes are called **Centaurs**. Figure 9.10 plots the semimajor axes and eccentricities of TNOs. The complex structure seen in this plot is usually organized by dividing the TNOs into several distinct groups.

Classical belt This is the large population of TNOs between the 3:2 and 2:1 Neptune resonances with eccentricities $\lesssim 0.2$. Most classical TNOs do not cross or come close to Neptune's orbit (i.e., they lie well below the

[12] The belt is sometimes called the Kuiper belt and its members Kuiper belt objects (KBOs).

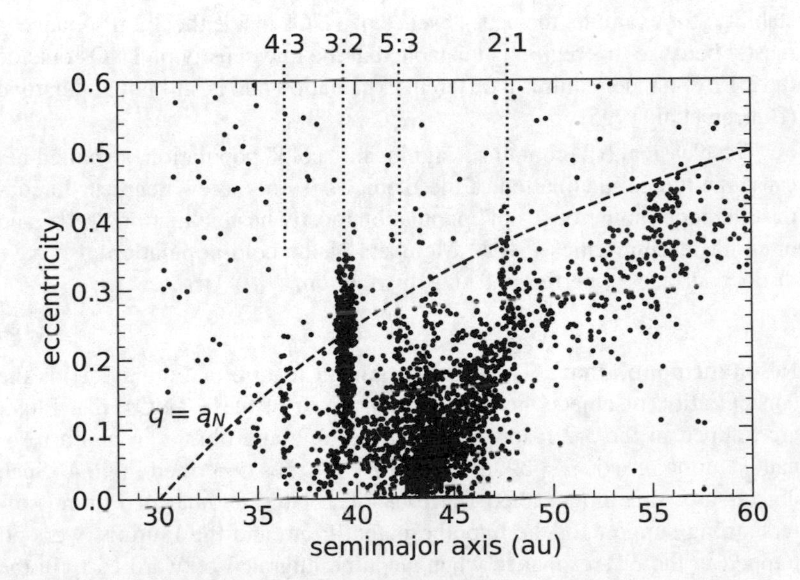

Figure 9.10: Semimajor axis versus eccentricity for ~ 3 000 TNOs. The vertical dotted lines denote the semimajor axes corresponding to several mean-motion resonances with Neptune; for example, objects in the 3:2 resonance have a mean motion that is $\frac{2}{3}$ of Neptune's. The curved dashed line traces the location where the perihelion $q = a(1 - e)$ equals Neptune's semimajor axis, $a_{\rm Nep} = 30.07$ au. The data are from https://www.minorplanetcenter.net/iau/lists/MPLists.html, and the plot shows all objects with $a > a_{\rm Nep}$ from their lists of trans-Neptunian objects, Centaurs, and scattered-disk objects.

dashed line in Figure 9.10 that traces orbits with perihelion $q = a_{\rm Nep}$) and are dynamically stable for the lifetime of the solar system—of course, if this were not so, we would not expect to see these objects today. Indeed, the distribution of TNOs in phase space is partly determined by the demands of stability; for example, the near-absence of TNOs inside the 3:2 resonance is mostly because this region is unstable, but the low density of TNOs outside the 2:1 resonance requires a different explanation and is still not understood (Duncan et al. 1995).

The classical belt contains a significant "cold" population, so called because the typical inclination of these objects is only a few degrees. In contrast the more numerous "hot" population mostly has inclinations $\gtrsim 10°$ and often has eccentricities $\gtrsim 0.2$. Members of the cold population also have distinct surface properties and size distribution.

Resonant population The most prominent feature of Figure 9.10 is the concentration of objects near $a = 39\,$au. Most of these TNOs, like Pluto, are trapped in the 3:2 resonance with Neptune and librate around a resonant semimajor axis $a = 39.40\,\text{au} = (\frac{3}{2})^{2/3} a_{\rm Nep}$ as described in §6.4. Such objects are sometimes called **Plutinos**. The large number of Plutinos offers strong support for the hypothesis that Pluto and the Plutinos were all trapped in the 3:2 resonance when Neptune migrated outward early in the history of the solar system. Populations of TNOs are also trapped in several other Neptune resonances, such as the 2:1 resonance.

Scattered disk This population is visible in Figure 9.10 as a broad band parallel to but somewhat below the dashed line $q = a_{\rm Nep}$. TNOs in the scattered disk come close enough to Neptune that they undergo a random walk in energy, as described in §9.3. Over the lifetime of the solar system, some evolve to less tightly bound orbits and will eventually escape or join the Oort cloud along the band seen in the top panel of Figure 9.9, while others evolve to more tightly bound orbits and become Centaurs. Because of these processes, the population of the scattered disk is steadily eroding.

There is also the detached disk described in footnote 6.

About one-third of Centaurs evolve to semimajor axes less than 7.37 au, at which point the orbital period is less than 20 yr so they are classified as Jupiter-family comets (Di Sisto & Rossignoli 2020). The lifetimes of Centaurs are much less than the age of the solar system—typically ~ 10 Myr with a range from < 1 Myr to > 100 Myr (Tiscareno & Malhotra 2003)—which explains why the number of known Centaurs is much smaller than the number of TNOs, even though the Centaurs are typically closer and therefore easier to discover.

These arguments strongly suggest that the trans-Neptunian belt is the reservoir in which the Jupiter-family comets have been stored for the past several Gyr. The small inclinations of the Jupiter-family comets reflect the small inclinations of TNOs, just as the isotropic distribution of comets with longer periods reflects the spherical shape of the Oort cloud.

The details of the evolutionary path from TNOs to Centaurs to Jupiter-family comets are not yet fully understood. Most likely the Centaurs come mainly from the scattered-disk population (Duncan & Levison 1997). One difficulty in interpreting the observations is that the typical observed Jupiter-family comet is much smaller than the typical Centaur or TNO, so connecting the flux of Jupiter-family comets to the properties of the trans-Neptunian belt requires an uncertain extrapolation of the size distribution of TNOs.

Binary TNO systems are common, particularly in the cold classical belt, where ~ 30% of TNOs are binary (Noll et al. 2020). The components of TNO binaries have almost the same colors despite the wide range of colors found among TNOs, which suggests that the systems were formed as binaries (rather than, say, by capture), as does the prevalence of prograde binary orbits. The survival of binaries in the cold classical belt sets strong constraints on its history (Morbidelli & Nesvorný 2020).

There are puzzling aspects of the current configuration of the trans-Neptunian belt. (i) The total mass is much less than the mass that we would expect in the region 30–50 au from an extrapolation of the solid masses in the outer planets.[13] (ii) The present quantity of material in the belt is too small for the observed TNOs to have grown by collision and accumulation

[13] For example, assume that the surface density varies with radius as $r^{-1.5}$ as in the minimum-mass solar nebula (eq. 6.111), and that all the mass between 15 au and 35 au was incorporated in Uranus and Neptune. Then the mass between 35 au and 50 au should be $\sim 20 M_\oplus$,

of smaller bodies. (iii) Given the eccentricities and inclinations of the classical TNOs, the typical collision velocities are high enough ($\sim 1\,\mathrm{km\,s^{-1}}$) that most collisions will result in shattering or erosion, so the objects cannot grow.

A possible resolution to these puzzles is that originally the belt had much more mass and much smaller eccentricities and inclinations; that some event excited the eccentricities and inclinations of the TNOs after they were fully formed; and that since then most of the mass of the belt has been ground down to dust in erosive collisions and then lost through radiation pressure or Poynting–Robertson drag. A second, more plausible resolution is that most objects in the belt were formed closer to the Sun and then transported to their current locations during the migration of the giant planets, and that only the cold classical belt was formed with its current properties (Morbidelli & Nesvorný 2020; Gladman & Volk 2021).[14] However, no single proposed migration model has yet been able to explain all of the properties of the TNO population. Perhaps some ingredients are missing, such as interactions with the Sun's birth cluster or additional planets that were subsequently scattered into much larger orbits.

The rich dynamical structure of the thousands of objects in the trans-Neptunian region offers a unique, though still ambiguous, fossil record of the early history of the outer solar system.

9.7 Earth-crossing asteroids

Most of the objects in the asteroid belt have semimajor axes between 2 au and 3.3 au and eccentricities $e \lesssim 0.3$. The corresponding perihelion distances $q = a(1-e)$ exceed 1.4 au, a distance that is well outside the orbit of Earth. However, a small fraction of asteroids come much closer. The **near-Earth asteroids**, or NEAs, are conventionally defined to be those with

far larger than the observed belt mass of ~ 0.05–$0.1 M_\oplus$.

[14] In these models the density of the massive planetesimal disk declines sharply outside $\sim$ 30 au, consistent with the requirement to stop Neptune's outward migration at its current semimajor axis. Thus, ironically, of the two explanations for the absence of planets beyond Neptune described at the beginning of this section, the one that does *not* predict the trans-Neptunian belt is more nearly correct.

$q < 1.3\,\mathrm{au}$. A subclass of NEAs is the **Earth-crossing asteroids**, those with orbits having a radial extent—from perihelion $q = a(1 - e)$ to aphelion $Q = a(1 + e)$—that overlaps the radial extent of the Earth's orbit, from $q_\oplus = 0.983\,\mathrm{au}$ to $Q_\oplus = 1.017\,\mathrm{au}$. The Earth-crossing asteroids are further subdivided into **Apollo asteroids**, which have semimajor axes $a > 1\,\mathrm{au}$ but perihelion distances $q < Q_\oplus$, and **Aten asteroids**, which have $a < 1\,\mathrm{au}$ but aphelion distance $Q > q_\oplus$.

Earth-crossing asteroids have the potential to collide with Earth. For example, the meteor that exploded over the city of Chelyabinsk on February 15, 2013, with an estimated energy 400–500 kilotons of TNT,[15] was an Apollo asteroid with an estimated radius of about 10 m. The 150-km diameter Chicxulub crater in the Yucatan peninsula of Mexico is the remnant of an impact that is believed to have caused the mass extinction at the Cretaceous–Tertiary boundary 65 Myr ago. The impact frequency as a function of mass, on the upper atmosphere and at the Earth's surface, is given by Bland & Artemieva (2006).

Small colliding bodies can be found on the Earth's surface as meteorites, and these provide invaluable insights into the early physical and chemical history of the solar system.

Not all Earth-crossing asteroids come close to the Earth, even over many thousands of orbits. For example, an Earth-crossing asteroid on an inclined orbit can only have a close encounter with Earth if it crosses the ecliptic plane at a radius between $q_\oplus$ and $Q_\oplus$—in terms of orbital elements this requires $q_\oplus < a(1 - e^2)/(1 \pm e\cos\omega) < Q_\oplus$. A more sensitive measure of the likelihood of collision is the **minimum orbit intersection distance** (MOID), the distance between the closest points of the osculating orbits of the Earth and the asteroid. Asteroids with an MOID $\leq 0.05\,\mathrm{au}$ are defined to be **potentially hazardous asteroids** if their radius exceeds about 70 m.[16]

Normally, it takes many centuries before an asteroid with a MOID of (say) 0.05 au actually comes within 0.05 au of the Earth, since the asteroid and Earth must arrive at the location of the MOID at the same time.

---

[15] 1 kiloton of TNT equals 4.184×10^{12} J.

[16] In practice the radius is estimated from the observed brightness of the asteroid and an assumption about its albedo.

And even when this happens, the chance of a collision is only of order $(R_\oplus/0.05\,\mathrm{au})^2 \simeq 10^{-6}$. Nevertheless, the danger of collisions with Earth is a good reason to discover as many potentially hazardous asteroids as possible and to determine their orbits accurately.

The typical lifetime of an asteroid on an Earth-crossing orbit is a few Myr (Table 9.1). Only a few percent of Earth-crossing asteroids end up colliding with Earth; most of the rest impact the Sun or are ejected from the solar system by Jupiter (Morbidelli & Gladman 1998). Because the lifetimes are only a small fraction of the lifetime of the solar system, some mechanism is needed to resupply the Earth-crossing orbits from the main part of the asteroid belt.

One of the most important components in this supply chain is the zones in phase space associated with resonances, particular the 3:1 mean-motion resonance with Jupiter and the ν_6 secular resonance (§6.6). Numerical and analytic studies show that particles in these zones can have their eccentricities excited within a few Myr to the point where the orbits become Mars-crossing or even Earth-crossing. Once this happens, one or more close encounters with Mars or Earth can extract the particle from the resonance and place it in an orbit similar to those we see in the current population of Earth-crossing asteroids (Morbidelli et al. 2002).

We still need a mechanism to re-supply the resonance zones. One possibility is collisions between asteroids. Collisions lead to cratering and fragmentation, and the velocities of the fragments may differ from the velocity of their parent body by several hundred $\mathrm{m\,s^{-1}}$. Thus collisions can inject small bodies into the resonance zones. A second possibility is that asteroids drift into the resonance zones by the Yarkovsky effect (§7.5). The Yarkovsky effect is probably more important than collisions for this process. First, the cosmic-ray exposure ages of meteorites measure the time since the most recent fragmentation of the meteorite's parent body. If collisions trigger the process of transferring an asteroid from the main belt to an Earth-crossing orbit, then many meteorites should have cosmic-ray exposure ages of only a few Myr, but most have ages much longer than this. Second, the size distribution of the NEAs is not compatible with collisional debris. Finally, the estimated collision rate in the main belt appears to be too small to maintain the Earth-crossing asteroid population in a steady state.

Appendix A

Physical, astronomical, and solar-system constants

Numbers in parentheses indicate one standard deviation uncertainty in the last digit of the preceding number. Values labeled "exact" are defined by convention; values labeled "nominal" are used to simplify comparisons of different studies but may not be precisely correct; and values labeled "theory" come from theoretical models constrained by other observations. For additional details see §1.5.

Orbital elements and obliquities are given for the year J2000.0.

The mean radius $\overline{R}$ is the volumetric mean radius, the radius of a sphere containing the same volume as the body. For a spheroidal body $\overline{R} = (R_{\rm eq}^2 R_{\rm pol})^{1/3}$, where $R_{\rm eq}$ is the equatorial radius and $R_{\rm pol}$ is the polar radius. The reference radius $R_{\rm p}$ is the radius used in defining quantities such as the moment of inertia factor $C/(MR^2)$ and the quadrupole moment J_2.

The gravitational fields of solar-system bodies are often expressed in terms of mass coefficients $C_{n,m}$, $S_{n,m}$ or normalized mass coefficients $\overline{C}_{n,m}$, $\overline{S}_{n,m}$. These are related to the quadrupole moment J_2 and the moments of inertia $A \le B \le C$ (eq. D.87) by the formulas:

$$J_2 = \frac{C - \frac{1}{2}(A+B)}{MR_{\rm p}^2} = -C_{2,0} = -5^{1/2}\overline{C}_{2,0},$$
$$\frac{B-A}{MR_{\rm p}^2} = 4(C_{2,2}^2 + S_{2,2}^2)^{1/2} = (\tfrac{20}{3})^{1/2}(\overline{C}_{2,2}^2 + \overline{S}_{2,2}^2)^{1/2}. \qquad \text{(A.1)}$$

The main sources for the constants below include:

- Particle Data Group, https://pdg.lbl.gov/
- JPL Solar System Dynamics, https://ssd.jpl.nasa.gov/
- NASA Planetary Fact Sheets, https://nssdc.gsfc.nasa.gov/planetary/planetfact.html
- Report of the IAU Working Group on Cartographic Coordinates and Rotational Elements: 2015 (Archinal et al. 2018)
- International Earth Rotation and Reference Systems Service (IERS) Conventions (2010), Technical Note No. 36, version 1.3.0, https://www.iers.org/IERS/EN/DataProducts/Conventions/conventions.html
- The Astronomical Almanac for the year 2020 (Washington, D.C.: U.S. Government Publishing Office)
- Explanatory Supplement to the Astronomical Almanac (Urban & Seidelmann 2013)

Other sources are referenced in the tables.

Physical and astronomical constants

gravitational constant	$\mathbb{G} =$	$6.674\,30(15) \times 10^{-11}$ m^3 kg^{-1} s^{-2}
	$g =$	$\mathbb{G}/(6.674\,30 \times 10^{-11}$ m^3 kg^{-1} s$^{-2})$
speed of light	$c =$	$299\,792\,458$ m s^{-1} (exact)
Stefan–Boltzmann constant	$\sigma =$	$5.670\,374\,419\ldots \times 10^{-8}$ W m^{-2} K^{-4} (exact)
astronomical unit	au =	149 597 870 700 m (exact)
parsec	pc =	au $\times$ 648 000/π
	=	$(3.085\,677\,58\ldots) \times 10^{16}$ m (exact)
Julian year	yr =	365.25 days = 31 557 600 s (exact)
age of the Universe[1]	$t_0 =$	13.787(20) Gyr
age of the solar system[2]	$t_{ss} =$	4.5673(2) Gyr

[1] Planck Collaboration et al. (2020).

[2] Connelly et al. (2012).

Solar constants

solar mass parameter	$\mathbb{G}M_\odot = 1.327\,124\,4 \times 10^{20}\ \mathrm{m^3\,s^{-2}}$ (nominal)
solar mass	$M_\odot = 1.988\,41/g \times 10^{30}$ kg
obliquity	$\epsilon = 7.25°$
moment of inertia[3]	$C/(M_\odot R_\odot^2) = 0.070$
solar quadrupole moment[4]	$J_2 = 2.25(2) \times 10^{-7}$
gravitational Love number[3]	$k_2 = 0.030$
solar radius	$R_\odot = 695\,700$ km (nominal)
solar luminosity	$L_\odot = 3.828 \times 10^{26}$ W (nominal)

Planetary constants

Mercury

semimajor axis	$a = 0.387\,10$ au
eccentricity	$e = 0.205\,64$
inclination	$I = 7.00°$
orbital period	$P = 87.969\ \mathrm{d} = 0.240\,85$ yr
mass parameter[5]	$\mathbb{G}M = 2.203\,19 \times 10^{13}\ \mathrm{m^3\,s^{-2}}$
mass	$M = 3.301\,00/g \times 10^{23}$ kg
Sun/planet mass ratio	$M_\odot/M = 6.023\,66 \times 10^6$
spin period	$P_s = 58.6462$ d
obliquity[5]	$\epsilon = 0.03°$
moments of inertia[5]	$C/(MR_\mathrm{p}^2) = 0.333(2)$
	$J_2 = [C - \frac{1}{2}(A+B)]/(MR_\mathrm{p}^2) = 5.032(1) \times 10^{-5}$
	$(B-A)/(MR_\mathrm{p}^2) = 3.216(2) \times 10^{-5}$
gravitational Love number[5]	$k_2 = 0.569(8)$
equatorial radius[6]	$R_\mathrm{eq} = 2\,440.53$ km
polar radius[6]	$R_\mathrm{pol} = 2\,438.26$ km
reference radius	$R_\mathrm{p} = 2\,440$ km

3 Claret (2019). Note that the symbol k_2 in this paper refers to the apsidal motion constant, which is a factor of two smaller than the Love number.
4 Genova et al. (2018).
5 Genova et al. (2019).
6 Archinal et al. (2018).

Venus

semimajor axis	$a = 0.723\,34\,\mathrm{au}$
eccentricity	$e = 0.006\,78$
inclination	$I = 3.39^\circ$
orbital period	$P = 224.701\,\mathrm{d} = 0.615\,20\,\mathrm{yr}$
mass parameter[7]	$\mathbb{G}M = 3.248\,59 \times 10^{14}\,\mathrm{m^3\,s^{-2}}$
mass	$M = 4.867\,31/g \times 10^{24}\,\mathrm{kg}$
Sun/planet mass ratio	$M_\odot/M = 408\,523.7$
spin period	$P_s = 243.018\,\mathrm{d}$
obliquity	$\epsilon = 177.36^\circ$
moments of inertia[7,8]	$C/(MR_\mathrm{p}^2) = 0.34(2)$
	$J_2 = [C - \frac{1}{2}(A+B)]/(MR_\mathrm{p}^2) = 4.46(1) \times 10^{-6}$
	$(B-A)/(MR_\mathrm{p}^2) = 2.17(1) \times 10^{-6}$
gravitational Love number[9]	$k_2 = 0.30(7)$
mean radius[6]	$\overline{R} = 6\,052\,\mathrm{km}$
reference radius	$R_\mathrm{p} = 6\,051\,\mathrm{km}$

Earth

semimajor axis	$a = 1.000\,00\,\mathrm{au}$
eccentricity	$e = 0.016\,71$
inclination	$I = 0.00^\circ$
orbital period	$P = 365.256\,\mathrm{d} = 1.000\,02\,\mathrm{yr}$
mass parameter	$\mathbb{G}M_\oplus = 3.986\,004 \times 10^{14}\,\mathrm{m^3\,s^{-2}}$ (nominal)
mass	$M_\oplus = 5.972\,17/g \times 10^{24}\,\mathrm{kg}$
Sun/planet mass ratio	$M_\odot/M_\oplus = 332\,946.1$
Sun/system mass ratio	$M_\odot/(M_\oplus + M_☽) = 328\,900.6$
spin period	$P_s = 0.997\,27\,\mathrm{d}$
obliquity	$\epsilon = 23.44^\circ$
moment of inertia	$C/(M_\oplus R_\oplus^2) = 0.330\,70$
quadrupole moment	$J_2 = 0.001\,082\,64$
gravitational Love number	$k_2 = 0.295$
displacement Love number	$h_2 = 0.608$

[7] Konopliv et al. (1999).

[8] Margot et al. (2021).

[9] Konopliv & Yoder (1996).

equatorial radius	$R_{\mathrm{eq}} = 6\,378.1$ km (nominal)
polar radius	$R_{\mathrm{pol}} = 6\,356.8$ km (nominal)
reference radius	$R_\oplus = 6\,378.136$ km

Moon

semimajor axis	$a = 384\,400$ km
eccentricity	$e = 0.0549$
inclination	$I = 5.15°$
orbital period	$P = 27.322$ d
mass parameter[10]	$\mathbb{G}M_{☽} = 4.902\,80 \times 10^{12}$ m^3 s^{-2}
mass	$M_{☽} = 7.345\,79/g \times 10^{22}$ kg
Earth/Moon mass ratio	$M_\oplus/M_{☽} = 81.3005$
spin period	$P_s = 27.322$ d
obliquity	$\epsilon = 6.68°$
moment of inertia[10]	$C/(M_{☽}R_{☽}^2) = 0.392\,73$
quadrupole moment	$J_2 = 0.000\,203\,22$
gravitational Love number[10]	$k_2 = 0.0242(2)$
displacement Love number[10]	$h_2 = 0.0424(1)$
mean radius[10]	$\overline{R} = 1\,737.15$ km
reference radius	$R_{☽} = 1\,738$ km

Mars

semimajor axis	$a = 1.523\,71$ au
eccentricity	$e = 0.093\,39$
inclination	$I = 1.85°$
orbital period	$P = 686.980$ d $= 1.880\,85$ yr
mass parameter[11]	$\mathbb{G}M = 4.282\,84 \times 10^{13}$ m^3 s^{-2}
mass	$M = 6.416\,91/g \times 10^{23}$ kg
Sun/planet mass ratio	$M_\odot/M = 3.098\,71 \times 10^6$
spin period	$P_s = 1.025\,96$ d
obliquity	$\epsilon = 25.19°$

[10] Williams et al. (2014).
[11] Genova et al. (2016).

moments of inertia[11,12]	$C/(MR_{\rm p}^2) =$	$0.3644(5)$
	$J_2 = [C - \frac{1}{2}(A+B)]/(MR_{\rm p}^2) =$	$0.001\,956\,61$
	$(B-A)/(MR_{\rm p}^2) =$	$0.000\,252\,43$
gravitational Love number[11]	$k_2 =$	$0.170(1)$
equatorial radius[6]	$R_{\rm eq} =$	$3\,396.2$ km
polar radius[6]	$R_{\rm pol} =$	$3\,376$ km
reference radius	$R_{\rm p} =$	$3\,396$ km

Jupiter

semimajor axis	$a =$	5.2029 au
eccentricity	$e =$	$0.048\,39$
inclination	$I =$	$1.30°$
orbital period	$P =$	$4\,332.820$ d $= 11.8626$ yr
mass parameter[13]	$\mathbb{G}M =$	$1.266\,865\,3 \times 10^{17}$ m^3 s^{-2} (nominal)
mass	$M =$	$1.898\,12/g \times 10^{27}$ kg
Sun/planet mass ratio	$M_\odot/M =$	$1\,047.57$
Sun/system mass ratio	$M_\odot/M' =$	$1\,047.35$
spin period[14]	$P_s =$	$0.413\,54$ d
obliquity	$\epsilon =$	$3.13°$
moment of inertia[15]	$C/(MR_{\rm p}^2) =$	0.28 (theory)
quadrupole moment[13]	$J_2 =$	$0.014\,697$
gravitational Love number[13]	$k_2 =$	$0.57(2)$
equatorial radius[16]	$R_{\rm eq} =$	$71\,492$ km (nominal)
polar radius[16]	$R_{\rm pol} =$	$66\,854$ km (nominal)
reference radius	$R_{\rm p} =$	$71\,492$ km

Saturn

semimajor axis	$a =$	9.5367 au

12 Konopliv et al. (2011).

13 Durante et al. (2020).

14 The spin period refers to the rotation of the magnetic field of the planet.

15 Ni (2018).

16 At a pressure of 1 bar, from Archinal et al. (2018).

eccentricity	$e = 0.053\,86$
inclination	$I = 2.49°$
orbital period	$P = 10\,755.7$ d $= 29.4475$ yr
mass parameter	$\mathbb{G}M = 3.793\,12 \times 10^{16}$ m^3 s^{-2}
mass	$M = 5.683\,17/g \times 10^{26}$ kg
Sun/planet mass ratio	$M_\odot/M = 3\,498.77$
Sun/system mass ratio	$M_\odot/M' = 3\,497.90$
spin period[14]	$P_s = 0.444\,01$ d
obliquity	$\epsilon = 26.73°$
moment of inertia	$C/(MR_\mathrm{p}^2) = 0.22$ (theory)
quadrupole moment	$J_2 = 0.016\,291$
gravitational Love number[17]	$k_2 = 0.39(2)$
equatorial radius[16]	$R_\mathrm{eq} = 60\,268$ km
polar radius[16]	$R_\mathrm{pol} = 54\,364$ km
reference radius	$R_\mathrm{p} = 60\,330$ km

Uranus

semimajor axis	$a = 19.1892$ au
eccentricity	$e = 0.047\,26$
inclination	$I = 0.77°$
orbital period	$P = 30\,687.2$ d $= 84.0168$ yr
mass parameter	$\mathbb{G}M = 5.793\,95 \times 10^{15}$ m^3 s^{-2}
mass	$M = 8.680\,99/g \times 10^{25}$ kg
Sun/planet mass ratio	$M_\odot/M = 22\,905.3$
Sun/system mass ratio	$M_\odot/M' = 22\,903.0$
spin period[14]	$P_s = 0.718\,33$ d
obliquity	$\epsilon = 97.77°$
moment of inertia[18]	$C/(MR_\mathrm{p}^2) = 0.22$ (theory)
quadrupole moment	$J_2 = 0.003\,511$
gravitational Love number[19]	$k_2 = 0.10$ (theory)
equatorial radius[16]	$R_\mathrm{eq} = 25\,559$ km
polar radius[16]	$R_\mathrm{pol} = 24\,973$ km
reference radius	$R_\mathrm{p} = 25\,559$ km

---

17 Lainey et al. (2017).

18 Nettelmann et al. (2013).

19 Gavrilov & Zharkov (1977).

Neptune

semimajor axis	$a =$	30.0699 au
eccentricity	$e =$	0.008 59
inclination	$I =$	1.77°
orbital period	$P =$	60 190.0 d = 164.7913 yr
mass parameter	$\mathbb{G}M =$	$6.835\,10 \times 10^{15}$ m^3 s^{-2}
mass	$M =$	$1.024\,09/g \times 10^{26}$ kg
Sun/planet mass ratio	$M_\odot/M =$	19 416.3
Sun/system mass ratio	$M_\odot/M' =$	19 412.2
spin period	$P_s =$	0.671 25 d
obliquity	$\epsilon =$	28.32°
moment of inertia[18]	$C/(MR_\mathrm{p}^2) =$	0.26 (theory)
quadrupole moment	$J_2 =$	0.003 536
gravitational Love number[19]	$k_2 =$	0.13 (theory)
equatorial radius[16]	$R_\mathrm{eq} =$	24 764 km
polar radius[16]	$R_\mathrm{pol} =$	24 341 km
reference radius	$R_\mathrm{p} =$	24 764 km

Appendix B

Mathematical background

B.1 Vectors

The position of the point with Cartesian coordinates (x, y, z) may be described by a position vector,

$$\mathbf{r} = x\hat{\mathbf{x}} + y\hat{\mathbf{y}} + z\hat{\mathbf{z}}, \tag{B.1}$$

where $\hat{\mathbf{x}}$, $\hat{\mathbf{y}}$, and $\hat{\mathbf{z}}$ are fixed unit vectors that point along the x, y, and z axes. The distance of the point from the origin is written r or $|\mathbf{r}|$ and is equal to $(x^2+y^2+z^2)^{1/2}$.

Similarly, we may represent an arbitrary vector $\mathbf{A}$ in component form as

$$\mathbf{A} = A_x\hat{\mathbf{x}} + A_y\hat{\mathbf{y}} + A_z\hat{\mathbf{z}}. \tag{B.2}$$

The magnitude of a vector $\mathbf{A}$ is $A \equiv |\mathbf{A}| \equiv (A_x^2 + A_y^2 + A_z^2)^{1/2}$.

The **scalar** or **dot product** of two vectors $\mathbf{A}$ and $\mathbf{B}$ is

$$\mathbf{A} \cdot \mathbf{B} \equiv |\mathbf{A}||\mathbf{B}| \cos\psi, \tag{B.3}$$

where ψ is the angle between the two vectors, placed tail to tail. Note that $\mathbf{A} \cdot \mathbf{B} = \mathbf{B} \cdot \mathbf{A}$ and $\mathbf{A} \cdot \mathbf{A} = |\mathbf{A}|^2$. Since $\hat{\mathbf{x}} \cdot \hat{\mathbf{x}} = \hat{\mathbf{y}} \cdot \hat{\mathbf{y}} = \hat{\mathbf{z}} \cdot \hat{\mathbf{z}} = 1$, and $\hat{\mathbf{x}} \cdot \hat{\mathbf{y}} = \hat{\mathbf{x}} \cdot \hat{\mathbf{z}} = \hat{\mathbf{y}} \cdot \hat{\mathbf{z}} = 0$, we may write the dot product in component form as

$$\mathbf{A} \cdot \mathbf{B} = \sum_{i=1}^{3} A_i B_i, \tag{B.4}$$

where the subscripts 1, 2, and 3 stand for x, y, and z, respectively. We often adopt the **summation convention**, in which we automatically sum from 1 to 3 over any dummy subscript that appears twice in one term of an equation. Thus the preceding equation simplifies to

$$\mathbf{A} \cdot \mathbf{B} = A_i B_i. \tag{B.5}$$

Two obvious but useful identities are

$$A^2 = (\hat{\mathbf{A}} \cdot \hat{\mathbf{x}})^2 + (\hat{\mathbf{A}} \cdot \hat{\mathbf{y}})^2 + (\hat{\mathbf{A}} \cdot \hat{\mathbf{z}})^2, \tag{B.6a}$$

$$\mathbf{A} = (\hat{\mathbf{x}} \cdot \mathbf{A})\hat{\mathbf{x}} + (\hat{\mathbf{y}} \cdot \mathbf{A})\hat{\mathbf{y}} + (\hat{\mathbf{z}} \cdot \mathbf{A})\hat{\mathbf{z}} = (\hat{\mathbf{n}}_i \cdot \mathbf{A})\hat{\mathbf{n}}_i; \tag{B.6b}$$

here $\hat{\mathbf{n}}_i$ is the unit vector along coordinate axis $i \in \{1, 2, 3\}$ in a Cartesian coordinate system.

The **vector** or **cross product** of two vectors is

$$\mathbf{A} \times \mathbf{B} \equiv AB \sin \psi \, \hat{\mathbf{p}}, \tag{B.7}$$

where $\hat{\mathbf{p}}$ is a unit vector that is perpendicular to the plane containing $\mathbf{A}$ and $\mathbf{B}$ and points in the direction of movement of the right thumb as the fingers rotate $\mathbf{A}$ into $\mathbf{B}$ around their common tail. Note that $\mathbf{A} \times \mathbf{B} = -\mathbf{B} \times \mathbf{A}$, $\mathbf{A} \times \mathbf{A} = \mathbf{0}$, and $\hat{\mathbf{x}} \times \hat{\mathbf{y}} = \hat{\mathbf{z}}$, $\hat{\mathbf{y}} \times \hat{\mathbf{z}} = \hat{\mathbf{x}}$, $\hat{\mathbf{z}} \times \hat{\mathbf{x}} = \hat{\mathbf{y}}$. In component form the cross product may be written

$$\mathbf{A} \times \mathbf{B} = \sum_{ijk=1}^{3} \epsilon_{ijk} \hat{\mathbf{n}}_i A_j B_k = \epsilon_{ijk} \hat{\mathbf{n}}_i A_j B_k. \tag{B.8}$$

Here ϵ_{ijk} is the permutation symbol defined in Appendix C.1, and the last equality uses the summation convention.

Some identities that involve the dot and cross product include:

$$\mathbf{A} \cdot (\mathbf{B} \times \mathbf{C}) = \mathbf{C} \cdot (\mathbf{A} \times \mathbf{B}) = \mathbf{B} \cdot (\mathbf{C} \times \mathbf{A}), \tag{B.9a}$$

$$\mathbf{A} \times (\mathbf{B} \times \mathbf{C}) = (\mathbf{A} \cdot \mathbf{C})\mathbf{B} - (\mathbf{A} \cdot \mathbf{B})\mathbf{C}, \tag{B.9b}$$

$$(\mathbf{A} \times \mathbf{B}) \cdot (\mathbf{C} \times \mathbf{D}) = (\mathbf{A} \cdot \mathbf{C})(\mathbf{B} \cdot \mathbf{D}) - (\mathbf{A} \cdot \mathbf{D})(\mathbf{B} \cdot \mathbf{C}), \tag{B.9c}$$

$$\begin{aligned}(\mathbf{A} \times \mathbf{B}) \times (\mathbf{C} \times \mathbf{D}) &= [\mathbf{A} \cdot (\mathbf{B} \times \mathbf{D})]\mathbf{C} - [\mathbf{A} \cdot (\mathbf{B} \times \mathbf{C})]\mathbf{D} \\ &= [\mathbf{C} \cdot (\mathbf{D} \times \mathbf{A})]\mathbf{B} - [\mathbf{C} \cdot (\mathbf{D} \times \mathbf{B})]\mathbf{A}.\end{aligned} \tag{B.9d}$$

In proving these identities, it is helpful to use the relation (C.2) between the permutation symbol and the Kronecker delta.

The velocity and acceleration of a particle may be written in Cartesian components as

$$\mathbf{v} \equiv \dot{\mathbf{r}} \equiv \frac{\mathrm{d}\mathbf{r}}{\mathrm{d}t} = \dot{x}\hat{\mathbf{x}} + \dot{y}\hat{\mathbf{y}} + \dot{z}\hat{\mathbf{z}}, \quad \mathbf{a} \equiv \ddot{\mathbf{r}} \equiv \frac{\mathrm{d}^2\mathbf{r}}{\mathrm{d}t^2} = \ddot{x}\hat{\mathbf{x}} + \ddot{y}\hat{\mathbf{y}} + \ddot{z}\hat{\mathbf{z}}. \tag{B.10}$$

B.2 Coordinate systems

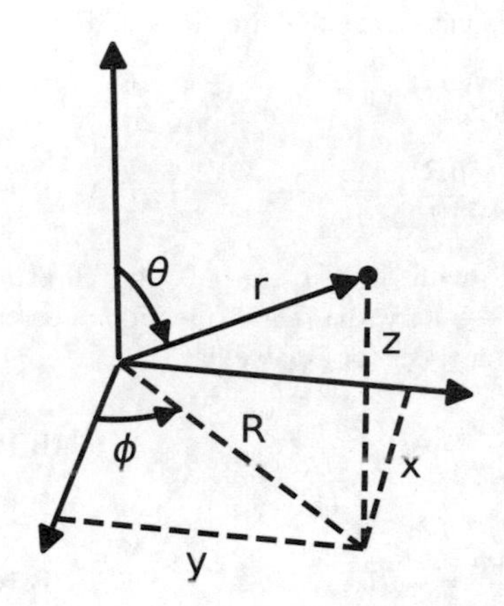

Figure B.1: The three main coordinate systems: Cartesian (x, y, z), cylindrical (R, ϕ, z), and spherical (r, θ, ϕ).

B.2.1 Cylindrical and polar coordinates

In cylindrical coordinates the location of a particle is denoted by (R, ϕ, z), where R is the perpendicular distance from the z-axis to the particle, and ϕ is the azimuthal angle between the x-axis and the projection of the position vector onto the (x, y) plane (Figure B.1). Thus the relation to Cartesian coordinates is

$$x = R\cos\phi, \quad y = R\sin\phi, \quad z = z. \tag{B.11}$$

Polar coordinates are simply the restriction of cylindrical coordinates to the $z = 0$ plane; in some cases we use ψ instead of ϕ to denote the azimuthal angle in polar coordinates.

In cylindrical coordinates the position vector is

$$\mathbf{r} = R\hat{\mathbf{R}} + z\hat{\mathbf{z}}. \tag{B.12}$$

An arbitrary vector may be written $\mathbf{A} = A_R\hat{\mathbf{R}} + A_\phi\hat{\boldsymbol{\phi}} + A_z\hat{\mathbf{z}}$, where $\hat{\boldsymbol{\phi}} = \hat{\mathbf{z}} \times \hat{\mathbf{R}}$ and

$$A_x = A_R\cos\phi - A_\phi\sin\phi, \quad A_y = A_R\sin\phi + A_\phi\cos\phi, \quad A_z = A_z. \tag{B.13}$$

The expressions for dot and cross products in cylindrical coordinates are simply equations (B.4) and (B.8), with the subscripts $(1,2,3)$ denoting (R,ϕ,z) instead of (x,y,z). Note that the decomposition into components must be carried out at the same position for both vectors in the product, since the directions of $\hat{\mathbf{R}}$ and $\hat{\boldsymbol{\phi}}$ depend on position.

The velocity in cylindrical coordinates is

$$\mathbf{v} = \frac{\mathrm{d}\mathbf{r}}{\mathrm{d}t} = \frac{\mathrm{d}R}{\mathrm{d}t}\hat{\mathbf{R}} + R\frac{\mathrm{d}\hat{\mathbf{R}}}{\mathrm{d}t} + \frac{\mathrm{d}z}{\mathrm{d}t}\hat{\mathbf{z}}. \tag{B.14}$$

To compute $\mathrm{d}\hat{\mathbf{R}}/\mathrm{d}t$, we use equations (B.13) with $A_R = 1$, $A_\phi = 0$, $A_z = 0$. Thus $\hat{\mathbf{R}} = \cos\phi\hat{\mathbf{x}} + \sin\phi\hat{\mathbf{y}}$, and $\mathrm{d}\hat{\mathbf{R}} = (-\sin\phi\hat{\mathbf{x}} + \cos\phi\hat{\mathbf{y}})\mathrm{d}\phi$. The expression in parentheses is just $\hat{\boldsymbol{\phi}}$. After carrying out a similar analysis for $\mathrm{d}\hat{\boldsymbol{\phi}}$, we have

$$\frac{\mathrm{d}\hat{\mathbf{R}}}{\mathrm{d}\phi} = \hat{\boldsymbol{\phi}}, \quad \frac{\mathrm{d}\hat{\boldsymbol{\phi}}}{\mathrm{d}\phi} = -\hat{\mathbf{R}}. \tag{B.15}$$

Thus

$$\frac{\mathrm{d}\hat{\mathbf{R}}}{\mathrm{d}t} = \dot{\phi}\hat{\boldsymbol{\phi}}, \quad \frac{\mathrm{d}\hat{\boldsymbol{\phi}}}{\mathrm{d}t} = -\dot{\phi}\hat{\mathbf{R}}. \tag{B.16}$$

The velocity is

$$\mathbf{v} = \dot{R}\hat{\mathbf{R}} + R\dot{\phi}\hat{\boldsymbol{\phi}} + \dot{z}\hat{\mathbf{z}}, \tag{B.17}$$

and the acceleration is

$$\mathbf{a} = \frac{\mathrm{d}^2\mathbf{r}}{\mathrm{d}t^2} = (\ddot{R} - R\dot{\phi}^2)\hat{\mathbf{R}} + (2\dot{R}\dot{\phi} + R\ddot{\phi})\hat{\boldsymbol{\phi}} + \ddot{z}\hat{\mathbf{z}}. \tag{B.18}$$

.

B.2.2 Spherical coordinates

The position of a particle is denoted by (r,θ,ϕ). The coordinate r is the radial distance from the origin to the particle; θ is the angle between the position vector and the z-axis, sometimes called the **polar angle** or the **colatitude**; and ϕ is the same azimuthal angle used in cylindrical coordinates (Figure B.1). The relation to Cartesian coordinates is

$$x = r\sin\theta\cos\phi, \quad y = r\sin\theta\sin\phi, \quad z = r\cos\theta. \tag{B.19}$$

In spherical coordinates the position vector is simply

$$\mathbf{r} = r\hat{\mathbf{r}}. \tag{B.20}$$

An arbitrary vector may be written $\mathbf{A} = A_r\hat{\mathbf{r}} + A_\theta\hat{\boldsymbol{\theta}} + A_\phi\hat{\boldsymbol{\phi}}$, where $\hat{\boldsymbol{\theta}} = \hat{\boldsymbol{\phi}} \times \hat{\mathbf{r}}$ and

$$\begin{aligned} A_x &= A_r \sin\theta\cos\phi + A_\theta\cos\theta\cos\phi - A_\phi\sin\phi, \\ A_y &= A_r \sin\theta\sin\phi + A_\theta\cos\theta\sin\phi + A_\phi\cos\phi, \\ A_z &= A_r\cos\theta - A_\theta\sin\theta. \end{aligned} \tag{B.21}$$

Once again, the expressions for dot and cross products in spherical coordinates are simply equations (B.4) and (B.8), with the subscripts $(1, 2, 3)$ denoting (r, θ, ϕ) instead of (x, y, z), and with the understanding that the decomposition into components must be carried out at the same position for both vectors in the product.

The rate of change of the unit vectors in spherical coordinates is

$$\frac{\mathrm{d}\hat{\mathbf{r}}}{\mathrm{d}t} = \dot{\theta}\hat{\boldsymbol{\theta}} + \dot{\phi}\sin\theta\hat{\boldsymbol{\phi}}, \quad \frac{\mathrm{d}\hat{\boldsymbol{\theta}}}{\mathrm{d}t} = -\dot{\theta}\hat{\mathbf{r}} + \dot{\phi}\cos\theta\hat{\boldsymbol{\phi}}, \quad \frac{\mathrm{d}\hat{\boldsymbol{\phi}}}{\mathrm{d}t} = -\dot{\phi}\sin\theta\hat{\mathbf{r}} - \dot{\phi}\cos\theta\hat{\boldsymbol{\theta}}. \tag{B.22}$$

Thus the velocity is

$$\mathbf{v} = \dot{\mathbf{r}} = \dot{r}\hat{\mathbf{r}} + r\dot{\theta}\hat{\boldsymbol{\theta}} + r\sin\theta\dot{\phi}\hat{\boldsymbol{\phi}}, \tag{B.23}$$

and the acceleration is

$$\begin{aligned} \mathbf{a} = \frac{\mathrm{d}\mathbf{v}}{\mathrm{d}t} &= (\ddot{r} - r\dot{\theta}^2 - r\sin^2\theta\dot{\phi}^2)\hat{\mathbf{r}} + (2\dot{r}\dot{\theta} + r\ddot{\theta} - r\sin\theta\cos\theta\dot{\phi}^2)\hat{\boldsymbol{\theta}} \\ &\quad + (r\sin\theta\ddot{\phi} + 2\dot{r}\sin\theta\dot{\phi} + 2r\cos\theta\dot{\theta}\dot{\phi})\hat{\boldsymbol{\phi}}. \end{aligned} \tag{B.24}$$

B.3 Vector calculus

Gradient In Cartesian coordinates we define the **gradient** of a scalar function of position $f(\mathbf{x})$ to be the vector

$$\nabla f \equiv \hat{\mathbf{x}}\frac{\partial f}{\partial x} + \hat{\mathbf{y}}\frac{\partial f}{\partial y} + \hat{\mathbf{z}}\frac{\partial f}{\partial z} = \hat{\mathbf{n}}_i\frac{\partial f}{\partial x_i}, \tag{B.25}$$

where in the last equality $(x_1, x_2, x_3) = (x, y, z)$ and we have used the summation convention. The symbol ∇ is called grad, del, or nabla. An alternative notation is

$$\nabla f = \frac{\partial f}{\partial \mathbf{r}}. \tag{B.26}$$

The direction of the gradient is the direction in which $f(\mathbf{r})$ increases most rapidly, and $|\nabla f|$ is the rate of increase in that direction.

The change in the value of f between the points $\mathbf{r}$ and $\mathbf{r} + \mathrm{d}\mathbf{r}$ is

$$\mathrm{d}f = \frac{\partial f}{\partial x_i}\mathrm{d}x_i = \nabla f \cdot \mathrm{d}\mathbf{r}. \tag{B.27}$$

If $\mathbf{r}$ and $\mathbf{r}+\mathrm{d}\mathbf{r}$ lie on a surface S on which f is constant then $\mathrm{d}f = 0$ and $\nabla f \cdot \mathrm{d}\mathbf{r} = 0$, so ∇f is orthogonal to $\mathrm{d}\mathbf{r}$ and hence to the surface S itself. In other words the gradient of f is normal to surfaces of constant f.

In cylindrical coordinates $\mathrm{d}\mathbf{r} = \mathrm{d}R\hat{\mathbf{R}} + R\mathrm{d}\phi\hat{\boldsymbol{\phi}} + \mathrm{d}z\hat{\mathbf{z}}$ (note that the coefficient of $\hat{\boldsymbol{\phi}}$ is $R\mathrm{d}\phi$, not $\mathrm{d}\phi$). Hence for consistency with equation (B.27) we must have

$$\nabla f = \hat{\mathbf{R}}\frac{\partial f}{\partial R} + \frac{\hat{\boldsymbol{\phi}}}{R}\frac{\partial f}{\partial \phi} + \hat{\mathbf{z}}\frac{\partial f}{\partial z}. \tag{B.28}$$

Similarly, in spherical coordinates

$$\nabla f = \hat{\mathbf{r}}\frac{\partial f}{\partial r} + \frac{\hat{\boldsymbol{\theta}}}{r}\frac{\partial f}{\partial \theta} + \frac{\hat{\boldsymbol{\phi}}}{r\sin\theta}\frac{\partial f}{\partial \phi}. \tag{B.29}$$

Divergence In Cartesian coordinates we define the **divergence** of a vector function $\mathbf{F}(\mathbf{r})$ to be the scalar function

$$\nabla \cdot \mathbf{F} \equiv \frac{\partial F_x}{\partial x} + \frac{\partial F_y}{\partial y} + \frac{\partial F_z}{\partial z} = \frac{\partial F_i}{\partial x_i}. \tag{B.30}$$

To illuminate the physical meaning of this expression, consider a cubical volume V occupying the region $x_{ia} \leq x_i \leq x_{ib}$, $i = 1, 2, 3$. Then

$$\begin{aligned}\int_V \mathrm{d}\mathbf{r}\,\nabla\cdot\mathbf{F} &= \int_{x_{1a}}^{x_{1b}}\mathrm{d}x_1\int_{x_{2a}}^{x_{2b}}\mathrm{d}x_2\int_{x_{3a}}^{x_{3b}}\mathrm{d}x_3\left(\frac{\partial F_1}{\partial x_1} + \frac{\partial F_2}{\partial x_2} + \frac{\partial F_3}{\partial x_3}\right)\\ &= \int_{x_{sa}}^{x_{2b}}\mathrm{d}x_2\int_{x_{3a}}^{x_{3b}}\mathrm{d}x_3\,[F_1(x_{1b}, x_2, x_3) - F_1(x_{1a}, x_2, x_3)]\\ &\quad + \text{two similar terms}.\end{aligned} \tag{B.31}$$

This expression can be written more compactly as $\int_S \mathrm{d}A\,\hat{\mathbf{n}}\cdot\mathbf{F}$, where the integral is over the surface S surrounding the volume V, $\mathrm{d}A$ is a small element of area on the surface, and $\hat{\mathbf{n}}$ is a unit vector normal to the surface and pointing outward. We may generalize this result to an arbitrary volume by dividing the volume into many small cubes and noting that the surface integrals from the inside faces of the cubes cancel. Hence for an arbitrary volume V,

$$\int_V \mathrm{d}\mathbf{r}\,\nabla\cdot\mathbf{F} = \int_S \mathrm{d}A\,\hat{\mathbf{n}}\cdot\mathbf{F}. \tag{B.32}$$

This result is known as the **divergence theorem**. The theorem gives a physical meaning to the divergence: the divergence of a vector field in a small volume measures the net flux of that field through a closed surface enclosing the volume.

In cylindrical coordinates

$$\nabla \cdot \mathbf{F} = \frac{1}{R}\frac{\partial}{\partial R}(RF_R) + \frac{1}{R}\frac{\partial F_\phi}{\partial \phi} + \frac{\partial F_z}{\partial z}. \tag{B.33}$$

This result can be derived by writing $\nabla \cdot \mathbf{F} = (\hat{\mathbf{R}}\partial/\partial R + \hat{\boldsymbol{\phi}}\partial/\partial\phi + \hat{\mathbf{z}}\partial/\partial z) \cdot (F_R\hat{\mathbf{R}} + F_\phi\hat{\boldsymbol{\phi}} + F_z\hat{\mathbf{z}})$ and evaluating the expression using equations (B.15). In spherical coordinates

$$\nabla \cdot \mathbf{F} = \frac{1}{r^2}\frac{\partial}{\partial r}(r^2 F_r) + \frac{1}{r\sin\theta}\frac{\partial}{\partial\theta}(\sin\theta F_\theta) + \frac{1}{r\sin\theta}\frac{\partial F_\phi}{\partial\phi}. \tag{B.34}$$

Curl In Cartesian coordinates we define the **curl** of a vector function $\mathbf{F}(\mathbf{r})$ to be the vector function

$$\nabla \times \mathbf{F} \equiv \hat{\mathbf{r}}\left(\frac{\partial F_z}{\partial y} - \frac{\partial F_y}{\partial z}\right) + \hat{\mathbf{y}}\left(\frac{\partial F_x}{\partial z} - \frac{\partial F_z}{\partial x}\right) + \hat{\mathbf{z}}\left(\frac{\partial F_y}{\partial x} - \frac{\partial F_x}{\partial y}\right). \tag{B.35}$$

Note that this definition is the same as the cross product of the vector operator ∇ with the vector function $\mathbf{F}$.

Some useful identities are

$$\nabla \times \nabla f = \mathbf{0}, \tag{B.36a}$$
$$\nabla \cdot (\nabla \times \mathbf{F}) = 0, \tag{B.36b}$$
$$\nabla \times (f\mathbf{F}) = \nabla f \times \mathbf{F} + f\nabla \times \mathbf{F}, \tag{B.36c}$$
$$\nabla \cdot (\mathbf{G} \times \mathbf{F}) = \mathbf{F} \cdot \nabla \times \mathbf{G} - \mathbf{G} \cdot \nabla \times \mathbf{F}. \tag{B.36d}$$

If we replace $\mathbf{F}$ by $\mathbf{G} \times \mathbf{F}$ in the divergence theorem (B.32), where $\mathbf{G}$ is a constant vector, then using equations (B.9a) and (B.36d) we find $\mathbf{G}\cdot\int \mathrm{d}\mathbf{r}\,\nabla\times\mathbf{F} = \mathbf{G}\cdot\int \mathrm{d}A\,\hat{\mathbf{n}}\times\mathbf{F}$. Since this result must hold for any constant vector $\mathbf{G}$, we conclude that

$$\int_V \mathrm{d}\mathbf{r}\,\nabla \times \mathbf{F} = \int_S \mathrm{d}A\,\hat{\mathbf{n}} \times \mathbf{F} \tag{B.37}$$

In cylindrical and spherical coordinates

$$\nabla \times \mathbf{F} = \left(\frac{1}{R}\frac{\partial F_z}{\partial\phi} - \frac{\partial F_\phi}{\partial z}\right)\hat{\mathbf{R}} + \left(\frac{\partial F_R}{\partial z} - \frac{\partial F_z}{\partial R}\right)\hat{\boldsymbol{\phi}} + \left(\frac{\partial F_\phi}{\partial R} + \frac{F_\phi}{R} - \frac{1}{R}\frac{\partial F_R}{\partial\phi}\right)\hat{\mathbf{z}}$$

$$= \left(\frac{1}{r}\frac{\partial F_\phi}{\partial \theta} + \frac{\cos\theta}{r\sin\theta}F_\phi - \frac{1}{r\sin\theta}\frac{\partial F_\theta}{\partial \phi}\right)\hat{\mathbf{r}} \tag{B.38}$$
$$+\left(\frac{1}{r\sin\theta}\frac{\partial F_r}{\partial \phi} - \frac{\partial F_\phi}{\partial r} - \frac{F_\phi}{r}\right)\hat{\boldsymbol{\theta}} + \left(\frac{\partial F_\theta}{\partial r} + \frac{F_\theta}{r} - \frac{1}{r}\frac{\partial F_r}{\partial \theta}\right)\hat{\boldsymbol{\phi}}.$$

Laplacian The divergence of the gradient of a scalar function is called the **Laplacian** of that function. Thus the Laplacian of $F(\mathbf{r})$ is

$$\nabla^2 F \equiv \nabla \cdot \nabla F. \tag{B.39}$$

Applying the divergence theorem (B.32) to ∇F, we have

$$\int_V \mathrm{d}\mathbf{r}\, \nabla^2 F = \int_S \mathrm{d}A\, \hat{\mathbf{n}} \cdot \nabla F. \tag{B.40}$$

In Cartesian, cylindrical, and spherical coordinates we have

$$\nabla^2 F = \frac{\partial^2 F}{\partial x^2} + \frac{\partial^2 F}{\partial y^2} + \frac{\partial^2 F}{\partial z^2}$$
$$= \frac{1}{R}\frac{\partial}{\partial R}\left(R\frac{\partial F}{\partial R}\right) + \frac{1}{R^2}\frac{\partial^2 F}{\partial \phi^2} + \frac{\partial^2 F}{\partial z^2}$$
$$= \frac{1}{r^2}\frac{\partial}{\partial r}\left(r^2\frac{\partial F}{\partial r}\right) + \frac{1}{r^2\sin\theta}\frac{\partial}{\partial \theta}\left(\sin\theta\frac{\partial F}{\partial \theta}\right) + \frac{1}{r^2\sin^2\theta}\frac{\partial^2 F}{\partial \phi^2}. \tag{B.41}$$

An important special case is the Laplacian of $F(\mathbf{r}) = 1/r$. Using the last of equations (B.41), we find $\nabla^2 F = 0$ for $r \neq 0$. The behavior of $\nabla^2 F$ near $r = 0$ needs to be handled more carefully because of the singularity. In equation (B.40), let V be a spherical volume of radius r_0 centered on the origin. Since $\nabla F = -\hat{\mathbf{r}}/r^2$ the integral on the right side is -4π, independent of r_0. Thus $\int_V \mathrm{d}\mathbf{r}\, \nabla^2 F = -4\pi$. Since $\nabla^2 F = 0$ for $r \neq 0$, this result must hold for *any* volume V that contains the origin, whatever its shape and size. These results imply that in three dimensions

$$\nabla^2(1/r) = -4\pi\delta(\mathbf{r}) \quad \text{or} \quad \nabla^2\frac{1}{|\mathbf{r}-\mathbf{r}'|} = -4\pi\delta(\mathbf{r}-\mathbf{r}'), \tag{B.42}$$

where $\delta(\mathbf{r})$ is the 3-dimensional delta function (Appendix C.2).

The gravitational potential from a point mass m at $\mathbf{r}'$ is $\Phi(\mathbf{r}) = -\mathbb{G}m/|\mathbf{r}-\mathbf{r}'|$. Therefore the gravitational potential from a mass density distribution $\rho(\mathbf{r}')$ is

$$\Phi(\mathbf{r}) = -\mathbb{G}\int \frac{\mathrm{d}\mathbf{r}'\rho(\mathbf{r}')}{|\mathbf{r}-\mathbf{r}'|}. \tag{B.43}$$

Thus

$$\nabla^2\Phi(\mathbf{r}) = -\mathbb{G}\int d\mathbf{r}'\rho(\mathbf{r}')\nabla^2\frac{1}{|\mathbf{r}-\mathbf{r}'|} = 4\pi\mathbb{G}\int d\mathbf{r}'\rho(\mathbf{r}')\delta(\mathbf{r}-\mathbf{r}') = 4\pi\mathbb{G}\rho(\mathbf{r}), \tag{B.44}$$

which is **Poisson's equation**. In a vacuum the right side is zero, so the potential must satisfy **Laplace's equation,** $\nabla^2\Phi = 0$.

B.4 Fourier series

If a function $f(x)$ is periodic with period h, that is, $f(x+h) = f(x)$ for all x, then it can be represented as a **Fourier series**,

$$f(x) = \sum_{n=0}^{\infty}\left(a_n \cos\frac{2\pi nx}{h} + b_n \sin\frac{2\pi nx}{h}\right). \tag{B.45}$$

We will use the more compact expression

$$f(x) = \sum_{n=-\infty}^{\infty} c_n \exp\left(\frac{2\pi i nx}{h}\right). \tag{B.46}$$

To determine the coefficients c_n we multiply (B.46) by $\exp(-2\pi i mx/h)$, with m an integer, and integrate from $x = 0$ to $x = h$. We use the identity

$$\int_0^h dx \exp\left[\frac{2\pi i(n-m)x}{h}\right] = h\,\delta_{mn}, \tag{B.47}$$

where δ_{mn} is the Kronecker delta (eq. C.1). Then

$$c_m = \frac{1}{h}\int_0^h dx \exp\left(-\frac{2\pi i mx}{h}\right) f(x). \tag{B.48}$$

B.5 Spherical trigonometry

Spherical triangles are defined by three intersecting great circles on the unit sphere. The lengths of the sides are conventionally labeled a, b, c and the angles opposite these sides are labeled A, B, C; by convention these angles are less than π (Figure B.2).

The **spherical cosine law** states that

$$\cos a = \cos b \cos c + \sin b \sin c \cos A,$$

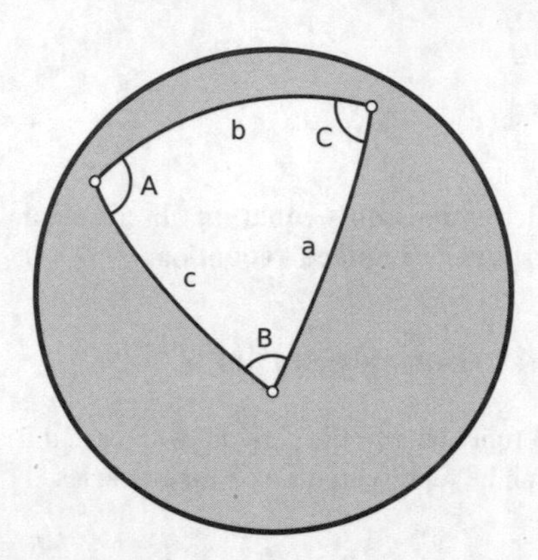

Figure B.2: A triangle on the unit sphere with sides a, b, c and angles A, B, C.

$$
\begin{aligned}
\cos b &= \cos c \cos a + \sin c \sin a \cos B, \\
\cos c &= \cos a \cos b + \sin a \sin b \cos C.
\end{aligned}
\tag{B.49}
$$

We now prove the first of these. Let O denote the center of the sphere, and let $\hat{\mathbf{A}}$, $\hat{\mathbf{B}}$, $\hat{\mathbf{C}}$ denote the unit vectors from O to the points A, B, C on the surface of the sphere. The normal to the plane through A, O and B is $\hat{\mathbf{A}} \times \hat{\mathbf{B}}$ and this vector has magnitude $\sin c$. Similarly the normal to the plane through A, O, and C is $\hat{\mathbf{A}} \times \hat{\mathbf{C}}$ and this has magnitude $\sin b$. The angle between these normals is A, so we have

$$(\hat{\mathbf{A}} \times \hat{\mathbf{B}}) \cdot (\hat{\mathbf{A}} \times \hat{\mathbf{C}}) = \sin b \sin c \cos A. \tag{B.50}$$

Using the identity (B.9c),

$$(\hat{\mathbf{A}} \times \hat{\mathbf{B}}) \cdot (\hat{\mathbf{A}} \times \hat{\mathbf{C}}) = (\hat{\mathbf{A}} \cdot \hat{\mathbf{A}})(\hat{\mathbf{B}} \cdot \hat{\mathbf{C}}) - (\hat{\mathbf{A}} \cdot \hat{\mathbf{C}})(\hat{\mathbf{A}} \cdot \hat{\mathbf{B}}) = \cos a - \cos b \cos c, \tag{B.51}$$

and the first of equations (B.49) follows from eliminating the left sides of (B.50) and (B.51). The second and third of equations (B.49) follow from the first by re-labeling the sides and angles.

Using the steps that led to equation (B.50), we can also write

$$|(\hat{\mathbf{A}} \times \hat{\mathbf{B}}) \times (\hat{\mathbf{A}} \times \hat{\mathbf{C}})| = \sin b \sin c \sin A. \tag{B.52}$$

Using the identity (B.9d),

$$|(\hat{\mathbf{A}} \times \hat{\mathbf{B}}) \times (\hat{\mathbf{A}} \times \hat{\mathbf{C}})| = |\mathbf{A} \cdot (\mathbf{B} \times \mathbf{C})|. \tag{B.53}$$

Eliminating the left side of equations (B.52) and (B.53) gives

$$\frac{\sin A}{\sin a} = \frac{|\mathbf{A} \cdot (\mathbf{B} \times \mathbf{C})|}{\sin a \sin b \sin c}. \tag{B.54}$$

Because of the identities (B.9a), the right side is unchanged if we exchange the label pairs $\{a, A\}$, $\{b, B\}$, and $\{c, C\}$, so the left side must be as well. Therefore we arrive at the **spherical sine law**,

$$\frac{\sin A}{\sin a} = \frac{\sin B}{\sin b} = \frac{\sin C}{\sin c}. \tag{B.55}$$

We also have

$$\begin{aligned}
\cos A &= -\cos B \cos C + \sin B \sin C \cos a, \\
\cos B &= -\cos C \cos A + \sin C \sin A \cos b, \\
\cos C &= -\cos A \cos B + \sin A \sin B \cos c.
\end{aligned} \tag{B.56}$$

B.6 Euler angles

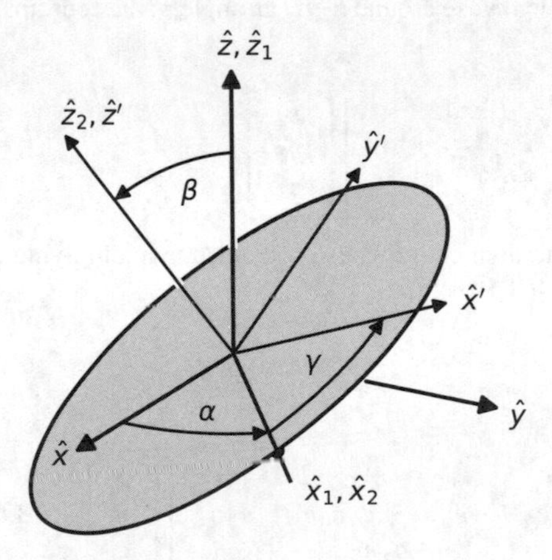

Figure B.3: The Euler angles (α, β, γ) that specify a rotation from coordinate axes (x, y, z) to (x', y', z').

The Euler angles (α, β, γ) shown in Figure B.3 describe the rotation of one Cartesian coordinate frame into another. Many different conventions are used to define the Euler angles, but the following is probably the most common in physics and astronomy.

Let the coordinates of a point be (x, y, z) in the initial frame and (x', y', z') in the final frame. We convert from the initial to the final frame by three rotations of the coordinate axes. We begin by rotating the original axes counterclockwise around $\hat{\mathbf{z}}$ by an angle α; the coordinates in the new frame are labeled (x_1, y_1, z_1) and are related to the original coordinates by

$$\begin{bmatrix} x_1 \\ y_1 \\ z_1 \end{bmatrix} = \begin{bmatrix} c_\alpha & s_\alpha & 0 \\ -s_\alpha & c_\alpha & 0 \\ 0 & 0 & 1 \end{bmatrix} \begin{bmatrix} x \\ y \\ z \end{bmatrix}. \tag{B.57}$$

Here and below we abbreviate $\cos\alpha$ by c_α, $\sin\alpha$ by s_α, and so forth. Next we rotate the axes counterclockwise around $\hat{\mathbf{x}}_1$ by an angle β; the coordinates in this frame are labeled (x_2, y_2, z_2) and are given by

$$\begin{bmatrix} x_2 \\ y_2 \\ z_2 \end{bmatrix} = \begin{bmatrix} 1 & 0 & 0 \\ 0 & c_\beta & s_\beta \\ 0 & -s_\beta & c_\beta \end{bmatrix} \begin{bmatrix} x_1 \\ y_1 \\ z_1 \end{bmatrix}. \tag{B.58}$$

Finally we rotate the axes counterclockwise around $\hat{\mathbf{z}}_2$ by an angle γ; the coordinates (x', y', z') in this frame are

$$\begin{bmatrix} x' \\ y' \\ z' \end{bmatrix} = \begin{bmatrix} c_\gamma & s_\gamma & 0 \\ -s_\gamma & c_\gamma & 0 \\ 0 & 0 & 1 \end{bmatrix} \begin{bmatrix} x_2 \\ y_2 \\ z_2 \end{bmatrix}. \tag{B.59}$$

The matrix relating the original and final coordinates is obtained by multiplying the three matrices in equations (B.57)–(B.59):

$$\begin{aligned} \mathbf{r}' = \begin{bmatrix} x' \\ y' \\ z' \end{bmatrix} &= \mathbf{R}(\alpha, \beta, \gamma)\mathbf{r} \\ &= \begin{bmatrix} c_\alpha c_\gamma - s_\alpha c_\beta s_\gamma & s_\alpha c_\gamma + c_\alpha c_\beta s_\gamma & s_\beta s_\gamma \\ -c_\alpha s_\gamma - s_\alpha c_\beta c_\gamma & -s_\alpha s_\gamma + c_\alpha c_\beta c_\gamma & s_\beta c_\gamma \\ s_\alpha s_\beta & -c_\alpha s_\beta & c_\beta \end{bmatrix} \begin{bmatrix} x \\ y \\ z \end{bmatrix}. \end{aligned} \tag{B.60}$$

Any matrix describing a rotation of the coordinate axes is orthogonal, and hence its inverse is equal to its transpose. Thus the transformation from the final to the original coordinates is

$$\mathbf{r} = \begin{bmatrix} x \\ y \\ z \end{bmatrix} = \mathbf{R}^{\mathrm{T}}(\alpha, \beta, \gamma)\mathbf{r}'$$
$$= \begin{bmatrix} c_\alpha c_\gamma - s_\alpha c_\beta s_\gamma & -c_\alpha s_\gamma - s_\alpha c_\beta c_\gamma & s_\alpha s_\beta \\ s_\alpha c_\gamma + c_\alpha c_\beta s_\gamma & -s_\alpha s_\gamma + c_\alpha c_\beta c_\gamma & -c_\alpha s_\beta \\ s_\beta s_\gamma & s_\beta c_\gamma & c_\beta \end{bmatrix} \begin{bmatrix} x' \\ y' \\ z' \end{bmatrix}. \quad \text{(B.61)}$$

The position of a particle can be specified by its radius r and the three Euler angles if the particle is assumed to be located at $(x', y', z') = (r, 0, 0)$.[1] From equation (B.61),

$$\mathbf{r} = (x, y, z) = r(c_\alpha c_\gamma - s_\alpha c_\beta s_\gamma, s_\alpha c_\gamma + c_\alpha c_\beta s_\gamma, s_\beta s_\gamma). \quad \text{(B.62)}$$

If the Euler angles of the particle are changing with time, then its angular speed is

$$\boldsymbol{\Omega} = \dot{\alpha}\hat{\mathbf{z}} + \dot{\beta}\hat{\mathbf{x}}_1 + \dot{\gamma}\hat{\mathbf{z}}_2, \quad \text{(B.63)}$$

where the unit vectors $\hat{\mathbf{z}}$, $\hat{\mathbf{x}}_1$, and $\hat{\mathbf{z}}_2$ point in the directions shown in Figure B.3. The velocity due to a rotation at angular velocity $\boldsymbol{\Omega}$ is $\boldsymbol{\Omega} \times \mathbf{r}$ (eq. D.17). Thus

$$\dot{\mathbf{r}} = \dot{r}\hat{\mathbf{r}} + \boldsymbol{\Omega} \times \mathbf{r} = \dot{r}\hat{\mathbf{r}} + (\dot{\alpha}\hat{\mathbf{z}} + \dot{\beta}\hat{\mathbf{x}}_1 + \dot{\gamma}\hat{\mathbf{z}}_2) \times \mathbf{r}. \quad \text{(B.64)}$$

In terms of the unit vectors of the original coordinates (x, y, z) or the final coordinates (x', y', z'), we have $\hat{\mathbf{z}} = s_\beta s_\gamma \hat{\mathbf{x}}' + s_\beta c_\gamma \hat{\mathbf{y}}' + c_\beta \hat{\mathbf{z}}'$, $\hat{\mathbf{x}}_1 = c_\alpha \hat{\mathbf{x}} + s_\alpha \hat{\mathbf{y}} = c_\gamma \hat{\mathbf{x}}' - s_\gamma \hat{\mathbf{y}}'$, and $\hat{\mathbf{z}}_2 = \hat{\mathbf{z}}' = s_\alpha s_\beta \hat{\mathbf{x}} - c_\alpha s_\beta \hat{\mathbf{y}} + c_\beta \hat{\mathbf{z}}$. Then

$$\boldsymbol{\Omega} = (c_\alpha \dot{\beta} + s_\alpha s_\beta \dot{\gamma})\hat{\mathbf{x}} + (s_\alpha \dot{\beta} - c_\alpha s_\beta \dot{\gamma})\hat{\mathbf{y}} + (\dot{\alpha} + c_\beta \dot{\gamma})\hat{\mathbf{z}} \quad \text{(B.65a)}$$
$$= (s_\beta s_\gamma \dot{\alpha} + c_\gamma \dot{\beta})\hat{\mathbf{x}}' + (s_\beta c_\gamma \dot{\alpha} - s_\gamma \dot{\beta})\hat{\mathbf{y}}' + (c_\beta \dot{\alpha} + \dot{\gamma})\hat{\mathbf{z}}'. \quad \text{(B.65b)}$$

[1] This representation is not unique as it uses four variables $(r, \alpha, \beta, \gamma)$ to specify the 3-dimensional vector $\mathbf{r}$.

B.7 Calculus of variations

Let $\mathbf{r} = \mathbf{r}(t)$, $t_0 \leq t \leq t_1$, be an arbitrary smooth curve, which we label by γ. We define a functional

$$S[\gamma] \equiv \int_{t_0}^{t_1} \mathrm{d}t\, L[\mathbf{r}(t), \dot{\mathbf{r}}(t), t]. \tag{B.66}$$

Now consider a nearby curve γ' defined by $\mathbf{r} = \mathbf{r}(t) + \epsilon\mathbf{h}(t)$, where $\mathbf{h}(t_0) = \mathbf{h}(t_1) = \mathbf{0}$. As $\epsilon \to 0$ we have

$$\begin{aligned} S[\gamma'] - S[\gamma] &= \int_{t_0}^{t_1} \mathrm{d}t\, [L(\mathbf{r} + \epsilon\mathbf{h}, \dot{\mathbf{r}} + \epsilon\dot{\mathbf{h}}, t) - L(\mathbf{r}, \dot{\mathbf{r}}, t)] \\ &= \epsilon \int_{t_0}^{t_1} \mathrm{d}t \left(\mathbf{h} \cdot \frac{\partial L}{\partial \mathbf{r}} + \dot{\mathbf{h}} \cdot \frac{\partial L}{\partial \dot{\mathbf{r}}} \right) + \mathrm{O}(\epsilon^2), \end{aligned} \tag{B.67}$$

where the integral on the second line is evaluated along the unperturbed curve γ. On this curve L can be considered to be a function only of time, $L(t) = L[\mathbf{r}(t), \dot{\mathbf{r}}(t), t]$. Hence we may integrate by parts to obtain

$$S[\gamma'] - S[\gamma] = -\epsilon \int_{t_0}^{t_1} \mathrm{d}t\, \mathbf{h} \cdot \left[\frac{\mathrm{d}}{\mathrm{d}t}\left(\frac{\partial L}{\partial \dot{\mathbf{r}}} \right) - \frac{\partial L}{\partial \mathbf{r}} \right] + \mathrm{O}(\epsilon^2). \tag{B.68}$$

The boundary term arising from the integration by parts vanishes since $\mathbf{h}(t_0) = \mathbf{h}(t_1) = \mathbf{0}$.

The curve γ is an extremum of $S[\gamma]$ if $I(\gamma) - I(\gamma') = \mathrm{O}(\epsilon^2)$ for all variations $\mathbf{h}$. On an extremal curve, the integral in equation (B.68) must vanish for all variations $\mathbf{h}(t)$ for which $\mathbf{h}(t_0) = \mathbf{h}(t_1) = \mathbf{0}$; thus the condition for an extremal curve is

$$\frac{\mathrm{d}}{\mathrm{d}t}\left(\frac{\partial L}{\partial \dot{\mathbf{r}}} \right) - \frac{\partial L}{\partial \mathbf{r}} = \mathbf{0}. \tag{B.69}$$

This is the **Euler–Lagrange equation**.

Appendix C

Special functions

In general we follow the conventions of the NIST Digital Library of Mathematical Functions, https://dlmf.nist.gov/, or Olver et al. (2010). Other standard references include Erdélyi (1953–1955) and Gradshteyn & Ryzhik (2015). Algorithms for evaluating many of these functions are described by Press et al. (2007).

C.1 Kronecker delta and permutation symbol

The **Kronecker delta** is a function δ_{mn} of two integers m and n, defined by

$$\delta_{mn} = \begin{cases} 0 & \text{if } m \neq n, \\ 1 & \text{if } m = n. \end{cases} \tag{C.1}$$

The 3-dimensional **permutation symbol** ϵ_{ijk} is defined to be zero if two or more of the indices i, j, and k are equal, $+1$ if (i, j, k) is an even permutation of $(1, 2, 3)$ [the even permutations are $(2, 3, 1)$ and $(3, 1, 2)$], and -1 if (i, j, k) is an odd permutation of $(1, 2, 3)$.

A useful relation is

$$\sum_{k=1}^{3} \epsilon_{ijk}\epsilon_{klm} = \delta_{il}\delta_{jm} - \delta_{im}\delta_{jl}. \tag{C.2}$$

C.2 Delta function

The **delta function** or **Dirac delta function** is a singular function defined by the properties

$$\delta(x) = \begin{cases} \infty & x = 0, \\ 0 & x \neq 0, \end{cases} \qquad \int_{-\infty}^{\infty} \mathrm{d}x\, \delta(x) = 1. \tag{C.3}$$

Thus

$$\int_{-\infty}^{\infty} \mathrm{d}x\, f(x)\delta(x-a) = f(a) \tag{C.4}$$

if $f(x)$ is an arbitrary continuous function. We also have

$$\delta[g(x)] = \sum_i \frac{\delta(x - x_i)}{|g'(x_i)|}, \tag{C.5}$$

where $g(x)$ is differentiable and $\{x_i\}$ is the set of all roots of $g(x)$, that is, solutions of $g(x_i) = 0$.

The delta function can be written as

$$\delta(x) = \frac{1}{\pi} \lim_{\epsilon \to 0} \frac{\epsilon}{\epsilon^2 + x^2} = \lim_{a \to \infty} \left(\frac{a}{\pi}\right)^{1/2} \mathrm{e}^{-ax^2} = \lim_{a \to \infty} \frac{\sin ax}{\pi x} = \frac{1}{2\pi} \int_{-\infty}^{\infty} \mathrm{d}x\, \mathrm{e}^{ixt}. \tag{C.6}$$

The 3-dimensional delta function is

$$\delta(\mathbf{r}) = \delta(x)\delta(y)\delta(z), \tag{C.7}$$

where $\mathbf{r} = (x, y, z)$ in Cartesian coordinates. The analog of equations (C.3) is

$$\delta(\mathbf{r}) = \begin{cases} \infty & \mathbf{r} = \mathbf{0}, \\ 0 & \mathbf{r} \neq \mathbf{0}, \end{cases} \qquad \int_V \mathrm{d}\mathbf{r}\, \delta(\mathbf{r}) = 1 \tag{C.8}$$

if the volume V contains the origin.

The delta function on the surface of a sphere is related to the spherical harmonics through equation (C.49).

The **periodic delta function** with period h is a sum of delta functions at $x = \ldots, -2h, -h, 0, h, 2h, \ldots$:

$$\delta_h(t) = h \sum_{k=-\infty}^{\infty} \delta(t - kh); \tag{C.9}$$

the normalizing factor of h is chosen such that the average value of $\delta_h(t)$ over any nonzero interval approaches unity as $h \to 0$.

The Fourier series for the periodic delta function is derived from equations (B.46) and (B.48):

$$\delta_h(x) = \sum_{n=-\infty}^{\infty} \exp\left(\frac{2\pi i n x}{h}\right) = \sum_{n=-\infty}^{\infty} \cos\left(\frac{2\pi n x}{h}\right) = \sum_{n=0}^{\infty}(2 - \delta_{n0})\cos\left(\frac{2\pi n x}{h}\right). \tag{C.10}$$

C.3 Gamma function

The **gamma function** is defined by

$$\Gamma(z) \equiv \int_0^\infty dt\, t^{z-1} e^{-t}, \quad (\mathrm{Re}\, z > 0). \tag{C.11}$$

For $\mathrm{Re}\, z \le 0$, $\Gamma(z)$ can be defined by analytic continuation for all complex numbers z except $0, -1, -2, \ldots$. Special values of $\Gamma(z)$ include

$$\Gamma(1) = 1, \quad \Gamma(\tfrac{1}{2}) = \pi^{1/2} = 1.772\,45, \quad \Gamma(\tfrac{3}{2}) = \tfrac{1}{2}\pi^{1/2} = 0.886\,23. \tag{C.12}$$

For non-negative integers n, the factorial function is defined as

$$n! \equiv \Gamma(n+1) = 1 \cdot 2 \cdot 3 \cdots n. \tag{C.13}$$

Useful relations include

$$\Gamma(z+1) = z\Gamma(z), \tag{C.14a}$$

$$\Gamma(z)\Gamma(1-z) = \frac{\pi}{\sin \pi z}, \tag{C.14b}$$

$$\Gamma(2z) = \frac{2^{2z-1}}{\pi^{1/2}}\Gamma(z)\Gamma(z+\tfrac{1}{2}). \tag{C.14c}$$

As $|z| \to \infty$ with $|\arg(z)| < \pi$,

$$\frac{\Gamma(z+a)}{\Gamma(z+b)} \to z^{a-b}\left[1 + \mathrm{O}(|z|^{-1})\right], \tag{C.15a}$$

$$\Gamma(z) \to (2\pi)^{1/2} z^{z-1/2} e^{-z}\left[1 + \mathrm{O}(|z|^{-1})\right]. \tag{C.15b}$$

The second of these results is **Stirling's approximation**.

C.4 Elliptic integrals

The **elliptic integrals** are defined by[1]

$$K(k) \equiv \int_0^{\pi/2} \frac{\mathrm{d}\phi}{(1-k^2\sin^2\phi)^{1/2}} = \int_0^1 \frac{\mathrm{d}t}{(1-t^2)^{1/2}(1-k^2t^2)^{1/2}},$$

$$E(k) \equiv \int_0^{\pi/2} \mathrm{d}\phi\,(1-k^2\sin^2\phi)^{1/2} = \int_0^1 \frac{\mathrm{d}t(1-k^2t^2)^{1/2}}{(1-t^2)^{1/2}}. \tag{C.16}$$

The **incomplete elliptic integrals** are

$$F(\phi,k) \equiv \int_0^{\phi} \frac{\mathrm{d}\phi}{(1-k^2\sin^2\phi)^{1/2}} = \int_0^{\sin\phi} \frac{\mathrm{d}t}{(1-t^2)^{1/2}(1-k^2t^2)^{1/2}},$$

$$E(\phi,k) \equiv \int_0^{\phi} \mathrm{d}\phi\,(1-k^2\sin^2\phi)^{1/2} = \int_0^{\sin\phi} \frac{\mathrm{d}t(1-k^2t^2)^{1/2}}{(1-t^2)^{1/2}}. \tag{C.17}$$

Thus $F(\frac{1}{2}\pi,k) = K(k)$ and $E(\frac{1}{2}\pi,k) = E(k)$. A complete description of elliptic integrals is given by Byrd & Friedman (1971).

For our purposes $0 \le k < 1$. As $k \to 1$, $E(k) \to 1$ and

$$K(k) \to \tfrac{1}{2}\log\left(\frac{16}{1-k^2}\right)[1 + \mathrm{O}(1-k)]. \tag{C.18}$$

The derivatives of the elliptic integrals are

$$\frac{\mathrm{d}K(k)}{\mathrm{d}k} = \frac{E(k)-(1-k^2)K(k)}{k(1-k^2)}, \qquad \frac{\mathrm{d}E(k)}{\mathrm{d}k} = \frac{E(k)-K(k)}{k}. \tag{C.19}$$

If the factor k^2 in the definition of the elliptic integrals is replaced by $-k^2$, we have

$$\int_0^{\pi/2} \frac{\mathrm{d}\phi}{(1+k^2\sin^2\phi)^{1/2}} = \frac{1}{(1+k^2)^{1/2}} K\left[\frac{k}{(1+k^2)^{1/2}}\right],$$

$$\int_0^{\pi/2} \mathrm{d}\phi\,(1+k^2\sin^2\phi)^{1/2} = (1+k^2)^{1/2} E\left[\frac{k}{(1+k^2)^{1/2}}\right]. \tag{C.20}$$

[1] Unfortunately, in some references and software, the argument of the elliptic integral is k^2 rather than k; thus, for example, $K(m) \equiv \int_0^{\pi/2} \mathrm{d}\phi/(1-m\sin^2\phi)^{1/2}$.

The elliptic integrals can also be defined in the Carlson or symmetric forms

$$R_F(x,y,z) \equiv \tfrac{1}{2}\int_0^\infty \frac{\mathrm{d}t}{s(t)}, \quad R_D(x,y,z) \equiv \tfrac{3}{2}\int_0^\infty \frac{\mathrm{d}t}{(t+z)s(t)}, \tag{C.21}$$

where $s^2(t) = (t+x)(t+y)(t+z)$ and for our purposes $x, y, z \geq 0$. The relation between the forms is

$$K(k) = R_F(0, 1-k^2, 1), \quad E(k) = R_F(0, 1-k^2, 1) - \tfrac{1}{3}k^2 R_D(0, 1-k^2, 1). \tag{C.22}$$

C.5 Bessel functions

The **Bessel functions** of the first and second kind, $J_\nu(z)$ and $Y_\nu(z)$, are linearly independent solutions of the differential equation

$$\frac{1}{z}\frac{\mathrm{d}}{\mathrm{d}z}\Big(z\frac{\mathrm{d}w}{\mathrm{d}z}\Big) + \Big(1 - \frac{\nu^2}{z^2}\Big)w = 0. \tag{C.23}$$

In series form,

$$J_\nu(z) = \sum_{k=0}^\infty \frac{(-1)^k}{k!\Gamma(\nu+k+1)}\big(\tfrac{1}{2}z\big)^{\nu+2k}. \tag{C.24}$$

The function $Y_\nu(z)$, which diverges as $z \to 0$, is defined by the relation

$$Y_\nu(z) = \frac{\cos\nu\pi J_\nu(z) - J_{-\nu}(z)}{\sin\nu\pi}, \tag{C.25}$$

or by its limiting value if ν is an integer. The most comprehensive description of Bessel functions is given by Watson (1944).

If $\nu \equiv n$ is an integer,

$$J_{-n}(z) = (-1)^n J_n(z) = J_n(-z), \quad Y_{-n}(z) = (-1)^n Y_n(z), \tag{C.26}$$

$$J_n(z) = \frac{1}{\pi}\int_0^\pi \mathrm{d}\theta\, \cos(z\sin\theta - n\theta). \tag{C.27}$$

If C_ν denotes either J_ν or Y_ν,

$$C_{\nu-1}(z) + C_{\nu+1}(z) = \frac{2\nu}{z}C_\nu(z), \quad C_{\nu-1}(z) - C_{\nu+1}(z) = 2\frac{\mathrm{d}C_\nu(z)}{\mathrm{d}z}. \tag{C.28}$$

We shall also use the identity

$$\exp(\mathrm{i}z\sin u) = \sum_{k=-\infty}^{\infty} J_k(z)\exp(\mathrm{i}ku). \tag{C.29}$$

The **modified Bessel functions** are

$$I_\nu(z) = \mathrm{i}^{-\nu} J_\nu(\mathrm{i}z), \quad K_\nu(z) = K_{-\nu}(z) = \frac{\pi}{2}\frac{I_{-\nu}(z) - I_\nu(z)}{\sin\nu\pi}; \tag{C.30}$$

the second equation is replaced by its limiting value if ν is an integer. For small $|z|$,

$$\begin{aligned} I_\nu(z) &\to \frac{1}{\Gamma(\nu+1)}\left(\tfrac{1}{2}z\right)^\nu \left[1 + \mathrm{O}(|z|^2)\right], \quad (\nu \neq -1, -2, \ldots), \\ K_\nu(z) &\to \begin{cases} \tfrac{1}{2}\Gamma(\nu)\left(\tfrac{1}{2}z\right)^{-\nu}\left[1 + \mathrm{O}(|z|^2)\right], & (\mathrm{Re}\,\nu > 0), \\ -\gamma - \log(\tfrac{1}{2}z) + \mathrm{O}(|z|^2\log|z|), & (\nu = 0), \end{cases} \end{aligned} \tag{C.31}$$

where $\gamma = 0.577\,216\cdots$ is Euler's constant. As $|z| \to \infty$ with $|\arg(z)| < \frac{1}{2}\pi$,

$$I_\nu(z) \to \frac{\mathrm{e}^z}{(2\pi z)^{1/2}}\left[1 + \mathrm{O}(|z|^{-1})\right], \quad K_\nu(z) \to \left(\frac{\pi}{2z}\right)^{1/2} \mathrm{e}^{-z}\left[1 + \mathrm{O}(|z|^{-1})\right]. \tag{C.32}$$

If Z_ν denotes either I_ν or $\mathrm{e}^{\mathrm{i}\pi\nu}K_\nu$,

$$Z_{\nu-1}(z) - Z_{\nu+1}(z) = \frac{2\nu}{z}Z_\nu(z), \quad Z_{\nu-1}(z) + Z_{\nu+1}(z) = 2\frac{\mathrm{d}Z_\nu(z)}{\mathrm{d}z}; \tag{C.33}$$

for $\nu = 0$ these imply

$$I_0'(z) = I_1(z), \quad K_0'(z) = -K_1(z). \tag{C.34}$$

We shall use the result (Erdélyi 1954, Table 4.16)

$$\int_0^\infty \mathrm{d}x\, x^{1-\nu}\mathrm{e}^{-px^2} I_\nu(ax) = \frac{(2p)^{\nu-1}}{\Gamma(\nu)a^\nu}\exp[a^2/(4p)]\gamma[\nu, a^2/(4p)], \quad a, p > 0, \tag{C.35}$$

where $\gamma(\nu, x)$ is an incomplete gamma function.

C.6 Legendre functions

The **associated Legendre functions**, $\mathsf{P}_l^m(x)$ and $\mathsf{Q}_l^m(x)$, are linearly independent solutions of the differential equation

$$\frac{\mathrm{d}}{\mathrm{d}x}\left[(1-x^2)\frac{\mathrm{d}w}{\mathrm{d}x}\right]-\frac{m^2}{1-x^2}w+l(l+1)w=0. \tag{C.36}$$

For our purposes we need only consider functions of the first kind, $\mathsf{P}_l^m(x)$, for real arguments in the interval $-1 \leq x \leq 1$, with $l = 0, 1, 2, \ldots$ and m an integer in the range $|m| \leq l$. Then

$$\mathsf{P}_l^m(x)=\frac{(-1)^m}{2^l l!}(1-x^2)^{m/2}\frac{\mathrm{d}^{l+m}}{\mathrm{d}x^{l+m}}(x^2-1)^l. \tag{C.37}$$

When m is even, $\mathsf{P}_l^m(x)$ is a polynomial of degree l, and when $m = 0$, these are called **Legendre polynomials** $\mathsf{P}_l(x) \equiv \mathsf{P}_l^0(x)$.

The associated Legendre functions satisfy the relations

$$\mathsf{P}_l^m(-x)=(-1)^{l-m}\mathsf{P}_l^m(x), \tag{C.38}$$

$$\mathsf{P}_l^{-m}(x)=(-1)^m\frac{(l-m)!}{(l+m)!}\mathsf{P}_l^m(x), \tag{C.39}$$

$$\int_{-1}^{1}\mathrm{d}x\,\mathsf{P}_l^m(x)\mathsf{P}_{l'}^m(x)=\frac{(l+m)!}{(l-m)!}\frac{2}{2l+1}\delta_{ll'}, \tag{C.40}$$

$$\mathsf{P}_l^m(1)=\delta_{m0}, \tag{C.41}$$

$$\mathsf{P}_l^m(0)=\begin{cases}\dfrac{2^m\pi^{1/2}}{\Gamma(\frac{1}{2}l-\frac{1}{2}m+1)\Gamma(\frac{1}{2}-\frac{1}{2}l-\frac{1}{2}m)}, & (l-m\text{ even}),\\[2ex] 0, & (l-m\text{ odd}).\end{cases} \tag{C.42}$$

The Legendre polynomials are generated by the relation

$$\frac{1}{(1-2xt+t^2)^{1/2}}=\sum_{l=0}^{\infty}\mathsf{P}_l(x)t^l, \quad |t|<1,\ |x|\leq 1. \tag{C.43}$$

Thus the inverse distance between the points $\mathbf{x}$ and $\mathbf{x}'$ can be written

$$\frac{1}{|\mathbf{x}-\mathbf{x}'|}=\sum_{l=0}^{\infty}\frac{r_<^l}{r_>^{l+1}}\mathsf{P}_l(\cos\gamma), \tag{C.44}$$

where $r_<$ and $r_>$ are the smaller and larger of $|\mathbf{x}|$ and $|\mathbf{x}'|$, and γ is the angle between the two vectors.

The associated Legendre functions can be written most compactly using the substitution $x = \cos\theta$; since $-1 \le x \le 1$, we may take $0 \le \theta \le \pi$ and let $c = \cos\theta$, $s = \sin\theta$:

$$\begin{aligned} &\mathsf{P}_0(c) = 1, \\ &\mathsf{P}_1(c) = c, \qquad \mathsf{P}_1^1(c) = -s, \\ &\mathsf{P}_2(c) = \tfrac{3}{2}c^2 - \tfrac{1}{2}, \qquad \mathsf{P}_2^1(c) = -3cs, \qquad \mathsf{P}_2^2(c) = 3s^2, \\ &\mathsf{P}_3(c) = \tfrac{5}{2}c^3 - \tfrac{3}{2}c, \qquad \mathsf{P}_3^1(c) = s(\tfrac{3}{2} - \tfrac{15}{2}c^2), \qquad \mathsf{P}_3^2(c) = 15cs^2, \qquad \mathsf{P}_3^3(c) = -15s^3. \end{aligned} \tag{C.45}$$

C.7 Spherical harmonics

A **spherical harmonic** is defined by the expression

$$Y_{lm}(\theta,\phi) = \left[\frac{2l+1}{4\pi}\frac{(l-m)!}{(l+m)!}\right]^{1/2} \mathsf{P}_l^m(\cos\theta)\mathrm{e}^{im\phi}. \tag{C.46}$$

The variables lie in the range $0 \le \theta \le \pi$ and $0 \le \phi \le 2\pi$ and usually represent the angular coordinates in a spherical coordinate system (Appendix B.2.2). The indices are usually restricted to be integers with $l = 0, 1, 2, \ldots$, and $|m| \le l$. With this definition

$$Y_{l,-m}(\theta,\phi) = (-1)^m Y_{lm}^*(\theta,\phi), \tag{C.47}$$

where the asterisk denotes complex conjugation.

The most important feature of the spherical harmonics, easily proved using equation (C.40), is that they are orthonormal in the sense that

$$\int_0^\pi \mathrm{d}\theta \sin\theta \int_0^{2\pi} \mathrm{d}\phi\, Y_{kn}^*(\theta,\phi) Y_{lm}(\theta,\phi) = \int \mathrm{d}\mathbf{\Omega}\, Y_{kn}^*(\theta,\phi) Y_{lm}(\theta,\phi) = \delta_{kl}\delta_{nm}, \tag{C.48}$$

where δ_{kl} is the Kronecker delta (Appendix C.1), and $\mathrm{d}\mathbf{\Omega} \equiv \sin\theta \mathrm{d}\theta \mathrm{d}\phi$ represents an element of solid angle. We also have

$$\sum_{l=0}^{\infty} \sum_{m=-l}^{l} Y_{lm}^*(\theta',\phi') Y_{lm}(\theta,\phi) = \delta(\cos\theta - \cos\theta')\delta(\phi - \phi'), \tag{C.49}$$

where $\delta(x)$ is the delta function (Appendix C.2).

The spherical harmonics satisfy the partial differential equation

$$\left[\frac{1}{\sin\theta}\frac{\partial}{\partial\theta}\left(\sin\theta\frac{\partial}{\partial\theta}\right)+\frac{1}{\sin^2\theta}\frac{\partial^2}{\partial\phi^2}+l(l+1)\right]Y_{lm}(\theta,\phi)=0. \tag{C.50}$$

An arbitrary scalar function of position $\mathbf{r}$ can be written in spherical coordinates as a series of spherical harmonics,

$$f(\mathbf{r})=f(r,\theta,\phi)=\sum_{l=0}^{\infty}\sum_{m=-l}^{l}f_{lm}(r)Y_{lm}(\theta,\phi). \tag{C.51}$$

Multiplying by $Y^*_{kn}(\theta,\phi)$, integrating over solid angle and using equation (C.48), we find

$$f_{lm}(r)=\int \mathrm{d}\boldsymbol{\Omega}\, Y^*_{lm}(\theta,\phi)f(\mathbf{r}), \tag{C.52}$$

where the integral is over the unit sphere.

If the directions (θ,ϕ) and (θ',ϕ') in spherical coordinates are separated by an angle γ, then

$$\cos\gamma=\cos\theta\cos\theta'+\sin\theta\sin\theta'\cos(\phi-\phi'), \tag{C.53}$$

and the addition theorem for spherical harmonics states that

$$\mathsf{P}_l(\cos\gamma)=\frac{4\pi}{2l+1}\sum_{m=-l}^{l}Y^*_{lm}(\theta',\phi')Y_{lm}(\theta,\phi). \tag{C.54}$$

Together with equation (C.44), this leads to an expression for the inverse distance between the points $\mathbf{x}=(r,\theta,\phi)$ and $\mathbf{x}'=(r',\theta',\phi')$:

$$\frac{1}{|\mathbf{x}-\mathbf{x}'|}=\sum_{l=0}^{\infty}\frac{4\pi}{2l+1}\frac{r_<^l}{r_>^{l+1}}\sum_{m=-l}^{l}Y^*_{lm}(\theta',\phi')Y_{lm}(\theta,\phi). \tag{C.55}$$

The first few spherical harmonics are:

$$\begin{aligned}
&Y_{00}(\theta,\phi)=\left(\tfrac{1}{4\pi}\right)^{1/2}, \\
&Y_{10}(\theta,\phi)=\left(\tfrac{3}{4\pi}\right)^{1/2}\cos\theta, && Y_{1,\pm1}(\theta,\phi)=\mp\left(\tfrac{3}{8\pi}\right)^{1/2}\sin\theta\,\mathrm{e}^{\pm\mathrm{i}\phi}, \\
&Y_{20}(\theta,\phi)=\left(\tfrac{5}{16\pi}\right)^{1/2}(3\cos^2\theta-1), && Y_{2,\pm1}(\theta,\phi)=\mp\left(\tfrac{15}{8\pi}\right)^{1/2}\sin\theta\cos\theta\,\mathrm{e}^{\pm\mathrm{i}\phi}, \\
& && Y_{2,\pm2}(\theta,\phi)=\left(\tfrac{15}{32\pi}\right)^{1/2}\sin^2\theta\,\mathrm{e}^{\pm2\mathrm{i}\phi}.
\end{aligned} \tag{C.56}$$

C.8 Vector spherical harmonics

The following discussion uses the notation and proofs of Barrera et al. (1985). We define three vector-valued functions of position,

$$\mathbf{Y}_{lm}(\theta,\phi) \equiv Y_{lm}\hat{\mathbf{r}}, \quad \mathbf{\Psi}_{lm}(\theta,\phi) \equiv r\nabla Y_{lm}, \quad \mathbf{\Phi}_{lm}(\theta,\phi) \equiv \mathbf{r}\times\nabla Y_{lm}, \tag{C.57}$$

where $Y_{lm}(\theta,\phi)$ is a spherical harmonic, and the gradient operator ∇ in spherical coordinates is given by equation (B.29).

An arbitrary vector field $\mathbf{A}(\mathbf{r})$ can be written in spherical coordinates as a series of vector spherical harmonics (cf. eq. C.51),

$$\mathbf{A}(\mathbf{r}) = \sum_{l=0}^{\infty}\sum_{m=-l}^{l}\left[f_{lm}(r)\mathbf{Y}_{lm}(\theta,\phi) + g_{lm}(r)\mathbf{\Psi}_{lm}(\theta,\phi) + h_{lm}(r)\mathbf{\Phi}_{lm}(\theta,\phi)\right]. \tag{C.58}$$

The vector spherical harmonics are orthogonal in two senses: first,

$$\mathbf{Y}_{lm}(\theta,\phi)\cdot\mathbf{\Psi}_{lm}(\theta,\phi) = \mathbf{Y}_{lm}(\theta,\phi)\cdot\mathbf{\Phi}_{lm}(\theta,\phi) = \mathbf{\Psi}_{lm}(\theta,\phi)\cdot\mathbf{\Phi}_{lm}(\theta,\phi) = 0; \tag{C.59}$$

and second,

$$\begin{aligned}\int \mathrm{d}\mathbf{\Omega}\,\mathbf{Y}^*_{lm}(\theta,\phi)\cdot\mathbf{\Psi}_{l'm'}(\theta,\phi) &= \int \mathrm{d}\mathbf{\Omega}\,\mathbf{Y}^*_{lm}(\theta,\phi)\cdot\mathbf{\Phi}_{l'm'}(\theta,\phi)\\ = \int \mathrm{d}\mathbf{\Omega}\,\mathbf{\Psi}^*_{lm}(\theta,\phi)\cdot\mathbf{\Phi}_{l'm'}(\theta,\phi) &= 0\end{aligned} \tag{C.60}$$

for all l, m, l', m'. We also have

$$\begin{aligned}\int \mathrm{d}\mathbf{\Omega}\,\mathbf{Y}^*_{lm}(\theta,\phi)\cdot\mathbf{Y}_{l'm'}(\theta,\phi) &= \delta_{l,l'}\delta_{m,m'},\\ \int \mathrm{d}\mathbf{\Omega}\,\mathbf{\Psi}^*_{lm}(\theta,\phi)\cdot\mathbf{\Psi}_{l'm'}(\theta,\phi) &= l(l+1)\delta_{l,l'}\delta_{m,m'},\\ \int \mathrm{d}\mathbf{\Omega}\,\mathbf{\Phi}^*_{lm}(\theta,\phi)\cdot\mathbf{\Phi}_{l'm'}(\theta,\phi) &= l(l+1)\delta_{l,l'}\delta_{m,m'}.\end{aligned} \tag{C.61}$$

The divergence is

$$\begin{aligned}\nabla\cdot[f(r)\mathbf{Y}_{lm}] &= \left(\frac{\mathrm{d}f}{\mathrm{d}r} + \frac{2f}{r}\right)Y_{lm},\\ \nabla\cdot[f(r)\mathbf{\Psi}_{lm}] &= -\frac{l(l+1)f}{r}Y_{lm},\\ \nabla\cdot[f(r)\mathbf{\Phi}_{lm}] &= 0.\end{aligned} \tag{C.62}$$

The curl is

$$\begin{aligned}
\nabla \times [f(r)\mathbf{Y}_{lm}] &= -\frac{f}{r}\mathbf{\Phi}_{lm},\\
\nabla \times [f(r)\mathbf{\Psi}_{lm}] &= \left(\frac{\mathrm{d}f}{\mathrm{d}r} + \frac{f}{r}\right)\mathbf{\Phi}_{lm},\\
\nabla \times [f(r)\mathbf{\Phi}_{lm}] &= -\frac{l(l+1)f}{r}\mathbf{Y}_{lm} - \left(\frac{\mathrm{d}f}{\mathrm{d}r} + \frac{f}{r}\right)\mathbf{\Psi}_{lm}.
\end{aligned} \tag{C.63}$$

The Laplacian is

$$\begin{aligned}
\nabla^2[f(r)\mathbf{Y}_{lm}] &= \left(\frac{1}{r^2}\frac{\mathrm{d}}{\mathrm{d}r}r^2\frac{\mathrm{d}f}{\mathrm{d}r}\right)\mathbf{Y}_{lm} - \frac{2+l+l^2}{r^2}f\mathbf{Y}_{lm} + \frac{2f}{r^2}\mathbf{\Psi}_{lm},\\
\nabla^2[f(r)\mathbf{\Psi}_{lm}] &= \left(\frac{1}{r^2}\frac{\mathrm{d}}{\mathrm{d}r}r^2\frac{\mathrm{d}f}{\mathrm{d}r}\right)\mathbf{\Psi}_{lm} + \frac{2l(l+1)}{r^2}f\mathbf{Y}_{lm} - \frac{l(l+1)}{r^2}f\mathbf{\Psi}_{lm},\\
\nabla^2[f(r)\mathbf{\Phi}_{lm}] &= \left(\frac{1}{r^2}\frac{\mathrm{d}}{\mathrm{d}r}r^2\frac{\mathrm{d}f}{\mathrm{d}r}\right)\mathbf{\Phi}_{lm} - \frac{l(l+1)}{r^2}f\mathbf{\Phi}_{lm}.
\end{aligned} \tag{C.64}$$

Appendix D

Lagrangian and Hamiltonian dynamics

The goal of this appendix is to provide a review and summary of the results in this subject that are employed in this book. We assume a background in classical mechanics at the advanced undergraduate level, including basic Hamiltonian mechanics (Tabor 1989; Lichtenberg & Lieberman 1992; José & Saletan 1998; Sussman & Wisdom 2001; Goldstein et al. 2002).

Let $\mathbf{r} = (x, y, z)$ denote the Cartesian coordinates of a particle moving in the force field arising from the potential $\Phi(\mathbf{r}, t)$. We define the **Lagrangian**

$$L(\mathbf{r}, \dot{\mathbf{r}}, t) \equiv \tfrac{1}{2}m|\dot{\mathbf{r}}|^2 - m\Phi(\mathbf{r}, t), \tag{D.1}$$

where $|\dot{\mathbf{r}}|^2 = \dot{x}^2 + \dot{y}^2 + \dot{z}^2$ is the square of the speed of the particle and m is its mass. The **action** maps a trajectory $\mathbf{r}(t)$ into a scalar S, given by

$$S \equiv \int_{t_0}^{t_1} \mathrm{d}t\, L[\mathbf{r}(t), \dot{\mathbf{r}}(t), t]. \tag{D.2}$$

Hamilton's principle states that the motion of the particle from time t_0 to t_1 is along a trajectory $\mathbf{r}(t)$ that is an extremum of the action for fixed end points $\mathbf{r}(t_0)$ and $\mathbf{r}(t_1)$. The proof is straightforward. According to the Euler–Lagrange equation (B.69), the trajectory is an extremum of S if and only if

$$\mathbf{0} = \frac{\mathrm{d}}{\mathrm{d}t}\left(\frac{\partial L}{\partial \dot{\mathbf{r}}}\right) - \frac{\partial L}{\partial \mathbf{r}} = m\ddot{\mathbf{r}} + m\frac{\partial \Phi}{\partial \mathbf{r}} \tag{D.3}$$

along the trajectory, and this is simply Newton's second law.

Hamilton's principle is also called the **principle of least action**, since in most cases the extremum is a minimum of the action, rather than a maximum or a saddle point.

The power of this approach is that the Lagrangian is a *scalar* function. Hence it is straightforward to compute as a function[1] $L(\mathbf{q}, \dot{\mathbf{q}}, t)$ of arbitrary or **generalized coordinates** $\mathbf{q}$ and their time derivatives $\dot{\mathbf{q}}$. Extremizing the action with L expressed in this form yields Lagrange's equations

$$\frac{\mathrm{d}}{\mathrm{d}t}\left(\frac{\partial L}{\partial \dot{\mathbf{q}}}\right) - \frac{\partial L}{\partial \mathbf{q}} = \mathbf{0}, \tag{D.4}$$

which are the equations of motion in the generalized coordinates. This approach evades the lengthy algebra that is often required to express vector differential equations, such as Newton's second law, in non-Cartesian coordinates.

For a given set of generalized coordinates $\mathbf{q}$, we define the **generalized momenta** $\mathbf{p}$ to be

$$\mathbf{p} \equiv \left(\frac{\partial L}{\partial \dot{\mathbf{q}}}\right)_{\mathbf{q},t}. \tag{D.5}$$

In this approach the momentum $\mathbf{p}$ depends on the choice of the coordinate $\mathbf{q}$, so $\mathbf{q}$ and $\mathbf{p}$ are sometimes called a **conjugate coordinate-momentum pair** or a **canonical coordinate-momentum pair**.

The **Hamiltonian** is

$$H(\mathbf{q}, \mathbf{p}, t) \equiv \mathbf{p} \cdot \dot{\mathbf{q}} - L(\mathbf{q}, \dot{\mathbf{q}}, t), \tag{D.6}$$

where it is understood that $\dot{\mathbf{q}}$ is to be eliminated in favor of $\mathbf{q}$, $\mathbf{p}$, and t using equation (D.5).

As an example, in spherical coordinates $\mathbf{q} = (r, \theta, \phi)$ the Lagrangian (D.1) is

$$L(\mathbf{q}, \dot{\mathbf{q}}, t) = \tfrac{1}{2}m(\dot{r}^2 + r^2\dot{\theta}^2 + r^2\sin^2\theta\,\dot{\phi}^2) - m\Phi(r, \theta, \phi, t), \tag{D.7}$$

where the expression for $|\dot{\mathbf{r}}|^2$ is taken from equation (B.23). The corresponding momenta are

$$p_r = \frac{\partial L}{\partial \dot{r}} = m\dot{r}, \quad p_\theta = \frac{\partial L}{\partial \dot{\theta}} = mr^2\dot{\theta}, \quad p_\phi = \frac{\partial L}{\partial \dot{\phi}} = mr^2\sin^2\theta\,\dot{\phi}. \tag{D.8}$$

[1] For notational simplicity, in this appendix we adopt the convention that symbols such as $\Phi(\mathbf{r}, t)$, $L(\mathbf{q}, \dot{\mathbf{q}}, t)$, and $H(\mathbf{q}, \mathbf{p}, t)$ denote functions of position and velocity in phase space rather than functions of the coordinates. Thus $\Phi(\mathbf{r}, t)$ and $\Phi(\mathbf{q}, t)$ have the same value if $\mathbf{r}$ and $\mathbf{q}$ are coordinates of the same point in different coordinate systems.

Notice that p_ϕ is simply the z-component of the angular momentum. The Hamiltonian is

$$H(\mathbf{q}, \mathbf{p}, t) = \frac{p_r^2}{2m} + \frac{p_\theta^2}{2mr^2} + \frac{p_\phi^2}{2mr^2 \sin^2 \theta} + m\Phi(r, \theta, \phi, t). \tag{D.9}$$

The analogous expressions for cylindrical coordinates are given in equations (D.25)–(D.27).

D.1 Hamilton's equations

The total derivative of each side of equation (D.6) is

$$\begin{aligned} \mathrm{d}H &= \left(\frac{\partial H}{\partial \mathbf{q}}\right)_{\mathbf{p},t} \cdot \mathrm{d}\mathbf{q} + \left(\frac{\partial H}{\partial \mathbf{p}}\right)_{\mathbf{q},t} \cdot \mathrm{d}\mathbf{p} + \left(\frac{\partial H}{\partial t}\right)_{\mathbf{q},\mathbf{p}} \mathrm{d}t && \text{(left side)} \\ &= \mathrm{d}\mathbf{p} \cdot \dot{\mathbf{q}} + \mathbf{p} \cdot \mathrm{d}\dot{\mathbf{q}} - \left(\frac{\partial L}{\partial \mathbf{q}}\right)_{\dot{\mathbf{q}},t} \cdot \mathrm{d}\mathbf{q} - \left(\frac{\partial L}{\partial \dot{\mathbf{q}}}\right)_{\mathbf{q},t} \cdot \mathrm{d}\dot{\mathbf{q}} - \left(\frac{\partial L}{\partial t}\right)_{\mathbf{q},\dot{\mathbf{q}}} \mathrm{d}t && \text{(right side)} \\ &= \mathrm{d}\mathbf{p} \cdot \dot{\mathbf{q}} - \left(\frac{\partial L}{\partial \mathbf{q}}\right)_{\dot{\mathbf{q}},t} \cdot \mathrm{d}\mathbf{q} - \left(\frac{\partial L}{\partial t}\right)_{\mathbf{q},\dot{\mathbf{q}}} \mathrm{d}t, && \text{(D.10)} \end{aligned}$$

where the second and fourth terms in the middle line have canceled because of equation (D.5). Since the first and third line must be the same, we conclude that

$$\dot{\mathbf{q}} = \left(\frac{\partial H}{\partial \mathbf{p}}\right)_{\mathbf{q},t}, \quad \left(\frac{\partial H}{\partial \mathbf{q}}\right)_{\mathbf{p},t} = -\left(\frac{\partial L}{\partial \mathbf{q}}\right)_{\dot{\mathbf{q}},t}, \quad \left(\frac{\partial H}{\partial t}\right)_{\mathbf{q},\mathbf{p}} = -\left(\frac{\partial L}{\partial t}\right)_{\mathbf{q},\dot{\mathbf{q}}}. \tag{D.11}$$

We may combine the first two of these equations with equations (D.4) and (D.5) and simplify the notation to obtain **Hamilton's equations**,[2]

$$\dot{\mathbf{q}} = \frac{\partial H}{\partial \mathbf{p}}, \quad \dot{\mathbf{p}} = -\frac{\partial H}{\partial \mathbf{q}}. \tag{D.12}$$

The **configuration space** of a system is the n-dimensional space with coordinates $\mathbf{q} = (q_1, \ldots, q_n)$. The corresponding **momentum space** has coordinates $(p_1, \ldots, p_n)$. A system with n-dimensional configuration and momentum spaces is

[2] The vectors on the right side of Hamilton's equations define a direction in phase space, $(\partial H/\partial \mathbf{p}, -\partial H/\partial \mathbf{q})$, just as the gradient defines a direction $(\partial H/\partial \mathbf{q}, \partial H/\partial \mathbf{p})$. The phase-space directions defined by these two vectors are orthogonal: the first vector points along the direction in which the rate of change of H is zero, and the second points along the direction in which the rate of change is as large as possible.

said to have n **degrees of freedom**. **Phase space** is the $2n$-dimensional space with coordinates $(q_1, \ldots, q_n, p_1, \ldots, p_n) \equiv (\mathbf{q}, \mathbf{p}) \equiv \mathbf{z}$; that is, $\mathbf{z}$ denotes a collection of $2n$ variables, with the first half being the values of the q_i's and the second half the values of the p_i's. The position of a system in phase space describes its dynamical state completely. Thus, through each point $\mathbf{z}_0$ in phase space there passes a unique phase-space trajectory $\mathbf{z}(t)$, which gives the future and past phase-space coordinates[3] of the particle that has coordinates $\mathbf{z}_0$ at $t = t_0$. No two distinct trajectories can ever intersect.

In terms of the $2n$-dimensional phase-space vector $\mathbf{z} \equiv (\mathbf{q}, \mathbf{p})$, Hamilton's equations can be written compactly as

$$\frac{\mathrm{d}\mathbf{z}}{\mathrm{d}t} = \mathbf{J}\frac{\partial H}{\partial \mathbf{z}}, \tag{D.13}$$

where the $2n \times 2n$ **symplectic matrix** is

$$\mathbf{J} \equiv \begin{bmatrix} \mathbf{0} & \mathbf{I} \\ -\mathbf{I} & \mathbf{0} \end{bmatrix}, \tag{D.14}$$

with $\mathbf{0}$ and $\mathbf{I}$ the $n \times n$ zero and unit matrix respectively. The symplectic matrix has the properties

$$\mathbf{J}^{-1} = \mathbf{J}^{\mathrm{T}} = -\mathbf{J}, \quad \mathbf{J}^2 = -\mathbf{I}, \quad \det(\mathbf{J}) = 1, \tag{D.15}$$

where $\mathbf{J}^{\mathrm{T}}$ is the transpose of $\mathbf{J}$, and $\det(\mathbf{J})$ is its determinant.

D.2 Rotating reference frame

We can use Lagrange's and Hamilton's equations to find the equation of motion of a particle in a rotating frame of reference.

Let $\mathbf{r}$ be the vector of Cartesian coordinates of the particle in an inertial reference frame, that is, a frame in which an isolated body does not accelerate. The equation of motion in the inertial frame is determined by the Lagrangian $L(\mathbf{r}, \dot{\mathbf{r}}, t)$ given by equation (D.1).

Now let $\mathbf{q}$ denote Cartesian coordinates of the particle in a frame that rotates with angular velocity $\mathbf{\Omega}(t)$ around the origin. We want to determine the Lagrangian in the new coordinates, $L(\mathbf{q}, \dot{\mathbf{q}}, t)$.

---

[3] Notice the imprecise use of the term "coordinates," which refers both to the generalized coordinates $\mathbf{q}$ that are conjugate to $\mathbf{p}$, and to the phase-space coordinates $(\mathbf{q}, \mathbf{p})$.

The rate of change of any physical vector $\mathbf{u}$ as observed in the rotating frame is related to its rate of change in the inertial frame by

$$\left(\frac{\mathrm{d}\mathbf{u}}{\mathrm{d}t}\right)_{\mathrm{in}} = \left(\frac{\mathrm{d}\mathbf{u}}{\mathrm{d}t}\right)_{\mathrm{rot}} + \boldsymbol{\Omega} \times \mathbf{u}. \tag{D.16}$$

Thus the velocities observed in the inertial and rotating frames are related by

$$\dot{\mathbf{r}} = \dot{\mathbf{q}} + \boldsymbol{\Omega} \times \mathbf{q} \quad \text{or} \quad \dot{\mathbf{q}} = \dot{\mathbf{r}} - \boldsymbol{\Omega} \times \mathbf{r}. \tag{D.17}$$

The kinetic energy per unit mass is

$$\begin{aligned} \tfrac{1}{2}|\dot{\mathbf{r}}|^2 &= \tfrac{1}{2}|\dot{\mathbf{q}} + \boldsymbol{\Omega} \times \mathbf{q}|^2 = \tfrac{1}{2}|\dot{\mathbf{q}}|^2 + \dot{\mathbf{q}} \cdot (\boldsymbol{\Omega} \times \mathbf{q}) + \tfrac{1}{2}|\boldsymbol{\Omega} \times \mathbf{q}|^2 \\ &= \tfrac{1}{2}|\dot{\mathbf{q}}|^2 + \boldsymbol{\Omega} \cdot (\mathbf{q} \times \dot{\mathbf{q}}) + \tfrac{1}{2}\Omega^2|\mathbf{q}|^2 - \tfrac{1}{2}(\boldsymbol{\Omega} \cdot \mathbf{q})^2; \end{aligned} \tag{D.18}$$

in deriving the last expression we have used the vector identities (B.9a) and (B.9c). In the rotating coordinates, the Lagrangian (D.1) becomes

$$L(\mathbf{q}, \dot{\mathbf{q}}, t) = \tfrac{1}{2}m|\dot{\mathbf{q}}|^2 + m\boldsymbol{\Omega}\cdot(\mathbf{q}\times\dot{\mathbf{q}}) + \tfrac{1}{2}m\Omega^2|\mathbf{q}|^2 - \tfrac{1}{2}m(\boldsymbol{\Omega}\cdot\mathbf{q})^2 - m\Phi(\mathbf{q}, t). \tag{D.19}$$

Substituting this result into Lagrange's equation (D.4) yields

$$\begin{aligned} \ddot{\mathbf{q}} &= -\frac{\partial \Phi}{\partial \mathbf{q}} - 2\boldsymbol{\Omega} \times \dot{\mathbf{q}} - \dot{\boldsymbol{\Omega}} \times \mathbf{q} + \Omega^2 \mathbf{q} - (\boldsymbol{\Omega} \cdot \mathbf{q})\boldsymbol{\Omega} \\ &= -\frac{\partial \Phi}{\partial \mathbf{q}} - 2\boldsymbol{\Omega} \times \dot{\mathbf{q}} - \dot{\boldsymbol{\Omega}} \times \mathbf{q} - \boldsymbol{\Omega} \times (\boldsymbol{\Omega} \times \mathbf{q}). \end{aligned} \tag{D.20}$$

The second, third and remaining terms on the right side are **fictitious forces** arising from the rotation of the frame of reference; respectively, the **Coriolis**, **Euler**, and **centrifugal force**. The centrifugal force per unit mass can be written as the negative gradient of the **centrifugal potential**,

$$\Phi_{\mathrm{cent}}(\mathbf{r}) = \tfrac{1}{2}(\boldsymbol{\Omega} \cdot \mathbf{q})^2 - \tfrac{1}{2}\Omega^2|\mathbf{q}|^2 = -\tfrac{1}{2}|\boldsymbol{\Omega} \times \mathbf{q}|^2. \tag{D.21}$$

The canonical momentum in the rotating frame is

$$\mathbf{p} = \frac{\partial L}{\partial \dot{\mathbf{q}}} = m\dot{\mathbf{q}} + m\boldsymbol{\Omega} \times \mathbf{q}. \tag{D.22}$$

Notice that the canonical momentum in the rotating frame is the same as the Newtonian momentum in the inertial frame, $m\dot{\mathbf{r}}$ (by the first of eqs. D.17), and is *not* equal to $m\dot{\mathbf{q}}$. The Hamiltonian is

$$H(\mathbf{q}, \mathbf{p}, t) = \frac{\mathbf{p}^2}{2m} + m\Phi(\mathbf{q}, t) - \boldsymbol{\Omega} \cdot (\mathbf{q} \times \mathbf{p}). \tag{D.23}$$

Since $\mathbf{r}$ and $\mathbf{q}$ are the same vector and $\mathbf{p} = m\dot{\mathbf{r}}$, the factor $\mathbf{q} \times \mathbf{p}$ is simply the angular momentum per unit mass $\mathbf{L}$ in the inertial frame. Therefore

$$H(\mathbf{q}, \mathbf{p}, t) = \frac{\mathbf{p}^2}{2m} + m\Phi(\mathbf{q}, t) - \mathbf{\Omega} \cdot \mathbf{L}. \tag{D.24}$$

Thus, the rotation of the frame adds an extra term $-\mathbf{\Omega} \cdot \mathbf{L}$ to the Hamiltonian of the orbiting particle.

This analysis can also be carried out in cylindrical polar coordinates (Appendix B.2.1). Here we assume for simplicity that the angular speed of the rotating frame is constant and parallel to the positive z-axis, so $\mathbf{\Omega} = \Omega\hat{\mathbf{z}}$. If (R, ϕ, z) are cylindrical coordinates in the inertial frame, then the azimuthal coordinate in the rotating frame is $\varphi = \phi - \Omega t$ and the Lagrangian can be written (cf. eq. D.7)

$$L(\mathbf{q}, \dot{\mathbf{q}}, t) = \tfrac{1}{2}m[\dot{R}^2 + R^2(\dot{\varphi} + \Omega)^2 + \dot{z}^2] - m\Phi(R, \varpi, z, t). \tag{D.25}$$

The corresponding momenta are

$$p_R = \frac{\partial L}{\partial \dot{R}} = m\dot{R}, \quad p_\varphi = \frac{\partial L}{\partial \dot{\varphi}} = mR^2(\dot{\varphi} + \Omega), \quad p_z = \frac{\partial L}{\partial \dot{z}} = m\dot{z}, \tag{D.26}$$

and the Hamiltonian is

$$H(\mathbf{q}, \mathbf{p}, t) = \frac{p_R^2}{2m} + \frac{p_\varphi^2}{2mR^2} + \frac{p_z^2}{2m} + m\Phi(R, \varphi, z, t) - \Omega p_\varphi. \tag{D.27}$$

Note that p_φ is simply the z-component of the angular momentum, so the last term is the same as the last term of equation (D.24).

D.3 Poisson brackets

Let $f(\mathbf{z})$ and $g(\mathbf{z})$ be any two scalar functions of a set of phase-space coordinates $\mathbf{z} = (\mathbf{q}, \mathbf{p})$. The **Poisson bracket** is another scalar function of $\mathbf{z}$, defined by

$$\{f, g\} \equiv \frac{\partial f}{\partial \mathbf{q}} \cdot \frac{\partial g}{\partial \mathbf{p}} - \frac{\partial f}{\partial \mathbf{p}} \cdot \frac{\partial g}{\partial \mathbf{q}}. \tag{D.28}$$

An equivalent definition is

$$\{f, g\} = \sum_{i,j=1}^{2n} J_{ij} \frac{\partial f}{\partial z_i} \frac{\partial g}{\partial z_j} = \left(\frac{\partial f}{\partial \mathbf{z}}\right)^{\mathrm{T}} \mathbf{J} \frac{\partial g}{\partial \mathbf{z}}. \tag{D.29}$$

Here "T" denotes the transpose of a vector or matrix, and the symplectic matrix $\mathbf{J}$ is defined in equation (D.14).

The Poisson bracket of the phase-space coordinates themselves is

$$\{z_i, z_j\} = J_{ij}, \tag{D.30}$$

that is,

$$\{q_i, q_j\} = 0, \quad \{p_i, p_j\} = 0, \quad \{q_i, p_j\} = -\{p_i, q_j\} = \delta_{ij}, \tag{D.31}$$

where δ_{ij} is the Kronecker delta (eq. C.1). For any functions $f(\mathbf{z})$, $g(\mathbf{z})$, and $h(\mathbf{z})$:

$$\{f, g\} = -\{g, f\}, \tag{D.32a}$$

$$\{f, f\} = 0, \tag{D.32b}$$

$$\{fg, h\} = f\{g, h\} + g\{f, h\}, \tag{D.32c}$$

$$\{fg, fh\} = fh\{g, f\} + fg\{f, h\} + f^2\{g, h\}, \tag{D.32d}$$

$$0 = \{\{f, g\}, h\} + \{\{g, h\}, f\} + \{\{h, f\}, g\}. \tag{D.32e}$$

The last of these, the only one that needs significant algebra to prove, is called the **Jacobi identity**.

D.4 The propagator

The Poisson bracket can be regarded as an operator $\mathbf{L}_g$, sometimes called the **Lie operator**, that depends on the function $g(\mathbf{z})$ and is defined by

$$\mathbf{L}_g f \equiv \{f, g\}. \tag{D.33}$$

Thus $\mathbf{L}_f g = -\mathbf{L}_g f$. Powers of this operator are

$$\mathbf{L}_g^n f = \underbrace{\mathbf{L}_g \mathbf{L}_g \cdots \mathbf{L}_g}_{n \text{ times}} f = \{\cdots\{\{f, g\}, g\}, \cdots, g\}, \tag{D.34}$$

with $L_g^0 f \equiv f$. We can define the exponential of $\mathbf{L}_g$ using the Taylor series for the exponential of a scalar:

$$\exp(\mathbf{L}_g) \equiv \sum_{n=0}^{\infty} \frac{\mathbf{L}_g^n}{n!}. \tag{D.35}$$

It is straightforward to show from the Jacobi identity (D.32e) that the **commutator** of $\mathbf{L}$ is

$$[\mathbf{L}_f, \mathbf{L}_g] \equiv \mathbf{L}_f\mathbf{L}_g - \mathbf{L}_g\mathbf{L}_f = \mathbf{L}_{\{g,f\}}. \tag{D.36}$$

Hamilton's equations (D.13) may be written

$$\dot{\mathbf{z}} = \{\mathbf{z}, H\} \quad \text{or} \quad \dot{\mathbf{z}} = \mathbf{L}_H\mathbf{z}; \tag{D.37}$$

here the Poisson bracket of a vector $\mathbf{z} = z_i\hat{\mathbf{n}}_i$ with components z_i along fixed unit vectors $\hat{\mathbf{n}}_i$ is understood to be $\{\mathbf{z}, H\} = \sum_i \hat{\mathbf{n}}_i\{z_i, H\}$.

The rate of change of any function $f(\mathbf{z}, t)$ along a trajectory determined by Hamilton's equations is

$$\frac{\mathrm{d}}{\mathrm{d}t}f[\mathbf{z}(t), t] = \{f, H\} + \frac{\partial f}{\partial t} = \mathbf{L}_H f + \frac{\partial f}{\partial t}. \tag{D.38}$$

We now show that Hamilton's equations have a simple formal solution. For brevity we assume that the Hamiltonian is autonomous or time-independent, that is, $H(\mathbf{q}, \mathbf{p}, t) = H(\mathbf{q}, \mathbf{p})$. This assumption is not restrictive, because any time-dependent Hamiltonian can be converted to a time-independent one in an extended phase space, as described in Box 2.1. The solution is motivated by the observation that the differential equation $\dot{z} = \lambda z$ has the solution $z(t) = \exp(\lambda t)z(0)$. The solution for Hamilton's equations has the analogous form

$$\mathbf{z}(t) = \mathbf{G}_t\mathbf{z}(0), \quad \text{where} \quad \mathbf{G}_t \equiv \exp(t\mathbf{L}_H) \tag{D.39}$$

is the **propagator**. To prove this, use equation (D.35) to write

$$\exp\left(t\mathbf{L}_H\right) = \sum_{n=0}^{\infty} \frac{t^n}{n!}\mathbf{L}_H^n. \tag{D.40}$$

Then

$$\begin{aligned}\dot{\mathbf{z}}(t) &= \frac{\mathrm{d}\mathbf{G}_t}{\mathrm{d}t}\mathbf{z}(0) = \frac{\mathrm{d}}{\mathrm{d}t}\sum_{n=0}^{\infty}\frac{t^n}{n!}\mathbf{L}_H^n\mathbf{z}(0) = \sum_{n=1}^{\infty}\frac{t^{n-1}}{(n-1)!}\mathbf{L}_H^n\mathbf{z}(0)\\ &= \mathbf{L}_H\sum_{m=0}^{\infty}\frac{t^m}{m!}\mathbf{L}_H^m\mathbf{z}(0) = \mathbf{L}_H\mathbf{z}(t),\end{aligned} \tag{D.41}$$

consistent with equation (D.37). In going from the first line to the second, we have made the replacement $n \to m + 1$.

D.5 Symplectic maps

The propagator G_t acts on a phase-space position $\mathbf{z}$ at time t_0 to produce a new position $\mathbf{Z} = \mathsf{G}_t\mathbf{z}$, the image of $\mathbf{z}$ under the Hamiltonian flow after time t. The **Jacobian matrix** $\mathbf{G}$ of the propagator is defined such that $G_{ij}(t, \mathbf{z})$ is the derivative of the ith component of $\mathsf{G}_t\mathbf{z}$ with respect to the jth component of $\mathbf{z}$:

$$G_{ij}(t, \mathbf{z}) = \frac{\partial Z_i}{\partial z_j} = \frac{\partial}{\partial z_j}[\mathsf{G}_t(\mathbf{z})]_i. \tag{D.42}$$

Using Hamilton's equations (D.13), the time derivative of the Jacobian matrix is

$$\begin{aligned}\frac{\partial G_{ij}}{\partial t} &= \frac{\partial^2 Z_i}{\partial t \partial z_j} = \frac{\partial}{\partial z_j}\sum_{k=1}^{2n} J_{ik}\frac{\partial H}{\partial Z_k} \\ &= \sum_{k,m=1}^{2n} J_{ik}\frac{\partial^2 H}{\partial Z_k \partial Z_m}\frac{\partial Z_m}{\partial z_j} = \sum_{k,m=1}^{2n} J_{ik}\frac{\partial^2 H}{\partial Z_k \partial Z_m}G_{mj}.\end{aligned} \tag{D.43}$$

This can be written more compactly in matrix notation as

$$\frac{\partial \mathbf{G}}{\partial t} = \mathbf{JHG}, \quad \text{where} \quad H_{km} \equiv \frac{\partial^2 H}{\partial Z_k \partial Z_m} \tag{D.44}$$

is the Hessian matrix of the Hamiltonian. Notice that $\mathbf{G}$ is a function of t and $\mathbf{z}$ while $\mathbf{H}$ is a function of $\mathbf{Z}$.

Now consider

$$\frac{\partial}{\partial t}\mathbf{G}^{\mathrm{T}}\mathbf{J}\mathbf{G} = \frac{\partial \mathbf{G}^{\mathrm{T}}}{\partial t}\mathbf{J}\mathbf{G} + \mathbf{G}^{\mathrm{T}}\mathbf{J}\frac{\partial \mathbf{G}}{\partial t} = (\mathbf{JHG})^{\mathrm{T}}\mathbf{JG} + \mathbf{G}^{\mathrm{T}}\mathbf{J}^2\mathbf{HG}. \tag{D.45}$$

Using the relations $(\mathbf{ABC})^{\mathrm{T}} = \mathbf{C}^{\mathrm{T}}\mathbf{B}^{\mathrm{T}}\mathbf{A}^{\mathrm{T}}$ and $\mathbf{J}^{\mathrm{T}}\mathbf{J} = -\mathbf{J}^2 = \mathbf{I}$ (eq. D.15) and observing that the Hessian is symmetric so $\mathbf{H}^{\mathrm{T}} = \mathbf{H}$, we conclude that the expression (D.45) vanishes. Since $\mathbf{G} = \mathbf{I}$ when $t = t_0$ we have[4]

$$\mathbf{G}^{\mathrm{T}}\mathbf{J}\mathbf{G} = \mathbf{J} \quad \text{or} \quad \mathbf{G}\mathbf{J}\mathbf{G}^{\mathrm{T}} = \mathbf{J}. \tag{D.46}$$

Any matrix satisfying these constraints, and by extension the operator G that gives rise to $\mathbf{G}$, is said to be **symplectic**. Thus the propagator for a phase-space flow governed by Hamilton's equations is symplectic.

[4] The second of these expressions is proved by left-multiplying the first by $\mathbf{GJ}$ and then right-multiplying by $\mathbf{G}^{-1}\mathbf{J}$ and using the relation $\mathbf{J}^2 = -\mathbf{I}$ (eq. D.15).

A small volume element $\mathrm{d}\mathbf{z}$ in phase space at time $t = 0$ is carried by the Hamiltonian flow into a new volume $\mathrm{d}\mathbf{Z}$ at time t. The two volumes are related by

$$\mathrm{d}\mathbf{Z} = |\det(\mathbf{G})|\mathrm{d}\mathbf{z}, \tag{D.47}$$

where $\det(\mathbf{G})$ is the determinant of the Jacobian matrix $\mathbf{G}$. From equation (D.46), $|\det(\mathbf{G}^{\mathrm{T}}\mathbf{J}\mathbf{G})| = |\det(\mathbf{J})| = 1$. Since $\det(\mathbf{AB}) = \det(\mathbf{A})\det(\mathbf{B})$ and $\det(\mathbf{A}^{\mathrm{T}}) = \det(\mathbf{A})$ for any square matrices $\mathbf{A}$ and $\mathbf{B}$, we conclude that $|\det(\mathbf{G})| = 1$, so $\mathrm{d}\mathbf{Z} = \mathrm{d}\mathbf{z}$. Thus volumes in canonical coordinates are conserved by a Hamiltonian flow (**Liouville's theorem**).

Note that if the operator $\mathsf{G} = \mathsf{G}_2\mathsf{G}_1$ is the composition of two operators, G_1 and G_2, then its Jacobian matrix is the product of their two Jacobian matrices, $\mathbf{G} = \mathbf{G}_2\mathbf{G}_1$. From this result it is straightforward to show that the composition of two symplectic operators is also symplectic.

For most purposes, the adjectives "symplectic" and "Hamiltonian" can be applied interchangeably to describe propagators in dynamical systems. The difference is that "symplectic" describes a local property of a phase-space flow because it depends only on the Jacobian matrix, while "Hamiltonian" describes a global property, the existence of a function $H(\mathbf{q}, \mathbf{p})$ that governs the flow through Hamilton's equations.

Informally, a symplectic map is a generalization of an area-preserving map to a space of $2n$ dimensions where $n > 1$.

D.6 Canonical transformations and coordinates

The advantage of Lagrange's equations is that they retain their form under any coordinate transformation $\mathbf{Q} = \mathbf{Q}(\mathbf{q}, t)$. Hamilton's equations allow an even wider range of transformations, of the form

$$\mathbf{Q} = \mathbf{Q}(\mathbf{q}, \mathbf{p}, t), \quad \mathbf{P} = \mathbf{P}(\mathbf{q}, \mathbf{p}, t). \tag{D.48}$$

Not all such transformations preserve the form of Hamilton's equations. Nevertheless, those that do—called **canonical transformations**—are sufficiently general to make Hamilton's equations a more powerful tool than Lagrange's for most problems in classical mechanics.

Consider a transformation from the original phase-space coordinates $\mathbf{z} = (\mathbf{q}, \mathbf{p})$ to some new set of phase-space coordinates $\mathbf{Z}$. Hamilton's equations (D.12) become

$$\frac{\mathrm{d}Z_i}{\mathrm{d}t} = \sum_j \frac{\partial Z_i}{\partial z_j}\frac{\mathrm{d}z_j}{\mathrm{d}t} = \sum_{jk} G_{ij}J_{jk}\frac{\partial H}{\partial z_k}$$

$$= \sum_{jkm} G_{ij} J_{jk} \frac{\partial H}{\partial Z_m} \frac{\partial Z_m}{\partial z_k} = \sum_m (\mathbf{G}\mathbf{J}\mathbf{G}^{\mathrm{T}})_{im} \frac{\partial H}{\partial Z_m}, \tag{D.49}$$

where $\mathbf{G}$ is the matrix with components G_{ij} defined in equation (D.42), although now $\mathbf{G}$ is derived from an operator **G** that describes a coordinate transformation at fixed time rather than time evolution in a fixed set of coordinates. Thus the form of Hamilton's equations is preserved if and only if equation (D.46) is satisfied; in other words canonical transformations are symplectic. The reverse is also true: all symplectic transformations are canonical. It is remarkable that these results are independent of the form of the Hamiltonian.

Any set of phase-space coordinates in which Hamilton's equations are valid is said to be a set of **canonical coordinates** or **variables**. The position $\mathbf{r}$ and Newtonian momentum $m\dot{\mathbf{r}}$ in Cartesian coordinates are canonical and so is any set of coordinates derived from these by a canonical transformation.

Phase-space volumes are conserved by canonical transformations; this is an obvious consequence of the proof of Liouville's theorem following equation (D.47).

The Poisson bracket of any two of the new coordinates is (eq. D.29)

$$\{Z_k, Z_l\} = \sum_{ij} J_{ij} \frac{\partial Z_k}{\partial z_i} \frac{\partial Z_l}{\partial z_j} = \sum_{ij} J_{ij} G_{ki} G_{lj} = (\mathbf{G}\mathbf{J}\mathbf{G}^{\mathrm{T}})_{kl} = J_{kl}, \tag{D.50}$$

where the last equation follows from (D.46) if the transformation is canonical and thus symplectic. Thus the Poisson bracket relation (D.30) is preserved in canonical transformations.

Similarly, let f and g be any two functions and write $\{f, g\}_{\mathbf{z}}$ for their Poisson bracket with respect to the variables $\mathbf{z} = (\mathbf{q}, \mathbf{p})$. Then if $\mathbf{Z}$ is a set of canonical coordinates,

$$\begin{aligned}
\{f, g\}_{\mathbf{z}} &= \sum_{ij} J_{ij} \frac{\partial f}{\partial z_i} \frac{\partial g}{\partial z_j} = \sum_{ijkl} J_{ij} \frac{\partial f}{\partial Z_k} \frac{\partial Z_k}{\partial z_i} \frac{\partial g}{\partial Z_l} \frac{\partial Z_l}{\partial z_j} \\
&= \sum_{ijkl} J_{ij} G_{ki} G_{lj} \frac{\partial f}{\partial Z_k} \frac{\partial g}{\partial Z_l} = \sum_{kl} J_{kl} \frac{\partial f}{\partial Z_k} \frac{\partial g}{\partial Z_l} = \{f, g\}_{\mathbf{Z}}.
\end{aligned} \tag{D.51}$$

Thus the Poisson bracket can be taken with respect to *any* set of canonical variables.

We now show that a transformation from $\mathbf{z} = (\mathbf{q}, \mathbf{p})$ to $\mathbf{Z} = (\mathbf{Q}, \mathbf{P})$ is canonical if

$$\mathbf{P} \cdot \mathrm{d}\mathbf{Q} = \mathbf{p} \cdot \mathrm{d}\mathbf{q} \quad \text{or} \quad \mathbf{Q} \cdot \mathrm{d}\mathbf{P} = \mathbf{q} \cdot \mathrm{d}\mathbf{p}. \tag{D.52}$$

We only prove the first of these results since the proof of the second is almost identical. In index notation, the first of equations (D.52) is

$$\sum_{ij} Z_i K_{ij} \mathrm{d}Z_j = \sum_{ij} z_i K_{ij} \mathrm{d}z_j, \quad \text{where} \quad \mathbf{K} \equiv \begin{bmatrix} \mathbf{0} & \mathbf{0} \\ \mathbf{I} & \mathbf{0} \end{bmatrix}. \tag{D.53}$$

We consider the new coordinates to be a function of the old, $\mathbf{Z}(\mathbf{z})$, with Jacobian matrix $G_{ij} = \partial Z_i / \partial z_j$. Then equation (D.53) can be written

$$\sum_{ijk} Z_i K_{ij} G_{jk} \mathrm{d}z_k = \sum_{ij} z_i K_{ij} \mathrm{d}z_j. \tag{D.54}$$

This is satisfied for arbitrary $\mathrm{d}\mathbf{z}$ if and only if

$$\sum_{ij} Z_i K_{ij} G_{jk} = \sum_{i} z_i K_{ik}. \tag{D.55}$$

Differentiating this result with respect to z_m gives

$$\sum_{ij} G_{im} K_{ij} G_{jk} + \sum_{ij} Z_i K_{ij} \frac{\partial^2 Z_j}{\partial z_k \partial z_m} = K_{mk}. \tag{D.56}$$

Now switch the indices m and k, and in the first sum switch the dummy indices i and j:

$$\sum_{ij} G_{im} K_{ji} G_{jk} + \sum_{ij} Z_i K_{ij} \frac{\partial^2 Z_j}{\partial z_k \partial z_m} = K_{km}. \tag{D.57}$$

Subtract (D.57) from (D.56) and use the result $K_{ij} - K_{ji} = -J_{ij}$, where $\mathbf{J}$ is the symplectic matrix. The result is $\mathbf{G}^{\mathrm{T}}\mathbf{J}\mathbf{G} = \mathbf{J}$, which implies that the transformation is symplectic and therefore canonical.

D.6.1 Generating functions

Let $(\mathbf{q}, \mathbf{p})$ and $(\mathbf{Q}, \mathbf{P})$ be canonical variables ("old" and "new," respectively), with $H(\mathbf{q}, \mathbf{p}, t) = H'(\mathbf{Q}, \mathbf{P}, t)$ the corresponding Hamiltonians. The trajectories in both sets of coordinates must be extrema of the action (D.2). Using equation (D.6), the action can be written

$$\int_{t_0}^{t_1} \mathrm{d}t \, [\mathbf{p} \cdot \dot{\mathbf{q}} - H(\mathbf{q}, \mathbf{p}, t)] = \int_{t_0}^{t_1} \mathrm{d}t \, [\mathbf{P} \cdot \dot{\mathbf{Q}} - H'(\mathbf{Q}, \mathbf{P}, t)]. \tag{D.58}$$

The two extremal trajectories must be the same, a requirement that can be satisfied if the two integrands satisfy the relation

$$\mathbf{p}\cdot\dot{\mathbf{q}} - H(\mathbf{q},\mathbf{p},t) = \mathbf{P}\cdot\dot{\mathbf{Q}} - H'(\mathbf{Q},\mathbf{P},t) + \frac{\mathrm{d}S}{\mathrm{d}t}, \tag{D.59}$$

where the **generating function** S is a function of the time and phase-space coordinates.

The generating function can be regarded as a function of $(\mathbf{q},\mathbf{p},t)$ or $(\mathbf{Q},\mathbf{P},t)$, or a mixture of the old and new phase-space coordinates. First suppose that $S \equiv S_1(\mathbf{q},\mathbf{Q},t)$. Then equation (D.59) becomes

$$\mathbf{p}\cdot\dot{\mathbf{q}} - H(\mathbf{q},\mathbf{p},t) = \mathbf{P}\cdot\dot{\mathbf{Q}} - H'(\mathbf{Q},\mathbf{P},t) + \frac{\partial S_1}{\partial t} + \frac{\partial S_1}{\partial \mathbf{q}}\cdot\dot{\mathbf{q}} + \frac{\partial S_1}{\partial \mathbf{Q}}\cdot\dot{\mathbf{Q}}. \tag{D.60}$$

This can be rewritten as

$$\left[\mathbf{p} - \frac{\partial S_1}{\partial \mathbf{q}}\right]\cdot\dot{\mathbf{q}} - \left[\mathbf{P} + \frac{\partial S_1}{\partial \mathbf{Q}}\right]\cdot\dot{\mathbf{Q}} - \left[H(\mathbf{q},\mathbf{p},t) - H'(\mathbf{Q},\mathbf{P},t) + \frac{\partial S_1}{\partial t}\right] = 0. \tag{D.61}$$

Since $\mathbf{q}$ and $\mathbf{Q}$ can be varied independently, this equation holds identically only if all of the terms in square brackets vanish, which requires

$$\mathbf{p} = \frac{\partial S_1}{\partial \mathbf{q}}, \quad \mathbf{P} = -\frac{\partial S_1}{\partial \mathbf{Q}}, \quad H'(\mathbf{Q},\mathbf{P},t) = H(\mathbf{q},\mathbf{p},t) + \frac{\partial S_1}{\partial t}. \tag{D.62}$$

Every well behaved function $S_1(\mathbf{q},\mathbf{Q})$ defines a canonical transformation through these relations. The definition is implicit since the generating function depends on both old and new coordinates, which is an inconvenience except in the simplest transformations.

Now let $S = S_2(\mathbf{q},\mathbf{P},t) - \mathbf{Q}\cdot\mathbf{P}$. The analog of equations (D.62) is

$$\mathbf{p} = \frac{\partial S_2}{\partial \mathbf{q}}, \quad \mathbf{Q} = \frac{\partial S_2}{\partial \mathbf{P}}, \quad H'(\mathbf{Q},\mathbf{P},t) = H(\mathbf{q},\mathbf{p},t) + \frac{\partial S_2}{\partial t}. \tag{D.63}$$

Let $S = S_3(\mathbf{Q},\mathbf{p},t) + \mathbf{q}\cdot\mathbf{p}$. Then (D.62) becomes

$$\mathbf{q} = -\frac{\partial S_3}{\partial \mathbf{p}}, \quad \mathbf{P} = -\frac{\partial S_3}{\partial \mathbf{Q}}, \quad H'(\mathbf{Q},\mathbf{P},t) = H(\mathbf{q},\mathbf{p},t) + \frac{\partial S_3}{\partial t}. \tag{D.64}$$

Finally, let $S = S_4(\mathbf{P},\mathbf{p},t) + \mathbf{q}\cdot\mathbf{p} - \mathbf{Q}\cdot\mathbf{P}$. Then (D.62) becomes

$$\mathbf{q} = -\frac{\partial S_4}{\partial \mathbf{p}}, \quad \mathbf{Q} = \frac{\partial S_4}{\partial \mathbf{P}}, \quad H'(\mathbf{Q},\mathbf{P},t) = H(\mathbf{q},\mathbf{p},t) + \frac{\partial S_4}{\partial t}. \tag{D.65}$$

D.7 Angle-action variables

In rare but important cases, the trajectories in a Hamiltonian system with n degrees of freedom (i.e., a $2n$-dimensional phase space) admit n independent **integrals of motion**, that is, phase-space functions $C_i(\mathbf{q},\mathbf{p})$, $i = 1,\ldots,n$ that are constant along all trajectories. Such systems are said to be **integrable** (see Appendix D.8). Integrable systems include all autonomous Hamiltonians with one degree of freedom, a test particle orbiting in a spherically symmetric time-independent potential, and the gravitational two-body problem, but not the gravitational N-body problem for $N \geq 3$.

When n integrals of motion are present, the trajectory of the system is restricted to an n-dimensional manifold in the $2n$-dimensional phase space. It can be shown that this manifold is an n-dimensional torus (e.g., Arnold 1989).

Much of the power of Hamiltonian mechanics lies in the ability to choose coordinates and momenta that simplify the dynamics. The dynamics is simplest if the momenta are integrals of motion, and because the manifold on which the integrals are fixed is a torus, it is natural for the coordinates to range from 0 to 2π. Such pairs of coordinates and momenta are called **angle-action variables**. If the Hamiltonian has dimensions of energy (mass times velocity squared) then the actions must have dimensions of angular momentum, or energy divided by frequency, to satisfy Hamilton's equations.

If we denote the angle-action variables by $(\boldsymbol{\theta},\mathbf{J})$, Hamilton's equations read

$$\frac{\mathrm{d}\mathbf{J}}{\mathrm{d}t} = -\frac{\partial H}{\partial \boldsymbol{\theta}}, \quad \frac{\mathrm{d}\boldsymbol{\theta}}{\mathrm{d}t} = \frac{\partial H}{\partial \mathbf{J}}. \tag{D.66}$$

Since the actions are integrals of motion, $\partial H/\partial\boldsymbol{\theta}$ must vanish, so the Hamiltonian depends only on the actions, $H = H(\mathbf{J})$. The equation of motion for the angles is then easy to solve,

$$\boldsymbol{\theta}(t) = \boldsymbol{\theta}_0 + \boldsymbol{\Omega}t, \quad \text{where} \quad \boldsymbol{\Omega}(\mathbf{J}) = \frac{\partial H}{\partial \mathbf{J}}. \tag{D.67}$$

Thus knowing the n integrals (the actions) is sufficient to allow integrating the $2n$ Hamilton's equations exactly.

To find the canonical transformation from variables $(\mathbf{q},\mathbf{p})$ to angle-action coordinates $(\boldsymbol{\theta},\mathbf{J})$, we assume that the transformation is described by a generating function $S_2(\mathbf{q},\mathbf{J})$. Then from equations (D.63),

$$\mathbf{p} = \frac{\partial S_2(\mathbf{q},\mathbf{J})}{\partial \mathbf{q}}, \quad \boldsymbol{\theta} = \frac{\partial S_2(\mathbf{q},\mathbf{J})}{\partial \mathbf{J}}. \tag{D.68}$$

On the n-torus there are n topologically distinct closed curves or cycles γ_i, that is, curves that cannot be continuously deformed into one another. The first of equations (D.68) shows that along one of these curves

$$S_2(\mathbf{q}, \mathbf{J}) = \int_{\mathbf{q}_0}^{\mathbf{q}} \mathrm{d}\mathbf{q}' \cdot \mathbf{p}(\mathbf{J}, \mathbf{q}') + c(\mathbf{J}), \tag{D.69}$$

where $\mathbf{q}_0$ is an arbitrary point on the torus, $c(\mathbf{J})$ is an arbitrary function, and the integral is unchanged by continuous deformations of the path from $\mathbf{q}_0$ to $\mathbf{q}$. After a circuit of the closed curve γ_i the generating function has changed by

$$\Delta S_2(\mathbf{J}) = \oint_{\gamma_i} \mathrm{d}\mathbf{q}' \cdot \mathbf{p}(\mathbf{J}, \mathbf{q}'). \tag{D.70}$$

Thus the generating function is multi-valued on the torus. The increment in S_2 in one circuit of the curve γ_i is the area in phase space enclosed by the curve, and this area is independent of deformations of the curve. Because S_2 is multi-valued, $\boldsymbol{\theta} = \partial S_2 / \partial \mathbf{J}$ is also multi-valued. To satisfy the condition that the angle variables vary from 0 to 2π during one circuit of the closed curve γ_i we need

$$2\pi = \frac{\partial \Delta S_2}{\partial J_i}; \tag{D.71}$$

here we have associated the ith component of the action vector $\mathbf{J}$ with the area enclosed by the curve γ_i. This equation is satisfied if we choose

$$J_i = \frac{1}{2\pi} \oint_{\gamma_i} \mathrm{d}\mathbf{q}' \cdot \mathbf{p}(\mathbf{J}, \mathbf{q}'). \tag{D.72}$$

D.8 Integrable and non-integrable systems

In this section we summarize some geometric features of trajectories in Hamiltonian systems, mostly without providing formal proofs—for more detail, see books such as Arnold (1989), Tabor (1989), Lichtenberg & Lieberman (1992) and Dumas (2014). We consider only autonomous Hamiltonians since all Hamiltonians can be converted to this form by extending the phase space (Box 2.1). We assume that the system has n degrees of freedom so the phase space has $2n$ dimensions.

Systems with one degree of freedom are straightforward to analyze, at least in principle. The Hamiltonian $H(q, p) = E$ where the energy E is a constant or integral of motion, so this equation can be inverted to give the momentum as a function of the coordinate and energy,

$$p = p(q, E). \tag{D.73}$$

Using this relation, the function $\partial H/\partial p$ can be rewritten as a function of q and E, which we call $g(q, E)$; then Hamilton's equation $\dot{q} = \partial H/\partial p = g(q, E)$, which can be integrated to give

$$t = \int_{q_0}^{q} \frac{\mathrm{d}q'}{g(q', E)}, \tag{D.74}$$

and this can be inverted to give $q(t, E)$. Thus the determination of the trajectory has been reduced to the determination of inverse functions and a quadrature (the evaluation of an indefinite integral). This process is analytic in simple cases and can always be done numerically.

Systems with two or more degrees of freedom can sometimes be solved similarly. Consider a canonical transformation to new variables $(\mathbf{Q}, \mathbf{P})$, defined by a generating function $S_2(\mathbf{q}, \mathbf{P})$. We seek variables such that the new Hamiltonian $H'(\mathbf{Q}, \mathbf{P})$ depends only on the momenta $\mathbf{P}$. Using equations (D.63) this requirement can be written

$$H\left[\mathbf{q}, \frac{\partial S_2}{\partial \mathbf{q}}(\mathbf{q}, \mathbf{P})\right] = H'(\mathbf{P}). \tag{D.75}$$

This is the **Hamilton–Jacobi equation**, a first-order partial differential equation for the generating function $S_2(\mathbf{q}, \mathbf{P})$. Since $H'(\mathbf{P})$ is independent of the coordinates $\mathbf{Q}$, Hamilton's equation $\dot{\mathbf{P}} = -\partial H'/\partial \mathbf{Q}$ implies that each of the n momenta P_i is an integral of motion. Moreover Hamilton's equation $\dot{\mathbf{Q}} = \partial H'/\partial \mathbf{P}$ implies that $\mathbf{Q}$ increases linearly with time, $\mathbf{Q} = \boldsymbol{\Omega} t + \text{const}$, where $\boldsymbol{\Omega} = \partial H'/\partial \mathbf{P}$ is constant because it depends only on the momenta. Therefore we have solved the equations of motion if we can find the generating function $S_2(\mathbf{q}, \mathbf{P})$.

In general the Hamilton–Jacobi equation is no easier to solve than Hamilton's equations of motion in the original coordinates. However, in some systems the Hamilton–Jacobi equation is separable, that is,

$$H\left[\mathbf{q}, \frac{\partial S_2}{\partial \mathbf{q}}(\mathbf{q}, \mathbf{P})\right] = \sum_{i=1}^{n} H_i\left[q_i, \frac{\partial S_{2i}}{\partial q_i}(q_i, \mathbf{P})\right]. \tag{D.76}$$

Since each coordinate q_i is independent, equation (D.75) is equivalent to the n separate Hamilton–Jacobi equations

$$H_i\left[q_i, \frac{\partial S_{2i}}{\partial q_i}(q_i, \mathbf{P})\right] = H_i'(\mathbf{P}), \quad \text{where} \quad \sum_{i=1}^{n} H_i'(\mathbf{P}) = H'(\mathbf{P}). \tag{D.77}$$

Since the momenta $\mathbf{P}$ are constants along any trajectory, each of these equations can be solved by a quadrature to yield the generating function $S_{2i}(q_i, \mathbf{P})$ and thus the relation between the old and new coordinates and momenta.

Hamiltonians of this kind are said to be **separable** or **integrable**. Any integrable system with n degrees of freedom has n integrals of motion, the n components of the momentum $\mathbf{P}$. The most important examples are the Hamiltonian for the n-dimensional harmonic oscillator and the Hamiltonian for motion of a test particle in a central potential, including the potential $-\mathbb{G}M/r$ that governs the Kepler problem.

In the special case of angle-action variables (Appendix D.7), the conserved momenta are the actions, and the phase space is periodic in the angle coordinates with period 2π. Each trajectory lies on an n-dimensional torus; the surface of the torus is defined by the n actions, and the position on the surface is defined by the n angles. If the frequencies $\mathrm{d}\theta_i/\mathrm{d}t = \Omega_i$ are incommensurate, the trajectory eventually densely covers the whole surface of the torus, while if they are commensurate—as in the Kepler problem—the trajectory traces a 1-dimensional curve on the torus. The tori are sometimes called **invariant tori** since a trajectory that is on the torus at any time can never leave it.

Most Hamiltonians are not integrable. However, many of the systems that are important in celestial mechanics are **near-integrable** in the sense that their Hamiltonians can be written in the form

$$H(\boldsymbol{\theta}, \mathbf{J}) = H_0(\mathbf{J}) + \epsilon H_1(\boldsymbol{\theta}, \mathbf{J}), \tag{D.78}$$

where H_0 is integrable, $(\boldsymbol{\theta}, \mathbf{J})$ are angle-action variables for H_0, and ϵ is a small parameter. The most important example from our perspective is a multi-planet system, in which $H_0(\mathbf{J})$ is the sum of the Kepler Hamiltonians for each planet, and $\epsilon H_1(\boldsymbol{\theta}, \mathbf{J})$ is the gravitational potential due to interactions among the planets. The system is near-integrable because planetary masses are much smaller than stellar masses.

The centerpiece of the study of near-integrable Hamiltonian systems is the **Kolmogorov–Arnold–Moser** or **KAM theorem**. Loosely speaking, the theorem states that if (i) $H_0(\mathbf{J})$ is a nonlinear function of $\mathbf{J}$; (ii) $H_1(\boldsymbol{\theta}, \mathbf{J})$ is sufficiently smooth; and (iii) ϵ is sufficiently small, then most of phase space remains filled with invariant tori.[5] In other words most of the invariant tori are distorted by the perturbation but not destroyed. Orbits on the surviving tori are called **regular orbits**.

---

[5] "Most" means that the measure of the phase space in which the tori do not survive is small and goes to zero as $\epsilon \to 0$.

Trajectories that do not lie on an invariant torus are extremely sensitive to small changes in the initial conditions or the parameters of the Hamiltonian. Specifically, if $\mathbf{z}(t)$ is an orbit that does not lie on an invariant torus, and $\delta\mathbf{z}(t)$ is its separation from a nearby orbit, then as $|t| \to \infty$ the separation grows exponentially, $|\delta\mathbf{z}(t)| \propto \exp(|t|/t_{\rm L})$ so long as $|\delta\mathbf{z}(t)| \ll |\mathbf{z}(t)|$. Trajectories with this property are said to be **chaotic**. The growth time $t_{\rm L}$ is called the **Liapunov time**. This exponential divergence implies that, in practice, the future or past of the trajectory cannot be computed over timescales $\gg t_{\rm L}$ even though the trajectory is deterministic.

The behavior of chaotic trajectories in systems with 2 degrees of freedom is fundamentally different from the behavior in phase spaces of higher dimension. With 2 degrees of freedom, the phase space has 4 dimensions, and conservation of energy restricts the motion to a 3-dimensional manifold in phase space, sometimes called the energy surface. A 2-torus has 2 dimensions, so an invariant torus divides the 3-dimensional space into volumes inside and outside the torus. Chaotic trajectories inside the torus are therefore permanently trapped inside, just as air molecules inside an inner tube are trapped inside the tube. In regions of phase space with a significant density of invariant tori, the chaotic trajectories are trapped between two tori, so their actions always remain close to their initial values. In other words the exponential divergence of nearby chaotic orbits is seen only in the angles.

In contrast, when the number of degrees of freedom $n \geq 3$, the energy surface has $2n - 1$ dimensions, the tori have n dimensions, and the difference between the dimensionality of the energy surface and the tori is $n - 1 > 1$. Thus the tori do not divide the energy surface, so any chaotic trajectory is free to wander throughout the available phase space. The actions will therefore diffuse away from their initial values, a process usually called **Arnold diffusion**, although this is not diffusion in the usual sense of statistical mechanics. In regions where the invariant tori occupy a large fraction of phase space, the rate of Arnold diffusion may be extremely slow[6] and impractical to calculate except by numerical experiments.

All of these results are strictly valid only when the strength of the perturbing Hamiltonian $\epsilon H_1(\boldsymbol{\theta}, \mathbf{J})$ is extremely small—far smaller than implied by typical ratios of planetary to stellar masses. Nevertheless the qualitative picture that emerges from the KAM theorem appears to remain valid for relatively large values of the perturbation parameter ϵ. Numerical experiments show that as ϵ grows, the density of invariant tori declines until eventually the tori disappear in macroscopically large regions of phase space. The disappearance of the last torus in a region of phase space

[6] Typically the instability time is of order $\exp[\alpha\epsilon^{-\beta}]$, where α and β are positive constants and $\beta < 1$ (see Yalinewich & Petrovich 2020).

allows orbits to wander rapidly throughout that region, a phenomenon called **global chaos**. The generic structure of phase space is often described as "islands" of nested invariant tori surrounded by a "sea" of chaotic orbits. The sea can also be thought of as a *single* chaotic orbit, since any one orbit will eventually pass arbitrarily close to every point in the sea. For examples of this process, see Figure F.1 for the standard map, and Figures 3.3 and 3.4 for the circular, restricted three-body problem. The value of the perturbation parameter ϵ required for the emergence of a chaotic sea can be predicted approximately by the resonance overlap criterion, described in Appendix F.

D.9 The averaging principle

We wish to study the behavior of trajectories in the near-integrable Hamiltonian (D.78), having n degrees of freedom. We assume that the unperturbed Hamiltonian $H_0(\mathbf{J})$ is independent of k of the n actions, so we may write the actions as $\mathbf{J} \equiv (J_{s,1}, \ldots, J_{s,k}, J_{f,1}, \ldots, J_{f,n-k})$, where $\partial H_0/\partial J_{s,i} = 0$ and $\partial H_0/\partial J_{f,i} \neq 0$. The angles conjugate to $\mathbf{J}_s$ and $\mathbf{J}_f$ are $\boldsymbol{\theta}_s$ and $\boldsymbol{\theta}_f$. The subscript "s" stands for "slow," since the rate of change of $\boldsymbol{\theta}_s$ under the influence of H_0 is $\dot{\boldsymbol{\theta}}_s = \partial H_0/\partial \mathbf{J}_s = \mathbf{0}$, while the rate of change of the components of $\boldsymbol{\theta}_f$ (the "fast" angles) is nonzero.

The **averaging principle** consists of the replacement of the Hamiltonian (D.78) by the averaged Hamiltonian

$$\overline{H}(\boldsymbol{\theta}_s, \mathbf{J}) = H_0(\mathbf{J}) + \epsilon \overline{H}_1(\boldsymbol{\theta}_s, \mathbf{J}), \tag{D.79}$$

where

$$\overline{H}_1(\boldsymbol{\theta}_s, \mathbf{J}) \equiv \frac{1}{(2\pi)^{n-k}} \int \mathrm{d}\boldsymbol{\theta}_f \, H_1(\boldsymbol{\theta}_s, \boldsymbol{\theta}_f, \mathbf{J}); \tag{D.80}$$

in words, $\overline{H}(\boldsymbol{\theta}_s, \mathbf{J})$ is the average of $H(\boldsymbol{\theta}, \mathbf{J})$ over the fast angles. Since $\overline{H}(\boldsymbol{\theta}_s, \mathbf{J})$ is independent of the fast angles, their conjugate actions $\mathbf{J}_f$ are conserved, so the averaged Hamiltonian can be considered to have only k degrees of freedom.

The averaging principle can also be applied either to Hamiltonians in which the slow frequencies $\partial H_0/\partial J_s$ are small but not zero, or to time-dependent Hamiltonians of the form

$$H(\boldsymbol{\theta}, \mathbf{J}, t) = H_0(\mathbf{J}, t) + \epsilon H_1(\boldsymbol{\theta}, \mathbf{J}, t), \tag{D.81}$$

so long as the changes in the Hamiltonian are slow compared to the orbital frequencies $\boldsymbol{\Omega} = \partial H_0/\partial \mathbf{J}_f$. Note that the averaging principle may fail if there is a resonance or near-resonance between two of the fast frequencies.

Arnold (1989, p. 292) has commented: "this principle is neither a theorem, an axiom, nor a definition, but rather a physical proposition, i.e., a vaguely formulated and, strictly speaking, untrue assertion." For example, if the original Hamiltonian has $n = 2$ degrees of freedom and $k = 1$ slow action, a significant fraction of phase space may be covered by chaotic orbits, but the averaged Hamiltonian will have only one degree of freedom so it is always integrable. Despite these concerns, there is strong analytic and numerical evidence that the averaging principle provides an accurate description of the slow evolution of dynamical systems that are close enough to being integrable.

D.10 Adiabatic invariants

Informally, an **adiabatic invariant** of a dynamical system is a function of the phase-space position that remains nearly constant along a trajectory when the parameters of the system are changed slowly.

In more detail, suppose that the Hamiltonian of the system is $H(\mathbf{z}, \lambda)$, where $\mathbf{z} = (\mathbf{q}, \mathbf{p})$ and λ is a parameter (e.g., the mass of the central star in a planetary system). For simplicity we focus on the case of a single parameter, although the arguments below can be generalized to several parameters.

During the interval $0 \leq t \leq \tau$, the parameter varies as $\lambda(t) = f(t/\tau)$ with τ much larger than the characteristic dynamical time in the system (e.g., the orbital period P of a planet). Over a short period of time near $t = 0$, the trajectory of the system will not differ significantly from a trajectory of the "frozen" system in which λ is fixed at $\lambda_0 = f(0)$, and we call this frozen trajectory $\mathbf{z}_0(t)$. Similarly, near $t = \tau$ the trajectory will be close to a trajectory in the frozen system with $\lambda_1 = f(1)$, which we call $\mathbf{z}_1(t)$. An adiabatic invariant is a scalar function of λ and the frozen trajectory, $A[\mathbf{z}(t), \lambda]$, with the property that $A[\mathbf{z}_1(t), \lambda_1] - A[\mathbf{z}_0(t), \lambda_0] \to 0$ as $\tau \to \infty$.

The existence of adiabatic invariants is remarkable, because it implies that some properties of the trajectory at $t = \tau$ are independent of the history of the evolution of the parameter from λ_0 to λ_1. In other words the adiabatic invariant is independent of the function $f(x)$ for $0 < x < 1$, so long as $f(0)$ and $f(1)$ are fixed, $f(x)$ is sufficiently smooth, and there are no resonances.

To find the adiabatic invariants of an integrable system, we revisit the derivation of angle-action variables in Appendix D.7. Let $S_2(\mathbf{q}, \mathbf{J}, \lambda)$ be the generating function for the angles and actions, now time-dependent because of the time dependence of λ. Then equation (D.68) is generalized to

$$\mathbf{p} = \frac{\partial S_2(\mathbf{q}, \mathbf{J}, \lambda)}{\partial \mathbf{q}}, \quad \boldsymbol{\theta} = \frac{\partial S_2(\mathbf{q}, \mathbf{J}, \lambda)}{\partial \mathbf{J}}, \quad H'(\boldsymbol{\theta}, \mathbf{J}, t) = H(\mathbf{J}, \lambda) + \frac{\partial S_2}{\partial \lambda}\frac{\mathrm{d}\lambda}{\mathrm{d}t}. \tag{D.82}$$

The generating function S_2 is multi-valued and increases by $\Delta S_2 = 2\pi J_i$ when θ_i increases from 0 to 2π. Therefore $S^* \equiv S_2 - 2\pi\mathbf{J}\cdot\boldsymbol{\theta}$ is periodic in $\boldsymbol{\theta}$ with zero average over any angle, and we can write

$$H'(\boldsymbol{\theta}, \mathbf{J}, t) = H(\mathbf{J}, \lambda) + \frac{\partial S^*}{\partial \lambda}\frac{\mathrm{d}\lambda}{\mathrm{d}t}. \tag{D.83}$$

Since $\mathrm{d}\lambda/\mathrm{d}t$ is of order λ/τ, with τ large, this Hamiltonian is near-integrable and has the form (D.81). Therefore we can apply the averaging principle, which states that the trajectory can be approximated by

$$\overline{H}'(\boldsymbol{\theta}, \mathbf{J}, t) = H(\mathbf{J}, \lambda) + \frac{\partial \overline{S}^*}{\partial \lambda}\frac{\mathrm{d}\lambda}{\mathrm{d}t}. \tag{D.84}$$

Since the average of S^* over any angle vanishes, $\overline{S}^* = 0$. Therefore $\overline{H}'(\boldsymbol{\theta}, \mathbf{J}, t) = H(\mathbf{J}, \lambda)$, so the averaged Hamiltonian is independent of all the angles. Thus the actions are constant in the averaged Hamiltonian, which implies that they are adiabatic invariants in the original (unaveraged) Hamiltonian. Since the action is proportional to the area enclosed by the orbit in phase space (eq. D.72), an equivalent statement is that the area enclosed by an orbit is an adiabatic invariant.

If the change in parameters is sufficiently smooth (i.e., if $f(x)$ is analytic) then the actions are conserved to remarkably high accuracy, with an error of order $\exp(-k\tau/P)$ where k is a positive constant. See Lichtenberg & Lieberman (1992) or Henrard (1993) for a more thorough discussion of adiabatic invariants in classical mechanics.

D.11 Rigid bodies

The **inertia tensor** I of a body with density distribution $\rho(\mathbf{r})$ is defined by

$$I_{ij} \equiv \int \mathrm{d}\mathbf{r}\, \rho(\mathbf{r})(r^2\delta_{ij} - r_i r_j), \tag{D.85}$$

where $\mathbf{r} = (r_1, r_2, r_3) = (x, y, z)$ is the position relative to the center of mass. An alternative formula is

$$\mathbf{I} \equiv \int \mathrm{d}\mathbf{r}' \rho(\mathbf{r}') \begin{bmatrix} y^2 + z^2 & -xy & -xz \\ -xy & z^2 + x^2 & -yz \\ -xz & -yz & x^2 + y^2 \end{bmatrix}. \tag{D.86}$$

The **principal axes** of the body are those in which the inertia tensor is diagonal; it is always possible to find such axes because the inertia tensor is symmetric ($I_{ij} = I_{ji}$). In this **principal-axis frame**,

$$\begin{aligned} I_{xx} = A &\equiv \int \mathrm{d}\mathbf{r}\, \rho(\mathbf{r})(y^2 + z^2), \\ I_{yy} = B &\equiv \int \mathrm{d}\mathbf{r}\, \rho(\mathbf{r})(x^2 + z^2), \\ I_{zz} = C &\equiv \int \mathrm{d}\mathbf{r}\, \rho(\mathbf{r})(x^2 + y^2), \end{aligned} \tag{D.87}$$

and $I_{xy} = I_{xz} = I_{yz} = 0$.

D.11.1 Rotation of a rigid body

Consider a rigid body that rotates about its center of mass at $\mathbf{r} = \mathbf{0}$. Its kinetic energy and spin angular momentum are

$$T = \tfrac{1}{2} \int \mathrm{d}\mathbf{r}\, \rho(\mathbf{r}) |\mathbf{v}(\mathbf{r})|^2, \quad \mathbf{S} = \int \mathrm{d}\mathbf{r}\, \rho(\mathbf{r}) \mathbf{r} \times \mathbf{v}, \tag{D.88}$$

where $\mathbf{v}(\mathbf{r})$ is the velocity at $\mathbf{r}$. If the angular velocity of the body is $\boldsymbol{\omega}$, then $\mathbf{v}(\mathbf{r}) = \boldsymbol{\omega} \times \mathbf{r}$ (eq. D.17). Using the vector identity (B.9a), we have

$$T = \tfrac{1}{2} \int \mathrm{d}\mathbf{r}\, \rho(\mathbf{r}) \mathbf{v} \cdot (\boldsymbol{\omega} \times \mathbf{r}) = \tfrac{1}{2} \int \mathrm{d}\mathbf{r}\, \rho(\mathbf{r}) \boldsymbol{\omega} \cdot (\mathbf{r} \times \mathbf{v}) = \tfrac{1}{2} \boldsymbol{\omega} \cdot \mathbf{S}. \tag{D.89}$$

Moreover using the identity (B.9b),

$$\mathbf{S} = \int \mathrm{d}\mathbf{r}\, \rho(\mathbf{r}) \mathbf{r} \times (\boldsymbol{\omega} \times \mathbf{r}) = \int \mathrm{d}\mathbf{r}\, \rho(\mathbf{r}) [r^2 \boldsymbol{\omega} - \mathbf{r}(\boldsymbol{\omega} \cdot \mathbf{r})]; \tag{D.90}$$

thus

$$S_j = \sum_{k=1}^{3} I_{jk} \omega_k \quad \text{or} \quad \mathbf{S} = \mathbf{I} \cdot \boldsymbol{\omega}, \tag{D.91}$$

where $\mathbf{I}$ is the inertia tensor defined in equation (D.85). Combining (D.89) and (D.91), we have

$$T = \tfrac{1}{2} \sum_{j,k=1}^{3} \omega_j I_{jk} \omega_k = \tfrac{1}{2} \boldsymbol{\omega} \cdot \mathbf{I} \cdot \boldsymbol{\omega}. \tag{D.92}$$

D.11.2 Euler's equations

If the body is subjected to an external torque $\mathbf{N}$, then the rate of change of the spin angular momentum in an inertial frame is $(\mathrm{d}\mathbf{S}/\mathrm{d}t)_{\text{in}} = \mathbf{N}$. We transform to the rotating principal-axis frame; in this frame the rate of change of angular momentum is (eq. D.16)

$$\left(\frac{\mathrm{d}\mathbf{S}}{\mathrm{d}t}\right)_{\text{rot}} = \left(\frac{\mathrm{d}\mathbf{S}}{\mathrm{d}t}\right)_{\text{in}} - \boldsymbol{\omega} \times \mathbf{S} = \mathbf{N} - \boldsymbol{\omega} \times \mathbf{S}. \tag{D.93}$$

Using equation (D.91),

$$\left(\frac{\mathrm{d}\mathbf{I}\cdot\boldsymbol{\omega}}{\mathrm{d}t}\right)_{\text{rot}} + \boldsymbol{\omega} \times (\mathbf{I}\cdot\boldsymbol{\omega}) = \mathbf{N}. \tag{D.94}$$

In the principal-axis system, the inertia tensor is diagonal and time-independent, so we arrive at **Euler's equations**

$$\begin{aligned} I_{11}\dot{\omega}_1 + (I_{33} - I_{22})\omega_2\omega_3 &= N_1, \\ I_{22}\dot{\omega}_2 + (I_{11} - I_{33})\omega_3\omega_1 &= N_2, \\ I_{33}\dot{\omega}_3 + (I_{22} - I_{11})\omega_1\omega_2 &= N_3. \end{aligned} \tag{D.95}$$

D.11.3 Hamilton's equations for a rotating rigid body

The Euler angles α, β, γ, described in Appendix B.6, can be used as generalized coordinates that relate an inertial reference frame to a frame fixed in the body. We choose these such that the body frame (x', y', z') is the principal-axis frame. Then from equations (B.65b), (D.87), and (D.92), the kinetic energy is

$$\begin{aligned} T &= \tfrac{1}{2}A\omega_{x'}^2 + \tfrac{1}{2}B\omega_{y'}^2 + \tfrac{1}{2}C\omega_{z'}^2 \\ &= \tfrac{1}{2}A(\dot{\alpha}\sin\beta\sin\gamma + \dot{\beta}\cos\gamma)^2 + \tfrac{1}{2}B(\dot{\alpha}\sin\beta\cos\gamma - \dot{\beta}\sin\gamma)^2 \\ &\quad + \tfrac{1}{2}C(\dot{\alpha}\cos\beta + \dot{\gamma})^2. \end{aligned} \tag{D.96}$$

The Lagrangian can be written $L(\alpha, \beta, \gamma, \dot{\alpha}, \dot{\beta}, \dot{\gamma}) = T - \Phi(\alpha, \beta, \gamma)$, where Φ is the potential energy as a function of the orientation of the body. The generalized momenta are

$$\begin{aligned} p_\alpha &= \frac{\partial L}{\partial \dot{\alpha}} = A\omega_{x'}\sin\beta\sin\gamma + B\omega_{y'}\sin\beta\cos\gamma + C\omega_{z'}\cos\beta, \\ p_\beta &= \frac{\partial L}{\partial \dot{\beta}} = A\omega_{x'}\cos\gamma - B\omega_{y'}\sin\gamma, \end{aligned}$$

$$p_\gamma = \frac{\partial L}{\partial \dot{\gamma}} = C\omega_{z'}. \tag{D.97}$$

It is straightforward to solve these for $\boldsymbol{\omega}$ in terms of the momenta:

$$\begin{aligned}
\omega_{x'} &= \frac{1}{A \sin\beta}(p_\alpha \sin\gamma + p_\beta \sin\beta\cos\gamma - p_\gamma \cos\beta\sin\gamma), \\
\omega_{y'} &= \frac{1}{B \sin\beta}(p_\alpha \cos\gamma - p_\beta \sin\beta\sin\gamma - p_\gamma \cos\beta\cos\gamma), \\
\omega_{z'} &= \frac{p_\gamma}{C}.
\end{aligned} \tag{D.98}$$

Since $S_{z'} = C\omega_{z'}$, the momentum p_γ is simply the projection of the total spin angular momentum along the body-fixed z'-axis. It is straightforward to show, using either equation (B.61) or equation (7.39), that p_α is the projection of the spin angular momentum along the z-axis of the original (x, y, z) reference frame.

The Hamiltonian $H_{\mathrm{rot}}(\alpha, \beta, \gamma, p_\alpha, p_\beta, p_\gamma) = \mathbf{p} \cdot \dot{\mathbf{q}} - L = \dot{\mathbf{q}} \cdot (\partial L/\partial \dot{\mathbf{q}}) - L$. Now $L = T - \Phi$, Φ is independent of $\dot{\mathbf{q}}$, and T is a quadratic function of $\dot{\mathbf{q}}$, so it is straightforward to show that $H = T + \Phi$. Thus

$$\begin{aligned}
H_{\mathrm{rot}} = &\frac{1}{2A\sin^2\beta}(p_\alpha \sin\gamma + p_\beta \sin\beta\cos\gamma - p_\gamma \cos\beta\sin\gamma)^2 \\
&+ \frac{1}{2B\sin^2\beta}(p_\alpha \cos\gamma - p_\beta \sin\beta\sin\gamma - p_\gamma \cos\beta\cos\gamma)^2 \\
&+ \frac{1}{2C}p_\gamma^2 + \Phi(\alpha, \beta, \gamma).
\end{aligned} \tag{D.99}$$

If the potential Φ is zero or independent of α, the Hamiltonian is independent of the coordinate α, so the momentum p_α is conserved.

The Hamiltonian can be simplified by a canonical transformation to the Andoyer variables described in §7.3.

Appendix E

Hill and Delaunay variables

A basic task in celestial mechanics is to develop canonical coordinates and momenta in which trajectories in the Kepler potential can be described as simply as possible. This can be done by finding angle-action variables (Appendix D.7) for the Kepler Hamiltonian (1.80). An additional benefit of angle-action variables is that they are adiabatic invariants (Appendix D.10) and therefore are conserved during slow variations in the parameters of the Hamiltonian.

E.1 Hill variables

We begin with the usual Cartesian coordinates $\mathbf{r}$ and $\mathbf{v} = \dot{\mathbf{r}}$ for a test particle in 6-dimensional phase space.[1] We define the **Hill variables**

$$\mathbf{q} = (r, w, \Omega), \quad \mathbf{p} = (\dot{r}, L, L_z). \tag{E.1}$$

Here r and $\dot{r}$ are the radius and radial velocity, $\mathbf{L} = \mathbf{r} \times \mathbf{v}$ is the angular momentum per unit mass, $L = |\mathbf{L}|$, $L_z = L \cos I$, $w = \omega + f$, and ω, f, Ω and I are the usual orbital elements seen in Figure 1.2: argument of periapsis, true anomaly, longitude of the ascending node, and inclination.

We now show that the Hill variables are a canonical coordinate momentum pair. The position vector $\mathbf{r}$ is completely determined by the orbital elements r, w, Ω and I. A small change $\mathrm{d}r$ produces a change in position $\mathrm{d}r\hat{\mathbf{r}}$. A change $\mathrm{d}w$ corresponds

[1] This derivation partly follows Laskar (2017).

to an infinitesimal rotation about the orbit normal $\hat{\mathbf{L}}$, which gives rise to a change in position $\mathrm{d}\mathbf{r} = \mathrm{d}w\,\hat{\mathbf{L}} \times \mathbf{r}$. Similarly, a change $\mathrm{d}I$ corresponds to a rotation about the line of nodes, whose direction is defined by a unit vector $\hat{\mathbf{n}} = (\cos\Omega, \sin\Omega, 0)$ (Figure 1.2), and gives rise to a change in position $\mathrm{d}\mathbf{r} = \mathrm{d}I\,\hat{\mathbf{n}} \times \mathbf{r}$. Finally, a change $\mathrm{d}\Omega$ corresponds to an infinitesimal rotation about the z-axis, which gives rise to a change in position $\mathrm{d}\mathbf{r} = \mathrm{d}\Omega\,\hat{\mathbf{z}} \times \mathbf{r}$. Thus

$$\mathrm{d}\mathbf{r} = \mathrm{d}r\,\hat{\mathbf{r}} + \mathrm{d}w\,\hat{\mathbf{L}} \times \mathbf{r} + \mathrm{d}I\,\hat{\mathbf{n}} \times \mathbf{r} + \mathrm{d}\Omega\,\hat{\mathbf{z}} \times \mathbf{r}. \tag{E.2}$$

Taking the dot product with the velocity $\mathbf{v}$ results in

$$\begin{aligned}\mathbf{v}\cdot\mathrm{d}\mathbf{r} &= \mathrm{d}r\,\mathbf{v}\cdot\hat{\mathbf{r}} + \mathrm{d}w\,\mathbf{v}\cdot(\hat{\mathbf{L}}\times\mathbf{r}) + \mathrm{d}I\,\mathbf{v}\cdot(\hat{\mathbf{n}}\times\mathbf{r}) + \mathrm{d}\Omega\,\mathbf{v}\cdot(\hat{\mathbf{z}}\times\mathbf{r}) \\ &= \mathrm{d}r\,\dot{r} + \mathrm{d}w\,\hat{\mathbf{L}}\cdot(\mathbf{r}\times\mathbf{v}) + \mathrm{d}I\,\hat{\mathbf{n}}\cdot(\mathbf{r}\times\mathbf{v}) + \mathrm{d}\Omega\,\hat{\mathbf{z}}\cdot(\mathbf{r}\times\mathbf{v});\end{aligned} \tag{E.3}$$

in the second line we have used the vector identity (B.9a) to rearrange each of the last three terms. Since $\mathbf{r}\times\mathbf{v} = \mathbf{L}$, the angular momentum, and $\hat{\mathbf{n}}\cdot\mathbf{L} = 0$, this expression simplifies to

$$\mathbf{v}\cdot\mathrm{d}\mathbf{r} = \dot{r}\,\mathrm{d}r + L\,\mathrm{d}w + L_z\,\mathrm{d}\Omega = \mathbf{p}\cdot\mathrm{d}\mathbf{q}, \tag{E.4}$$

where the last equation follows from the definition (E.1) of the Hill variables. According to equation (D.52), the conservation of this dot product proves that the transformation from Cartesian coordinates and momenta to Hill variables is canonical.

E.2 Delaunay variables

Derivations of these or closely related variables are given in many textbooks (Landau & Lifshitz 1976; Goldstein et al. 2002; Binney & Tremaine 2008); however, we may simplify those derivations by starting with the Hill variables.

We consider a canonical transformation from the Hill variables (E.1) to new coordinates and momenta $(\mathbf{Q}, \mathbf{P})$ defined by the generating function

$$S_2(\mathbf{q}, \mathbf{P}) = P_2 w + P_3\Omega + s(r, P_1, P_2). \tag{E.5}$$

Then from equations (D.63) we have

$$L = \frac{\partial S_2}{\partial w} = P_2, \quad L_z = \frac{\partial S_2}{\partial \Omega} = P_3, \quad Q_3 = \frac{\partial S_2}{\partial P_3} = \Omega; \tag{E.6}$$

thus the second and third new momenta are the same as the old, as is the third coordinate. We also have

$$\dot{r} = \frac{\partial S_2}{\partial r} = \frac{\partial s}{\partial r}, \tag{E.7}$$

which is satisfied if we define (cf. eq. 1.20)

$$s(r, P_1, P_2) = \int^r \mathrm{d}r'\,\sigma\left(2E + \frac{2\,\mathbb{G}M}{r'} - \frac{P_2^2}{r'^2}\right)^{1/2}. \tag{E.8}$$

Here E is the energy per unit mass of the Kepler orbit, considered to be a function of P_1 and $P_2 = L$. The integral starts at periapsis and follows the orbit in and out; $\sigma = +1$ as the radius grows from periapsis to apoapsis, -1 as it shrinks back to periapsis, and so forth.

The first coordinate is then

$$Q_1 = \frac{\partial s(r, P_1, L)}{\partial P_1} = \frac{\partial E(P_1, L)}{\partial P_1}\int^r \mathrm{d}r'\,\sigma\left(2E + \frac{2\,\mathbb{G}M}{r'} - \frac{L^2}{r'^2}\right)^{-1/2}. \tag{E.9}$$

At this point we replace $E = -\frac{1}{2}\,\mathbb{G}M/a$ and $L = [\,\mathbb{G}Ma(1-e^2)]^{1/2}$ by the semimajor axis a and eccentricity e, and change the dummy variable from radius to eccentric anomaly, setting $r' = a(1 - e\cos u')$ (eqs. 1.28, 1.32, 1.46). The integral simplifies to

$$Q_1 = \frac{(\,\mathbb{G}M)^{1/2}}{2a^{1/2}}\frac{\partial a}{\partial P_1}\int \mathrm{d}u'\,(1 - e\cos u')\sigma\,\mathrm{sgn}(\sin u'). \tag{E.10}$$

Since both σ and $\mathrm{sgn}(\sin u')$ are $+1$ when the radius is increasing and -1 when it is decreasing, the factor $\sigma\,\mathrm{sgn}(\sin u')$ is always unity; and since the integral starts at periapsis, where $u = 0$,

$$Q_1 = \frac{(\,\mathbb{G}M)^{1/2}}{2a^{1/2}}\frac{\partial a}{\partial P_1}(u - e\sin u). \tag{E.11}$$

We are free to choose $\partial P_1/\partial a = \frac{1}{2}(\,\mathbb{G}M)^{1/2}/a^{1/2}$. Then $P_1 = (\,\mathbb{G}Ma)^{1/2}$, a quantity we call Λ throughout this book, and Kepler's equation (1.49) implies that $Q_1 = \ell$, the mean anomaly.

A similar analysis for Q_2 yields

$$Q_2 = w - (1-e^2)^{1/2}\int_0^u \frac{\mathrm{d}u'}{1 - e\cos u'} = w - 2\tan^{-1}\left[\left(\frac{1+e}{1-e}\right)^{1/2}\tan\tfrac{1}{2}u\right] = w - f, \tag{E.12}$$

where the last equality follows from equation (1.51b). Since $w = f + \omega$, we conclude that $Q_2 = \omega$, the argument of periapsis.

In summary, the new canonical coordinates and momenta $(\mathbf{Q}, \mathbf{P})$, called the **Delaunay variables**, are

$$\ell, \qquad\qquad \Lambda = (\,\mathbb{G}Ma)^{1/2},$$

$$\begin{aligned} &\omega, \qquad && L = [\,\mathbb{G}Ma(1-e^2)]^{1/2}, \\ &\Omega, \qquad && L_z = L\cos I. \end{aligned} \tag{E.13}$$

The coordinates (ℓ, ω, Ω) vary in the range 0 to 2π, and the Kepler Hamiltonian (1.85) depends only on the momenta, not the coordinates (in fact it depends on only one of the three momenta). Therefore the Delaunay variables are also angle-action variables for the Kepler Hamiltonian. Several canonical coordinate-momentum pairs derived from Delaunay variables are described in §1.4.

Appendix F

The standard map

The standard map, also known as the **Chirikov–Taylor map**, is a map from the (x, y) plane into itself. The map is

$$y_{n+1} = y_n + K \sin x_n, \quad x_{n+1} = x_n + y_{n+1} = x_n + y_n + K \sin x_n. \tag{F.1}$$

The map depends on one parameter K, which can be chosen to be positive (if $K < 0$, simply replace x_n by $x_n+\pi$ to obtain a map with $K > 0$). The map is area-preserving since the Jacobian determinant is unity:

$$\begin{vmatrix} \dfrac{\partial x_{n+1}}{\partial x_n} & \dfrac{\partial x_{n+1}}{\partial y_n} \\ \dfrac{\partial y_{n+1}}{\partial x_n} & \dfrac{\partial y_{n+1}}{\partial y_n} \end{vmatrix} = \begin{vmatrix} 1 + K \cos x_n & 1 \\ K \cos x_n & 1 \end{vmatrix} = 1. \tag{F.2}$$

If the sequence $\{x_n, y_n\}$ is a solution of equations (F.1), then so is $\{x_n + 2\pi p, y_n\}$, where p is an arbitrary integer. Thus the dynamics can be restricted to a cylinder by treating x_n as the azimuthal coordinate of the cylinder and taking the modulus of x relative to 2π after every step. Moreover $\{x_n, y_n + 2\pi q\}$ is also a solution, where q is an arbitrary integer, so the dynamics can be restricted to a torus by taking the modulus of x *and* y relative to 2π at the end of every step. However, the restriction of the phase space to a torus does obscure some important information: in particular, we must do extra bookkeeping if we want to answer the important question of whether y_n can grow or decay without limit over many steps.

When x and y are interpreted as a coordinate and a momentum respectively, the standard map is generated by the time-dependent Hamiltonian

$$H(x,y,t) = \tfrac{1}{2}y^2 + K\delta_1(t)\cos x, \qquad \text{(F.3)}$$

where $\delta_1(t)$ is the periodic delta function with unit period (eq. C.9), x_n is the coordinate at time $t = n$, and y_n is the momentum just before the kick at $t = n$. Physically, this Hamiltonian describes the **kicked rotor**: a simple pendulum of unit mass (cf. eq. 6.1), with the gravitational field switched on in short pulses at integer times.

Plots of the standard map are shown in Figure F.1 for $K = 0.1$, 0.5, 1.0 and 2.0. For small K, the trajectories in the map follow 1-dimensional curves in the 2-dimensional phase space. As K increases, significant chaotic regions emerge, mapped out by single trajectories that cover 2-dimensional regions of phase space rather than a 1-dimensional curve. For $K = 2$ most of the phase space is covered by a single chaotic trajectory, and chaos is **global** in the sense that the momentum y_n can grow or decay without limit; in effect the chaotic orbit diffuses in momentum with the characteristic long-time behavior $|y_{n+k} - y_n| \sim K^{1/2}k^{1/2}$. Numerical experiments show that global chaos is present for all $K > K_{\rm crit} = 0.9716$.

Much of the behavior of the standard map is generic for area-preserving maps or Hamiltonian flows: as described in Appendix D.8, the phase space exhibits "islands" of regular orbits embedded in a chaotic "sea," and the fraction of phase space occupied by the chaotic sea grows with the amplitude of the perturbation parameter K.

F.1 Resonance overlap

Using the Fourier series for the periodic delta function, equation (C.10), the Hamiltonian for the standard map can be written

$$H(x,y,t) = \tfrac{1}{2}y^2 + K\cos x \sum_{n=-\infty}^{\infty} \cos(2\pi nt) = \tfrac{1}{2}y^2 + K \sum_{n=-\infty}^{\infty} \cos(x - 2\pi nt). \qquad \text{(F.4)}$$

If the parameter K is small enough, the resonances associated with each term will be narrow, and we can examine the effects of each term separately. Thus we consider

$$H_n(x,y,t) = \tfrac{1}{2}y^2 + K\cos(x - 2\pi nt). \qquad \text{(F.5)}$$

We apply a canonical transformation to a new coordinate and momentum (X, Y) defined by the generating function

$$S_2(x,Y,t) = (Y + 2\pi n)(x + \pi - 2\pi nt); \qquad \text{(F.6)}$$

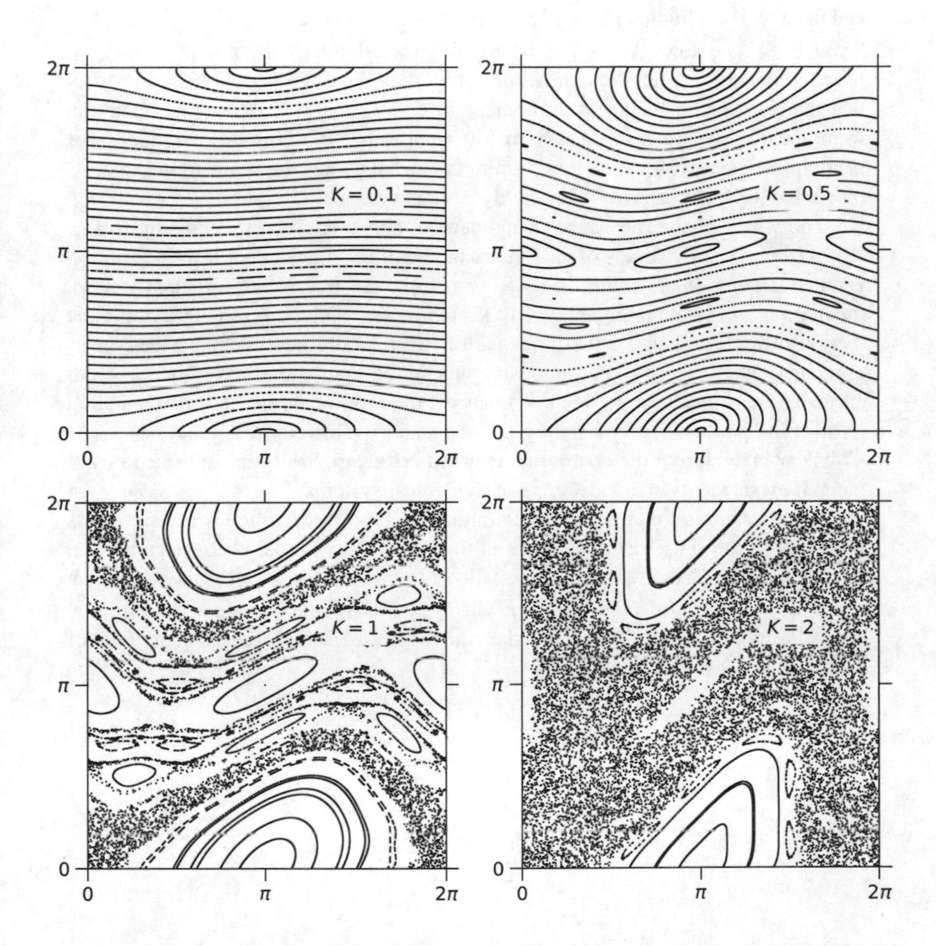

Figure F.1: Trajectories in the standard map (F.1) for $K = 0.1$, 0.5, 1 and 2. The coordinates are x_n and y_n.

from equations (D.63) we have $X = \partial S_2/\partial Y = x+\pi-2\pi nt$, $y = \partial S_2/\partial x = Y+2\pi n$, and the new Hamiltonian $H'_n = H_n + \partial S_2/\partial t = \frac{1}{2}(Y+2\pi n)^2 - K\cos X - 2\pi n(Y+2\pi n) = \frac{1}{2}Y^2 - K\cos X - 2\pi^2 n^2$. Dropping the unimportant constant $2\pi^2 n^2$, we have the pendulum Hamiltonian, whose properties are described in §6.1. In particular the resonance is centered at momentum $Y = 0$ or $y = 2\pi n$, so the separation in momentum between resonances of different n is 2π. This linear sequence of pendulum-type resonances of equal strength is what makes the standard map a good model for the local behavior of many dynamical systems.

The width of the resonance in momentum space is given by equation (6.13), $w = 4K^{1/2}$. The dynamics of each resonance will be independent if the separation is much larger than the width, or $K \ll \frac{1}{4}\pi^2$. If the resonances are sufficiently close, the motion will be more complicated. There are few analytic results describing the motion of a particle under the influence of two or more nearby resonances; however, in 1959 Boris Chirikov famously conjectured that when the resonance width exceeded the separation between resonances, the motion would become chaotic in most of the phase space covered by the resonances. This conjecture, known as the **Chirikov criterion** or the **resonance-overlap criterion**, has been confirmed by numerical experiments in a wide range of dynamical systems.

For the standard map a natural definition for "mostly chaotic" is that the chaos is global, in the sense defined earlier in this appendix. Global chaos occurs when $K > K_{\rm crit} = 0.9716$; at this point the ratio of the resonance width to the separation is $w/(2\pi) = 2K_{\rm crit}^{1/2}/\pi = 0.63$. This result suggests a sharper version of the Chirikov criterion: motion is mostly chaotic when the resonance width exceeds about 60% of the separation between resonances.

Appendix G

Hill stability

The goal of this appendix is to derive sufficient conditions for Hill stability of an arbitrary three-body system. See Golubev (1968), Marchal & Bozis (1982) and Gladman (1993) for more details.

A general three-body system is Hill stable if none of the three bodies can have a close encounter. Hill stability does not preclude the escape of one of the bodies to infinity.[1]

Let the masses of the three bodies be m_0, m_1 and m_2, with positions and velocities in the barycentric frame denoted by $\mathbf{r}_i$, $\mathbf{v}_i$, $i = 0, 1, 2$. The kinetic energy, potential energy and angular momentum of the system are

$$T = \tfrac{1}{2} \sum_{i=0}^{2} m_i v_i^2, \quad W = -\mathbb{G} \sum_{\substack{i,j=0 \\ i>j}}^{2} \frac{m_i m_j}{|\mathbf{r}_i - \mathbf{r}_j|}, \quad \mathbf{L} = \sum_{i=0}^{2} m_i \mathbf{r}_i \times \mathbf{v}_i. \tag{G.1}$$

We shall also define

$$I = \sum_{i=0}^{2} m_i r_i^2; \tag{G.2}$$

this quantity—or sometimes half of it—is often referred to as the moment of inertia, since it is related to the moments of inertia (D.87) in the principal-axis frame by $2I = A + B + C$.

---

[1] In a planetary system, Hill stability is often interpreted more liberally, to mean that the planets cannot have a close encounter. In this usage Hill stability does not preclude the collision of one of the planets with the host star.

We may write

$$\mathbf{L} = \sum_{i=0}^{2} a_i \mathbf{b}_i \tag{G.3}$$

where

$$a_i \equiv m_i^{1/2} r_i, \quad \mathbf{b}_i \equiv m_i^{1/2} \frac{\mathbf{r}_i \times \mathbf{v}_i}{r_i}. \tag{G.4}$$

Then

$$L^2 = \Big(\sum_{i=0}^{2} a_i b_{ix}\Big)^2 + \Big(\sum_{i=0}^{2} a_i b_{iy}\Big)^2 + \Big(\sum_{i=0}^{2} a_i b_{iz}\Big)^2, \tag{G.5}$$

where b_{ix}, b_{iy}, b_{iz} are the components of $\mathbf{b}$ along three Cartesian axes x, y, z. According to the Cauchy–Schwarz inequality, $(\sum_i a_i b_{ix})^2 \le \sum_i a_i^2 \sum_i b_{ix}^2$ with similar relations for the y and z components. Thus

$$\begin{aligned} L^2 &\le \sum_{i=0}^{2} a_i^2 \left(\sum_{i=0}^{2} b_{ix}^2 + \sum_{i=0}^{2} b_{iy}^2 + \sum_{i=0}^{2} b_{iz}^2 \right) = \sum_{i=0}^{2} a_i^2 \sum_{i=0}^{2} |\mathbf{b}_i|^2 \\ &= I \sum_{i=0}^{2} m_i \frac{|\mathbf{r}_i \times \mathbf{v}_i|^2}{r_i^2}. \end{aligned} \tag{G.6}$$

Using the vector identity (B.9c), we can write $|\mathbf{r}_i \times \mathbf{v}_i|^2 = r_i^2(v_i^2 - \dot{r}_i^2)$, so we arrive at the inequality[2]

$$L^2 \le I \sum_{i=0}^{2} m_i (v_i^2 - \dot{r}_i^2) \le 2IT. \tag{G.7}$$

Now set $T = E - W$, where E is the total energy. We shall assume that the system is bound, so $E < 0$. Then we can rewrite equation (G.7) as

$$-I^{1/2} W \ge \frac{L^2}{2I^{1/2}} + I^{1/2}|E|. \tag{G.8}$$

If we vary the moment of inertia I at fixed energy E and angular momentum L, the right side is minimized at $I = \frac{1}{2}L^2/|E|$ where it is equal to $2^{1/2} L|E|^{1/2}$. Hence

$$I^{1/2} W \le -2^{1/2} L|E|^{1/2}. \tag{G.9}$$

The right side of this equation depends only on the conserved energy E and angular momentum L, while the left side depends only on the positions of the three masses.

---

[2] This is a weak form of the **Sundman inequality**, which states that $L^2 + \frac{1}{4}(\mathrm{d}I/\mathrm{d}t)^2 \le 2IT$ (e.g., Saari 1971).

Moreover it is straightforward to show that

$$I = \frac{1}{2(m_0 + m_1 + m_2)} \sum_{i=0}^{2}\sum_{j=0}^{2} m_i m_j (\mathbf{r}_i - \mathbf{r}_j)^2 \tag{G.10}$$

(recall that we are in the barycentric frame, so $\sum_i m_i \mathbf{r}_i = \mathbf{0}$), so $I^{1/2}|W|$ depends only on the relative positions $\mathbf{r}_i - \mathbf{r}_j$ of the three masses. We then have

$$\begin{aligned} I^{1/2}W = &-\frac{\mathbb{G}}{M^{1/2}}(m_0 m_1 r_{01}^2 + m_0 m_2 r_{02}^2 + m_1 m_2 r_{12}^2)^{1/2} \\ &\times \left(\frac{m_0 m_1}{r_{01}} + \frac{m_0 m_2}{r_{02}} + \frac{m_1 m_2}{r_{12}}\right), \end{aligned} \tag{G.11}$$

where $M \equiv m_0 + m_1 + m_2$ and $r_{ij} \equiv |\mathbf{r}_i - \mathbf{r}_j|$.

If we rescale the vectors $\mathbf{r}_i$ to $\lambda \mathbf{r}_i$ with λ a fixed real number, then $I \propto \lambda^2$ and $W \propto \lambda^{-1}$ so $I^{1/2}W$ is independent of λ. Therefore we can plot contours of $I^{1/2}W$ as a function of the position of m_2 in a figure in which we assume without loss of generality that (i) the three masses lie in the (x, y) plane; (ii) m_0 and m_1 lie along the x-axis; (iii) the unit of length is r_{01}.

Figure G.1 shows contours of $I^{1/2}W$ for the case $m_1 = m_2 = 0.1m_0$. The geometry is remarkably similar to that of the zero-velocity curves in the circular restricted three-body problem (Figure 3.1). In particular, (i) there are five extrema—maxima or saddle points—of $I^{1/2}W$, which we may call Lagrange points by analogy to the discussion in §3.1; (ii) the maxima at the Lagrange points L4 and L5 occur when the three masses form an equilateral triangle; and (iii) when $m_2 \to 0$ the contours of $I^{1/2}W$ are the same as the zero-velocity curves in the planar, circular, restricted three-body problem (Problem 3.14).

Let $(I^{1/2}W)_\mathrm{L}$ denote the minimum value of $I^{1/2}W$ at the three collinear Lagrange points. The inequality (G.9) shows that if

$$(I^{1/2}W)_\mathrm{L} > -2^{1/2}L|E|^{1/2} \tag{G.12}$$

then the region of possible motions contains three disconnected parts: either (i) m_2 always orbits m_0 and can never have a close encounter with m_1, (ii) m_2 always orbits m_1 and never has a close encounter with m_0, or (iii) m_2 never has a close encounter with either m_0 or m_1. Whether or not this inequality is satisfied is straightforward to determine: the left side can be determined given the three masses m_0, m_1 and m_2, and the right side is an integral of motion that is determined by the initial conditions.

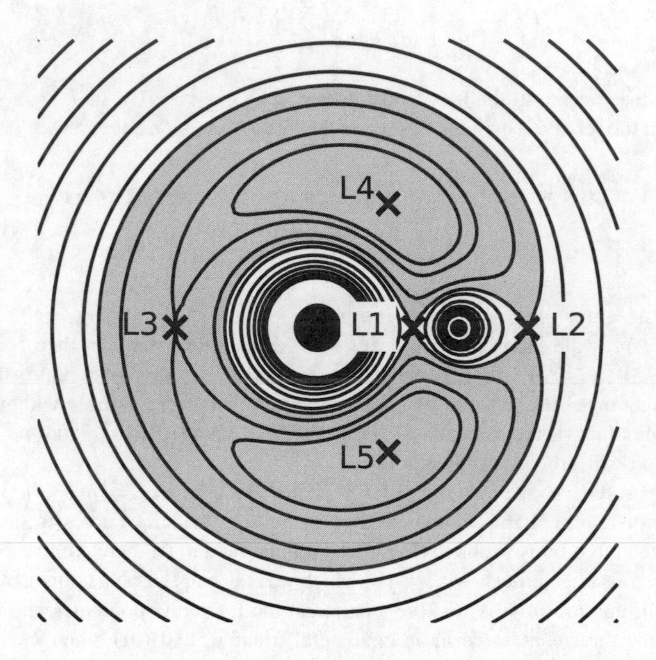

Figure G.1: Contours of $I^{1/2}W$ (eq. G.11) for the three-body problem. Here I is the moment of inertia in the barycentric frame, and W is the potential energy (eqs. G.1). The plot shows $I^{1/2}W$ as a function of the position of body 2, in coordinates where bodies 0 and 1 are separated by unit distance along the x-axis. The Lagrange points L1, ..., L5 are marked by crosses. Shading denotes regions in which $I^{1/2}W$ is greater than the minimum of the values at the collinear Lagrange points L1, L2, L3. The bodies have masses $m_1 = m_2 = 0.1m_0$. This figure can be compared to Figure 3.1, which shows the zero-velocity surfaces in the planar, circular, restricted three-body problem.

It is worthwhile to rewrite the inequality (G.12) for the case of a two-planet system with $m_1, m_2 \ll m_0$. We introduce a small parameter ϵ and assume that m_1 and m_2 are $\mathrm{O}(\epsilon)$. To find the location of the collinear Lagrange points L1 and L2, we write $r_{01} = 1$, $r_{02} = 1 + x$, and $r_{12} = |x|$ and assume $|x| = \mathrm{O}(\epsilon^{1/3})$. We expand $I^{1/2}W$ in powers of ϵ and find the extremum in x of the non-constant term with the smallest power of ϵ. We find $x = \pm r_\mathrm{H}$, where

$$r_\mathrm{H} = |\mathbf{r}_0 - \mathbf{r}_1| \left(\frac{m_1 + m_2}{3m_0} \right)^{1/3} + \mathrm{O}(m_1, m_2/m_0)^{2/3}. \tag{G.13}$$

This result is similar to the definition of the Hill radius in equation (3.112) except that here $|\mathbf{r}_0 - \mathbf{r}_1|$ is not necessarily constant, so r_H varies during the evolution of the three-body system. The value of $I^{1/2}W$ at L1 and L2 is

$$\begin{aligned}(I^{1/2}W)_{\mathrm{L1,L2}} = &-\mathbb{G}m_0(m_1 + m_2)^{3/2} \\ &\times \left[1 + \frac{3^{4/3} m_1 m_2}{2m_0^{2/3}(m_1 + m_2)^{4/3}} + \mathrm{O}(m_1, m_2/m_0) \right].\end{aligned} \tag{G.14}$$

In these formulas the distances to L1 and L2 and the values of $I^{1/2}W$ at L1 and L2 are the same, but differences between the two values arise at higher order in $(m_1, m_2)/m_0$.

In the limit where $m_1, m_2 \ll m_0$, the dominant contributions to the energy and angular momentum are

$$\begin{aligned} E &= -\frac{\mathbb{G}m_0 m_1}{2a_1} - \frac{\mathbb{G}m_0 m_2}{2a_2}, \\ L &= (L_1^2 + L_2^2 + 2L_1 L_2 \cos i)^{1/2}, \\ L_1 &= m_1[\mathbb{G}m_0 a_1(1 - e_1^2)]^{1/2}, \end{aligned} \tag{G.15}$$

with a similar expression for L_2. Here i is the inclination between the orbits of m_1 and m_2. We substitute these expressions into the right side of (G.12) and simplify by introducing a small parameter δ and assuming that $|a_1 - a_2|$, e_1, e_2 and I are all $\mathrm{O}(\delta)$. Then we find that the planets are Hill stable if

$$\begin{aligned}|a_1 - a_2| \geq\ & 2 \cdot 3^{1/6} \overline{a} \\ &\times \left[\left(\frac{m_1 + m_2}{m_0} \right)^{2/3} + \frac{i^2}{3^{4/3}} + \frac{m_1 + m_2}{3^{4/3} m_1 m_2} (m_1 e_1^2 + m_2 e_2^2) \right]^{1/2},\end{aligned} \tag{G.16}$$

where $\overline{a} = \frac{1}{2}(a_1 + a_2)$ is the mean semimajor axis. When the orbits are circular, $e_1 = e_2 = 0$, this result reduces to the stability criterion (3.137) derived using Hill's approximation. When the orbits are coplanar but eccentric, $i = 0$, the result is less sharp than (3.136); that is, systems that are guaranteed to be Hill stable by equation (3.136) are not guaranteed to be stable by (G.16).

Appendix H

The Yarkovsky effect

Our goal is to determine the Yarkovsky force on a spherical body traveling on a circular orbit around its host star. The following derivations can be skipped without affecting the reader's understanding of the rest of the book.

The spin axis of the body is assumed to be normal to its orbital plane. To be definite, we refer to the body as an asteroid and the host star as the Sun. We first examine a patch of the asteroid surface that is small enough to be considered flat. Let z denote the height relative to this surface, so the asteroid body occupies the volume $z < 0$. The diffusion of heat through the asteroid is described by the heat-conduction equation (e.g., Landau & Lifshitz 1987)

$$\frac{\partial T}{\partial t} = \frac{\kappa}{C_V} \frac{\partial^2 T}{\partial z^2}, \quad z \leq 0, \tag{H.1}$$

in which $T(z,t)$ is the temperature and κ and C_V are the thermal conductivity and the heat capacity per unit volume, assumed constant within the asteroid. We denote the surface temperature by $T_s(t) = T(z = 0, t)$.

The surface temperature is determined by three processes:

Absorption of solar radiation Let $f(t)$ be the incident solar flux, that is, the solar energy impacting unit surface area of the asteroid in unit time. Then the rate of energy absorption per unit area is αf where $\alpha < 1$ is the absorption coefficient.[1]

[1] A black body has $\alpha = 1$. The **Bond albedo**, the ratio of the energy incident on the body to the energy that is immediately re-emitted, is $A_B = 1 - \alpha$.

Thermal emission The asteroid emits thermal radiation at a rate $\epsilon\sigma T_\mathrm{s}^4$ per unit area, where ϵ is the emissivity and the Stefan–Boltzmann constant is $\sigma = 5.670\,374\ldots \times 10^{-8}\,\mathrm{W\ m^{-2}\,K^{-4}}$. A black body has $\epsilon = 1$.

Thermal conduction The heat flux—the rate at which energy is conducted outward through the asteroid per unit area—is $-\kappa\partial T/\partial z$.

The rate of energy gain at the surface from solar radiation and conduction from the interior must balance the rate of loss from emission, so

$$\alpha f - \kappa \frac{\partial T}{\partial z}\bigg|_{z=0} = \epsilon\sigma T_\mathrm{s}^4. \tag{H.2}$$

We now separate this equation into time-averaged and variable components, denoting the time average by an overline. The time average of the heat flow at any depth must vanish, so the time-averaged temperature $\overline{T}$ must be independent of depth. Since $\partial\overline{T}/\partial z = 0$, the time average of equation (H.2) is

$$\alpha\overline{f} = \epsilon\sigma\overline{T_\mathrm{s}^4}. \tag{H.3}$$

In most circumstances the temporal variation of the surface temperature is small, even at the surface, so we may write $T_\mathrm{s}(z,t) = \overline{T} + \Delta T(0,t)$, where $|\Delta T| \ll \overline{T}$. The thermal emission $\epsilon\sigma T_\mathrm{s}^4$ can then be approximated as $\epsilon\sigma[\overline{T}^4 + 4\overline{T}^3\Delta T(0,t)]$, and the average $\overline{T_\mathrm{s}^4}$ can be approximated as $\overline{T}^4$. Then we can subtract (H.3) from (H.2) to obtain

$$\alpha(f - \overline{f}) - \kappa \frac{\partial T}{\partial z}\bigg|_{z=0} = 4\epsilon\sigma\overline{T}^3\Delta T|_{z=0}. \tag{H.4}$$

Now suppose that $f - \overline{f}$ is equal to the real part of $A\exp(\mathrm{i}\omega t)$, where the frequency ω is real and the amplitude A may be complex. Then the time-varying component of the temperature $\Delta T(z,t)$ must equal the real part of some function of the form $B(z)\exp(\mathrm{i}\omega t)$; and the heat-conduction equation (H.1) implies that $B(z)$ must be proportional to $\exp(kz)$, where $k^2 = \mathrm{i}\omega C_V/\kappa$. Since the time-varying component of the temperature must die away deep in the body of the asteroid, the real part of k must be positive, which implies that

$$k = [1 + \mathrm{i}\,\mathrm{sgn}(\omega)]\left(\frac{|\omega|C_V}{2\kappa}\right)^{1/2}. \tag{H.5}$$

Combining these results with equation (H.4), we find that the time-varying component of the temperature in the asteroid interior is the real part of

$$\Delta T(z,t) = \frac{\alpha A}{4\epsilon\sigma\overline{T}^3 + \kappa k}\exp(kz + \mathrm{i}\omega t), \quad z \le 0. \tag{H.6}$$

The variable component of the rate of emission per unit area is the real part of

$$4\epsilon\sigma\overline{T}^3\Delta T(0,t) = \frac{4\epsilon\sigma\overline{T}^3\alpha A}{4\epsilon\sigma\overline{T}^3 + \kappa k}\exp(\mathrm{i}\omega t) = \frac{\alpha A}{1 + \Lambda + \mathrm{i}\,\mathrm{sgn}(\omega)\Lambda}\exp(\mathrm{i}\omega t), \tag{H.7}$$

where

$$\Lambda(\omega) \equiv \frac{|\omega|^{1/2} I_{\mathrm{th}}}{2^{5/2}\epsilon\sigma\overline{T}^3}, \quad \text{and} \quad I_{\mathrm{th}} \equiv (C_V\kappa)^{1/2} \tag{H.8}$$

is the **thermal inertia**, which depends only on the bulk—not surface—properties of the asteroid material.

The factor $[1 + \Lambda + \mathrm{i}\,\mathrm{sgn}(\omega)\Lambda]^{-1}$ in equation (H.7) can be written as $(1 + 2\Lambda + 2\Lambda^2)^{-1/2}\exp(-\mathrm{i}\phi_{\mathrm{th}})$, where

$$\cos\phi_{\mathrm{th}} = \frac{1 + \Lambda}{(1 + 2\Lambda + 2\Lambda^2)^{1/2}}, \quad \sin\phi_{\mathrm{th}} = \frac{\mathrm{sgn}(\omega)\Lambda}{(1 + 2\Lambda + 2\Lambda^2)^{1/2}}. \tag{H.9}$$

Then

$$4\epsilon\sigma\overline{T}^3\Delta T(0,t) = \frac{\alpha A}{(1 + 2\Lambda + 2\Lambda^2)^{1/2}}\exp[\mathrm{i}(\omega t - \phi_{\mathrm{th}})]. \tag{H.10}$$

Thus the peak emission lags the peak solar flux by a time $\tau = \phi_{\mathrm{th}}/\omega$.

We may now generate a map of the thermal emission on the asteroid surface. We work in the frame rotating with the asteroid's circular orbit and erect a Cartesian coordinate system (X, Y, Z) centered on the asteroid, with the Z-axis parallel to the orbital angular momentum, the X-Y plane coinciding with the orbital plane, and the X-axis pointing away from the Sun. Thus the trailing face of the asteroid has coordinates $Y < 0$. The corresponding spherical coordinates are (θ, ϕ). The asteroid's spin rate in this rotating frame is the synodic frequency ω_{syn}, which we assume to be nonzero.[2]

[2] If the asteroid's spin frequency in an inertial frame is ω_{in}, with $\omega_{\mathrm{in}} > 0$ for prograde spin and $\omega_{\mathrm{in}} < 0$ for retrograde spin, then $\omega_{\mathrm{syn}} = \omega_{\mathrm{in}} - n$, where n is the mean motion. The synodic period is $2\pi/|\omega_{\mathrm{syn}}|$. See footnote 13 in §3.4 for more on the synodic frequency and period.

In this coordinate system, the solar flux and the thermal emission at a given point are time-independent, and the time variability at a given point on the surface arises because of the asteroid's spin. The normal to the surface at (θ, ϕ) is $\hat{\mathbf{n}} = \sin\theta\cos\phi\hat{\mathbf{X}} + \sin\theta\sin\phi\hat{\mathbf{Y}} + \cos\theta\hat{\mathbf{Z}}$. The incident solar flux is then

$$f(\theta, \phi) = \frac{L_\odot}{4\pi a^2} W(-\hat{\mathbf{X}} \cdot \hat{\mathbf{n}}) = \frac{L_\odot}{4\pi a^2} W(-\sin\theta\cos\phi), \tag{H.11}$$

where $W(x) = x$ if $x > 0$ (day side) and 0 if $x < 0$ (night side). Here $L_\odot = 3.828 \times 10^{26}$ W is the solar luminosity and a is the asteroid's semimajor axis—recall that we assume that the orbit is circular. The flux can be written as a Fourier series (Appendix B.4)

$$f(\theta, \phi) = \frac{L_\odot}{4\pi a^2} \sin\theta \sum_{m=-\infty}^{\infty} c_m \exp(im\phi), \tag{H.12}$$

where (eq. B.48)

$$c_m = -\frac{1}{2\pi} \int_{\pi/2}^{3\pi/2} \mathrm{d}\phi \cos m\phi \cos\phi. \tag{H.13}$$

For example, $c_0 = 1/\pi$, $c_1 = c_{-1} = -\frac{1}{4}$.

A given point attached to the asteroid surface (call it P) has polar angle θ = constant and azimuthal angle $\phi = \omega_{\rm syn} t$ + const. Therefore the incident flux at P corresponding to the term with index m in equation (H.12) varies with time as $\exp(\mathrm{i}\omega t)$ with $\omega = m\omega_{\rm syn}$. The term with $m = 0$ gives the mean flux,

$$\overline{f} = \frac{L_\odot c_0}{4\pi a^2} \sin\theta = \frac{L_\odot}{4\pi^2 a^2} \sin\theta, \tag{H.14}$$

and this gives the mean temperature $\overline{T}(\theta)$ through equation (H.3). The terms with $m \neq 0$ give rise to temperature variations that are time-independent in the (X, Y, Z) frame but time-variable at the point P. From equation (H.10) the emission is

$$4\epsilon\sigma\overline{T}^3 \Delta T(\theta, \phi) = \frac{\alpha L_\odot}{4\pi a^2} \sin\theta \sum_{\substack{m=-\infty \\ m\neq 0}}^{\infty} \frac{c_m}{(1 + 2\Lambda + 2\Lambda^2)^{1/2}} \exp[\mathrm{i}(m\phi - \phi_{\rm th})], \tag{H.15}$$

where Λ and $\phi_{\rm th}$ are functions of m through their dependence on the forcing frequency $\omega = m\omega_{\rm syn}$. We show below that only terms with $m = \pm 1$ contribute to the total force on the asteroid, so we may simplify this result to

$$4\epsilon\sigma\overline{T}^3 \Delta T(\theta, \phi) = -\frac{\alpha L_\odot}{8\pi a^2} \frac{\sin\theta}{(1 + 2\Lambda + 2\Lambda^2)^{1/2}} \cos(\phi - \phi_{\rm th}), \tag{H.16}$$

where Λ and $\phi_{\rm th}$ are now evaluated at $\omega = \omega_{\rm syn}$.

To determine the force exerted by this emission, we need to know the angular distribution of the radiation emitted from an element of its surface. The simplest realistic model is **Lambert's law** (e.g., Seager 2010). In spherical polar coordinates with the polar axis parallel to the surface normal $\hat{\mathbf{n}}$, Lambert's law states that the probability that an emitted photon will be directed into a small solid angle $\mathrm{d}\mathbf{\Omega} = \sin\vartheta \mathrm{d}\vartheta \mathrm{d}\varphi$ is $\mathrm{d}p(\vartheta) = \pi^{-1}\cos\vartheta\,\mathrm{d}\mathbf{\Omega}$ for $\vartheta < \frac{1}{2}\pi$, and zero otherwise since the body is opaque. By symmetry, the net force per unit surface area $\mathrm{d}\mathbf{F}/\mathrm{d}A$ must be along the $\hat{\mathbf{n}}$-axis. A photon with energy E carries momentum E/c, where c is the speed of light, and if the photon is emitted at an angle (ϑ, φ) then the component of its momentum along $\hat{\mathbf{n}}$ is $(E/c)\cos\vartheta$. By Newton's third law the net force on the surface is equal and opposite to the rate of emission of momentum, so

$$\begin{aligned} \frac{\mathrm{d}\mathbf{F}}{\mathrm{d}A} &= -\frac{\epsilon}{c}\sigma T_{\rm s}^4\hat{\mathbf{n}}\int \mathrm{d}p(\vartheta)\cos\vartheta = -\frac{2\epsilon}{c}\sigma T_{\rm s}^4\hat{\mathbf{n}}\int_0^{\pi/2}\mathrm{d}\vartheta\,\sin\vartheta\cos^2\vartheta \\ &= -\frac{2\epsilon}{3c}\hat{\mathbf{n}}\sigma T_{\rm s}^4 \simeq -\frac{2\epsilon}{3c}\hat{\mathbf{n}}\sigma(\overline{T}^4 + 4\overline{T}^3\Delta T), \end{aligned} \tag{H.17}$$

where we have assumed as usual that the temperature variations are small.

The force in the Z direction, perpendicular to the orbital plane, must vanish after integrating over the entire surface of the spherical asteroid, so we will not consider it further. The force in the Y direction, $\mathrm{d}F_Y/\mathrm{d}A$, is proportional to $(\hat{\mathbf{Y}}\cdot\hat{\mathbf{n}})\Delta T = \sin\theta\sin\phi\Delta T$; similarly $\mathrm{d}F_X/\mathrm{d}A$ is proportional to $\sin\theta\cos\phi\Delta T$. When we integrate these forces over the azimuthal angle ϕ, any component of ΔT proportional to $\exp(\mathrm{i}m\phi)$ with $m \neq \pm 1$ integrates to zero, which justifies the restriction of equation (H.16) to terms with $m = \pm 1$. Using equations (H.16) and (H.17) we obtain an explicit expression for the force per unit polar angle,

$$\frac{\mathrm{d}F_X}{\mathrm{d}\theta} = R^2\sin\theta\int_0^{2\pi}\mathrm{d}\phi\,\frac{\mathrm{d}F_X}{\mathrm{d}A}(\theta,\phi) = \frac{\alpha L_\odot R^2\sin^3\theta}{12ca^2(1+2\Lambda+2\Lambda^2)^{1/2}}\cos\phi_{\rm th}; \tag{H.18}$$

in the expression for $\mathrm{d}F_Y/\mathrm{d}\theta$, $\cos\phi_{\rm th}$ is replaced by $\sin\phi_{\rm th}$.

The final step is to integrate the forces over the polar angle θ of the asteroid surface. A complication is that Λ and the lag angle $\phi_{\rm th}$ depend on polar angle through the mean temperature, $\overline{T}(\theta) \propto \sin^{1/4}\theta$ (eqs. H.3 and H.14). Because of this dependence the integrals for $F_{X,Y} = \int_0^\pi \mathrm{d}\theta\,\mathrm{d}F_{X,Y}/\mathrm{d}\theta$ are not analytic. We adopt a simple fix that leads to only a small loss of accuracy. We replace Λ by Λ_*, defined by replacing $\overline{T}(\theta)$ by $T_* \equiv \overline{T}(\theta_*)$ in equation (H.8). Similarly we replace $\phi_{\rm th}$ by $\phi_{\rm th}^*$, defined by replacing Λ with Λ_* in equations (H.9). We choose $\theta_* = \frac{1}{3}\pi$.

The fractional errors in F_X and F_Y due to this approximation are $\lesssim 1.5\%$. Using the result $\int_0^\pi \mathrm{d}\theta \sin^3\theta = \frac{4}{3}$ we then have

$$F_X = \frac{\alpha L_\odot R^2}{9ca^2}\frac{1+\Lambda_*}{1+2\Lambda_*+2\Lambda_*^2}, \tag{H.19a}$$

$$F_Y = \mathrm{sgn}(\omega_{\mathrm{syn}})\frac{\alpha L_\odot R^2}{9ca^2}\frac{\Lambda_*}{1+2\Lambda_*+2\Lambda_*^2}, \tag{H.19b}$$

where

$$\Lambda_* \equiv \frac{(|\omega_{\mathrm{syn}}|C_V\kappa)^{1/2}}{2^{5/2}\epsilon\sigma T_*^3}, \quad \epsilon\sigma T_*^4 = \frac{3^{1/2}\alpha L_\odot}{8\pi^2 a^2}. \tag{H.20}$$

The azimuthal force F_Y is positive if the force is in the direction of orbital motion; thus if the spin is prograde ($\omega_{\mathrm{syn}} > 0$) the Yarkovsky effect adds angular momentum to the orbit.

In the limiting case where the thermal inertia is negligible, Λ_* is much less than unity so

$$F_X = \frac{\alpha L_\odot R^2}{9ca^2}, \quad F_Y = 0. \tag{H.21}$$

The expression for the radial force F_X can be compared to the radiation pressure force derived in Box 1.5, which in the current notation is $F_X = L_\odot R^2/(4ca^2)$. The coefficient is different because the assumptions are different in the two calculations: in Box 1.5 we assumed that the body was a perfect absorber ($\alpha = 1$) and that the absorbed energy was re-radiated isotropically; while here we assume that the absorption coefficient $\alpha \leq 1$, we neglect the additional pressure due to incident photons that are absorbed or scattered, and we assume that the absorbed energy is re-radiated according to Lambert's law. The azimuthal force F_Y vanishes in equation (H.21); there is no analog to Poynting–Robertson drag, because we are only working to $\mathrm{O}(1/c)$ whereas the Poynting–Robertson force is $\mathrm{O}(1/c^2)$.

Appendix I

Tidal response of rigid bodies

The response of a body to tidal forces depends on its internal structure. Giant planets and stars are mostly or entirely composed of gas. Smaller planets such as the Earth can be approximated as fluids for some purposes, since the gravitational forces within them are much larger than the strength of the solid material of which they are composed. Even smaller bodies, including many planetary satellites in the solar system, should instead be treated as elastic solids (Love 1944; Landau & Lifshitz 1986; Wisdom & Meyer 2016), and these respond quite differently to tidal forces. As an extreme example, a completely rigid body would not be distorted at all by a tidal field, so its Love numbers k_l (eq. 8.21) would be zero.

The following derivations can be skipped without affecting the reader's understanding of the rest of the book.

We model the solid body as a uniform, incompressible sphere having the linear stress-strain relation (Hooke's law)

$$\sigma_{ij} = -p\delta_{ij} + \mu\left(\frac{\partial u_i}{\partial x_j} + \frac{\partial u_j}{\partial x_i}\right). \tag{I.1}$$

Here $\mathbf{u}(\mathbf{r})$ is the displacement vector, the change in position of the mass element originally at position $\mathbf{r}$ as a result of the tidal force, $\sigma_{ij}(\mathbf{r})$ is the stress tensor, $p(\mathbf{r})$ is the pressure, δ_{ij} is the Kronecker delta (Appendix C.1) and μ is the rigidity of the solid, assumed independent of position. Hooke's law is valid only so long as the displacements are small, $|\partial u_i/\partial x_j| \ll 1$.

Since the body is incompressible we must have

$$\nabla \cdot \mathbf{u} = \frac{\partial u_i}{\partial x_i} = 0; \tag{I.2}$$

here and throughout this appendix we have adopted the summation convention of Appendix B.1. Equation (I.2) implies that the displacement field can be written in terms of a vector potential $\mathbf{A}(\mathbf{r})$,

$$\mathbf{u}(\mathbf{r}) = \nabla \times \mathbf{A}(\mathbf{r}). \tag{I.3}$$

If the displacement is caused by a static gravitational potential $\Phi(\mathbf{r})$, then force balance requires

$$\frac{\partial \sigma_{ij}}{\partial x_j} = \rho \frac{\partial \Phi}{\partial x_i}. \tag{I.4}$$

Equations (I.1) and (I.4) can be combined to give

$$\mu \nabla^2 \mathbf{u} = \nabla(p + \rho\Phi) \quad \text{or} \quad \mu \frac{\partial^2 u_i}{\partial x_j^2} = \frac{\partial (p + \rho\Phi)}{\partial x_i}. \tag{I.5}$$

The function $p + \rho\Phi$ can be expanded as a series in spherical harmonics, as in equation (C.51). Taking the divergence of equation (I.5) and using equation (I.2), we find that $\nabla^2(p + \rho\Phi) = 0$. Therefore $p + \rho\Phi$ is a solution of Laplace's equation, which in turn requires that the radial function associated with the spherical harmonic $Y_{lm}(\theta, \phi)$ must be proportional to r^l or r^{-l-1}. The latter function is singular at the center of the body and therefore is not allowed. Since equation (I.5) is linear, we can determine its solution for an assumed form

$$p(\mathbf{r}) + \rho\Phi(\mathbf{r}) = a_{lm} r^l Y_{lm}(\theta, \phi), \tag{I.6}$$

and then add the results for different harmonics. Similarly, we write the tidal potential and self-gravitational potential at the surface, the radial displacement of the surface due to the tidal field, and the pressure at the surface as

$$\Phi_{\text{tide}}(R, \theta, \phi) = f_{lm}^{\text{tide}} Y_{lm}(\theta, \phi), \tag{I.7a}$$

$$\Phi_{\text{self}}(R, \theta, \phi) = f_{lm}^{\text{self}} Y_{lm}(\theta, \phi), \tag{I.7b}$$

$$\Delta R(\theta, \phi) = d_{lm} Y_{lm}(\theta, \phi), \tag{I.7c}$$

$$p(\theta, \phi) = p_{lm} Y_{lm}(\theta, \phi). \tag{I.7d}$$

The gradient of equation (I.6) can be written in terms of vector spherical harmonics (Appendix C.8),

$$\nabla(p + \rho\Phi) = a_{lm}\nabla r^l Y_{lm}(\theta,\phi) = a_{lm} r^{l-1}\left[l\mathbf{Y}_{lm}(\theta,\phi) + \mathbf{\Psi}_{lm}(\theta,\phi)\right]. \quad \text{(I.8)}$$

Similarly, the vector potential $\mathbf{A}(\mathbf{r})$ can be expanded in vector spherical harmonics according to equation (C.58):

$$\mathbf{A}(\mathbf{r}) = \sum_{l'=0}^{\infty}\sum_{m'=-l'}^{l'}\left[f_{l'm'}(r)\mathbf{Y}_{l'm'}(\theta,\phi) + g_{l'm'}(r)\mathbf{\Psi}_{l'm'}(\theta,\phi)\right.$$
$$\left. + h_{l'm'}(r)\mathbf{\Phi}_{l'm'}(\theta,\phi)\right]. \quad \text{(I.9)}$$

According to equations (I.3) and (C.63), the terms involving $f(r)$ and $g(r)$ give rise to a displacement field proportional to $\mathbf{\Phi}_{lm}$. According to equations (C.64), the Laplacian of this field is also proportional to $\mathbf{\Phi}_{lm}$, but this is orthogonal to the right side of equation (I.5) according to equations (I.8) and (C.60). Therefore we may set $f_{l'm'}(r) = g_{l'm'}(r) = 0$. Moreover according to equations (C.61), the terms in the Laplacian with index l', m' are orthogonal to (I.8) unless $l = l'$ and $m = m'$. Therefore

$$\mathbf{u}(\mathbf{r}) = \nabla\times\mathbf{A} = -\frac{l(l+1)}{r}h_{lm}(r)\mathbf{Y}_{lm}(\theta,\phi) - \left[h'_{lm}(r) + \frac{h_{lm}(r)}{r}\right]\mathbf{\Psi}_{lm}(\theta,\phi). \quad \text{(I.10)}$$

Using equations (C.64) we can now reduce equation (I.5) to

$$\frac{\mu l(l+1)}{r^3}\left[l(l+1)h_{lm} - 2rh'_{lm} - r^2h''_{lm}\right]\mathbf{Y}_{lm} + \frac{\mu}{r^3}\left[-l(l+1)h_{lm}\right.$$
$$\left. + l(l+1)rh'_{lm} - 3r^2h''_{lm} - r^3h'''_{lm}\right]\mathbf{\Psi}_{lm} = a_{lm}r^{l-1}(l\mathbf{Y}_{lm} + \mathbf{\Psi}_{lm}). \quad \text{(I.11)}$$

Since $\mathbf{Y}_{lm}$ and $\mathbf{\Psi}_{lm}$ are orthogonal, this equation can only be satisfied if the coefficients of both vanish. This is true if

$$h_{lm}(r) = -a_{lm}\frac{r^{l+2} + b_{lm}r^l}{2\mu(l+1)(2l+3)}, \quad \text{(I.12)}$$

where b_{lm} is arbitrary; in addition there can be a term $\propto r^{-l-1}$ which we discard because it is singular at the center of the body.

The radial distortion $\Delta R(\theta,\phi)$ is simply the radial component of the displacement field at $r = R$:

$$\Delta R = \hat{\mathbf{r}}\cdot\mathbf{u}|_{r=R} = -\frac{l(l+1)}{R}h_{lm}(R)Y_{lm}(\theta,\phi). \quad \text{(I.13)}$$

Then equations (I.7c) and (I.12) yield

$$d_{lm} = \frac{la_{lm}}{2\mu(2l+3)}(R^{l+1} + b_{lm}R^{l-1}). \tag{I.14}$$

We next determine the self-gravitational potential arising from the tidal distortion. Since the body has constant density, the only changes to the density distribution are at the surface, and their effect on the potential can be approximated as that of a razor-thin surface-density distribution $\Sigma(\theta,\phi) = \rho\Delta R(\theta,\phi)$ that is painted onto the sphere of radius R. The potentials internal and external to R satisfy Laplace's equation, and so they are the sums of terms of the form

$$\Phi_{\text{self,int}}(\mathbf{r}) = f_{lm}^{\text{self}}\left(\frac{r}{R}\right)^l Y_{lm}(\theta,\phi), \quad \Phi_{\text{self,ext}}(\mathbf{r}) = f_{lm}^{\text{self}}\left(\frac{R}{r}\right)^{l+1} Y_{lm}(\theta,\phi); \tag{I.15}$$

the coefficients f_{lm}^{self} are the same because the potential must be continuous at $r = R$. Poisson's equation (B.44), $4\pi\,\mathbb{G}\rho = \nabla^2\Phi$, can be integrated across the boundary at R at constant angular position (θ,ϕ) to give

$$4\pi\,\mathbb{G}\Sigma = \left.\frac{\partial\Phi_{\text{self,ext}}}{\partial r}\right|_{r=R} - \left.\frac{\partial\Phi_{\text{self,int}}}{\partial r}\right|_{r=R}, \tag{I.16}$$

and evaluating this result gives $f_{lm}^{\text{self}} = -4\pi\,\mathbb{G}\rho d_{lm}R/(2l+1)$. Thus the potential due to self-gravity inside the body is

$$\Phi_{\text{self,int}}(\mathbf{r}) = f_{lm}^{\text{self}}\left(\frac{r}{R}\right)^l Y_{lm}(\theta,\phi), \tag{I.17}$$

where

$$f_{lm}^{\text{self}} = -\frac{2\pi\,\mathbb{G}\rho l R^{l+2} a_{lm}}{\mu(2l+1)(2l+3)}(1 + b_{lm}/R^2). \tag{I.18}$$

The final condition is that the stress acting outward across the surface $r = R$ must be balanced by the gravitational force from the weight of material above it, which requires

$$[\hat{\mathbf{x}}_i\sigma_{ij}x_j + (1-\gamma)g\rho\mathbf{r}\Delta R]_{(R,\theta,\phi)} = \mathbf{0}; \tag{I.19}$$

here $g = \mathbb{G}M/R^2 = \frac{4}{3}\pi\,\mathbb{G}\rho R$ is the gravitational acceleration at the surface of the body. The quantity $\gamma = 0$, but we shall use other values of γ in the following subsection to account for the effects of rotation, so we retain it for now. Using equation (I.1), we have

$$\hat{\mathbf{x}}_i\sigma_{ij}x_j = -p\mathbf{r} + \mu[(\mathbf{r}\cdot\nabla)\mathbf{u} + \nabla(\mathbf{r}\cdot\mathbf{u}) - \mathbf{u}]. \tag{I.20}$$

We use equation (I.10) as well as the identities $(\mathbf{r}\cdot\nabla)f(r)\mathbf{Y}_{lm} = rf'\mathbf{Y}_{lm}$, $(\mathbf{r}\cdot\nabla)f(r)\mathbf{\Psi}_{lm} = rf'\mathbf{\Psi}_{lm}$, $\nabla[f(r)\mathbf{r}\cdot\mathbf{Y}_{lm}] = (rf)'\mathbf{Y}_{lm}+f\mathbf{\Psi}_{lm}$, $\nabla[f(r)\mathbf{r}\cdot\mathbf{\Psi}_{lm}] = \mathbf{0}$ to obtain

$$\hat{\mathbf{x}}_i\sigma_{ij}x_j = -p\mathbf{r} + \frac{2\mu l(l+1)}{R}(h_{lm} - rh'_{lm})_{r=R}\mathbf{Y}_{lm} + \frac{\mu}{R}[(2-l-l^2)h_{lm} - r^2 h''_{lm}]_{r=R}\mathbf{\Psi}_{lm}. \tag{I.21}$$

To satisfy the component of equation (I.19) perpendicular to $\mathbf{r}$, we must have $(2-l-l^2)h_{lm} - r^2h''_{lm} = 0$ at $r = R$, which requires[1]

$$b_{lm} = -\frac{l(l+2)}{l^2-1}R^2, \tag{I.22}$$

which implies from (I.14) that

$$d_{lm} = -\frac{l(2l+1)a_{lm}R^{l+1}}{2\mu(2l+3)(l^2-1)}. \tag{I.23}$$

To satisfy the component parallel to $\mathbf{r}$ we require

$$p_{lm} = \frac{2\mu l(l+1)}{R^2}(h_{lm} - rh'_{lm})_{r=R} + (1-\gamma)g\rho d_{lm}, \tag{I.24}$$

and this yields

$$p_{lm} = a_{lm}R^l\left[\alpha_l - (1-\gamma)\frac{\beta_l}{\nu}\right], \tag{I.25}$$

where

$$\alpha_l \equiv \frac{l}{(2l+3)(l+1)}, \quad \beta_l \equiv \frac{l(2l+1)}{2(2l+3)(l^2-1)}, \tag{I.26}$$

and

$$\nu \equiv \frac{\mu}{g\rho R} \tag{I.27}$$

is a dimensionless measure of the relative importance of rigidity and self-gravity. Applying equation (I.22) to (I.18) and using the relation $g = \mathbb{G}M/R^2 = \frac{4}{3}\pi\mathbb{G}\rho R$, we have

$$f_{lm}^{\rm self} = \frac{3a_{lm}R^l}{2\nu\rho}\delta_l, \quad \text{where} \quad \delta_l \equiv \frac{l}{(2l+3)(l^2-1)}, \tag{I.28}$$

[1] This expression and others in the derivation diverge when $l = 1$. The divergence arises because an isolated body is neutrally stable to a rigid-body displacement, which can be written in spherical coordinates as an $l = 1$ displacement field.

Now equation (I.6) can be rewritten $a_{lm}R^l = p_{lm} + \rho f_{lm}^{\rm self} + \rho f_{lm}^{\rm tide}$. Moreover $f_{lm}^{\rm self}/f_{lm}^{\rm tide} = k_l$, the gravitational Love number, so we have

$$a_{lm}R^l = a_m R^l \left[\alpha_l - (1-\gamma)\frac{\beta_l}{\nu}\right] + \frac{3a_{lm}R^l}{2\nu}\delta_l\left(1+\frac{1}{k_l}\right). \tag{I.29}$$

Eliminating the common factor of $a_{lm}R^l$, setting $\gamma = 0$ since we neglect rotation, and rearranging the terms, we arrive at an expression for the Love number,

$$k_l = \frac{3}{2l-2}\frac{1}{1+\widetilde{\nu}_l}, \quad \text{where} \quad \widetilde{\nu}_l = \frac{2l^2+4l+3}{l}\nu = \frac{2l^2+4l+3}{l}\frac{\mu}{g\rho R}. \tag{I.30}$$

The displacement Love number h_l is obtained from the definitions (8.22) and (8.21), which imply that

$$h_l = -k_l\frac{g\Delta R}{\Phi_{\rm self}} = -k_l\frac{g d_{lm}}{f_{lm}^{\rm self}} = \frac{2l+1}{3}k_l = \frac{2l+1}{2l-2}\frac{1}{1+\widetilde{\nu}_l}. \tag{I.31}$$

I.1 Tidal disruption of a rigid body

We examine the stresses on a homogeneous spherical satellite in a circular orbit of radius $r_{\rm p}$ around a planet of mass $m_{\rm p}$ (Aggarwal & Oberbeck 1974). The satellite is assumed to be synchronously rotating, and we work in rotating coordinates centered on the satellite, with the positive x-axis pointing away from the host planet and the z-axis normal to the orbit. The satellite is subject to two potentials in addition to its own self-gravity: a centrifugal potential $\Phi_{\rm cent}(\mathbf{r}) = -\frac{1}{2}n^2(x^2+y^2)$ (eq. D.21) with $n^2 = \mathbb{G}m_{\rm p}/r_{\rm p}^3$, and the dominant $l = 2$ component of the tidal field, given by equation (8.75). The sum of these potentials can be written

$$\begin{aligned}\Phi_{\rm cent}(\mathbf{r}) + \Phi_{\rm tide}(\mathbf{r}) &= -\frac{\mathbb{G}m_{\rm p}}{2r_{\rm p}^3}(x^2+y^2) + \frac{\mathbb{G}m_{\rm p}}{2r_{\rm p}^3}(y^2+z^2-2x^2)\\ &= \frac{\mathbb{G}m_{\rm p}}{2r_{\rm p}^3}(z^2-3x^2)\\ &= \Phi_1(\mathbf{r}) + \Phi_2(\mathbf{r}),\end{aligned} \tag{I.32}$$

where

$$\Phi_1(\mathbf{r}) = -\frac{\mathbb{G}m_{\rm p}}{3r_{\rm p}^3}r^2, \quad \Phi_2(\mathbf{r}) = \frac{\mathbb{G}m_{\rm p}}{6r_{\rm p}^3}(-7x^2+2y^2+5z^2). \tag{I.33}$$

The potential $\Phi_1(\mathbf{r})$ is spherically symmetric. Thus it can be combined with the potential due to the self-gravity of the unperturbed satellite, $\Phi_0(r) = \frac{2}{3}\pi\mathbb{G}\rho r^2$. We simply replace $\Phi_0(r)$ by $\Phi_0(r) + \Phi_1(r) = \frac{2}{3}\pi(1-\gamma)\mathbb{G}\rho r^2$, where

$$\gamma \equiv \frac{m_\mathrm{p}}{2\pi\rho r_\mathrm{p}^3}. \tag{I.34}$$

The parameter γ, already seen in equation (8.72), is a dimensionless measure of the relative importance of tides and self-gravity; far from the planet $\gamma \to 0$, and when the satellite radius equals the Hill radius $\gamma = \frac{2}{9}$.

The pressure $p_0(r)$ in the unperturbed satellite is obtained by solving the equation of hydrostatic equilibrium $\mathrm{d}p_0/\mathrm{d}r = \rho\,\mathrm{d}(\Phi_0 + \Phi_1)/\mathrm{d}r$ with the boundary condition $p_0(R) = 0$. Thus we find

$$p_0(r) = \tfrac{2}{3}\pi(1-\gamma)\mathbb{G}\rho^2(R^2 - r^2). \tag{I.35}$$

The second potential $\Phi_2(\mathbf{r})$ is a solution of Laplace's equation, $\nabla^2\Phi_2 = 0$, and thus can be written in spherical coordinates as a sum of terms of the form $f_{lm}^\mathrm{tide}(r/R)^l Y_{lm}(\theta,\phi)$ with $l = 2$. The strain field induced by such a potential was studied earlier in this appendix. The total potential $\Phi_\mathrm{self} + \Phi_\mathrm{tide}$ is a similar sum with terms $f_{lm}^\mathrm{tide}(1+k_l)(r/R)^l Y_{lm}(\theta,\phi)$, where k_l is the Love number. Then the coefficients a_{lm} defined in equation (I.6) satisfy $a_{lm}R^l = p_{lm} + \rho f_{lm}^\mathrm{tide}(1+k_l)$ with p_{lm} given by equation (I.25). With these relations we can determine the coefficients a_{lm} as functions of f_{lm}^tide. Now the displacement field is determined by the vector potential $\mathbf{A} = h_{lm}(r)\mathbf{\Phi}_{lm}$, with $h_{lm}(r)$ related to a_{lm} by (I.12) and (I.22). To reduce the complexity of the formulas, we restrict ourselves to $l = 2$ and a perfectly rigid satellite, $\mu \to \infty$. Thus the Love number $k_l = 1$ from (I.30). Then the pressure field and vector potential induced by the tide are

$$\begin{aligned} p_\mathrm{tide}(\mathbf{r}) &= p_{2m}(r/R)^l Y_{2m}(\theta,\phi) = \tfrac{2}{19}\rho f_{2m}^\mathrm{tide}(r/R)^l Y_{2m}(\theta,\phi), \\ \mathbf{A}(\mathbf{r}) &= \frac{\rho f_{2m}^\mathrm{tide}}{114\mu R^2} r^2(8R^2 - 3r^2)\mathbf{\Phi}_{2m}(\theta,\phi). \end{aligned} \tag{I.36}$$

The corresponding strain field is

$$\begin{aligned} \mathbf{u}(\mathbf{r}) = \nabla\times\mathbf{A} &= \frac{\rho f_{2m}^\mathrm{tide}}{38\mu R^2}\left[(6r^3 - 16R^2 r)\mathbf{Y}_{2m}(\theta,\phi) + (5r^3 - 8R^2 r)\mathbf{\Psi}(\theta,\phi)\right] \\ &= \frac{\rho f_{2m}^\mathrm{tide}}{38\mu R^2}\left\{-14\mathbf{r}r^2 Y_{2m}(\theta,\phi) + \nabla\left[(5r^2 - 8R^2)r^2 Y_{2m}(\theta,\phi)\right]\right\}. \end{aligned} \tag{I.37}$$

To express the strain field in Cartesian coordinates we just replace $f_{2m}^{\rm tide} r^2 Y_{2m}(\theta,\phi)$ in the second of equations (I.37) by the quadratic potential $\Phi_2(\mathbf{r})$ from equation (I.33), and replace r^2 by $x^2 + y^2 + z^2$. The stress field then follows from equation (I.1), with the pressure p replaced by the sum of the equilibrium pressure $p_0(r)$ from equation (I.35) and the pressure induced by the tidal field $p_{\rm tide}(\mathbf{r})$ from equation (I.36). For example,

$$\begin{aligned}\sigma_{xx}(x,y,z) &= \tfrac{2}{57}\pi\,\mathbb{G}\rho^2\big[(75\gamma-19)R^2 + (19-75\gamma)x^2 + (19-60\gamma)y^2 \\ &\quad + (19-69\gamma)z^2\big], \\ \sigma_{xy}(x,y,z) &= -\tfrac{10}{19}\pi\,\mathbb{G}\rho^2\gamma xy. \end{aligned} \tag{I.38}$$

The expressions for the stresses show that the principal planes on which the shear stresses $\sigma_{xy} = \sigma_{xz} = \sigma_{yz} = 0$ are the coordinate planes, and that the stresses in the coordinate planes are linear functions of x^2, y^2 and z^2. Therefore the maximum stresses in these planets will occur either at the center of the body, $\mathbf{r} = \mathbf{0}$, or at the intersection of one of the coordinate axes with the surface. At these locations

$$\begin{aligned}\sigma_{xx}(0,0,0) &= \frac{2\pi}{57}(75\gamma-19)\,\mathbb{G}\rho^2R^2, \\ \sigma_{yy}(0,0,0) &= \frac{2\pi}{57}(3\gamma-19)\,\mathbb{G}\rho^2R^2, \\ \sigma_{zz}(0,0,0) &= -\frac{2\pi}{57}(21\gamma+19)\,\mathbb{G}\rho^2R^2, \\ \sigma_{xx}(R,0,0) = \sigma_{yy}(0,R,0) &= \sigma_{zz}(0,0,R) = 0, \\ \sigma_{xx}(0,R,0) = \sigma_{yy}(R,0,0) &= \frac{10\pi}{19}\gamma\,\mathbb{G}\rho^2R^2, \\ \sigma_{xx}(0,0,R) = \sigma_{zz}(R,0,0) &= \frac{4\pi}{19}\gamma\,\mathbb{G}\rho^2R^2, \\ \sigma_{yy}(0,0,R) = \sigma_{zz}(0,R,0) &= -\frac{14\pi}{19}\gamma\,\mathbb{G}\rho^2R^2. \end{aligned} \tag{I.39}$$

Most materials are much stronger in compression than in tension, so positive stresses are more dangerous than negative ones. For $\gamma < \frac{19}{60}$ the largest positive stresses are at the edge, $\sigma_{xx}(0,R,0)$ and $\sigma_{yy}(R,0,0)$. For $\gamma > \frac{19}{60}$ the largest positive stress is at the center, $\sigma_{xx}(0,0,0)$. These results suggest that the tidal disruption may occur either as a fracture at the edge that propagates to the center or a fracture at the center that propagates to the edge, depending on the distance $r_{\rm p}$, which is related to γ by equation (I.34). These arguments lead directly to the tidal disruption criterion (8.79).

Appendix J

Relativistic effects

By far the most important effect of general relativity on planetary orbits is the relativistic precession of the line of apsides. In the Newtonian two-body problem the line of apsides is fixed, so even the slow precession induced by general relativity can have important consequences for celestial mechanics (see for example §5.4.1).

If a test particle orbits a point mass M with semimajor axis a and eccentricity e, the orbit-averaged relativistic precession rate of the line of apsides is[1]

$$\frac{\mathrm{d}\omega}{\mathrm{d}t} = \frac{3(\mathbb{G}M)^{3/2}}{c^2 a^{5/2}(1-e^2)}, \tag{J.1}$$

where c is the speed of light. The fractional error in this expression is $\mathrm{O}(v^2/c^2)$. Here we sketch the derivation of the Hamiltonian that describes the dominant relativistic corrections to the Kepler Hamiltonian, from which the result (J.1) follows. Our main goal is to describe the equations of motion in forms that can be used for numerical integration. The derivation is self-contained, but it will be easier to follow for readers with some background in general relativity (Weinberg 1972; Hartle 2003; Poisson & Will 2014; Carroll 2019).

The motion of a freely falling particle between two points A and B in curved spacetime is along a geodesic, a path on which the elapsed proper time τ is an

[1] Because of spherical symmetry, relativity has no effect on the longitude of the ascending node, Ω, so the argument of periapsis ω precesses at the same rate as the longitude of periapsis $\varpi = \omega + \Omega$.

extremum:

$$0 = \delta \int_A^B \mathrm{d}\tau = \delta \int_A^B \left(g_{\mu\nu}\mathrm{d}x^\mu \mathrm{d}x^\nu\right)^{1/2}. \tag{J.2}$$

Here $(x^0, x^1, x^2, x^3) \equiv (t, \mathbf{r})$, where t is time and $\mathbf{r}$ is position, $\{g_{\mu\nu}\}$ are the components of the metric tensor, and μ and ν are integer indices in the set $\{0, 1, 2, 3\}$. We have adopted the summation convention of Appendix B, except that here the summation runs over integers from 0 to 3 instead of 1 to 3.

The metric tensor in the gravitational field of a point mass M was derived by Karl Schwarzschild in 1916. The point mass is assumed to be at rest at the origin, and the coordinates are chosen such that the relation between proper time and coordinates approaches the Minkowski or flat-space metric far from the point mass,

$$\mathrm{d}\tau^2 = c^2\mathrm{d}t^2 - \mathrm{d}\mathbf{r}^2. \tag{J.3}$$

There are several commonly used coordinates consistent with these requirements, and in each of these the relativistic corrections have a different appearance. In isotropic Schwarzschild coordinates the metric is given by

$$\mathrm{d}\tau^2 = \frac{[1 - \frac{1}{2}\psi(r)]^2}{[1 + \frac{1}{2}\psi(r)]^2} c^2 \mathrm{d}t^2 - [1 + \tfrac{1}{2}\psi(r)]^4 \mathrm{d}\mathbf{r}^2, \quad \text{where} \quad \psi(r) \equiv \frac{\mathbb{G}M}{c^2 r}; \tag{J.4}$$

in standard Schwarzschild coordinates

$$\mathrm{d}\tau^2 = [1 - 2\psi(r)]c^2\mathrm{d}t^2 - \frac{2\psi(r)}{1 - 2\psi(r)} \frac{(\mathbf{r}\cdot\mathrm{d}\mathbf{r})^2}{r^2} - \mathrm{d}\mathbf{r}^2; \tag{J.5}$$

and in harmonic Schwarzschild coordinates

$$\mathrm{d}\tau^2 = \frac{1 - \psi(r)}{1 + \psi(r)} c^2\mathrm{d}t^2 - \frac{1 + \psi(r)}{1 - \psi(r)} \frac{\psi^2(r)}{r^2} (\mathbf{r}\cdot\mathrm{d}\mathbf{r})^2 - \left[1 + \psi(r)\right]^2 \mathrm{d}\mathbf{r}^2. \tag{J.6}$$

We parametrize the spacetime trajectory by t so we may rewrite (J.2) in isotropic coordinates as

$$0 = \delta \int_A^B \mathrm{d}t \frac{\mathrm{d}\tau}{\mathrm{d}t} = \delta \int_A^B \mathrm{d}t \left(\frac{[1 - \frac{1}{2}\psi(r)]^2}{[1 + \frac{1}{2}\psi(r)]^2} c^2 - [1 + \tfrac{1}{2}\psi(r)]^4 |\dot{\mathbf{r}}|^2 \right)^{1/2}, \tag{J.7}$$

where $\dot{\mathbf{r}} = \mathrm{d}\mathbf{r}/\mathrm{d}t$ is the velocity. There are similar expressions in the other coordinate systems.

The classical result analogous to equation (J.7) is Hamilton's principle: the trajectory of the particle is an extremum of the action, $\int_A^B \mathrm{d}t\, L(\mathbf{r}, \dot{\mathbf{r}})$, where $L(\mathbf{r}, \dot{\mathbf{r}})$ is the Lagrangian (eq. D.2). Thus we can *define* the Lagrangian in isotropic coordinates to be any linear function of the integrand of equation (J.7),

$$L(\mathbf{r},\dot{\mathbf{r}}) = a + b\left(\frac{[1-\frac{1}{2}\psi(r)]^2}{[1+\frac{1}{2}\psi(r)]^2}c^2 - [1+\tfrac{1}{2}\psi(r)]^4|\dot{\mathbf{r}}|^2\right)^{1/2}, \tag{J.8}$$

where a and b are constants.

The velocities in planetary systems are much smaller than the speed of light and the gravitational field is weak, in the sense that $|\dot{\mathbf{r}}|/c \ll 1$ and $\psi(r) = \mathbb{G}M/(c^2r) \ll 1$. Therefore to isolate the strongest relativistic effects, we can expand the Lagrangian in powers of $1/c$:

$$L(\mathbf{r},\dot{\mathbf{r}}) = a+bc-\frac{b}{c}\left(\tfrac{1}{2}|\dot{\mathbf{r}}|^2+\frac{\mathbb{G}M}{r}\right)-\frac{b}{c^3}\left(\frac{3\mathbb{G}M}{2r}|\dot{\mathbf{r}}|^2-\frac{\mathbb{G}^2M^2}{2r^2}+\tfrac{1}{8}|\dot{\mathbf{r}}|^4\right)+\mathrm{O}\left(\frac{b}{c^5}\right). \tag{J.9}$$

We now set $a = c^2$ and $b = -c$ to ensure that the Lagrangian approaches its Newtonian form $L(\mathbf{r},\dot{\mathbf{r}}) = \frac{1}{2}|\dot{\mathbf{r}}|^2 + \mathbb{G}M/r$ as $c \to \infty$. Carrying out the same analysis in all three coordinate systems, we find

$$L(\mathbf{r},\dot{\mathbf{r}}) = \tfrac{1}{2}|\dot{\mathbf{r}}|^2 + \frac{\mathbb{G}M}{r} + \frac{1}{c^2}\Bigg[\left(\tfrac{3}{2}-\alpha\right)\frac{\mathbb{G}M}{r}|\dot{\mathbf{r}}|^2 + \left(\alpha-\tfrac{1}{2}\right)\frac{\mathbb{G}^2M^2}{r^2} + \alpha\frac{\mathbb{G}M}{r^3}(\mathbf{r}\cdot\dot{\mathbf{r}})^2 + \tfrac{1}{8}|\dot{\mathbf{r}}|^4\Bigg] + \mathrm{O}(c^{-4}), \tag{J.10}$$

where $\alpha = 0$ for isotropic or harmonic coordinates and $\alpha = 1$ for standard coordinates.

The momentum is

$$\mathbf{p} = \frac{\partial L}{\partial \dot{\mathbf{r}}} = \dot{\mathbf{r}}\left[1 + (3-2\alpha)\frac{\mathbb{G}M}{c^2r} + \frac{|\dot{\mathbf{r}}|^2}{2c^2}\right] + 2\alpha\frac{\mathbb{G}M}{c^2r^3}(\mathbf{r}\cdot\dot{\mathbf{r}})\mathbf{r} + \mathrm{O}(c^{-4}). \tag{J.11}$$

The Hamiltonian is

$$\begin{aligned} H(\mathbf{r},\mathbf{p}) &= \mathbf{p}\cdot\dot{\mathbf{r}} - L(\mathbf{r},\dot{\mathbf{r}}) \\ &= \tfrac{1}{2}p^2 - \frac{\mathbb{G}M}{r} + \frac{1}{c^2}\Bigg[\left(\alpha-\tfrac{3}{2}\right)\frac{\mathbb{G}M}{r}p^2 + \left(\tfrac{1}{2}-\alpha\right)\frac{\mathbb{G}^2M^2}{r^2} \\ &\qquad - \alpha\frac{\mathbb{G}M}{r^3}(\mathbf{r}\cdot\mathbf{p})^2 - \tfrac{1}{8}p^4\Bigg] + \mathrm{O}(c^{-4}). \end{aligned} \tag{J.12}$$

The first two terms are, of course, the usual Kepler Hamiltonian for a test particle (eq. 1.80), and the terms proportional to $1/c^2$ are the dominant relativistic correction.

Hamilton's equations of motion become

$$\frac{\mathrm{d}\mathbf{r}}{\mathrm{d}t} = \frac{\partial H}{\partial \mathbf{p}} = \mathbf{p}\left[1 + (2\alpha - 3)\frac{\mathbb{G}M}{c^2 r} - \frac{p^2}{2c^2}\right] - 2\alpha\frac{\mathbb{G}M}{c^2 r^3}(\mathbf{r}\cdot\mathbf{p})\mathbf{r} + \mathrm{O}(c^{-4}),$$

$$\frac{\mathrm{d}\mathbf{p}}{\mathrm{d}t} = -\frac{\partial H}{\partial \mathbf{r}} = -\frac{\mathbb{G}M}{r^3}\mathbf{r}\left[1 + (\tfrac{3}{2} - \alpha)\frac{p^2}{c^2} + (2\alpha - 1)\frac{\mathbb{G}M}{c^2 r} + 3\alpha\frac{(\mathbf{r}\cdot\mathbf{p})^2}{c^2 r^2}\right] + 2\alpha\frac{\mathbb{G}M}{c^2 r^3}(\mathbf{r}\cdot\mathbf{p})\mathbf{p} + \mathrm{O}(c^{-4}). \qquad \text{(J.13)}$$

These can be combined into a second-order differential equation for the acceleration,

$$\frac{\mathrm{d}^2\mathbf{r}}{\mathrm{d}t^2} = -\frac{\mathbb{G}M}{r^3}\mathbf{r} + (4 - 2\alpha)\frac{\mathbb{G}^2 M^2}{c^2 r^4}\mathbf{r} - (1 + \alpha)\frac{\mathbb{G}M|\dot{\mathbf{r}}|^2}{c^2 r^3}\mathbf{r} + (4 - 2\alpha)\frac{\mathbb{G}M}{r^3 c^2}(\mathbf{r}\cdot\dot{\mathbf{r}})\dot{\mathbf{r}} + 3\alpha\frac{\mathbb{G}M}{c^2 r^5}(\mathbf{r}\cdot\dot{\mathbf{r}})^2\mathbf{r} + \mathrm{O}(c^{-4}). \qquad \text{(J.14)}$$

If the relativistic corrections are small, then for many purposes we can orbit average their effects on the Hamiltonian. The first two terms of equation (J.12) are equal to the Kepler Hamiltonian $-\frac{1}{2}\mathbb{G}M/a$ (eq. 1.85). Since the remaining terms are already smaller by $\mathrm{O}(c^{-2})$, their orbit averages can be evaluated under the assumption that the orbits are Kepler orbits and that $\mathbf{p} = \dot{\mathbf{r}}$ (eq. J.11). Then using equation (1.66a) and the results from Problem 1.2, we obtain

$$\begin{aligned}\langle H\rangle &= -\frac{\mathbb{G}M}{2a} - \frac{3\,\mathbb{G}^2 M^2}{c^2 a^2(1 - e^2)^{1/2}} + \frac{15\,\mathbb{G}^2 M^2}{8c^2 a^2} + \mathrm{O}(c^{-4}) \\ &= -\frac{\mathbb{G}^2 M^2}{2\Lambda^2} - \frac{3\,\mathbb{G}^4 M^4}{c^2 \Lambda^3 L} + \frac{15\,\mathbb{G}^4 M^4}{8c^2 \Lambda^4} + \mathrm{O}(c^{-4}).\end{aligned} \qquad \text{(J.15)}$$

In the second line we have expressed the result in terms of Delaunay elements (eq. 1.84). Note that the terms proportional to α have canceled, so the orbit-averaged post-Newtonian Hamiltonian is the same for all three coordinate systems.

The second term in the Hamiltonian yields the relativistic precession rate (J.1) through the relation $\dot{\omega} = \partial H/\partial L$. The third term does not contribute to the precession since it is independent of L.

J.1 The Einstein–Infeld–Hoffmann equations

The results so far in this appendix describe the motion of a test particle around a point mass. The more general case of N bodies of masses $m_1, \ldots, m_N$, with positions and velocities $\mathbf{r}_i$, $\mathbf{v}_i$, is described by the **Einstein–Infeld–Hoffmann equations of motion**. In the barycentric frame and harmonic coordinates,

$$
\begin{aligned}
\frac{\mathrm{d}^2\mathbf{r}_i}{\mathrm{d}t^2} = &-\sum_{j\neq i}\frac{\mathbb{G}m_j\mathbf{r}_{ij}}{r_{ij}^3} \\
&+\frac{1}{c^2}\sum_{j\neq i}\frac{\mathbb{G}m_j\mathbf{r}_{ij}}{r_{ij}^3}\Bigg[-v_i^2-2v_j^2+4\mathbf{v}_i\cdot\mathbf{v}_j+\frac{3(\mathbf{r}_{ij}\cdot\mathbf{v}_j)^2}{2r_{ij}^2} \\
&+4\sum_{k\neq i}\frac{\mathbb{G}m_k}{r_{ik}}+\sum_{k\neq j}\frac{\mathbb{G}m_k}{r_{jk}}-\sum_{k\neq j}\frac{\mathbb{G}m_k}{2r_{jk}^3}\mathbf{r}_{ij}\cdot\mathbf{r}_{jk}\Bigg] \\
&+\frac{1}{c^2}\sum_{j\neq i}\frac{\mathbb{G}m_j}{r_{ij}^3}\mathbf{r}_{ij}\cdot(4\mathbf{v}_i-3\mathbf{v}_j)(\mathbf{v}_i-\mathbf{v}_j) \\
&-\frac{7}{2c^2}\sum_{j\neq i}\sum_{k\neq j}\frac{\mathbb{G}^2m_jm_k\mathbf{r}_{jk}}{r_{ij}r_{jk}^3}+\mathrm{O}(c^{-4}).
\end{aligned}
\tag{J.16}
$$

Here $\mathbf{r}_{ij} = \mathbf{r}_i - \mathbf{r}_j$ and $r_{ij} = |\mathbf{r}_{ij}|$. When there is only one massive body, the equations of motion for a test particle reduce to (J.14) with $\alpha = 0$. Thorough discussions of these equations are given by Misner et al. (1973), Will (1993), and Poisson & Will (2014).

Problems

Problems are marked [1], [2], [3] in order of increasing difficulty.

1.1 [1] A **geosynchronous orbit** is an orbit with period equal to the spin period of the Earth. Show that a circular geosynchronous orbit has radius $r_{\text{sync}} = 42\,164$ km. You may approximate the Earth as spherical and ignore the effects of the Moon and Sun. Hint: 42 241 km is not correct.

1.2 [1] Prove the following formulas for time averages over a bound Kepler orbit of semimajor axis a and eccentricity e:

$$\langle v^4 \rangle = \left(\frac{\mathbb{G}M}{a}\right)^2 \left[4(1-e^2)^{-1/2} - 3\right], \tag{P.1a}$$

$$\langle v^2/r \rangle = \frac{\mathbb{G}M}{a^2}\left[2(1-e^2)^{-1/2} - 1\right], \tag{P.1b}$$

$$\langle (\mathbf{r}\cdot\mathbf{v})^2/r^3 \rangle = \frac{\mathbb{G}M}{a^2}\left[(1-e^2)^{-1/2} - 1\right]. \tag{P.1c}$$

1.3 [1] Prove the following formulas for time averages over a bound Kepler orbit of semimajor axis a and eccentricity e:

$$\langle (a/r)^4 \cos f \rangle = \frac{e}{(1-e^2)^{5/2}}, \tag{P.2a}$$

$$\langle (a/r)^4 \cos^3 f \rangle = \frac{3e}{4(1-e^2)^{5/2}}, \tag{P.2b}$$

$$\langle (a/r)^4 \cos f \sin^2 f \rangle = \frac{e}{4(1-e^2)^{5/2}}, \tag{P.2c}$$

$$\langle (r/a)^3 \cos f \rangle = -\tfrac{5}{8}e(4+3e^2), \tag{P.2d}$$

$$\langle (r/a)^3 \cos^3 f \rangle = -\tfrac{5}{8}e(3+4e^2), \tag{P.2e}$$

$$\langle (r/a)^3 \cos f \sin^2 f \rangle = -\tfrac{5}{8} e(1 - e^2). \tag{P.2f}$$

1.4 [1] Many computing languages provide the function atan2(y, x), which yields the angle in radians between the positive x-axis and the vector from the origin to the point (x, y). Find an expression for the true anomaly of a bound orbit in terms of the eccentric anomaly using this function.

1.5 [1] Find the maximum time that a comet on a parabolic orbit can spend inside the Earth's orbit ($r < 1$ au) during a single perihelion passage. You may ignore perturbations from the planets.

1.6 [1] A spacecraft travels around the Earth on a circular orbit with an altitude of 300 km. The spacecraft engine can give it one or more velocity impulses Δv. (a) What Δv (in km s^{-1}) is needed to place the spacecraft on an orbit that escapes from the Earth's gravitational field? (b) The mission designers want to place the spacecraft in a geosynchronous orbit, that is, a circular orbit of radius $r_{\text{sync}} = 42\,164$ km (Problem 1.1). This can be done by two impulses, the first to place the spacecraft on an eccentric orbit with apoapsis equal to r_{sync} (a **Hohmann transfer orbit**) and the second at apoapsis to circularize the orbit. What is the total Δv required? (c) Suppose that the transfer to geosynchronous orbit is accomplished by a slow steady burn, so the spacecraft spirals out on a nearly circular orbit. What is the total Δv required?

1.7 [2] A spacecraft travels around a star with a mass of $1M_\odot$, in a circular orbit with a semimajor axis of 1 au. The spacecraft engine can give it one or more velocity impulses Δv. What is the minimum total Δv needed to crash the spacecraft into the star? You may assume that the radius of the star is very small. Hint: a single impulse $\Delta v = 29.8$ km s^{-1} is not the best solution.

1.8 [2] A body on a Kepler orbit suffers an instantaneous velocity impulse or kick when it passes through periapsis. Show that (a) the periapsis distance of the orbit after the kick must be the same as or smaller than the original periapsis distance; (b) if the velocity kick is $\mathrm{O}(\epsilon)$, then the change in periapsis is $\mathrm{O}(\epsilon^2)$.

1.9 [1] In July 2015 the New Horizons spacecraft encountered Pluto. The impact parameter of the encounter was 13 700 km and the relative velocity was 13.8 km s^{-1}. By what angle was the spacecraft's trajectory deflected during the encounter? The mass of Pluto is 1.303×10^{22} kg.

1.10 [2] A test particle orbits a central mass. Erect a Cartesian coordinate system (x_1, x_2, x_3) with origin at the central mass, the positive x_1-axis pointing toward the periapsis of the orbit, the positive x_2-axis pointing toward true anomaly $f = \frac{1}{2}\pi$, and the positive x_3-axis parallel to the angular-momentum vector. Thus the unit vectors $(\hat{\mathbf{x}}_1, \hat{\mathbf{x}}_2, \hat{\mathbf{x}}_3)$ form a right-handed orthonormal triad. The position of the test particle can be written $\mathbf{r} = r(\hat{\mathbf{x}}_1 \cos f + \hat{\mathbf{x}}_2 \sin f)$. Let $\hat{\mathbf{n}}$ be an arbitrary unit vector. Show that the orbit average of the square of the projection of $\mathbf{r}$ along $\hat{\mathbf{n}}$ is

$$\langle (\mathbf{r}\cdot\hat{\mathbf{n}})^2 \rangle = \tfrac{1}{2}a^2\left[j^2 + 5(\mathbf{e}\cdot\hat{\mathbf{n}})^2 - (\mathbf{j}\cdot\hat{\mathbf{n}})^2\right], \tag{P.3}$$

where $\mathbf{e}$ is the eccentricity vector (Box 1.1) and $\mathbf{j}$ is the dimensionless angular momentum (eq. 1.33). Hint: use equations (1.65d)–(1.65f) and (B.6a).

1.11 [3] Write code to solve Kepler's equation (1.49) for the eccentric anomaly u. The algorithm should converge quickly for all values of the mean longitude ℓ and eccentricity $e < 1$.

1.12 [2] Find Gauss's f and g functions for unbound orbits, in terms of the true anomaly and eccentricity.

1.13 [2] A test particle orbits in an axisymmetric, near-Kepler potential of the form $\Phi(R,z) = -\mathbb{G}M/(R^2+z^2)^{1/2} + \epsilon\phi(R,z)$, where (R,z) are cylindrical coordinates and $\phi(R,z)$ is an even function of z. If the orbit is nearly circular and lies near the equatorial plane $z = 0$, show that the apsidal and nodal precession rates are

$$\frac{\mathrm{d}\varpi}{\mathrm{d}t} = -\epsilon\left(\frac{R_g^3}{\mathbb{G}M}\right)^{1/2}\left(\frac{1}{R}\frac{\partial\phi}{\partial R} + \frac{1}{2}\frac{\partial^2\phi}{\partial R^2}\right)_{(R_g,0)} + \mathrm{O}(\epsilon^2),$$

$$\frac{\mathrm{d}\Omega}{\mathrm{d}t} = \tfrac{1}{2}\epsilon\left(\frac{R_g^3}{\mathbb{G}M}\right)^{1/2}\left(\frac{1}{R}\frac{\partial\phi}{\partial R} - \frac{\partial^2\phi}{\partial z^2}\right)_{(R_g,0)} + \mathrm{O}(\epsilon^2), \tag{P.4}$$

where R_g is the guiding center of the orbit.

1.14 [1] A **collision orbit** is an orbit with negligible angular momentum. Suppose that a test particle approaches a point mass M on a collision orbit. Show that as the angular momentum approaches zero, (a) the test particle approaches and recedes from the central body in the same direction; (b) the radius of the test particle before and after the encounter is

$$r(t) = \left(\frac{9\mathbb{G}M}{2}\right)^{1/3} |t-t_0|^{2/3} + \mathrm{O}(|t-t_0|^{4/3}), \tag{P.5}$$

where t_0 is the collision time.

1.15 [2] Show that the following set of coordinates (left column) and momenta (right column) are canonical:

$$\begin{aligned} &\lambda = \ell + \omega + \Omega, && \Lambda, \\ &-[2(\Lambda - L)]^{1/2} \sin \varpi, && [2(\Lambda - L)]^{1/2} \cos \varpi, \\ &-[2(L - L_z)]^{1/2} \sin \Omega, && [2(L - L_z)]^{1/2} \cos \Omega. \end{aligned} \tag{P.6}$$

This is an alternative form of the Poincaré variables (1.91).

1.16 [2] Show that the following set of coordinates (left column) and momenta (right column) are canonical:

$$\begin{aligned} &\lambda = \ell + \omega + \Omega, && \Lambda, \\ &(\Lambda - L)^{1/2} e^{i\varpi}, && i(\Lambda - L)^{1/2} e^{-i\varpi}, \\ &(L - L_z)^{1/2} e^{i\Omega}, && i(L - L_z)^{1/2} e^{-i\Omega}. \end{aligned} \tag{P.7}$$

1.17 [3] A planet of radius R_p transits a star of radius R_*. Assume that the stellar disk has uniform surface brightness (i.e., there is no limb darkening). Let $r \equiv R_\mathrm{p}/R_* < 1$ and let $R_* d$ be the distance between the centers of the star and planet. Show that between first and second contact ($1 - r < d < 1 + r$), the flux from the star is a fraction $1 - f$ of its unobscured value, where (Mandel & Agol 2002)

$$\pi f(r, d) = r^2 \cos^{-1} \frac{d^2 + r^2 - 1}{2rd} + \cos^{-1} \frac{d^2 - r^2 + 1}{2d} - \left[d^2 - \tfrac{1}{4}(d^2 - r^2 + 1)^2\right]^{1/2}. \tag{P.8}$$

1.18 [1] An imaged planet is detected and tracked, but only for a small fraction of its orbit. The observations are used to estimate the orbital elements of the planet, following the procedure in §1.6.4. It is found that the semimajor axis a and orbital period P are poorly determined but the ratio a^3/P^2 is relatively reliable. Why is this so?

1.19 [2] Show that $r \exp(\mathrm{i}f)$ can be written in terms of the mean anomaly ℓ as a power series in the eccentricity,

$$\begin{aligned} \frac{r}{a} \exp(\mathrm{i}f) = {} & \exp(\mathrm{i}\ell) + \left[-\tfrac{3}{2} + \tfrac{1}{2}\exp(2\mathrm{i}\ell)\right]e + \left[-\tfrac{3}{8}\cos\ell - \tfrac{5}{8}\mathrm{i}\sin\ell \right. \\ & \left. + \tfrac{3}{8}\exp(3\mathrm{i}\ell)\right]e^2 + \left[-\tfrac{1}{3}\cos 2\ell - \tfrac{5}{12}\mathrm{i}\sin 2\ell + \tfrac{1}{3}\exp(4\mathrm{i}\ell)\right]e^3 + \mathrm{O}(e^4), \end{aligned} \tag{P.9}$$

and evaluate the term proportional to e^4.

1.20 [2] Equations (1.176), (1.177) and (1.179) imply that around an oblate planet

$$\kappa_R^2(R) + \kappa_z^2(R) - 2\kappa_\phi^2(R) = 0. \tag{P.10}$$

(a) Without using a multipole expansion, prove that this result holds in *any* axisymmetric potential $\Phi(R, z)$, so long as the potential is an even function of z and the local density of the mass generating the potential vanishes. Hint: if the density is zero, the potential must satisfy Laplace's equation

$$\nabla^2\Phi = \frac{1}{R}\frac{\partial}{\partial R}R\frac{\partial\Phi}{\partial R} + \frac{\partial^2\Phi}{\partial z^2} = 0. \tag{P.11}$$

(b) Show that if equation (P.10) holds and the apsidal and nodal precession rates $\dot{\varpi}$ and $\dot{\Omega}$ are small, then

$$\dot{\varpi} = -\dot{\Omega} \tag{P.12}$$

with an error $\mathrm{O}(\dot{\varpi}^2, \dot{\Omega}^2)$. Hint: use equations (1.174) and (1.175).

1.21 [3] If a particle orbiting in a Kepler potential is subjected to an external force per unit mass $\mathbf{F}_{\mathrm{ext}}$, its energy E and angular momentum $\mathbf{L}$ change at a rate $\mathrm{d}E/\mathrm{d}t = \mathbf{F}_{\mathrm{ext}} \cdot \mathbf{v}$ and $\mathrm{d}\mathbf{L}/\mathrm{d}t = \mathbf{r} \times \mathbf{F}_{\mathrm{ext}}$. Starting from these results prove Gauss's equations (1.200) for the rate of change of semimajor axis, eccentricity, inclination and longitude of the ascending node.

1.22 [1] Prove that the general solution to the differential equations (1.203) for the evolution of an orbit under weak radiation forces is given implicitly by

$$t = t_0 + \frac{2ca_0^2}{5k_{\mathrm{rad}}A^2}\int_e^{e_0} \mathrm{d}e \frac{e^{3/5}}{(1-e^2)^{3/2}}, \quad a = \frac{a_0}{A}\frac{e^{4/5}}{1-e^2}; \tag{P.13}$$

here a_0 and e_0 are the semimajor axis and eccentricity at the initial time t_0, and the constant $A = e_0^{4/5}/(1-e_0^2)$.

1.23 [3] A small body orbiting a planet is subject to radiation pressure from the planet's host star. Approximate the radiation pressure by its lowest order term in an expansion in powers of v/c. Thus, from equation (a) of Box 1.5, the force per unit mass on a body of mass m and radius R is $k\hat{\mathbf{r}}_*$ where $k = L R^2/(4mr_*^2c)$, L is the luminosity of the star, and the vector from the star to the planet is $\mathbf{r}_* = r_*\hat{\mathbf{r}}_*$. Find the differential equations for the evolution of the small body's semimajor axis, eccentricity and inclination. You should average over the orbit of the small body but not over the orbit of the planet, and ignore the possibility that the small body is occulted by the planet. Express the result in terms of $\hat{\mathbf{x}} \cdot \hat{\mathbf{r}}_*$, $\hat{\mathbf{y}} \cdot \hat{\mathbf{r}}_*$ and $\hat{\mathbf{z}} \cdot \hat{\mathbf{r}}_*$, where $\hat{\mathbf{x}}$ points from the planet to the periapsis of the small body, $\hat{\mathbf{z}}$ is parallel to the angular momentum of the small body, and $\hat{\mathbf{y}} = \hat{\mathbf{z}} \times \hat{\mathbf{x}}$.

2.1 [2] (a) Analyze the behavior of the drift-kick-drift leapfrog integrator in a harmonic potential, following the steps in equations (2.12)–(2.17). Show that the scale factor $\kappa_\pm$ is given by equation (2.23), the same as for the modified Euler method. (b) If the scale factor is the same for modified Euler and leapfrog, why is the former first-order and the latter second-order in accuracy?

2.2 [1] Prove that the implicit first-order integrator (2.56) is symplectic.

2.3 [3] Prove that the trapezoidal rule (2.75) and the implicit midpoint method (2.76) are reversible when applied to reversible systems, and that the implicit midpoint method is symplectic when applied to Hamiltonian systems.

2.4 [2] Let $\mathbf{E}_h$ denote Euler's method with timestep h. Prove that the implicit midpoint method (2.76) can be written as $\mathbf{E}_{h/2}\mathbf{E}^{-1}_{-h/2}$.

2.5 [1] In a central potential, where $\nabla\Phi(\mathbf{r})$ is parallel or anti-parallel to $\mathbf{r}$, the angular momentum $\mathbf{r} \times \mathbf{v}$ of a test particle is conserved. Prove that the drift and kick operators conserve angular momentum exactly in a central potential, and hence that the modified Euler and leapfrog integrators conserve angular momentum as well.

2.6 [2] Let $\mathbf{P}_h$ be an integrator of order k with timestep h. Then $\mathbf{P}_{ah}\mathbf{P}_{bh}\mathbf{P}_{ah}$ is an integrator with timestep h if $2a + b = 1$. Prove that this integrator has order $k + 1$ or higher if k is even and $a^{-1} = 2 - 2^{1/(k+1)}$.

2.7 [3] Evaluate the coefficients $\{\beta_m\}$ for a Störmer integrator of order $k = 8$.

2.8 [3] At time $t \to -\infty$ a test particle is in a bound orbit around a star of mass M. The orbit has semimajor axis a and eccentricity e. The star loses mass in a spherically symmetric wind at a rate

$$\frac{\mathrm{d}M}{\mathrm{d}t} = -\frac{\Delta M}{(2\pi\sigma^2)^{1/2}} \exp\left[-\frac{t^2}{2\sigma^2}\right]. \tag{P.14}$$

This exercise uses numerical orbit integrations to find the eccentricity of the planet as $t \to \infty$. Assume $\Delta M = 0.5M$ and consider initial eccentricities $e = 0$ and 0.6. For each of these values, plot the difference between the initial and final eccentricity as a function of the dimensionless ratio σ/P, where $P = 2\pi a^{3/2}(\mathbb{G}M)^{-1/2}$ is the initial orbital period. Hint: when $e \neq 0$ the answer depends on the orbital phase of the particle at $t = 0$.

2.9 [3] Write code that can follow a test particle on a highly eccentric Kepler orbit, and use it to integrate an orbit with an eccentricity $e = 0.9999$ for 1 000 orbital periods. You may use public or commercial software in the code. The goal is to achieve a fractional energy error less than 10^{-6} at the end of the integration. How many evaluations of the force per orbit are needed to reach this accuracy?

2.10 [2] In the numerical integration of a Kepler orbit, roundoff error produces a relative energy error at step j given by $\Delta E/E = 2^{-p} f_j$, where p is the precision and f_j is of order unity. The corresponding error in the mean motion is given by $\Delta n/n = \frac{3}{2}\Delta E/E$ if $|\Delta E/E| \ll 1$. (a) Show that the error in mean longitude after N steps is $\Delta \ell = h \sum_{j=1}^{N} \Delta n_j (N - j)$, where h is the timestep. (b) Assume that the errors f_j are statistically independent, with mean $\overline{f}$ and standard deviation σ. Then prove equations (2.159) for $N \gg 1$.

3.1 [1] Plot the eccentricities and inclinations of the Jupiter Trojan asteroids, using data from the JPL Small-Body Database Search Engine https://ssd.jpl.nasa.gov/sbdb_query.cgi.

3.2 [1] What is the libration period in years for a Trojan asteroid? You may model the Sun–Jupiter–asteroid system using the circular restricted three-body problem, and assume that the libration amplitude is small.

3.3 [2] Spacecraft located near the Sun–Earth Lagrange points L1 or L2 are unstable, in the sense that their mean distance from the Lagrange point grows with time as $\exp(\lambda t)$. What is the e-folding time λ^{-1} in days?

3.4 [2] This problem investigates the behavior of spacecraft orbits around the Sun–Earth Lagrange points L1 or L2. (a) In the limit $m_1 \ll m_0$, show that the quantity ν_L^2 in equations (3.31) is equal to $4\Omega^2$, where Ω is the mean motion of m_1 around m_0. (b) Show that a solution of the linearized equations of motion (3.25) is

$$x_a = A\cos(\omega t + \phi), \quad x_b = -bA\sin(\omega t + \phi), \quad x_c = B\cos(\omega_c t + \phi_c), \tag{P.15}$$

where A, B, ϕ and ϕ_c are arbitrary constants and $\omega, \omega_c > 0$. Find ω, ω_c and b. (c) Why does this motion appear to be stable when we have shown that the collinear Lagrange points are unstable? Comment: the frequencies ω and ω_c are found to differ by less than 4%. By varying the amplitudes A and B, we also vary the frequencies ω and ω_c through nonlinear terms in the equations of motion. If the amplitudes and phases are chosen such that $\omega = \omega_c$ and $\phi = \phi_c$, we have a periodic orbit called a **halo orbit**. Such orbits never intersect the Sun–Earth line $x_b = x_c = 0$, so the Sun does not interfere with spacecraft communications. Many spacecraft are launched into halo orbits around L1 or L2.

3.5 [2] The gravitational force exerted on a body of mass m_1 at position $\mathbf{r}_1$ by a body of mass m_2 at $\mathbf{r}_2$ is $\mathbb{G}m_1m_2(\mathbf{r}_2-\mathbf{r}_1)/|\mathbf{r}_2-\mathbf{r}_1|^3$. Suppose that this force is modified to $\mathbb{G}m_1m_2(\mathbf{r}_2-\mathbf{r}_1)/|\mathbf{r}_2-\mathbf{r}_1|^\alpha$. How does the position of the triangular Lagrange point in the circular restricted three-body problem depend on the parameter α?

3.6 [1] A satellite travels around the Earth in a circular orbit. (a) At what semimajor axis is the nodal precession due to the Earth's quadrupole moment equal to the average nodal precession rate due to the Sun? You may neglect the obliquity of the Earth relative to the ecliptic and the eccentricity of the Earth's orbit. (b) Explain the relation of this radius to the Laplace radius defined in equation (5.74).

3.7 [2] This exercise derives the variational ellipse (3.95) without using Hamilton's equations. (a) For simplicity, assume that the Moon orbits the Earth in the plane of the ecliptic and that the Earth–Sun orbit is circular. Show that the quadrupole potential due to the Sun can be written in polar coordinates as

$$\Phi_\odot(r,\phi,t) = -\frac{\mathbb{G}m_\odot r^2}{4r_\odot^3} - \frac{3\,\mathbb{G}m_\odot r^2}{4r_\odot^3}\cos 2(\phi-\phi_\odot), \qquad \text{(P.16)}$$

where $r_\odot$, $m_\odot$ and $\phi_\odot$ are the distance, mass and azimuthal angle of the Sun. (b) Ignore the first, axisymmetric, term in equation (P.16) since it causes only a small fractional change in the mean motion (see Problems 3.8 and 3.9). The second term causes a fractional change in the potential that is of order $\epsilon = m_\odot\overline{r}^3/[(m_1+m_2)r_\odot^3]$, where m_1 and m_2 are the masses of the Earth and Moon and $\overline{r}$ is the mean radius of the lunar orbit. Write $r(t) = \overline{r} + \Delta r(t)$, $\phi(t) = \overline{\phi}(t) + \Delta\phi(t)$, where $\Delta r(t)$ and $\Delta\phi(t)$ are $\mathrm{O}(\epsilon)$ and $\mathrm{d}\overline{\phi}/\mathrm{d}t = \overline{n} = [\mathbb{G}(m_1+m_2)/\overline{r}^3]^{1/2}$. Then using equation (B.18), show that to the lowest order in ϵ the trajectory of the Moon can be written in the form

$$\begin{aligned}\frac{\mathrm{d}^2\Delta r}{\mathrm{d}t^2} - 2\overline{n}\,\overline{r}\frac{\mathrm{d}\Delta\phi}{\mathrm{d}t} - 3\overline{n}^2\Delta r &= \frac{3\,\mathbb{G}m_\odot\overline{r}}{2r_\odot^3}\cos 2(\overline{\phi}-\phi_\odot),\\ \frac{\mathrm{d}^2\Delta\phi}{\mathrm{d}t^2} + \frac{2\overline{n}}{\overline{r}}\frac{\mathrm{d}\Delta r}{\mathrm{d}t} &= -\frac{3\,\mathbb{G}m_\odot}{2r_\odot^3}\sin 2(\overline{\phi}-\phi_\odot).\end{aligned} \qquad \text{(P.17)}$$

(c) Let $n_0^2 = \mathbb{G}m_0/r_0^3$ be the squared mean motion of the Sun. Show that $n_0^2/n^2 = \epsilon$ and thus that terms like $\mathrm{d}(\overline{\phi} - \phi_0)/dt$ can be set equal to $\overline{n}$ with a fractional error that is $\mathrm{O}(\epsilon^{1/2})$. Using this approximation, show that the solution to equations (P.17) is

$$\Delta r = -\frac{n_0^2 \overline{r}}{\overline{n}^2} \cos 2(\overline{\phi} - \phi_0) + \alpha \overline{r} \cos(\overline{n}t + \beta),$$
$$\Delta \phi = \frac{11 n_0^2}{8\overline{n}^2} \sin 2(\overline{\phi} - \phi_0) - 2\alpha \sin(\overline{n}t + \beta), \tag{P.18}$$

where α and β are arbitrary constants. The terms independent of α describe the variational ellipse (3.94). (d) What is the interpretation of the terms proportional to α? (e) Show that the azimuthal term independent of α vanishes at lunar and solar eclipses, which is probably why this term was not known to Greek astronomers.

3.8 [2] Consider the term in the disturbing function (3.75)

$$H_{\mathrm{ann}} = -\frac{3\,\mathbb{G}m_0 a^2}{4a_0^3} e_0 \cos(\lambda_0 - \varpi_0). \tag{P.19}$$

(a) Using Lagrange's equations (1.187), show that this term leaves the semi-major axis, eccentricity, inclination, longitude of periapsis, and longitude of node constant, and gives rise to variations in the azimuthal angle

$$\Delta \phi = -\frac{3n_0}{n} e_0 \sin(\lambda_0 - \varpi_0). \tag{P.20}$$

This term is known as the **annual equation**. (b) Physically, the annual variation (P.20) arises because the angular speed of the Moon is determined by the net gravitational attraction from the Earth and Sun. The Sun's attraction tends to cancel the attraction from the Earth, and this cancellation is stronger when the Sun is at its periapsis. Using this physical picture, derive equation (P.20) without using Hamilton's equations. Hint: approximate the effect of the solar potential using the first term of equation (P.16), and observe that since this term is spherically symmetric, the angular momentum of the orbit remains constant. Comment: the amplitude of this periodic variation in the Moon's azimuth is $3n_0 e_0/n = 0.215^\circ$; the accurate estimate from Brown's lunar theory is 0.186°.

3.9 [2] Consider the term in the disturbing function (3.75)

$$H = \frac{\mathbb{G}m_0 a^2}{2a_0^3} e \cos(\lambda - \varpi) = \tfrac{1}{2} n_0^2 a^2 (k \cos \lambda + h \sin \lambda). \tag{P.21}$$

(a) Evaluate the effects of this Hamiltonian using the simplified Lagrange's equations (1.193), working to first order in the strength of the perturbation and assuming that the unperturbed eccentricity is zero. Show that the resulting perturbation in azimuth vanishes, and that the radius is constant and equal to $r = a + \frac{1}{2}a(n_0/n)^2$. Thus the radius of a circular orbit is not equal to the semimajor axis. Comment: a similar situation arises for orbits around an oblate planet; see Box 1.4. (b) Derive this relation between r and a without using Hamilton's equations. Hint: approximate the effect of the solar potential using the first term of equation (P.16).

3.10 [3] Examine the motion of two bodies in the planar Hill's problem. The bodies have masses m_1 and m_2. In the absence of forces other than the gravitational attraction of the host star, their positions relative to the reference vector $\overline{a}$ can be written (cf. eq. 3.104)

$$\Delta x_i = \alpha_i - k_{\mathrm{H},i} \cos \Omega t + h_{\mathrm{H},i} \sin \Omega t,$$
$$\Delta y_i = -\tfrac{3}{2}\alpha_i \Omega t + \gamma_i + 2k_{\mathrm{H},i} \sin \Omega t + 2h_{\mathrm{H},i} \cos \Omega t, \qquad \text{(P.22)}$$

where $k_{\mathrm{H},i}$ and $h_{\mathrm{H},i}$ are the Hill eccentricity components (3.128) and $i = 1, 2$. Now subject each body to some additional force $\mathbf{F}_i$. (a) Show that

$$m_i \frac{\mathrm{d}k_{\mathrm{H},i}}{\mathrm{d}t} = \frac{\sin \Omega t}{\Omega} F_{x,i} + \frac{2 \cos \Omega t}{\Omega} F_{y,i},$$
$$m_i \frac{\mathrm{d}h_{\mathrm{H},i}}{\mathrm{d}t} = \frac{\cos \Omega t}{\Omega} F_{x,i} - \frac{2 \sin \Omega t}{\Omega} F_{y,i}. \qquad \text{(P.23)}$$

Hint: cf. equations (3.129) and (3.130). (b) If the forces $\mathbf{F}_i$ arise from the mutual interaction of the two bodies, gravitational or otherwise, show that $m_1 k_{\mathrm{H},1} + m_2 k_{\mathrm{H},2}$ and $m_1 h_{\mathrm{H},1} + m_2 h_{\mathrm{H},2}$ are conserved. (c) In Hill's problem the eccentricity $\epsilon_i = (k_{\mathrm{H},i}^2 + h_{\mathrm{H},i}^2)^{1/2}$. Show that if the two bodies initially have circular orbits, $\epsilon_1 = \epsilon_2 = 0$, then $m_1\epsilon_1 = m_2\epsilon_2$ at all future times.

3.11 [3] A planet composed of a massive compact core and an extended, opaque, low-density atmosphere travels on a circular orbit around its host star. The planet fills its Roche lobe. Let (x, y, z) be Cartesian coordinates with origin at the center of mass of the planet, x-axis pointing along the star-planet line and z-axis normal to the orbital plane. (a) Show that the lengths of the principal axes of the planet (the intersections of the planetary surface with the coordinate axes) are r_{H}, $\frac{2}{3}r_{\mathrm{H}}$, and $(3^{2/3} - 3^{1/3})r_{\mathrm{H}}$, where r_{H} is the Hill radius. (b) Show that the area obscured by the planet when it transits its host star is $1.3349\, r_{\mathrm{H}}^2$.

3.12 [3] (a) Prove that Hill's equations (3.109) can be derived from the Hamiltonian

$$H = \tfrac{1}{2}(p_\xi^2 + p_\eta^2 + p_\zeta^2) + \eta p_\xi - \xi p_\eta - \xi^2 + \tfrac{1}{2}\eta^2 + \tfrac{1}{2}\zeta^2 - \frac{1}{\rho}, \quad \text{(P.24)}$$

where $\rho^2 = \xi^2 + \eta^2 + \zeta^2$. (b) For simplicity, ignore motion in ζ by setting $\zeta = p_\zeta = 0$. Transform to new canonical coordinates (λ, h) and momenta (p_λ, p_h) using the generating function

$$\begin{aligned} S_2(p_\lambda, p_h, \xi, \eta, \tau) &= (\tfrac{1}{2}\xi^2 + 2p_\lambda^2 + \tfrac{1}{2}p_h^2 - 2\xi p_\lambda)\cot\tau \\ &\quad + (\xi p_h - 2p_\lambda p_h)\csc\tau - \xi\eta + p_\lambda\eta, \end{aligned} \quad \text{(P.25)}$$

and show that the old coordinates and momenta can be written in terms of the new as

$$\begin{aligned} \xi &= 2p_\lambda + h\sin\tau - p_h\cos\tau, \\ \eta &= \lambda + 2h\cos\tau + 2p_h\sin\tau, \\ p_\xi &= -\lambda - h\cos\tau - p_h\sin\tau, \\ p_\eta &= -p_\lambda - h\sin\tau + p_h\cos\tau. \end{aligned} \quad \text{(P.26)}$$

(c) Show that the new Hamiltonian is

$$H' = H + \frac{\partial S_2}{\partial \tau} = -\tfrac{3}{2}p_\lambda^2 - \frac{1}{\rho}. \quad \text{(P.27)}$$

(d) Show that at large distances ($\rho \to \infty$) the coordinate-momentum pair (h, p_h) is equal to $(h_{\rm H}, k_{\rm H})$, where $k_{\rm H}$ and $h_{\rm H}$ are the Hill eccentricity components defined in equation (3.128).

3.13 [2] Consider Hill's problem when body 1 is on a circular orbit. Show that the Hill eccentricity components $k_{\rm H}$, $h_{\rm H}$ defined in equations (3.128) are related to the usual eccentricity components $k = e\cos\varpi$, $h = e\sin\varpi$ (eqs. 1.71) of body 2 by

$$k_{\rm H} + \mathrm{i}h_{\rm H} = \left(\frac{m_0}{m_1 + m_2}\right)^{1/3} \mathrm{e}^{\mathrm{i}\lambda_0}(k - \mathrm{i}h), \quad \text{(P.28)}$$

where λ_0 is the mean longitude at time $t = 0$. You may assume that $|k|, |h| \ll 1$.

3.14 [2] Equation (G.11) defines the function $I^{1/2}W$ in the general three-body problem. (a) Show that this function has an extremum when the three bodies form an equilateral triangle. (b) Show that as $m_2 \to 0$, the contours of $I^{1/2}W$ have the same shape as the zero-velocity surfaces in the orbital plane of the circular restricted three-body problem.

4.1 [2] Prove that the equations of motion resulting from the astrocentric Hamiltonian (4.14) are the same as equations (4.5).

4.2 [1] Let $(\mathbf{q}, \mathbf{p})$ be a set of canonical coordinates and momenta. Define a new set of coordinates $\mathbf{q}'$ by a linear transformation, $\mathbf{q}' = \mathbf{A}\mathbf{q}$, where $\mathbf{A}$ is a constant matrix. What are the new momenta $\mathbf{p}'$ canonically conjugate to $\mathbf{q}'$?

4.3 [2] Let $\{\mathbf{r}_i, \mathbf{p}_i\}$, $i = 0, \ldots, N-1$ be a set of canonical coordinates and momenta for an N-body system. Define a new set of coordinates by $\mathbf{r}'_i = \sum_{j=0}^{N-1} A_{jk}\mathbf{r}_k$, and let $\mathbf{p}'_i$ be the momentum conjugate to $\mathbf{r}'_i$. The total angular momentum of the system is $\mathbf{L} = \sum_{i=0}^{N-1} \mathbf{r}_i \times \mathbf{p}_i$. Prove that in the new coordinates the angular momentum has the same form, $\mathbf{L} = \sum_{i=0}^{N-1} \mathbf{r}'_i \times \mathbf{p}'_i$.

4.4 [3] The expansion of the disturbing function in equations (4.98), (4.99) and (4.104) is simplified in hierarchical three-body systems in which $\alpha = a_1/a_2 \ll 1$. Expand the disturbing function defined by these equations in powers of α and discard all terms that are $\mathrm{O}(\alpha^3)$ or higher. Also assume that the inclination of the outer body $I_2 = 0$. Show that the result is equivalent to the lunar disturbing function (3.75).

4.5 [3] Write code to evaluate the Laplace coefficient $b_s^m(\alpha)$ and its derivative $Db_s^m(\alpha) = \mathrm{d}b_s^m(\alpha)/\mathrm{d}\alpha$.

4.6 [2] Show by numerical examples that the Laplace coefficient $b_s^m(\alpha)$ can be evaluated by making crude estimates of $b_s^n(\alpha)$ and $b_s^{n-1}(\alpha)$ for $n \gg m$ and then repeatedly using the recursion relation (4.118) to evaluate the sequence $b_s^{n-2}(\alpha), b_s^{n-3}(\alpha), \ldots, b_s^0(\alpha)$. Hint: prove and use the sum rule $\frac{1}{2}b_s^0(\alpha) + \sum_{m=1}^{\infty} b_s^m(\alpha) = |1-\alpha|^{-2s}$.

4.7 [1] An ensemble of systems of equally spaced, equal-mass planets has a uniform distribution of separations $|\Delta a|$ over the range $0.5r_{1/4}$–$2.5r_{1/4}$ in which the relation (4.139) between separation and lifetime is approximately correct. Show that the number of systems that survive for a time $\tau \to \tau + \mathrm{d}\tau$ is $\mathrm{d}N \propto \mathrm{d}\tau/\tau$.

5.1 [1] Prove equations (5.42), (5.44) and (5.49).

5.2 [1] (a) Show that the quantities $\sum_{k=1}^{N} |Z_{e,k}|^2$ and $\sum_{k=1}^{N} |Z_{I,k}|^2$ are constants of motion for the Lagrange–Laplace equations (5.29). (b) The **angular-momentum deficit** is minus the difference between the z-component of the total angular momentum of a planetary system and the z-component of the total angular momentum of a system with the same planetary masses and semimajor axes but zero eccentricities and inclinations. Show that the angular-momentum deficit is

$$\mathrm{AMD} \equiv \sum_{k=1}^{N} m_k \big[\mathbb{G}(m_0 + m_k)a_k\big]^{1/2}\big[1 - (1 - e_k^2)^{1/2}\cos I_k\big], \tag{P.29}$$

where m_0 is the mass of the central star and m_k, a_k, e_k and I_k are the masses, semimajor axes, eccentricities and inclinations of the planets. (c) Show that the angular-momentum deficit is conserved in the secular dynamics of an isolated planetary system. (d) What is the relation between the angular-momentum deficit and the conserved quantities in part (a)?

5.3 [2] A satellite orbits a flattened planet. (a) Use the orbit-averaged quadrupole potential, equation (5.64), and Lagrange's equations (1.187) to show that the longitude of the ascending node Ω and the argument of periapsis $\omega = \varpi - \Omega$ precess at the rates

$$\frac{\mathrm{d}\Omega}{\mathrm{d}t} = -\frac{3(\mathbb{G}M_\mathrm{p})^{1/2}J_2R_\mathrm{p}^2}{2a^{7/2}(1-e^2)^2}\cos I,$$

$$\frac{\mathrm{d}\omega}{\mathrm{d}t} = \frac{3(\mathbb{G}M_\mathrm{p})^{1/2}J_2R_\mathrm{p}^2}{4a^{7/2}(1-e^2)^2}(5\cos^2 I - 1). \tag{P.30}$$

(b) Compare these results in the limit $e \ll 1$, $I \ll 1$ to equations (1.180). Why is the term proportional to J_2^2 in equation (1.180b) missing? (c) At the **critical inclination**, $I = \cos^{-1}(1/\sqrt{5}) = 63.4°$, the argument of periapsis does not precess. Why might an orbit of this kind be useful for an artificial satellite?

5.4 [2] This problem examines the same system as Problem 5.3, a satellite orbiting a flattened planet, using Milankovich's equations rather than Lagrange's equations. (a) Find the equations governing the rate of change of the dimensionless angular momentum $\mathbf{j}$ and the eccentricity vector $\mathbf{e}$. (b) Show that the scalar eccentricity e and the inclination I relative to the planet's equator are constants of motion. (c) Show that the angular-momentum vector precesses uniformly around the spin axis $\hat{\mathbf{n}}_\mathrm{p}$ of the planet at angular speed

$$\boldsymbol{\omega}_j = -\frac{3(\mathbb{G}M_\mathrm{p})^{1/2}J_2R_\mathrm{p}^2\cos I}{2a^{7/2}(1-e^2)^2}\hat{\mathbf{n}}_\mathrm{p}. \tag{P.31}$$

(d) In the frame rotating with the precession of the angular-momentum vector, the rate of change of eccentricity is $(\mathrm{d}\mathbf{e}/\mathrm{d}t)_\mathrm{rot} = \mathrm{d}\mathbf{e}/\mathrm{d}t - \boldsymbol{\omega}_j \times \mathbf{e}$ (eq. D.16).

Show that

$$\left(\frac{\mathrm{d}\mathbf{e}}{\mathrm{d}t}\right)_{\mathrm{rot}} = \frac{3(\mathbb{G}M_{\mathrm{p}})^{1/2}J_2R_{\mathrm{p}}^2}{4a^{7/2}j^7}\left[j^2 - 5(\mathbf{j}\cdot\hat{\mathbf{n}}_{\mathrm{p}})^2\right]\mathbf{e}\times\mathbf{j}. \tag{P.32}$$

(e) Show that in this frame the eccentricity vector precesses uniformly around the angular-momentum vector at angular speed

$$\boldsymbol{\omega}_e = \frac{3(\mathbb{G}M_{\mathrm{p}})^{1/2}J_2R_{\mathrm{p}}^2}{4a^{7/2}(1-e^2)^2}(5\cos^2 I - 1)\,\hat{\mathbf{j}}, \tag{P.33}$$

where $\hat{\mathbf{j}} = \mathbf{j}/j$ is the unit vector parallel to $\mathbf{j}$. (f) Compare these results to equations (P.30).

5.5 [3] Suppose that the orbit of the Moon is suddenly rotated such that its inclination relative to the ecliptic is 90°. The semimajor axis and the eccentricity are unchanged, and the line of apsides lies in the ecliptic. Use the secular equations of motion to estimate how long it will take for the Moon to collide with the Earth. You may solve the equations analytically or numerically. The elements of the lunar and solar orbits are given in Appendix A.

5.6 [2] A planet orbits a star of mass M_{h} in a binary-star system. The companion star has mass M_{c} and the binary orbit has semimajor axis a_{c} and eccentricity e_{c}. The planet executes small-amplitude ZLK librations around the equilibrium point $\omega = \pm\frac{1}{2}\pi$ and $e = e_0 \equiv [1 - (\frac{5}{3})^{1/2}|\cos I_0|]^{1/2}$ where $(\mathbb{G}M_{\mathrm{h}}a)^{1/2}\cos I_0$ is the conserved component of the planet's angular momentum normal to the binary orbit (see §5.4 for notation). Show that the libration period is

$$P_L = \frac{4\pi}{3^{3/2}}\frac{na_{\mathrm{c}}^3}{\mathbb{G}M_{\mathrm{c}}}\frac{(1-e_{\mathrm{c}}^2)^{3/2}}{e_0(2+3e_0^2)^{1/2}} = \frac{2}{3^{3/2}}\frac{P_{\mathrm{c}}^2}{P}\frac{M_{\mathrm{h}}+M_{\mathrm{c}}}{M_{\mathrm{c}}}\frac{(1-e_{\mathrm{c}}^2)^{3/2}}{e_0(2+3e_0^2)^{1/2}}. \tag{P.34}$$

Here $P = 2\pi/n = 2\pi a^{3/2}(\mathbb{G}M_{\mathrm{h}})^{-1/2}$ is the orbital period of the planet and $P_{\mathrm{c}} = 2\pi a_{\mathrm{c}}^{3/2}[\mathbb{G}(M_{\mathrm{h}}+M_{\mathrm{c}})]^{-1/2}$ is the orbital period of the companion. Comment: this result gives rise to the rule of thumb that the libration period is of order the square of the binary period divided by the planet period.

5.7 [2] A planet orbits a host star that belongs to a binary system. The semimajor axis a_{c} of the stellar companion is much greater than the semimajor axis a of the planet, so the effects of the companion can be investigated in the quadrupole approximation. The planet is initially on a circular orbit, close enough to the host star that relativistic precession is important. If the planetary orbit is unstable to ZLK oscillations, show that the maximum value of the cosine of the inclination to the companion orbit $(\cos I)_{\rm max}$ coincides with the maximum eccentricity $e_{\rm max}$, and the two are related by

$$(\cos I)^2_{\rm max} = \tfrac{3}{5} + \frac{8\epsilon_{\rm gr}}{5e^2_{\rm max}}\left[1 - (1 - e^2_{\rm max})^{-1/2}\right], \tag{P.35}$$

where $\epsilon_{\rm gr}$ is defined by equation (5.105).

6.1 [2] Prove that the angle variables for the pendulum Hamiltonian are given by equations (6.8) and (6.12).

6.2 [2] Prove that the trajectory on the separatrix of the pendulum Hamiltonian (6.1) with initial conditions $q = 0$, $\dot{q} > 0$ at $t = 0$ is $q(t) = 4\tan^{-1}\exp(\omega t) - \pi$.

6.3 [2] Saturn's satellites Mimas and Tethys are in a 2:1 mean-motion resonance. Call the inner satellite (Mimas) body 1 and the outer satellite (Tethys) body 2. Thus $n_1 \simeq 2n_2$. (a) Keeping terms in the disturbing function up to degree 2 (eqs. 4.98, 4.99 and 4.104), show that there are six slow angles associated with this resonance: $\phi_1 = \lambda_1 - 2\lambda_2 + \varpi_1$, $\phi_2 = \lambda_1 - 2\lambda_2 + \varpi_2$, $\phi_3 = 2\lambda_1 - 4\lambda_2 + \varpi_1 + \varpi_2$, $\phi_4 = 2\lambda_1 - 4\lambda_2 + \Omega_1 + \Omega_2$, $\phi_5 = 2\lambda_1 - 4\lambda_2 + 2\Omega_1$ and $\phi_6 = 2\lambda_1 - 4\lambda_2 + 2\Omega_2$. The indirect potential (4.104) modifies the coefficient of the term involving ϕ_2 but does not add any new slow angles. The associated sub-resonances are sometimes labeled e_1, e_2, e_1e_2, I_1I_2, I_1^2, and I_2^2 since these factors determine the strength of the corresponding terms in the disturbing function. The sub-resonances are separated in semimajor axis because of apsidal and nodal precession due to Saturn's quadrupole moment (see §1.8.3 and Appendix A). (b) Order the sub-resonances by increasing semimajor axis of Mimas and determine the separation of each sub-resonance from its neighbors (in km). (c) If Mimas and Tethys undergo convergent migration due to tidal friction from Saturn, which sub-resonance is encountered first? (d) Mimas and Tethys are found in the I_1I_2 sub-resonance. Using the pendulum model for resonances from §6.1.2, show that the width of the I_1I_2 sub-resonance is much less than the distance to its neighboring sub-resonances, so the dynamics of each sub-resonance can be treated in isolation. The semimajor axis, eccentricity, and inclination (relative to Saturn's equator) of Mimas are $a = 1.855 \times 10^5$ km, $e = 0.020$, $I = 1.6°$, and for Tethys $a = 2.947 \times 10^5$ km, $e = 0.001$, $I = 1.1°$. The mass of Tethys is 1.087×10^{-6} times the mass of Saturn and the mass of Mimas is only 6% of the mass of Tethys.

6.4 [3] (a) Show that the separatrices that exist in the Henrard–Lemaitre Hamiltonian (6.48) for $\Delta > 3$ can be written parametrically as $r = \beta + \gamma \cos\phi$, where (r, ϕ) are polar coordinates relative to a center at $(x, y) = (\alpha, 0)$, with $0 < \beta \leq 2$ and

$$\alpha = -\frac{4}{\beta^2}, \quad \gamma = \frac{8}{\beta^2}, \quad \Delta = \frac{16}{\beta^4} + \frac{\beta^2}{2}. \tag{P.36}$$

This curve is sometimes called the limaçon of Pascal. (b) Show that the areas of the interior and resonance zones are

$$A_{\text{int}} = (\beta^2 + 32/\beta^4)\cos^{-1}(\beta^3/8) - \frac{12}{\beta}(1 - \beta^6/64)^{1/2},$$

$$A_{\text{res}} = (\beta^2 + 32/\beta^4)[\pi - 2\cos^{-1}(\beta^3/8)] + \frac{24}{\beta}(1 - \beta^6/64)^{1/2}. \tag{P.37}$$

(c) What is the area inside the resonance zone at the critical point $\Delta = 3$?

6.5 [3] Equation (6.79), which describes the resonant component of Neptune's disturbing function at Pluto as a function of the slow angle $\phi_{\text{s}} = 3\lambda_{\text{P}} - 2\lambda_{\text{N}} - \varpi_{\text{P}}$, is valid only when the eccentricity and inclination of Pluto are small. (a) Show that without this restriction the resonant disturbing function can be written

$$H(\phi_{\text{s}}) = -\frac{\mathbb{G}m_{\text{N}}}{4\pi}\int_0^{4\pi} \mathrm{d}u_{\text{P}}\, \frac{1 - e\cos u_{\text{P}}}{\Delta}, \tag{P.38}$$

where

$$\Delta^2 = (x_{\text{P}} - a_{\text{N}}\cos\lambda_{\text{N}})^2 + (y_{\text{P}} - a_{\text{N}}\sin\lambda_{\text{N}})^2 + z_{\text{P}}^2. \tag{P.39}$$

Here $\lambda_{\text{N}} \equiv \frac{3}{2}(u_{\text{P}} - e_{\text{P}}\sin u_{\text{P}}) + \omega_{\text{P}} - \frac{1}{2}\phi_{\text{s}}$, $(x_{\text{P}}, y_{\text{P}}, z_{\text{P}})$ is the position of Pluto as a function of its orbital elements (eqs. 1.70); u_{P}, e_{P}, and ω_{P} are Pluto's eccentric anomaly, eccentricity, and argument of periapsis; a_{N} is the semimajor axis of Neptune; and we have assumed that the orbit of Neptune is circular and lies in the ecliptic plane $z = 0$. (b) Why does the indirect potential not contribute to $H(\phi_{\text{s}})$? (c) Why is the result independent of the longitude of Pluto's node? (d) By evaluating $H(\phi_{\text{s}})$ numerically, find the libration period when the libration amplitude is small. For Pluto's orbital parameters use $e_{\text{P}} = 0.2502$, $I_{\text{P}} = 17.09°$, $\omega_{\text{P}} = 112.6°$.

6.6 [2] Saturn's satellites Enceladus and Dione are locked in a 2:1 mean motion resonance, with the resonant argument $\phi_{\text{s}} = \lambda_{\text{E}} - 2\lambda_{\text{D}} + \varpi_{\text{E}}$. This argument librates around 0 with a very small amplitude. Assume that Enceladus is much less massive than Dione (the actual mass ratio is about 0.1). The eccentricity of Enceladus is $e_\mathrm{E} = 0.0047$, the mean motions are $\dot{\lambda}_\mathrm{E} = 262.732^\circ\,\mathrm{d}^{-1}$ and $\dot{\lambda}_\mathrm{D} = 131.535^\circ\,\mathrm{d}^{-1}$ (these include the effects of Saturn's quadrupole moment) and the apsidal precession rate of Enceladus due to the Saturn quadrupole is $\dot{\varpi}_\mathrm{E} = 0.416^\circ\,\mathrm{d}^{-1}$. Find the mass of Dione. Hint: use the relevant term from the disturbing function (4.98), Table 6.1 and Lagrange's equations (1.188).

6.7 [2] Prove that the coefficients F_m and G_m in equations (6.89) and Table 6.1 are related by

$$F_m = \left(\frac{m+1}{m}\right)^{2/3} G_{-m-1}, \quad m \neq 0, -1. \tag{P.40}$$

6.8 [2] Two planets travel around a star of mass M_* in coplanar and nearly circular orbits. The inner planet has mass, semimajor axis, mean longitude and longitude of periapsis M, a, λ and ϖ. The analogous quantities for the outer planet are M', a', λ' and ϖ'. The planets are close to a mean-motion resonance with resonant angle $m\lambda - (m+1)\lambda' + \varpi$, so their interactions are governed by the Hamiltonian (eq. 4.98)[2]

$$\begin{aligned} H_1 &= \frac{\mathbb{G}MM'}{a'}(m+1+\tfrac{1}{2}\alpha D)b_{1/2}^{m+1}(\alpha)e\cos[m\lambda-(m+1)\lambda'+\varpi] \\ &\equiv B_m e\cos[m\lambda-(m+1)\lambda'+\varpi], \end{aligned} \tag{P.41}$$

where $\alpha = a/a'$ and $b_{1/2}^{m+1}(\alpha)$ is a Laplace coefficient. We assume that the planet masses are small, $M, M' \ll M_*$, so the angle-action variables for the inner planet can be written (see eqs. 1.88 and Box 4.1)

$$\begin{aligned} J_1 &= \Lambda = M(\mathbb{G}M_*a)^{1/2}, & J_2 &= \Lambda - L = M(\mathbb{G}M_*a)^{1/2}[1-(1-e^2)^{1/2}], \\ \theta_1 &= \lambda = \ell + \varpi, & \theta_2 &= -\varpi, \end{aligned} \tag{P.42}$$

with analogous expressions for M'. The total Hamiltonian for the two planets is then

$$H_\mathrm{res} = -\frac{\mathbb{G}^2M_*^2M^3}{2\Lambda^2} - \frac{\mathbb{G}^2M_*^2M'^3}{2\Lambda'^2} + H_1. \tag{P.43}$$

[2] This formula assumes that $m \neq -2$ so no indirect term is present in the resonant Hamiltonian. We also ignore other terms associated with this resonance, such as the one proportional to $e'\cos[m\lambda-(m+1)\lambda'+\varpi']$. One reason why this simplification is often valid is that the apsidal precession rates $\dot{\varpi}$ and $\dot{\varpi}'$ are different, due to the quadrupole moment of the star, relativistic precession, or precession due to a third planet. See Problem 6.3.

(a) Show that there are two conserved fast actions,

$$J_\mathrm{f} = J_1 + mJ_2, \quad J'_\mathrm{f} = J'_1 - (m+1)J_2. \tag{P.44}$$

(b) The variation in semimajor axis is small if the planet masses are small, so we can treat B_m as a constant. Moreover, for small eccentricity $J_2 \simeq \frac{1}{2}M(\mathbb{G}M_*a)^{1/2}e^2$, so without additional loss of accuracy we can write $B_m e = (2J_2)^{1/2}\gamma$, where $\gamma = B_m M^{-1/2}(\mathbb{G}M_*a)^{-1/4}$ is assumed to be constant. Following the derivation in §6.2, show that the Hamiltonian can be written in the standard form (6.37), with parameters

$$\alpha = \frac{(\mathbb{G}M_*)^2 M^3 m}{J_\mathrm{f}^3} - \frac{(\mathbb{G}M_*)^2 M'^3 (m+1)}{J_\mathrm{f}'^3},$$

$$\beta = -\frac{3(\mathbb{G}M_*)^2 M^3 m^2}{2J_\mathrm{f}^4} - \frac{3(\mathbb{G}M_*)^2 M'^3 (m+1)^2}{2J_\mathrm{f}'^4}. \tag{P.45}$$

Thus the analysis in §6.2 applies not just to a test particle in resonance with a massive planet but also to two planets of comparable mass. (c) What is the physical interpretation of the factor α?

7.1 [1] Using Euler's equations (Appendix D.11.2), prove that rotational motion of an isolated rigid body around one of its three principal axes is unstable if and only if the principal axis is the one with the intermediate moment of inertia. This effect is sometimes called the "tennis racket instability."

7.2 [1] The Earth's spin angular velocity and angular momentum are not precisely aligned with its symmetry axis. (a) Assuming that the Earth is rigid and axisymmetric, calculate the rate of precession of the angular velocity and angular-momentum vectors relative to the surface of the Earth using Euler's equations. You may assume that there are no external torques. (b) Carry out the same calculation using Andoyer variables. Comment: the actual precession, known as **Chandler wobble** or **free nutation**, has a period about 40% longer than this estimate because the Earth is not rigid.

7.3 [1] A small satellite with inertia tensor $\mathbf{I}$ orbits a planet of mass M. The position vector of the satellite relative to the center of the planet is $\mathbf{r}$. Prove that the torque on the satellite is

$$\mathbf{N} = \frac{3\mathbb{G}M}{r^5}\mathbf{r} \times (\mathbf{I} \cdot \mathbf{r}); \tag{P.46}$$

here $\mathbf{I}\cdot\mathbf{r} \equiv \sum_{i,j=1}^{3} \hat{\mathbf{n}}_i I_{ij} r_j$, where $\hat{\mathbf{n}}_i$ is the unit vector along axis i.

7.4 [1] A triaxial satellite is in a circular orbit around its host planet. Because of tidal friction, the satellite is in synchronous rotation. The moments of inertia associated with the principal axes $\hat{\mathbf{n}}_1$, $\hat{\mathbf{n}}_2$ and $\hat{\mathbf{n}}_3$ are I_{11}, I_{22} and I_{33} with $I_{11} < I_{22} < I_{33}$ (eqs. D.87). In what directions do the axes point?

7.5 [2] Let $\mathbf{S} = (S_1, S_2, S_3)$ be the spin angular momentum in a reference frame with Cartesian unit vectors $\hat{\mathbf{n}}_i$, $i = 1, 2, 3$. (a) Show that in this frame $\mathbf{S}$ can be written in terms of Andoyer variables as $\mathbf{S} = [\sin h(S^2 - S_3^2)^{1/2}, -\cos h(S^2 - S_3^2)^{1/2}, S_3]$. (b) Show that the Poisson brackets of the spin components are (cf. eq. 5.46)

$$\{S_i, S_j\} = \epsilon_{ijk} S_k. \tag{P.47}$$

(c) The orientation-dependent gravitational potential energy H of an axisymmetric body in an external field depends only on the direction of its symmetry axis. If the spin angular momentum is parallel to the symmetry axis we can write $H = H(\mathbf{S})$. In this case, prove that (cf. eq. 5.57)

$$\frac{\mathrm{d}\mathbf{S}}{\mathrm{d}t} = -\mathbf{S} \times \frac{\partial H}{\partial \mathbf{S}}. \tag{P.48}$$

(d) Use this result to derive equation (7.10) for the precession rate of a planet.

7.6 [1] The precession rate of the unit spin vector of an axisymmetric satellite is $\mathrm{d}\hat{\mathbf{S}}/\mathrm{d}t = -\alpha(\hat{\mathbf{S}}\cdot\hat{\mathbf{L}})\hat{\mathbf{L}}\times\hat{\mathbf{S}}$, where $\hat{\mathbf{L}}$ is the unit normal to the orbit and (eq. 7.10)

$$\alpha = \frac{3\mathbb{G}M}{2a^3\omega(1-e^2)^{3/2}}\frac{C-A}{C}, \tag{P.49}$$

with M the planet mass, ω the satellite spin frequency, and $A = B$ and C the principal moments of inertia. We assume here that the eccentricity e of the satellite orbit is zero. If the orbit normal $\hat{\mathbf{L}}$ precesses with frequency $\mathbf{\Omega}_\mathrm{s}$, the rate of change of the spin vector in the precessing frame is (cf. eq. D.16)

$$\frac{\mathrm{d}\hat{\mathbf{S}}}{\mathrm{d}t} = -\mathbf{\Omega}_\mathrm{s}\times\hat{\mathbf{S}} - \alpha(\hat{\mathbf{S}}\cdot\hat{\mathbf{L}})\hat{\mathbf{L}}\times\hat{\mathbf{S}}. \tag{P.50}$$

A Cassini state is a stationary solution of this equation. Prove Cassini's third law and the equilibrium condition (7.64).

7.7 [2] A spherical asteroid has a surface that emits according to Lambert's law. Is there any distribution of Bond albedo $A_\mathrm{B}(\theta, \phi)$ on the surface that will produce a steady-state YORP torque?

8.1 [1] Prove that a synchronous equilibrium state of a two-body system is stable if and only if the orbital angular momentum is more than three times the spin angular momentum.

8.2 [2] (a) Assuming that the Moon's spin is negligible, show that the rate of energy dissipation in the Earth–Moon system is

$$\dot{E} = \frac{\mathbb{G}M_\oplus M_☽}{2a^2}\left(1 - \frac{\omega}{n}\right)\dot{a}, \tag{P.51}$$

where ω is the Earth's spin angular velocity, $n = [\mathbb{G}(M_\oplus + M_☽)/a^3]^{1/2}$ is the mean motion, and $\dot{a} = 3.83\,\mathrm{cm\,yr^{-1}}$ is the rate of change of the semimajor axis. (b) Evaluate $\dot{E}$ in watts. (c) What is the ratio of the energy lost from the Earth's spin to the total energy lost from the Earth–Moon system?

8.3 [2] Mars's satellite Phobos has a semimajor axis of 9 376 km, which is shrinking at $3.85\,\mathrm{cm\,yr^{-1}}$ due to tides from Mars (Jacobson 2010). (a) When will Phobos crash into Mars? (b) What was the initial semimajor axis of Phobos, assuming it was born at the time of formation of the solar system, and how does this compare to the synchronous radius? (c) The mass of Phobos is 1.06×10^{16} kg, and the gravitational Love number of Mars is $k_2 = 0.170$. What is Mars's quality factor Q? (d) The shrinkage of Phobos's orbit was first detected in 1945. How could this have been done using ground-based telescopes?

8.4 [1] The Moon is receding from the Earth at $3.83\,\mathrm{cm\,yr^{-1}}$. Assuming that the quality factor and other properties of the Earth and Moon have remained constant, when was the Moon at 1 Earth radius? Compare this result to the age of the Earth.

8.5 [2] A planet identical to Jupiter orbits a star identical to the Sun in a nearly circular orbit. The planet is in synchronous rotation. The ratio of the quality factor to the Love number in both the star and the planet may be written $Q/k_2 = 10^6 q_6$, where q_6 is a free parameter. (a) At what semimajor axis is the semimajor axis decay time $a/|\mathrm{d}a/\mathrm{d}t|$ equal to 10^{10} yr? (b) At what semimajor axis is the eccentricity decay time $e/|\mathrm{d}e/\mathrm{d}t|$ equal to 10^{10} yr? (c) Use a database such as the Extrasolar Planets Encyclopedia (http://exoplanet.eu/) or the NASA Exoplanet Archive (https://exoplanetarchive.ipac.caltech.edu/index.html) to plot eccentricity versus semimajor axis for known exoplanets. Is there a clear signature of eccentricity damping, and if not, can you suggest why?

8.6 [2] A rigid, homogeneous body is covered by an ocean with density q times the density of the body. Show that the gravitational and displacement Love numbers are

$$k_l = \frac{3q}{2l+1-3q}, \quad h_l = \frac{2l+1}{2l+1-3q}. \tag{P.52}$$

8.7 [1] The displacement Love number of a planet with an extended atmosphere is defined by the surfaces of constant pressure. Prove that in this case, the displacement Love number h_l is related to the gravitational Love number k_l by $h_l = 1 + k_l$.

8.8 [1] A satellite travels around a planet in an orbit with mean motion n and eccentricity $e \ll 1$. The satellite is small, in the sense that its mass is much less than the planet's mass and its Love number (8.23) is dominated by the factor proportional to the rigidity μ. The eccentricity is damped by tidal friction in the satellite, as described by equation (8.67). Argue that the orbital angular momentum is conserved and hence that the rate of tidal heating of the satellite is

$$\dot{E} = \frac{42\pi}{19} \frac{\rho^2 R^7 n^5}{\mu Q} e^2, \tag{P.53}$$

where ρ, R and Q are the density, radius, and quality factor of the satellite.

9.1 [2] At radius a, a planetesimal disk has surface number density Σ_N and a Rayleigh eccentricity distribution (eq. 9.13). Assuming that the mean eccentricity $\langle e \rangle \ll 1$, prove that (a) the total number of planetesimals on orbits that cross a is $4\pi a^2 \Sigma_N \langle e \rangle$; (b) the sum of the squared eccentricities of all the planetesimals on orbits that cross a is $24 a^2 \Sigma_N \langle e \rangle^3$.

9.2 [1] The **Fisher probability distribution** for the angle θ between two unit vectors, $0 \le \theta \le \pi$, is

$$p(\theta) = \frac{\kappa}{2 \sinh \kappa} \sin\theta \exp(\kappa \cos\theta). \tag{P.54}$$

As required for a probability distribution, $\int_0^\pi \mathrm{d}x\, p(x) = 1$. What is the relation between the Fisher distribution and the Rayleigh distribution for the inclination I (eq. 9.13)?

9.3 [2] The velocity distribution function $f(\mathbf{v})$ in a planetesimal disk is a Schwarzschild distribution, equation (9.17). Let

$$V_n \equiv \int \mathrm{d}\mathbf{v}_1 \mathrm{d}\mathbf{v}_2 f(\mathbf{v}_1) f(\mathbf{v}_2) |\mathbf{v}_2 - \mathbf{v}_1|^n. \tag{P.55}$$

(a) Argue that the mean collision velocity between planetesimals in the disk is V_2/V_1 (neglect gravitational focusing). (b) The trans-Neptunian belt has mean eccentricity and inclination $\langle e\rangle = 0.14$ and $\langle I\rangle = 11°$. What is the mean collision velocity at a heliocentric distance of 45 au? Hint: carry out the integrations over velocity space numerically.

9.4 [1] A comet enters the Sun's planetary system on a nearly parabolic orbit and has a close encounter with Jupiter that converts it to a Jupiter-family comet (orbital period $P < 20$ yr). Show that its Tisserand parameter (Box 3.1) cannot exceed $2^{3/2} = 2.828$. You may neglect the influence of planets other than Jupiter and assume that Jupiter is on a circular orbit.

9.5 [3] A distribution of comets on parabolic orbits has a Rayleigh distribution of inclinations, $\mathrm{d}n \propto I\exp(-\frac{1}{2}\lambda I^2)\mathrm{d}I$; you may assume that $\lambda \gg 1$, so the inclinations are small. The comets all have the same periapsis distance q, and the arguments of periapsis and longitudes of the node are randomly distributed. They orbit in a system containing a planet of radius R on a circular orbit in the equatorial plane, with semimajor axis $a_\mathrm{p} > q \gg R$. The probability that a comet will collide with the planet on a single periapsis passage is $f_\mathrm{c}(R/a_\mathrm{p})^2$. Find f_c as a function of the mean inclination $\langle I\rangle$. You may neglect gravitational focusing, and you may use numerical or analytic methods.

9.6 [2] How fast does the ecliptic precess due to the tidal field from the Galaxy? (a) First make an approximate estimate by modeling the solar system as consisting of a single planet on a circular orbit, with the semimajor axis of Jupiter. (b) For a more accurate estimate include all of the planets, assuming they are on circular, coplanar orbits and that the planetary system precesses as a rigid body. (c) Why is this last assumption accurate?

9.7 [2] (a) Show that circular orbits around a star are stable in the Galactic tidal field if and only if $|\cos I| > (\frac{4}{5})^{1/2}$, where I is the inclination relative to the Galactic midplane. Hint: use equations (9.65) and follow the derivation in equations (5.91)–(5.96). (b) Are circular orbits in the solar system stable (ignore perturbations from the planets)? (c) If the orbits are unstable, at what semimajor axis is the growth time of the instability shorter than the age of the solar system?

9.8 [2] Find the orbit-averaged torque exerted by the Galactic tidal potential (9.59) on an orbit with eccentricity $e = 1$, and thereby derive the first of equations (9.66) without using the Milankovich formalism.

9.9 [2] Comets in the outer Oort cloud are uniformly distributed in phase space at a given semimajor axis. (a) Prove that in this case, the probability that a comet has eccentricity and inclination in the interval $\mathrm{d}e\mathrm{d}I$ is $e \sin I \,\mathrm{d}e\mathrm{d}I$. (b) Prove equation (9.70). Hint: use Delaunay variables.

9.10 [2] TNO 26308 is a binary system with an orbital period of 130.16 d and a semimajor axis of 11 370 km. The ratio of the brightnesses of the two components is 0.11. (a) Assuming the components have the same albedo and a mean density of $0.5\,\mathrm{g\,cm^{-3}}$, what is the mass and radius of each component? (b) The system's heliocentric orbit has a semimajor axis of 47.998 au. What is its Hill radius (you may approximate the orbit as circular)? (c) The eccentricity of the binary orbit is $e = 0.473$, and the inclination relative to the heliocentric orbit is 75.4°. If the system undergoes ZLK oscillations, what is the maximum eccentricity achieved (eq. 5.98), and is there a danger of collision? (d) Using equation (P.34), estimate roughly the period of the ZLK oscillations. (e) Suggest a reason why the oscillations might be suppressed in this system. (f) Using equations (8.23) and (8.67), estimate the eccentricity damping time $\tau = e/|\dot{e}|$. You may assume that the smaller member of the binary is synchronously rotating, with quality factor $Q = 20$ and rigidity $\mu = 3\,\mathrm{GPa}$. (g) The typical collision velocity in the trans-Neptunian belt is about $1\,\mathrm{km\,s^{-1}}$ (Problem 9.3). Estimate roughly the minimum radius of a TNO that would disrupt the binary if it collided with the smaller member.

9.11 [1] An asteroid with a semimajor axis of 1.2 au and an eccentricity $e = 0.3$ collides with the Earth. The asteroid is approximately spherical, with a radius $R = 1\,\mathrm{km}$ and a density $\rho = 3\,\mathrm{g\,cm^{-3}}$. Estimate the energy of the impact in kilotons of TNT (1 kiloton of TNT equals 4.184×10^{12} J).

References

Most of the articles listed here are available on-line from the Smithsonian Astrophysical Observatory/NASA Astrophysics Data System (ADS), https://ui.adsabs.harvard.edu/. Newer articles are often available in pre-publication form on the free arXiv e-print service, https://arxiv.org/.
The following abbreviations are used for frequently cited journals:

A&A	Astronomy and Astrophysics
AJ	The Astronomical Journal
ApJ	The Astrophysical Journal
ApJL	The Astrophysical Journal Letters
ApJS	The Astrophysical Journal Supplement
ARAA	Annual Review of Astronomy and Astrophysics
CeMec	Celestial Mechanics
CeMDA	Celestial Mechanics and Dynamical Astronomy
GeoRL	Geophysical Research Letters
MNRAS	Monthly Notices of the Royal Astronomical Society

Aggarwal, H. R. & Oberbeck, V. R. 1974, ApJ, 191, 577

Agol, E. & Fabrycky, D. C. 2018, in H. J. Deeg & J. A. Belmonte, eds. Handbook of Exoplanets (Dordrecht: Springer), 797. arXiv:1706.09849

Agol, E., Steffen, J., Sari, R., et al. 2005, MNRAS, 359, 567

Alexander, M. E. 1973, Astrophysics and Space Science, 23, 459

Allan, R. R. & Ward, G. N. 1963, Mathematical Proceedings of the Cambridge Philosophical Society, 59, 669

Andrews, S. M., Huang, J., Pérez, L. M.,

et al. 2018, ApJL, 869, L41

Applegate, J. H., Douglas, M. R., Gürsel, Y., et al. 1986, AJ, 92, 176

Archinal, B. A., Acton, C. H., A'Hearn, M. F., et al. 2018, CeMDA, 130, 22

Arnold, V. I. 1984, in R. Z. Sagdeev, ed. Nonlinear and Turbulent Processes in Physics (Chur: Harwood), 1161

Arnold, V. I. 1989, Mathematical Methods of Classical Mechanics, 2nd ed. (New York: Springer–Verlag)

Barrera, R. G., Estévez, G. A. & Giraldo, J. 1985, European Journal of Physics, 6, 284

Barrow–Green, J. 1997, Poincaré and the Three-Body Problem (Providence, RI: American Mathematical Society)

Baruteau, C., Crida, A., Paardekooper, S.-J., et al. 2014, in H. Beuther et al., eds. Protostars and Planets VI (Tucson: University of Arizona Press), 667. arXiv:1312.4293

Beekman, G. 2005, Journal of the British Astronomical Association, 115, 207

Bender, C. M. & Orszag, S. A. 1999, Advanced Mathematical Methods for Scientists and Engineers (New York: Springer)

Bennett, M. & Bovy, J. 2019, MNRAS, 482, 1417

Binney, J. & Tremaine, S. 2008, Galactic Dynamics, 2nd ed. (Princeton, NJ: Princeton University Press)

Binney, J., Gerhard, O. E. & Hut, P. 1985, MNRAS, 215, 59

Biscani, F. & Izzo, D. 2021, MNRAS, 504, 2614

Bland, P. A. & Artemieva, N. A. 2006, Meteoritics and Planetary Science, 41, 607

Blanes, S. & Casas, F. 2016, A Concise Introduction to Geometric Numerical Integration (Boca Raton, LA: CRC Press)

Boquet, F. 1889, Annales de l'Observatoire de Paris, Mémoires, 19, B1

Borderies, N. & Goldreich, P. 1984, CeMec, 32, 127

Bottke, W. F., Vokrouhlický, D., Rubincam, D. P., et al. 2006, Annual Review of Earth and Planetary Sciences, 34, 157

Breiter, S. & Vokrouhlický, D. 2015, MNRAS, 449, 1691

Brouwer, D. 1937, AJ, 46, 149

Brouwer, D. & Clemence, G. M. 1961, Methods of Celestial Mechanics (New York: Academic Press)

Brown, E. W. 1897–1908, Memoirs of the Royal Astronomical Society, 53, 39; 53, 163; 54, 1; 57, 51; 59, 1

Brown, E. W. 1911, MNRAS, 71, 438

Brown, E. W. 1936, MNRAS, 97, 62

Brown, G. & Rein, H. 2020, Research Notes of the AAS, 4, 221

Burdet, C. A. 1967, Zeitschrift für angewandte Mathematik und Physik, 18, 434

Burns, J. A. 1976, American Journal of Physics, 44, 944

Burns, J. A., Lamy, P. L. & Soter, S. 1979, Icarus, 40, 1

Burrows, A., Marley, M., Hubbard, W. B., et al. 1997, ApJ, 491, 856

Byrd, P. F. & Friedman, M. D. 1971, Handbook of Elliptic Integrals for Engineers and Scientists, 2nd ed. (Berlin: Springer–Verlag)

Capitaine, N., Wallace, P. T. & Chapront, J. 2003, A&A, 412, 567

Carroll, S. M. 2019, Spacetime and Geometry (Cambridge: Cambridge University Press)

Cartwright, D. E. 1999, Tides: A Scientific History (Cambridge: Cambridge University Press)

Cary, J. R. 1981, Physics Reports, 79, 129

Chandler, J. F., Battat, J.B.R., Murphy, T. W., et al. 2021, AJ, 162, 78

Chandrasekhar, S. 1963, ApJ, 138, 1182

Chandrasekhar, S. 1969, Ellipsoidal Figures of Equilibrium (New Haven, CT: Yale University Press)

Claret, A. 2019, A&A, 628, A29

Claret, A. & Bloemen, S. 2011, A&A, 529, A75

Cochran, W. D., Fabrycky, D. C., Torres, G., et al. 2011, ApJS, 197, 7

Cohen, C. J. & Hubbard, E. C. 1965, AJ, 70, 10

Cohen, C. J., Hubbard, E. C. & Oesterwinter, C. 1973, Astronomical Papers prepared for the use of the American Ephemeris and Nautical Almanac, 22, Part I

Collins, B. F. & Sari, R. 2010, AJ, 140, 1306

Colombo, G. 1966, AJ, 71, 891

Connelly, J. N., Bizzarro, M., Krot, A. N., et al. 2012, Science, 338, 651

Cook, A. 2000, Astronomy & Geophysics, 41, 6.21

Correia, A.C.M. 2015, A&A, 582, A69

Dalrymple, G. B. 2001, in C.L.E. Lewis & S. J. Knell, eds. The Age of the Earth, Geological Society, London, Special Publications, 190, 205

Danby, J.M.A. 1964a, AJ, 69, 165

Danby, J.M.A. 1964b, AJ, 69, 294

Darwin, G. H. 1899, The Tides (Boston: Houghton Mifflin)

Deck, K. M., Holman, M. J., Agol, E., et al. 2012, ApJL, 755, L21

Deck, K. M., Payne, M. & Holman, M. J. 2013, ApJ, 774, 129

Dehnen, W. & Binney, J. J. 1998, MNRAS, 298, 387

Dekker, T. J. 1971, Numerische Mathematik, 18, 224

Delaunay, C. E. 1860, Mémoires de l'Académie des Sciences, 28, 1

Delaunay, C. E. 1867, Mémoires de l'Académie des Sciences, 29, 1

Deprit, A. 1967, American Journal of Physics, 35, 424

Deprit, A. 1969, CeMec, 1, 12

Di Sisto, R. P. & Rossignoli, N. L. 2020, CeMDA, 132, 36

Dones, L. & Tremaine, S. 1993, Icarus, 103, 67

Dones, L., Brasser, R., Kaib, N., et al. 2015, Space Science Reviews, 197, 191

Dumas, H. S. 2014, The KAM Story (Singapore: World Scientific)

Duncan, M. & Levison, H. F. 1997, Science, 276, 1670

Duncan, M., Levison, H. F. & Budd, S. M. 1995, AJ, 110, 3073

Duncan, M., Levison, H. F. & Lee, M. H. 1998, AJ, 116, 2067

Duncan, M., Quinn, T. & Tremaine, S. 1987, AJ, 94, 1330

Duncan, M., Quinn, T. & Tremaine, S. 1988, ApJL, 328, L69

Durante, D., Parisi, M., Serra, D., et al. 2020, GeoRL, 47, e2019GL086572

Earn, D.J.D. & Tremaine, S. 1992, Physica D, 56, 1

Efroimsky, M. & Makarov, V. V. 2013, ApJ, 764, 26

Erdélyi, A. 1953–1955, Higher Transcendental Functions, 3 volumes (New York: McGraw–Hill)

Erdélyi, A. 1954, Tables of Integral Transforms, 2 volumes (New York: McGraw–Hill)

Everhart, E. 1985, in A. Carusi & G. B. Valsecchi, eds. IAU Colloquium 83: Dynamics of Comets: Their Origin and Evolution (Dordrecht: Reidel), 185

Fernández, J. A. 1980, MNRAS, 192, 481

Fernández, J. A. 2005, Comets (Dordrecht: Springer)

Fernández, J. A. & Ip, W.-H. 1984, Icarus, 58, 109

Folkner, W. M. 2010, in S. A. Klioner et al., eds. IAU Symposium 261: Relativity in Fundamental Astronomy: Dynamics, Reference Frames, and Data Analysis (Cambridge: Cambridge University Press), 155

Ford, E. B., Kozinsky, B. & Rasio, F. A. 2000, ApJ, 535, 385

Forest, E. & Ruth, R. D. 1990, Physica D, 43, 105

Friedland, L. 2001, ApJL, 547, L75

Fukushima, T. 1999, in J. Svoren, E. M. Pittich & H. Rickman, eds. IAU Colloquium 173: Evolution and Source Regions of Asteroids and Comets (Tatranská Lomnica, Slovakia: Astronomical Institute of the Slovak Academy of Sciences), 309

Fuller, J., Luan, J. & Quataert, E. 2016, MNRAS, 458, 3867

Gaudi, B. S. 2011, in S. Seager, ed. Exoplanets (Tucson: University of Arizona Press), 79. arXiv:1002.0332

Gavrilov, S. V. & Zharkov, V. N. 1977, Icarus, 32, 443

Gear, C. W. 1971, Numerical Initial Value Problems in Ordinary Differential Equations (Englewood Cliffs, NJ: Prentice–Hall)

Genova, A., Goossens, S., Lemoine, F. G., et al. 2016, Icarus, 272, 228

Genova, A., Mazarico, E., Goossens, S., et al. 2018, Nature Communications, 9, 289

Genova, A., Goossens, S., Mazarico, E., et al. 2019, GeoRL, 46, 3625

Gibbs, J. W. & Wilson, E. B. 1901, Vector Analysis (New Haven, CT: Yale University Press)

Gladman, B. 1993, Icarus, 106, 247

Gladman, B. & Volk, K. 2021, ARAA, 59, 203

Glowinski, R., Osher, S. J. & Yin, W., eds. 2016, Splitting Methods in Communication, Imaging, Science, and Engineering (Cham: Springer)

Goldreich, P. 1965, AJ, 70, 5

Goldreich, P. 1966, Reviews of Geophysics and Space Physics, 4, 411

Goldreich, P. & Soter, S. 1966, Icarus, 5, 375

Goldstein, H. 1975–1976, American Journal of Physics, 43, 737; 44, 1123

Goldstein, H., Poole, C. & Safko, J. 2002, Classical Mechanics, 3rd ed. (San Francisco: Addison–Wesley)

Golubev, V. G. 1968, Soviet Physics Doklady, 13, 373

Gradshteyn, I. S. & Ryzhik, I. M. 2015, Tables of Integrals, Series, and Products, 8th ed., D. Zwillinger & V. Moll, eds. (Amsterdam: Academic Press)

Grazier, K. R., Newman, W. I., Hyman, J. M., et al. 2005, ANZIAM Journal, 46, C786 and C1086

Greenberg, A. H., Margot, J.-L., Verma, A. K., et al. 2020, AJ, 159, 92

Greenzweig, Y. & Lissauer, J. J. 1992, Icarus, 100, 440

Guo, R., Liu, C., Mao, S., et al. 2020, MNRAS, 495, 4828

Gutzwiller, M. C. 1998, Reviews of Modern Physics, 70, 589

Hairer, E., Nørsett, S. P. & Wanner, G. 1993, Solving Ordinary Differential Equations I. Nonstiff Problems (Berlin: Springer)

Hairer, E., Lubich, C. & Wanner, G. 2006, Geometric Numerical Integration, 2nd ed. (Berlin: Springer)

Hamilton, C. & Rafikov, R. R. 2019, MNRAS, 488, 5489

Hartle, J. B. 2003, Gravity: an Introduction to Einstein's General Relativity (San Francisco: Addison–Wesley)

Hayashi, C. 1981, Progress of Theoretical Physics Supplements, 70, 35

Hayes, W. B. 2008, MNRAS, 386, 295

Heggie, D. C. 1973, in B. D. Tapley & V. Szebehely, eds. Recent Advances in Dynamical Astronomy (Dordrecht: Reidel), 34

Heggie, D. C. & Rasio, F. A. 1996, MNRAS, 282, 1064

Heisler, J. & Tremaine, S. 1986, Icarus, 65, 13

Hénon, M. 1965, Annales d'Astrophysique, 28, 992

Hénon, M. 1969, A&A, 1, 223

Hénon, M. 1970, A&A, 9, 24

Hénon, M. 1974, A&A, 30, 317

Hénon, M. & Petit, J.-M. 1986, CeMec, 38, 67

Henrard, J. 1982, CeMec, 27, 3

Henrard, J. 1989, CeMec, 45, 245

Henrard, J. 1993, in C.K.R.T. Jones et al., eds. Dynamics Reported, 2, (Berlin: Springer–Verlag), 117

Henrard, J. & Lemaitre, A. 1983, CeMec, 30, 197

Henrard, J. & Murigande, C. 1987, CeMec, 40, 345

Henrici, P. 1962, Discrete Variable Methods in Ordinary Differential Equations (New York: Wiley)

Hill, G. W. 1894, Annals of Mathematics, 9, 31

Holman, M. J. & Murray, N. W. 2005, Science, 307, 1288

Holman, M. J. & Wiegert, P. A. 1999, AJ, 117, 621

Holmberg, J. & Flynn, C. 2000, MNRAS, 313, 209

Hunter, J. D. 2007, Computing in Science and Engineering, 9, 90

Hut, P. 1980, A&A, 92, 167

Hut, P. 1981, A&A, 99, 126

Ida, S., Kokubo, E. & Makino, J. 1993, MNRAS, 263, 875

Iliffe, R. 2017, Priest of Nature (Oxford: Oxford University Press)

Ito, T. & Ohtsuka, K. 2020, Monographs on Environment, Earth and Planets, 7, 1. arXiv:1911.03984

Ito, T. & Tanikawa, K. 2002, MNRAS, 336, 483

Jacobson, R. A. 2010, AJ, 139, 668

Jacobson, R. A., Spitale, J., Porco, C. C., et al. 2008, AJ, 135, 261

Jewitt, D. & Luu, J. 1993, Nature, 362, 730

José, J. V. & Saletan, E. J. 1998, Dynamics: A Contemporary Approach (Cambridge: Cambridge University Press)

Julian, W. H. & Toomre, A. 1966, ApJ, 146, 810

Kaplan, G. H. 2005, United States Naval Observatory Circular No. 179. arXiv:astro-ph/0602086

Kevorkian, J. & Cole, J. D. 1996, Multiple Scale and Singular Perturbation Methods (New York: Springer)

Kinoshita, H. 1972, Publications of the Astronomical Society of Japan, 24, 423

Kinoshita, H. & Nakai, H. 1984, CeMec, 34, 203

Kinoshita, H. & Nakai, H. 1996, Earth, Moon, and Planets, 72, 165

Kinoshita, H. & Nakai, H. 2007, CeMDA, 98, 67

Knežević, Z., Milani, A., Farinella, P., et al. 1991, Icarus, 93, 316

Knuth, D. E. 1981, The Art of Computer Programming. Vol. 2: Seminumerical Algorithms, 2nd ed. (Reading, MA: Addison–Wesley)

Konopliv, A. S. & Yoder, C. F. 1996, GeoRL, 23, 1857

Konopliv, A. S., Banerdt, W. B. & Sjogren, W. L. 1999, Icarus, 139, 3

Konopliv, A. S., Asmar, S. W., Folkner, W. M., et al. 2011, Icarus, 211, 401

Kozai, Y. 1962, AJ, 67, 591

Królikowska, M. & Dybczyński, P. A. 2020, A&A, 640, A97

Lainey, V. 2016, CeMDA, 126, 145

Lainey, V., Jacobson, R. A., Tajeddine, R., et al. 2017, Icarus, 281, 286

Lambert, J. D. & Watson, I. A. 1976, IMA Journal of Applied Mathematics, 18, 189

Landau, L. D. & Lifshitz, E. M. 1976, Mechanics, 3rd ed. (Amsterdam: Elsevier)

Landau, L. D. & Lifshitz, E. M. 1986, Theory of Elasticity, 3rd ed. (Oxford: Butterworth–Heinemann)

Landau, L. D. & Lifshitz, E. M. 1987, Fluid Mechanics, 2nd ed. (Oxford: Pergamon)

Laplace, P. S. 1799–1825, Traité de mécanique céleste, 5 volumes (Paris: Chez J.B.M. Duprat). English translation by N. Bowditch 1829–1839, Celestial Mechanics (Bronx, NY: Chelsea Publishing)

Laskar, J. 1986, A&A, 157, 59

Laskar, J. 1988, A&A, 198, 341

Laskar, J. 1989, Nature, 338, 237

Laskar, J. 2013, in B. Duplantier et al., eds. Chaos: Poincaré Seminar 2010 (Basel: Springer), 239. arXiv:1209.5996

Laskar, J. 2017, CeMDA, 128, 475

Laskar, J. & Gastineau, M. 2009, Nature, 459, 817

Laskar, J. & Robutel, P. 1993, Nature, 361, 608

Laskar, J., Correia, A.C.M., Gastineau, M., et al. 2004a, Icarus, 170, 343

Laskar, J., Robutel, P., Joutel, F., et al. 2004b, A&A, 428, 261

Le Verrier, U.J.J., 1855, Annales de l'Observatoire Impérial de Paris, 1, 258

Levesque, D. & Verlet, L. 1993, Journal of Statistical Physics, 72, 519

Lichtenberg, A. J. & Lieberman, M. A. 1992, Regular and Chaotic Dynamics (New York: Springer–Verlag)

Lidov, M. L. 1961, in Problems of Motion of Artificial Celestial Bodies (Moscow: Russian Academy of Sciences), 119 (in Russian). Original article and English translation available at https://ui.adsabs.harvard.edu/abs/1963pmac.book..119L/abstract.

Lindblom, L. 1992, Philosophical Transactions of the Royal Society A, 340, 353

Lithwick, Y., Xie, J. & Wu, Y. 2012, ApJ, 761, 122

Love, A.E.H. 1944, A Treatise on the Mathematical Theory of Elasticity, 4th ed. (New York: Dover)

Luger, R., Sestovic, M., Kruse, E., et al. 2017, Nature Astronomy, 1, 0129

Luo, L., Katz, B. & Dong, S. 2016, MNRAS, 458, 3060

Ma, L. & Fuller, J. 2021, ApJ, 918, 16

Malhotra, R. 1993, Nature, 365, 819

Malhotra, R. & Williams, J. G. 1997, in S. A. Stern & D. J. Tholen, eds. Pluto and Charon (Tucson: University of Arizona Press), 127

Mandel, K. & Agol, E. 2002, ApJL, 580, L171

Marchal, C. 1990, The Three-Body Problem (Amsterdam: Elsevier)

Marchal, C. & Bozis, G. 1982, CeMec, 26, 311

Margot, J.-L., Campbell, D. B., Giorgini, J. D., et al. 2021, Nature Astronomy, 5, 676

Matsumura, S., Peale, S. J. & Rasio, F. A. 2010, ApJ, 725, 1995

Maury, J. L. & Segal, G. P. 1969, NASA Technical Report TM-X-63542, X-553-69-46. https://ntrs.nasa.gov/citations/19690017325

Meibom, S. & Mathieu, R. D. 2005, ApJ, 620, 970

Mikkola, S. 2020, Gravitational Few-Body Dynamics (Cambridge: Cambridge University Press)

Mikkola, S. & Tanikawa, K. 1999, MNRAS, 310, 745

Milankovich, M. 1939, Bulletin de l'Académie Royal Serbe, Académie des Sciences Mathématiques et Naturelles A, Sciences Mathématiques et Physiques, 6, 1

Mills, S. M., Fabrycky, D. C., Migaszewski, C., et al. 2016, Nature, 533, 509

Misner, C. W., Thorne, K. S. & Wheeler, J. A. 1973, Gravitation (San Francisco: W. H. Freeman)

Morbidelli, A. & Gladman, B. 1998, Meteoritics and Planetary Science, 33, 999

Morbidelli, A. & Nesvorný, D. 2020, in D. Prialnik, M. A. Barucci & L. Young, eds. The Trans-Neptunian Solar System (Amsterdam: Elsevier), 25. arXiv:1904.02980

Morbidelli, A., Bottke, W. F., Froeschlé, C., et al. 2002, in W. F. Bottke, A. Cellino, P. Paolicchi, et al., eds. Asteroids III (Tucson: University of Arizona Press), 409

Morrison, S. & Malhotra, R. 2015, ApJ, 799, 41

Müller, J., Murphy, T. W., Schreiber, U., et al. 2019, Journal of Geodesy, 93, 2195

Muller, J.-M., Brisebarre, N., de Dinechin, F., et al. 2010, Handbook of Floating-Point Arithmetic (Boston: Birkhaäuser)

Murray, C. D. 1985, CeMec, 36, 163

Murray, C. D. & Dermott, S. F. 1999,

Solar System Dynamics (Cambridge: Cambridge University Press)

Musielak, Z. E. & Quarles, B. 2014, Reports on Progress in Physics, 77, 065901

Naoz, S. 2016, ARAA, 54, 441

Nettelmann, N., Helled, R., Fortney, J. J., et al. 2013, Planetary and Space Science, 77, 143

Newcomb, S. 1899, Astronomische Nachrichten, 148, 321

Newhall, X X, Standish, E. M. & Williams, J. G. 1983, A&A, 125, 150

Ni, D. 2018, A&A, 613, A32

Nicholson, P. D., Hamilton, D. P., Matthews, K., et al. 1992, Icarus, 100, 464

Noll, K., Grundy, W. M., Nesvorný, D., et al. 2020, in D. Prialnik, M. A. Barucci & L. Young, eds. The Trans-Neptunian Solar System (Amsterdam: Elsevier), 201. arXiv:2002.04075

Obertas, A., Van Laerhoven, C. & Tamayo, D. 2017, Icarus, 293, 52

Ogilvie, G. I. 2014, ARAA, 52, 171

O'Keefe, J. A. 1976, Tektites and Their Origin (New York: Elsevier)

Olver, F.W.J., Lozier, D. W., Boisvert, R. F. & Clark, C. W., eds. 2010, NIST Handbook of Mathematical Functions (Cambridge: Cambridge University Press). See also https://dlmf.nist.gov/

Oort, J. H. 1950, Bulletin of the Astronomical Institutes of the Netherlands, 11, 91

Paddack, S. J. 1969, Journal of Geophysical Research, 74, 4379

Park, R. S., Folkner, W. M., Williams, J. G., et al. 2021, AJ, 161, 105

Peale, S. J. 1999, ARAA, 37, 533

Peale, S. J. 2003, CeMDA, 87, 129

Peirce, B. 1849, AJ, 1, 31

Pendse, C. G. 1935, Phil. Trans. Roy. Soc. A 234, 145

Petit, A. C., Pichierri, G., Davies, M. B., et al. 2020, A&A, 641, A176

Petit, J.-M. & Hénon, M. 1986, Icarus, 66, 536

Pitjeva, E. V. & Pitjev, N. P. 2014, CeMDA, 119, 237

Pitjeva, E. V., Pitjev, N. P., Pavlov, D. A., et al. 2021, A&A, 647, A141

Planck Collaboration, Aghanim, N., Akrami, Y., et al. 2020, A&A, 641, A6

Plummer, H. C. 1918, An Introductory Treatise on Dynamical Astronomy (Cambridge: Cambridge University Press)

Poincaré, H. 1892–1897, Les méthodes nouvelles de la mécanique céleste, 3 volumes (Paris: Gauthier–Villars). English translation by D. L. Goroff 1993, New Methods of Celestial Mechanics (New York: American Institute of Physics)

Poincaré, H. 1896, Comptes Rendus, 123, 1031

Poisson, E. & Will, C. M. 2014, Gravity: Newtonian, Post-Newtonian, Relativistic (Cambridge: Cambridge University Press)

Poisson, S. D. 1809, Journal de l' École Polytechnique, Cahier 8, 1

Press, W. H., Teukolsky, S. A., Vetterling, W. T. & Flannery, B. P. 2007, Numerical Recipes: the Art of Scientific Computing, 3rd ed. (Cambridge: Cambridge University Press)

Preto, M. & Tremaine, S. 1999, AJ, 118, 2532

Quillen, A. C. 2006, MNRAS, 365, 1367

Quillen, A. C. 2011, MNRAS, 418, 1043

Quinlan, G. D. 1999, arXiv:astro-ph/9901136

Quinlan, G. D. & Tremaine, S. 1990, AJ, 100, 1694

Quinn, T. R., Tremaine, S. & Duncan, M. 1991, AJ, 101, 2287

Radzievskii, V. V. 1954, Doklady Akademii Nauk SSSR, 97, 49

Rafikov, R. R. 2003, AJ, 125, 922

Rannou, F. 1974, A&A, 31, 289

Rein, H. & Liu, S.-F. 2012, A&A, 537, A128

Rein, H. & Spiegel, D. S. 2015, MNRAS, 446, 1424

Rein, H. & Tamayo, D. 2017, MNRAS, 467, 2377

Rein, H. & Tamayo, D. 2018, MNRAS, 473, 3351

Rein, H. & Tamayo, D. 2019, Research Notes of the AAS, 3, 16

Rein, H., Tamayo, D. & Brown, G. 2019, MNRAS, 489, 4632

Roberts, J.A.G. & Quispel, G.R.W. 1992, Physics Reports, 216, 63

Roy, A. E., Walker, I. W., Macdonald, A. J., et al. 1988, Vistas in Astronomy, 32, 95

Rubincam, D. P. 2000, Icarus, 148, 2

Saari, D. G. 1971, Transactions of the American Mathematical Society, 156, 219

Sackett, P. D. 1999, in J.-M. Mariotti & D. Alloin, eds. Planets Outside the Solar System: Theory and Observations (Dordrecht: Kluwer), 189. arXiv:astro-ph/9811269

Saha, P. 2009, MNRAS, 400, 228

Salo, H. & Yoder, C. F. 1988, A&A, 205, 309

Sanz-Serna, J. M. 1988, BIT Numerical Mathematics, 28, 877

Schönrich, R., Binney, J. & Dehnen, W. 2010, MNRAS, 403, 1829

Schröder, K.-P. & Connon Smith, R. 2008, MNRAS, 386, 155

Seager, S. 2010, Exoplanet Atmospheres (Princeton, NJ: Princeton University Press)

Seager, S. & Mallén–Ornelas, G. 2003, ApJ, 585, 1038

Shampine, L. F. 1986, Mathematics of Computation, 46, 135

Shen, Y. & Tremaine, S. 2008, AJ, 136, 2453

Shevchenko, I. I. 2017, The Lidov–Kozai Effect—Applications in Exoplanet Research and Dynamical Astronomy (Cham: Springer)

Souchay, J., Mathis, S. & Tokieda, T., eds. 2013, Tides in Astronomy and Astrophysics (Heidelberg: Springer)

Spitale, J. N., Jacobson, R. A., Porco, C. C., et al. 2006, AJ, 132, 692

Spurzem, R., Giersz, M., Heggie, D. C., et al. 2009, ApJ, 697, 458

Sridhar, S. & Tremaine, S. 1992, Icarus, 95, 86

Stewart, G. R. & Ida, S. 2000, Icarus, 143, 28

Stiefel, E. L. & Scheifele, G. 1971, Linear and Regular Celestial Mechanics (Berlin: Springer–Verlag)

Strang, G. 1968, SIAM Journal on Numerical Analysis, 5, 506

Struve, O. 1952, The Observatory, 72, 199

Sussman, G. J. & Wisdom, J. 1988, Science, 241, 433

Sussman, G. J. & Wisdom, J. 1992, Science, 257, 56

Sussman, G. J. & Wisdom, J. 2001, Structure and Interpretation of Classical Mechanics (Cambridge, MA: MIT Press)

Suzuki, M. 1990, Physics Letters A, 146, 319

Szebehely, V. 1967, Theory of Orbits: the Restricted Problem of Three Bodies (New York: Academic Press)

Tabor, M. 1989, Chaos and Integrability in Nonlinear Dynamics (New York: Wiley)

Thomas, P. C., Burns, J. A., Hedman, M., et al. 2013, Icarus, 226, 999

Tiscareno, M. S. & Malhotra, R. 2003, AJ, 126, 3122

Tisserand, F. 1889–1896, Traité de mécanique céleste, 4 volumes (Paris: Gauthier–Villars)

Touma, J. & Wisdom, J. 1993, Science, 259, 1294

Touma, J. & Wisdom, J. 1994, AJ, 108, 1943

Touma, J. R., Tremaine, S. & Kazandjian, M. V. 2009, MNRAS, 394, 1085

Tremaine, S. 2001, CeMDA, 79, 231

Tremaine, S. & Yavetz, T. D. 2014, American Journal of Physics, 82, 769

Tremaine, S., Touma, J. & Namouni, F. 2009, AJ, 137, 3706

Urban, S. E. & Seidelmann, P. K. 2013, Explanatory Supplement to the Astronomical Almanac, 3rd ed. (Mill Valley, CA: University Science Books)

Valtonen, M. & Karttunen, H. 2005, The Three-Body Problem (Cambridge: Cambridge University Press)

Viswanathan, V., Fienga, A., Minazzoli, O., et al. 2018, MNRAS, 476, 1877

Vokrouhlický, D. 1998, A&A, 335, 1093

Vokrouhlický, D., Bottke, W. F., Chesley, S. R., et al. 2015, in P. Michel, F. E. DeMeo & W. F. Bottke, eds. Asteroids IV (Tucson: University of Arizona Press), 509. arXiv:1502.01249

Vokrouhlický, D., Nesvorný, D. & Dones, L. 2019, AJ, 157, 181

Waldvogel, J. 2008, CeMDA, 102, 149

Walsh, K. J. 2018, ARAA, 56, 593

Warner, B., Pravec, P. & Harris, A. P. (2019) Asteroid Lightcurve Data Base (LCDB) V3.0 urn:nasa:pds:ast-lightcurve-database::3.0. NASA Planetary Data System. https://sbn.psi.edu/pds/resource/lc.html

Watson, G. N. 1944, Theory of Bessel Functions (Cambridge: Cambridge University Press)

Weidenschilling, S. J. 1977, Astrophysics and Space Science, 51, 153

Weinberg, S. 1972, Gravitation and Cosmology (New York: Wiley)

Weiss, L. M., Marcy, G. W., Petigura, E. A., et al. 2018, AJ, 155, 48

Weisskopf, V. F. 1975, Science, 187, 605

Whiteside, D. T. 1976, Vistas in Astronomy, 19, 317

Whittaker, E. T. 1937, A Treatise on the Analytical Dynamics of Particles and Rigid Bodies, 4th ed. (Cambridge: Cambridge University Press)

Wiegert, P., Innanen, K. & Mikkola, S. 2000, AJ, 119, 1978

Will, C. M. 1993, Theory and Experiment in Gravitational Physics (Cambridge: Cambridge University Press)

Williams, J. G. & Faulkner, J. 1981, Icarus, 46, 390

Williams, J. G., Konopliv, A. S., Boggs, D. H., et al. 2014, Journal of Geophysical Research: Planets, 119, 1546

Williams, J. P. & Cieza, L. A. 2011, ARAA, 49, 67

Wilson, C. 1985, Archive for History of Exact Sciences, 33, 15

Winn, J. N. & Fabrycky, D. C. 2015, ARAA, 53, 409

Wisdom, J. 1980, AJ, 85, 1122

Wisdom, J. 2006, AJ, 131, 2294

Wisdom, J. & Holman, M. 1991, AJ, 102, 1528

Wisdom, J. & Meyer, J. 2016, CeMDA, 126, 1

Wisdom, J., Holman, M. & Touma, J. 1996, Fields Institute Communications, 10, 217

Wisdom, J., Peale, S. J. & Mignard, F. 1984, Icarus, 58, 137

Wolszczan, A. & Frail, D. A. 1992, Nature, 355, 145

Wu, Y. & Murray, N. 2003, ApJ, 589, 605

Wyatt, M. C. 2008, ARAA, 46, 339

Wyatt, M. C. 2020, in D. Prialnik, M. A. Barucci & L. Young, eds. The Trans-Neptunian Solar System (Amsterdam: Elsevier), 351. arXiv:1909.12312

Wyatt, M. C., Bonsor, A., Jackson, A. P., et al. 2017, MNRAS, 464, 3385

Wyatt, S. P. & Whipple, F. L. 1950, ApJ, 111, 134

Yabushita, S. 1980, A&A, 85, 77

Yalinewich, A. & Petrovich, C. 2020, ApJL, 892, L11

Yee, S. W., Winn, J. N., Knutson, H. A., et al. 2020, ApJL, 888, L5

Yoshida, H. 1982, CeMec, 28, 239

Yoshida, H. 1990, Physics Letters A, 150, 262

Yoshida, H. 1993, CeMDA, 56, 27

von Zeipel, H. 1910, Astronomische Nachrichten, 183, 345 (in French). See also Ito & Ohtsuka (2020).

von Zeipel, H. 1916, Arkiv för Matematik, Astronomi och Fysik, 11, 1

Index

Page numbers ending in "p" refer to problems.